LIFE AS WE KNOW IT

Cellular Origin, Life in Extreme Habitats and Astrobiology

Volume 10

Series Editor:

Joseph Seckbach
The Hebrew University of Jerusalem, Israel

The titles published in this series are listed at the end of this volume.

Life as We Know It

Edited by

J. Seckbach

The Hebrew University of Jerusalem, Israel

Springer

A C.I.P. Catalogue record for this book is available from the Library of Congress.

ISBN-10 1-4020-4394-5 (HB)
ISBN-10 1-4020-4403-8 (e-book)
ISBN-13 978-1-4020-4394-9 (HB)
ISBN-13 978-1-4020-4403-8 (e-book)

Published by Springer,
P.O. Box 17, 3300 AA Dordrecht, The Netherlands.

www.springer.com

LEGEND FOR THE COVER PICTURE
Title: "***Let everything that has breath praise God, Halleluyah***"
(Psalm 150).

This picture was designed and produced by the artists **Baruch and Issac Nachshon (info@nachshonart.com**, and: **www.nachshonart.com**). They contributed this illustration which was specially produced for our volume of "***Life As we Know It***" following the request of the editor. The Editor in Chief (**Joseph Seckbach**) would like to thank both of the artists for their contribution to the cover of this volume.

Printed on acid-free paper

TABLE OF CONTENTS[1]

[1] I thank Professor Julian Chela-Flores for practical and constructive advice in sorting this Table

The Distribution of Life in the Universe

The Destiny of Life in the Universe

PART 3:
DEEPER ISSUES OF THE PHENOMENON OF LIFE

Philosophical and Theological Aspects of Life

DEDICATION

This volume is dedicated to the prominent late scientist Professor **Joan Oró (1923–2004).** Born in Lleida, Spain and from 1960's he worked with NASA on the Viking missions which explored the planet Mars. His work was essential in the analysis of samples of Martian soil, and suggested that no life was detected on Mars. Dr. Joan Oró was a biochemist; he received his PhD from Baylor College of Medicine, Houston, TX in 1956. His scientific research has been centered on the experimental study of the prebiotic chemical reactions that took place on the primitive earth. Dr. Oró discovered the prebiotic synthesis of adenine from hydrogen cyanide. He proposed the theory on the key role of comets on the synthesis of biomolecules on the primitive earth. Dr. Oró is the author or co-author of more than thirty books and some three hundred publications related to the origin of Life (ISSOL) and organized its first meeting (ICOL) in Barcelona, Spain. In addition, he organized some thirty other International meetings in related fields of science. He participated in the Apollo mission to the Moon and the Viking project to the planet Mars. He enlightened us with his oral presentations in many conferences on chemical evolution and Astrobiology. Dr. Oró was a member of many NASA committees and a Holder of the Oparin Gold Medal Award and many other scientific recognitions and distinctions in the USA and Spain. Doctor Honoris Causa of the Universities of Granada, Houston and Lleida.

The late Professor J. Oro with Joseph Seckbach, editor of this volume and the ***Cellular Origins, Life in Extreme Habitats and Astrobiology*** series.

1

2

3

4

5

6

7

8

9

10

Legends for the Oró photos' Page

Fig. 1 and 4 courtesy of Jordi Llorca (Barcelona, Spain); Fig. 2, 3, 5, 9. from Joseph Seckbach photo collections. At fig. 5, Oró (center) behind him are Stanley Miller (left) and Julien Chela-Flores (right). Fig. 6 courtesy of photo archive of Julian Chela-Flores, 1998; from left to right; F. Drake, Christian de Duve, and J. Oró. Fig. 7, courtesy by Karen Meech (Honolulu, Hawaii, 2000). Fig. 8, courtesy of ICTP photo archives (2000) from left to right: T. Owen, J. Chela-Flores, S. Miller, J. Oró, F. Drake, and Bill Schopf. Fig. 10, courtesy of ICTP photo archives (1995) from left to right: J. Oró, J. Chela-Flores, S. Fox, and F. Raulin.

PREFACE

As pointed out by Carol Cleland in this book, the question "What is life?" has caused large quantities of ink to flow without eliciting a unanimously agreed answer. The issue is not just academic. Now that increasing efforts are being devoted to the detection of extraterrestrial life, it is obviously imperative that one should know what to look for and what tools to use.

My own answer to the question is: "Life is all that is common to all known living organisms". This definition is far from circular, being restricted to a small number of key properties, roughly those we share with the *colibacilli* in our gut; it agrees basically with that offered by Cleland. The problem, as she remarks, is that the definition does not necessarily apply to unknown forms of life and, therefore, is unsatisfactory both in theory and in practice. This consideration is, of course, rigorously correct, though not heuristically very useful. In my opinion, extraneous life has a good chance of resembling life "as we know it", except, possibly, for the sign of the chirality of some molecules.

This view is based on a number of arguments. First, there is the fact that the building blocks used in the development of Earth life were probably products of a universally occurring cosmic chemistry, which is likely to be the same everywhere. Furthermore, Earth life is most likely the outcome of a complex set of highly interrelated chemical and selective processes, linking, for example, nucleic acids to energy, information transfers to base pairing, proteins to RNA, and membranes to lipid bilayers. Chemistry obeys strictly deterministic rules and may be expected to give the same results under the same physical-chemical conditions. As to selection, there are good reasons to believe that it leads, much more frequently than is generally assumed, to an optimizing outcome reproducibly linked to the prevailing conditions.

Admittedly, these considerations leave entirely open the possibility that the products of cosmic chemistry would react differently under different conditions. One can even not rule out the existence of life forms based on a different kind of chemistry, including, as some would have it, carbon replaced by silicon, and water by ammonia. Personally, I recognize the validity of such open-mindedness in a philosophical or, even, scientific context. In practice, however, I see no usefulness in the reservations made and very much doubt their relevance to the detection of extraterrestrial life. Even restricted in the way I suggest, the problem is sufficiently formidable not to be complicated by additional hypothetical caveats.

For the time being and, no doubt, for some time to come, "life as we know it" is by far the best guide to follow in the quest for "other worlds." This book, with its informed and eclectic survey of our biosphere, represents a valuable contribution to this endeavor.

Institute of Cellular Phathology, Buxell, Belgium **Christian de Duve**
E-mail: **deduve_private@icp.ucl.ac.be**
February 20, 2006

Biodata of **Christian de Duve** author of the Preface

Dr. de Duve was born in England and educated in Belgium. He is emeritus professor at the Catholic University of Louvain (Belgium) and at the Rockefeller University in New York. He shared the Nobel Prize in Physiology or Medicine 1974 with Albert Claude and George Palade, "for their discoveries concerning the structural and functional organization of living cells." He is known for the discovery of lysosomes and peroxisomes and has, in recent years, become interested in the origin and evolution of life. Dr. de Duve is the author of several books, including "*Life Evolving: Molecules, Mind, and Meaning*" (Oxford University Press, 2002) and "*Singularities: Landmarks on the Pathways of Life*" (Cambridge University Press, 2005).

E-mail: **deduve_private@icp.ucl.ac.be**

FOREWORD

Life As We Know It: All of us in our own way have our understanding of what is life. Providing a simple definition, however, has never been easy. As Luisi (1998) pointed out there have been a number of definitions of Life. The reason for this perhaps can be laid at the door of the complexity of the life process and the myriad variations on the theme. Definitions are expected to be tidy and simple, lacking exceptions. In a recent attempt to provide a definition, Ruiz-Mirazo (2004) proposed defining living beings as "autonomous systems with open ended evolution capabilities". In echoing other attempts to provide a definition they designated four essential attributes: a semi-permeable membrane, an energy transduction system and two types of macro-molecular components, namely catalysts and records. In the same vein, Oró (2002) had previously provided ten attributes of living system followed by a list of ten different kinds of Life. These latter covered the full spectrum from complex and simple viruses to intelligent human life and designated groups, e.g. plant life, fungal life, eukaryotic unicellular and eukaryotic multicellular life, animal life. prokaryotic unicellular life, prokaryotic commensal and interdependent cellular life, subcellular and molecular life.

However we define or understand Life there is within that definition a vast panoply of forms, variations and themes. This volume presents a "birds-eye-view" of some of the simpler (or early forms) of life. But that adjective, "simpler", presents perhaps an anthropocentric view in how one defines simple. There are discussions of the range of habitats that these organisms inhabit, which would perhaps render the term "simple" as simplistic. Following chapters on some key landmarks in the evolution of Life, the origin of Life and possible extraterrestrial Life there are reviews devoted to the philosophical issues of life as exemplified in humankind. This volume treads new ground in providing this broad boundary of "life" and thus will introduce readers to new realms to which they may well have given little attention. In providing this framework, the volume takes the reader from the origin of life through the nature of various early forms, the inhabited environments and finally to the complexity of features that are attributable to, and driven by the views and evolution of mankind. By looking at this broad horizon, the volume almost certainly provides something new and challenging for all readers and helps us focus on the overriding question of what is life.

The views and descriptions are based on our understanding and knowledge of life on this planet. If there is life elsewhere, for which we are looking, we must ask whether such life, if found, will fall neatly within the boundaries, framework and attributes described by the various contributions in this volume and from earlier workers. If this turns out not to be the case perhaps we can anticipate another volume, years in the future, entitled "Life As Others Know It".

Luisi, P. L. (1998) About various definitions of Life. Origins of Life and Evolution of the Biosphere, **28**: 613-622.

Oró, J. (2002) Historical understanding of Life's beginnings. In: J. W. Schopf (ed.) *Life's Origins*. (Univ. California Press: pp. 7-45.

Ruiz-Mirazo, K., Pereto, J. and Moreno, A. (2004) A universal definition of Life: Autonomy and open-ended evolution. Origin of Life and evolution of the Biosphere **34**:323-346.

University of California
Santa Barbara, Ca. USA
E-mail: **chapman@lifesci.ucsb.edu**

David J. Chapman

February 20, 2006

Biodata of **David J. Chapman** author of the ***Foreword***

Dr. David J. Chapman is a Professor of Marine Biology in the Department of Ecology, Evolution and Marine Biology and additionally the Graduate Marine Science Program at the University of California Santa Barbara. He received his PhD from the Scripps Institution of Oceanography in 1965 and a D.Sc. from the University of Auckland (New Zealand) in 1979. His main research interests cover a broad spectrum of algal physiology and biochemistry, with an emphasis on pigments and natural products, applied uses of algae and the evolution of biochemical systems in the earliest algal life of the Archaean and Precambrian eras. In addition to his own published works, books and papers, he has served in a number of editorial positions. Among his books are: ***The Algae*** (with V.J. Chapman) (1973); ***Biochemical Phylogeny of the Protists*** (with M.A. Ragan) (1978); ***Seaweeds and their uses*** (with V.J. Chapman) (1980). He edited the series of ***Progress in Phycological Research*** (with F.E. Round) Volume 1-13 (1982-1999);and the ***Experimental Phycology. A Laboratory Manual***. (with C. Lobban and B. Kremer) (1988); ***Handbook of Protoctista*** (with L. Margulis, J.O. Corliss, M. Melkonian) (1990). Dr. Chapman has been awarded at UCLA (1981) the Distinguished Teaching Award in the Biological Sciences faculty. He is also a Fellow of the Linnean Society of London and a member of a number of national and international phycological and plant biological societies.

E-mail: **chapman@lifesci.ucsb.edu**

"Bless God, O my soul.....He established the earth upon its foundations, that it falter not forever and ever.......He causes vegetation to sprout for the cattle, and plants through man's labor, to bring forth bread from the earth; and wine that gladdens man's heart....He made the moon for festivals, the sun knows its destination......How abundant are Your works, God, with wisdom You make them all, the earth is full of Your possessions."

Psalm 104

BRIEF INTRODUCTION

This book, ***Life As we Know It*** is volume 10 of the ***Cellular Origins, Life in Extreme Habitats and Astrobiology*** series. The other nine volumes of this group are provided in the following website:
http://www.springeronline.com/sgw/cda/frontpage/0,10735,5-198-69-33111926-0,00.html.

We cannot define yet the concept of "Life", but we can nevertheless describe the characteristics and circumstances of living organisms which could be considered as physical systems with a high level of complexity. All organisms on Earth descended from a common evolutionary ancestor. Among them are the three domains of life, i.e., **Archaea, Bacteria** [both groups are Prokaryotes] and **Eukarya**. The Archaea contain some extremophiles, such as the halophiles and the thermophiles, while the Bacteria contain cyanobacteria and heterotrophic bacteria. In the Eukaryota are the nucleated cells such as, heterotrophic protists, algae, fungi, plants, and the animal kingdom. One might find life to be ubiquitously spread all over the earth's biosphere and find it almost in all environments from the microbial extremophilic environments to the human versatile habitats (e.g., the Eskimo igloos or the Bedouin tents, to the astronauts living in the Space Station).

These chapters cover many sections of the phenomenon of life from the Big Bang hypothesis to the search for extraterrestrial intelligent life in the Universe (SETI). This book deals also with the origin of life, the theory of microbial symbiotic association, fossil records, the prokaryotic world (and the extreme-philes), protists (e.g., fungi and algae), and several aspects of human culture, history, sport, philosophy (including the phenomenon of conscience), theology, the intriguing question of resuscitation after clinical death, and, finally, artificial intelligence. Other chapters discuss the emergence of life and its evolution in the Universe and the possibility of searching for extraterrestrial life on Mars, Europa and planets, including their satellites, in other solar systems.

Joseph Seckbach

The Hebrew University of Jerusalem
E-mail: **seckbach@huji.ac.il**
February 20, 2006

Biodata of **Joseph Seckbach** author of the chapter ***"Extremophiles"* and *"Introduction"* as well as *"What do we call Life"* (with co-authors: Raulin, Oren, Kolb and Chela-Flores.)**

Professor Joseph Seckbach is the initiator and chief editor of ***Cellular Origins, Life in Extreme Habitats and Astrobiology* (COLE)** book series, and author of several chapters and ***Introductions*** in this series. He earned his Ph.D. from the University of Chicago, Chicago, IL. (1965) and spent his postdoctoral years in the Division of Biology at Caltech (Pasadena, CA, USA). Then he headed at the University of California at Los Angeles (UCLA) a team for searching for extraterrestrial life. Dr. Seckbach has been appointed to the faculty of the Hebrew University (Jerusalem, Israel), performed algal research, and taught biological courses. He spent his sabbatical periods in Tübingen (Germany), UCLA and Harvard University, and served at Louisiana State University (LSU) (1997-1998) as the first selected occupant [of the John P. Laborde endowed] Chair for the Louisiana Sea Grant and Technology transfer, and as a visiting Professor in the Department of Life Sciences at LSU (Baton Rouge, LA, USA).

Among his publications are books, scientific articles concerning plant ferritin (phytoferritin), cellular evolution, acidothermophilic algae, and life in extreme environments. He also edited and translated several popular books. Dr. Seckbach is the co-author (with R. Ikan) of the ***Chemistry Lexicon*** (1991, 1999) published by Deveer (Tel Aviv) and of other volumes. He is also the co-editor of the *Proceeding of* ***Endocytobiology VII Conference*** (Freiburg, Germany, 1999) and the Proceedings of the ***Algae and Extreme Environments*** meeting (Trebon, Czech Republic, 2001) His recent interest is in the field of enigmatic microorganisms and life in extreme environments.

E-mail: **seckbach@huji.ac.il**

ACKNOWLEDGEMENTS

I owe so much to so many people in various ways in preparing this volume of "***Life As we Know It***" (Number 10 of the "***Cellular Origins, Life in Extreme Habitats and Astrobiology***" series). I thank to those who advised me and pointed out potential authors and reviewers for the chapters. I will not mention them for not skipping anyone of them. I appreciate our associate editors, Professors **David Chapman, Julian Chela-Flores, Aharon Oren, and Francois Raulin,** for constant interest, assistance, and readiness to help in many aspects of this and other volumes of the **(COLE)** series. Special gratitude is due to the free contribution of the artists **Baruch and Issac Nachshon** who designed following the editor's request the special attractive picture for this book's cover.

I thank to the following peer reviewers and external referees (Names sorted in alphabetical order, and those in bold fonts reviewed several manuscripts) who criticized and evaluated the chapters:
Alterman Wladyslow, Bahn Peter, Basiuk Vladimir, Blanc Michel, Buick Rager**, Castenholz Richard**, Chapman David, **Chela-Flores Julian**, Cleland Carol, Coyne George, Deamer Dave, Deming Jody, Despois Dedier, Doron Alex, Doruin J-M., Douglas Angela, Drew Endy, Ehrenfreund Pascale, Elitzur Avshalom, Falk Rafael, Fedo Christopher, Frankel Richard, Fry Iris, Gaidos Eric, Galili Yair, Gilad amos, Grandpierre Attila, Guillemin J.C., Harman Oren, Harvey Zeev, Hoham Ron, Jeon Kwang, Krumbein Wolfgang, Kugrens Paul, Lamarchand L., Larkum Tony, Lewin Ralph, Lineweaver Charley, Llorca Jordi, Maccone Claudio, **McKay Chris**, Montebugnoli Stelio, Matsuno Koichiro, Mercader Perez Juan, Messirotti Mauro, Molitoris H-P., Mojzsis S., Morgan Hugh, **Nealson Keneth**, **Oren Aharon**, Prieur Daniel, Pross Addy, Rainey Fred, Raulin FlorenceErard Chris, **Raulin François**, Reisser W., Rowe Gilbert, Schüler Dirk, Seckbach Elie, Shah Tahir, Spring Stefan, **Svoboda Josef**, **Teodorani Massimo**, Trigo Josep, **Umiel Nakdimon**, Vincent Warwick. Walsh Maud, Wasser Solomon, Watling Roy, Westall Frances, Whitehoause Martin, Wreschner Daniel, Yayanos Art, Zook Douglas.

I acknowledge the contributions of the photos by Gordon Beakes (UK), J. Chela-Flores (Italy), Jordi Llorca (Spain) and by Karen Meech (Hawaii), these illustrations are adding attractiveness to this volume.

Hebrew University of Jerusalem **Joseph Seckbach**
E-mail: **seckbach@huji.ac.il**
February 20, 2006

LIST OF AUTHORS FOR *LIFE AS WE KNOW IT*
All senior authors are underlined

AUMANN ROBERT JOHN
CENTER FOR THE STUDY OF RATIONALITY, AND DEPARTMENT OF MATHEMATICS AND ECONOMICS, THE HEBREW UNIVERSITY OF JERUSALEM, 91904, ISRAEL

BEAKES GORDON W.
DIVISION OF BIOLOGY, RIDLEY BUILDING, NEWCASTLE UNIVERSITY, NEW CASTLE UP ON TYNE, NE1 7RU, UNITED KINGDOM.

BELISHEVA NATALIA K.
POLAR-ALPINE BOTANICAL GARDEN-INSTITUTE, KOLA SCIENTIFIC CENTER, RUSSIAN ACADEMY OF SCIENCES, FERSMAN STR. 14, 189204, APATITY, RUSSIAN FEDERATION.

BELL PHILIP J. L.
MACQUARIE UNIVERSITY, SYDNEY, NEW SOUTH WALES, 2109, AUSTRALIA.

BLUMBERG BARUCH S.
FOX CHASE CANCER CENTER, 333 COTTMAN AVENUE, PHILADELPHIA, PA 19111, USA.

BONCH-OXMOLOVSKAYA ELIZAVETA
WINOGRADSKY INSTITUTE OF MICROBIOLOGY RAS, PROSPECT 60 LET OKTYABRYA 7/2, MOSCOW, 117312, RUSSIA.

BRASIER MARTIN D.
DEPARTMENT OF EARTH SCIENCES, PARKS ROAD, OXFORD, OX1 3PR, UK.

CARNERUP ANNA M.
DEPARTMENT OF APPLIED MATHEMATICS, RESEARCH SCHOOL OF PHYSICAL SCIENCES AND ENGINEERING, THE AUSTRALIAN NATIONAL UNIVERSITY, 0200 CANBERRA, ACT, AUSTRALIA.

CHAPMAN DAVID
DEPARTMENT OF ECOLOGY, EVOLUTION AND MARINE BIOLOGY, UNIVERSITY OF CALIFORNIA, SANTA BARBARA, SANTA BARBARA, CA 93106-9610 USA.

CHAPMAN RUSSELL L.
DEPARTMENT OF OCEANOGRAPHY AND COASTAL SCIENCES, SCHOOL OF THE COAST AND ENVIRONMENT. LOUISIANA STATE UNIVERSITY, BATON ROUGE, LA 70803 USA.

CHELA-FLORES JULIAN
THE ABDUS SALAM INTERNATIONAL CENTER FOR THEORETICAL PHYSICS, STRADA COSTIERA 11, 34014 TRIESTE, ITAL, AND INSTITUTO DE ESTUDIOS AVANZADOS, CARACAS 1015A, VENEZUELA.

CHRISTY ANDREW G.
DEPARTMENT OF APPLIED MATHEMATICS, RESEARCH SCHOOL OF PHYSICAL SCIENCES AND ENGINEERING, THE AUSTRALIAN NATIONAL UNIVERSITY, 0200 CANBERRA, ACT, AUSTRALIA, AND DEPARTMENT OF EARTH AND MARINE SCIENCES, AUSTRALIAN NATIONAL UNIVERSITY 0200 CANBERRA, ACT, AUSTRALIA.

CLELAND CAROL E.
PHILOSOPHY DEPARTMENT AND CENTER FOR ASTROBIOLOGY, UNIVERSITY OF COLORADO, BOULDER, CO. USA.

COYNE GEORGE V., S.J.
VATICAN OBSERVATORY, V-00120 VATICAN CITY STATE.

DE DUVE CHRISTIAN
CHRISTIAN DE DUVE INSTITUTE OF CELLULAR PATHOLOGY, 75.50 AVENUE HIPPOCRATE 75, B-1200 BRUXELLES, BELGIUM AND THE ROCKEFELLER UNIVERSITY, 1230 YORK AVENUE, NEW YORK , N.Y. 10021, USA.

DEAMER DAVID
DEPARTMENT OF CHEMISTRY AND BIOCHEMISTRY, UNIVERSITY OF CALIFORNIA, SANTA CRUZ, CA. 96064 USA.

EHRENFREUND PASCALE
LEIDEN INSTITUTE OF CHEMISTRY, ASTROBIOLOGY LABORATORY, P.O.BOX 9502, 2300 RA LEIDEN, THE NETHERLANDS.

ELITZUR AVSHALOM
UNIT FOR INTERDISCIPLINARY STUDIES, BAR-ILAN UNIVERSITY, RAMAT-GAN 52900, ISRAEL.

EVANS JOHN
DEPARTMENT OF APPLIED MATHEMATICS AND THEORETICAL PHYSICS. CAMBRIDGE UNIVERSITY, CAMBRIDGE, UK.

FERRI FRANCESCA
BIOCHEMISTRY DEPARTMENT, UNIVERSITY OF BOLOGNA, VIA IRNERIO 48, 40126 BOLOGNA, ITALY.

FUNDERBURKE LAWRENCE
12115 MAGNOLIA BLVD. #113, NORTH HOLLYWOOD, CA. 91607 USA.

GADOTH NATAN
DEPARTMENT OF NEUROLOGY, MEIR GENERAL HOSPITAL, KFAR-SABA AND THE SACKLER FACULTY OF MEDICINE, TEL-AVIV UNIVERSITY, ISRAEL.

GARCÍA-RUIZ JUAN MANUEL
LABORATORIO DE ESTUDIOS CRISTALOGRAFICOS, INSTITUTO ANDALUZ DE CIENCIAS DE LA TIERRA, CSIC -UNIVERSIDAD DE GRANADA, E 18002 GRANADA SPAIN.

GUNDE-CIMERMAN NINA
BIOLOGICAL DEPARTMENT, BIOOTECHNICAL FACULTY, UNIVERSITY OF LJUBLJANA, 1000 LJUBLJANA, SLOVENIA.

HYDE STEPHEN T.
DEPARTMENT OF APPLIED MATHEMATICS, RESEARCH SCHOOL OF PHYSICAL SCIENCES AND ENGINEERING, THE AUSTRALIAN NATIONAL UNIVERSITY, 0200 CANBERRA, ACT, AUSTRALIA.

KOLB VERA
DEPARTMENT OF CHEMISTY, UNIVERSITY OF WISCONSIN-PARKSIDE, KENOSHA, WISCONSIN 53141, USA.

KRUPIČKA JIRI
UNIVERSITY OF ALBERTA [1701]11200-53 AVE. EDMONTON, ALBERTA, T6H OS4, CANADA.

LAMMER HELMUT
SPACE RESEARCH INSTITUTE, AUSTRIAN ACADEMY OF SCIENCES, SCHMIEDLSTRASSE 6, A-8042 GRAZ, AUSTRIA.

LARAS GIUSEPPE
14, VIA PISACANE - 20129 MILAN, ITALY.

LARSSON ANN-KRISTIN
DEPARTMENT OF APPLIED MATHEMATICS, RESEARCH SCHOOL OF PHYSICAL SCIENCES AND ENGINEERING, THE AUSTRALIAN NATIONAL UNIVERSITY, 0200 CANBERRA, ACT, AUSTRALIA.

LEE JOHN J.
DEPARTMENT OF BIOLOGY, CITY COLLEGE OF CITY UNIVERSITY OF NEW YORK, CONVENT AVE @138 STREET, NEW YORK, NY 10031, USA.

LINEWEAVER CHARLES H.
PLANETARY SCIENCE INSTITUTE, AUSTRALIAN NATIONAL UNIVERSITY, CANBERRA, AUSTRALIA.

LLORCA JORDI
DEPARTMENT OF INORGANIC CHEMISTRY, UNIVERSITY OF BARCELONA, MARTÍ I FRANQUÈS 1-11, 08028, BARCELONA, SPAIN.

LUISI PIER LUIGI
BIOLOGY DEPARTMENT, THIRD UNIVERSITY OF ROMA, V.LE MARCONI, 446, 00146 ROMA, ITALY.

MACCONE CLAUDIO
SETI PERMANENT STUDY GROUP, INTERNTIONAL ACADEMY OF ASTRONAUTICS VIA MARTORELLI, 43 -TORINO (TO) 10155, ITALY.

MARTIN EDUARDO L.
INSTITUTO DE ASTROFISICA DE CANARIAS, LA LAGUNA, TENERIFE, E-38200, SPAIN, AND PHYSICS DEPARTMENT, UNIVERSITY OF CENTRAL FLORIDA, ORLANDO FL-32816, USA.

MATSUNO KOICHIRO
DEPARTMENT OF BIOENGINEERING, NAGAOKA UNIVERSITY OF TECHNOLOGY, NAGAOKA 940-2188, JAPAN.

MCKAY CHRISTOPHER P.
SPACE SCIENCE DIVISION, NASA-AMES RESEARCH CENTER, MOFFETT FIELD, CA. 94035-1000, USA.

MCLOUGHLIN NICOLA
DEPARTMENT OF EARTH SCIENCES, PARKS ROAD, OXFORD, OX1 3PR, UK.

NE'EMAN YUVAL
DEPARTMENT OF PHYSICS, TEL-AVIV UNIVERSITY, TEL-AVIV, 69978, ISRAEL.

OREN AHARON
THE INSTITUTE OF LIFE SCIENCES, HEBREW UNIVERSITY OF JERUSALEM, JERUSALEM 91904, ISRAEL.

PENZ THOMAS
SPACE RESEARCH INSTITUTE, AUSTRIAN ACADEMY OF SCIENCES, SCHMIEDLSTRASSE 6, A-8042 GRAZ, AUSTRIA, AND INSTITUTE OF PHYSICS, UNIVERSITY OF GRAZ, UNIVERSITÄTSPLATZ 5, A-8010 GRAZ, AUSTRIA.

RAULIN FRANCOIS
LISA, UNIVERSITÉS PARIS 12 AND PARIS 7, CRÉTEIL, FRANCE.

RISKIN SHLOMO
P.O.B. 1037, EFRAT 90435, ISRAEL.

ROSE EBEN C.
DEPARTMENT OF GEOLOGY AND GEOPHYSICS,
YALE UNIVERSITY, P.O.BOX 208109, NEW HAVEN,
CT 06520-8109 USA.

ROSS IAN K.
MOLECULAR, CELLULAR AND DEVELOPMENTAL BIOLOGY, UNIVERSITY OF CALIFORNIA, SANTA BARBARA, CA. 93106-9610, USA.

RUSSELL ROBERT JOHN
THE CENTER FOR THEOLOGY AND THE NATURAL SCIENCES; THE GRADUATE THEOLOGICAL UNION, 2400 RIDGE ROAD, BERKELEY, CA. 94709, USA.

SAFFO MARY BETH
MARINE BIOLOGICAL LABORATORY (WOOD HOLE, MA) AND HARVARD UNIVERSITY. MCZ LABS 408, HARVARD UNIVERSITY, 26 OXFORD ST., CAMBRIDGE, MA 02138-2902 USA.

SCHRADER MALCOLM E.
DEPARTMENT OF INORGANIC AND ANALYTICAL CHEMISTRY, THE HEBREW UNIVERSITY OF JERUSALEM, JERUSALEM 91904, ISRAEL.

SECKBACH JOSEPH
HEBREW UNIVERSITY OF JERUSALEM, HOME: P.O.B. 1132, EFRAT 9043, ISRAEL.

SHAH TAHIR
FACULTY OF ENGINEERING, UNIVERSITY OF TRIESTE, ITALY

SHOSTACK SETH
SETI INSTITUTE, 515 N. WHISMAN ROAD, MOUNTAIN VIEW, CA. 94043, USA.

SIPPER MOSHE
DEPARTMENT OF COMPUTER SCIENCE, BEN-GURION UNIVERSITY, P.O.B. 653, BEER-SHEVA 84105, ISRAEL.

STANO PASQUALE
CENTRO STUDI "E. FERMI" COMPENDIO VIMINALE 00184 ROMA, ITALY. AND BIOLOGY DEPARTMENT, THIRD UNIVERSITY OF ROMA, V.LE MARCONI, 446, 00146 ROMA, ITALY.

SVOBODA JOSEF
DEPARTMENT OF BIOLOGY, UNIVERSITY OF TORONTO AT MISSISSAUGA, MISSISSAUGA, ONTARIO, L5L 1C6, CANADA.

TAMBURINI CHRISTIAN
LABORATOIRE DE MICROBIOLOGIE, GÉOCHIMIE ET ECOLOGIE MARINES (LMGEM), UMR 6117, CNRS – INSUE, UNIVERSITÉ LA MÉDITERRANÉE, COM, CAMPUS DE LUMINY, CASE 907, 13288 MARSEILLE, CEDEX 9, FRANCE.

TEODORANI MASSIMO
INAF - ISTITUTO DI RADIOASTRONOMIA, RADIOTELESCOPI DI MEDICINA, VIA FIORENTINA- 40060 VILLAFONTANA (BO), ITALY.

TRIGO-RODRÍGUEZ JOSEP M.
ASTROBIOLOGY CENTER, INSTITUTE OF GEOPHYSICS AND PLANETARY PHYSICS, UNIVERSITY OF CALIFORNIA AT LOS-ANGELES, LOS ANGELES, CA, USA.

UMIEL NAKDIMON
DIVISION OF ORNAMENTAL HORTICULTURE, THE INSTITUTE OF HORTICULTURE, AGRICULTURAL RESEARCH ORGANIZATION, THE VOLCANI CENTER, BET-DAGAN, ISRAEL.

WATERS DEBRA A.
DEPARTMENT OF OCEANOGRAPHY AND COASTAL SCIENCES, SCHOOL OF THE COAST AND ENVIRONMENT. LOUISIANA STATE UNIVERSITY, BATON ROUGE, LA 70803 USA.

ZALAR POLONA
UNIVERSITY OF LJUBLJANA, BIOTECHNICAL FACULTY, DEPARTMENT BIOLOGY, VEČNA POT 111, 1000 LJUBLJANA, SLOVENIA.

PART 1:
LIFE AS WE KNOW IT

Biodiversity and Extremophiles

Seckbach
Oren
Chapman, Waters
Ross
Oren
Bonch-Osmolorskaya
Gunde-Cimerman
Tamburini

Biodata of **Joseph Seckbach**, author of the chapter ***"Extremophiles"* and *"Introduction"* as well as *"What do will call Life"* (with co-authors: Raulin, Oren, Kolb and Chela-Flores.)**

Professor Joseph Seckbach is the initiator and chief editor of ***Cellular Origins, Life in Extreme Habitats and Astrobiology* (COLE)** book series, and author of several chapters and ***Introductions*** in this series. He earned his Ph.D. from the University of Chicago, Chicago, IL. (1965) and spent his postdoctoral years in the Division of Biology at Caltech (Pasadena, CA, USA). Then he headed at the University of California at Los Angeles (UCLA) a team for searching for extraterrestrial life. Dr. Seckbach has been appointed to the faculty of the Hebrew University (Jerusalem, Israel), performed algal research, and taught biological courses. He spent his sabbatical periods in Tübingen (Germany), UCLA and Harvard University, and served at Louisiana State University (LSU) (1997-1998) as the first selected occupant [of the John P. Laborde endowed] Chair for the Louisiana Sea Grant and Technology transfer, and as a visiting Professor in the Department of Life Sciences at LSU (Baton Rouge, LA, USA).

Among his publications are books, scientific articles concerning plant ferritin (phytoferritin), cellular evolution, acidothermophilic algae, and life in extreme environments. He also edited and translated several popular books. Dr. Seckbach is the co-author (with R. Ikan) of the ***Chemistry Lexicon*** (1991, 1999) published by Deveer (Tel Aviv) and of other volumes. He is also the co-editor of the *Proceeding of* ***Endocytobiology VII Conference*** (Freiburg, Germany, 1999) and the Proceedings of the ***Algae and Extreme Environments*** meeting (Trebon, Czech Republic, 2001) His recent interest is in the field of enigmatic microorganisms and life in extreme environments.

E-mail: **seckbach@huji.ac.il**

J. Seckbach (ed.), Life as We Know It, 3-20.

THE EXTREMOPHILES:
Diversity of Life Environments

JOSEPH SECKBACH
P.O.B. 1132 Efrat, 90435, Israel

1. Introduction

Life in various forms appears almost all over Earth, and most organisms live in environments considered "normal" from our anthropocentric point of view. There are other organisms, however, that are exceptional, living in very harsh conditions – they are termed extremophiles. The extremophiles live in extraordinary habitats; some of them are not able to grow in more moderate conditions or even to tolerate them for short periods. Such harsh environments are lethal for most other organisms. The majority of known extremophiles are unicellular – members of the domains Archaea and the Bacteria; there are also a few members of the Eukaryotes (protists and multicellular forms) among these exceptional organisms. The study of extremophiles and their habitats increases our understanding of the limits of life and the nature of the first organisms on early Earth, and in addition gives us an insight into the possibilities of extraterrestrial life: one of the goals of the field of astrobiology (DesMarais et al. 2003).

Astrobiology, an interdisciplinary research field, is the new scientific branch of studies that includes the origin of life, various parameters of terrestrial environmental conditions and their inhabitants (extremophiles), astronomy, astrophysics and the possibilities for extraterrestrial life. In essence, astrobiology is all encompassing. From comparison of the physical and atmospheric conditions on some celestial bodies with the inhospitable niches and environments on Earth, we find that some extremophiles could serve as candidates or analogues for extraterrestrial life in the Solar System and beyond. Thus, the study and understanding of extremophiles is valuable for astrobiology. Among the current celestial places suitable as candidates for finding past and perhaps present life are Mars, the Jovian satellites (i.e., Europa, Callisto, and Ganymede), and – for exploring initial stages of life as they may have occurred on young Earth – Saturn's moon Titan.

2. General Overview on the Extremophiles

Most extremophiles are (micro)organisms that thrive in habitats which for other terrestrial life-forms are intolerably hostile or even lethal environments. They are classified (see below) according to the conditions in which they grow as

TABLE 1. The extremophiles on Earth and possibly on extraterrestrial bodies

Environmental parameter	Type	Location and Conditions	Examples
Temperature	Psychrophiles Thermophiles Hyperthermophiles	-20°C - <15°C 50°C - 80°C 80°C - >100°C	*Psychrobacter*, *Polaromonas*, cyanobacteria, diatoms and icy green algae, some insects and *Cyanidium*, Prokaryotes, *Matsigocladus, Pyrolobus fumarii* (113°C)
pH	Acidophiles Alkaliphiles	pH 0 - 4 some hot springs pH >9 in African soda lakes	*Picrophilus oshimae, Cyanidium caldarium, Dunaliella acidophila, Natronobacterium, Spirulina* sp., fungi
Salinity	Halophiles	In salt lakes (neutral, and alkaline pH). In salt solutions up to saturation. Synthesis of glycerol to balance the external osmotic pressure.	Cyanobacteria, diatoms, flagellates, *Dunaliella; and Halobacteriaceae.*
Desiccation	Xerophiles	Anhydrobiotic condition Low water activity	Prokaryotes, extreme halophiles, endoliths, fungi and lichens (a symbioitic system), nematodes.
Hydrostatic pressure	Barophiles Piezophiles	Weight lovers Pressure lovers	Prokaryotic strain MT41 tolerates >100 MPa.
Radiation, Ultraviolet and ionizing radiation		High level of Radiation	Cyanobacteria, desert plants *Deinococcus radiodurans*
Thriving in organic Solvents and under various atmospheres	CO_2 gas Organic toxic solutions	Thriving at pure gas solutions of damaging agents.	Cyanidiaceae (grown under pure CO_2)
Gravity	Hyper-gravity Hypo-gravity	>1 g < 1 g	Such microorganisms are not known on Earth (this factor may be related to extraterrestrial environments).

Based on Oren and Seckbach (2001), Seckbach and Oren (2004), and other sources.

thermophiles, hyperthermophiles, psychrophiles, halophiles, acidophiles, alkaliphiles, barophiles, etc. (Table 1). In addition to the various extremophilic factors one may add that microbial spores can withstand a variety of different hostile conditions in their dormant stage for long periods of time. There are some limits in long-term dormancy, namely, thermal decomposition and ionizing radiation (McKay, 2000). Extremophiles thrive in hot niches, ice, salt solutions, acid and alkaline media; some may grow in toxic waste, organic solvents, heavy metals while extremophiles are found as well in several other habitats that were previously considered inhospitable for life and toxic for organisms (see Table 1). especially if nutritional extremophily as discussed below is added.

Many of the extremophiles cope with more than one environmental parameter, and may be termed as multi- or poly-extremophiles (Rothschild and Mancinelli, 2001). For example, the unicellular rhodophytan *Cyanidium caldarium* (Seckbach, 1994, 1999; Seckbach and Walsh, 1999) and its cohorts are thermo-acidophiles growing in an acid media (pH 0.5 to 3.5) at elevated temperatures (maximal temperature 57°C); they tolerate salinity range from 3 to 10% (Pinto et al., 1994), 1 N H_2SO_4 (Allen, 1959) and thrive under pure CO_2 (Seckbach, 1994). Likewise the barophiles/piezophiles, the weight/pressure lovers are also either thermophiles, living near hydrothermal vents at the bottom of the oceans, or psychrophiles (cold lovers) in the depth of the seas and oceans.

Finally, the metabolism of many prokaryotes themselves renders them "extreme" in comparison to their metabolically limited eukaryotic relatives (Gaidos et al., 1999; Nealson, 2001; Nealson and Berelson. 2003; Nealson and Conrad, 1999). Thus, organisms that are otherwise quite normal, living in hospitable places with regard to the variables mentioned above, become extreme for their remarkable abilities to be nourished by the inedible and breathe the unbreathable (Table II). While eukaryotes metabolize a limited range of carbon sources (usually things that can be easily converted into glucose, acetyl-coA, or other compounds that feed into the TCA cycle, the prokaryotes metabolize a range of carbon compounds ranging from one carbon (methane, methanol, carbon monoxide) to complex polycyclic compounds. This distinction, however, pales in comparison to the diversity with regard to inorganic substrates. The ability to use inorganic energy sources is, as far as is known, a property of the prokaryotic world – microbes utilizing all manner of inorganic energy sources (Table II): things that are totally unavailable to the eukaryotic world. Similarly, many prokaryotes are able to live very well in the absence of molecular oxygen, using a wide array of other electron acceptors as "oxygen substitutes" (Table II). This process of anaerobic respiration is occasionally found among a few fungal species that can use nitrate, but for the most part is a prokaryotic trait. This being said, it should be stressed that the prokaryotic world is vast, and not all organisms do all these things. In fact, the ability to perform many of these reactions is often an identifying trait used to classify certain groups of prokaryotes. The other point of interest is that under extreme conditions (i.e., Temperature, pH, salinity), it is often the case that oxygen or organic nutrients become limiting, thus making this nutritional extremophily a commonly encountered adaptation in harsh environments (Nealson and Stahl 1997).

Among the extremophiles are representatives of all domains of life [Prokarya (Archaea, Bacteria) and Eukarya]. Most extremophiles are prokaryotes (Archaea and Bacteria), but this group also includes eukaryotic cells (Roberts, 1999). Among the Eukaryotes are protists (e.g., algae, fungi, protozoa) and multicellular forms living in some of these harsh habitats. Some extremophiles are relegated to a particular habitat and are not able to survive milder conditions, while others are more flexible and can grow also in "normal" environments.

A number of these drastic environments may represent the primeval conditions that existed for the pioneering microbes evolving during the early periods in young Earth. For example, hyperthermophiles live in very warm or hot conditions and obtain their energy by heterotrophic and/or chemotrophic (with inorganic chemicals) metabolic pathways, while thermophiles of less extreme temperatures (i.e., below 74°C) may even be photosynthetic. The ability of thermophiles and hyperthermophiles to cope with such extreme environments may agree with the idea proposed by several scientists that the first organisms evolved in warm zones. The thermophiles are also placed at the roots of the archaeal branch in the tree of life based on molecular sequences. They may have evolved in the primordial soup of little ponds, in hot springs, inside rocks or even at the hydrothermal vents at the openings found around the black smokers at the bottom of the oceans. The hydrothermal vents are located at mid-ocean ridges where plates spread apart and spew out hot water containing

TABLE 2. Nutritional Extremophiles

(1)Energy Sources **(2) Oxidants**	**Metabolic types of Organisms**	**Environments**	**Group(s) of organisms**
(1) Complex Organic polymers Cellulose Lignin chitin	Heterotrophs[a]	Many	Bacteria Archaea Fungi
(1) Inorganics H_2 H_2S Fe(II) CO CH_4 CH_3OH NH_3	Lithotrophs Aerobic[b] Anaerobic[c]	Redox Interfaces[d] Hydrothermal Vents[e] Layered Systems[f]	Bacteria Archaea
(1) Light Anaerobic Rhodopsin	Photosynthetic Bacteria[d] Halophiles[f]		Bacteria[e] Archaea[g]
(2) none	Fermentors	Many	Fungi Bacteria Archaea
(2) Inorganics CO_2 $SO_4^{=}$ Fe(III) NO_3^- Mn(IV)	Anaerobic Respirers	Redox Interfaces Deep Subsurface And vents	Bacteria Archaea Fungi[h]

a. The ability to metabolize complex organics is characteristic of many groups of prokaryotes, and with the exception of many fungi, not a general property of eukaryotes. b. Many lithotrophs are obligate oxygen utilizers. c. In almost every case of inorganic energy sources, microbes, or microbial consortia exist that can oxidize the substrate in the absence of molecular oxygen. d. The photosynthetic bacteria use inorganic electron donors, and produce no oxygen during photosynthesis. e. This process is found only in the Bacteria. f. Rhodopsin-based photosynthesis involves the use of light to produce a membrane gradient directly, and use. this to synthesize ATP. g. It is now known to occur in both Bacteria and archaeal groups, but not eukaryotes. h. A few fungi are known that are capable of anaerobic respiration of nitrate. Other than these, this is a property that is restricted to the prokaryotes.

dissolved minerals providing material and energy for a rich diversity of extremophilic life. Other extremophiles may have evolved at great depth of beneath the surface environments. Living in such places, these microorganisms (if they existed) may have been protected from the impact of the asteroid bombardment on the planet and shielded from high doses of UV irradiation during the early stages of Earth's evolution. Huge extraterrestrial asteroids and boulders from space hit young Earth between 4.5 Ga and 3.8 Ga (bya = billion years ago).

Actually the first appearance of Life is still a highly controversial and argued area. Much before 3.8 Ga, the Earth was so hot that life as we know it would not have been possible. As the heat was coming from the interior, there was not much escape available by going down. Around 4 Ga, it is thought to have cooled enough to have a habitable solid crust. It is generally accepted that the origin of Life took place when the impact ceased as early as 3.5 to 3.8 Ga (Glausiusz, 2004), while others reported upon evidence for Life on Earth before 3.8 Ga (Mojzsis et al., 1996; Rosing, M.T. and Frei, R. 2004).

The conditions below Earth's surface were suitable then for thermophiles deriving their energy via chemoautotrophic oxidation of inorganic chemicals. Among these microbes where perhaps CO fermenting microbes like those that grow with CO as the sole source of carbon and energy, generating hydrogen as a by-product (Table II), or methanotrophic or acetogenic microbes, capable of growth on hydrogen and CO_2 (Table II). In contrast, other extremophiles, like *Acidithiobacillus thiooxidans,* which oxidizes sulfur or sulfide to sulfate, and *A. ferrooxidans*, which oxidizes ferrous to the ferric iron, as well as others that oxidize hydrogen to water, nitrites to nitrates, are incompatible with earliest life because of their requirement for oxygen, or oxygen-derived oxidants. The fact that there is almost no limit in the habitats on Earth brings the astrobiologists to propose that extremophiles may serve as models for candidates for extraterrestrial life in planets and moons of the Solar System and beyond. In fact the whole field of extremophily has modified to some degree, the concept of the habitable zone, as originally enunciated by Kasting et al. (1993). Clearly, knowing what we know about extremophiles on Earth allows one to expand the habitable zones to take into account subsurface and subglacial zones not previously imagined in the planetary models. Some enzymes from extremophiles are tolerant of extreme parameters such as, heat, salinity, acidity, alkalinity, and high or low pressers. They have been introduced recently into applied biotechnology systems and industry. We can find the extremophiles and their cellular productions in food and dairy products, health food, medical and pharmaceutical lines, antibiotics, detergents, in fermentation processes, and so on.

Many extremophiles thrive under anaerobic conditions, as outlined below, and many of the most extreme environments are anoxic or contain very low levels of oxygen. These environments may be similar to the primordial Earth, where the atmosphere was anoxic (Nealson and Conrad, 1999; Knoll, 2003). The oxidative atmosphere evolved after the appearance of the oxygenic photosynthesis process by cyanoprokaryota (Avers, 1989; Shapiro, 1986). While this chapter covers most extremophiles, it does not deal with microorganisms thriving under anaerobic conditions. The anoxic atmospheres appeared as the original natural environment, while the oxygenated atmosphere is a secondary evolution.

Additional reviews on extremophiles can be found in various chapters in the series *Cellular Origins, Life in Extreme Habitats and Astrobiology*; see: www.springer online.com and in Horikoshi and Grant (1998), Roberts (1999), Rothschild and Mancinelli (2001), and Seckbach (1999, 2000a, 2000b, 2003) and Seckbach and Walsh (1999).

2.1. LIVING AT THE END RANGES OF TEMPERATURE

2.1.1. Thermophiles and Hyperthermophiles

The organisms living in warm or hot habitats are the thermophiles and hyperthermophiles, respectively. On the tree of life these microbes are placed close to the root of the prokaryotic position. It has been reported that among organisms there are some differences in heat tolerance (Table 1). For example, the upper temperature limit for eukaryotes is 62°C, while the thermo-cyanoprokaryotes can occur at < 70°C, thermophilic bacteria may thrive at 70°C to <100°C and hyperthermophilic Archaea can be detected at a maximum of 113°C observed in the archaeon *Pyrolobus fumarii* (Blöchl et al., 1997, see also Seckbach and Walsh, 1999; and tables therein). This latter temperature point is considered the highest current temperature for life and might support the source where life has originated. Under special treatments and conditions the maximum temperature of life and the edge of stability can reach higher values, such as 121-125°C (Blöchl et al., 1997, Kashefi and Lovley, 2003; Seckbach and Oren, 2004). Such archaeal anaerobes gain their energy from reducing Fe^{+3} into Fe^{+2}, forming other minerals, such as magnetite, in the process.

The thermophiles thrive in geothermic sources, such as in hot springs, hot geysers, or at hydrothermal vents at the ocean floors. Black smokers on the sea floor expel hot water at 350°C, gases, minerals, and other materials from the inside of the Earth. One of the microbes associated with the walls of black smokers is *Methanopyrus* that produces methane (CH_4) gas. For more information see chapter by Bonch-Osmolovskaya in this volume).

2.1.2 . The Psychrophiles like Cold and Chilly Environments

Life cannot grow where liquid water does not exist. On the opposite scale from the spectrum of the thermophiles are the psychrophiles (cold-loving organisms) that grow at very cold ranges of temperature, even below freezing some of these may live in icy layers. In snow and ice layers one can find eukaryotic diatoms, green algae (e.g., *Chlamydomonas nivalis*) that "paint" snow with their pigmentation (green, red, orange). Psychrophilic microorganisms are in permafrost of Siberia, in the Arctic zones, and in the icy depth close to the Vostok Lake (Antarctica). The Antarctic sea is is populated with Bacteria, algae, and diatoms. In Antarctica the endolithic microbes – among them Bacteria, cyanobacteria, algae, and the symbiotic association-lichens live a few millimeters below the surface of rocks at very cold temperatures. In addition, at the ocean floor are worms dwelling in methane-enriched environments called cold seeps.

The psychrophiles can tolerate severe cold environments at the level of –20°C (Rivkina et al., 2000; Junge et al., 2004). Furthermore, lichens survive Antarctica winters at –65°C. Many psychrophiles are also barophiles (poly-extremophiles) and such microorganisms possibly could survive on Mars when the air temperature reaches minus 10°C during daytime throughout the warmest part of the year.

Recently, a psychrophile in the group *Psychrobacter* was isolated from a subsurface water lens in Siberia, that was capable of growth at –10°C, and respiration at –20°C. The minus 20°C conditions were maintained because of the very high salt concentration in the water lens. Such a situation demonstrates again the issue of multiple extremes so often encountered in the real world (Bakermans et al., 2003). For additional reference see chapter by Gunde-Cimerman in this volume.

2.2. MICROORGANISMS LIVING AT THE EDGE OF pH RANGES

2.2.1. Acidophiles

Acidophilic members are among all three domains of life. Some may be phototrophic others heterotrophic – all require acid media. Seckbach (2000a) presented a table of the lower pH limits for different groups of organisms. Most of the thermo-acidophiles are prokaryotes (Archaea and Bacteria) with the exception of some fungi and the algal members. Their habitats are in warm, hot (such as occur at volcanic fields, sulfatara fields, hot springs), ambient and cold areas. Some acidic habitats are formed by microorganisms, such as after oxidation of elemental sulfur or ferrous ions and other compounds by bacteria (e.g., *Acidithiobacillus thiooxidans*, *A. ferrooxidans*), which leads to accumulation of sulfuric acid and ferric ions that precipitate $Fe(OH)_3$ with release of protons in various environments, according to the following equation:

$$2\ Fe^{2+} + 0.5\ O_2 + 5\ H_2O \rightarrow 2\ Fe(OH)_3 + 4\ H^+$$

Thus, acidic environments can be expected wherever reduced sulfur compounds (and especially iron sulfides) are exposed to molecular oxygen, allowing the production of excess protons and sulfuric acid.

Several poly-extremophiles thrive in various ranges of pH, from milder solutions to very acidic media. Among the prokaryotes are thermo-acidic organisms (Oren and Seckbach, 2001; Seckbach and Walsh, 1999; Weiss Bizzoco, 1999; see tables therein). The archaeal hyperacidophilic *Picrophilus oshimae* and *Picrophilus torridus* grow down to pH –0.06, the lowest level of pH known to support life. Another archaeon, *Thermoplasma acidophilum,* grows at pH 1.8-2 (min. at 0.4). The eukaryotic thermoacidophilic red alga *Cyanidium caldarium* thrives at pH 0-3.5 (Seckbach, 1994, 1999) and tolerates 1 N sulfuric acid (Allen, 1959; Seckbach, 1994, 2000b). Some green algae are acidophiles, such as *Dunaliella acidophila* at pH 0.5-3.0 (Pick, 1999), *Chlamydomonas acidophila* (pH 1-2) or *Euglena mutabilis* (pH 1-5). Among diatoms,

Pinnularia braunii grows at pH 0-4. Cyanobacteria are known to thrive at neutral to alkaline pH. Yet there are some exceptions, such as the *Oscillatoria* and *Spirulina* discovered in Bavarian lakes at acidic condition measured a pH of 0.06 to 2.9 (Steinberg et al., 1998) but not shown to grow at these levels (Castenholz, personal communication). Hundreds of lakes in Sweden and Canada showed cyanobacteria present even in the most acidic lakes (Steinberg et al., 1998) which are very oligotrophic. There are three species of fungi growing at pH near 0 (Schleper et al., 1995): *Acontium cylatium*, *Cephalosporium* spp., and *Trichosporon ceribriae*, which grow at pH–0.06 while their optimum growth is at pH ~0.5.

These acidic pH values are only external to the cells; the intracellular cyto-solution (cytosol) keeps its pH level around neutral pH (Beardall and Entwisle, 1984; Oren and Seckbach, 2001; Seckbach, 2000a). The cell membrane rejects and prevents the H^+ from entering the cell by using a variety of mechanisms for also pumping the protons out. Strong cyto-proton pump or low proton membrane permeability regulates the H^+ ions and stops them from entering the plasmalemma. The internal neutral pH has to protect some cyto-constituents like DNA, ATP, and chlorophyll from the acidification damage. Sasaki et al. (1999) reported about intracellular extract solutions from four species of Dictyotales to be highly acidic (at pH 0.5 to 0.9), similar values of pH were observed in some species of Desmarestiales (both groups are brown algae, Phaeophyta). Their intracellular pH measured to be 0.5-0.9, because their vacuoles contain sulfuric acid. Apparent these low values are due to the vacuolar content, but the true cytoplasmic pH is supposedly much higher.

Further data on acidophiles are reported in volumes 1, 2, and 6 of Cellular Origins, Life in Extreme Habitats and Astrobiology (COLE) and elsewhere. See website: www.spingeronline.com

2.2.2. Alkaliphiles

In some soils and soda lakes in east Africa and under dry climates (such as in Mojave desert), the soil is very alkaline. There are microorganisms (Archaea, Bacteria, cyanobacteria, and protists) that prefer to live on the high scale of the pH (alkaline levels). Such high pH levels were found together with high salt solutions. In highly alkaline-saline lakes, as in Mono Lake, CA, are diatoms, and green algae such as *Nanochloris, Chlamydomonas* and *Dunaliella.* The cyanobacterium *Spirulina platensis* is an obligate alkaliphile, and reaches a high population at pH values of 11 and above. *Plectonema nostocorum* has been reported to grow up to pH 13 which is considered the highest level at which life has been found. The diatoms are prominent components in the biota of many alkaline lakes.

These alkaliphiles keep the high alkali pH outside the cell and pump ions in and out the cells (similar to the acidophiles) to keep a lower and milder intracellular milieu which might reach up pH 9 (Horikoshi, 1998; see table 6.2 there) while the external pH level is much higher. For further data on alkaliphiles see Horikoshi (1998), Oren and Seckbach (2001), Horikoshi and Grant (1998), and Seckbach and Oren (2004).

Another setting for alkaline environments is that in which anaerobic water interacts with ultramafic (high Mg and Fe content) magmatic rocks, producing hydrogen and ultra high pH waters as by-products (Sleep et al., 2004). The pH of such environments is often around 12, and the conditions so harsh that cell numbers are quite low. Little is known of the microbial flora of such environments, although a marine isolate was recently described from a serpentinization site in the deep sea (Takai et al., 2005).

Given that the early Earth may have been very active volcanically, such serpentinization environments may have been very common, and thus todays sites may serve as interesting analogs of early Earth. Furthermore, one can imagine that as Mars lost its water, evaporitic environments similar to todays alkaline hypersaline lakes and their inhabitants may serve as analogues for ancient microbial habitats on Mars (see below).

2.3. HALOPHILES – THE SALT LOVERS

In the Solar System our Earth is considered the sole water-endowed planet, some 71% of its area are seas and oceans, mostly salt water. Similar water bodies have not yet been proven to exist outside the Solar System. The dawn of the appearance of the first organisms is unknown, although it is generally agreed that life was present and leaving signs of its past existence from 3.5 Ga onwards (Knoll, 2003). They were most probably simple cells similar to our present day Prokaryotes and ruled on Earth for 2–2.5 Ga (until the evolution of the eukaryotic cell). In this "Prokaryotic world" the Archaea and Bacteria were the sole organisms on Earth and they occupied almost every niche on the surface and subsurface of Earth.

While salt appears to be a universal requirement of life, those organisms adapted to very high salt concentrations (halophiles) are regarded as extreme. Brines are the medium for halophilic microorganisms, and such environments occur ubiquitously on Earth. Among the hypersaline-tolerant halophiles are photosynthetic green algae (such as *Dunaliella* species), diatoms, and others. The enigmatic Cyanidiaceae, a group of thermo-acidophilic red algae, including *Cyanidioschyzon merolae*, *Cyanidium caldarium* and *Galdieria sulphuraria* tolerate NaCl solutions from 3 to 10% of NaCl (Pinto et al., 1994; Seckbach, 1994, and see above). Among the halophiles at different ranges of pH are also cyanoprokaryota, green and purple Bacteria, sulfur-reducing Bacteria, and methanogenic Archaea.

Moderate halophiles live in the marine environments (containing water of 3% salt), others thrive in inland lakes within bodies of salty water, while the hyper-halophiles grow in very high concentration salt solutions. Rod shaped halobacteria strains can grow up to concentrations of ~30% salt, levels that are present in the Dead Sea (Israel-Jordan) and in the Great Salt Lake (Utah, USA). Other may tolerate saturated salt solutions of a wide variety of different salts. As the water evaporates from salty aqueous bodies, high concentrations of salt solutions remain, then only halophiles will remain and obtain these areas for themselves. Vreeland et al., (2000) revived bacteria that were enclosed in salt crystals for 250 million years. Assuming this report is authentic and fully reliable, we may have a new record of halophilic life in dry condition for vast periods.

The halophiles have to react against the high gradient gaps in osmotic pressure of their external solution vs. their internal cytosol medium. They have to avoid plasmolysis – the process where cell aqueous solution escapes due to the hypertonic external solution that causes the cytoplasm shrinkage. To avoid the escape (plasmolysis) of cell liquid, the halophiles balance these concentration differences of the solutes inside and outside the cell membrane by synthesizing intracellular organic compounds (e.g., glycerol, betaines, sucrose, K^+ ions, etc.). These compounds are synthesized and/or accumulated in response to osmotic stress at high salinity. Halophiles may regulate the movement of solutions by other means as well (Oren, 2002). The cell membrane may also regulate the osmotic pressure on both sides of the plasmalemma.

Some Antarctic lakes are salty at different concentration and some can reach saline concentration close to the Dead Sea level. Such saline lakes do not freeze during the Antarctic winter even when the temperature declines to –50°C. For further literature on the halophiles, see Oren (1999, 2002) and the chapters in this volume.

As a final point, it should be noted that many hypersaline environments are so because they are also evaporative sites (because of high temperature), where nutrients can be limiting, and where pH can become very high. Thus, many alkaline lakes have proven to be a source of extremely interesting multi-extremophiles (for salinity, temperature, pH, and often nutrients see Sorokin et al., 2002). More information is available at the chapter of Oren (Life at high concentration of salt) in this volume.

2.4. XEROPHILES – THE DRY LIVING AND DESICCATED ORGANISMS

Some microorganisms have the ability to grow in low water activity and tolerate water stress. Low water availability can cause high concentration of salt and other solutes. Deserts and dry valleys of Antarctica or the Siberian permafrost support a variety of microorganisms in dry conditions. A survey (Davis, 1972) has shown that cyanobacteria (e.g., *Nostoc* sp.) survive desiccation for a record of 107 years. The diatom *Nitzschia palea*, and the chlorophytes *Pleurococcus* sp. and *Cystococcus* sp. were shown to be viable after 98 years of dry storage in a herbarium.

Spores have a low, if any, rate of metabolism, and remain in long-term periods of dormancy before being revived. Well known are the cases claiming of viable bacterial spores resuscitated from the stomachs of insects imbedded in amber dated at 25–40 million years (Cano and Borucki, 1995). If the report (Vreeland et al., 2000) of the restoration to life of bacteria that had been encapsulated for 250 million years in a salt crystal is proven to be true, it will provide strong evidence in support of the long period during which life can be persevered.

2.5. ORGANISMS THRIVING AT HIGH PRESSURE

Barophiles (weight loving) and piezophiles (pressure loving) microorganisms both in the subsurface of Earth and in the depths of the ocean are obligated to live heterotrophically, chemoautotrophically or chemolithotrophically, in darkness and at high pressure. This elevated hydrostatic pressure can reach over a thousand atmospheres (at the bottom of the ocean). For every 10 m depth in the water column the hydrostatic pressure increases by about 1 atmosphere (0.1 MPa). The pressure at the deepest point in the ocean (at the bottom of Mariana Trench near the Philippines) is about 110 MPa. The barophilic bacteria are poly-extremophiles being psychrophilic in addition. Among the Archaea are the baro-thermophiles observed in the hydrothermal vents; *Themococcus barophilus* grows at temperatures up to 100°C. Our knowledge about barophiles must be expanded for us to understand more about this fascinating group. For additional discussion on deep-sea Bacteria, see Yayanos (2000) and Tamburini's paper in this volume.

2.6. MORE EXTREME PARAMETERS FOR MICROORGANISMS

There are other factors of extreme environments that support life under various atmospheres. Some algae thrive under pure CO_2 (Seckbach, 1994), and other micro-organisms grow under ammonia (Siegel, 1999) or on hydrogen gas (e.g., the methanogenic Archaea). Microorganisms can grow in environments with chemicals, heavy metals, organic solvents that are lethal for most other organisms. Another parameter is gravity, which might be relevant more to life in space or in celestial places, rather than to earthling organisms.

3. Astrobiology

In recent years the new multidisciplinary field of astrobiology has been established. Currently, in this area there are numerous books, publication of a plethora of articles in various journals and proceedings of conferences, meetings, workshops and a great deal of research being conducted (DesMarais et al., 2003).

The wide variety of astrobiology-related disciplines include prebiotic chemistry, astronomy, astrophysics, origin of life, microfossil records, evolution, diversity of microorganisms, extremophiles, geobiology, molecular biology, extremophiles and extra-terrestrial possibilities of life. It also includes life in the Universe, discovery of new stars, planets and satellites and the search for extraterrestrial intelligent life (SETI). Astrobiology has been referred also as bioastronomy, exobiology, extrater-restrial life and the Russian term cosmobiology.

Many of the extremophiles that we have described here are ubiquitous on Earth, and may be considered as candidates for inhabiting exterritoral sites. Furthermore, if there is life to based on silicon instead of carbon as well in non-aqueous solvents, (thoughts far beyond the scope of this chapter!) then life may perhaps exist on planets and satellites. The most popular extraterrestrial bodies within the Solar System with regard to the possibility of providing habitats for life (as we know it), or perhaps for the precursors of life, are Mars and the Jovian moons such as Europa. Titan, the satellite of Saturn, may serve as a prebiotic laboratory resembling the Earth's prebiotic periods. Europa has a very cold environment (minus 162°C) and from its image one can distinguish huge cracked shells of ice floating and moving on water. These ice sheets may cover and float over an ocean of warm water originating from Europa's hydrothermal vents that may reach a couple of hundred degrees Celsius. One problem such a habitat would have, of course, is that with kilometers of ice over this ocean, there would be no ability to generate oxidants photosynthetically, and the ecosystem would have limited capacity for metabolism in the absence of good oxidants (Gaidos et al., 1999). Thus the issue of life here may well be one of nutritional, rather than physical or chemical, extremes. The surface appearance of Europa resembles Alaska and Antarctica with their frozen seas. If Prokaryotes can survive such severe conditions on Earth, why not there in the heavenly bodies? One can ask the question: How unlikely is it that life does not exist elsewhere?

Similar to Jupiter, Saturn is a gas planet surrounded by hydrogen and helium (with traces of methane and ammonia). Its largest moon Titan is covered by a reddish fog over the icy continent and contains organic molecules including hydrocarbon oceans (while no oxygen is present in its atmosphere). Similar prebiotic organic chemistry may have led to life on Earth. Recently a special issue of Science magazine covered the Cassini reports on the atmosphere and surface of Titan (Science, 2005).

3.1. MARS

Our knowledge about Mars has increased but still not complete. Wherever there is free liquid water and other certain elements, there should be life. Water allows the solution chemistry that is essential for enzymatic reactions, natural processes and other functions in all cells. There are still debates among the astrobiologists whether Mars contains briny water on the upper layers of the surface (Möhlmann, 2004) or whether the Martian water is within layers present deeper in the subsurface. Such water might be accumulated in salty oceans warmed up by volcanic sources, while it could be also located as hydrated minerals. On Mars, similar to Earth, liquid water should go together with life since it is vital for cell metabolism, enzymatic reactions, processes as photosynthesis, etc.

The recently discovered methane in the Martian atmosphere provides encouragement for the scientists searching for life (Formisano et al., 2004). This detection of Martian methane may not originated from chemical reaction but rather it is believed that it may have evolved from biological sources

The images of Mars, as well as abundant geological data from the recent MER expedition (Lemmon et al., 2004) indicate that in the past it was warmer and wetter, with abundant water in rivers, canyons, lakes and perhaps even large oceans. The bulk of the Martian water might be tied up in the subsurface, either as permafrost in the craters in the polar latitudes, or in deep aquifers as reservoirs of water (McKay and Marinova, 2001). Possibly some forms of life may have flourished there in the past. Perhaps spores from the ancient "green Martian Planet" are still present as dormant spores and might be revived when environmental conditions will be improved by terraforming (Graham, 2004). If Martian surface conditions are at present non-inhabitable for living forms, the subsurface (lithologies in subterranean areas) may offer the only dwelling place for extant life forms. Therefore, finding of liquid water on Mars is vital for the presence of contemporary life on this planet.

Today the red planet is dry, cold (–50°C to +12°C); the Martian salty solutions are lowering the freezing point of its water so that microbes might grow in liquid water. To this end, it is interesting to note that the recently discovered *Psychrobacter cryopegella* (Bakermans et al., 2003) might well be able to survive on certain environments on Mars, as pointed out recently Jakosky et al. (2003).

Mars also has a very low atmospheric pressure (~6 to 10 mbars) and the oxygen in the atmosphere is very low (<0.15%). Its atmosphere contains 95% CO_2, 2.7% N_2, 1.6% Ar and other minute ingredients. The soil of Mars contains mineral ions such as: Na^+, Mg^{2+}, Ca^{2+}, Cl^- and K^+. Organisms require sulfur, potassium, sodium, chloride, potassium, iron, and trace elements, all thought to be present on Mars.

Since it would be difficult for organisms to grow at the surface regions in the cryosphere and under UV radiation of Mars, if life does exist on Mars today, it would by necessity be located in the subsurface where warmer conditions and perhaps liquid ground water may be present. The thin Martian ozone layer does not significantly block UV rays, but the radiations of very short wavelengths penetrate all the way to the surface. If some species of Bacteria or Archaea may overcome Martian conditions (similar to those Bacteria existing in permafrost – see Möhlmann 2004), then this planet might perhaps harbor psychrophilic life. Also, microbial dormant spores may have life preserved there in 3.5 to 4 Ga old sediments that were left over from when the planet was warmer and had liquid water. Earthling Arctic and Antarctic life forms wait for the icy coverage to melt into liquid water (conditions that might be similar on Mars).

Mars (and other planets' moons) might carry life in hydrothermal lakes under the surface or in ice-rich regions similar to the way microbial life is harbored in the icy layer and in subsurface warm water which is assumed to exist in Vostok Lake (Antarctica).

One imagines that as the early water evaporated from Mars, a high concentration of dissolved salt was left so that only halophiles remained and had the planet for themselves. By understanding osmophily (halophily) and extremophiles on Earth we may detect life that lives or has lived on other planets. Such osmophiles are best suited for life in brines and evaporates. Halophiles may possibly occur in the Martian icy soil, as they thrive in the hypersaline ponds of Antarctica. Therefore, a preparatory search for extant and extinct life of Mars should be a focused on osmophiles or on their fossilized remnants. All the answers about Martian life can be found only after samples from Mars will be brought to Earth. See also chapter by Mancinelli in *Adaptation to Life at High Salt Concentration* (volume 9 of Cellular Origins, Life in Extreme Habitats and Astrobiology series).

Recently there have been some proposals to create a greenhouse effect for warming Mars towards restoring habitable conditions on this planet. Such terraforming ideas for Mars have been published (Friedmann and Ocampo-Friedmann, 1995; McKay and Marinova, 2001). Only in time will we know if such proposals are realistic or just wishful dreams.

Is the future export or transfer of microbes from Earth to Mars practical? Some scientists believe that the red planet is currently too cold and dry to allow the survival of any terrestrial organisms. Terrestrial life forms could perhaps survive there only after warming up the planet to a minimal level. Such warming of Mars can take place by using artificial greenhouse gases (Marinova et al., 2004). Mars appears to have sufficient water, if warmed, to form large bodies, and to once again be capable of harboring life. There might be a possibility that genetic engineering may "create" and convert organisms and equip them with special features to resist extraterrestrial environments such as those existing on Mars. If introduced on Mars they may assist in the terraforming there.

4. Acknowledgements

I thank Prof. Richard Castenholz, Prof. David J. Chapman, Dr. Rocco Mancinelli, and Dr. Christopher McKay for their efficient reviews on this chapter and for providing practical suggestions. Last but least, lots of thanks are due to Professor Kenneth Nealson (USC, Los-Angeles) who proofread this manuscript and suggested several practical additions and variations for improving this chapter.

5. References

Avers, C.J. (1989) *Process and Pattern in Evolution.* Oxford University Press, Oxford.

Allen, M.B. (1959) Studies with *Cyanidium caldarium*, an anomalously pigmented chlorophyte. Arch. f. Mikrobiol. **32,** 270-277.

Bakermans, C., Tsapin, A.I., Souza-Egipsy, V., Glichinsky, D.A. and Nealson, K.H. (2003) Reproduction and metabolism at –10°C of bacteria isolated from Siberian permafrost. Envion. Microbiol. **5,** 321-326.

Beardall, J. and Entwisle, L. (1984) Internal pH of the obligate acidophile *Cyanidium caldarium* Geitler (Rhodophyta?) Phycologia **23,** 397-399.

Blöchl, E., Rachel, R., Burggraf, S., Hafenbradl, D., Jannasch, H.W. and Stetter, K.O. *Pyrolobus funarii,* gen. and sp. nov. represents a novel group of archaea, extending the upper temperature limit for life to 113°C. Extremophiles **1,** 14021.

Cano, R.J. and Borucki, M. (1995) Revival and identification of bacterial spores in 25- to 40-million-year-old Dominican amber. Science **268,** 1060-1064.

Davis, J.S. (1972) Survival records in the algae, and the survival role of certain algal pigments, fat and mucilaginous substances. The Biologist **54,** 52-93.

DesMarais, D.J., Allamandola, L. J., Benner, S., Boss, A.P., Deamer, D. Falkowski, P.G., Farmer, J.D., Hedges, S.B., Jakowsky, B.M., Knoll, A., H. Liskowsky, D.R., Meadows, V.S., Meyer, M.A., Pilcher, C.B., Nealson, K.H., Spormann, A.M., Trent, J.D., Turner, W.W., Woolfe, N.J., and Yorke, H.W. (2003) Astrobiology Roadmap, Roadmap for NASA's Office of Space Science.. Astrobiol. J. **3,** 219-235.

Formisano, V., Atreya, S., Encrenaz, T, Ignatiev, N. and Giuranna, M. (2004) Detection of methane in the atmosphere of Mars. Science **306,** 1758-1761.

Friedmann, E.I. and Ocampo-Friedmann, R. (1995) A primitive cyanobacterium as pioneer microorganism for terraforming, Mars. Adv. Space Res. **15,** 243-246.

Gaidos, E.J., Nealson, K.H. and Kirschvink, J.L. (1999) Life in ice-covered oceans. Science **284,** 1631-1633.

Glausiusz, J. (2004) When Life was Hell, Discover **23,** 10.

Graham, J.M. (2004) The biological terraforming of Mars: planetary ecosynthesis as ecological succession on a global scale. Astrobiology **4,** 168-195.
Horikoshi, K. (1998) Alkaliphiles, In: K. Horikoshi and W.D. Grant (eds.), *Extremophiles: Microbial Life in Extreme Environments.* Wiley-Liss, New York, pp. 155-179.
Horikoshi, K. and Grant W.D. (eds.) (1998) *Extremophiles. Microbial life in Extreme Environments.* Wiley-Liss, New York.
Jakosky, B.M., Nealson, K.H., Bakermans, I.C., Ley, R.E. and Mellon, M.T. (2003) Sub-freezing activity of microorganisms and the potential habitability of Mars' polar regions. Astrobiol. J. **3,** 343-350.
Junge, K., Eicken, H. and Deming J.W. (2004) Bacterial activity at –2 to –20°C in Arctic wintertime sea ice. Appl. Environ. Microbiol. **70,** 550-557.
Kashefi, K. and Lovley, D.R. (2003) Extending the upper temperature limit for life. Science **301,** 934.
Kasting, J.F., Whitmire, D.P. and Reynolds, R.T. (1993) Habitable zones around main sequence stars. Icarus, **101,** 108-124.
Knoll, A.H. (2003) Life on a Young Planet. Princeton University Press.
Lemmon, M.T., Wolff, M.J. Smith, M.D. Clancy R.T., Banfield, D. Landis, G.A. Ghosh A., Smith, P H. Spanovich, N. Whitney, B. Whelley, P. Greeley, R. Thompson, S. Bell, J.F. and Squyres. III, S.W. (2004) Atmospheric imaging results from the Mars Exploration Rovers: Spirit and Opportunity. Sci. **306** (Issue 5702), 1753-1756.
Marinova, M.M., McKay, C.P. and Hahimoto, H. (2005) Radiative-convective model of warming Mars with artificaial greenhouse gases. J. Geophys. Res. **110,** xxxx .
McKay, Ch.P. (2000) Life in the cold and dry limits: Earth, Mars and beyond, In: J. Seckbach (ed.), *Journey to Diverse Microbial Worlds.* Kluwer Academic Publishers, Dordrecht, pp. 377-386.
McKay, Ch.P. and Marinova M.M. (2001) The physics, biology, and environmental ethics of making Mars habitable. Astrobiology **1,** 89-109.
Möhlmann, D.T.F. (2004) Adsorption water-related potential chemical and biological processes in the upper Martian surface. Astrobiology **4,** 280.
Mojzsis, S., Arrenius, G., McKeegan, K.D., Harrison , T.M., Nuitman, A.P. and Friend C.R.L. (1996) Evidence for life on Earth before 3,800 Million years ago. Nature **384,** 55-59.
Nealson, K.H. (2001) Searching for life in the universe: Lessons from the Earth. Ann. New York Acad. Sci. **950,** 241-258.
Nealson, K.H. and Berelson. (2003) Layered microbial communities and the searh for life in the universe. Geomicrobiol. J. **20,** 451-462.
Nealson, K.H. and Conrad, P.G. (1999) Life, past, present and future. Phil. Trans. Roy. Soc. Lond. B **354,** 1-17.
Nealson, K.H. and Stahl. (1997) Microorganisms and biogeochemical cycles: what can we learn from layered microbial communities? In: J. Banfield and K.H. Nealson (eds.) *Geomicrobiology,* Mineral. Soc. America, Washington, D.C. pp. 1-34.
Oren, A. (ed.) (1999) *Microbiology and Biogeochemistry of Hypersaline Environments.* CRC Press, Boca Raton.
Oren, A. (2002) *Halophilic Microorganism and their Environments.* volume 5 of Cellular Origins, Life in Extreme Habitats and Astrobiology (series editor J. Seckbach) Kluwer Academic Publishers, Dordrecht.
Oren, A. and Seckbach, J. (2001) Oxygenic photosynthesis microorganisms in extreme environments, In: J. Elster, J. Seckbach, W.F. Vincent and O. Lhotsky (eds.), *Algae and Extreme Environments.* Nova Hedwigia **Beiheft 123,** pp. 13-31.
Pick, U. (1999) *Dunaliella acidophila* – a most extreme acidophilic alga, In: J. Seckbach (ed.), *Enigmatic Microorganisms and Life in Extreme Environments.* Kluwer Academic Publishers, Dordrecht, pp. 467-478.
Pinto, G., Albertano, P. and Pollio, A. (2004) Italy's contribution to the systematics of *Cyanidium caldarium "sensu lato"*, In: J. Seckbach (ed.), *Evolutionary Pathways and Enigmatic Algae: Cyanidium caldarium (Rhodophyta) and Related Cells.* Kluwer Academic Publishers, Dordrecht, pp. 157-166.
Rivkina, E.M., Friedmann, E.I., McKay, C.P. and Glichinsky, D. (2000) Metabolic activity of permafrost bacteria below the freezing point. Appl. Environ. Microbiol. **66,** 3230-3233.
Roberts, D.M.L. (1999) Eukaryotic cells under extreme conditions, In: J. Seckbach (ed.), *Enigmatic Microorganisms and Life in Extreme Environments.* Kluwer Academic Publishers, Dordrecht, pp. 163-173.

Rosing, M.T. and Frei, R. (2004). U-rich Archean sea-floor sediments from Greenland – indications of >3700 Ma oxygenic photosynthesis. Earth Planet. Sci. Lett. **217,** 237-244.

Rothschild, L.J. and Mancinelli, R.L. (2001) Life in extreme environments. Nature **409,** 1092-1101.

Sasaki, H., Kataoka, H., Kamiya, M. and Kawai, H. (1999) Accumulation of sulfuric acid in Dictyotales (Phaeophyceae): Taxonomic distribution and ion chromatography of cell extracts. J. Phycol. **35,** 732-739.

Schleper, C., Pühler, G., Kühlmorgen, B. and Zillig, W. (1995) Life at extremely low pH. Nature **375,** 741-742.

Science magazine 13 May (2005) [DOI: 10.1126/science.308.5724.968] **308 (Issue 5724)** 968-995.

Seckbach, J. (ed.) (1994) *Evolutionary Pathways and Enigmatic Algae: Cyanidium caldarium (Rhodophyta) and Related Cells.* Kluwer Academic Publishers, Dordrecht.

Seckbach, J. (1999) The Cyanidiophyceae: Hot spring algae, In: J. Seckbach (ed.), *Enigmatic Microorganisms and Life in Extreme Environments.* Kluwer Academic Publishers, Dordrecht, pp. 425-435.

Seckbach, J. (2000a) Acidophilic microorganisms, In: J. Seckbach (ed.), *Journey to Diverse Microbial Worlds.* Kluwer Academic Publishers, Dordrecht, pp. 107-116.

Seckbach, J. (2000b) Extremophiles as models for extraterrestrial life, In: G. Lemarchand and K. Meech (eds.), *A New Era in Bioastronomy.* ASP Conference Series **213,** 379-386.

Seckbach, J. (2003) Life on the edge and beyond, In: W.E. Krumbein, D.M. Paterson and G.A. Zavarzin (eds.), *Fossil and Recent Biofilms: A Natural History of Life on Earth.* Kluwer Academic Publishers, Dordrecht, pp. 401-414.

Seckbach, J. and Walsh, M.M. (1999) Ubiquity of life as related to certain extremophiles: Mini review on Cyanidia, In: E. Wagner, J. Normann, H. Greppin, J.H.P. Hackstein, R.G. Herrmann, K.V. Kowallik, H.E.A. Schenk and J. Seckbach (eds.), *From Symbiosis to Eukaryotism: Endocytobiology VII.* University of Geneva. pp. 85-104.

Seckbach, J. and Oren, A. (2004) Introduction to the extremophiles, In: J. Seckbach (ed.), *Origins: Genesis, Evolution and Diversity of Life.* Kluwer Academic Publishers, Dordrecht, pp. 373-393.

Shapiro, R. (1986) *Origins: A Skeptic's Guide to the Creation of Life on Earth.* Summit Books, NY.

Siegel, B.Z. (1999) Growth of organisms in ammonia: *Kakabekia*, a microbial enigma, In: J. Seckbach (ed.), *Enigmatic Microorganisms and Life in Extreme Environments.* Kluwer Academic Publishers, Dordrecht, pp. 89-96.

Sleep, N.H., Meibom, A., Fridriksson Th., Coleman R.G. and Bird D.K. (2004) H_2-rich fluids from serpentinization: Geochemical and biotic implications. PNAS **101,** 12818-12823.

Sorokin, D.Y., Gorlenko, V.M., Tourova, T.P., Kalganova, T.V., Tsapin, A.I. , Nealson, K.H. and Kuenen, G.J. (2002) *Thioalkalimicrobium cyclicum* sp. Nov., and *Thioalkalivibrio jannaschii* sp. Nov., new species of haloalkaliphilic obligately chemolithoautotrophic sulfur-oxidizing bacteria from hypersaline alkaline Mono Lake (California). Int. J. Syst. Env. Microbiol. **52,** 913-920.

Steinberg, C.E.W., Schäfer, H. and Beisker, W. (1998) Do acid-tolerant cyanobacteria exist? Acta Hydrochim. Hydrobiol. **26,** 13-16.

Takai, K., Moyer, C.L., Miyazaki, M. Nogi, Y., Hirayama, H., Nealson, K.H., and Horikoshi, K. (2005) *Marinobacter alkaliphilus* sp. nov., a novel alkaliphilic bacterium isolated from subseafloor alkaline serpentine muid from Ocean Drilling Program Site 1200 at South Chamorro Seamount, Mariana Forearc. Extremophiles **9,** 17-27.

Vreeland, R.H., Rosenzweig, W.D. and Powers, D.W. (2000) Isolation of a 250 million-year old halotolerant bacterium from a primary salt crystal. Nature **407,** 897-900.

Weiss, Bizzoco, R.L. (1999) Life in the extreme: New prokaryotes living in high temperature low pH environments, In: J. Seckbach (ed.), *Enigmatic Microorganisms and Life in Extreme Environments.* Kluwer Academic Publishers, Dordrecht, pp. 305-314.

Yayanos, A.A. (2000) Deep sea bacteria, In: J. Seckbach (ed.), *Journey to Diverse Microbial Worlds: Adaptation to Exotic Environments.* Kluwer Academic Publishers, Dordrecht, pp. 161-174.

Biodata of **Aharon Oren** author of "***Prokaryotic Life***" of "***Life In Saline and Hypersaline Environments***" and co-author (with Seckbach et al.) of "***What do we call Life? A brief Outlook on Life***"

Aharon Oren is professor of microbial ecology at the Institute of Life Sciences, The Hebrew University of Jerusalem. He earned his Ph.D. at the Hebrew University in 1978, and spent a post-doctoral period at the University of Illinois at Urbana-Champaign. His research interests are in ecology, physiology, biochemistry and taxonomy of halophilic microorganisms and the microbiology of hypersaline environments. He has authored over 250 scientific publications, and he has written or edited five books, including "*Microbiology and Biogeochemistry of Hypersaline Environments*" (CRC Press, Boca Raton, 1999) and "*Halophilic Microorganisms and their Environments*", published in 2002 by Kluwer Scientific Publishers as Volume 5 in the series *"Cellular Origin and Life in Extreme Habitats and Astrobiology*.

E-mail: **orena@cc.huji.ac.il**

J. Seckbach (ed), Life as We Know It, 21-35.

PROKARYOTIC LIFE

AHARON OREN
The Institute of Life Sciences, and The Moshe Shilo Minerva Centre for Marine Biogeochemistry, The Hebrew University of Jerusalem, Jerusalem 91904, Israel

1. Introduction

When in the 1940s and the 1950s electron microscopes became available for the study of the structure of cells, it became clear that the organisms living on Earth could be divided into two groups on the basis of the organization of their cells. The first, designated eukaryotes, includes the plant and animal world, the algae, the fungi, ciliate, flagellate and amoeboid protozoa. All these organisms share a common basic structure of their cells. Each cell contains a nucleus that contains the cell's DNA, and a double membrane surrounds this nucleus. During cell division the DNA condenses into chromosomes, which are distributed evenly over the daughter cells. The eukaryotic cell contains other membrane-surrounded organelles such as mitochondria (the sites of energy generation during respiration) and chloroplasts (the sites of conversion of light energy into cellular energy and reducing power during photosynthesis in green plants and algae), as well as a complex system of intracellular membranes. The second type of cells, found in all bacteria, shows a far simpler structure in the electron microscope. The cells are generally much smaller than those of the eukaryotes, the genetic material is not surrounded by a membrane, and organelles such as mitochondria and chloroplasts are absent. Such cells that lack a true nucleus are named prokaryotes (Madigan et al., 2003). The prokaryotes encompass all bacteria, as well as the cyanobacteria – a group previously classified in the plant world as 'blue-green algae' (Stanier and van Niel, 1962).

The prokaryotes are by far the most abundant living organisms on planet Earth. It may also be argued that they are also the most important organisms: without the prokaryotes the degradation of organic material and the global biogeochemical cycles of matter would quickly come to a halt. However, due to their small size their presence is not directly obvious in most cases.

This short chapter highlights a number of aspects of the life of the prokaryotes. It first examines the structure of the prokaryotic cell and its mode of growth and multiplication. It then discusses the abundance of prokaryotes in nature, made possible by their ability to colonize nearly every environment and to adapt to varied environmental conditions, including extreme environments hostile to even the hardiest eukaryotes. Afterwards it explores the phylogenetic and physiological diversity of the prokaryotes and the role they play in the biogeochemical cycles of carbon, nitrogen, sulfur and other elements. Finally, a few paragraphs are devoted to different types of interrelationships between the prokaryotes and the eukaryotic world. These include symbiotic associations as well as pathogen-host interactions.

2. What are prokaryotes?

The first obvious difference between a typical prokaryote cell and a typical eukaryote is its size. Prokaryotic cells generally are shaped like small spheres, small straight or curved rods, or may be spirally shaped, and most are in the range of 0.5 μm to 5 μm in size (1 μm = 0.001 mm). Most eukaryotic cells are much larger than that: most plant and animal cells are between 10 and 100 μm in diameter. There are a few exceptions: we know giant bacteria that measure about 500 μm (= 0.5 mm), and the smallest eukaryotic protozoa are less than 1 μm in size.

Comparison of the ultrastructure of a typical prokaryote with that of a typical eukaryote shows the far greater complexity of the latter. The prokaryotic cell lacks organelles such as mitochondria, chloroplasts, Golgi apparatus, endoplasmatic reticulum and other intracellular membrane structures found in eukaryotes. No membrane-surrounded nucleus is found in any prokaryote. Instead, the prokaryotic 'chromosome' is a single circular strand of DNA that contains between less than 500,000 base pairs (*Nanoarchaeum equitans*) to over 9 million base pairs (*Myxococcus xanthus* and *Nostoc punctiforme*, the largest known prokaryotic genomes) (Madigan et al., 2003). This DNA is visible in thin sections in the electron microscope as a distinct area within the cytoplasm.

The cell wall of prokaryotes has some unique chemical properties. Most contain peptidoglycan, a complex polymer that consists of chains of two unusual sugar molecules, cross-linked by bridges formed by short peptides (chains of amino acids). It is interesting to note that penicillin and related antibiotics interfere with the biosynthesis of these cross-links. As no eukaryotic cell has peptidoglycan in its wall, such antibiotics specifically attack prokaryotes. Presence of peptidoglycan is limited to the domain Bacteria; a modified form of this wall polymer is found in some members of the domain Archaea (see section 6).

Many prokaryotes are motile. A cell may possess one or more flagella, long screw-shaped appendages that rotate like propellers and push the cell forward in the medium. This mode of motility is very different from that of ciliate and flagellate eukaryotic protozoa, in which cilia or flagella are much thicker, have a more complex ultrastructure, and show a flexing rather than a rotating movement.

3. Growth and multiplication of prokaryotes

Prokaryotes generally live as single cells. In some species the daughter cells do not separate following cell division, so that filaments, chains or clusters may be formed. Thus, bacteria of the genus *Streptococcus* typically occur in chains, while the genus *Staphylococcus* is characterized by growth in irregular clusters. Growth as multicellular filaments is common among the cyanobacteria.

Prokaryotic cells divide by binary fission. When sufficient nutrients are available, the cell grows, it elongates, and a cross-wall is formed, and finally the two daughter cells are separated. The single chromosome has replicated as well, so that each daughter cell obtains an identical copy of the genetic material. Structurally the process is much simpler than the complex phenomena occurring during the mitosis of the eukaryotic cell.

Although the life cycle of the prokaryotes does not include a sexual reproduction stage, genetic material is extensively exchanged between prokaryotes by processes such as transformation (uptake of DNA from the environment), conjugation (transfer of genetic material during physical contact and fusion between two related prokaryotes), and transduction (introduction of fragments of foreign DNA during infection with bacteriophages – viruses that attack prokaryotes).

The process of cell growth and division can be extremely rapid in many prokaryotes. The high activity of the cells is enabled by the large surface/volume ratio, which allows efficient and rapid uptake of nutrients through the cell membrane. *Escherichia coli*, the facultative anaerobic inhabitant of the human intestine which has become the most extensively studied of all prokaryotic organisms, can complete its cell cycle and divide within 20 minutes. That is also one of the reasons why this organism has become such a popular object of research. *E. coli* is by no means the fastest grower of all: the marine *Vibrio natriegens* reportedly has a generation time as short as 9.8 minutes under optimal conditions (Eagon, 1962). Not all bacteria grow that fast: most species of prokaryotes studied have doubling times of a number of hours, and species that need several days to grow and divide to yield two new cells have been described as well.

Such rapid growth, however, is seldom found in nature, and if it occurs at all, it is sustained for a short period only. A simple calculation makes this clear: if a single *E. coli* cell would grow with a generation time of 20 minutes for two days, the mass of the cells produced would many times exceed the mass of planet Earth! In reality such organisms will only seldom find the large amounts of nutrients and other favorable conditions to sustain such rapid growth. In its natural habitat, the human colon, *E. coli* has an average doubling time of 12 hours rather than the minimum of 20 minutes in laboratory culture. In the aquatic environment and in the soil, most bacteria probably experience starvation conditions, to resume growth as soon as sufficient nutrients may become available (defined sometimes as a 'feast and famine' existence) (Koch, 1971).

4. Where do we find prokaryotes?

There are hardly any environments on Earth where prokaryotes do not occur in high numbers. Even in the most oligotrophic (low in nutrients) ocean water bacteria are present at number between 10^4-10^5 per milliliter, and in more eutrophic (nutrient-rich) seawater and freshwaters prokaryote numbers, as determined by microscopic enumeration, range between 10^5 and 10^6 per milliliter or higher. Agricultural soils harbor around 10^8-10^9 bacteria per gram. Even higher densities are found in specialized environments such as the large intestine of mammals and the rumen of ruminant animals (Whitman et al., 1998). Here counts of 10^{10}-10^{11} per milliliter and higher are common. Not all these organisms have been cultivated and studied in the laboratory. In fact, most species of prokaryotes that exist in nature are still awaiting isolation and characterization (see section 5).

Thanks to their tremendous ability to utilize a wide variety of energy sources and nutrients (see section 8) and their adaptability to diverse environmental conditions, prokaryotes can make a living under conditions under which no eukaryote can survive. The prokaryotes are thus the true specialists for the colonization of extreme environments. Many environments on Earth are characterized by extremes of temperature, pH, salt concentration, and pressure. Although there are also a few eukaryotes that can withstand and even thrive at environmental extremes, the prokaryotes are in most cases superior in their ability to grow under extreme conditions (Horikoshi and Grant, 1991; Madigan, 2000; Madigan and Marrs, 1997; Madigan and Oren, 1997; Seckbach, 1999, 2000). The highest temperature conducive for eukaryotic life is about 62°C. Some thermophilic prokaryote, especially species of the domain Archaea (see section 6) will only start to grow at temperatures above 80°C. The species *Pyrolobus fumarii* has its optimal growth temperature at 106°C and grows up to 113°C. A recently discovered organism designated as 'strain 121', was even reported to grow up to 121°C (Kashefi and Lovley, 2003). *Polaromonas vacuolata* is one of the most psychrophilic (cold-loving) organisms, with a minimum growth temperature of 0°C and an optimum of 4°C. The barophilic or piezophilic (pressure-loving) strain MT41 isolated from the depths of the Mariana Trench grows optimally at a hydrostatic pressure of 700 atmospheres, and is still able to grow at pressures over 1000 atmospheres such as encountered at the deepest places in the ocean. The extreme halophile *Halobacterium salinarum*, which can be isolated from salt lakes, salted hides, and salted fish, requires at least 250 g/l salt and grows up to NaCl saturation. Growth at the extremes of pH extends from the extremely acidophilic thermophile *Picrophilus oshimae* from acidic hot springs, which grows optimally at pH 0.7 and can still maintain growth at a pH of -0.06, to the alkaliphilic halophile *Natronobacterium gregoryi*, which lives in soda lakes, has its growth optimum at pH 10, and tolerates up to pH 12 (Seckbach and Oren, 2004).

Many extremophiles are optimally adapted to life under multiple stress factors (Rothschild and Mancinelli, 2001). Thus we know 'polyextremophiles' that grow simultaneously at high temperature and at low pH (the above-mentioned *Picrophilus* is an example), at high pH and high salt concentration (e.g., *Natronobacterium*), at high temperature and high pH, at high temperature and high pressure, and at low temperature and high pressure.

Extensive information is available about the mechanisms enabling microorganisms to withstand such environmental extremes. In some cases the organism establishes barriers between the hostile outside world and the cell's cytoplasm. The intracellular enzymatic machinery can then function without special adaptations. Thus, acidophilic and alkaliphilic prokaryotes all maintain a near-neutral intracellular pH. Many halophilic and halotolerant prokaryotes that live in high salt environments keep the salt out. They accumulate intracellular organic solutes to provide the necessary osmotic balance, and no special salt-adapted intracellular enzymes are needed (see the chapter on "Life in saline and hypersaline environments" by Oren in this volume). In all these cases the cytoplasmic membrane performs a key function in maintaining the intracellular conditions 'non-extreme'. In other cases, however, the cells are unable to prevent the extreme conditions from affecting the cytoplasm. All intracellular components of thermophilic and psychrophilic microorganisms need to function at the temperature of the environment, and barophiles cannot maintain a low hydrostatic pressure inside the cells. Some halophiles (salt lovers) accumulate KCl at concentrations at least as high as the salt concentration in their medium, and their intracellular enzymes are modified so that they are functional at high salt, at the expense of their ability to function at low salt (Seckbach and Oren, 2004).

5. How many species of prokaryotes are there on Earth?

To answer the question how many species of prokaryotes are found on our planet, one needs first to agree on a species definition. This is not straightforward for organisms that do not reproduce sexually and show little morphological diversity. Today the species for prokaryotes is mainly defined on the basis of DNA similarity. Comparison of certain genes, notably the gene that encodes the RNA of the small subunit of the ribosome ('16S rRNA') has proven useful in the characterization of prokaryote species (Oren, 2004; Rosselló-Mora and Amann, 2001).

The number of prokaryotic species whose names have been validly published (as of September 25, 2004) is 6,693; these belong to 1,272 genera and 171 families (for updates see http://www.bacterio.cict.fr) (Euzéby, 1997). Extensive information on the phylogenetic as well as on the physiological diversity within the prokaryotes can be found in two handbooks. The first is 'Bergey's Manual of Systematic Bacteriology' (Garrity, 2001, Vol. 1), a manual that provides a systematic treatise on the Archaea and certain groups of the Bacteria, using a 16S rRNA-based classification. Volume 2 is to appear in 2005, and subsequent volumes will follow soon. The second is 'The Prokaryotes' (Dworkin et al., 1999). This on-line resource provides in-depth information on all aspects of prokaryote biology.

The number of prokaryote species described is very small compared to the around a million recognized insect species and the more than 260,000 vascular plants described and named. One reason is that most prokaryote species have not yet been brought into culture. The 6,693 prokaryote species described thus form only a small fraction of the true number of species on Earth. That not all bacteria can easily be cultured has been known for several decades. Comparison of the number of colonies developing on agar plates inoculated with natural water or soil samples with the number of bacteria recognizable microscopically in the same sample generally shows a discrepancy of several orders of magnitude. Thus, only between 0.001-0.1 percent of the bacteria from samples of oligotrophic seawater typically yield colonies. For soil samples the value is about 0.3%, and for freshwater and aquatic sediment samples values of about 0.25% have been reported (Amann et al., 1995). This phenomenon is known as 'the great plate count anomaly' (Staley and Konopka, 1985).

Use of techniques developed in the last 10-15 years derived from molecular biology has led to the insight that most of these non-cultivated cells belong to yet-to-be described species. When the genes encoding 16S rRNA are isolated from the DNA extracted from the biomass collected from the environment, sequenced, and compared with the database of 16S rRNA gene sequences of the established prokaryote species or environmental clones, it only seldom happens that a sequence recovered from the environment fully matches that of a known species or sequence (Amann et al., 1995). The extent of the differences generally suggests that we deal with new species, new genera, and often even with new families, orders, classes, and even phyla, of which no cultured representatives are known as yet (Hugenholtz et al., 1998). It is now a major challenge to obtain these organisms in culture and to study them, so that their properties and their function in the ecosystem will become clear. Based on the large number of novel 16S rRNA gene sequences it may be concluded that the prokaryote species described and named thus far represent no more than 1-2%, and possibly even no more than 0.1%, of the true number of existing bacterial species (Amann et al., 1995; DeLong and Pace, 2001; Dykhuizen, 1998).

Use of such molecular techniques has led to the realization that we often even do not know the identity of the numerically dominant prokaryotes in common environments such as water and soil. Thus, characterization of 16S rRNA gene sequences from marine bacterioplankton has led to the recognition that an organism still designated with the code name SAR11 (Giovannoni et al., 1990) is the most abundant bacterium in the oceans, and probably even the most abundant bacterium on Earth. This organism accounts for a third and in some regions as much as half of the cells present in sea surface waters and nearly a fifth of the cells present in the mesopelagic (middle area of open ocean) zone, and it contributes about 12% of the total marine prokaryote biomass (Morris et al., 2002). The organisms that harbors the SAR11 16S rRNA sequence is a small (0.4 x 0.2 μm) cell. The SAR11 bacterium has only recently been brought into culture, but cell densities obtained are very low, and little is known as yet about its properties (Rappé et al., 2002). SAR11 is just one case in which the truly important prokaryotic components of an ecosystem were identified only in recent years. There are many more such cases, and major efforts are now being made to obtain more information on the identity of such formerly uncultured organisms to elucidate their role in the ecosystem (Oren, 2004).

6. Phylogenetic diversity among prokaryotes: the domains Bacteria and Archaea

Prokaryotic life has existed on our planet for over 3.6 billion years. The first eukaryotes appeared only about 1.5-2.2 billion years ago. It is therefore understandable that the prokaryotic world has had accordingly much time to develop a great extent of diversity, enabling the colonization of all available niches that enable organisms to make a living (see also section 5).

How great the true diversity within the prokaryote world really is, was established after molecular sequence information became available from the 1970s onwards. The use of small subunit ribosomal RNA (16S rRNA for the prokaryotes,

18S rRNA for the eukaryotes) as a molecular chronometer to determine the phylogeny of the different life forms has revolutionized our views on prokaryote evolution. Analysis of ribosomal RNA sequence data has made it clear that rather than being a single, phylogenetically coherent group as was assumed until that time, the prokaryotes consist of two fundamentally different groups. Therefore, the forms of life inhabiting Earth are now divided into three domains, two of prokaryotes – the Bacteria and the Archaea, and one of eukaryotes – the Eukarya (Pace, 1997; Woese, 1987; Woese et al., 1990). The prokaryotes thus represent a large proportion of life's genetic diversity.

Many members of the domain Archaea that have been isolated in culture are extremophiles, adapted to high temperatures, whether or not combined with low pH, or high salt concentrations (see section 4). The Archaea also include the methanogens that inhabit anaerobic environments. Morphologically and (electron) microscopically the Archaea resemble the Bacteria. However, the Archaea do not have peptidoglycan in their cell wall (some have the structurally related pseudopeptidoglycan), and they differ from the Bacteria in many other aspects, including the structure of their membrane lipids, their sensitivity to antibiotics, and others.

It is becoming increasingly clear that the domain Archaea not only consists of hyperthermophiles, thermoacidophiles, halophiles and haloalkaliphiles. Molecular characterization of the prokaryotes based on 16S rRNA sequence analysis (see also section 5) has shown that different, yet uncultured Archaea represent a significant fraction of the bacterioplankton in the open sea (DeLong, 1992; Fuhrman et al., 1992; Karner et al., 2001). According to the current estimate, the world oceans harbor $1.3x10^{28}$ archaeal and $3.1x10^{28}$ bacterial cells. Unfortunately, none of these common types of marine Archaea has yet been cultured, and therefore hardly anything is known about their physiology and their possible function in the ocean ecosystem.

7. The diverse modes of dissimilatory metabolism of the prokaryotes drive the global biogeochemial cycles

Besides their small size and their simple structural organization, the most prominent difference between the prokaryotic and the eukaryotic world is the tremendous metabolic diversity displayed by the prokaryotes (Lengeler et al., 1999; Madigan et al., 2003). Within the eukaryotic world we find only two main modes of energy generation: aerobic respiration and (in plants and algae) oxygenic photosynthesis (use of water as electron donor with formation of molecular oxygen). One may even argue that aerobic respiration of animals and plants and oxygenic photosynthesis of plants and algae are in fact performed not by the eukaryotic organisms themselves but by their mitochondria (respiration) and chloroplasts (photosynthesis) which evolutionary may have derived from the domain Bacteria of the prokaryotes (see section 8). Other types of metabolism, such as the alcohol fermentation of yeasts, are rarely found among other eukaryotes.

Within the world of the prokaryotes we find a much greater variety of metabolic types, both among the Bacteria and the Archaea. This diversity is expressed at the level of assimilatory metabolism (the nutrients taken up to be incorporated to build new cell material) as well as at the level of dissimilatory metabolism (production of the energy necessary to support the processes of life). The variety of carbon and nitrogen-containing compounds that can be assimilated and support growth of prokaryotes is almost unlimited. The prokaryote world can use many compounds that cannot be exploited by eukaryotes, including substances that are toxic to eukaryotes. With regard to energy generation, it is not exaggerated to state that nearly any process that is thermodynamically feasible can be utilized by at least some types of prokaryotes in their dissimilatory metabolism. Light is used as energy source not only in 'plant-type' oxygenic photosynthesis (by members of the phylum Cyanobacteria), but in different modes of anoxygenic photosynthesis (photosynthesis without evolution of oxygen) as well. Here electron donors other than water supply the electrons for autotrophic CO_2 fixation. We thus know purple and green sulfur bacteria that oxidize sulfide and other reduced sulfur compounds in the light, while other bacterial phototrophs can use hydrogen or reduced iron as electron donors. Not all anoxygenic phototrophic bacteria lead an autotrophic mode of life: many use light as the energy source while taking up organic compounds as carbon source (photoheterotrophy). Different bacteriochlorophylls present as light-harvesting and reaction center pigments in anoxygenic phototrophs enable the use of near-infrared radiation of wavelengths between 700 and 1020 nm, wavelengths not used by eukaryotic phototrophs. Chlorophyll or bacteriochlorophyll-based photosynthesis has never yet been shown to occur in the archaeal domain, but some halophilic Archaea can use light energy to generate energy (in the form of a gradient of protons across the cell membrane). The pigment involved is bacteriorhodopsin, a protein that shows a high degree of resemblance with rhodopsin, the visual pigment of the human eye. Recently the same type of light energy utilization was documented in yet uncultured marine members of the domain Bacteria (Béjà et al., 2001).

Many heterotrophic prokaryotes obtain their energy from aerobic respiration, and some (e.g., representatives of the genus *Pseudomonas* and related organisms) are able to oxidize a tremendous variety of organic compounds that are not used by any eukaryote. Oxygen is by no means the only electron acceptor available for respiration: many prokaryotes use nitrate, sulfate, elemental sulfur, trivalent iron, and a range of additional oxidized compounds as electron acceptors in processes termed anaerobic respiration. In the absence of oxygen, nitrate is reduced to gaseous nitrogen ('denitrification'), and sulfide is formed anaerobically from sulfate – two major processes in the biogeochemical cycles of nitrogen and sulfur. When no suitable electron acceptor for respiration is available, fermentative prokaryotes can degrade complex organic material to simple organic fermentation products such as organic acids and alcohols and gain energy from the process. As stated above, yeasts and a few other eukaryotes have a limited potential for fermentative growth as well, but the range of fermentable substrates and the variety of the products formed is much greater among the prokaryotes.

Certain prokaryotes can live in the presence as well as in the absence of oxygen. Examples are *E. coli*, which inhabits the human colon and can switch from aerobic respiration to fermentation, and denitrifying bacteria. Others (e.g., methanogens, many fermentative bacteria) are obligate anaerobes for which oxygen is toxic.

A mode of life altogether unknown in the eukaryotic world but displayed by many prokaryotes is chemoautotrophy or chemolithotrophy. Here the energy necessary for autotrophic CO_2 fixation is generated by the oxidation of inorganic electron donors. In many cases, molecular oxygen is the electron acceptor. Thus, nitrification, in which ammonium ions are oxidized to nitrite and further to nitrate, is based on the action of two groups of aerobic chemoautotrophic bacteria. Chemoautotrophic prokaryotes also oxidize reduced sulfur compounds, molecular hydrogen, divalent iron, and other reduced inorganic compounds. Oxidation of sulfide or elemental sulfur can also be coupled to the reduction of nitrate to gaseous nitrogen. Another well-known case of anaerobic chemoautotrophic growth is the oxidation of molecular hydrogen with CO_2 as electron acceptor, one of the processes performed by the methanogenic Archaea.

No single prokaryote can perform all the above-mentioned processes, but some are markedly versatile indeed. Thus, bacteria of the genus *Rhodobacter* can grow photoheterotrophically in the light (using light as energy source and organic compounds as carbon source) or photoautotrophically (using molecular hydrogen or low concentrations of H_2S as electron donor). They can also grow chemoheterotrophically in the dark by oxidizing certain organic compounds with oxygen as electron acceptor, and in the absence of suitable electron acceptors they can live by fermentation. Moreover, they can grow chemoautotrophically by using hydrogen as electron donor and oxygen as electron acceptor. Such versatility may be advantageous in environments in which conditions frequently change. However, *Rhodobacter*-like organisms seldom develop to high densities, as more specialized bacteria tend to dominate in any single niche. Another versatile prokaryote is *E. coli*, which is generally grown in the laboratory as an aerobe, but can grow anaerobically as well, either leading a fermentative life as it does in its natural habitat, the human colon, or living by anaerobic respiration using compounds such as nitrate, dimethylsulfoxide, or fumarate as electron acceptor.

The biogeochemical cycles of elements such as carbon, nitrogen, and sulfur are based on the cooperation between different groups of prokaryotes. Organic matter, produced mainly by the photoautotrophs (higher plants, algae) is oxidized back to CO_2 either by aerobic respiration or, when oxygen becomes depleted, by anaerobic degradation processes in which nitrate or sulfate serve as electron acceptors (anaerobic respiration). In many cases, fermentative degradation of complex organic compounds to simpler organic molecules such as short-chain fatty acids and alcohols is an intermediate stage in the process. When no suitable electron acceptors are available, methane is produced besides CO_2. In oxic environments this methane can be oxidized to CO_2 by specialized methanotrophic members of the Bacteria. Anaerobic oxidation of methane is possible as well, in a recently discovered process in which sulfate serves as electron acceptor. This process yields little energy, and the small amount of energy generated is shared by two cooperating prokaryotes: a methane-oxidizing member of the Archaea and a sulfate-reducing member of the Bacteria (Boetius et al., 2000). Such cases of 'syntrophy', in which different types of prokaryotes cooperate to enable degradation of substrates, are very common during the anaerobic degradation of organic matter.

Many important steps in the nitrogen cycle are mediated entirely by prokaryotes. These include autotrophic nitrification, i.e., the oxidiation of ammonium ions via nitrite to nitrate (see above), denitrification, i.e., anaerobic respiration in which nitrate serves as electron acceptor and is reduced to gaseous nitrogen (N_2), and the reduction of nitrogen to yield ammonium ions (nitrogen fixation). Nitrogen fixation is essential to warrant the continued availability of nitrogen to biological processes. The process is energetically expensive, and is performed by a limited number of prokaryotes only, some of which live in symbiosis with higher plants and animals (see section 8). The recent discovery of anaerobic oxidation of ammonia to gaseous nitrogen with nitrite as electron acceptor (Jetten et al., 1998) shows that our understanding of the basic processes that drive the biogeochemical cycles on Earth may still be far from complete.

The prokaryotes also play a central role in the sulfur cycle which is based on oxidation and reduction reactions involving different forms of sulfur. Sulfate can serve as electron acceptor in anaerobic respiration when oxygen is not available. The sulfide formed can be reoxidized by anoxygenic phototrophic sulfur bacteria and/or by chemolithotrophic sulfur-oxidizing bacteria. The prokaryotes participate in a similar way in the biogeochemical cycles of iron, manganese, and other elements.

8. Interactions between prokaryotes and eukaryotes

During the approximately two billion years that prokaryotes and eukaryotes have coexisted on our planet, these two types organisms have not evolved independently, but they have often developed often very complex interactions.

When hearing the word 'bacteria', most people first of all think about nasty pathogens that cause unpleasant diseases (i.e., germs). Indeed, host-pathogen interactions are among the best known cases in which prokaryotes and eukaryotes interact. During millions of years of evolution the bacteria have learned to invade the bodies of humans, other animals and plants and to use these as a rich source of nutrients for growth. At the same time the eukaryotic hosts have evolved often extremely complex mechanisms to defend themselves to these attacks. However, it must also be stated that the prokaryotes have also become an integral part of the human and the animal body. The intestinal system is densely colonized by bacteria, to the extent that the human colon contains more bacterial cells than there are human cells in the body. These bacteria make a significant contribution to the digestion of food in the intestine. An even more dense and complex community of prokaryotes is found in the rumen of ruminant animals, and the activities of this community are absolutely essential for the ability of cows, sheep, and other ruminants to digest their feed. The animals alone cannot degrade cellulose, the main component of their feed, and they rely on the bacteria to break down cellulose to simpler compounds.

There are countless similar symbiotic associations between prokaryotes and eukaryotes (Seckbach, 2002). To give just one more well-known example: the symbiosis between bacteria of the genus *Rhizobium* and leguminous plants. The bacteria multiply within the root nodules of the plant, and there they fix gaseous nitrogen to ammonium ions which are used by the plant as a nitrogen source.

Finally it may be commented that it is most likely that the mitochondria and chloroplasts found as intracellular organelles within eukaryotic cells may have originated from prokaryotes that had entered a symbiotic association with primitive eukaryotic cells. Both mitochondria and chloroplasts have the size characteristic of prokaryotic cells, they divide by binary fission just as bacteria do, they contain their own DNA and their own ribosomes, which in size and structure are unlike those of the animal or plant cells that harbor the organelles, but instead resemble those of prokaryotes. On the basis of the nucleotide sequence of the 16S rRNA found in these organelles (see section 5) the plant chloroplast appears to be phylogenetically related to the cyanobacteria, which not surprising as both perform oxygenic photosynthesis. Similarly, the mitochondria are related to certain aerobic prokaryotes of the domain Bacteria.

9. Conclusions

As shown in the sections above, the prokaryotes are an extremely versatile group. Thanks to their small size, their fast growth rates and their metabolic versatility they rapidly colonize nearly any environment where suitable sources of energy, carbon and other nutrients are available. Nearly any process that is thermodynamically feasible will support growth of at least some types of prokaryotes. Such processes include conversions of organic compounds, whether or not coupled to the reduction of external electron acceptors, oxidation of reduced inorganic compounds, or use of light as energy source. All these processes can occur in the presence as well as in the absence of molecular oxygen.

During the past decades we have gained the insight that the less than seven thousand species described thus far of Bacteria and Archaea combined, represent only a small fraction of the true diversity within the prokaryote world. The realization that over 99 percent of the prokaryote species inhabiting our planet has not yet been isolated and characterized presents a major challenge to microbial ecologists, taxonomists and physiologists to fill the gaps in our knowledge and to provide an improved picture of the diversity within the prokaryotic world, representing a major part of diversity of life itself.

10. References

Amann, R.I., Ludwig, W., and Schleifer, K.-H. (1995) Phylogenetic identification and in situ detection of individual microbial cells without cultivation. Microbiol. Rev. **59,** 143-169.

Béjà, O., Arvind, L., Koonin, E.V., Suzuki, M.T., Hadd, A., Nguyen, L.P., Jovanovich, S., Gates, C.M., Feldman, R.A., Spudich, J.L., Spudich, E.N., and DeLong, E.F. (2001) Bacterial rhodopsin: evidence for a new type of photosynthesis in the ocean. Science **289,** 1902-1906.

Boetius, A., Ravenschlag, K., Schubert, C.J., Rickert, D., Widdel, F., Gieseke, A., Amann, R., Jørgensen, B.B., Witte, U., and Pfannkuche, O. (2000) A marine microbial consortium apparently mediating anaerobic oxidation of methane. Nature **407,** 623-626.

DeLong, E.F. (1992) Archaea in coastal marine environments. Proc. Natl. Acad. Sci. USA **89,** 5685-5689.

DeLong, E.F., and Pace, N.R. (2001) Environmental diversity of bacteria and archaea. Syst. Biol. **50,** 470-478.

Dworkin, M., Falkow, S., Rosenberg, E., Schleifer, K.-H., and Stackebrandt, E. (eds.) (1999) *The Prokaryotes: An Evolving Electronic Resource for the Microbiological Community*, 3rd ed., Springer-Verlag, New York, http://link.springer-ny.com/link/service/books/10125/.
Dykhuizen, D.E. (1998) Santa Rosalia revisited: Why are there so many species of bacteria? Antonie van Leeuwenhoek **73,** 25-33.
Eagon, R.G. (1962) *Pseudomonas natriegens*, a marine bacterium with a generation time of less than 10 minutes. J. Bacteriol. **83,** 736-737.
Euzéby, J. (1997) List of bacterial names with standing in nomenclature: a folder available on the internet (URL: http://www.bacterio.cict.fr). Int. J. Syst. Bacteriol. **47,** 590-592.
Fuhrman, J.A., McCallum, K., and Davis, A.A. (1992) Novel major archaebacterial group from marine plankton. Nature **356,** 148-149.
Garrity, G.M. (editor-in-chief) (2001) *Bergey's Manual of Systematic Bacteriology, 2nd ed., Vol. 1. The Archaea and the Deeply Branching and Phototrophic Bacteria* (D.R. Boone and R.W. Castenholz, eds). Springer-Verlag, New York.
Giovannoni, S.J., Britschgi, T.B., Moyer, C.L., and Field, K.G. (1990) Genetic diversity in Sargasso Sea bacterioplankton. Nature **345,** 60-63.
Horikoshi, K., and Grant, W.D. (eds.) (1991) *Superbugs. Microorganisms in Extreme Environments.* Japan Scientific Societies Press, Tokyo; Springer-Verlag, Berlin.
Hugenholtz, P., Goebel, B.M., and Pace, N.R. (1998) Impact of culture-independent studies on the emerging phylogenetic view of bacterial diversity. J. Bacteriol. **180,** 4765-4774.
Jetten, M.S.M., Strous, M., van de Pas-Schoonen, K.T., Schalk, J., van Dongen, U.G.J.M., van de Graaf, A.A., Logemann, S., Muyzer, G., van Loosdrecht, M.C.M., and Kuenen, J.G. (1998) The anaerobic oxidation of ammonium. FEMS Microbiol. Rev. **22,** 421-437.
Karner, M.B., DeLong, E.F., and Karl, D.M. (2001) Archaeal dominance in the mesopelagic zone of the Pacific Ocean. Nature. **409,** 507-510.
Kashefi, K., and Lovley, D.R. (2003). Extending the upper temperature limit for life. Science **301,** 934.
Koch, A.L. (1971) The adaptive responses of *Escherichia coli* to a feast and famine existence. Adv. Microb. Physiol. **6,** 147-217.
Lengeler, J.W., Drews, G., and Schlegel, H.G. (eds.) (1999) *Biology of the Prokaryotes.* Thieme, Stuttgart.
Madigan, M.T. (2000) Bacterial habitats in extreme environments, In: J. Seckbach (ed.) *Journey to Diverse Microbial Worlds. Adaptation to Exotic Environments.* Kluwer Academic Publishers, Dordrecht, pp. 63-72.
Madigan, M.T., and Marrs, B.L. (1997) Extremophiles. Sci. Am. **276 (4),** 66-71.
Madigan, M.T., and Oren, A. (1999) Thermophilic and halophilic extremophiles. Curr. Opinion Microbiol. **2,** 265-269.
Madigan, M.T., Martinko, J.M., and Parker, J. (2003) *Brock Biology of Microorganisms.* 10th ed. Prentice Hall, Upper Saddle River.
Morris, R.M., Rappé, M.S., Connon, S.A., Vergin, K.L., Siebold, W.A., Carlson, C.A., and Giovannoni, S.J. (2002) SAR11 clade dominates ocean surface bacterioplankton communities. Nature **420,** 806-810.
Oren, A. (2004) Prokaryote diversity and taxonomy: present status and future challenges. Phil. Trans. R. Soc. B **359,** 623-638.
Pace, N.R. (1997) A molecular view of microbial diversity and the biosphere. Science **276,** 734-740.
Rappé, M.S., Connon, S.A., Vergin, K.L. and Giovannoni, S.J. (2002) Cultivation of the ubiquitous SAR11 marine bacterioplankton clade. Nature **418,** 630-633.
Rosselló-Mora, R., and Amann, R. (2001) The species concept for prokaryotes. FEMS Microbiol. Rev. **25,** 39-67.
Rothschild, L.J., and Mancinelli, R.L. (2001) Life in extreme environments. Nature **409,** 1092-1101.
Seckbach, J. (ed.) (1999) *Enigmatic Microorganisms and Life in Extreme Environments.* Kluwer Academic Publishers, Dordrecht.
Seckbach, J. (ed.) (2000) *Journey to Diverse Microbial Worlds. Adaptation to Exotic Environments.* Kluwer Academic Publishers, Dordrecht.
Seckbach, J. (ed.) (2002) *Symbiosis: Methods and Model Systems.* Kluwer Academic Publishers, Dordrecht.
Seckbach, J., and Oren, A. (2004) Introduction to the extremophiles. In: J. Seckbach. (ed.), *Origins.* Kluwer Academic Publishers, Dordrecht, pp. 373-393.

Staley, J.T., and Konopka, A. (1985) Measurement of in situ activities of nonphotosynthetic microorganisms in aquatic and terrestrial habitats. Ann. Rev. Microbiol. **39,** 1379-1384.

Stanier, R.Y., and van Niel, C.B. (1962) The concept of a bacterium. Arch. Mikrobiol. **42,** 17-35.

Whitman, W.B., Coleman, D.C. and Wiebe, W.J. (1998) Prokaryotes: the unseen majority. Proc. Natl. Acad. Sci. USA **95,** 6578-6583.

Woese, C.R. (1987) Bacterial evolution. Microbiol. Rev. **51,** 221-271.

Woese, C.R., Kandler, O., and Wheelis, M.L. (1990) Towards a natural system of organisms: proposal of the domains Archaea, Bacteria, and Eucarya. Proc. Natl. Acad. Sci. USA **87,** 4576-4579.

Biodata of **Russell L. Chapman** author (with coauthor Debra A. Waters) of ***"The Algae—Diverse Life Forms and Global Importance."***

Dr. Russell L. Chapman is Professor of the Department of Oceanography and Coastal Sciences in the School of the Coast and Environment at Louisiana State University. He earned his Ph.D. at the University of California at Davis in 1973 (under the direction of Dr. Norma J. Lang). He is the author of numerous journal articles and book chapters dealing with his research interests, which include algal ultrastructure and phylogeny, biology of Trentepohliaceae (Chlorophyta), and molecular evolution in algae and bryophytes. He has served both as president and as a board of trustees member for the Phycological Society of America.

E-mail: **chapman@lsu.edu**

J. Seckbach (ed.), Life as We Know It, 37-51.

THE ALGAE—DIVERSE LIFE FORMS AND GLOBAL IMPORTANCE

RUSSELL L. CHAPMAN AND DEBRA A. WATERS
Department of Oceanography and Coastal Sciences
School of the Coast and Environment
Louisiana State University, Baton Rouge, LA 70803 USA

1. Introduction

The very first concepts for this chapter focused narrowly on algal life cycles. We quickly realized, however, that a detailed review of the multitude of life cycles within the group of organisms known as algae would be overwhelming and under-intriguing to the nonspecialist and perhaps boring to some phycologists. On the other hand, a vivid portrait of the diverse array of organisms known as algae could introduce a critically important component of Earth's biota to those who are unfamiliar with these amazing life forms. Since we believe algae are the most important plants on earth, we think this approach is worthwhile.

Finally, lest we be accused of unfairly ranking the algae above the undeniably essential land plants, we hasten to point out that the land plants would not even be here were it not for the green algae. Thus, the tremendous importance of the land plants simply amplifies that of the algae from which they arose.

2. Overlapping Kingdoms?

Beginning an analysis with the big picture provides a context for the more detailed observations to follow. When it comes to describing living organisms, the big picture is framed by a system of nomenclatural boundaries known as kingdoms–the Animal Kingdom, the Plant Kingdom, the Protist Kingdom, and the Prokaryote Kingdom. There are systems that recognize only two or three of these kingdoms or add a fifth or sixth, but the specific approach to kingdoms does not fundamentally affect our story (Cavalier–Smith 2004, Ward et al. 2004).

The algal big picture depicts a group of organisms so fascinatingly diverse that they exceed the boundaries of a normal framework. The blue green algae, or cyanobacteria, are prokaryotic organisms and hence members of the Prokaryote Kingdom. Many of the larger and more complex green, brown, and red algae are routinely classified as members of the Plant Kingdom, reflecting their photosynthetic capacity and complex morphology. The beautiful and highly motile algae called euglenoids were historically treated in the taxonomic code as being in the Animal Kingdom, even though they have chloroplasts. But more modern approaches place them in the

Protist Kingdom. (For an update on the current view of the systematics of algae, the reader is directed to the *American Journal of Botany* volume 91, issue 10 [2004], which is devoted to an exhaustive overview of the Plant Tree of Life.)

Even taking into account the varying nomenclatural schemes, the algae include members from at least three kingdoms. Without getting into a heated discussion of current taxonomic and nomenclatural concepts about the classification of living things, one can appreciate that any group of life forms, however loosely connected, that spans three or more kingdoms will consist of very diverse members, to say the least.

3. Playing with Semantics

The humorless skeptic can be heard to opine, If the blue green algae are cyanobacteria and therefore not algae, and the protist-like algae are protists and therefore not algae, then that leaves algae in just one kingdom, so they are not really such an interesting and diverse group of organisms after all. We acknowledge that one could say the algae aren't necessarily algae in the modern sense and direct the reader to an apropos poem by Ralph Lewin, Ph.D. (Lewin 1987) titled "The Algae Aren't Algae Anymore." To avoid further debate, we continue this chapter by agreeing that when the term algae is used, it means those organisms traditionally known as algae for hundreds of years. This approach should suffice for all but the most belligerent of readers.

4. Extreme Lifestyles

4.1. SIZE EXTREMES

4.1.1. The Tiny

Although most algae comprising the phytoplankton are microscopic, the cyanobacteria and eukaryotic algae known as the picoplankton rank among the smallest living cells, ranging in size from 0.2 to 2.0 micrometers (Moon-van der Staay et al. 2000). Despite their minuteness, the picoplankton are the powerhouses of primary productivity of the world's oceans. The foundation of the marine food web that provides many consumers with the cod, bass, sole, shrimp, and lobster that they so love is built upon the microscopic backs of the picoplankton.

More important even than their role as food for the fish that feed us is the picoplankton's role as oxygen producers. And, these little organisms in large numbers do a great job. Some estimates indicate that the algae produce more than 50 percent of the world's oxygen, and the picoplankton may be the single most important group of algae contributing to global oxygen production. Additionally, in this era of global warming and great concern about greenhouse gases such as carbon dioxide, the value of the picoplankton as carbon dioxide consumers cannot be overestimated.

4.1.2. The Mighty

For all who have seen the underwater forests of giant kelp off the coast of southern California–either in person or in the underwater scenes of many movies and television shows–there is no question that *Macrocystis pyrifera* is a giant. In fact, at more than 50 meters (164 ft) in length, individual kelp plants can grow one-and-a-half times taller than the tallest oak trees. And the giant kelp grows not only large, but also very fast, easily adding 18-24 inches a day (45.72-60.96 cm/day)–a growth rate that ranks this alga ahead of terrestrial fast–growers like sugar cane (12.6 in/day or 32 cm/day) and bamboo (>12 in/day or >30.48 cm/day).

When thinking big and thinking about algae, the coralline red algae might be considered a very special and interesting case. These algae, growing by themselves or with corals, can form reefs and islands, and thus collectively represent a large–scale algal presence. Another interesting, albeit unfortunate, case involves the invasive seaweed *Caulerpa*, which can grow in large, grassland–like carpets covering extensive submarine areas with amazing coenocyctic cells, a continuous cytoplasm lacking cross walls and filled with a multitude of nuclei.

4.2. HABITAT EXTREMES

The mention of coral islands summons visions of pleasant tropical environments, and certainly many algae are found in tropical or temperate habitats. Some algae, however, have physiological and biochemical adaptations that allow them not only to survive, but even thrive in extreme environments. Thermophilic algae grow in the temperature and pH extremes of hot thermal pools, whereas psychrophilic algae spend their lives in the extreme cold and intense UV radiation of persistent snowfields.

In the caldera of Yellowstone National Park in Wyoming, USA, the boundaries of lifestyle extremes are vividly and colorfully outlined in areas of zonation around alkaline and acidic thermal pools. *Cyanidium caldarium*, a eukaryotic, photosynthetic red alga, can grow in temperatures up to 56 degrees C (113 degrees F) in water 100 times more acidic than lemon juice. Photosynthetic cyanobacteria, on the other hand, do not thrive in acid pools, but can withstand even higher temperatures of 70-73 degrees C (158-163 degrees F).

Another challenge that some algae have adapted to is extreme saline conditions. For example, the green algal flagellates *Dunaliella salina* and *D. parva* grow in sites like the Great Salt Lake in Utah, USA, and the Dead Sea in Israel (Giordano and Bowes 1997, Oren and BenYosef 1997, Bageshwar et al. 2004). LANDSAT satellite images of the Dead Sea providing spatial and temporal information on a bloom of *Dunaliella parva* (Oren and BemYosef 1997) indicate that these halotolerant algae not only survive, but even thrive in that extreme environment.

Blood–red footprints in the snow might conjure omens or inspire tales about ghosts but in fact are produced by the spores of dense blooms of one of the many species of snow–loving algae, such as the unicellular green alga, *Chlamydomonas nivalis* (Bidigare et al. 1993, Duval and Hoham 2000, Thomas and Duval 1995). The red carotenoid pigments of *C. nivalis* are concentrated in the compressed snow of footprints. In many cases snow algae are so abundant that even without the concentrating mechanism of a footprint, the snow appears pink or red to the naked eye. Algae–laden snow can also be green or other colors, depending on the species of snow algae living there.

Some marine algae are surf–loving organisms. The sea palm, *Postelsia palmaeformis*, grows at its best in the extreme intertidal habitat of coastal California, where the waves hit the rocks with incredible pounding force.

The cold waters of the northern Atlantic provide a home to many species of phytoplankton, some of which form massive blooms in water cold enough to kill people after brief exposure. Some micro–algae even grow on sea ice in Antarctica, where nearby forests of brown macroalgae such as *Desmarestia anceps* can be found in the 1–2 degree C water. Red algae such as *Iridaea cordata* and *Phyllophora antarctica* also thrive there.

The algae need light for photosynthesis. Nevertheless some algae grow with very limited amounts of light, such as at great depths in the ocean, in aquatic and terrestrial caves, and even in the soil down to a depth of 20 mm or more.

Some green algae in the charophyte lineage conquered the land more than 600 million years ago and gave rise to the land plants, but other green algae that became terrestrial remained algae. There are green algae that grow on leaves and tree trunks, rooftops, walls, and sidewalks. Some of these algae are simply little round green things, but others, like those in the order Trentepohliales, have more complex morphologies and have evolved specific dispersal mechanisms that take advantage of their terrestrial habitat and lifestyle.

Algae can even be hitchhikers. Algae living on white polar bears turn their fur partly green. The Central and South American sloths provide slow moving, but apparently very suitable homes for algae. Cold–water diatoms hitch a ride on the largest mammal on Earth–the blue whale. Adhering to a whale's skin, they can give the whale's belly a yellow tinge, leading to the nickname "sulfur bottom." Algae can even inhabit the skin of immuno–compromised humans.

5. Back to Basics

With the exception of a few comments in section 3 above, we have so far avoided answering the basic question, What are the algae? From the comments already made, it is clear that the algae are not a natural monophyletic group: that is, there is no single group of organisms known as algae that includes the most recent common ancestor of all those organisms and all of the descendants of that common ancestor. Within the mixed assemblage of organisms known as algae, there are many natural or monophyletic groups such as the brown algae, the diatoms, and the red algae. But, although the algae as a whole comprise a mixed assemblage of not necessarily closely related organisms, the term algae remains useful.

Graham & Wilcox (2000) defined algae this way: "...we can say that algae are (with numerous exceptions) aquatic organisms that (with frequent exceptions) are photosynthetic, oxygenic autotrophs that are (except for the kelps) typically smaller and less structurally complex than land plants. This rather inelegant definition allows us to include the cyanobacteria..." They go on to say, "In addition to cyanobacteria, the algae consist of several groups of remotely related protists, many members of which have acquired plastids through endosymbiosis. Although it is possible to clearly distinguish algae from plants, animals, and fungi, they cannot, as a whole, be separated from the rest of the protists."

Given this definition, it should be clear why many phycologists have no difficulty in including in their phycology courses the blue–green algae, which are now recognized as bacteria because of their prokaryotic condition (Sanders 2004). As molecular phylogenetic studies have helped us understand more precisely where the different groups of algae belong in the tree of life, our appreciation of the biology and diversity of these fascinating organisms has not diminished.

6. Sex and Life

The presence or absence of sex and the variation in life cycles among different types of algae are intriguing points of comparison. Sex is good, or so it would seem since the green, brown, and red algae, the diatoms, and most other algal groups do it. The absence of sex among eukaryotic algae is perhaps more interesting. The blue–green algae are prokaryotes and are thus basically sexless. Although euglenoid algae are also sexless, because they are eukaryotic organisms it is a likely assumption that the group lost sexual reproduction early in its evolution.

Algae that reproduce sexually can have gametes that are motile, amoeboid, or nonmotile. Variations on the theme of motile gametes include isogamy (gametes similar, as in the unicellular green *Chlamydomonas*), anisogamy (gametes different, as in the green alga *Bryopsis*), and oogamy (small, motile sperm and large, nonmotile egg, as in the morphologically distinctive green alga *Chara*). The beautiful desmids and related conjugating green filaments (e.g. *Spirogyra*) have amoeboid gametes. The red algae never form flagellated cells, so their nonmotile sperm seems to be directionless when it comes to mating. Their unique sexual lifestyle is discussed in more detail below.

Observing sexual processes in microscopic planktonic algae such as the dinoflagellates is very problematic. Thus the absence of observations of sexual reproduction in any particular group of algae may not necessarily mean that the group does not undergo sexual reproduction but rather that it has just never been observed.

The product of sexual reproduction among algae may be a diploid zygote that becomes a long–lived nonmotile spore. Spores often have special adaptations making them resistant to desiccation and chemical and biological onslaughts. This feature is particularly useful for freshwater algae that may occur in vernal pools or small lakes that dry up periodically. Similarly, special pigments in snow algal spores protect them from intense cold and UV radiation.

In other cases there is no resistant diploid zygote, but there are alternating haploid and diploid generations. In some cases the alternate generations look similar (i.e., alternation of isomorphic generations, as in the brown alga *Ectocarpus*), whereas in others, such as the giant kelps (brown algae), the sporophyte is huge and the gametophyte is microscopic (i.e., alternation of dramatically dissimilar generations). To further complicate the scene, some brown algae (e.g. *Fucus*) appear to have only a macroscopic sporophyte generation, because they have retained the gametophytic generation within the specialized chambers in the sporophyte thallus.

The red algae have done their own amazing thing. In addition to both a sporophyte and a gametophyte generation, many red algae have evolved a third generation, the carposporophyte, which is formed on the gametophyte generation. The carposporophyte gives rise to diploid spores (carpospores) that produce tetrasporophyte generations. This complex variation on a theme might compensate for the fact that the nonmotile red alga sperm cells may not be too effective in finding and fertilizing the egg cells. Thus, amplifying each successful mating generates many diploid offspring from each fusion rather than just one sporophyte per successful mating.

Sexual reproduction in the terrestrial green alga *Cephaleuros* is not well documented, but because the algae grow on leaves rather than in the water, the gametes from a given thallus cannot swim to gametes from other thalli. Thus, if gametes mate and fuse they do so with genetically identical gametes from the same gametangium, which defeats the whole concept of sexual reproduction as a mechanism to achieve genetic diversity. It is reported that the biflagellate gametes can forego sex altogether and simply germinate directly to give rise to haploid thalli (Chapman 1984).

7. In the Absence of Sex

Algae that lack sexual reproduction must have alternative methods of reproducing. Asexual reproduction can be an extremely effective form of propagation; even those algae that do have sex may propagate very effectively through asexual means. Asexual aplanospores, akinetes, filament fragments, and zoospores can serve as propagules, and simple cell fission is an important mode of asexual reproduction.

The filamentous green alga *Pithophora* forms many akinetes per filament, and each akinete can give rise to another filament. Ponds and canals can be choked by the masses of *Pithophora* filaments that under favorable conditions are created asexually.

Asexual reproduction is apparently very important to the successful lifestyle of the terrestrial green alga *Cephaleuros*. In addition to the thwarted sexual process mentioned above, *Cephaleuros* reproduces asexually by means of zoosporangia that are filled with quadriflagellate zoospores. This alga and its related genera have evolved a unique abscission process that releases the entire zoosporangium for dispersal by wind, rain, and insects to some other suitable host, thus achieving good dispersal of zoospores despite the absence of an aquatic milieu in which to swim. If the zoosporangia end up in a suitable new environment, they then release the zoospores, each of which can settle down and give rise to a thallus. Thousands upon thousands of zoospores are produced and released, and this process can quite successfully lead not only to nearly complete colonization of the leaves of the host tree, but also to colonization of new hosts in the region, not necessarily close to the initial tree.

Whether algal reproduction is sexual, asexual, or both, successful reproduction can lead to an amazing number of algae. In favorable conditions, some algae can quickly form a bloom that may overwhelm a given habitat.

8. Structure and Form

8.1. STRUCTURE

Most algae do not build houses, but there are exceptions. Perhaps the most beautiful and fascinating algal homes are the glass houses, or frustules, built by the diatoms. Many consider the siliceous skeletons of diatoms to be among the most beautiful biological objects one can find; Victorians created microscopic art from diatom frustules. On a more pragmatic level geologic sediments composed of massive deposits of well–preserved diatom frustules are harvested for use in filters and as abrasives.

The micro–architectural efforts of other algae result in less–long–lasting homes. Many groups develop extra–cellular matrices, known as loricae, which may be cup–shaped or obovoid and clear or colored. The euglenoid alga *Trachelomonas* is a classic example of a flagellated, unicellular alga living in a lorica. The important phytoplankton algae known as dinoflagellates are divided into two major groups based on the structure of their walls. In the armored dinoflagellates the rigid, plate–like components of the cell walls are obvious even at the light–microscopic level. The so–called naked dinoflagellates are not really naked but have delicate wall structures devoid of obvious plate–like components.

Some algae find their housing in a symbiotic association with other organisms. The lichens are a fine example of this phenomenon.

Dinoflagellates known as zooxanthellae live in association with corals, which rely on the photosynthetic activities of the algae for some nutrients. This productive symbiosis is at risk from sea temperature rise, increasing pollution, and physical damage. A large percentage of the world's coral has already been killed, and some estimate that at least half of the remaining corals are threatened (Attrill et al. 2004; Hoegh–Guldberg 2004; Lirman and Miller 2003).

The green alga *Chlorella* presents one example of an alga living amicably within protists such as *Paramecium*. Prokaryotic, autotrophic blue–green algae living in such endosymbiotic arrangements are thought to have given rise to the chloroplasts of eukaryotic organisms. The evidence for this theory includes the fact that chloroplasts have prokaryotic–like genomes separate from the nuclear genome of the eukaryotic host, and so they can carry out their own protein synthesis.

An even more complicated scenario is that of secondary endosymbiosis, wherein a nonphotosynthetic eukaryote engulfed a photosynthetic eukaryote rather than a prokaryote. The unicellular, motile cryptomonads appear to be the evolutionary result of such an event. The evidence for this fascinating biological development is the fine structure of the cryptomonad itself, a motile, autotrophic cell containing a nucleus, mitochondria, and an endoplasmic reticulum that contains the endosymbiont. The endosymbiont has its own plasma membrane, cytoplasm, chloroplast, ribosomes, and a nucleomorph that, although only a vestige of a nucleus, is surrounded by a nuclear envelope and contains a functioning genome.

8.2. FORM

The diverse organisms known as algae exhibit a wide array of forms, or morphology. Algae can be unicellular and either motile (e.g. *Chlamydomonas*) or nonmotile (e.g. *Chlorella*). They can form multicellular colonies with the number of cells per colony fixed (which makes the colony a coenobium e.g. *Pediastrum*), or variable (e.g. *Tetraspora*). Colonies can consist of only a few cells (e.g. *Scenedesmus*) or hundreds of cells (e.g. *Volvox*), and the colonies can be either motile (e.g. *Synura*) or nonmotile (e.g. *Pediastrum*). In some cases each cell of the colony can form a new colony, and thus, a colony with 100 cells will go from one colony to 100 colonies in one reproductive cycle (e.g. *Hydrodictyon*).

The algae can also form filaments, which can be unbranched (e.g. *Spirogyra*) or branched (e.g. *Chaetophora*). All branches may be similar in size and morphology (e.g. *Cladophora*), or there can be differentiation, such as large horizontal branches with erect branches arising from them (e.g. *Stigeoclonium*). This condition, called heterotrichy, quite clearly represents more differentiation of form and function and hence more advanced morphology.

Algae can form flattened blades that can be one cell layer thick (e.g. *Monostroma*), or develop thick tissue-like blades of several layers that result from meristematic growth (e.g. the tough blades of the kelps, discussed below).

Perhaps the most elaborate morphology belongs to the giant kelps, brown algae that have specialized holdfasts, or haptera; massive, strong stems with vascular tissue that conducts photosynthates from the alga's sunlit upper regions to the lower, dimly lit regions where new growth is occurring; and extensive blades with air bladders, or pneumatocysts, that float on the surface of the sea.

In addition to this broad array of forms and morphologies, the coral–like calcium carbonate structures of the coralline red algae and chlorophytan dasycladaleans add another whole realm of unique forms.

No discussion of algal form, however brief, would be complete without at least a few words on multinucleate algae. Most algae large enough to see with the naked eye contain cells–hundreds or thousands of them–each with a single nucleus. In contrast, some algae, including taxa within the red and green algal groups in particular, have large cells, each with many nuclei. These cells are called coenocytic cells. In siphonous algae, the number of cell cross walls is even further reduced, resulting in a multinucleate plant that can be very large (even meters long), yet basically single celled. One might assume that a large alga composed of a single cell rather than 30,000 individual cells would be particularly fragile. The success of the large, siphonous marine green algae *Codium* and *Caulerpa* indicate that, apparently, nothing could be farther from the truth. Both organisms have given new meaning to the term "seaweed," because in multiple locations they have spread over vast territories, wherein they have eliminated many of the indigenous algae and animals. Two examples are the spread of *Codium* northward along the Atlantic coast of the USA and elsewhere (Chapman 1998, Goff et al. 1992) and the catastrophic spread of *Caulerpa* around the Mediterranean (Meinesz et al. 1993, Verlaque et al. 2003, 2004). In 2000, a clone of the Mediterranean *Caulerpa* was identified as having invaded coastal California, USA. Aggressive efforts to control this invasion before it reaches the dimension of the Mediterranean disaster are underway (Jousson et al. 2000).

9. The Importance of Plankton

The freshwater and marine phytoplankton play a vital role in the global ecology of our planet. Conservative estimates indicate that at least half of all the oxygen produced on Earth comes from algae, mainly the phytoplanktonic algae inhabiting the water that covers two–thirds of the surface of the planet. The picoplankton, the very tiny algal forms mentioned in section 4 above, are not the only important component of the phytoplankton. Indeed, three types of planktonic algae, all abundant in lakes and oceans, have rather poetically been called "the pasturage of the seas." The three groups are the diatoms, dinoflagellates, and coccolithophorids, all microscopic but, along with the green algae, still giants in global oxygen production and carbon dioxide uptake. As primary producers in the aquatic food chain, as oxygen generating powerhouses, and as sinks for carbon dioxide, the phytoplankton and their beneficial roles cannot be over–appreciated. The benefits they provide far outweigh the problems caused when good algae go bad and form massive harmful algal blooms (HABs).

9.1. THE PROBLEM OF HABS

As wonderful as the algae are for their positive roles in providing oxygen and as the starting point for the aquatic food web, some species sometimes form HABs. Reports of red tide fish kills have been recorded since the mid–1800s. Red tides occur worldwide, caused by several species of marine phytoplankton. Red tide blooms in the Gulf of Mexico are generally caused by the dinoflagellate *Karenia brevis*. The potent neurotoxins produced in red tide blooms not only kill fish, but can also contaminate shellfish and affect the human respiratory system.

The diatoms are important as food for other aquatic organisms and as producers of the oxygen we and other animals need. Until recently, blooms of diatoms were not considered to be of major concern. Alas, it turns out that not all diatoms are benign, and now people must be alert for blooms of *Pseudonitzschia pungens*. Shellfish feeding on *Pseudonitzschia* can accumulate the toxin the alga produces, called domoic acid. Eating shellfish with high levels of this toxin can cause amnesic shellfish poisoning.

Harmful algal blooms caused by dinoflagellates, diatoms, and some other phytoplankton algae are occurring in more places, more often, and more extensively than ever before (Boesch et al. 1997; Gilbert et al. 2001; Hoagland et al. 2002; Van Dolah 2000). Some of the increase may simply be the result of more detection and reporting; however, increased pollution of coastal waters, global change in climate, and increased transportation of foreign species in ships' ballast waters are clearly parts of the story.

9.2. NOT ALL BLOOMS ARE BAD

Some algal blooms are decidedly more benign than others. The coccolithophorids, the "coccolith bearers," can form massive blooms visible even through satellite imagery. Although recent evidence (Houdan et al. 2004; Jordan and Chamberlain 1997) indicates these algae can have important, sometimes harmful ecological effects, they are generally not HABs, but rather positive indicators of the productivity of an area. The tiny calcium carbonate discs borne around the surface of the motile unicells are the coccoliths. Large coccolithophorids can be easily seen with a light microscope, although their morphology is far more impressive when viewed with the scanning electron microscope. The concept of phytoplankton abundance takes on a certain dramatic quality when one considers that the famed White Cliffs of Dover are chalk consisting mainly of billions upon billions of minute coccoliths accumulated over millions of years as coccolithophorid blooms came and went.

9.3. DEAD ZONES

Leaving aside HABs, the sheer abundance of phytoplankton can create a special problem under certain conditions. For the past 30 years, a dead zone caused by hypoxic water has become an annual summer phenomenon in the Gulf of Mexico, extending west from the mouth of the Mississippi River and averaging about 5,000 sq. miles (12,950 km^2), ranging to 7,700 sq. miles (19,943 km^2) in 1999. Dead zones are abandoned by any organism that can move out of the lethal area, but sessile organisms die and many square miles of lifeless sea bottom can be the result.

The lack of oxygen is caused by bacteria that are efficiently oxidizing the remains of the phytoplankton that were previously blooming in great abundance near the surface in the area. After depletion of the nutrients, the algae die, sink, and provide the substrate for the oxygen–consuming process carried out by the bacteria. If the water column is stratified because of temperature and/or salinity conditions, the lower level of the water column becomes hypoxic and remains so until fresh oxygenated waters are brought into the zone, often as a result of a major storm.

Although phytoplankton abundance could be blamed as the starting point of dead zones, ill will toward the algae is misplaced. In fairness to the algae, the real starting point for the problem is an excess of nutrients in the water–way above what would be created by natural causes alone. And those nutrients are the result of human activity. In the case of the Gulf of Mexico, the nutrients come from the American heartland via the mighty Mississippi River. The hypoxic zones occur to the west of the mouth of the river where the nutrients are carried by prevailing currents.

Since the mid–1960s reports of hypoxic zones have increased worldwide. As reported by Joyce (2000), "dead zones exist in areas such as Japan's Seto Island Sea and its harbors, the northern Adriatic Sea, the Baltic Sea (with more than 38,000 square miles of hypoxic water, about 33 percent of the sea), the Black Sea (90 percent hypoxic), and many others."

10. Of Myth and Legend

The notion that algae are the primary culprits in the formation of dead zones can be taken as a myth, but some algae provide real grist for the legend mill.

Brown algae that form extensive beds floating in the sea are very different in habitat from those that grow along rocky coastlines, although very similar in other respects. These brown algae are *Sargassum* spp., and there is rich lore about the Sargasso Sea–a 2–million square mile oasis of algal life in the North Atlantic Ocean. Columbus mistakenly thought this algal mass meant his ships must be nearing land. His crew may have feared entrapment in the seemingly endless algal mass, but *Sargassum* does no harm and provides a floating habitat that benefits many unusual organisms in what has otherwise been described as a biological desert in the ocean.

Science fiction might possibly describe a mutant *Sargassum* that, triggered by an experiment gone wrong on a ship on a mysterious mission in the Sargasso Sea, grows outrageously fast, catches the ship, traps it, grows over it, and sinks it. But, like many myths and legends, that algal story is just simply fiction. Now, what is true is that the Bermuda Triangle falls within the Sargasso Sea; whether or not there is some correlation between the two…that is another question.

11. Diverse Organisms–Diverse Impacts

As previously discussed, excess nutrients released through human activity can lead to an unwanted abundance of algae, but human activity can also reduce or eliminate beneficial algal growth. The forests of biological diversity formed by giant kelp beds off the coast of California are only a fraction of the size they were before 17th and 18th century hunters slaughtered the sea otters for their fur. The sea otters help kelp forests flourish by eating the sea urchins that damage the kelp through grazing and by cutting the giant plants free of their holdfasts, causing them to wash up on the shore and die (Graham 2004). Although there are some encouraging results that indicate efforts to protect the sea otter population and reduce the sea urchin population are working, evidence from El Niño events indicates that global warming could also destroy forests of giant kelp (Edwards 2004), which need cold water to survive.

Microbiology and molecular biology have generated a great market for agar, which comes from certain species of red algae that grow in clean coastal waters (e.g. *Gracilaria),* Yet even as the demand for agar has grown, the areas of clean coastal water where these red algae grow has been diminished because of pollution and coastal development. The corals that are home to millions of algal cells are dying at a dramatic level worldwide because of pollution and increasing water temperatures caused at least partially by human activity.

As the basis for the food web that provides seafood that nourishes the Earth's human inhabitants and our economy and as the source of much of the oxygen we breathe, the algae are essential to life on this planet. The algae exhibit a grand diversity that we can study and appreciate, but the bottom line is that we must not do anything that will diminish these diverse, fascinating organisms. It is not hyperbole to say that if the algae do not survive and thrive on Earth, neither will we.

12. References

Attrill, M. J., Kelmo, F. and Jones, M. B. (2004) Impact of the 1997–98 El Niño event on the coral reef–associated echinoderm assemblage from northern Bahia, northeastern Brazil. Climate Research **26(2):** 151–158.

Bageshwar, U. K., Premkumar, L., Gokhman, I., Savchenko, T., Sussman, J. L., and Zamir, A. (2004) Natural protein engineering: a uniquely salt-tolerant, but not halophilic, alpha-type carbonic anhydrase from algae proliferating in low- to hyper-saline environments. Protein Engineering Design & Selection **17(2):** 191–200.

Bidigare, R. R., Ondrusek, M. E., Kennicutt, M. C., Iturriaga, R., Harvey, H. R., Hoham, R. W. and Macko, S. A. (1993) Evidence for a photoprotective function for secondary carotenoids of snow algae. Journal of Phycology **29(4):** 427–434.

Boesch, D. F., Anderson, D. M., Horner, R. A., Shumway, S. E., Tester, P. A. and Whitledge, T. E. (1997) Harmful algal blooms in coastal waters: options for prevention, control, and mitigation. An assessment conducted for the National Fish and Wildlife Foundation and the National Oceanic and Atmospheric Administration Coastal Ocean Program. NOAA Coastal Ocean Office.

Cavalier–Smith, T. (2004). Only six kingdoms of life. Proceedings of the Royal Society of London, series b-biological sciences **271(1545):** 1251–1262.

Chapman, A. S. (1998) From introduced species to invader: what determines variation in the success of Codium fragile ssp tomentosoides (Chlorophyta) in the North Atlantic Ocean? Helgolander Meeresuntersuchungen **52(3–4):** 277–289.

Chapman, R. L. (1984) An assessment of the current state of our knowledge of the Trentepohliaceae. In: D.E.G. Irvine and D. M. John (eds.) *Systematic Association Special Volume* No. **27,** Systematics of the Green Algae. Academic Press, London, pp. 233-250.

Duval, B., and Hoham, R. W. (2000) Snow algae in the northeastern U.S.: photomicrographs, observations, and distribution of *Chloromonas* spp. (Chlorophyta). Rhodora **102(911):** 365–372.

Edwards, M. S. (2004) Estimating scale–dependency in disturbance impacts: El Niños and giant kelp forsts in the northeast Pacific. Oecologia **138(3):** 436–447.

Gilbert, P. M., Magnien, R., Lomas, M. W., Alexander, J., Fan, C.L., Haramoto, E., Trice, M. and Kana, T. M. (2001) Harmful algal blooms in the Chesapeake and coastal bays of Maryland, USA: comparison of 1997, 1998, and 1999 events. Estuaries **24(6A):** 875–883.

Giordano, M. and Bowes, G. (1997) Gas exchange and C allocation in Dunaliella salina cells in response to the N source and CO2 concentration used for growth. Plant Physiology **115(3):** 1049–1056.

Goff, L. J., Liddle, L., Silva, P. C., Voytek, M., and Coleman, A. W. (1992) Tracing species invasion in Codium, a siphonous green alga, using molecular tools. American Jounral of Botany **79(11):** 1279–1285.

Graham, L. E. and Wilcox, L. W. (2000*) Algae*. Prentice–Hall, Inc. Upper Saddle River, New Jersey.

Graham, M. H. (2004) Effects of local deforestation on the diversity and structure of Southern California giant kelp forest food webs. Ecosystems **7(4):** 341–357.

Hoagland, P., Anderson, D. M., Kaoru, Y., and White, A. W. (2002) The economic effects of harmful algal blooms in the United States: estimates, assessment issues, and information needs. Estuaries **25(4B):** 819–837.

Hoegh–Guldberg, O. (2004) Coral reefs in a century of rapid environmental change. Symbiosis **37(1–3):** 1–31.

Houdan, A., Bonnard, A., Fresnel, J., Fouchard, S., Billard, C., and Probert, I. (2004) Toxicity of coastal coccolithophores (Prymenesiophyceae, Haptophyta). Journal of Plankton Research **26(8):** 875–883.

Jordan, R. W., and Chamberlain, A. H. L. (1997) Biodiversity among haptophyte algae. Biodiversity and Conservation **6(1):** 131–152.

Jousson, O., Pawlowski, J., Zaninetti, L., Zechman, F. W., Dini, F., Di Guiseppe, G., Woodfield, R., Millar, A., and Meinesz, A. (2000) Invasive alga reaches California – The alga has been identified that threatens to smother Californian coastal ecosystems. Nature **408(6809):** 157–158.

Joyce, S. (2000) The dead zones: oxygen–starved coastal waters. Environmental Health Perspectives **108(3):** A120–A125.

Lewin, R. A. (1987) *Biology of Algae and Diverse other Verses*. Boxwood Press.

Lirman, D. and Miller, M. W. (2003) Modeling and monitoring tools to assess recovery status and convergence rates between restored and undisturbed coral reef habitats. Restoration Ecology **11(4):** 448–456.

Meinesz, A., Devaugelas, J., Hesse, B., and Mari, X. (1993) Spread of the introduced tropical green-alga Caulerpa taxifolia in northern Mediterranean waters. Journal of Applied Phycology **5(2):** 141–147.

Moon–van der Staay, S. Y., van der Staay, G. W. M., Guillou, L., Vaulot, D., Claustre, H., and Medlin, L. K. (2000) Abundance and diversity of prymnesiophytes in the picoplankton community from the equatorial Pacific Ocean inferred from 18S rDNA sequences. Limnology and Oceanography **45(1):** 98-109.

Oren, A. and Ben Yosef, N. (1997) Development and spatial distribution of an algal bloom in the Dead Sea: a remote sensing study. Aquatic Microbial Ecology **13(2):** 219–223.

Sanders, W. B. (2004) Bacteria, algae, and phycobionts: maintaining useful concepts and terminology. Lichenologist **36:** 269-275.

Thomas, W. H. and Duval, B. (1995) Sierra-Nevada, California, USA, snow algae–snow albedo changes, algal bacterial interrelationships, and ultraviolet-radiation effects. Arctic and Alpine Research **27(4):** 389-399.

Van Dolah, F. M. (2000) Marine algal toxins: origins, health effects, and their increased occurrence. Environmental Health Perspectives **108** (Suppl. 1): 133–141.

Verlaque, M., Afonso–Carrillo, J., Gil–Rodriguez, M. C., Durand, C., Boudouresque, C. F., and Le Parco, Y. (2004) Biological Invasions **6(3):** 269–281.

Verlaque, M., Durand, C., Huisman, J. M., Boudouresque, C. F., and Le Parco, Y. (2003) On the identity and origin of the Mediterranean invasive Caulerpa racemosa (Caulerpales, Chlorophyta). European Journal of Phycology **38(4):** 325–339.

Ward, J. J., Sodhi, J. S., McGuffin, L. J., Buxton, B. F. and Jones, D. T. (2004) Prediction and functional analysis of native disorder in proteins from the three kingdoms of life. Journal of Molecular Biology **337(3):** 635–645.

Biodata of **Ian K. Ross** author of "***Fungal Lives***"

Professor Ross is currently Research Professor Emeritus of Microbiology. He received his B.S. degree from George Washington University in 1952 and the Ph.D. degree in Mycology from McGill University, Montreal in 1957. His post-doctoral research was with Kenneth B. Raper at the University of Wisconsin (1957-1958) on the cellular slime molds. From 1958 to 1959 he was on the faculty of the Department of Botany, Yale University, while working on the genetics and physiology of mating interactions and morphogenesis in plasmodial slime molds. He moved to UCSB in 1964, where he continued his work on the physiology and genetics of slime mold development for several years before turning to the molecular biology of developmental regulation in filamentous fungi - which led to a current interest in the role of mitochondria in aging in fungi. He has served on the editorial boards of several journals and on the national executive committees of major mycological societies. He has served as reviewer for the National Science Foundation, the National Institutes of Health, the United States Department of Agriculture, and various other international agencies. He received the Mycological Society of America Weston Award for Teaching Excellence in 1987 and the MSA Distinguished Mycologist Award in 1999. He was a BP Venture Research Fellow from 1987 to 1991. His 1979 text, Biology of the Fungi, was an innovative approach to the study of fungi that influenced a generation of young experimental mycologists. His most recent book, Aging of Cells, Humans and Societies (Wm. C. Brown, pub., 1995), reflects his interest in research on the underlying causes of the aging processes in all organisms. In 2001 Dr. Ross received the the Chancellors award for excellence in undergraduate research mentoring.

E-mail: **ross@lifesci.ucsb.edu**

J. Seckbach (ed.), Life as We Know It, 53-73.

FUNGAL LIVES

IAN K. ROSS
Molecular, Cellular and Developmental Biology,
University of California,
Santa Barbara, CA 93106-9610

1. Introduction

According to recent reports (Hawkworth, 2001), there are around 1.5 million species of fungi on earth, but what is a fungus? The word fungus encompasses a bewildering array of forms and kinds of organisms: uninucleate, multinucleate, acellular, coenocytic, cellular, spherical, filamentous, simple, highly complex, motile, non-motile, parasitic, saprobic, microscopic, macroscopic, through almost any combinations of extremes desired. There are, in other words, no simple definitions that can be used to delimit fungi. There are five major characteristics of fungi, three of which are found in all animals and some plants. They are eukaryotic and heterotrophic and lack chlorophyll. The other two characteristics are not normally found in other eukaryotic organisms: fungi are absorptive, that is they obtain all their food by absorbing soluble nutrients from their environment and neither photosynthesize (autotrophism) nor surround and engulf their food (ingestion), and fungi are basically multinucleate. It is this ability to possess more than one nucleus in a common cytoplasm, along with an extraordinary flexibility of somatic and sexual couplings that contributes to gene-cytoplasm interactions not normally found outside the fungal kingdom and that form the basis of the variability of fungi and thus to their fungal lives.

One aspect of fungi not always discussed in general reviews but nevertheless an important aspect of any consideration of fungal lives is the attitude of humans towards fungi. If you ask a class of undergraduates if they would walk over a stranger's lawn to kick a dandelion, they look at you with a touch of pity at asking such a ridiculous question, but if you ask how many would take a walk to kick a mushroom or puffball, many of them would raise their hands. In recent years fungi – those called molds – are regarded by large segments of the population to be horrendous things that must be destroyed. Homes infested with 'toxic mold' have been burned to the ground with all the owner's possessions inside – multi-million dollar settlements have been obtained from juries in moldy home cases because of the public distain and disgust with what they think are 'toxic molds'. This aspect of fungal impact on human lives is an important part of how fungal lives affect all organisms on earth. Just how the fungi have gained this notoriety may become clearer as you read on.

The place of fungi in the list of organisms has long been controversial. Even now there are several different classifications of the various groups that have been either traditionally or properly regarded as fungi. To avoid too much confusion, I have elected to follow the classification in Kendrick's The Fifth Kingdom (2001), one reason being that this is the most accessible reference to everyone with access to the Internet. Fungi, once regarded as 'lower plants', are now clearly recognized as possessing their own Kingdom – the Eumycota – and have been phylogenetically placed as closer to animals than they are to plants. Older texts often lump into the Fungi several other groups of organism – including the slime molds, Myxomycota, and the water molds, the Oomycota – that are now known not to belong to the Eumycota. The fungal kingdom as now recognized consists of a monophyletic group composed of 4 phyla, the Chytridiomycota, the Zygomycota, the Glomeromycota and the Dikaryomycota (the latter phylum combines the Ascomycotina and the Basidiomycotina that earlier had been regarded as phyla in their own right and the old group Deuteromycota has been dropped since molecular studies have shown that its members virtually all belong to the Dikaryomycota). Although members of these phyla have their own distinctive characteristics, they have in common that multinuclearity mentioned above and a cell wall that surrounds the cytoplasm. This cell wall, composed in the Eumycota mainly of chitin and glucans, provides a tough envelope that not only protects the cytoplasm, but enables the fungus to penetrate and utilize a vast array of substrates. Although the modern systematics provides the formal names for the various major fungal taxa, the Ascomycotina and the Basidiomycotina are often referred to informally by older terms, the ascomycetes and basidiomycetes. I will be using the latter terms when discussing general aspects of either group.

2. Fungal Structure

The main body of most fungi (the mycelium) is composed of thin filaments, the hyphae (sing. hypha) (Figures 1 and 2). These filaments contain all the organelles of typical

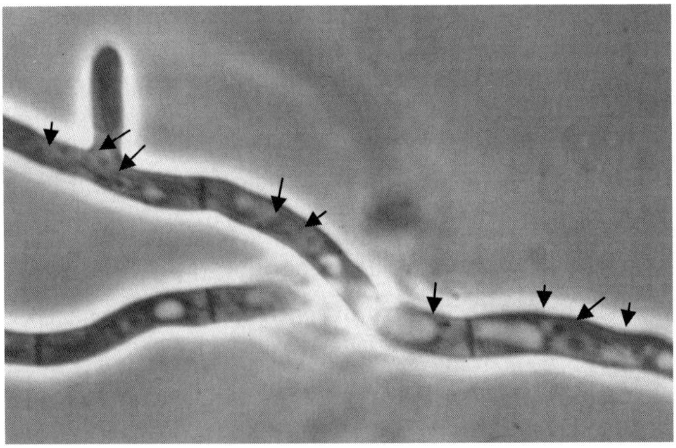

Figure 1. Hypha with nuclei indicated by arrows.

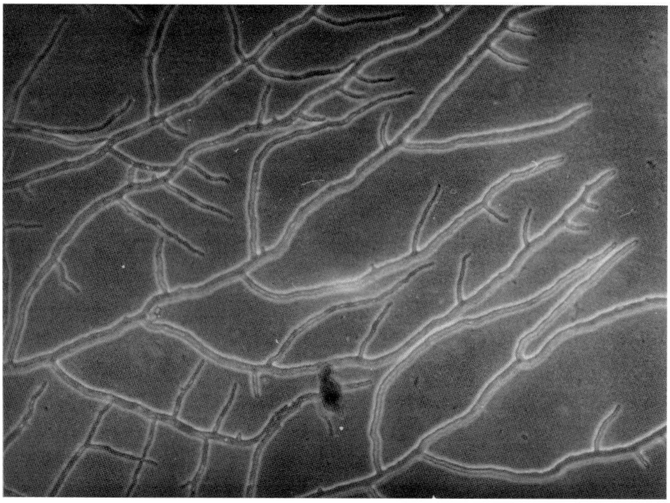

Figure 2. A mycelium showing branching and hyphal tips.

eukaryotes – nuclei,mitochondria, Golgi apparatus, etc., with nuclei multiplying by mitosis (Roberson and Fuller, 1988). The nuclei are usually haploid throughout the majority of the life cycle. The hyphae of most fungi contain cross walls, septa, that provide rigidity to the walls, but do not divide the filaments into separate cells. Unlike most cross walls in plants and algae, the septa of fungi are usually perforate with a hole that allows cell organelles and nutrients to pass through. In many hyphae the nuclei can pass through as well, consequently the whole mycelium is essentially a large open plumbing system in which nuclei, mitochondria and other organelles and materials can move throughout. This free passage can be restricted in some hyphae, particularly in the Basidiomycotina, after mating (Ross, 1979). The structure of fungal hyphae makes it very difficult to define categorically a fungal "cell". Since the contents of a hypha can move through the septa (in most cases), what part of the hypha is a cell? Several expressions have been used to describe individual components of a hypha, such as cell, compartment, segment, but none of these really successfully describes the contents of a hypha. Some fungi have no septa in their hypha, so can the whole mycelium be called a single cell? I will use the term 'cell' when referring to normal cell contents – cell membrane, nuclei, ER, Golgi, mitochondria, etc., or when referring to functions usually associated with normal cells that are not affected by the unique hyphal structure.

Fungal filaments grow by apical extension – this means that as the tip of the filament grows it is constantly entering new environments and receiving new information about the nature of that substrate. The cell membrane at the tip contains a variety of receptors, for sugars, amino acids, etc., that can transduce such reception into biochemical information that can regulate gene expression in the nuclei. Such flow of information permits the fungal filament to respond rapidly to changing environments to a degree not found in uninucleate cells. The apex of the hypha is the region where not only are most of the signal receptors located, but where the fungus secretes enzymes that digest complex compounds in the substrate to release the sugars, amino acids, etc., that are then absorbed by the hyphal tip. As the tip elongates branch hyphae start emerging further back in the hypha and form more hyphal tips. The hyphae of a mycelium may actually fuse with each other as they meet to form a complex network of cytoplasm and nuclei with great potential for receiving and responding to environmental signals. This often makes it difficult to define an individual (Rayner, 1996).

The salient difference between the uni-nucleate cells of animals and plants and the multinucleate filaments of fungi lies in the possibility of storing a much larger number of different alleles. Cells with single nuclei have only one (haploid) or two (diploid) alleles of any one gene per cell – cells or filaments with many nuclei can store any mutations that occur, and even if those mutations could be detrimental or lethal to a haploid or diploid cell, they may exist in the filaments because there are many other alleles present that can counter any potential adverse effects. This is important because what may be a detrimental mutation under some conditions could be beneficial under others – if such a mutation occurs in a haploid uni-nucleate cell in the wrong conditions, that cell would probably die, but if in a multi-nucleate filament, it can be retained in a living organism until conditions change. Multinuclearity and the ability to store numerous alleles that might be activated by different environmental signals permits fungi to enter into a tremendous array of interactions with other fungi and with all forms of life on earth – resulting in a tremendous array of lives.

3. Fungal Life Cycles and Genetic Variability

Most animals and plants have a relatively simple life cycle – haploid gametes – diploid zygote – diploid adult – haploid gametes. Fungi, however, have several complicated life cycles. Very few fungi have the simple animal/plant type of cycle, and these are mainly single celled yeasts. Many fungi, especially those in the Chytridiomycota and Zygomycota, have basically haploid, multi-nucleate thalli that can produce sexual structures either constitutively or when induced by environmental signals (since fungi may be single-celled, filamentous, or some other form, the vegetative body of fungi is often called a thallus). Gametes from these structures fuse to form a diploid cell in which the nuclei undergo meiosis to form haploid spores that develop into the mature fungus. Some of the Chytridiomycota add a complication – the diploid nucleus resulting from gamete fusion is able to multiply by mitosis and the diploid cell or filament grows into a second mature thallus. On this diploid thallus eventually develop special structures, meiosporangia, in which meiosis occurs to permit the formation of haploid spores that develop into the haploid thallus (Kendrick, 2001). Most of these fungi are limited in their potential for variability because the only fusions that occur are between gametes, specialized cells with, normally, only one function – to mate.

It is in the life cycles of the Dikaryota that we find radical changes in behavior not found in other organisms – changes that blur the distinction between sexual and non-sexual structures and permit the complicated cycles unique to these fungi. One of the major changes is that the fusion of cells or filaments does not take place only among specialized sexual cells – purely vegetative filaments can fuse with each other. What are the implications of this? Consider a substrate upon which 100 spores of a zygomycete land. Each germinates and starts to grow, but growth is as individuals – each competing for the resources in that substrate – those with genes that permit the most rapid and efficient utilization of that substrate will out-compete the others, leaving them to die off with the loss of whatever genetic potential they may have had on a different substrate. If, however, 100 spores of almost any member of the ascomycetes and many of the basidiomycetes land on the same substrate the young developing filaments will soon contact others and non-sexual fusions, anastamoses, will usually occur. This vegetative fusion joins the cytoplasm of the fusing hyphae into a continuum through which can move nutrients, nuclei and other organelles as well as information received from any part of that continuum. As such anastamoses continue, eventually the nuclei from all 100 spores will become incorporated into a single mosaic utilizing the substrate resources and permitting essentially all of the different genetic potential to survive and to be utilized as the substrate is modified and changes (Ritz and Crawford, 1999). These cytoplasmic and nuclear mosaics are called heterokaryons (different nuclei) and heteroplasmons (different cytoplasmic organelles, particularly mitochondria).

But there is yet another attribute – asexual spores. Most spores of plants such as ferns and mosses are the result of former sexual action – most spores of the ascomycetes are not, as are those of many of the basidiomycetes. This means that an entire vegetative thallus, which may encompass literally kilometers of hyphae and millions of nuclei, may produce asexual, vegetative, spores from almost any part of its thallus – thus any nucleus or its mitotic progeny has the capability of winding up in a spore. Since such thalli are warehouses of many mutations, this means that the asexually produced, mitotic, progeny of a fungal thallus can be highly variable, without the necessity of passing through a sexual and meiotic cycle. In addition, fungi play the numbers game – reproducing asexually not in pairs, or dozens, or hundreds, or thousands or millions, but in billions – a single rotten orange may have trillions of spores ready to be wafted around on the breeze, with many millions of mutational varieties included.

Thus here we have to add another 'life' of the fungi, the ability of asexual organisms to produce highly variable progeny. The mutational aspect that drives part of this diversity of progeny is merely the normal effect to having multinuclear filaments where any nucleus can wind up in a spore. But in addition to this, many of these asexual forms have evolved a system of gene recombination that does not involve mating and meiosis – parasexuality (essentially 'getting around sex') (Clutterbuck, 1974). The large number of nuclei residing in a common cytoplasm permits rare and random nuclear fusion. Next, rare and random crossing over of arms of a pair of homologous chromosomes may occur in those rare and random diploid nuclei, leading to a recombination of genes on those particular chromosomes. This may be followed by rare and random reversal of such recombinant diploid nuclei to a haploid state by gradual loss of chromosomes at successive mitoses until the haploid state is achieved. These events may be rare and random in the percentage of nuclei involved, but when the numbers are in the billions this means that a significant number of diploid nuclei, diploid nuclei with recombinant genes and haploid nuclei with different combinations of genes may co-exist with the vast majority of unchanged haploid nuclei in the filaments. This adds another level of potential for genetic diversity and the ability of fungi to be highly opportunistic since they possess this reservoir of different gene combinations that may be induced to activate as conditions and environments change – this becomes an important issue when we start to consider the lives of fungi impacting on humans. It is important to recognize that in many ascomycetes the vegetative stage with its asexual ascospores is the dominant and most dispersive stage. Such ascomycetes are among the most common members of the fungal populations in field and man-made structures. Many of the ascomycetes that lack such prolific asexual means of multiplication are still very common, but not as much as the asexual forms – in fact so successful are these asexual forms that many, if not the majority, of them have either lost the sexual stage completely, or use it so rarely that it has not been reported. These fungi made up the old taxon, the Deuteromycota.

Relatively few fungal life cycles have been mentioned so far, but space limits further discussion to two highly complex cycles of some of the basidiomycetes and ascomycetes. Although basidiomycetes may produce copious quantities of asexual spores, these more advanced forms depend mainly on sexual spores for numbers and genetic diversity. But they are not dependent on a two-sex way of life. If sexes are defined as the ability of one to mate successfully with another sex, and animals and plants have therefore two, then some basidiomycetes exceed all such banality by having thousands of sexes, each capable of successfully mating with any and all of the others. I say 'all' advisedly because mating in such fungi may indeed involve more, many more, than just two sexes – a multi-allele system. The basidiomycetes have evolved a system that promotes outbreeding to the nth degree while enjoying all the attributes of asexual fusion and a mosaic thallus – this is the homokaryon-dikaryon system (Chiu and Moore, 1999).

A haploid spore of such basidiomycetes germinates into a filamentous thallus that, although septate, may contain from one to many nuclei per septate segment or compartment. The septa contain perforations that allow nutrients, etc, including organelles, to pass through from one segment to another – another continuum of cytoplasm. These haploid thalli are essentially filamentous, indefinitely growing gametothalli. They may grow and live for a long time – for decades, at least in the laboratory, but they are not the main vegetative body of the fungus. When two haploid mycelia (homokaryons – because all the nuclei in each mycelium contain one of the genetic loci that regulate mating – mating type loci) come in contact, the filaments of one will anastamose with those of the other. This somatic fusion usually occurs regardless of the mating type loci in the mycelia – in other words, two mycelia of the same or different mating type compatibility can undergo somatic fusion. Only after cell fusion do the filaments appear to recognize what kind of partner they have. If the mating type locus in the nuclei of both filaments is the same, we have somatic fusion leading to a cytoplasmic continuum and no change in development or morphogenesis. If, however, the mating type loci are different – and since some of these fungi have over 20,000 different and compatible mating type loci this difference is the norm – recognition at the molecular level occurs between the cytoplasms and nuclei of the two fusing filaments. A series of events then occurs – the nuclei in the fusing filaments multiply and the daughter nuclei begin to migrate through the fusion sites into each mycelium, reciprocally. This leads to the infiltration of both thalli with nuclei from each other. Now, instead of either many nuclei or just one per septate segment or compartment, every segment that subsequently grows contains but two nuclei, one from each original homokaryon, and every mitosis is conjugate so that all future filaments will be composed of a series of bi-nucleate segments – the dikaryon. If this sequence of events follows the initial somatic fusion, we now term the fusion to be somatogamy – a confusing term meaning non-sexual sex (Figure 3).

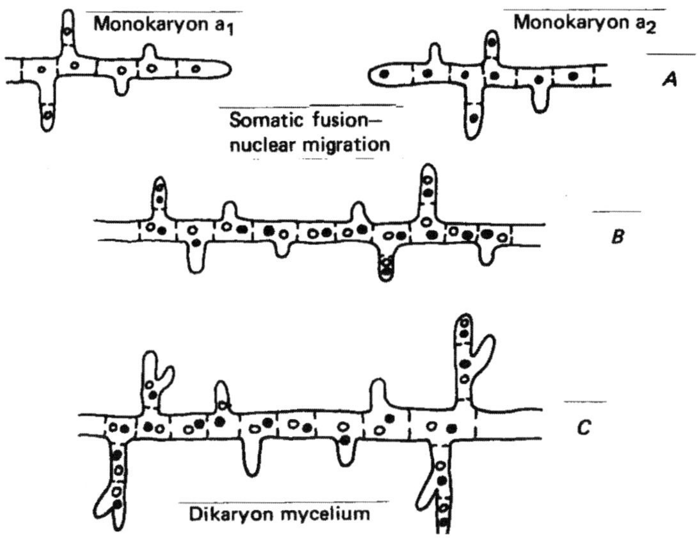

Figure 3. Nuclear migration after somatic fusion (somatogamy) and formation of the dikaryon.

These haploid, conjugate nuclei will not normally fuse in the vegetative hyphae – thus there are no diploid nuclei in this stage – but the dikaryon is functionally a normal diploid with all the attributes of gene cooperation, recessiveness and dominance found in diploids, with another fundamental difference. They can fuse with other dikaryons and other homokayons. The diploid zygotes of animals and plants, indeed the zygotes of almost every organism that has sex, cannot fuse with any other cell and is thus individually at the mercy of any environmental change. The dikaryons of basidiomycetes can undergo somatic fusion with any other dikaryon – in which case there is no nuclear migration and a cytoplasmic continuum is established – or with any other homokaryon. In that case, another unusual series of events occurs – either or both of the nuclei in the dikaryon will commence dividing and passing through the fusion sites to transform the homokaryon into a dikaryon – but not one necessarily containing the same nuclear combination.

The dikaryon is a long-lived entity – some of the largest living organisms are not blue whales, but fungal mycelia in forests of the North West and Central states of the U. S. A., with one reported from Oregon to cover over 2,200 acres and possibly up to 8,000 years old (Smith et al., 1992) . Although the dikaryon is the major vegetative stage of most members of the Basidiomycota, it does not necessarily seem to be obligatory. There are species in this fungal group in which the haploid nuclei actually fuse after somatogamy so that the resulting mycelium is indeed a normal diploid organism. Such species are rare, but their successful existence makes the dominance of the dikaryon over the diploid state an intriguing evolutionary selection problem.

There are two other aspects of vegetative fusion that deserve attention. One is the fact that it is often noted that fusions of hyphae between closely related species or even strains of the same fungus will result in the death of one partner. This almost resembles the immune rejection behavior found in animals and appears to be governed by complex genetic systems. The second interesting aspect is that in the Basidiomycota so far examined mitochondria do not migrate out of their original hyphae (Baptista-Ferreira, et al., 1983, Jin and Horgen, 1994) This means that any dikaryon mosaic is also a mosaic of mitochondria, containing areas of identical nuclei but in different mitochondrial backgrounds – another degree of complexity leading to even more diverse physiological and developmental potential (Ross, 2004).

With an appropriate supply of nutrients and receiving the proper environmental signals, dikaryons will sporulate. In most of the basidiomycete families sporulation occurs by the development of highly complex structures, the fruitbodies – mushrooms (Figure 4). Without going into the technical details of induction and morphogenesis

Figure 4. A group of mature mushrooms, showing the gills that bear the basidia.

here (Ross, 1985), suffice it to say that the mushroom is composed of dikaryotic filaments that form distinct tissues with different functions (Moore, 1998). One tissue is segregated out as the meiotic tissue. The filaments of this tissue grow with the developing mushroom and eventually differentiate at their tips into club-shaped meiotic cells – the basidia – in which the two nuclei of the dikaryon in that cell fuse to form a diploid nucleus that immediately undergoes meiosis to form 4 haploid nuclei. These haploid nuclei are segregated into usually 4 spores – the basidiospores – although basidia with 2, 8, 16 or more spores can be found in some species (Figure 5). Because it is suspected that this fertile tissue of the mushroom develops originally from one dikaryotic segment of the mycelium, the only genetic variation among the spores of a single mushroom comes from meiosis. But, since dikaryons can be genetic mosaics, the different mushrooms on a fruiting mycelium in nature may originate from segments containing different combinations of nuclei – another degree of variability potential.

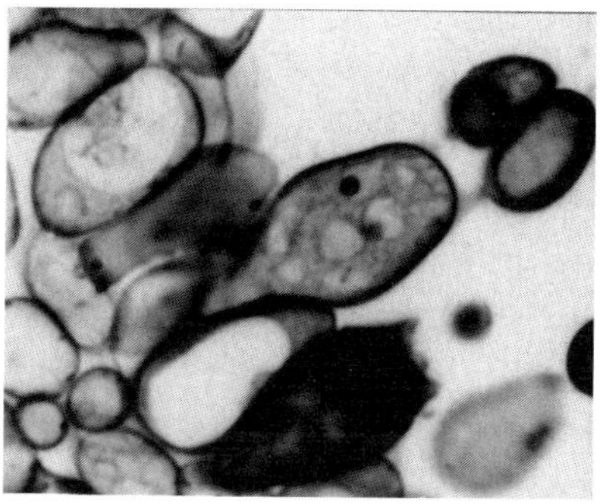

Figure 5. Developing basidium of the common cultivated mushroom, *Agaricus bisporus,* with only two spores, with two of the haploid meiotic products still within the basidium.

I have managed to skip over the dikaryon stage of the Ascomycotina because, even though that stage in the Basidiomycotina is complicated, it is, to some degree simpler than that of the Ascomycotina. Since both Ascomycotina and Basidiomycotina are in the Dikaryomycota, there has to be a dikaryon stage in the Ascomycotina. Indeed there is, but it plays no role in vegetative growth and when it does appear it is essentially parasitic on the homokaryon vegetative phase. The vegetative stage of most ascomycetes is a heterokaryon formed either by the accumulation of mutations or from the somatic fusion of two or more vegetative homokaryotic thalli. The heterokaryon of most filamentous ascomycetes is the multiplication stage in that asexual spore production is abundant. The sexual stage is initiated when specialized structures are formed on the hyphae – ascogonia, or female sexual structures. These are single celled or multicellular structures that are specialized to receive a male element. Notice the use of male and female – ascomycetes do not have the multi-allele system of basidiomycetes but have only two mating type loci – A and a, or α and a depending on the fungus. The male and female terminology comes from the presence of morphologically distinct sexual elements and that the female is usually large and fixed in place while the male can be a specialized hypha or a simple single cell released from the hyphae or even a normal asexual spore. They can be either of the two mating types – that is male can be either A or a, and female can be either A or a. The rules of mating here are both structural and genetic – only structural males can mate with females and the mating types of both must be different. Usually the sexual structures are inside in a closed or eventually open multi-filamentous (multi-cellular) complex macroscopic fruiting body or carpophore (Figure 6).

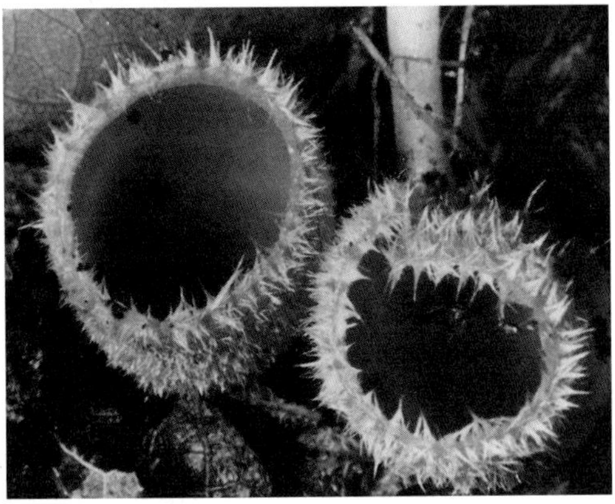

Figure 6. Ascomycete fruiting bodies (carpophores) called cup fungi, showing the interior lining of asci.

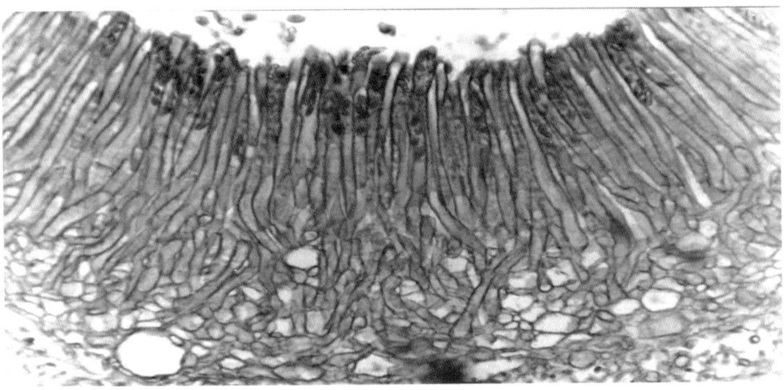

Figure 7. Cylindrical asci form a layer inside cup fungi.

The female element may be uni- or multicellular. If composed of only one cell, it is also the ascogonium, the cell that will eventually generate asci, but if multicellular, there may be several supporting cells around the pre-ascogonium. The male element either lands on or grows to the female structure and fuses with it. The 'male' nuclei migrate into the ascogonium, where a process of dikaryon formation begins. The nuclei in the ascogonium undergo many mitoses and the ascogonium begins to produce buds that elongate into filaments. Meanwhile, the nuclei from the male pair up with nuclei from the female and move as conjugate pairs into the buds and the resulting filaments. There the nuclei undergo conjugate mitosis and the filaments become dikaryotic. After a period of dikaryon growth, during which the dikaryon is dependent on the vegetative phase for sustenance, and depending on the particular fruiting body structure being produced, the tips of the dikaryon hyphae develop into a large, elongated cell in which nuclear fusion and then meiosis occurs – this cell is the ascus (means sac) (Kendrick, 2001). For reasons not yet evaluated, in the majority of ascomycetes the 4 haploid nuclear progeny of meiosis often undergo a mitosis to form 8 nuclei. The ascus cytoplasm around each nucleus is encased with a spore wall and the mature ascus contains 8 ascospores. The exceptions to this may have anywhere from 1 or a few ascospores to dozens or hundreds (Figure 8).

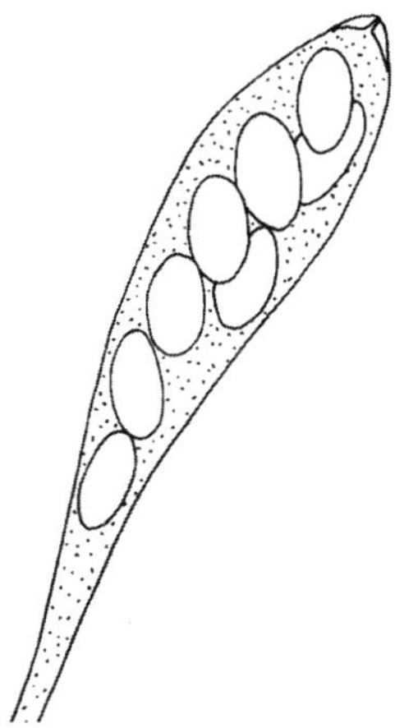

Figure 8. An ascus containing 8 ascospores.

3. Physiologial Variation

Just as the life cycles of fungi are varied and complex, so is the physiology of fungi (Griffin, 1994). All organisms have a primary metabolism – that which produces and regulates the formation of nucleic acids, proteins, essential carbohydrates, that are required for cellular growth, development and multiplication. But, like plants, the vegetative body of fungi is non-motile – subject to all the slings and arrows of the environment without the animal's ability to get up and move. It is not surprising then to find that there has been an evolutionary selection for genes and signal transduction pathways that regulate the production of many compounds that may not be required for normal cell growth and division, but may play important roles in modifying or protecting from changes in the environment. Such compounds are usually termed secondary metabolites or natural products. Fungi and plants are well known to be able to produce thousands of such metabolites – and even though the value of their production may be suspected in several cases, we just do not know what the value of most of these secondary metabolites may be to the producing organism – fungus or plant. Indeed, mutants lacking the ability to form these compounds often show no deleterious phenotype – they continue living with no sign of distress in the laboratory. These metabolites are usually produced towards the end of a growth phase, or in association with sporulation and can be very valuable to humans – antibiotics like penicillin are examples of 'good' metabolites. Other metabolites, the 'ugly' will be discussed below. I will come back to this metabolic versatility later.

As noted above, fungi produce digestive enzymes from the hyphal tips that break down complex compounds in the substrate through which the hypha is growing. In other words, fungi grow by destroying their neighbourhood. If the neighbourhood is barley malt or grape juice, we get beer, wine and pleasure, if your big toe, athletes foot and disgust, if food plants, rot and decay and financial loss, if fallen leaves and branches, humus and fertility. This short list of what may occur when fungi grow indicates that there must be a variety of fungal associations matching their variable physiology.

4. Fungal Associations

Fungi exist in all possible interactions with their environments, including other organisms. Many fungi are almost totally independent of other forms of life, except for access to organic compounds (dead or decayed forms of life), others are found only in environments that contain specific nutrients, and others are found in ephemeral or constant interactions with other living organisms (Figure 9).

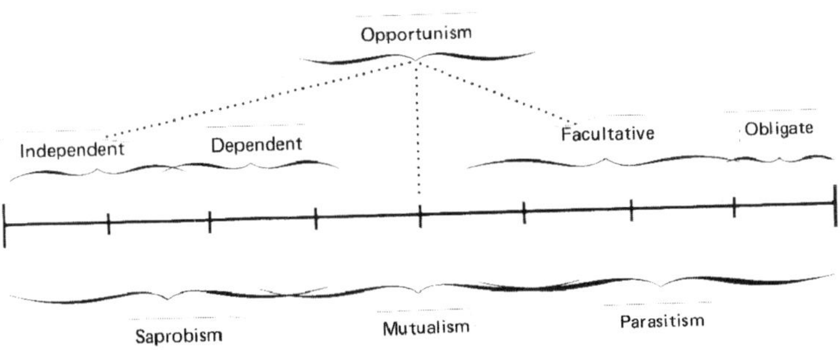

Figure 9. The various degrees of interaction of fungi with the environment.

Fungal associations with other organisms range from superficial (using the other organism merely as a support, as in lichens on trees), through varying degrees of symbiosis (described below) to ravaging predator. The term symbiosis has had varying definitions since coined by de Bary in 1879 (cf de Bary 1887) to describe any condition in which two organisms lived together (the literal translation of symbiosis) with some physiological dependence (if fungi had emotions, we would today call it a "relationship"). A symbiotic relationship can be benign in that one or both partners gain from it and neither gets hurt, or detrimental in that one partner usually has an advantage that damages the other. The first is mutualism, the second parasitism. Fungi are masters at both but it must be recognized that there is a fine line between the two states that will be discussed below after describing two of the most common mutualistic interactions between fungi and photosynthetic organisms and one with animals.

Mutualistic associations of fungi with animals are numerous and varied, but space does not allow more than a brief description of the association between fungi and ants to illustrate the hidden complexities in these relationships. Certain species of ants (now ca 210 species) began cultivating fungi as food around 50-60 million years ago in South America. These fungus – growing (attine) ants are obligately dependent on this symbiosis. So dependent are these ants on the fungal symbiont that virgin queens carry their mother's symbiont when leaving their parent colony. However, although this vertical transmission of the symbiont ensures that the new colonies will start with the original strain, there is always the possibility that alien fungi could enter the colony and spread. What is surprising is that ant colonies rear only a single fungus clone – a remarkable ability in a wild natural environment with no sterile transfer hoods or other means of ensuring such pure culture. An exciting paper by Poulen and Boomsa (2005) has shown that the fungus itself imposes this cultural purity. The fungus is believed to produce incompatibility compounds that are inhibitory to other strains of the same species. These compounds pass through the digestive tract of the ants and are incorporated into the fungal growth substrate when the fecal droppings of the ants are used as manure. This ensures that the fungus is growing on a subtrate that will inhibit any alien strain. Such symbiont regulating behavior is also suspected in fungus-growing termite associations.

The first of the plant-fungus associations is the interaction between fungi and certain green algae and/or cyanobacteria to form lichens (Ahmadjian, 1967, Honegger,1998). The vast majority of lichen fungi belong to the Ascomycota (about 20,000 species) along with a few members of the Basidiomycota. Lichens are found all over the globe but are more conspicuous in areas of more extreme climate conditions, the Antarctic, tundra, mountains above tree line, as well as being abundant in forests and rocky areas anywhere – among which can be included the "rocks" of city buildings.

The two major components of a lichen are the fungus, a non-photosynthesis organism, and the photosynthesetic organism (or organisms, since there may be more than one alga or cyanobacterium involved in the same lichen – I will refer to the photosynthetic partner simply as 'the alga'). Neither component seems to be able to succeed without the other, although the photosynthetic partner(s) may be found growing alone, but slowly and poorly in nature, the fungal partner is not. Attempts at separating and growing lichen partners on agar medium show that the fungi grow exceedingly slowly and the green partner almost as slowly. Some intact lichens are also very slow growers, 1-2mm per year! But the important difference between intact lichens and the separated components is the ability of the intact lichen to undergo complex and specific differentiation to form large macroscopic structures and to develop equally complex physiological abilities lacking in either separated component. In other words, the successful interaction of the fungus and the alga induces the expression of genes that enable the fungus to form those structures we call lichens and to produce chemicals that permit them to live in such extreme environments as exposed bare rock on high mountains – a classic case of the sum being far more than the individual entities.

This symbiotic relationship is certainly mutual, in that both partners can live in areas where neither one could live alone, but, in both anthropomorphic and Orwellian terms, the fungus appears to be "more equal" than the alga. The alga is essentially entrapped in hyphae and induced to give up most of its photosynthesis products to the fungus in exchange for water and minerals supplied by the fungus which gains the ability to grow and reproduce and spread. Of interest to the discussion to come, is that the physical interaction of the fungal hyphae and the algal cells closely resembles that of parasitic fungal interactions. The hyphae contact the algal cells and penetrate the cell wall, but not the algal cell membrane. This allows a very close and physiological association of algal cell cytoplasm and fungal hyphal cell membranes that permits the establishment of molecular systems that regulate the passage of algal photosynthate from the algal cells to the fungal hyphae. I will return to this in the section on parasitism. I have not the space here to dwell on the importance of lichens to soil production, to animals as food, to humans as both food and source of major natural dyes, or to discuss the recent use of lichens to track pollution in industrial areas – but these are all areas of great interest.

A second very important life of fungi, although not one well known at all to most non-biologists, is that of their mutualistic associations with plants. About 90% of land plants have a fungal-root association (mycorrhiza) that is not only vital to the health and welfare of present day plants, but may also have been highly instrumental in permitting plants to leave the aquatic habitats and start to grow on land about 400-500 million years ago. There are many kinds of mycorrhiza – of which the Ectomycorrhiza (EM), the Endomycorrhiza or Arbuscular mycorrhiza (AM), and the Ericaceous mycorrhiza (ErM) are the most common and probably the most important (Kendrick, 2001). The EM grow mainly on the outside of the root with hyphae extending inwards between cortical cells of the root but not penetrating the cell walls. They extend hyphae for some distance into the soil and can form large networks of mycelia. The AM grow mainly inside the root tissue and penetrate the cell walls, but not the cell membranes, a situation similar to that described for lichen above. The hyphae that penetrate the wall invaginate the cell membrane and form a bulbous compartment permitting very close contact between the hyphae and the cell. In ErM, the fungal hyphae loosely surround the roots but do not form the complex mantle of the EM.

All mycorrhiza have a very complex physiological relationship with the host plant that has evolved from an interaction of the fungal ability to absorb nutrients, especially minerals such as phosphorus, from their environments and of the hyphae to penetrate different substrates, such as soils and roots, and the defenses of the host plants. Fungi are more efficient at obtaining phosphorus from soils than are roots, and plants are very efficient at manufacturing sugars and other compounds. Nutrients and water absorbed from soils are transported from the external hyphae to the hyphae inside the root tissues. There, by various mechanisms, water, minerals, and nutrients are released and picked up by the plant cells at the same time as sugars in the root cells are released and picked up by the hyphae. It is a mutual exchange of materials vital to the successful growth of both partners. One mechanism for mineral exchange appears to be the death and digestion of some of the invading hyphal structures by the plant cells which would lead to the diffusion of the minerals in the hyphal structures into the plant cells and tissues.

Although most plants can be grown without mycorrhiza under controlled conditions, their success in the field really does depend on the presence of mycorrhiza. Most EM fungi belong to the basidiomycetes, with a few ascomycetes. The ErM all belong to the ascomycetes. AM fungi all belong to the Glomeromycota, a taxon of fungi that are obligatory mycorrhizal. Although this formal nomenclature is relatively recent, it has been shown that members of this group have been found in the earliest of land plants with morphology virtually identical to present day species. There is now a considerable industry that has developed to produce mycorrhizal fungi capable of being used to inoculate seedlings of commercial plants to ensure success when transplanting them to the field (Brundrett, 2004).

It is probable that the mutualistic association of symbiotic fungi evolved from the natural tendency of fungi to invade and destroy their neighborhood and from the natural tendency of that neighborhood, if another living organism, to defend itself. In the case of lichens it has been shown that if the individual components of a separated lichen are brought together again in a Petri dish the normal reaction is for the fungus to attack and kill the alga unless conditions are exactly right to allow both to defend and adapt to the other (Ahmadjian, 1967). When conditions are unfavorable for either partner, such as those found in extreme environments, the overgrowth of the alga by the fungus may protect it from extremes of light or desiccation and enable it to synthesize defensive substances, while the fungus finds a source of nutrients sufficient to survive, but, in the presence of the algal defenses, not enough to induce fatal parasitism – a highly simplified scenario, but possibly indicative of what may have occurred. The same scenario could be applied to mycorrhiza because they also have hyphal structures that penetrate the cell walls of the plant root cells that may be digested by the plant cells.

It is known that many fungi can penetrate and enter leaves, roots and other tissues of plants – and by growing there and destroying as they go can cause damage and disease of those plants. Some fungi may have highly complex relationships with their host plants that never lead to widespread destruction, others may co-exist with their hosts for some time before becoming lethal infestations, and others may merely enter and kill plant cells so that the plant dies without any physiological interaction. In general, all plants have defenses against attack by fungi, insects, nematodes, etc., and most pathogenic fungi have biochemical abilities that enable them to overcome such defenses. All such interactions are genetically controlled and usually for every plant defense gene there is a fungal gene that can overcome that defense. This constant warfare has led to a vast array of secondary metabolites produced by both plants and fungi that are involved in plants resisting infection and fungi attempting to overcome them. It is the ability of fungi to store mutations, as well as the tremendous numbers of nuclei present in even small fungal thalli, as described above, that enables them to so rapidly gain the capability of attacking formerly resistant plants and to cause widespread epidemics of plant disease.

Plant pathogenic fungi are some of the most economically important fungi because of their ability to devastate large areas of commercially important crops and other vital plants. They have evolved a large number of ways to travel from plant to plant, to enter the plants, to react to plant defenses, and to obtain nutrients from their hosts. There are far too many different life cycles and strategies found among the plant pathogens to be described here – the life cycle of the common wheat rust, with 5 different spore types and two different hosts is one such complex cycle (Deacon, 1997). Many pathogens are affected and regulated by hormones produced by the host, such as those that may enter a plant and grow very slowly as the host develops. Then, when the host begins reproduction and produces plant hormones regulating flowering and seed formation, those same hormones can trigger massive growth of the fungus, as is seen in corn smut (Figure 10), and other fungi actually produce plant hormones themselves that may induce the host to undergo unusual growth or modified flowering patterns.

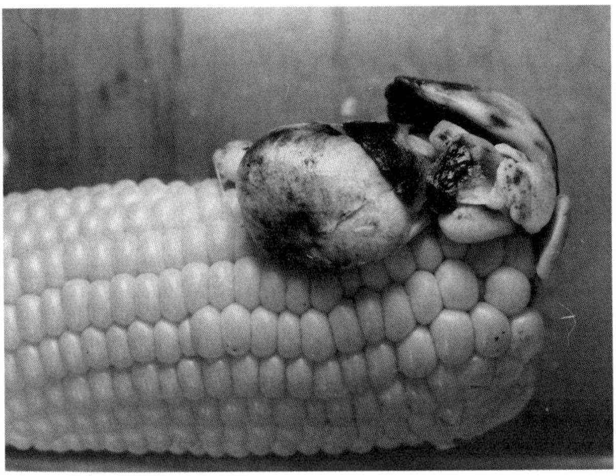

Figure 10. Corn smut. The large growths are single kernels that have been induced to hypertrophy by the internal fungal mycelium. This mycelium had been present in the plant from just after seed germination to the final maturation and flowering of the plant. The fungal mycelium was itself induced to produce hypertrophy inducing hormones by the plants own hormones.

It may have been noted that I have kept emphasizing the physiological versatility of fungi with their production of a vast array of different compounds and their ability to react rapidly to a change of environment – this means that many fungi are very opportunistic, that is they do not have single or few substrates on which they can grow, nor narrow and specific environmental conditions, such as temperature, within which they have to grow. Such opportunistic fungi can not only grow on a vast variety of substrates and in extreme conditions, but they can also grow under conditions that humans find most unpleasant – they can live in and on humans and other animals (Rippon, 1988). Many, if not most, of the fungal diseases of humans are caused by opportunistic fungi that normally live as saprobes on decaying vegetation. Some of the more serious fungal diseases stem from fungi that are found in warm habitats – such as compost piles (*Aspergillus fumigatus* – aspergillosis of the lungs), or on birds and their nests (*Histoplasma*). These fungi therefore find the 37°C temperature of humans to their liking and are not repressed by this temperature. In the same vein, most of these fungi have evolved to survive in highly competitive environments and are genetically capable of producing quite an array of secondary metabolites that kill or damage bacteria, other molds, insects, nematodes etc. This ability enables them to overcome many of the animal or human hosts' defense mechanisms and permits the fungus to at least grow and at worst flourish and kill the host. Fortunately, the human immune defense is adequate to prevent this in most cases and the major incidence of fungal disease is in immuno-compromised people, such as those with AIDS, organ transplants requiring immune suppression drugs, or on steroid therapy (Casadevall, 1995).

Not only can fungi grow in the relatively unfavorable environment of a human body, they can also survive under far more extreme conditions. After the nuclear mishap at Chernobyl the buildings that survived were the most sterile place on the whole earth, yet when it became possible for humans to enter, under extreme protection, fungi were found growing on the walls. These fungi were exposed to thousands of Sieverts yet survived and grew (the total radiation dose limit for individual members of the public as defined by the Code of Federal Regulations (10 CFR 20.1301) is 1 mSv/y). Over 2000 strains representing 200 species from 98 genera have been isolated around Chernobyl. Some of these fungi have developed the ability to grow towards the source of ionizing radiation (Zhdanova et al., 2004) Even outer space is not free of fungi – brought there as contaminant travelers on space vehicles. During the establishment of the International Space Station in 2000 it was found that fungi and bacteria were growing across the glass of portholes and obstructing the view! In addition, wiring and aluminum screens were found to be corroded by fungi (Makimura et al., 2001) Fungi grow well in the Dead sea (Kis-Papo, 2003) and in many other hostile environments of high osmotic potential – one of the reasons that pickled vegetables, jams, and other foods theoretically preserved from decay by the high concentration of salts or sugars may often be found to be mouldy. It is also known that fungi can not only grow under extreme conditions, they can survive such conditions for eons. A recent paper describes the rescue of fungi from cores bored from under-ocean sediments that date back 450,000 years – these fungi had to survive under the pressures and lack of oxygen faced in those sediments.

Finally in this listing of how fungi have many lives is one that has come to the fore in very recent years, and is itself nothing more than a further expression of the opportunistic way of life. As mentioned above, many secondary metabolites of fungi are useable in some way by other organisms and humans, but some are not useable, they are "sueable". This refers, as mentioned in the beginning of this chapter, to the many law suits being brought about because certain secondary metabolites, the mycotoxins, have been publicized as highly toxic and deadly aspects of mold growing in buildings. The legal profession and the media have been active in propagating often inaccurate and very unscientific accounts of mold activity with the result that the layman (which includes much of the legal profession and media personnel) has become so paranoid that the finding of minute amounts of certain molds is used as justification to burn the house down (cf http://www.mp group.com/el_escorial.htm). This ugly aspect of fungal lives is a reflection of the general ignorance about fungi, what they are and how they live accompanied with an equal ignorance of just how valuable and how essential to humans and all other life are these many fungal lives.

5. References

Ahmadjian, V. (1967). *The Lichen Symbiosis*. Blaisdell, Waltham, Mass.

Baptista-Ferreira, J. L. C., Economou, A., and Casselton, L. A. (1983) Mitochondrial genetics of Coprinus: recombination of mitochondrial genomes. Curr. Genet. **7,** 405-407.

Brundrett, M. (2004). Diversity and classification of mycorrhizal associations. Biological Rev. **78,** 473-495.

Casadevall, A. (1995). Antibody immunity and invasive fungal infections. Infection and Immunity **63,** 4211-4218.

Chiu, S. - W. and Moore, D. (1999). Sexual development of higher fungi. In: R. Oliver and M. Schweizer (eds.) *Molecular Fungal Biology*. Cambridge: Cambridge University Press, pp. 231-271.

Clutterbuck, J. A. (1974). *Aspergillus nidulans*. In: R. C. King (ed.) *Handbook of Genetics*. Plenum Press, New York, pp. 447-510.

Deacon, J. W. (1997). *Modern Mycology*, 3rd ed. Blackwell Science Ltd, Oxford.

de Bary, A. (1887). *Comparative Morphology and Biology of the Fungi, Mycetozoa and Bacteria*. Oxford University (Clarandon), New York and London.

Griffin, D. H. (1994). *Fungal Physiology*. New York, USA: Wiley.

Hawksworth, D. L. (2001). The magnitude of fungal diversity: the 1.5 million species estimate revisited. Mycol. Res. **105,** 1422-1432.

Honegger, R. (1998). The lichen symbiosis-what is so spectacular about it? The Lichenologist, **30,** 193-212(20).

Jin, T., and Horgen, P. A. (1994). Uniparental mitochondrial transmission in the cultivated button mushroom, *Agaricus bisporus*. Appl. Environ. Microbiol. **60,** 4456-4460.

Kendrick, B. (2001). *The Fifth Kingdom*, 3rd Ed., Mycologue Publications, Canada (http://www.mycolog.com/fifthtoc.html).

Kis-Papo, T., Oren A., Wasser S. P. and Nevo E. (2003). Survival of filamentous fungi in hypersaline Dead Sea water. Microb Ecol. **2003,**183-90.

Makimura, K., Hanazawa R., Takatori K., Tamura Y., Fujisaki R., Nishiyama Y., Abe S., Uchida K., Kawamura Y. Ezaki and Yamaguchi H. (2001). Fungal flora on board the Mir-Space Station, identification by morphological features and ribosomal DNA sequences. Microbial Immunol. **45,** 357-363.

Moore, D. (1998). *Fungal Morphogenesis*. Cambridge: Cambridge University Press. 469.

Poulsen, M. and Boomsa, J. J. (2005). Mutualistic fungi control crop diversity in fungus-growing ants. Science **307,** 741-744.
Rayner, A. D. M. (1996). Interconnectedness and individualism in fungal mycelia. In: B. C. Sutton (ed.) *A Century of Mycology*, Cambridge: Cambridge University Press. pp. 193-232.
Rippon, J. W. (1988). *Medical Mycology, The Pathogenic Fungi and the Pathogenic Actinomycetes.* Philadelphia: WB Saunders.
Ritz, K. and Crawford, J. W. (1999). Colony development in nutritionally heterogeneous environments. In: A. R. Gow, G. D. Robson and G. M. Gadd (eds.) *The Fungal Colony*, NCambridge: Cambridge University Press. pp. 49-74.
Roberson, R. W. and Fuller, M. S. (1988). Ultrastructural aspects of the hyphal tip of *Sclerotium rolfsii* preserved by freeze substitution. Protoplasma, **146,** 143-149.
Ross, I. K. (1979). Biology *of the Fungi.* McGraw-Hill, New York.
Ross, I. K. (1985). Determination of the initial steps in differentiation in *Coprinus congregatus.* In: D. Moore, D. Wood, and L. Casselton (eds) *Developmental biology of higher fungi*, Cambridge: Cambridge U. Press. pp. 353-373.
Ross, I. K. (2004). Sex, Mitochondria and mortality. Ann NY Acad Sci*.,* **1019,** 581-584.
Smith, M. L., Bruhn, J. N., and Anderson, J. B. (1992). The fungus *Armillaria bulbosa* is among the largest and oldest living organisms Nature, **356,** 428-431.
Zhdanova, N. N., Tugay, T., Dighton, J., Zheltonozhsky, V. and McDermott, P. (2004). Ionizing radiation attracts soil fungi. Mycol. Res. **108,** 1089-1096.

Biodata of **Aharon Oren** author of "***Prokaryotic Life***" of "***Life In Saline and Hypersaline Environments***" and co-author (with Seckbach et al.) of "***What do we call Life? A brief Outlook on Life***"

Aharon Oren is professor of microbial ecology at the Institute of Life Sciences, The Hebrew University of Jerusalem. He earned his Ph.D. at the Hebrew University in 1978, and spent a post-doctoral period at the University of Illinois at Urbana-Champaign. His research interests are in ecology, physiology, biochemistry and taxonomy of halophilic microorganisms and the microbiology of hypersaline environments. He has authored over 250 scientific publications, and he has written or edited five books, including "*Microbiology and Biogeochemistry of Hypersaline Environments*" (CRC Press, Boca Raton, 1999) and "*Halophilic Microorganisms and their Environments*", published in 2002 by Kluwer Scientific Publishers as Volume 5 in the series *"Cellular Origin and Life in Extreme Habitats and Astrobiology*.

E-mail: **orena@cc.huji.ac.il**

J. Seckbach (ed.), Life as We Know It, 75-87.

LIFE IN SALINE AND HYPERSALINE ENVIRONMENTS

AHARON OREN
The Institute of Life Sciences, and The Moshe Shilo Minerva Center for Marine Biogeochemistry, The Hebrew University of Jerusalem, Jerusalem 91904, Israel

1. Introduction

With a surface of 71 percent of the globe and with an average depth of 3.8 km, the oceans are by far the largest ecosystem on planet Earth. Seawater contains about 35 gram of dissolved salts per liter on the average, 78 percent of which is sodium chloride or common salt. All microbial, plant and animal life in the marine habitat has therefore to be adapted to life in a salty environment.

A salt concentration of 35 g/l is by no means the highest at which life can occur. We know many hypersaline environments, both natural and man-made, with far higher values (Table 1). Some of these are salt-saturated, such as the Dead Sea (around 340 g/l total dissolved salts) and the crystallizer ponds of solar salterns in which seawater is evaporated to produce common salt. Such environments are inhabited by microorganisms, and in some cases macroorganisms as well, that are adapted to life at high salt concentrations.

Table 1. Examples of hypersaline environments inhabited by halophilic and halotolerant microorganisms.

Type of environment	Examples
Natural salt lakes of near-neutral pH	The Dead Sea Great Salt Lake, Utah
Alkaline soda lakes	Mono Lake, California East-African soda lakes
Man-made saline and hypersaline lakes	Salt pans for the production of salt by evaporation of sea water
Salt crystals	Crude solar salt Rock salt
Salted foods	Salted fish Pickled food products

Life at high salt concentrations requires special adaptations of the cell's physiology. As a result, the diversity of microbes encountered generally decreases with increasing salinity. However, the community density of salt-adapted microorganisms in salt lakes and other high-salt environments is often very high, especially at the highest salt concentrations, at which predatory organisms are absent. The waters of hypersaline lakes such as the Great Salt Lake, the Dead Sea, hypersaline soda lakes, and saltern crystallizer ponds are often colored pink-red by dense communities of pigmented microorganisms (Grant and Tindall, 1986; Javor, 1989, 2002; Litchfield, 1991; Oren, 1988, 1997, 2000, 2002a; Post, 1977, 1981). Salt-loving (halophilic) microorganisms are also found on salted food products, and they even can survive for long periods within salt crystals.

This chapter intends to provide an overview of the diversity of life at high salt concentrations and of the special adaptations that salt-loving and salt-tolerant microorganisms have developed enabling them to grow in salt-saturated brines.

2. Microbial life at high salt concentrations – a diverse and colorful world

Microscopic examination of samples taken from the water or from the surface sediments of salt lakes, solar salterns, and other hypersaline environments can a highly rewarding experience. As an example, Fig. 1 shows a representative selection of microphotographs of organisms found in saltern evaporation and crystallizer ponds. These include prokaryotic as well as eukaryotic microorganisms, and many of these are pigmented red, orange, or green.

Unicellular green algae of the genus *Dunaliella* are found worldwide in environments ranging from seawater salinity up to the highest salt concentrations, including in saltern crystallizer ponds (Fig. 1A) and in the Dead Sea (Borowitzka, 1981; Javor, 1989; Oren, 1988, 1997, 2002a; Post, 1977, 1981). *Dunaliella* is the main primary producer in such environments. Depending on the species and the environmental conditions, the cells can be green (chlorophyll *a* and *b* are the main photosynthetic pigments) or red-orange (in cells that in addition accumulate large amounts of β-carotene). Other genera of unicellular eukaryotic algae are occasionally encountered in hypersaline environments. Other unicellular eukaryotes often found at high salt concentrations include ciliate and flagellate protozoa (Fig. 1C, D).

Shallow sediments of salt lakes and saltern evaporation ponds are often covered by microbial mats dominated by cyanobacteria. The high salt concentrations prevent massive development of predatory animals, and as a result the cyanobacteria and associated heterotrophic microorganisms may develop to high densities. Both unicellular (Fig. 1F) and filamentous types of cyanobacteria (Fig. 1E, G) can be found (Caumette et al., 1994; Javor, 1989; Oren, 2000, 2002a; Oren et al., 1995). Unicellular cyanobacteria that grow near the surface of the sediment often have an orange-brown color due to a high content of carotenoid pigments, while the somewhat deeper occurring filamentous types are colored dark green (Oren et al., 1995). Below the cyanobacterial layers a red-purple layer of salt-adapted purple sulfur bacteria is often found (Caumette et al., 1994; Oren et al., 1995).

In hypersaline environments at or approaching salt saturation (i.e., salt concentrations of 300 g/l and above), red blooms of extremely halophilic *Archaea* (family *Halobacteriaceae*) commonly occur, generally accompanied by green or red *Dunaliella* cells. The brines of crystallizer ponds of solar salterns worldwide show often intense red-pink colors (Javor, 1989, 2002; Oren, 1994, 2002a), and red blooms have also been reported from the hypersaline north arm of the Great Salt Lake (Post, 1977, 1981), and they occasionally occur in the Dead Sea as well (Oren, 1988, 1997). Red extremely halophilic *Archaea* may also colonize salted fish and other highly salted food products.

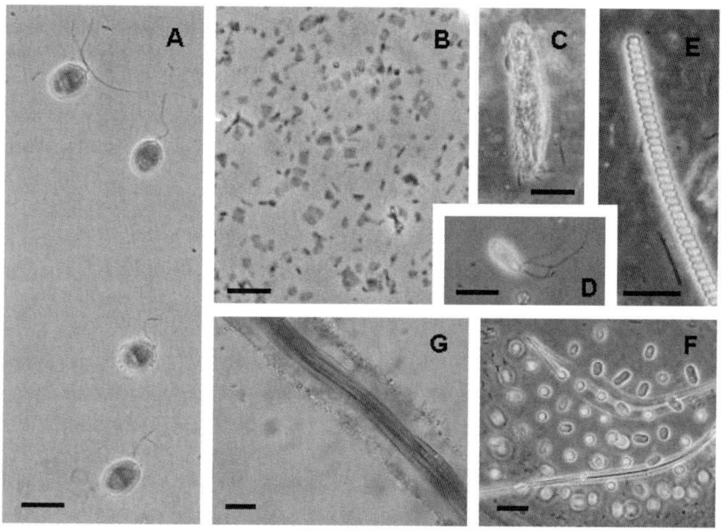

Figure 1. Salt-loving and salt-tolerant microorganisms: (A), orange β-carotene-rich cells of the unicellular green alga *Dunaliella salina* from a crystallizer pond of the salterns of Eilat, Israel (salt concentration 350 g/l); (B), square *Archaea* from the Eilat crystallizer ponds; (C) and (D), unidentified ciliate and flagellate protozoa from an evaporation pond of the Eilat salterns (salt concentration 204 g/l); (E-G): different types of halophilic cyanobacteria: *Halospirulina tapeticola* (E) and *Halothece*-like cells (F) from evaporation ponds of the Eilat salterns (salt concentration 120 and 204 g/l, respectively), and *Microcoleus chthonoplastes* within a common polysaccharide sheath, from the bottom sediments of a salt-saturated salt pan in Sečovlje, Slovenia (G). The scale bars represent 20 μm.

Red halophilic *Archaea* come in a variety of shapes. Some show rod or coccoid morphology, but others may appear as flat pleomorphic disks (e.g., *Haloferax volcanii* from the Dead Sea) (Mullakhanbhai and Larsen, 1975) and even extremely thin (0.1-0.2 μm) flat perfectly square-shaped or rectangular cells. These square prokaryotes were first recognized during microscopic examination of a sample from a brine pool on the coast of the Sinai peninsula, Egypt (Walsby, 1980). Now they are known to be the most abundant component of the biota of most saltern crystallizer ponds, where they may occur in numbers of 10^7-10^8 per milliliter and higher (Fig. 1B). These interestingly shaped *Archaea* have only recently been isolation in pure culture (Burns et al., 2004).

The pink-red color of the *Halobacteriaceae* is mainly due to carotenoid pigments in their cell membrane. These pigments protect the cells against damage by high light intensities (Dundas and Larsen, 1962; Larsen, 1973). Some species may in addition contain the purple pigments bacteriorhodopsin and halorhodopsin. Bacteriorhodopsin is a membrane-bound protein, structurally very similar to rhodopsin, the visual pigment of the human eye. When excited by light it catalyzes the transfer of a proton (H^+) from the cell's cytoplasm to the outer medium. The proton gradient thus established can be used to generate energy in the form of ATP. Light absorption by bacteriorhodopsin thus enables the cells to use sunlight as an energy source (DasSarma et al., 2001; Lanyi, 1998, 1999; Stoeckenius, 1976). It should be noted, however, that the halophilic *Archaea* cannot grow as photoautotrophs, but require organic carbon sources for growth. Halorhodopsin, a related protein, catalyzes the light-driven transport of chloride ions from the outer medium to the cytoplasm (Lanyi, 1990).

Many representatives of the domain *Bacteria* are also able to grow at every high salinities. Colorless bacteria of the genus *Halomonas* can be found at a wide range of salt concentrations, and grow up to salt saturation (Ventosa et al., 1998; Vreeland et al., 1980). The recent discovery of *Salinibacter ruber*, a red pigmented extremely halophilic member of the *Bacteria* that resembles the *Halobacteriaceae* in many properties (Antón et al., 2002), shows that the ability to grow in salt-saturated environments is more widespread than previously assumed.

When NaCl (halite) crystals are formed in salt-saturated brines, bacterial cells are often trapped within the crystals. Such bacteria may retain their viability for many years (Norton and Grant, 1988). Viable halophilic prokaryotes, *Archaea* as well as *Bacteria*, have been recovered from salt crystals extracted from up to 250 million years old halite crystals from underground salt deposits (McGenity et al., 2000; Vreeland et al., 2000). While the true age of these 'ancient' prokaryotes has been contested, it is well established that halophilic microorganisms can survive within salt crystals for very long periods.

3. What adaptations does a microorganism require to live at high salt?

To investigate what the special features of a microorganism should be that grows at high salt concentrations up to saturation, let's imagine what would happen when a cell previously grown without salt suddenly finds itself surrounded by salt-saturated brine.

Every cell is surrounded by a lipid membrane which is highly impermeable to ions, but is permeable to water (Fig. 2A). Ions only enter and leave the cell through special carrier molecules located within the membrane, and the activity of there carriers is closely regulated, so that the cell can control its internal ion concentrations. This is also true for microorganisms that live at high salt concentrations (van de Vosseberg et al., 1999). Regulation of the composition of the cells's cytoplasm thus depends primarily on the processes occurring within the cell membrane. The cell wall, found exterior to the membrane, confers physical strength to the cell, but does present a barrier to water or to ions.

Cells generally maintain a concentration of solutes that is somewhat higher than that found in their surrounding medium. Accordingly, the concentration of water inside the cell is slightly lower than outside, and water therefore tends to accumulate through the semipermeable membrane. This results in the swelling of the cytoplasm, which pushes the cell membrane toward the cell wall (the so-called turgor pressure). When such a cell is now transferred to a high salt medium, i.e., a medium in which the concentration of water is strongly reduced, water will rapidly leave the cell through the membrane to the outer medium. The volume of the cytoplasm will shrink, the turgor pressure, essential for growth of the cell, will be abolished, and the cell membrane may even become detached from the cell wall (a phenomenon known as 'plasmolysis') (Fig. 2B).

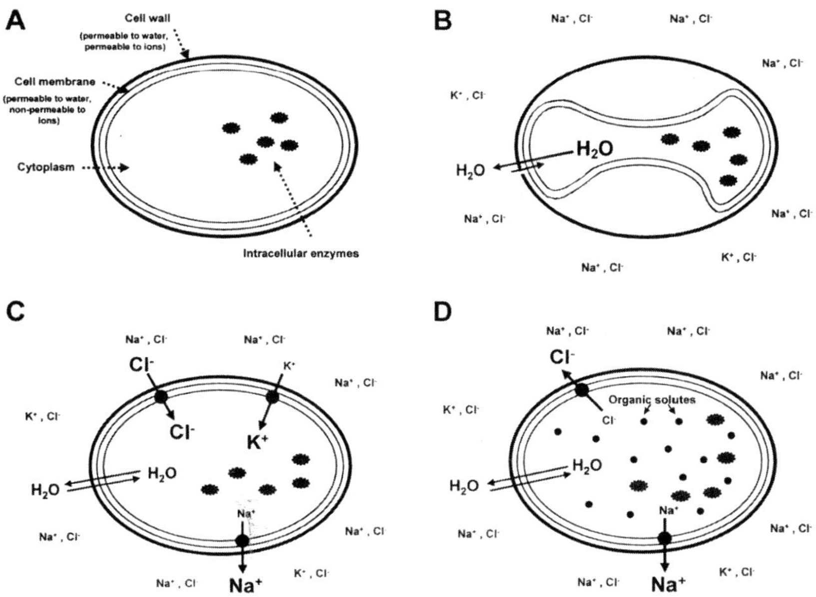

Figure 2. Osmotic adaptation in microorganisms. (A), water and ion movement in microorganisms living in a low-salt environment; (B), loss of water when a non-salt-adapted microorganism is transferred to a high-salt medium; (C), osmotic adaptation based on accumulation of KCl as osmotic solute, as found in halophilic *Archaea* of the family *Halobacteriaceae* and halophilic *Bacteria* of the genus *Salinibacter*; (D), osmotic adaptation based on the biosynthesis or accumulation of organic osmotic solutes while maintaining a low intracellular ionic concentration.

It is therefore clear that in order to function at high salt, a microorganism must maintain an intracellular water concentration (more generally termed 'water activity') no higher than that of the brine in which it lives (Brown, 1976, 1990; Kushner, 1978). This implies that the concentration of solutes within the cell cannot be significantly lower than that of the salt in the surrounding medium. When a turgor pressure is to be maintained, the concentration of solutes within the cell even has to be higher than outside. Nature has devised two fundamentally different strategies how such osmotic balance can be achieved between the cytoplasm and the saline and hypersaline environment. The first is based on the accumulation of salts within the cell, the second on the exclusion of salts and the accumulation of organic solutes (Kushner, 1978; Oren, 2000, 2002a).

The first possibility, realized in a few specialized groups of prokaryotes only, is to balance the salt outside the cell with equivalent concentrations of salts inside. This 'salt-in' strategy is used by the red extremely halophilic aerobic representatives of the domain *Archaea* (the family *Halobacteriales*), by a group of anaerobic fermentative members of the domain *Bacteria* belonging to the order *Halanaerobiales*, and by a recently discovered red extremely halophilic member of the domain *Bacteria* named *Salinibacter* (Antón et al., 2002). In these organisms the dominant intracellular ions are never Na^+ and Cl^- as in most seawater-derived brines, but K^+ and Cl^- (Christian and Waltho, 1962) (Fig. 2C). The maintenance of such an intracellular ionic composition requires the establishment of gradients of Na^+ and K^+ at the expense of energy. The distribution of Cl^- over the cell membrane is not in thermodynamic equilibrium, and active inward transport of chloride is required as well. Adaptation to life at high salt concentration can therefore only be achieved at the expense of considerable amounts of energy to pump ions in and out against the thermodynamic equilibrium.

The 'salt-in' strategy has other far-reaching consequences for those organisms that use this mode of life to achieve osmotic equilibrium with their surroundings. The activity of enzymes is strongly influenced by salt, and most enzymes from non-halophilic microorganisms function poorly, if at all in the presence of molar concentrations of salt. Moreover, at high salt concentrations the solubility of enzymes and other proteins is reduced. Therefore an organism such as *Halobacterium* that maintains molar concentrations of KCl in its cytoplasm needs to possess salt-adapted enzymes. Examination of the properties of the enzymes and structural proteins of such 'salt-in' strategists shows a large excess of acidic amino acids (glutamate, aspartate) over basic amino acids (lysine, arginine), and generally a relatively low content of hydrophobic amino acid residues (Dennis and Shimmin, 1997; Lanyi, 1974; Ng et al., 2000; Reistad, 1970). Such proteins have therefore a large excess of negative charges on their surface, and the hydrophobic interactions that stabilize the protein structure are very weak. Molar concentrations of salt are needed to stabilize these proteins, and when suspended in a low salt environment such proteins will denature, and all enzymatic activity is lost (Eisenberg et al., 1992; Madigan and Oren, 1999; Mevarech et al., 2000). *Halobacterium*, *Salinibacter*, and other microorganisms that accumulate high salt intracellularly are depend therefore on the constant presence of high salt concentrations in their medium, and their far-going adaptation to life at high salt is offset by their lack of flexibility and their inability to survive in the absence of salt.

The second strategy, much more widespread in the microbial world, is to use small organic molecules as so-called 'osmotic solutes' or 'compatible solutes' to provide osmotic balance (Brown, 1976, 1990). This strategy is used by most halophilic and halotolerant representatives of the domain *Bacteria* (except *Salinibacter* and the anaerobic *Halanaerobiaceae*, as mentioned above). Furthermore it is found in halophilic methanogenic *Archaea*, as well as in salt-loving and salt-tolerant eukaryotic microorganisms such as algae and fungi. Figure 3 shows a representative selection of such organic osmotic solutes used by different groups of microorganisms. Most of these solutes are derivatives of amino acids (e.g., glycine betaine, ectoine, Nε-acetyllysine) or sugars and sugar alcohols (e.g., glycerol, glucosylglycerol, trehalose) (Galinski, 1993, 1995). All these compounds are highly soluble in water, uncharged or zwitterionic at the physiological pH, and they do not inhibit enzymatic activities of non-salt-adapted enzymes. Cells either synthesize such compatible solutes or, when available in the medium, can take these up by means of active transport through the membrane. Cells may often contain more than one different compatible solute, and cocktails of three or more different solutes are not unusual. When the osmotic balance is provided by organic solutes there is no need for intracellular accumulation of ions such as K^+, Na^+ and Cl^-, and these are excluded from the cells by the highly impermeable membrane and if necessary by active removal by ion pumps located in the membrane (Fig. 2D).

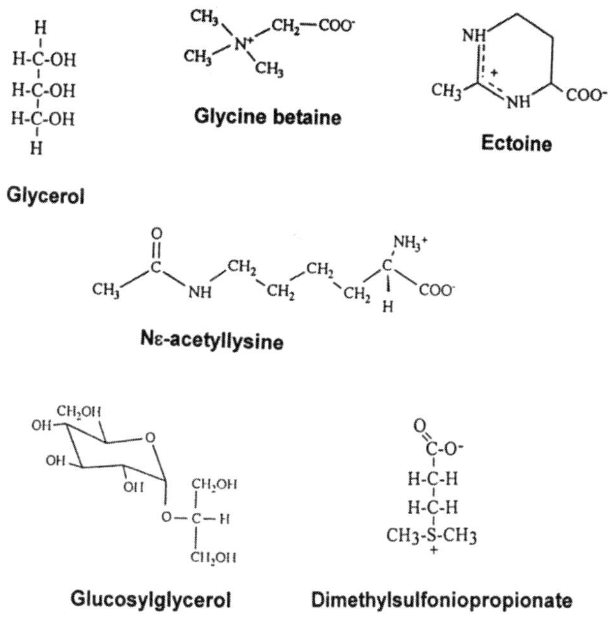

Figure 3. Some organic osmotic solutes: glycerol (found e.g., in unicellular green algae of the genus *Dunaliella*), glycine betaine (produced by halophilic cyanobacteria and other phototrophic prokaryotes), ectoine (synthesized by many heterotrophic as well as by some phototrophic prokaryotes), Nε-acetyllysine (produced by some Gram-positive *Bacteria*), glucosylglycerol (produced by cyanobacteria living at intermediate salt concentrations), and dimethylsulfoniopropionate (DMSP), synthesized by marine algae.

Eukaryotic microorganisms such as the unicellular green alga *Dunaliella* and many yeasts produce glycerol as osmotic solute (Ben-Amotz and Avron, 1973, 1983; Brown, 1990). Glycerol may be considered as the ideal compatible solute: it is a small molecule, easy to synthesize, and it can be mixed with water at any desired ratio. However, its distribution is limited to the eukaryotic world, as the cell membrane of prokaryotes is highly permeable to glycerol. The compatible solutes most widespread in the prokaryotic world are glycine betaine (a derivative of the amino acid glycine) and ectoine (1,4,5,6-tetrahydro-2-methyl-4-pyrimidine carboxylic acid, produced from intermediates in the biosynthetic pathway of lysine and threonine). Another osmotic compound worth mentioning is dimethylsulfoniopropionate (DMSP, Fig. 3), which is commonly found in eukaryotic marine macro- and microalgae. DMSP is synthesized from the amino acid methionine, and yields upon degradation dimethylsulfide (DMS), the compound that gives the characteristic smell to marine algae.

The fundamental difference between the 'salt-in' strategy and the 'salt-out, organic solutes in' mode of osmotic adaptation is the great degree of flexibility that the use of organic compatible solutes confers upon the cells. Changes in the salinity of the outside medium rapidly lead to adjustment of the intracellular concentrations of organic osmotic solutes by synthesis or uptake of more solute when the salt concentration increases, or by degradation of excess solute or its excretion out of the cells following a decrease in salinity of the cell's environment. No special modification of the intracellular proteins is required to enable their function in high concentrations of organic compatible solutes, and intracellular enzymes thus retain their activity over a wide range of extracellular salt concentrations. A possible drawback of the 'organic solute in' strategy is the large amount of energy the cell has to spend to synthesize the large amounts of compatible solutes necessary; energetically this mode of adaptation to life at high salt is more expensive than the 'salt-in' strategy of *Halobacterium* and related organisms (Oren, 1999).

4. Adaptation to high salt concentration and the tree of life

The previous sections show that the world of halophilic microorganisms is very diverse, both with respect to the range of salt concentrations in which the organisms can grow, their morphology, their pigmentation, their mode of energy generation, and the strategy they use to achieve osmotic balance between the cell cytoplasm and the high salt concentration in the environment.

Halophilic microorganisms are diverse also from the evolutionary point of view. Halophiles are found all over the phylogenetic tree of life: we find halophilic microorganisms in each of the three domains of life – *Archaea*, *Bacteria*, and *Eukarya* (Oren, 2000, 2002a, 2002b). Within each domain we encounter salt-adapted microbes in several branches (Fig. 4).

The best-known and most widespread organisms found worldwide in salt-saturated habitats are the red-pigmented heterotrophic prokaryotes of the order *Halobacteriales*. When the *Archaea* were recognized as the third domain of life in the late 1970s, the archaeal nature of these extreme halophiles was quickly recognized (Magrum et al., 1978). The archaeal domain also contains methanogenic anaerobic microorganisms, not closely related to the *Halobacteriales*, and these synthesize organic osmotic solutes rather than accumulating KCl as the *Halobacteriales* do. Within the domain *Bacteria* we encounter halophilic and halotolerant microorganisms in many of the major branches. These include aerobic and anaerobic phototrophs, aerobic heterotrophs, heterotrophic bacteria that live in the absence of oxygen by fermentation or by anaerobic respiration, and chemoautotrophs (Oren, 2000, 2002a; Ventosa et al., 1998; Vreeland et al., 1980). Most synthesize and/or accumulate organic compatible solutes, but the 'salt-in' strategy is found within the domain *Bacteria* as well, namely in the obligatory anaerobic fermentative *Halanaerobiales* (low G+C branch of the Gram-positive bacteria) and also in the recently discovered red extremely halophilic *Salinibacter* (*Flavobacterium-Cytophaga* branch) (Antón et al., 2002). We find halophiles among the eukaryotes as well: the unicellular green alga *Dunaliella* is ubiquitously found up to the highest salt concentrations, and we also know halophilic fungi. Finally, the occurrence of a few high-salt-adapted macroorganisms may be noted, such as the brine shrimp *Artemia*.

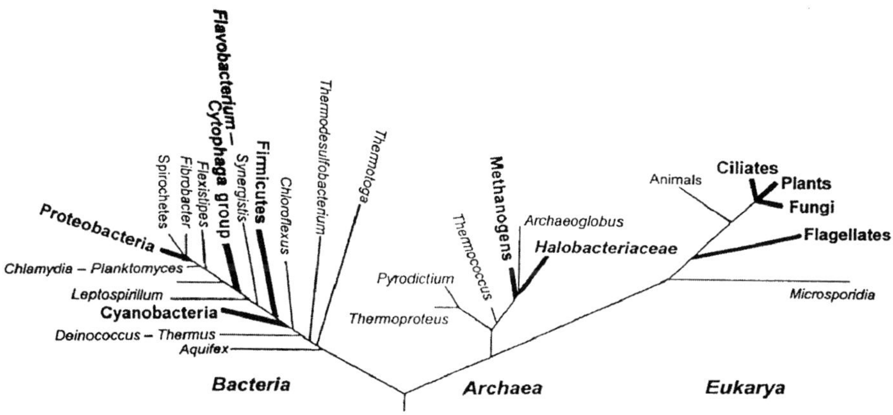

Figure 4. The three-domain tree of life, showing the distribution of micro- and macroorganisms adapted to life at high salt concentrations (from 100 g/l salt up to saturation). Branches containing salt-loving and/or highly salt-tolerant organisms are indicated with bold lines.

It may thus be assumed that the potential for growth at high salt concentrations has originated repeatedly during the evolution of life on Earth. Each group of halophiles appears to have solved the problems related to life at high salt in different ways. A survey of the forms of life adapted to high salt concentrations thus shows a highly diverse and often brighty colored world, a world that provides us with many questions about the adaptation of life to adverse environmental conditions, and ultimately about the nature of life itself.

5. References

Antón, J., Oren, A., Benlloch, S., Rodríguez-Valera, F., Amann, R., and Rosselló-Mora, R. (2002) *Salinibacter ruber* gen. nov., sp. nov., a novel extreme halophilic bacterium from saltern crystallizer ponds. Int. J. Syst. Evol. Microbiol. **52,** 485-491.

Ben-Amotz, A., and Avron, M. (1973) The role of glycerol in the osmotic regulation of the halophilic alga *Dunaliella parva*. Plant Physiol. **51,** 875-878.

Ben-Amotz, A., and Avron, M. (1983) Accumulation of metabolites by halotolerant algae and its industrial potential. Ann. Rev. Microbiol. **37,** 95-119.

Borowitzka, L.J. (1981) The microflora. Adaptations to life in extremely saline lakes. Hydrobiologia **81,** 33-46.

Brown, A.D. (1976) *Microbial water stress.* Bacteriol. Rev. **40,** 803-846.

Brown, A.D. (1990) *Microbial water stress physiology. Principles and perspectives.* John Wiley & Sons, Chichester.

Burns, D.G., Camakaris, H.M., Janssen, P.H., and Dyall-Smith, M.L. (2004) Cultivation of Walsby's square haloarchaeon. FEMS Microbiol. Lett. **238,** 469-473.

Caumette, P., Matheron, R., Raymond, N., and Relexans, J.-C. (1994) Microbial mats in the hypersaline ponds of Mediterranean salterns (Salins-de-Giraud, France). FEMS Microbiol. Ecol. **13,** 273-286.

Christian, J.H.B., and Waltho, J.A. (1962) Solute concentrations within cells of halophilic and non-halophilic bacteria. Biochim. Biophys. Acta **65,** 506-508.

DasSarma, S., Kennedy, S.P., Berquist, B., Ng, W.V., Baliga, N.S., Spudich, J.L., Krebs, M.P., Eisen, J.A., Johnson, C.H., and Hood, L. (2001) Genomic perspective on the photobiology of *Halobacterium* species NRC-1, a phototrophic, phototactic, and UV-tolerant haloarchaeon. Photosynth. Res. **70,** 3-17.

Dennis, P.P., and Shimmin, L.C. (1997) Evolutionary divergence and salinity-mediated selection in halophilic archaea. Microbiol. Mol. Biol. Rev. **61,** 90-104.

Dundas, I.D., and Larsen, H. (1962) The physiological role of the carotenoid pigments of *Halobacterium salinarium*. Arch. f. Mikrobiol. **44,** 233-239.

Eisenberg, H., Mevarech, M., and Zaccai, G. (1992) Biochemical, structural, and molecular genetic aspects of halophilism. Adv. Prot. Chem. **43,** 1-62.

Galinski, E.A. (1993) Compatible solutes of halophilic eubacteria: molecular principles, water-solute interactions, stress protection. Experientia **49,** 487-496.

Galinski, E.A. (1995) Osmoadaptation in bacteria. Adv. Microb. Physiol. **37,** 273-328.

Grant, W.D., and Tindall, B.J. (1986) The alkaline saline environment, pp. 25-54 In: Herbert, R.A., and Codd, G.A. (Eds.), *Microbes in extreme environments.* Academic Press, London.

Javor, B. (1989) *Hypersaline Environments. Microbiology and Biogeochemistry.* Springer-Verlag, Berlin.

Javor, B.J. (2002) Industrial microbiology of solar salt production. J. Ind. Microbiol. Biotechnol. **28,** 42-47.

Kamekura, M. (1998) Diversity of extremely halophilic bacteria. Extremophiles **2,** 289-295.

Kushner, D.J. (1978) Life in high salt and solute concentrations, pp. 317-368 In: Kushner, D.J. (Ed.), *Microbial life in extreme environments.* Academic Press, London.

Lanyi, J.K. (1974) Salt-dependent properties of proteins from extremely halophilic bacteria. Bacteriol. Rev. **38,** 272-290.

Lanyi, J.K. (1990) Halorhodopsin, a light-driven electrogenic chloride-transport system. Physiol. Rev. **70,** 319-330.

Lanyi, J.K. (1998) Understanding structure and function in the light-driven proton pump bacteriorhodopsin. J. Struct. Biol. **124,** 164-178.

Lanyi, J.K. (1999) Bacteriorhodopsin. Int. Rev. Cytol. **187,** 161-202.

Larsen, H. (1973) The halobacteria's confusion to biology. Antonie van Leeuwenhoek **39,** 383-396.

Litchfield, C.D. (1991) Red – the magic color for solar salt production, pp. 403-412 In: Hocquet, J.-C., and Palme, R. (Eds.), *Das Salz in der Rechts- und Handelsgeschichte.* Berenkamp, Hall in Tirol.

Madigan, M.T., and Oren, A. (1999) Thermophilic and halophilic extremophiles. Curr. Opin. Microbiol. **2,** 265-269.

Magrum, L.J., Luehrsen, K.R., and Woese, C.R. (1978) Are extreme halophiles actually "bacteria"? J. Mol. Evol. **11,** 1-8.

McGenity, T.J., Gemmell, R.T., Grant, W.D., and Stan-Lotter, H. (2000) Origins of halophilic microorganisms in ancient salt deposits. Environ. Microbiol. **2,** 243-260.

Mevarech, M., Frolow, F., and Gloss, L.M. (2000) Halophilic enzymes: proteins with a grain of salt. Biophys. Chem. **86,** 155-164.

Mullakhanbhai, M.F., and Larsen, H. (1975) *Halobacterium volcanii* spec. nov., a Dead Sea halobacterium with a moderate salt requirement. Arch. Microbiol. **104,** 207-214.

Ng, W.V., Kennedy, S.P., Mahairas, G.G., Berquist, B., Pan, M., Shukla, H.D., Lasky, S.R., Baliga, N.S., Thorsson, V., Sbrogna, J., Swartzell, S., Weir, D., Hall, J., Dahl, T.A., Welti, R., Goo, Y.A., Leithauser, B., Keller, K., Cruz, R., Danson, M.J., Hough, D.W., Maddocks, D.G., Jablonski, P.E., Krebs, M.P., Angevine, C.M., Dale, H., Isenberger, T.A., Peck, R.F., Pohlschroder, M., Spudich, J.L., Jong, K.-H., Alam, M., Freitas, T., Hou, S., Daniels, C.J., Dennis, P.P., Omer, A.D., Ebhardt, H., Lowe, T.M., Liang, P., Riley, M., Hood, L., and DasSarma, S. (2000) Genome sequence of *Halobacterium* species NRC-1. Proc. Natl. Acad. Sci. USA **97**, 12176-12181.

Norton, C.F., and Grant, W.D. (1988) Survival of halobacteria within fluid inclusions in salt crystals. J. Gen. Microbiol. **134,** 1365-1373.

Oren, A. (1988) The microbial ecology of the Dead Sea, pp. 193-229 In: Marshall, K.C. (Ed.), *Advances in microbial ecology*, Vol. 10. Plenum Publishing Company, New York.

Oren, A. (1994) The ecology of the extremely halophilic archaea. FEMS Microbiol. Rev. **13,** 415-440.

Oren, A. (1997) Microbiological studies in the Dead Sea: 1892-1992, pp. 205-213 In: Niemi, T., Ben-Avraham, Z., and Gat, J.R. (Eds.), *The Dead Sea - the lake and its setting.* Oxford University Press, Oxford.

Oren, A. (1999) Bioenergetic aspects of halophilism. Microbiol. Mol. Biol. Rev. **63,** 334-348.

Oren, A. (2000) Life at high salt concentrations, In: M. Dworkin, S. Falkow, E. Rosenberg, K.-H. Schleifer and E. Stackebrandt (Eds.), *The Prokaryotes: An Evolving Electronic Resource for the Microbiological Community*, 3rd ed., release 3.1, 20 January 2000. Springer-Verlag, New York.

Oren, A. (2002a) *Halophilic Microorganisms and their Environments.* Kluwer Academic Publishers, Dordrecht.

Oren, A. (2002b) Diversity of halophilic microorganisms: environments, phylogeny, physiology, and applications. J. Ind. Microbiol. Biotechnol. **28,** 56-63.

Oren, A., Kühl, M., and Karsten, U. (1995) An e ndoevaporitic microbial mat within a gypsum crust: zonation of phototrophs, photopigments, and light penetration. Mar. Ecol. Prog. Ser. **128,** 151-159.

Post, F.J. (1977) The microbial ecology of the Great Salt Lake. Microb. Ecol. **3,** 143-165.

Post, F.J. (1981) Microbiology of the Great Salt Lake north arm. Hydrobiologia **81,** 59-69.

Reistad, R. (1970) On the composition and nature of the bulk protein of extremely halophilic bacteria. Arch. Microbiol. **71,** 353-360.

Stoeckenius, W. (1976) The purple membrane of salt-loving bacteria. Sci. Am. June 1966: 38-46.

van de Vosseberg, J.L.C.M., Driessen, A.J.M., Grant, W.D., and Konings, W.N. (1999) Lipid membranes from halophilic and alkali-halophilic Archaea have a low H^+ and Na^+ permeability at high salt concentration. Extremophiles **3,** 253-257.

Ventosa, A., Nieto, J.J., and Oren, A. (1998) Biology of moderately halophilic aerobic bacteria. Microbiol. Mol. Biol. Rev. **62,** 504-544.

Vreeland, R.H., Litchfield, C.D., Martin, E.L., and Elliot, E. (1980) *Halomonas elongata*, a new genus and species of extremely salt-tolerant bacteria. Int. J. Syst. Bacteriol. **30,** 485-495.

Vreeland, R.H., Rosenzweig, W.D., and Powers, D.W. (2000) Isolation of a 250 million-year-old halotolerant bacterium from a primary salt crystal. Nature **407,** 897-900.

Walsby, A.E. (1980) A square bacterium. Nature **283,** 69-71.

Biodata of **Elizaveta Bonch-Osmolovskaya** author of ***"Thermophilic Communities as Autonomous Ecosystems"***

Dr. Elizaveta Bonch-Osmolovskaya is currently the Head of the Laboratory of Hyperthermophilic Microbial Communities in Winogradsky Institute of Microbiology, Russian Academy of Sciences, Moscow, Russia. She obtained her Ph.D. from the Institute of Microbiology in 1979 and continued her research in the Laboratory of Prof. G.A. Zavarzin at the same Institute. The Laboratory of Hyperthermophilic Microbial Communities was organized in 1996. Scientific interests of Dr. Bonch-Osmolovskaya are in the areas of phylogeny, metabolism and ecology of thermophilic prokaryotes.

E-mail: **lbo@mail.ru**

J. Seckbach (ed.), Life as We Know It, 89-105.

THERMOPHILIC COMMUNITIES AS AUTONOMOUS ECOSYSTEMS

ELIZAVETA BONCH-OSMOLOVSKAYA
Winogradsky Institute of Microbiology RAS, Prospect 60 Let Oktyabrya 7/2, Moscow, 117312 Russia

1. Introduction

Interest in ecosystems that could be regarded as autonomous, i.e., not depending on the modern biosphere of the Earth, has been aroused in the past decade in connection with modeling of the most ancient biocoenoses of our planet that developed in the absence of an oxygen atmosphere and the energy pool of organic compounds having a photosynthetic origin. The interest in such ecosystems on the Earth is also fueled by numerous attempts to envisage corresponding environments on other planets or their satellites (Mars, Europa), where liquid water can exist only in subsurface environments, or around zones of hydrothermal activity (Boston et al., 1992; McCollom, 1999). Thermophilic microbial communities are recognized as analogues of hypothetical ecosystems: those of early Earth and extraterrestrial ones. The natural habitats of thermophilic microorganisms – terrestrial and submarine hydrothermal areas – contain reduced products of volcanic origin, which can serve as energy substrates for chemolithotrophic thermophilic prokaryotes. There are a few other reasons that make microbial communities of thermal habitats good candidates to consider as the analogs of ancient or extraterrestrial ecosystems: elevated temperatures on the surface of the Archean Earth and the possible availability of liquid water on other planets in volcanic zones. This hypothesis is also in agreement with the phylogeny of thermophilic prokaryotes, many of which represent deep lineages on the tree of life. In this paper, we discuss the possible existence on Earth of thermophilic microbial communities that might be considered autonomous and not dependent on the modern photosynthesis-based biosphere in terms of energy substrates, electron acceptors and carbon sources.

Thermophilic microorganisms characterized by elevated optimal temperatures for growth as compared to other forms of life were known for a long time but until the last quarter of the 20th century they were believed to be the variations of known mesophilic species. The pioneering works of T.D. Brock in Yellowstone National Park (Brock, 1978) showed the diversity of thermophilic microorganisms inhabiting natural thermal habitats – hot springs and pools of volcanic origin. New organisms discovered by Brock differed both taxonomically and metabolically, the genera *Thermus* and *Sulfolobus* being the best examples. The latter – extremely thermophilic – organism was found to belong to a newly introduced *Archaea* domain (Woese and Fox, 1977); and the subsequent search for new thermophilic archaea undertaken by Wolfram Zillig and Karl Stetter resulted in the discovery of hyperthermophilic prokaryotes growing optimally at 80°C and above and the description of many novel archaeal and bacterial taxa represented by metabolically new organisms (Stetter, 1996).

The next breakthrough in the investigation of thermophilic microbial communities stemmed from the discovery of deep-sea hot vents. It was understood that complex ecosystems could exist by deriving their energy from hydrothermal hydrogen sulfide utilized by lithoautotrophic mesophilic bacterial symbionts of endemic invertebrate animals (Jannasch and Mottl, 1985). Such ecosystems, however, are known to occur in the marginal area where hot, sulfide-rich hydrothermal fluid mixes with cold, oxygen-rich oceanic water, and oxygen acts as the terminal electron acceptor in both symbiotic bacteria and their invertebrate hosts. This makes such ecosystems dependent on the modern biosphere for a continual supply of molecular oxygen as the electron acceptor.

In what follows, we shall review the biodiversity of anaerobic chemolithoautotrophic thermophilic prokaryotes, the available evidence on their activity in natural habitats, and existing examples of autonomous thermophilic microbial communities.

2. Biodiversity of Anaerobic Lithoautotrophic Thermophilic Prokaryotes

2.1. ARCHAEA

Anaerobic thermophilic prokaryotes capable of lithoautotropic growth with inorganic energy and carbon sources are diverse both phylogenetically and metabolically and include archaea of both kingdoms, bacteria of different phylogenetic lineages, hyperthermophiles, and extreme and moderate thermophiles (Table 1). Such organisms were isolated from different thermal sources: regions of post-volcanic activity (terrestrial hot springs and shallow-water and deep-sea hot vents) and the hot deep subsurface biosphere. The most common substrate for anaerobic lithoautotrophs inhabiting thermal environments is molecular hydrogen, a common component of gases evolving in the process of Earth's degasation. CO_2, another common component, serves as a carbon source in autotrophic growth.

Oxidation of hydrogen in the absence of oxygen can be coupled with reduction of different electron acceptors (Fig. 1). The first thermophilic lithoautotrophic isolates were methanogens reducing CO_2 to methane:

$$4H_2 + CO_2 \rightarrow CH_4 + 2H_2O. \qquad (1)$$

In terrestrial hot springs, this reaction is performed by rod-shaped methanogens. Moderately thermophilic *Methanobacterium thermoautotrophicus* (now *Methanothermobacter*) was initially isolated from hot springs of Yellowstone National Park (Zeikus et al., 1972; Wasserfallen et al., 2000), while members of the hyperthermophilic genus *Methanofervidus* were found in Iceland's thermal habitats (Stetter et al., 1981). In submarine hot vents both shallow-water and deep-sea, the same reaction is performed by microorganisms with coccoid cells, moderate to hyperthermophilic, first identified as members of the genus *Methanococcus*, but later reclassified as *Methanothermococcus* and *Methanocaldococcus*, respectively (Huber et al., 1982;

TABLE 1. Anaerobic thermophilic prokaryotes using molecular hydrogen as the energy substrate for autotrophic growth.

Genus	Opt growth T°, C°	Habitat(s)	Electron acceptor(s)	Reduced product(s)
		Archaea – *Euryarchaeota*		
Methanothermobacter[1]	55-65	T, DS	CO_2	CH_4
Methanothermococcus[2, 3]	65	SM, DM, DS	CO_2	CH_4
Methanocaldococcus[3, 4]	80-88	SM, DM	CO_2	CH_4
Methanothermus[5]	83-88	T	CO_2	CH_4
Methanopyrus[6]	98	DM	CO_2	CH_4
Ferroglobus[7]	85	SM	NO_3^-, Fe(III)	NO, NO_2^-, Fe(II)
Geoglobus[8]	90	DM	Fe(III)	Fe(II)
		Archaea – *Crenarchaeota*		
Thermoproteus[9]	90	T	S°	H_2S
Pyrobaculum[10, 11]	100-102	T, SM	S°, NO_3^-, Fe(III), O_2	H_2S, N_2, Fe(II), H_2O
Pyrodictium[12]	100	SM	S°, $S_2O_3^{-2}$	H_2S
Ignicoccus[13]	95	SM, DM	S°	H_2S
Acidianus[14]	70-90	T, SM	S°, O_2	H_2S
Stigiolobus[15]	80	T, SM	S°	H_2S
Pyrolobus[16]	105	DM	NO_3^-, O_2	NO_2^-, H_2O
Strain 121[17]		DM	Fe(III)	Fe(II)
		Gram-negative Bacteria – *Aquificales*		
Aquifex[18]	85		NO_3^-, O_2	N_2
Persephonella[19]	70-73	DM	NO_3^-, O_2	N_2
Desulfurobacterium[20]	70	DM	S°, NO_3^-	H_2S, NH4+
Thermovibrio[21]			NO_3^-	NH4+
Balnearium[22]	70-75	DM	S°	H_2S
		Gram-negative Bacteria – deep lineages		
Thermodesulfobacterium[23, 24]	75	DM	SO_4^{-2}	H_2S
Thermodesulfatator[25]	70	DM		
Desulfacinum[26]	60	DS, SM	SO_4^{-2}, SO_3^{-2}, $S_2O_3^{-2}$	H_2S
Deferribacter[27]	55	DS, DM	Fe(III), Mn(IV) So, NO_3^-	Fe(II), Mn(II) H_2S, NO_2^-
Desulfurella[28]	55	T	S°, $S_2O_3^{-2}$	H_2S
Geothermobacterium[29]	85-90	T	Fe(III)	Fe(II)
		Gram-negative Bacteria – epsilon-Proteobacteria		
Nautilia[30]	55	DM	S°	H_2S
Caminibacter[31]	55	DM	S°, NO_3^-, O_2	H_2S, NH_4^+, H_2O
Hydrogenimonas[32]	55	DM	S°, NO_3^-, O_2	H_2S, NH_4^+, H_2O
Nitratiruptor[33]	55	DM	NO_3^-, O_2	N_2, H_2O
Lebetimonas[34]	50	DM	S°	H_2S
		Gram-positive Bacteria		
Moorella (Clostridium)[35, 36]	58-60	T	CO_2	Acetate
Thermoanaerobacter (Acetogenium)[37, 38]	58	T	CO_2	Acetate
Thermoterrabacterium[39]	55	T	Fe(III)	Fe(II)
Desulfotomaculum[40]	50-68	T, DS	S°, $S_2O_3^{-2}$	H_2S
Ammonifex[41]	70	T	NO_3^-, S°, SO_4^{-2}	NH_4^+, H_2S

[1]Zeikus et al., 1972; [2]Huber et al., 1981; [3]Whitman et al., 2001; [4]Jones et al., 1983; [5]Stetter et al., 1981; [6]Kurr et al., 1991; [7]Haffenbradl et al., 1996; [8]Kashefi et al., 2002; [9]Zillig et al., 1981; [10]Huber et al., 1987; [11]Völkl et al., 1993; [12]Stetter et al., 1983; [13]Huber et al., 2000; [14]Segerer et al., 1986; [15]Segerer et al., 1991; [16]Blöchl

Jones et al., 1983; Whitman et al., 2001). The only known marine methanogen with rod-shaped cells is *Methanopyrus kandleri* (Kurr et al., 1987). This unique organism, found until now only in deep-sea hydrothermal habitats, grows at temperatures ranging from 84 to 110°C and optimally at 98°C, which are among the highest known for living organisms and correspond to characteristics of its habitat where the temperature of water exceeds significantly the boiling point because of hydrostatic pressure.

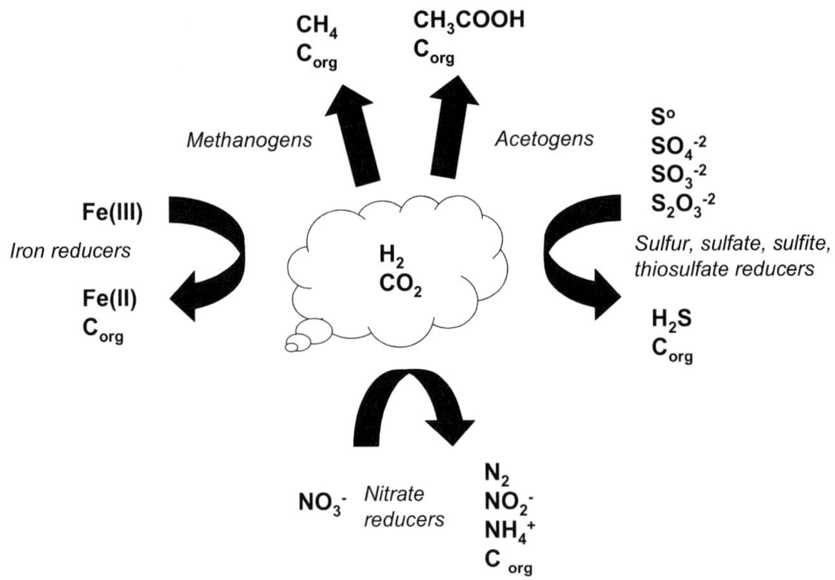

Figure 1. Anaerobic hydrogen oxidation by thermophilic lithoautotrophic prokaryotes inhabiting thermal environments.

---

et al., 1997; [17]Kashefi and Lovley, 2003; [18]Huber et al., 1992; [19]Götz et al., 2001; [20]L'Haridon et al., 1998; [21]Huber et al., 2002; [22]Takai et al., 2003a; [23]Zeikus et al., 1983; [24]Jeanthon et al., 2002; [25]Moussard et al., 2004; [26]Rees et al., 1995; [27]Greene et al., 1997; [28]Bonch-Osmolovskaya et al., 1990; [29]Kashefi et al., 2002; [30]Miroshnichenko et al., 2002; [31]Alain et al., 2002; [32]Takai et al., 2003b; [33]Nakagawa et al., 2005; [34]Takai et al., 2005; [35]Wiegel et al., 1981; [36]Collins et al., 1994; [37]Leigh et al., 1981; [38] Collins et al., 1994; [39]Slobodkin et al., 1997; [40]Widdel and Pfennig, 1984; [41]Huber et al., 1996.

Sources of isolation: T - terrestrial hot springs; SM - shallow-water submarine hot vents; DM - deep-sea hot vents; DS – deep subsurface thermal environment.

In addition to methanogenesis, hyperthermophilic archaea of the *Euryarchaeota* kingdom were found to perform other lithotrophic catabolic reactions. Hyperthermophilic archaea of the genus *Archaeoglobus* (Stetter, 1988), although requiring organic sources of carbon, can derive energy for growth from anaerobic oxidation of hydrogen coupled with reduction of sulfate:

$$4H_2+SO_4^{2-} \rightarrow S^{-2}+4H_2O. \quad (2)$$

Representatives of this genus were found in submarine and deep subsurface thermal habitats, but until now none were found in terrestrial ones.

Another lithoautotrophic process performed by anaerobic hyperthermophilic *Euryarchaeota* inhabiting deep-sea hot vents is oxidation of molecular hydrogen concurrently with reduction of ferric iron to ferrous:

$$H_2+Fe(III) \rightarrow 2H^{+}+Fe(II) \quad (3)$$

This reaction is carried out by *Geoglobus ahangari* (Kashefi and Lovley, 2002) and by *Ferroglobus placidus* (Tor et al., 2001), initially described as an organism using ferric iron as electron donor for denitrification (Haffenbradl et al., 1996).

Reduction of ferric iron with molecular hydrogen is so far the only lithoautotrophic process found in both kingdoms of hyperthermophilic archaea. It was recently found in the new deep-sea isolate – Strain 121 belonging to *Crenarchaeota* and having the highest temperature range for growth with an optimum at 110°C and maximum at 121°C (Kashefi and Lovley, 2003).

All other hyperthermophilic lithoautotrophic archaea of the *Crenarchaeota* kingdom use a completely different set of inorganic electron acceptors than those of *Euryarchaeota*. Reduction of elemental sulfur with molecular hydrogen:

$$H_2+S^{o} \rightarrow H_2S \quad (4)$$

found in the first anaerobic hyperthermophilic *Crenarchaeota* (Zillig et al., 1981) was claimed to represent the most ancient respiration process (Fischer et al., 1983). Hyperthermophilic sulfur-reducing *Crenarchaeota* include neutrophiles and acidophiles, obligate anaerobes (Zillig et al., 1981; Stetter et al., 1983; Segerer et al., 1991; Huber et al., 2000), and facultative anaerobes capable also of aerobic lithoautotrophic growth by oxidizing elemental sulfur with oxygen (Segerer et al., 1986). Other facultatively anaerobic representatives of hyperthermophilic *Crenarchaeota* are capable of lithoautotrophic growth with molecular hydrogen and nitrate, reduced to molecular nitrogen (Völkl et al., 1993) or ammonium (Blöchl et al., 1997):

$$6H_2+2NO_3^{-} \rightarrow N_2+6H_2O \quad (5)$$
$$5H_2+NO_3^{-} \rightarrow NH_4^{+}+3H_2O \quad (6)$$

The latter reaction is performed by *Pyrolobus fumarii,* an organism isolated from a deep-sea habitat and growing optimally at 113°C (Blöchl et al., 1997).

It follows that different types of anaerobic lithotrophic metabolism occur widely among thermophilic and hyperthermophilic archaea. Most of them have the necessary enzymatic apparatus to utilize CO_2 as a source of carbon (Schönheit and Schäfer, 1995). Lithoautotrophic archaea have the highest temperatures of growth known up to now; and those on top of the list were isolated from deep-sea hydrothermal habitats.

2.2. BACTERIA

The metabolic and phylogenetic diversity among Bacteria capable of anaerobic lithoautotrophic growth is no less than among Archaea. Thermophilic bacteria perform all kinds of reactions found in hyperthermophilic archaea except for methanogenesis (Fig. 1). While most of the thermophilic lithoautotrophic Archaea are hyperthermophiles, lithoautotrophic anaerobic Bacteria differ significantly in their temperature characteristics. The order *Aquificales*, one of the deepest lineages in the domain Bacteria (Huber et al., 1991), includes lithotrophic hyperthermophiles that share ecological niches with Archaea. These organisms are facultative or obligate anaerobes isolated from terrestrial or submarine sources. They reduce elemental sulfur or nitrate and most of them can also grow microaerobically. Phylogenetic analyses of thermal microbial communities showed the predominance of *Aquificales* in thermal habitats (Reysenbach and Cady, 2001).

The majority of other thermophilic bacteria capable of lithoautotrophic growth are Gram-negative anaerobes. Some of them represent deep lineages, like sulfate-reducing *Thermodesulfobacterium* (Zeikus et al., 1973; Jeanthon et al., 1999), iron-, manganese-, sulfur-, and nitrate-reducing *Deferribacter* (Greene et al., 1995), and obligately sulfur-reducing *Desulfurella* (Bonch-Osmolovskaya et al., 1990). The presence of new phylogenetic lineages in the microbial community of Obsidian Pool, Yellowstone National Park, was detected by means of 16SrDNA clone library analyses (Hugenholz et al., 1998). The unknown organism was thought to be a sulfate reducer. However, the first cultivated representative of the new lineage, *Geothermobacterium hydrogeniphilum* (Kashefi et al., 2003) turned out to be an obligately lithoautotrophic hydrogen-utilizing, iron-reducing anaerobe growing at the highest temperature range known for bacteria (optimum at 85–90°C, maximum at 100°C).

Numerous investigations of microbial communities in deep-sea hot vents by molecular methods revealed the presence of abundant populations of epsilon-Proteobacteria of the epsilon subclass associated with colonies of *Alvinella* spp., which are endemic polychaeta worms (Reysenbach et al., 2000, etc.). However, the attempts to isolate such organisms under mesophilic conditions, typical of known epsilon-Proteobacteria, were not successful. In 2002–2004, several new moderately thermophilic bacteria were isolated, which proved to be the closest relatives to the deep-sea organisms detected as environmental clones, and a new order *Nautiliales* was proposed (Miroshnichenko et al., 2004). All new isolates are obligately or facultatively anaerobic lithoautotrophs, utilizing molecular hydrogen and CO_2 as the energy and carbon sources and reducing elemental sulfur to hydrogen sulfide or nitrate to ammonium or growing microaerobially.

The number of Gram-positive thermophilic bacteria capable of anaerobic lithoautotrophic growth is fairly small. Most isolates belong to the genus *Desulfotomaculum* and, in addition to diverse organic compounds, can use molecular hydrogen for lithoautotrophic growth, reducing sulfur compounds to hydrogen sulfide (Widdel and Pfennig, 1984). *Ammonifex degensii* (Huber et al., 1996), another Gram-positive thermophilic bacterium, is obligately dependent on molecular hydrogen oxidized in the course of reduction of nitrate to ammonium or sulfate to H_2S.

Acetogens, another group of thermophilic anaerobic Gram-positive lithoautotrophs, perform the only reaction that is known in bacteria but has never been found in archaea: anaerobic oxidation of molecular hydrogen coupled with CO_2 reduction to acetate:

$$4H_2+2CO_2 \rightarrow CH_3COOH+2H_2O \quad (7)$$

Thermophilic lithotrophic acetogens are represented by the genus *Moorella* – formerly *Clostridium* (Wiegel et al., 1981; Collins et al., 1994) – and *Thermoanaerobacter kivui* – formerly *Acetogenium kivui* (Leigh et al., 1981).

Thermophilic iron-reducers constitute another group of lithoautotrophic Gram-positive thermophiles represented by the genus *Thermoterrabacterium* (Slobodkin et al., 1997) and a new organism *Thermincola ferriacetica* (Zavarzina et al., manuscript in preparation) isolated from terrestrial hot springs of Yellowstone and Kunashiri Island, respectively.

There are few inorganic energy substrates other than molecular hydrogen that can support anaerobic growth of both Archaea and Bacteria. Some of nitrate-reducing organisms can use sulfur compounds as electron donors and a hyperthermophilic archaeon *Ferroglobus placidus* mentioned above is able to oxidize ferrous iron to ferric via nitrate reduction (Haffenbradl et al., 1996). A larger group of thermophilic organisms is able to utilize CO as a sole energy and carbon source (Fig. 2). CO is a typical component of volcanic exhalations and can serve as an energy substrate for thermophilic methanogens or acetogens (Daniels et al., 1977; Daniel et al., 1990). Anaerobic oxidation of CO to CO_2 accompanied by production of molecular hydrogen from water:

$$CO+H_2O \rightarrow CO_2+H_2, \quad (8)$$

was initially discovered in mesophilic phototrophic bacteria. The first thermophilic organism anaerobically oxidizing CO with concurrent production of H_2 was *Carboxydothermus hydrogenoformans* (Svetlichny et al., 1991); then several more thermophilic prokaryotes capable of this reaction, so-called "hydrogenogenic" CO-trophs (Svetlichny et al., 2001), were found (Sokolova et al., 2000, 2002, 2004a). Bacterial isolates were moderate to extreme thermophiles belonging to Gram-positive bacteria with low G+C content where they did not form a single phylogenetic cluster. Recently, the same reaction was shown in a hyperthermophilic archaeon of the genus *Thermococcus* isolated from deep-sea hot vents of the East Pacific Rise (Sokolova et al., 2004b). All known so far hydrogenogenic CO-trophs are obligate anaerobes; some are capable of fermentation or anaerobic respiration with different electron accceptors; others are obligately dependent on CO as the energy substrate. Several thermophilic hydrogenogenic CO-trophs – *C. hydrogenoformans, Carboxydocella thermautotropha* (Sokolova et al., 2002), *Thermincola ferrireducens* (Zavarzina et al., manuscript in preparation) – are capable of lithoautotrophic growth and thus can act as primary producers in thermophilic microbial communities.

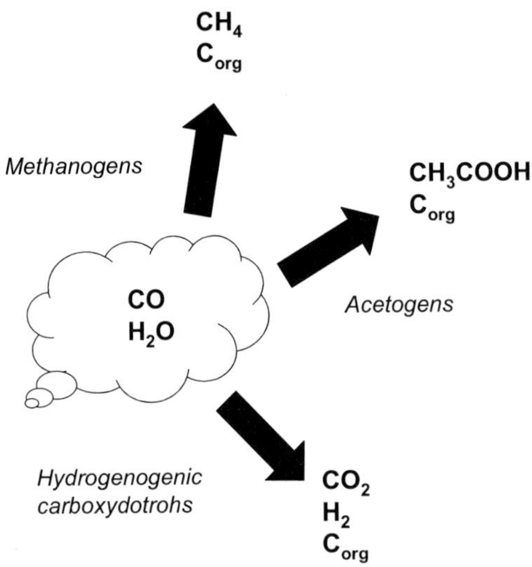

Figure 2. Anaerobic CO utilization by thermophilic prokaryotes inhabiting thermal environments.

3. *In Situ* Assays of Chemolithoautotrophic Activity in Natural Thermal Habitats

The radioisotopic method makes possible *in situ* studies of microbial processes: methanogenesis, acetogenesis, sulfate reduction, and assimilation of inorganic carbon in microbial cells. According to this method, radio-labeled substrates are added to sediment or water samples that are then incubated *in situ* or, if impossible, under similar conditions, and the radioactivity of the formed products is determined. Controls for abiotic processes are essential under this approach. Such experiments, allowing approximate rates of microbiological processes to be estimated, were conducted in natural thermal habitats: hot springs of Kamchatka

(Bonch-Osmolovskaya et al., 1999), shallow-water hot vents in the Western Pacific Ocean (Namsaraev et al., 1994), deep-sea hot vents of the East Pacific Rise (Bonch-Osmolovskaya et al., manuscript in preparation), and a high-temperature oil reservoir of Samotlor, Western Siberia (Bonch-Osmolovskaya et al., 2003). Some of the data obtained is presented in Table 2. Inorganic carbon assimilation in microbial cells was detected in all volcanic habitats studied. This process could not be attributed to lithotrophic methanogenesis, for, according to our measurements, the rate of the latter was lower, in certain cases, by several orders of magnitude. In some tests, to compensate for the loss of hydrogen after decompression, the headspace of bottles with deep-sea samples was filled with molecular hydrogen, and that increased significantly the rate of lithotrophic methanogenesis. Even so, its rate remained much lower than the rate of carbon assimilation (Table 2). Sulfate reduction in the deep-sea samples studied was not detectable because of a high background, but increased drastically in the presence of additional molecular hydrogen (Bonch-Osmolovskaya et al., manuscript in preparation).

TABLE 2. Rates of microbial processes in natural thermal environments.

Habitat	Methanogenesis from CO_2 without and with extra H_2, mg C l^{-1} day^{-1}		Aceto-genesis, mg C l^{-1} day^{-1}	Sulfate reduction without and with extra H_2, mg S $l^{-1}day^{-1}$		Carbon assimilation without and with extra H_2, mg C l^{-1} day^{-1}		Acetate minera-lization, mg C l^{-1} day^{-1}
	-	H_2		-	H_2	-	H_2	
Kamchatka hot spring, 80°C, pH 6.5[1]	0.0001	n.d.	0.002	n.d.	n.d.	0.07	n.d.	1.0
New Guinea, coastal hot vents, 85°C[2]	0.0035	n.d.	n.d.	0.28	n.d.	0.0053	n.d.	0.0297
Genesis Field, East Pacific Rise, 80°C[3]	0.001	0.0136	n.d.	0	247.0	4.8	0.23	0.0025
Samotlor oil field, depth 2600 m, 84°C[4]	0.0008	0.0006	n.d.	0.0001	0.0001	n.d.	n.d.	n.d.

[1]incubation *in situ* (Bonch-Osmolovskaya et al., 1999)
[2]incubation *in situ* (Namsaraev et al., 1994)
[3]incubation aboard ship at *in situ* temperature and ambient pressure (Bonch-Osmolovskaya et al., manuscript in preparation)
[4]incubation in laboratory at *in situ* temperature and ambient pressure (Bonch-Osmolovskaya et al., 2003)
"-" - no addition n.d. - no data

By contrast, addition of molecular hydrogen failed to stimulate methanogenesis or sulfate reduction in samples of formation water from a high-temperature oil reservoir.

Addition of molecular hydrogen to the headspace of bottles with deep-sea samples supplied with ^{14}C-labeled bicarbonate had an unexpected impact on the process of carbon assimilation, which was inhibited rather than stimulated (Table 2). A possible explanation could lie in the aerobic nature of the predominant lithoautotrophic processes: the addition of hydrogen caused rapid formation of hydrogen sulfide and subsequent binding of free oxygen present in samples.

It is worth mentioning that the rate of acetogenesis in Kamchatka hot springs was as high as 0.015–0.2 mg C l^{-1} day^{-1} in habitats with temperatures from 60 to 80°C and pH 3.0–8.5 (Bonch-Osmolovskaya et al., 1999). High rates of acetogenesis at extreme temperature and pH values suggest the presence of microorganisms not known in laboratory cultures that could act as primary producers of organic matter in this system.

Measurements of the rate of ^{14}C-2 labeled acetate oxidation to CO_2 allowed the rate of organic matter mineralization to be estimated in some of the habitats studied. In terrestrial and shallow-water thermal habitats, this process was found to be about ten times faster than organic matter production. These ecosystems, therefore, are strongly influenced by surrounding areas as sources of allochthonous organic matter.

4. Looking for Autonomous Thermophilic Microbial Communities

Microbial communities utilizing the energy of hydrogen produced from basalt-ground water interactions were proposed as an example of an autonomous ecosystem (Stevens and McKinley, 1995). This work was criticized later by claims that the amount of hydrogen produced was insufficient to support the microbial community existing in that environment (Anderson et al., 1998). Autonomous microbial communities are most likely to be found in anaerobic environments lacking a supply of allochthonous organic matter and having sources of geologically produced inorganic energy substrate. Such conditions are known to occur in Lidy Hot Springs, Idaho, USA, with water temperature of 58°C, anoxic conditions and a negligible content of organic matter (Chapelle et al., 2002). The ratio of bacterial and archaeal cells in this habitat was estimated by different molecular methods. It was found that 95–99% of the microbial population was represented by Archaea, and the dominating organisms were closely related to known methanogens. However, no biogeochemical or cultivation experiments were undertaken at this site.

Another hydrogen-driven microbial ecosystem could exist beneath the active deep-sea hydrothermal seafloor – in the so-called sub-vent biosphere (Deming and Baross, 1993; Takai and Horikoshi, 1999, etc.). However, the cultivation approach applied to microorganisms present in hydrothermal emissions and potentially representing the sub-vent microbial communities recorded the predominance of organotrophic thermophiles (Summit and Baross, 1998). In a recent investigation of the Kairei hydrothermal field in the Central Indian Ridge, cultivation methods also revealed the predominance of *Thermococcales* – a hyperthermophilic anaerobic organotrophic *Euryarchaeota*. Meanwhile, analyses of rDNA clone libraries showed the predominance of *Methanococcales* in superheated hydrothermal emissions (Takai et al., 2004). These data were supported by a light isotopic composition of methane, suggesting its biological origin. The bacterial part of the community consisted of members of *Aquificales* and Epsilonproteobacteria - also represented by lithoautotrophs.

5. Summary

The investigations of the last two decades have revealed significant diversity of thermophilic anaerobes capable of lithoautotrophic growth. The existence of some of them was already predicted by molecular methods and, most probably, many anaerobic thermophilic chemolithoautotrophs are still unknown. All these microorganisms could play the role of primary producers in anaerobic thermal ecosystems by utilizing inorganic electron donors of volcanic origin and CO_2 as a carbon source. However, autonomous ecosystem must depend also on electron acceptors of nonphotosynthetic origin, and most of the acceptors used by thermophilic anaerobes are formed with direct or indirect participation of oxygen. For this reason, lithotrophic methanogenesis, lithotrophic acetogenesis and anaerobic CO oxidation can be regarded as the most likely catabolic processes to take place in an autonomous microbial system.

The application of radioisotopic methods makes possible the direct measurement of rates of these processes under *in situ* conditions and under laboratory conditions imitating those *in situ*. The results showed that high-temperature lithotrophic methanogenesis and acetogenesis do indeed occur *in situ* (quantitative characterization of anaerobic CO-oxidation is on the way). However, the rates of primary production were significantly higher than those of energy-giving processes studied; thus, inorganic carbon assimilation must be coupled with other lithoautotrophic processes in the frames of the same ecosystem. Radioisotopic experiments *in situ* showed how difficult it is to find an ecosystem that would not be connected with the photosynthesis-based biosphere: in the form of direct input of allochthonous organic matter, utilization of energy substrates from buried organic matter, or of traces of oxygen used by microaerophilic lithoautotrophic prokaryotes. Even so, recent investigations showed that the finding of habitats existing on the modern Earth and depending solely on the energy of inorganic compounds of volcanic origin is not an unfeasible task.

6. Acknowledgements

This work was supported by the “Molecular and Cell Biology” Program of the Russian Academy of Sciences.

7. References

Alain, K., Querellou, J., Lesongeur, F., Pignet, P., Crassons, P., Ragnenes, G., Cueff, V. and Cambon-Bonavita, M.-A. (2002) *Caminibacter hydrogenophilus* gen. nov. sp. nov., a novel thermophilic hydrogen-oxidizing bacterium isolated from an East-Pacific Rise hydrothermal vent. Int. J. Syst. Evol. Microbiol. **52,** 1317-1323.

Anderson, R.T., Chapelle, F.H. and Lovley, D.R. (1998) Evidence against hydrogen-based microbial ecosystems in basalt aquifiers. Science **281,** 976-977.

Blöchl, E., Rachel, R., Burggraf, S., Hafenbradl, D., Jannasch, H.W. and Stetter, K.O. (1997) *Pyrolobus fumarii* gen. nov., sp. nov., represents a novel group of archaea, extending the upper temperature limit for life to 113°C. Extremophiles **1,** 14-21.

Bonch-Osmolovskaya, E.A., Sokolova, N.G., Kostrikina, N.A., and Zavarzin, G.A. (1990) *Desulfurella acetivorans* gen. nov., sp. nov., a new thermophilic sulfur-reducing bacterium. Arch. Microbiol. **153,** 151-155.

Bonch-Osmolovskaya, E.A., Miroshnichenko, M.L., Slobodkin, A.I., Sokolova, T.G., Karpov, G.A., Kostrikina, N.A., Zavarzina, D.G., Prokofeva, M.I., Rusanov, I.I. and Pimenov, N.V. (1999) Distribution and diversity of anaerobic lithotrophic prokaryotes in terrestrial hydrotherms of Kamchatka. Mikrobiologiya, **68,** 398-406.

Bonch-Osmolovskaya, E.A., Miroshnichenko, M.L., Lebedinsky, A.V., Chernyh, N.A., Nazina, T.N., Ivoilov, V.S., Belyaev, S.S., Boulygina, E.S., Lysov, Yu.P., Perov, A.N., Mirzabekov, A.D., Hippe, H., Stackebrandt, E., L'Haridon, S. and Jeanthon, C. (2003) Radioisotopic, culture-based and oligonucleotide microchip analyses of thermophilic microbial communities in a continental high-temperature petroleum reservoir. Appl. Environ. Microbiol. **69,** 6143-6151.

Boston, P.M., Ivanov, M.V. and McKay, C.P. (1992) On the possibility of chemosynthetic ecosystem in subsurface habitats on Mars. Icarus **95,** 300-308.

Brock, T.D. Thermophilic microorganisms and life at high temperatures. 1978. New York: Springer-Verlag.

Chapelle, F.H., O'Neil, K., Bradley, P.M., Ciufo, S.A., Knobel, L.L. and Lovley, D.R. (2002) A hydrogen-based subsurface microbial community dominated by methanogens. Nature, **415,** 312-315.

Collins, M.D., Lawson, P.A., Willems, A., Cordoba, J.J., Fernandez-Garayzabal, J., Garsia, P., Cai, J., Hippe. H. and Farrow, J.A.E. (1994) The phylogeny of the genus *Clostridium:* proposal of five new genera and eleven new species combinations. Int. J. Syst. Bacteriol. **44,** 812-826.

Daniel, S.L., Hsu, T., Dean, S.I. and Drake, H.L. (1990) Characterization of the H_2- and CO-dependent chemolithoautotrophic potentials of the acetogens *Clostridium thermoaceticum* and *Acetogenium kivui*. J. Bacteriol. **172,** 4464-4471.

Daniels, L., Fuchs, G., Thauer, R. and Zeikus, J.G. (1977) Carbon monoxide oxidation by methanogenic bacteria. J. Bacteriol. **132,** 118-126.

Deming, J.W. and Baross, J.A. (1993) Deep-sea smokers: windows to a subsurface biosphere? Geochim. Cosmochim. Acta **57,** 3219-3230.

Fischer, F., Zillig, W., Stetter, K.O. and Schreiber, G. (1983) Chemolithotrophic metabolism of extremely thermophilic archaebacteria. Nature **301,** 511-513.

Götz, D., Banta, A., Beveridge, T.G., Rushdi, A.I., Simoneit, B.R.T. and Reysenbach, A.-L. (2001). *Persephonella marina* gen. nov., sp. nov. and *Persephonella guaymasensis* sp. nov., two novel thermophilic hydrogen-oxidizing microaerophiles from deep-sea hydrothermal vents. Int. J. Syst. Evol. Microbiol. **52,** 1349-1359.

Greene, A.C., Patel, B.K.C. and Sheehy, A. (1997) *Deferribacter thermophilus* gen. nov., sp. nov., a novel thermophilic manganese- and iron-reducing bacterium isolated from a petroleum reservoir. Int. J. Syst. Bacteriol. **47,** 505-509.

Haffenbradl, D., Keller, M., Dirmeier, R., Rachel, R., Rossnagel, P., Burggraf, S., Huber, H., and Stetter, K.O. (1996) *Ferroglobus placidus* gen. nov., sp. nov., a novel hyperthermophilic archaeum that oxidizes Fe^{2+} at neutral pH under anoxic conditions. Arch. Microbiol. **166,** 308-314.

Huber, H., Thomm, M., König, H., Ties, G. and Stetter, K.O. (1982) *Methanococcus thermolithotrophicus*, a novel thermophilic lithotrophic methanogen. Arch. Microbiol. **132,** 47-50.

Huber, R., Kristjansson, J.K. and Stetter K.O. (1987) *Pyrobaculum* gen. nov., a new genus of neutrophilic rod-shaped archaebacteria from continental solfataras growing optimally at 100°C. Arch. Microbiol. **145,** 95-101.

Huber, R., Wilharm, T., Huber, H., Trincone, A., Burggraf, S., König, H., Rachel, R., Rockenger, I., Fricke, H. and Stetter, K.O. (1992) *Aquifex pyrophilus* gen. nov.., sp. nov. represents a novel group of marine hydrogen oxidizing bacteria. Syst. Appl. Microbiol. **15,** 340-351.

Huber, R., Rossnagel, P., Woese, C.R., Rachel, R., Langworthy, T.A. and Stetter, K.O. (1996) Formation of ammonium from nitrate during chemolithotrophic growth of the extremely thermophilic bacterium *Ammonifex degensii* gen. nov., sp. nov. Syst. Appl. Microbiol. **19,** 40-49.

Huber, H., Burggraf, S., Mayer, T., Wyschkony, I., Rachel, R. and Stetter, K.O. (2000) *Ignicoccus* gen. nov., a novel genus of hyperthermophilic, chemolithotrophic Archaea, represented by two new species, *Ignicoccus islandicus* sp. Nov. and *Ignicoccus pacificus* sp. nov. Int. J. Syst. Evol. Microbiol. **50,** 2093-2100.

Huber, H., Diller, S., Horn, C. and Rachel, R. (2002) *Thermovibrio ruber* gen. nov., sp. nov., an extremely thermophilic, chemolithoautotrophic, nitrate-reducing bacterium that forms a deep branch within the phylum Aquificae. Int. J. Syst. Evol. Microbiol. **52,** 1859-1865.

Hugenholz, P., Pitulle, C., Hershberger, K.L. and Pace, N.R. (1998) Novel division level bacterial diversity in a Yellowstone hot spring. J. Bacteriol. **180,** 366-376.

Jannasch, H.W. and Mottl, M.J. (1985) Geomicrobiology of deep-sea hydrothermal vents. Science, **229,** 717-725.

Jeanthon, C., L'Haridon, S., Cueff V., Banta, A., Reysenbach, A.-L. and Prieur, D. (2002) *Thermodesulfobacterium hydrogenophilum* sp. nov., a thermophilic chemolithoautotrophic sulfate-reducing bacterium isolated from a deep-sea hydrothermal vent at Guaymas Basin, and emendation of the genus *Thermodesulfobacterium.* Int. J. Syst. Microbiol., **52,** 765-772.

Jones, W.J., Leigh, J.A., Mayer, F., Woese, C.R. and Wolfe, R.S. (1983) *Methanococcus jannaschii* sp. nov., an extremely thermophilic methanogen from a submarine hydrothermal vent. Arch. Microbiol. **136,** 254-261.

Kashefi, K., Tor, J.M., Holmes, D.E., Gaw Van Praagh, C.V., Reysenbach, A.-L. and Lovley, D.R. (2002) *Geoglobus ahangari* gen. nov., sp. nov., a novel hyperthermophilic archaeon capable of oxidizing organic acids and growing autotrophically on hydrogen with Fe(III) serving as the sole electron acceptor. Int. J. Syst. Evol. Microbiol. **52,** 719-728.

Kashefi, K. and Lovley, D.R. (2003) Extending the upper temperature limit for life. Science, **301,** 934.

Kashefi, K., Holmes, D.E., Reysenbach, A.-L. and Lovley, D.R. (2002) Use of Fe(III) as an electron acceptor to recover previously uncultured hyperthermophiles: isolation and characterization of *Geothermobacterium ferrireducens* gen. nov., sp. nov. Appl. Environm. Microbiol. **68,** 1735-1742.

Kurr, M., Huber, R., König, H., Jannasch, H.W., Fricke, H., Trincone, A., Krstjansson, J.K. and Stetter, K.O. (1991) *Methanopyrus kandleri,* gen. and sp. nov. represents a novel group of hyperthermophilic methanogens, growing at 110°C. Arch. Microbiol. **156,** 239-247.

L'Haridon, S., Cilia, V., Messner, P., Raguenes, G., Gambacorta, A., Sleytr, U.B., Prieur, D. and Jeanthon, C. (1998) *Desulfurobacterium thermolithotrophum* gen. nov., sp. nov., a novel autotrophic, sulfur-reducing bacterium isolated from a deep-sea hydrothermal vent. Int. J. Syst. Bacteriol. **48,** 701-711.

Leigh, J.A., Mayer, F. and Wolfe, R.S. (1981) *Acetogenium kivui,* a new thermophilic hydrogen-oxidizing, acetogenic bacterium. Arch. Microbiol. **129,** 275-280.

McColom, T.M. (1999) Methanogenesis as a potential source of chemical energy for primary biomass production by autotrophic organisms in hydrothermal systems on Europa. J. Geophys. Res. **104** (E12), 30729-39742.

Miroshnichenko, M.L., Kostrikina, N.A., L'Haridon, S., Jeanthon, S., Hippe, S., Stackebrandt, E., Bonch-Osmolovskaya, E.A. (2002) *Nautilia lithotrophica* gen. nov., sp. nov., a thermophilic sulfur-reducing epsilon proteobacterium isolated from a deep-sea hydrothermal vent. Int. J. Syst. Evol. Microbiol. **52,** 1299-1304

Miroshnichenko, M.L., L'Haridon, S., Schumann, P., Spring, S., Bonch-Osmolovskaya, E.A., Jeanthon, C. and Stackebrandt E. (2004) *Caminibacter profundus* sp. nov., isolated from a deep-sea hydrothermal vent represents a novel thermophilic bacterium of the *Nautiliales* ord. nov. within the class "*Epsilonproteobacteria*". Int. J. Syst. Evol. Microbiol. **54,** 41-45.

Moussard, H., L'Haridon, S., Tindall, B., Banta, A., Schumann, P., Stakebrandt, E., Reysenbach, A.-L. and Jeanthon, C. (2004) *Thermodesulfatator indicus* gen. nov., sp. nov., a novel thermophilic chemolithoautotrophic sulfate-reducing bacterium isolated from the Central Indian Ridge. Int. J. Syst. Evol. Microbiol., **54,** 227-233.

Nakagawa, T., Takai, K., Iganaki, F., Horikoshi, K. and Sako, Y. (2005) *Nitratiruptor tergarcus* gen. nov., sp. nov., and *Nitratifractor salsuginis* gen. nov., sp. nov., two nitrate-reducing chemolithoautotrophs within the epsilon-*Proteobacteria,* isolated from a deep-sea hydrothermal system in the Mid-Okinawa Trough. J. Syst. Evol. Microbiol., **55,** 925-933.

Namsaraev, B.B., Bonch-Osmolovskaya, E.A., Miroshnichenko, M.L., Pikuta, E.V., Kachalkin, V.I. , Miller, Yu. M., Propp, L.I. and Tarasov, V.G. (1994) Microbial processes of the carbon cycle in shallow-water hydrotherms of the Western margin of the Pacific Ocean. Microbiologiya **63,** 110-111.

Rees, G.N., Grassia, G.S., Sheehy, A.J., Dwivedi, P. and Patel, B.K.C. (1995) *Desulfacinum infernum* gen. nov., sp. nov., a thermophilic sulfate-reducing bacterium from a petroleum reservoir. Int. J. Syst. Bacteriol. **45,** 85-89.

Reysenbach, A-L., Longnecker, K. and Kirshtein, J. (2000) Novel bacterial and archaeal lineages from an in situ growth chamber deployed at a Mid-Atlantic Ridge hydrothermal vent. Appl. Environ. Microbiol. **66,** 3788-3797.

Reysenbach, A.-L. and Cady, S.L. (2001) Microbiology of ancient and modern hydrothermal systems. Trends in Microbiol. **3,** 79-87.

Schoenheit, P. and Schaefer, T. Metabolism of hyperthermophiles. (1995) World J. Microbiol. Biotechnol. 1995. **11,** 26-57.

Segerer, A., Neuner, A., Kristjansson, J.K. and Stetter, K.O. (1986) *Acidianus infernus* gen. nov., sp. nov. and *Acidianus brierley* comb. nov. facultatively aerobic, extremely acidophilic thermophilic sulfur-metabolising archaebacteria. Int. J. Syst. Bacteriol. **36,** 559-564.

Segerer, A., Trincone, A., Gahrtz, M. and Stetter, K.O. (1991) *Stigiolobus azoricus* gen. and sp. nov. represents a novel genus of anaerobic, extremely thermoacidophilic Archaea of the order *Sulfolobales.* J. Bacteriol. **41,** 495-501.

Slobodkin, A., Reysenbach, A.-L., Strutz, N., Dreier, M. and Wiegel, J. (1997) *Thermoterrabacterium ferrireducens* gen. nov. and sp. nov., a thermophilic anaerobic dissimilatory Fe(III)-reducing bacterium from a continental hot spring. Int. J. System. Bacteriol. **47,** 541-547.

Sokolova, T.G., Gonzales, J.M., Kostrikina, N.A., Chernyh, N.A., Tourova T.P., Kato C., Bonch-Osmolovskaya, E.A. and Robb, F.T. (2000) *Carboxydobrachium pacificum* gen. nov., sp. nov., a new anaerobic, thermophilic, CO-utilizing marine bacterium from Okinawa Trough. Int. J. Syst. Evol. Microbiol. **51,** 141-149.

Sokolova, T.G., Kostrikina, N.A., Chernyh, N.A., Tourova, T.P. and Bonch-Osmolovskaya, E.A. (2002) *Carboxydocella thermautotropha* gen. nov., sp. nov., a novel anaerobic CO-utilizing thermophile from a Kamchatkan hot spring. Int. J. Syst. Evol. Microbiol. **52,** 1961-1967.

Sokolova, T.G., González, J.M., Kostrikina, N.A., Chernyh, N.A., Slepova, T.V., Bonch-Osmolovskaya, E.A. and Robb, F.T. (2004a) *Thermosinus carboxydivorans* gen. nov., sp. nov., a new anaerobic thermophilic carbon monoxide oxidizing hydrogenogenic bacterium from a hot spring of Yellowstone National Park. Int. J. Syst. Evol. Microbiol. **54,** 2353-2359.

Sokolova, T.G., Jeanthon, C., Kostrikina, N.A., Chernyh, N.A., Lebedinsky, A.V., Stackebrandt, E. and Bonch-Osmolovskaya, E.A. (2004b) The first evidence of anaerobic CO oxidation coupled with H_2 production by a hyperthermophilic archaeon isolated from deep-sea hydrothermal vents. Extremophiles **8,** 317-323.

Stetter, K.O., Thomm, M., Winter, J., Wildgruber, G., Huber, H., Zillig, W., Janecovic, D., König, H., Palm, P., Wunderl, S. (1981) *Methanothermus fervidus* sp. nov., a novel extremely thermophilic methanogen isolated from Icelandic hot spring Zentralbl. Bakteriol. Hyg. I. Abt. Orig. **C 2,** 166-178.

Stetter, K.O., König, H. and Stackebrandt, E. (1983) *Pyrodictium* gen. nov., a new genus of submarine disc-shaped sulfur-reducing archaebacteria growing optimally at 105°C. Syst. Appl. Microbiol. **4,** 535-551.

Stetter, K.O. (1988) *Archaeoglobus fulgidus* gen. nov., sp. nov., a new taxon of extremely thermophilic archaebacteria. Syst. Appl. Microbiol. **10,** 172-173.

Stetter, K.O. (1996) Hyperthermophilic prokaryotes. FEMS Microbiol. Revs. **18,** 149-158.

Stevens, T.O. and McKinley, J.P. (1995) Lithoautotrophic microbial ecosystems in deep basalt aquifiers. Science **270,** 450-454.

Svetlichny, V.A., Sokolova, T.G., Gerhardt, M., Ringpfeil, M., Kostrikina, N.A. and Zavarzin, G.A. (1991) *Carboxydothermus hydrogenoformans* gen.nov., sp.nov., a CO-utilizing thermophilic anaerobic bacterium from hydrothermal environments of Kunashir Island. System. Appl. Microbiol. **14,** 254-260.

Svetlitchny, V.A., Peschel, C., Acker, G. and Meyer, O. (2001) Two membrane-associated NiFeS-carbon monoxide dehydrogenases from the anaerobic carbon-monoxide-utilizing eubacterium *Carboxydothermus hydrogenoformans*. J. Bacteriol. **183,** 5134-5144.

Takai, K., and Horikoshi., K. (1999) Genetic diversity of Archaea in deep-sea hydrothermal vent environments. Genetics **152,** 1285-1297.

Takai, K., Nakagawa, S., Sako, Y. and Horikoshi, K. (2003a) *Balnearium lithotrophicum* gen. nov., sp. nov., a novel thermophilic, strictly anaerobic, hydrogen-oxidizing chemolithoautotroph isolated from a black smoker chimney in the Suiyo Seamount hydrothermal system. Int. J. Syst. Evol. Microbiol. **53,** 1947-1954.

Takai, K., Nealson, K. and Horikoshi, K. (2003b) *Hydrogenimonas thermophila* gen. nov., sp. mov., a novel thermophilic hydrogen-oxidizing chemolithoautotroph within the ε-*Proteobacteria,* isolated from a black smoker in a Central Indian Ridge hydrothermal field. Int. J. Syst. Evol. Microbiol. **54,** 25-32.

Takai, K., Gamo, T., Tsunogai, U., Nakayama, N., Hirayama, H., Nealson, K.H. and Horikoshi, K. (2004) Geochemical and microbiological evidence for a hydrogen-based, hyperthermophilic subsurface lithoautotrophic microbial ecosystem (HyperSLiME) beneath an active deep-sea hydrothermal field. Extremophiles **8,** 269-283.

Takai, K., Hirayama, H., Nakagawa, T., Suzuki, Y., Nealson, K. and Horikoshi, K. (2005) *Lebetimonas acidiphila* gen. nov., sp. nov., a novel thermophilic, acidophilic, hydrogen-oxidizing chemolithoautotroph within the epsilon-*Proteobacteria,* isolated from a deep-sea hydrothermal fumarole in the Mariana Arc. Int. J. Syst. Evol. Microbiol., **55,** 183-189.

Tor, J.M., Kashefi, K. and Lovley, D.R. (2001) Acetate oxidation coupled to Fe(III) reduction in hyperthermophilic microorganisms. Appl. Environ. Microbiol. **67,** 1363-1365.

Völkl, P., Huber, R., Drobner, E., Rachel, R., Burggraf, S., Trincone, A. and Stetter, K.O. (1993) *Pyrobaculum aerophilum* sp. nov., a novel nitrate-reducing hyperthermophilic archaeum. Appl. Environm. Micorbiol. **59,** 2918-2926.

Wasserfallen, A., Nolling, J., Pfister P., Reeve J. and Conway de Macario, E. (2000) Phylogenetic analysis of 18 thermophilic *Methanobacterium* isolates supports the proposals to create a new genus, *Methanothermobacter* gen. nov., and to reclassify several isolates in three species, *Methanothermobacter thermautotrophicus* comb. nov., *Methanothermobacter wolfeii* comb. nov., and *Methanothermobacter marburgensis* sp. nov. Int. J. Syst. Evol. Microbiol. **50,** 43-53.

Whitman, W.B., Boone, D.R. and Koga, Y. (2001) *Methanococcales.* In: G.M. Garrity (ed.) *Bergey's Manual of SystematicBacteriology.* 2nd edn. Springer, Berlin, Heidelberg, New York, p. 236.

Widdel, F. and Pfennig, N. (1984) Dissimilatory sulfate- and sulfur-reducing bacteria. In: N.R. Krieg, and J.G. Holt (eds.) *Bergey's Manual of SystematicBacteriology.* The Williams & Wilkins Co, Baltimore, pp. 663-679.

Wiegel, J., Braun, M. and Gottschalk, G. (1981) *Clostridium thermoautotrophicum* species novum, a thermophile producing acetate from molecular hydrogen and carbon dioxide. Curr. Microbiol. **5,** 255-260.

Woese, C.R. and Fox, G.E. (1977) Phylogenetic structure of the prokaryotic domain: The primary kingdoms. Proc. Nat. Acad. Sci. **74,** 5088-5090.

Zeikus, J.G. and Wolfe, R.S. (1972) *Methanobacterium thermoautotrophicum* sp. nov.: an anaerobic, autotrophic extreme thermophile. J. Bacteriol. **109,** 707-713.

Zeikus, J.G., Dawson, M.A., Thompson, T.E., Ingvorsen, K. and Hatchikian, E.C. (1983) Microbial ecology of volcanic sulfidogenesis: isolation and characterization of *Thermodesulfobacterium commune* gen. nov. and sp. nov. J. Gen. Microbiol. **129,** 1159-1169.

Zillig, W., Stetter, K.O., Schaefer, W., Janekovic, D., Wunderl, S., Holz, I. and Palm, P. (1981) *Thermoproteales*: a novel type of extremely thermoacidophilic anaerobic archaebacteria isolated from Icelandic solfataras. Zbl. Bakt. Hyg. I. Abt. Orig. 1981. **C2,** 205-227.

Biodata of **Nina Gunde-Cimerman** author of "***Life in Ice***"

Dr. Nina Gunde-Cimerman is currently Associate Professor of Microbiology at the University of Ljubljana (Slovenia), Biotechnical Faculty, Department of Biology and principal investigator of the research microbiology group. She obtained her Ph.D. from the University of Ljubljana in 1993. In addition, she is in charge of the MZKI Fungal Culture Collection of the National Institute of Chemistry at her campus. Dr. Gunde-Cimerman scientific interests are in the areas of the microbiology, ecology, taxonomy and physiology of extremophilic fungi (primarily halophilic, xerophilic and psychrophilic) inhabiting solar salterns and hypersaline lakes as well as polar glacial and sub-glacial environments.

E-mail: **nina.gunde-cimerman@bf.uni-lj.si**

J. Seckbach (ed.), Life as We Know It, 107-124.

LIFE IN ICE

NINA GUNDE-CIMERMAN[1,2]
[1]*Biology Department, Biotechnical Faculty, University of Ljubljana, 1000 Ljubljana, Slovenia;* [2]*Laboratory of Biotechnology, National Institute of Chemistry, 1000 Ljubljana, Slovenia*

1. Introduction

Most of our planet is permanently cold. More than two thirds of Earth is covered by sea water, mostly deep oceans, with a constant low temperature of approx. 2°C. Besides, terrestrial and aquatic environments in polar and alpine regions, the upper atmosphere, plants and animals inhabiting cold regions and Antarctic rocks represent together more than 80% of the biosphere (Cavicchioli et al., 2002; Russell, 1990). Ice is characteristically common to all cold environments. For long it has been considered only as a life preserving medium, entrapping randomly deposited microbes that sometimes remain viable for extensive periods of time. Recent discoveries have unraveled that different types of ice, either as snow, sea ice, accretion or glacial ice, provide environments for active microbial life quite different from any other (Christner et al., 2000; Price and Sowers, 2004).

2. Different ice Environments

2.1. SEA-ICE

Sea-ice environments are geologically recent developments on Earth, with persisting polar ice caps having appeared only 60 millions years ago. Sea ice is temporally and spatially highly variable and is for the most part seasonally transient, constantly being broken up and reshaped by wind and ocean currents.

When temperatures seasonally drop, frozen seawater forms a semisolid matrix, permeated by a network of channels and pores, filled with brine formed from expelled salts as the ice crystals freeze. The salinity of sea ice brines within channels and cracks can become as high as 200 grams per liter and drop to below 10 grams per liter as the sea ice melts. Sea ice represents the coldest habitats on Earth, with temperatures ranging from –1°C to as low as –50°C in winter, with brine networks that remain liquid to –35°C. Changes in salinity and temperature are the dominant factors influencing the biological assemblages within sea ice. Seasonal differences in light further contribute to its complexity (Brown and Bowman, 2001; Thomas and Dieckman, 2002). It is within this labyrinth that sea-ice organisms live.

The nature of sea ice microbial communities is primarily determined by physical forces. These complex assemblages of organisms are adapted to swift transitions between water and ice and the subsequent rapid physical and chemical changes. They seasonally rise through the water when surface waters freeze (Thomas and Dieckmann, 2002). In winter most of the microbial biomass was detected in brine pockets within the central mass of the ice. Microorganisms stick to or are caught between ice crystals and move within its brine channels. In spring, a dense community sometimes develops at the bottom of the ice, where the ice algae can extract nutrients continuously from the underlying water. The highest concentrations of bacteria were found in this layer as well (Gosink et al., 1993). When snow and ice at the surface melt in spring, the bottom layers of ice sink in the water, providing the first pulse of primary production in the water column. In summer microorganisms in sea ice are released and seed the oceans with psychrotolerant and psychrophilic micro-organisms.

Sea ice microbial communities have been shown to have a major role in trophic food webs (Gosink et al., 1993). Well developed autotrophic algal communities usually support a microbial loop and a host of allochthonous and autochthonous protists, fungi, zooplankton and small metazoans, that are all playing an important role in secondary mineralization of dissolved and particulate organic matter.

Most bacteria isolated from sea ice have been found to be pigmented, with half of the taxa psychrophilic, while the rest were tolerant (Brown and Bowman, 2001). Many of the isolated bacterial strains were known for their attachment to particles and extracellular enzymes that degrade a wide variety of polymeric substances.

Nevertheless, the amount of specific data available on sea ice dwelling bacteria, a well as on heterotrophic Eukarya is still very limited (Brown and Bowman, 2001; Junge et al., 2002; Gunde – Cimerman et al., 2005). Conventional cultivation methods have yielded many bacterial isolates with novel 16S rRNA sequences, but they have also revealed that, unlike in other environments, cultivable organisms can comprise surprisingly large fractions of the total population (Deming, 2002). When DNA was extracted from seven Antarctic and one Arctic sea ice sample, no Arcahea, but bacterial phylotypes grouped within the alpha and gamma proteobacteria were found. The fact that all bacterial isolates from Arctic and Antarctic sea ice position at shallow branches within phylogenetic trees tends to support the idea of relative evolutionary youth (Thomas and Dieckmann, 2002). The majority of bacterial phylotypes were heterotrophic taxa, while the detected eukaryotic clones were determined as belonging to a variety of autotrophic and heterotrophic nanoplankton, including viruses, bacteria, algae, fungi, protists, flatworms and small crustaceans. Studies based on 16 S rRNA clone libraries indicated that certain sea ice dwelling bacterial communities are common at both poles (Brown and Bowman, 2001).

2.2. ICE SHELVES

Ice shelves consist of 20–100m thick sea ice that occurs in polar regions. Until recently, ice shelf ecosystems were thought to be restricted to Antarctica. However ice shelf cryo-ecosystem that began to form about 4500 years ago during Holocene cooling, were described as well in Arctic regions. Since that time they contracted by 90%, probably as a result of a warmer climate. They are still undergoing further fragmentation and loss (Vincent et al., 2004).

All ice shelves are characterized by undulating surface topographies in which elongate lakes and streams are oriented parallel to each other and to the prevailing wind. Large quantities of marine sediments can be brought up to the surface of the ice by basal freezing and surface ablation and redistributed by streams and wind across the ice shelves. These sediments at the base of the melt-water lakes and in melt-holes on the ice ridges contain highly pigmented microbial mats connected with the ice surfaces. Water temperatures are typically near 0°C, but may be slightly elevated in the more saline layers overlying some of the microbial mats.. The major ion composition resembles diluted seawater and is dominated by sodium, chloride, and high concentrations of sulfate. Inorganic carbon concentrations, phosphorous and dissolved organic nitrogen values are low in melt-water but can be order-of-magnitude higher in the interstitial waters of the microbial mats (Vincent, 2000; Vincent et al., 2004).

The microbial mat communities are dominated by cold-tolerant prokaryotes that provide local refuge for the growth and development of viruses and more complex organisms. The microbial mat consortia thus consists of viruses, bacteria, filamentous cyanobacteria, protists including diatoms, flagellates and ciliates, and micro-invertebrates including nematodes, rotifers and turbellaria (flat-worms). Although similar microbial communities are found in both polar regions, their species composition and mat architecture may differ significantly. These perennial communities remain frozen throughout almost all the year, punctuated by brief periods of thaw each summer. The persistent and large over-wintering biomass is then available to initiate photosynthesis and growth during the brief periods of thaw. Strategies that allow these communities to flourish include tolerance of freeze and thaw cycles, prolonged dormancy, growth at continuously low temperatures and the ability to survive extremely high solar irradiance as well as winter darkness (Vincent et al., 2004).

2.3. GLACIAL ICE

Glacial ice is formed from compacted snow at the poles as well as at lower latitude high altitude locations. Samples of the atmosphere and particulates prevailing at the time of snowfall are preserved chronologically. Thus ice core analyses have been used to document and date past climate changes, geological events and human activities. Direct sampling has repeatedly demonstrated the presence of contemporary and ancient microorganisms, characteristic of northern latitudes, temperate as well as tropical regions (Abyzov, 1993; Ma et al., 1999 a, b). Viable bacteria have been isolated and identified either by direct isolation or by DNA sequencing in glacial ice from the Andes, Himalayas, and Arctic and Antarctic regions. Some sequences were similar to contemporary species, while others exhibited little similarity to well-characterized present-day taxons. Species diversity varied among ice cores. They have been immured in ice for periods ranging from months up to 750.000 years (Reeve, 2002). The numbers of bacteria present in ice were generally low, with the highest numbers determined for non-polar glaciers. It seems that episodic deposition events rather than age primarily determined the numbers of recoverable bacteria (Christner et al., 2003)

By PCR amplification of fragments of the eukaryotic 18 S rRNA gene, extracted from 2000 and 4000 years old ice-core samples from North Greenland, a diversity of fungi, plants, algae and protists were identified (Price, 2000). Viable fungi were directly isolated as well from 38.000 years old ice (Abyzov, 1993; Ma et al., 1999 a, b).

2.4. CRYOCONITE HOLES

Cryoconite holes occur on glaciers in the Arctic and Antarctic as well as in alpine environments in the glacial ablation zone. They are formed when wind-blown particulates lodge on the surface in small depressions in the ice and cause vertical melting by absorbing more radiation than the surrounding ice. The result is a sediment inclusion, 90% immersed in water with the remaining head space filled with air surrounded by ice. These holes range from less than 1cm to 1m in width. They seldom grow deeper than 60 cm. They can coalesce into bigger holes or become interconnected by melt-water channels. Although the vast majority of cryoconite holes can be classified as ultra-oligotrophic, they nevertheless provide a habitat for microbial colonization and growth. The dominant organisms are typically filamentous mat-forming Cynobacteria of the order Oscillatoriales capable of nitrogen fixation. Photosynthesis by these cyanobacteria as well as algae can generate sufficient nutrients to support the development of complex communities. Light microscopy has documented the presence of pollen grains, bacteria, algae, diatoms, filamentous fungi, yeasts and occasionally micro-invertebrates such as rotifers, tardigrades and nematodes. Rarely psychrophiles have been detected amongst which psychrotrophs prevail (Mueller et al., 2001; Vincent et al., 2004).

As solar radiation decreases, an ice cover forms at the water surface and grows downwards. During the freezing solutes are rejected from the ice and therefore concentrate until the temperature drop stabilizes or the solutes precipitate. With faster freezing rates solutes can not fractionate and can therefore become trapped in bubbles or brine pockets with up to 5 times the salinity of sea-water. Such ponds freeze only after reaching brine temperatures of –12°C. Water that freezes in the cryoconite holes forms so-called black ice. Without solar irradiation winter glacial ablation can occur only via wind-induced sublimation and as a result cryoconite holes become shallower. As the solar input becomes stronger again, melting takes place at the sediment-ice contact zone, thereby melting downwards, away from the ice cover. Since the survival of cryoconite holes largely depends on the melt regime of the glacier surface, cryoconite holes can last from several days to weeks or entire seasons. On some stable glaciers they are thought to have remained for at least hundred years, serving as biological refuges during periods of extreme cold (Mueller et al., 2001; Reeve et al., 2002; Vincent et al., 2004).

2.5. SUBGLACIAL ENVIRONMENTS

Glacial ice has been until recently considered primarily as an extremely stable, frigid and static environment. However, recent investigations have shown that ice and glaciers are much more dynamic than previously assumed. Ice in glaciers can be permeated by a continuous network of aqueous veins, formed due to sea salts deposited as aerosols, essentially insoluble in ice crystals. The thickness of the quasi liquid water layer on the surface of an ice crystal has been calculated to be about 50 nm at a temperature of -10°C (Cavicchioli et al., 2002).

Besides, polythermal glaciers are particularly dynamic as well on the geomorphological level. Recently it has been shown that they are capable of previously unrecognized rapid movements. These events are produced when such glaciers slide abruptly along the beds causing displacements of a large mass of ice over a relatively short distance. Such large mass movements cause increased water pressure at the base of the glacier and short term loss of the connection between ice and underlying bed. These glaciers undergo rapid thinning in their low reaches, causing acceleration of ice discharge at the outlets near the ice edge. The scale for such changes in very large outlet glaciers, experiencing substantial surface melting, can be from 10,000 years to a few weeks or days and even within a single day (Ekstrom et al., 2003; Fohnestock, 2003).

Additionally, polythermal glaciers are characterized by cold (subfreezing) ice at the surface, margins, and terminus of the glacier and a warm core. The basal ice beneath the glacier can be at the pressure melting point and in this case basal melting occurs. Subglacial waters, flowing at the bed of the glacier, thus originate as supraglacial waters, which reach the bed through crevasses and moulins, and of groundwater generated by frictional and geothermal melting of ice at the glacier base. These subglacial waters interact with the rocks and sediments that underlie the ice and hence contain high solute concentrations and high suspended sediment concentrations. If refreezing, the ice contains significant sediment concentrations and thus provide a record of the geochemical and microbiological conditions at the glacier bed. The subglacial environment is not easily accessible for microbiological analyses because of its depth and difficulties in sampling aseptically. However, basal ice and sediments transported to and exposed at the glacier margins are relatively easily accessible and represent the local subglacial environment (Foght et al., 2004).

These environments were previously considered abiotic, but recent investigations have shown that this type of ice can provide a habitat for non-photosynthetic microbial life and a significant global reservoir of biological activity, with the potential to influence glacial melt-water chemistry. Subglacial bacterial populations have been recently found in Northern and Southern hemisphere and beneath high and low altitude glaciers (Lanoil, 2004).

A clear link between the presence of sediment and the level of microbial activity was observed. The majority of subglacial bacteria were associated with debris-rich ice from subglacial sediments, with smaller numbers in the overlying clean glacier ice. It seems that the sediment, present in the soils that the glacier overrides, is the primary source of bacteria. It provides nutrient-enriched microzones and there is also evidence that the bacteria are physically attached to the sediment particles. However, bacteria may also be washed into the subglacial environment from the glacial surface via crevasses or localized subglacial melting (Skidmore et al., 2000).

In the basal ice investigated so far numerous types of aerobic and anaerobic bacteria have been detected, including heterotrophs, nitrate and sulfate reducers, and methanogens. The presence of methanogens suggests that although the subglacial melt-waters are essentially aerobic, the underlying thawed subglacial sediments contain as well anaerobic microenvironments. The aerobic heterotrophs showed a high proportion of pigmented colonies. Total bacterial counts in thawed basal ice ranged from 10^5–10^7 cells/ml. The high degree of cultivability of the total bacterial counts was surprising, but in agreement with studies of microbial populations in sea ice. It was speculated that it may be a function of the relative simplicity of these cold environments (Lanoil, 2004).

2.6. POLAR LAKES

Most polar lakes are characterized by prolonged, sometimes perennial ice cover, typically 3-6 m thick, containing a layer of sand and organic matter of Aeolian origin. This layer represents a dynamic equilibrium between downward movement of sediments as a result of melting during the summer and upward movement of ice from ablation at the surface and freezing at the bottom. During summer, when solar radiation is continuous, liquid water inclusions are present in this layer. The ice melt-water supports a viable microbial assemblage associated with the sediment layer, thus permanent ice covers provide viable habitats for the microscale proliferation of life (Kriss et al., 1976; Priscu et al., 1998; Belzile et al., 2001; de Wit et al., 2003; Gaidos, 2004).

Antarctic lake Vostok represents a special example of polar lakes. It is comprised of two types of ice; the upper 3500 m layer of ice contains traces of nutrients of aeolian origin including diverse acids, sea salts and mineral grains. Ice below 3500 m comprises refrozen water from lake Vostok, accreted to the bottom of the upper glacial ice. This layer is at least 420.000 years old and has a temperature of a few degrees below freezing. Nutrients in the accretion ice include salts and dissolved organic carbon and nitrogen. Total levels are high enough to support growth of a heterotrophic microbial assemblage. Microbiological analyses have revealed incorporation and respiration of radioactively labeled organic substrates. The accretion ice represents a habitat of interconnected liquid veins in which motile bacteria, probably originating from deep glacial ice, were detected by direct scanning (Siegert et al., 2001; Cavicchioli et al., 2002).

2.7. PERMAFROST

Permafrost underlies about 20% of Earth's land surface. This permanently frozen soil is known to be fixed by ice, but also to contain un-frozen water. The proportion of water decreases from 5-6% by weight at –1.5°C, to 2–3% unfrozen water at –10°C and 1-2% between –15-20°C (Rivkina et al., 2000). Aerobic and anaerobic microorganisms (bacteia, cyanobacteria, green algae, yeast and fungi) were isolated from permafrost sediments which were maintained in a frozen state for extended periods. Viable bacteria and even yeasts were isolated from up to 3 million years old Siberian permafrost, although eucaryotes were preserved in considerably lower numbers than prokaryotic cells (Golubev, 1998). Upon thawing these micro-organisms were able to resume their metabolic activity (Takano, 2004).

Until recently it was assumed that microorganisms caught within ice inclusions are in a state of suspended animation (anabiosis). However, radioactively labeled substrate incorporation experiments demonstrated typically shaped growth curves, which correlated with temperature (Rivkina et al., 2000). It seems that in frozen soils both soil particles and bacterial cells are covered by thin films of water. Nutrients thus reach the cells and waste products can be eliminated by diffusion. It seems that these processes are limited by the thickness of the unfrozen films of water, which in turn depends on the temperature. The thickness of the film decreases from about 15 nm at –1.5°C to about 5 nm at –10°C (Rivkina et al., 2000).

2.8. SNOW

Until recently, the most investigated microorganisms living on the surface of snow and within it were the snow algae. In temperate regions they color snow in shades of red, green and orange depending on the irradiance levels of sunlight. Their vertical distribution is associated with the water surface surrounding the snow crystals and their appearance is more related to the melting process than to phototaxis. In polar regions red snow is characteristic of low nutrients areas whereas green snow is most often associated with higher nutrients areas. Snow is an environment that is severely limited in nutrients and higher local levels of nutrients in the snow-pack therefore strongly affects the spatial distribution of algae (Hoham and Ling, 2000).

The majority of the dominant snow algae belong to the green algal flagellate class Chlamydomonadaceae. The true snow algae grow optimally at temperatures below 10°C. However, photosynthesis was reported even from frozen samples of snow algae. They may often be subject to overnight freezes or repeated freeze-thaw events. From year to year populations of snow algae stay in approximately the same locations. The resting spores remain dormant during summer, may form daughter cells through division after the first freezes in autumn and become covered with new snow in fall, winter and spring. As light is known to penetrate 2 m of wet snow (Hoham and Ling, 2000) photoactive responses may occur at this distance from the surface (Hoham and Duval, 2001).

Besides algae, bacterial populations and associated metabolic activity was recently detected on the surface and in the upper 20 cm of snow and firn collected at the South pole. Direct counts by epifluorescence microscopy revealed population densities from 5. 10^2–10^3 cells per ml of snow-melt. The cultivation independent approach based on 16S rDNA sequencing revealed that most of them belong to the Deinococcus – Thermus group. Similar bacteria have been also observed in soils of the Antarctic dry Valleys (Carpenter et al., 2000).

2.9. EXTRATERRESTRIAL HABITATS

Extreme polar environments are being investigated as analogues of viable extraterrestrial habitats: Arctic tundra for soil covered Mars, subglacial environments for similar environments under polythermic ice formations on Mars and Lake Vostok for ice-covered Europa.

Most information so far exists on the conditions on Mars. Early in its history, Mars had an atmosphere with greenhouse effect effect that eventually melted the ice on the red planet. Some models and recent photos suggest that during times of high obliquity on Mars, superficial melting may have occurred in the north polar ice cap. Under these conditions the ice cap may have exhibited polythermic conditions and may have been more dynamic. Thus basal melting may have occurred, producing sediment-rich basal ice. As the atmosphere of Mars dissipated due to insufficient planetary mass, the temperatures dropped, freezing all water on the surface. Mars explorations discovered remnants of ancient Martian salty seas that drained away thousands of years ago. As the conditions on early Mars got colder and harsher, it lost liquid water through evaporation or sequestration into permafrost. Remaining bodies of water became increasingly salty, and then finally all liquid water disappeared and the salt deposits eventually lithified into the evaporate rocks (Nasa web site, 2004). However, liquid water may have persisted on the surface of Mars for a billion years, a time span that allowed microbial life to appear on planet Earth. If life evolved similarly on Mars before martian freezing, then microorganisms resembling extremophilic life on Earth may still be viable on Mars. (Christner et al., 2000; Deming, 2002; Onofri et al., 2003).

3. Viability and Diversity of Microorganisms in Ice

Ice provides a frigid and stable environment and is a natural site for the study of the prolonged anabiosis at low temperature. The immediate environment protects microorganisms from UV irradiation, oxidation and chemical damage. When microorganisms become desiccated, proteins and nucleic acids decay rates are drastically reduced and thus DNA damage is minimized (Ma et al., 1999). Some microorganisms remained viable for 750.000 years in ice and up to 3 million years in permafrost (Rivkina et al., 2000).

When both Bacteria and Archaea were surveyed across a range of polar and other cold environments, the Bacteria were observed to dominate and be present in greater diversity than Archaea (Deming, 2002; Cavicchioli et al., 2002). The important exceptions were the cold interior and the deep waters of the oceans, well below the upper mixed layer. Conventional quantitative cultivation methods have revealed that culturable organisms comprised surprisingly large fractions of the total population. Nevertheless, many isolates with novel 16S rRNA sequences that grew at –1°C were discovered. Bacterial isolates were asigned to the genera *Arthrobacter, Aureobacterium, Bacillus, Bradyrhizobium, Brevibacterium, Cellulomonas, Clavibacter, Flavobacterium, Frankia, Friedmaniella, Methylobacterium, Microbacterium, Micrococcus, Micromonospora, Mycobacterium, Nocardia, Nocardoides, Paenibacillus, Planococcus, Propioniferax, Sphingomonas, Staphylococcus* and *Stenotropomonas* (Christner et al., 2000). Most of the isolated bacteria were ubiquitous, while certain genera such as *Mycobacterium, Micrococcus, Brevibacterium, Planococcus, Arthrobacter, Clavibacter* and *Friedmaniella*, were isolated previously from Siberian permafrost and tundra soil, or sea ice (Christner et al., 2000).

Recently, ultra small, free living bacteria were discovered in Greenland ice. Many of these proteobacteria were smaller than 1 um and could belong to the group of ultramicrobacteria (Miteva, 2004).

In the extremely cold area of the world subglacial environments represent quite special habitats. They enable, on a geological scale, active propagation of micro-organisms, which are later again subjected to freezing and anabiosis. The few studies performed so far indicate that these habitats are dominated by aerobic heterotrophic beta proteobacteria. (Foght et al., 2004).

In cold polar regions the majority of studies have been oriented towards psychrophilic bacteria, while the occurrence and diversity of psychrophilic eukaryotic microorganisms remains largely unknown. Few studies exist on the biodiversity of algae in sea ice while the presence of fungi was investigated in Antarctic soil and Siberian permafrost sediments (Babjeva and Reshetova, 1998; Rivkina et al., 2000; Takano, 2004; Onofri et al., 2004). Contrary to soils in temperate zones, where filamentous fungi prevail, permafrost and Antarctic soils are dominated by yeasts. The most commonly detected yeast genera were *Cryptococcus, Leucosporidium, Sporobolomyces*, *Rhodotorula* and *Cystofilobasidium.* They were found with the highest frequency in the youngest layers of permafrost, less than 10.000 years old, although they were also detected in three millions years old Pliocene samples. In all cases the share of the yeasts represented 25% of all aerobic heterotrophs, independent of the organic matter content (Dmitriev et al., 1997a, 1997b; Rivkina et al., 2000; Soinam et al., 2000).

So far there were no investigations on the occurrence of fungi in sea ice, although their characteristic small subunit rRNA gene sequences were present in DNA extracted from diverse Antarctic and one Arctic sea ice sample (Brown and Bowman, 2001).

The presence of fungi was least investigated in polar glaciers. Low numbers of different viable filamentous fungi have been isolated from ice ranging in age from 10,000–38,600 on both poles (Ma et al., 1999a, 1999b; Christner et al., 2000, 2002; Abyzov, 1993). Viable yeasts, primarily of the genera *Cryptococcus* and *Rhodotorula* have been found only in the upper, younger ice-sheet horizons and surface layers of ice and snow. All findings of fungi in glacier ice were interpreted as the result of coincidental Aeolian deposits of spores or mycelium into the ice during its geological history. No fungi have yet been discovered in subglacial environments.

4. Cellular Responses to Life at Low Temperatures

The microorganisms which have learnt to cope with the stresses imposed by life at low temperatures are named psychrophilic and psychrotolerant. By definition psychrotolerant organisms grow well at temperatures close to the freezing point of water, but have fastests growth above 20°C, whereas psychrophilic organisms grow fastest at temperatures of 15°C or lower but are unable to grow above 20°C (Deegenaars and Watson, 1998).

A reduction in temperature slows down most physiological processes. It changes protein-protein interactions, reduces membrane fluidity, solute uptake rates and provokes an increased viscosity of water, which can be followed by the formation of intracellular crystalline ice. Moreover, it induces a reduction in salt solubility, an increase in gas solubility and a decrease in the pH of biological buffers, affecting both the protein solubility and the charge of amino acids. Thus, particularly enzymes showing a high degree of hydrophobicity and multimeric enzymes such as ATP-ases and fatty acid synthetases are subject to cold denaturation (Gerday et al., 2000). However, the most temperature sensitive process appears to be translation. The ribosomes and associated proteins play an important role in temperature sensing (Cheng, 1998; Mindock, 2001; Cavicchioli et al., 2002).

Microorganisms revived from ice are likely to have endured desiccation, solar irradiation, freezing, a period of frozen dormancy and thawing. The main reaction of a microbial cell to overcome the associated stresses are decreased cell size, prevention of physical cell rupture caused by freezing and thawing, synthesis of intracellular solutes that prevent desiccation damage and changed composition of cell membranes. Adaptive strategies range from the level of individual types of molecules to that of the whole organism. Many microrganisms form endospores, have thick, pigmented cell walls and/or extracellular polysaccharides around the cells (Christner et al., 2000). The most frequently observed photoprotective compounds are melanins, carotenoids and mycosporine-like amino acids.

A characteristic of psychrophilic microorganisms is their small size, in comparison with the size of cells at "normal" temperatures. One of the major physical threats that a cell needs to respond to at very low temperature is ice formation within the cell, since the volume increase of water, as ice is formed, can lead to cell lysis.

It is known that due to ordering effects water in confined spaces has very different properties from bulk water. When bulk water freezes it typically forms crystalline ice, but when it freezes on surfaces, it causes a net alignment of the water molecules resulting in polarization of ice and amorphous structure. As a result it has a lowered freezing point, altered ionic solubility, increased viscosity and reduced dielectric constants compared to bulk water.

The typical bacterial cell is only a few microns and therefore a highly confined space. Thus, intracellular water close to its surface has a depressed freezing point, while the water in the middle of the cell has a very high solute concentration, creating an internal salt concentration gradient (Mindock et al., 2001; Pomeroy,

Wiebe 2001). At low temperatures in the environment ice is formed by the separation of pure water (as ice) and a corresponding increase in salt concentration, leading to an osmotic gradient across the cell membrane. An increase of solute concentration inside of the cell of substances which are at the same time osmolytes and cryoprotectants, can counteract this phenomenon (Blackwell et al., 2001).

The most critical metabolic requirement at low temperatures is maintenance of functional lipid membranes The main adaptations include regulation of its fluidity, passive and active permeability, the regulation of ion channels, which in turn influences nutrient uptake, electron transport, environmental sensing and recognition processes (Zlatanov et al., 2001; Tanghe et al., 2002). Psychrophiles achieve this notably by increasing the proportion of unsaturated fatty acids and the modulation of the activity of the enzymes involved in fatty acids and lipid biosynthesis (Russell, 1997; Chattopadhyay and Jagannadham, 2001; Georlette et al., 2004). A novel enzyme family, the polyketide synthases, which is active at low temperatures and is required for the biosynthesis of PUFAs, has recently been found in several psychrophilic bacteria isolated from sea ice (Thomas and Dieckmann, 2002; Rodriguez-Vargas et al., 2002; Tanghe et al., 2002; Sakamoto and Murata, 2002).

Many sea ice diatoms and sea-ice algae release ice-active substances, mainly glycoproteins, which may modify the surfaces and optical properties of ice around the cells and/or increase the ability of cells to adhere to ice crystals. The viscous biofilms surrounding the cells have an important buffering and cryoprotectant role by altering the immediate physicochemical environment (Deming et al., 2002, Davies and Sykes, 1997, Krembs et al., 2002). High concentrations of these glycoproteins in winter ice represent also a previously unrecognized form of organic matter that may contribute significantly to polar ocean carbon cycles. Besides, such extracellular materials could be biotechnologicaly produced to be used as stabilizers, flocculants, gelling or emulsifying agents (Lee et al., 2003).

5. Conclusions

Investigations of extremophilic microorganisms have gained importance in the last decade. On one hand these microorganisms represent a largely unknown biodiversity that only lately became recognized. On the other hand they represent a source of mainly untapped biotechnological potential (Aguilar, 1996; Wang et al., 2003). However, not all areas of research on extremophiles have obtained the same degree of attention. Most investigations were dealing with extremely hot habitats and thermophilic microorganisms. Ice inhabiting microorganisms were until recently recognized mainly as living fossils, enabling insight into biological past. Only recently it became evident that their role and presence in extremely cold areas is much more significant than considered so far. They have been discovered as active communities in sea ice, microbial mats on sea ice-shelves, in glacial inclusions covered with thin films of ice, in permafrost, in brine networks of glacial ice, as ephemeral cryoconite holes and as biomass rich microbial communities in subglacial environments. In all cases they contribute to local geobiochemistry and seed the immediate environment, whether soil or sea, with so far unknown forms of life.

Their importance increased with the findings considerable global glaciations half a billion years ago. There is now mounting evidence that the Precambrian biosphere experienced extreme low temperature conditions at several intervals during the Paleo- and Neoproterozoic periods and perhaps even during the earliest steps in the emergence and evolution of life. The timing, duration and extent of cooling during Precambrian are subjects of considerable discussion and debate (Bodisetisch et al., 2005).

It has been argued that Earth became covered in ice by glaciers that extended from the poles to the equator as many as four times during the Neoproterozoic period (between 750 and 580 million years ago) in what is commonly referred to as Snowball earth periods. Estimates of the length of these episodes range from 3 to 30 million years each. Silicate weathering and photosynthesis, which are the major sinks for CO_2 at present, would have been inhibited by ice that covered both continents and oceans. Huge amounts of CO_2 could have accumulated in the atmosphere. Greenhouse gasses, particularly large amounts of CO_2 would have been needed to overcome the albedo effect caused by the glaciations. Evidence based on a new snowball marker, the element iridium that continually rains on Earth from outer space indicate that our planet most probably has been entirely frozen solid for 12 million consecutive years (Bodisetisch et al., 2005).

Research on polar microbiota may provide insight into how microorganisms survived, adapted and diversified during critical periods in the evolution of life on Earth. The entire world ocean was frozen by runaway ice-albedo events near the time of origin of the eukaryotic cell and the adaptive radiation of metazoa. The extensive microbial mat communities on and in ice in Antarctica and in the Arctic are dominated by similar organisms to those in the Proterozoic fossil record. Studies on present day equivalents show how mat forming species can provide protected microhabitats for the survival, growth and evolution of less tolerant biota, including multicellular eukaryotes, during periods of extensive glaciation. Antarctica developed its permanent ice cap some 14 million years ago and microbes may well have existed in its continental crust. It is hypothesized that their descendants may live in subglacial rock crevices, lakes and sediments and perhaps in glacial ice itself. These cyoenvironmnts and a variety of other habits such as cryoconite holes, thick ice covered lakes, nunataks, and subglacial environments may have played a role as refuges for microorganisms and other biota during periods of glaciation and therefore favored the development of endemic species. To which extent a local population of microorganisms can genetically diverge from the rest of the global gene pool is currently a much debated issue in microbial ecology. If "microbial endemism" (i.e. genotypes of bacteria, or other microorganisms specific to a geographic region) is at all possible, then polar regions should be amongst the first places to look for such organisms and to examine evolutionary processes that can give rise to microbial speciation. Some polar habitats have remained sealed for hundreds of thousands of years or longer, and pose unique opportunities for exploring microbial evolution in the absence of gene flow from outside biota. Many such habitats contain food webs that are reduced in complexity by comparison with latitudes elsewhere. They represent excellent systems for better understanding microbial processes in general, and for studying interactions of evolutionary significance in the absence of any confounding effects associated with higher plants or animals (Vincent, 2000).

Most so far detected psychrophilic and psychrotolerant microorganisms are culturable. Nevertheless, as revealed by 16 S and 18 S rRNA sequences, many more need to be isolated and characterized.

At what point does ice fail to support life? It seems that even the coldest forms of ice retain liquid water within their matrices, between individual ice crystals, wetting each crystal surface. Besides, solute-enriched water films can be formed on its surface or particulates can be entrapped in air bubbles. Even the coldest ice forms so far examined contained living bacteria (Deming, 2002). Incubation experiments using radioactively labeled substrates have shown that metabolism and macromolecular synthesis occur down to –20°C (Deming, 2002). The inevitable presence of some liquid even within the coldest of natural ice formations promises more surprises regarding the lower temperature limit for microbial life and the mechanisms supporting it.

6. References

Abyzov, S.S. (1993) Microorganisms in the Antarctic Ice, In: E.I. Friedmann (ed.), *Antarctic Microbiology*. Wiley-Liss, New York, pp. 265-285.

Aguilar, A. (1996) Extremophile research in the European Union: from fundamental aspects to industrial expectations. FEMS Microbiol. Rev. **18,** 89-92.

Babjeva, I. and Reshetova, I. (1998) Yeast resources in natural habitats at polar circle latitude. Food Technol. Biotechnol. **36,** 1-5.

Belzile, C., Vincent, F.W., Gibson, J.A.E. and Van Hove, P. (2001) Bio-optical characteristics of the snow, ice and water column of a perennially ice-covered lake in the High Arctic. Cand. J. Fish. Aquat. Sci. **58,** 2405-2418.

Blackwell, M., Longstretch, D.J., David, C. and Barker, S.A. (2001) Glycine betaine in fungi. Inoculum **52,** 22.

Bodisetisch, B., Koeberl, C., Master, S., Reimold, W.U. (2005) Estimating Duration and Intensity of Neoproterozoic Snowball Glaciation from Ir Anomalies. Science, **308,** 239-242.

Brown, M.V. and Bowman, J.P. (2001) A molecular phylogenetic survey of sea - ice microbial communities (SIMCO). FEMS Microbiol. Ecol. **35,** 267-275.

Cavicchioli, R., Siddiqui, S. K., Andrews, D., Sowers, K. (2002). Low-temperature extremophiles and their applications. Curr. Opinion Biotech. **13,** 253-261.

Carpenter, E. J., Lin, S. and Capone, D.G. (2000) Bacterial activity in South Pole Snow, Appl. Envir. Microbiol., 4514-4517.

Chattopadhyay, M.K. and Jagannadham, M.V. (2001) Maintenance of membrane fluidity in Antarctic bacteria. Polar Biol. **24,** 386-388.

Cheng, C.C. (1998) Evolution of diverse antifreeze proteins. Curr. Opin. Genet. Dev. **8,** 715-720.

Christner, B.C. (2002) Incorporation of DNA and protein precursors into macromolecules by bacteria at -15°C. Appl. Environ. Microbiol. **68,** 6435-6438.

Christner, B.C., Mosley-Thompson, E., Thompson, L.G. and Reeve, J. (2003) Bacterial recovery from ancient glacial ice. Environ. Microbiol. **5,** 433-436.

Christner, B.C., Mosley-Thompson, E., Thompson, L.G., Zagorodnov, V., Sandman, K. and Reeve, J.N. (2000) Recovery and identification of viable bacteria immured in glacial ice. Icarus **144,** 479-485.

Davies, P. L., Sykes, B. D. (1997) Antifreeze proteins. Curr. Opinion Struct. Biol. **7,** 828-834.

De Wit, R., Dyer, P., Genilloud, O., Goetlich, E., Hodgson, D., de Hoog, S., Jones, B., Laybourn-Parry, J., Marinelli, F., Stackebrandt, E., Swings, J., Tindall, B.J., Vyverman, W. and Wilmotte, A. (2003)

Antarctic lakes – 'hot spots' for microbial diversity and biotecnological screening, In: *1st FEMS Congress, June 29- July 3, Abstract book.* Ljubljana, Slovenia, p. 228.

Deegenaars, M.L. and Watson, K. (1998) Heat shock response in psychrophilic and psychrotrophic yeast from Antarctica. Extremophiles **2,** 41-49.

Deming, J.W. (2002) Psychrophiles and polar regions. Curr. Opin. Microbiol. **5,** 301-309.

Dmitriev, V.V., Gilichinsky, D.A., Faizutdinova, R.N., Shershunov, I.N., Golubev, W.I. and Duda, V.I. (1997a) Occurrence of viable yeasts in 3-million-year-old permafrost in Siberia. Mikrobiologiya **66,** 655-660.

Dmitriev, V.V., Gilichinsky, D.A., Faizutdinova, R.N., Ostroumova, N.V., Golubev, W.I. and Duda, V.I. (1997b) Yeasts in late Pleistocene-early Pleistocene Siberian permafrost. Cryosphere of the Earth **1,** 67-70.

Ekström, G., Nettles, M. and Abers, G.A. (2003) Glacial earthquakes. Science **302,** 622-624.

Fahnestock, M. (2003) Geophysics: glacial flow goes seismic. Science **302,** 578-579.

Foght, J., Aislabie, J., Turner, S., Brown, C.E., Ryburn, J., Saul, D.J. and Lawson, W. (2004) Culturable bacteria in subglacial sediments and ice from two southern hemisphere glaciers. Microb. Ecol. **47,** 329-340.

Gaidos, E., Lanoil, B., Thorsteinsson, T., Graham, A., Skidmore, M., Han, S.-K., Rust, T. and Popp, B. (2004) A viable microbial community in a subglacial volcanic crater lake, Iceland. Astrobiology **4,** 327-344.

Georlette, D., Blaise, V., Collins, T., D'Amico, S., Gratia, E., Hoyoux, A., Marx, J.C., Sonan, G., Feller, G. and Gerday, C. (2004) Some like it cold: biocatalysis at low temperatures. FEMS Microbiol. Rev. **28,** 25-42.

Gerday, C., Aittaleb, M., Bentahir, M., Chessa, J.P., Claverie, P., Collins, T., D`Amico, S., Dumont, J., Garsoux, G., Georlette, D., Hoyoux, A., Lonthienne, T., Meeuwis, M.A. and Feller, G. (2000) Cold adapted enzymes: from fundamentals to biotechnology. Tibtech **18,** 103-107.

Golubev, W.I. (1998) New species of basidiomycetous yeasts, *Rhodotorula creatinovora* and *R. yakutica*, isolated from permafrost soils of Eastern-Siberian Arctic. Mykologiya Phytopathologiya **32,** 8-13.

Gosink, J.J., Irgens, R.L., Staley, J.T. (1993). Vertical distribution of bacteria in arctic sea ice. FEMS Microbiol. Ecol. **102,** 85-90.

Gunde-Cimerman, N., Sonjak, S., Zalar, P., Frisvad, J.C., Diderichsen, B. and Plemenitaš, A. (2003) Extremophilic fungi in Arctic ice: a relationship between adaptation to low temperature and water activity. Phys. Chem. Earth **28,** 1273-1278.

Hoham, R.W. and Ling, H.U. (2000) Snow algae – the effects of Chemical and Physical factors on their life cycles and Populations In:*COLE, Journey to Diverse Microbial worlds, Adpations to exotic environmnts. Ed. J. Seckbach. Kluwer academic publishers.Dordrecht, Boston, London,* 133-145.

Hoham, R.W. and Duval B. (2001) Microbila ecology of snow and freshwater ice with emphasis on snow algae. *In: Snow Ecology: An Interdisciplinary Examination of Snow covered Ecosystems, (Jones, H.G., Pomeroy, J.W, Walker, D.A. and Hoham, R.W. eds.), Cambridge University Press, Cambridge, U.K.,* 168-228.

Junge, K., Imhoff, F., Staley, T. and Deming, J.W. (2002) Phylogenetic diversity of numerically important Arctic sea-ice bacteria cultured at subzero temperature. Microb. Ecol. **43,** 315-328.

Krembs, C., Eicken, H., Junge, K. and Deming, J.W. (2002) High concentrations of exopolymeric substances in Arctic winter sea ice: implications for the polar ocean carbon cycle and cryopeotection of diatoms. Deep-Sea Res. I **49,** 2163-2181.

Kriss, A.E., Mitskevich, I.N., Rozanova, E.P. and Osnitskaia, L.K. (1976) Microbiological studies of the Wanda Lake (Antarctica). Mikrobiologiia **45,** 1075-1081.

Lanoil, B. (2004) Microbiology of subglacial environments, In: *Extremophiles 2004, 5th International Conference on Extremophiles, September 19-23, Abstract Book.* Cambridge, Maryland, p. 19.

Lee, Y.K., Kim, H.W., Liu, C.L., Lee, H.K. (2003). A simple method for DNA extraction from marine bacteria that produce extracellular materials. J. Microbiol. Methods **52,** 245-250.

Libkind, D., Brizzio, S. and van Broock, M. (2004) *Rhodotorula mucilaginosa*, a carotenoid producing yeast strain from a Patagonian high-altitude lake. Folia Microbiol. **49,** 19-25.

López-García, P., Rodríguez-Valera, F., Pedrós-Alió, C. and Moreira, D. (2001) Unexpected diversity of small eukaryotes in deep-sea Antarctic plankton. Nature **409,** 603-607.

Ma, L.J., Rogers, S.O., Catranis, C.M. and Starmer, W.T. (1999a) Detection and characterization of ancient fungi entrapped in glacial ice. Mycologia **92,** 286-295.

Ma, L.J., Catranis, C.M., Starmer, W.T. and Rogers, S.O. (1999b) Revival and characterization of fungi from ancient polar ice. Mycologist **13,** 70-73.
Margesin, R., Zacke, G. and Schinner, F. (2002) Characterization of heterotrophic microorganisms in Alpine glacier cryoconite. Arctic, Antarctic and Alpine Res. **34,** 88-93.
McRae, C.F., Hocking, A.D. and Seppelt, R.D. (1999) *Penicillium* species from terrestrial habitats in the Windmill Islands, East Antarctica, including a new species, *Penicillium antarcticum*. Polar Biol. **21,** 97-111.
Mindock, C.A., Petrova, M.A. and Hollingsworth, R.I. (2001) Re-evaluation of osmotic effects as a general adaptive strategy for bacteria in sub-freezing conditions. Biophys. Chem. **89,** 13-24.
Miteva, V.I., Brenchley, J.E. (2004) Diversity of ultra-small microorganisms in greenland glacier ice. Extremophiles 2004; 5th International Conference on Extromophiles,Sept. 19-23: pp. 80. Cambridge, Maryland, USA.
Mueller, D.R., Vincent, W.F., Pollard, W.H., Fritsen, C.H. (2001) Glacial cryoconite ecosystems: A bipolar comparison of algal communities and habitats. Nova Hedwigia **123,** 173-197.
Onofri, S., Selbmann, L., Zucconi, L. and Pagano, S. (2004) Antarctic microfungi as models for exobiology. Planet. Space Sci. **52,** 229-237.
Pennisi, E. (2003) Microbiology. Neither cold nor snow stops tundra fungi. Science **301,** 1307.
Pomeroy, L.R., Wiebe, W.J. (2001) Temperature and substrates as interactive limiting factors for marine heterotrophic bacteria. Aquat Microb Ecol **23,** 187-204.
Price, P.B. (2000) A habitat for psychrophiles in deep Antarctic ice. Proc. Natl. Acad. Sci. USA **97,** 1247-1251.
Price, P.B. and Sowers, P. (2004) temperature dependence of metabolic rates for microbial growth, maintenance and survival, PNAS, **101, 13,** 4631-4636.
Priscu, J.C., Fritsen, C.H., Adams, E.E., Giovannoni, S.J., Paerl, H.W., McKay, C.P., Doran, P.T., Gordon, D.A., Lanoil, B.D. and Pinckney, J.L. (1998) Perennial Antarctic lake ice: an oasis for life in a polar desert. Science **280,** 2095-2098.
Reeve, J.N., Christner, B.C., Kvitko, B.H., Mosley-Thompson, E. and Thompson, L.G. (2002) Life in glacial ice, In: *Extremophiles 2002, 4th International Congress on Extremophiles, September 22-26, Abstract Book*. Naples, Italy, p. 27.
Rivkina, E.N., Friedmann, E.I., McKay, C.P. and Gilichinsky, D.A. (2000) Metabolic activity of permafrost bacteria below the freezing point. Appl. Environ. Microbiol. **66,** 3230-3233.
Rodriguez-Vargas, S., Estruch, F. and Randez-Gil, F. (2002) Gene expression analysis of cold and freeze stress in baker's yeast. Appl. Environ. Microbiol. **68,** 3024-3030.
Russell, N.J. (1990) Cold adaptation of microorganisms. Phil. Trans. R. Soc. Lond. B **326,** 595-611.
Russell, N.J. (1997). Psychrophilic bacteria-molecular adaptations of membrane lipids. Comp. Biochem. Physiol., Part A, Physiol. **118,** 489-493.
Sakamoto, T., Murata, N. (2002) Regulation of the desaturation of fatty acids and its role in tolerance to cold and salt stress. Current Opinion in Microbiology **5,** 206-210.
Soinam, V.S., Vorobyova, E.A., Suzina, N.E., Dimitriev, V.V., Faizutdinova, R.N., Duda, V.I., Vishnivetskaya, T.A., Ostroumova, N.V., Gilichinsky, D.A. Viable ancient permafrost yeast. Tenth International Symposium on Yeasts, 27th August - 1st September 2000, Arnhem, The Netherlands.
Siegert, M.J., Ellis-Evans, J.C., Tranter, M., Mayer, C., Petit, J.R., Salamatin, A. and Priscu, J.C. (2001) Physical, chemical and biological processes in Lake Vostok and other Antarctic subglacial lakes. Nature **414,** 603-609.
Skidmore, M.L., Foght, J.M. and Sharp, M.J. (2000) Microbial life beneath a high Arctic glacier. Appl. Environ. Microbiol. **66,** 3214-3220.
Takano, Y., Kobayashi, K., Marumo, K. and Ishikawa, Y. (2004) Biochemical indicators and enzymatic activity below permafrost environment, In: *Extremophiles 2004, 5th International Conference on Extremophiles, September 19-23, Abstract Book*. Cambridge, Maryland, p. 84.
Tanghe, A., Van Dijck, P.V., Dumortier, F., Teunissen, A., Hohmann, S. and Thevelein, J.M. (2002) Aquaporin expresion correlates with freez tolerance in baker's yeast, and overexpression improves freeze tolerance in industrial strains. Appl. Environ. Microbiol. **68,** 5981-5989.
Thomas, D.N. and Dieckmann, G.S. (2002) Antarctic sea ice-a habitat for extremophiles. Science **295,** 641-644.
Vincent, W.F. (2000) Evolutionary origins of Antarctic microbiota: invasion, selection and endemism. Antar. Sci. **12,** 374-385.

Vincent, W.F. Mueller, D.R. and Bonilla, S. (2004) Ecosystems on ice: the microbial ecology of Markham Ice Shelf in the high Arctic. Cryobiol. **48,** 103-12.

Wang, W., Vinocur, B. and Altman, A. (2003) Plant responses to drought, salinity and extreme temperatures: towards genetic engineering for stress tolerance. Planta **218,** 1-14.

Zlatanov, M., Pavlova, K. and Grigorova, D. (2001) Lipid composition of some yeast strains from Livingston Island, Antarctica. Folia Microbiol. **46,** 402-406.

Biodata of **Christian Tamburini** author of ***"Life under Pressure. Deep-sea microbial ecology"***

Dr. Christian Tamburini is currently is a research scientist (CR2) of the CNRS (since 2003). He obtained his Ph.D. ("Organic matter degradation by deep-sea bacteria: from the potential rates measurement to the real CO2 flux producted *in situ*") from the University of Marseilles in 2002 and continued his research at the IAMC (Instituto de l'Ambiente Marino Costiero, Sez. Messina) in 2003 to develop FISH and Micro-FISH techniques. Dr. Christian Tamburini scientific interests are in the areas of microbial ecology of mesopelagic and bathypelagic, effect of pressure on the modification of the structure and the metabolism of the prokaryotic community, effect of pressure on biodegradation of particles sinking throughout the water column developing innovative instruments and devices to collect and incubate deep-water samples without decompression and devices to simulate particle flow throughout the water column.
For more information, check www.com.univ-mrs.fr/ctambu_us

E-mail: **Christian.Tamburini@com.univ-mrs.fr**

J. Seckbach (ed.), Life as We Know It, 125-143.

LIFE UNDER PRESSURE:
Deep-sea microbial ecology

CHRISTIAN TAMBURINI
Laboratoire de Microbiologie, Géochimie et Ecologie Marines (LMGEM), UMR 6117 CNRS – INSUE, Université de la Méditerranée, COM, Campus de Luminy, Case 907, 13288 Marseille, Cedex 9, France

1. Introduction

Life under pressure concerns all organisms living in the largest habitat (by volume) of the Ocean: the deep sea. In spite of being the largest world ecosystem, the deep sea has been classically underestimated regarding its contribution to the Biosphere and represents one of the least known and most poorly understood environments on our planet.

Due to its peculiar features, namely the high hydrostatic pressure conditions, the low temperature (predominantly at 2–4°C except for hydrothermal vents – up to 460°C, the Sulu Sea at around 10°C and the Mediterranean Sea at around 13°C), the darkness, and the low concentration of readily utilizable organic resources, the deep marine environment represents "extreme" conditions for life. Indeed, life exists at temperatures at least to 110°C (Deming and Baross, 1993; Prieur et al., 1995; Deming and Baross, 2000), at pressures to at least 110 MPa (Kato et al., 1998), and in deep hypersaline anoxic basins with concentrations of $MgCl_2$ around 5M (Van Der Wielen et al., 2005).

In 1977, scientists made a stunning discovery on the bottom of the Pacific Ocean (Galápagos Islands) that forever changed understanding of our planet and life on it. They discovered deep-sea hydrothermal vents and – to their complete surprise – a lush community of exotic life thriving around them based on a chemosynthetic primary production (Corliss et al., 1979). The matter of life arising at "extreme conditions" (high temperature, high pressure, etc.) together with the biotechnological interest for the retrieval of new biological treats (e.g. genes, enzymes, metabolic pathways, cell components and products) have stimulated the scientific community to pursue efforts at these unusual deep-sea sites and have provided new impetus for deep-sea research in general. The Challenger Expedition (1873-1876) is commonly credited as the historical beginning of deep-sea biology. The finding of live specimens at great depths obliterated the azoic-zone theory, below 600-m depth, that had been suggested in the 1840s by Edward Forbes. A very good summary of the controversial concept of the azoic-zone is provided in Jannasch and Taylor (1984). In 1884, Certes (1884) examined sediment and water collected from depths to 5,000 m and cultured bacteria from almost every sample. In 1904, Portier used a sealed and autoclaved glass-tube device as a bacteriological sampler and reported counts of colonies from various depths and location (summarized in Jannasch and Wirsen, 1984). Research into the effects of high pressure on the physiology of deep-sea bacteria was developed during the last century. A synthesis of that work can be found in ZoBell (1970), pioneer biologist in the study of the effects of hydrostatic pressure on microbial activities (ZoBell and Johnson, 1949).

During the past decade, several mechanisms of bacterial adaptation to the deep-sea environment have been better elucidated by laboratory experimentation with pure cultures (Prieur et al., 1995; Kato and Qureshi, 1999; Kato et al., 2000; Bartlett, 2002). New molecular features as well as new metabolic pathways have been described that have important links to scientific endeavors and industrial fields (Colquhoun et al., 1998; Abe and Horikoshi, 2001). Nevertheless, the ecological roles of deep-sea bacteria and their relative contribution to global biogeochemical cycles remain to be fully explored.

In this chapter, we focus attention on the bathypelagic waters of the Ocean, the cold deep sea, with some attention to hydrothermal vents as epiphenomena in the deep sea. After a short section providing some definitions, special emphasis is given to the technological effort required to design devices that avoid decompression of deep samples in order to estimate pressure effects on the metabolic activities of prokaryotes. Finally, the role of prokaryotes in mineralizing organic matter in mesopelagic and bathypelagic waters will be considered.

2. Definitions

Hydrostatic pressure increases with depth according to the equation of Saunders and Fofonoff (1976) that accounts for the latitudinal and depth variation in *g*, the gravitational acceleration, and the variation in dynamic height. However, the old rule of thumb, well known by scuba divers, that pressure increases by 1 bar (0.1 MPa) per 10 meters of depth in the sea has proven accurate enough for biology. Several units are used to express pressure (1 atmosphere (atm) = 1.01325 bars = 1.01325×10^5 Pascals), but the accepted unit (in SI units) is Pascal. It is the pressure acting by a force of 1 Newton on a plan area of 1 m^2. At 3,800 m for example (the average depth of the Ocean), the pressure is equal to 38 MPa.

The deep sea is conveniently defined here as waters of 1,000 m and below. It underlies 88% of the global area covered by seawater and includes 75% of the Ocean's total volume (Jannasch and Taylor, 1984).

The term "barophilic" microorganism was first used by ZoBell and Johnson (1949) (from the Greek "baro" meaning weight) and recently replaced by the term "piezophilic" (from the Greek "piezo" meaning pressure; Yayanos, 1995). Today, piezophiles are defined as organisms with an optimal growth rate at a pressure higher

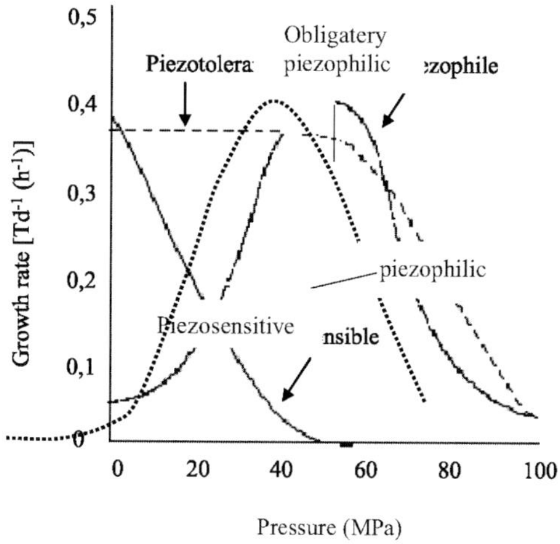

Figure 1. Growth behavior of different types of bacteria under pressure (modified from Abe and Horikoshi, 2001).

than 0.1 MPa or requiring increased pressure for growth. Figure 1 depicts the relationships between microbial growth rate and hydrostatic pressure (Abe and Horikoshi, 2001).

3. High Pressure Technologies

The main reasons for limited knowledge of the deep realm concern the difficulties, expense and time involved in sampling remote parts of the ocean.

For the majority of researchers who do not have access to pressure-retaining water samplers, previously decompressed samples and cultures can be studied under *in situ* conditions simulated in simple, stainless steel pressure vessels (see Fig. 1 and the review of Deming, 2001). Several authors (Schwarz and Colwell, 1975; Deming, 1985; Turley and Lochte, 1990; Meyer-Reil and Köster, 1992; Turley, 1993; Boetius and Lochte, 1994) have measured microbial activities in samples of deep-sea sediments or sinking particles decompressed during recovery and recompressed to *in situ* pressure conditions. Bacterial cultures are known to survive brief changes in pressure conditions, especially if temperature is unchanged (Zobell, 1970; Deming, 1993). However, the effects of successive pressure shocks on the metabolic rates of natural microbial populations is not known. Yayanos and Delong (1987) have shown that the cell division rate of an obligately piezophilic bacterial strain cultivated under copiotrophic conditions is not altered by repeated compression-decompression sequences. On the other hand, (Bianchi and Garcin, 1993) have shown that under oligotrophic conditions, the metabolic rates of deep-water microbial populations decompressed during retrieval and then recompressed are clearly lower than their undecompressed counterparts.

In the following, we discuss the important technological effort that researchers have made, and continue to make, to measure microbial activities without any change to in situ environmental conditions.

3.1. HIGH-PRESSURE SAMPLING FOR BATHYPELAGIC WATERS

Sterilizable pressure-retaining samplers for retrieving, recovering and sub-sampling undecompressed deep-sea water samples have been developed independently by several laboratories. Three laboratories have worked and/or are presently working on the effects of hydrostatic pressure on microbial measurements.

In the 1970s, Jannasch and collaborators (Jannasch et al., 1973) designed a multi-chambered stainless steel device deployed by the Alvin submarine. Limitations of the device included the pressure shock incurred when the valve was opened for filling and that only a single end-point measurement was possible. To overcome these limitations Jannasch and Wirsen (1977) designed a sampler to retrieve deep water at ambient pressure and to concentrate microbial populations without any loss of pressure. No pressure gradient occurred at any point during sampling due to a floating piston. During sampling, microorganisms from a 3-L sample were concentrated to a 13-ml volume by filtration over a 0.2-μm pore-size filter. Concentrated samples were kept in a storage vessel, independently of the sampling unit, under ambient pressure and temperature. Protection against pressure changes during sample recovery, concentration, and storage depended on gas pressure in a chamber separated from the sample by a piston floating in a buffer chamber of sterile water. The sampler could be sterilized and made ready for other deployments, while the concentrated sample was processed on board (or later in the land-based laboratory). Using these samplers, Jannasch and colleagues concluded that ambient (high hydrostatic) pressure resulted in microbial activities reduced relative to those measured under atmospheric pressure in decompressed samples (Jannasch and Wirsen, 1973; Jannasch et al., 1973; Jannasch and Wirsen, 1977; Jannasch and Wirsen, 1982). This work was criticized by several argues by Yayanos and Dietz (1982) and by Bianchi et al. (submitted) as for example : different incubation time for decompressed and undecompressed samples, change in sample's temperature, control done with different sampler.

Simultaneously, Colwell and collaborators developed two samplers (sample volumes of 350 and 400 ml) and a pressure-retaining transfer system (Tabor and Colwell, 1976). These operate entirely on the basis of hydrostatic pressure, generated by a hand pump. Use of hydrostatic pressure has the enormous advantage of greater safety during operation, with no need for the extra precautions require to work with gas (potentially explosive). Deming et al. (1980) reported measurements on samples taken at ambient pressure in the Puerto Rico Trench at 3450 to 7730-m depths, showing that deep-sea bacteria are adapted to ambient pressures. Using the same equipment, Tabor et al. (1981) provided data on substrate utilization by microbial populations, compared results from samples taken at ambient pressure and their decompressed counterparts, and confirmed the adaptation of deep-sea microorganisms to *in situ* pressure conditions in complete contradiction to the conclusions of Jannasch and colleagues (see above).

Later, Bianchi and collaborators developed several prototypes of high-pressure retaining samplers. With the first prototype, Bianchi and Garcin (1993) used a 1.5-L high-pressure sampler and pressure-retaining transfer bench to confirm that decompressed deep-sea samples underestimate microbial activity relative to those incubated at ambient pressure conditions. To conduct high-pressure growth experiments over a time course using a single, undivided bacteriological culture, Tholosan et al. (1999) designed a 5-l, high-pressure sampler (HP-5L) which they used to study 27 pairs of samples. Each pair included the water sample taken at ambient pressure and incubated in the HP-5L and a sub-sample withdrawn from the HP-5L then slowly decompressed showing that during stratified period of the water column, undecompressed samples are higher than their decompressed counterpart.

In order to obtain microbial activity data unbiased by decompression during retrieval, but having the same frequency of sampling used in measuring other hydrological parameters, Bianchi et al. (1999b) developed a high-pressure serial sampler (HPSS). This high-pressure device borrows from the designs of its predecessors, being simpler in operation than the first ones. It is based on a commercially available multi-sampling device that includes a CTD and can be equipped with 500 ml high-pressure bottles (HPBs). The HPSS allows collection of several ambient-pressure water samples at different depths during the same hydrocast down to a depth of 3500 m. Functioning of the HPBs depicted in Figure 2: when the filling valve is opened at depth by the operator via the deck unit connected to the electric wire, the natural hydrostatic pressure moves the floating piston downwards and seawater enters the upper chambers of two HPBs. The distilled water flushes from the lower chambers to the exhaust tanks through a nozzle that acts as a hydraulic brake avoiding decompression during filling. Samples are collected into two identical high-pressure bottles filled at the same time through the same two-way valve (Figure 2). One of these bottles retains the *in situ* pressure, while the other which lacks a check valve progressively decompresses as the sampler surfaces. The decompressed bottle, like a classical retrieving bottle (e.g. Niskin bottle), is used in comparative fashion to estimate the effect of decompression on measurements of deep-sea microbial activity. Decompressed and high-pressure (ambient) samples are treated in exactly the same way. High-pressure sampling, sub-sampling and transfer are performed using hydrostatic pressure (instead of gas pressure) thanks to a high pressure pump (similar to an HPLC pump) and/or special hydrostatic pressure bench in order to diminish risk during manipulation. Sub-sampling is performed by use of counter-pressure (a maximum of 0.05 MPa) from the bottom of the HP bottle; opening the sampling valve allows the counter-pressure to move the floating piston upwards. The primary sample within the HPB is maintained under *in situ* pressure while the secondary sample (decompressed) is analyzed directly or else fixed (by adding a preservative solution like formaldehyde), stored, and analyzed later.

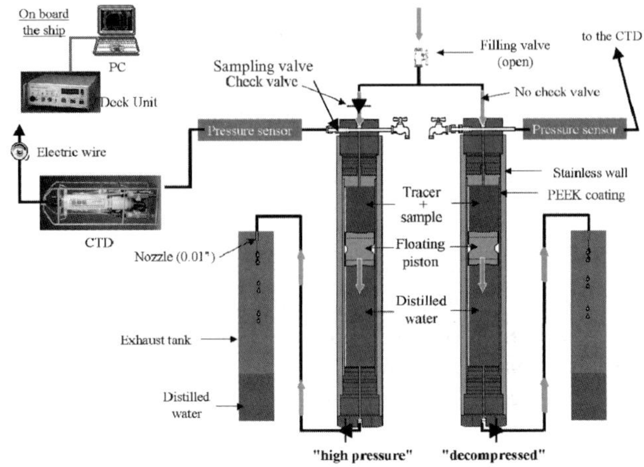

Figure 2. Diagrammatic representation of the high-pressure bottles (HPB) configured for sample filling (from Tamburini et al., 2003). See text for additional details.

3.2. SAMPLING THE DEEP WATER-SEDIMENT INTERFACE

Microbial activity at the deep-sea boundary layer (interface between ocean and sediments) has been studied using several different strategies of varying degrees of sophistication. The use of manned submersible, the most expensive and difficult to acquire for systematic measurements, has been adopted by some authors (Jannasch and Wirsen, 1973; Jannasch and Wirsen, 1980; Cahet and Sibuet, 1986; Cahet et al., 1990). For example, Jannasch and Wirsen (1973) performed a series of incubations at 1830-m depth, using the manned submersible Alvin. Sterile 120-mL incubation bottles were deposited on the seafloor and filled with water that included surface sediment particles.

An alternative approach to using manned submersibles that also avoids the double pressure shock of decompression and recompression inherent to conventionally recovered samples is the use of benthic landers (Picon, 2000; Viollier et al., 2003; Witte et al., 2003; Tamburini, 2002). Recently, we used an autonomous lander fitted with several chambers to incubate samples of the benthic interface *in situ* (Tamburini, 2002). A syringe injects a substrate (for example, ^{14}C-amino acids), while several syringes containing a solution of buffered formaldehyde are available to end activity at a desired time and thus permit a time–course experiment that is analyzed when the lander is recovered. To estimate the effect of pressure conditions on microbial activity measurements, decompressed samples are also collected by means of a multiple corer. Sub-samples are retrieved and processed under atmospheric pressure, following the same protocol as applied to those incubated *in situ* using the lander (Picon, 2000; Tamburini, 2002). Using an another benthic lander and similar approach, Witte et al. (2003) followed phytodetritus biodegradation on the deep seafloor by tracking oxygen consumption, change in organic carbon composition, and flux of $\delta^{13}C$ under *in situ* conditions.

3.3. SAMPLING DEEP-SEA SEDIMENTS

Microbial activity in deep-sea sediments has been measured most commonly in sediment slurries prepared from decompressed cores and sometimes repressurized to *in situ* conditions (Deming, 1993). Injection of substrates directly into undisturbed sediment cores has also been developed (Meyer-Reil, 1986) and used with (repressurized) – deep-sea sediments (Relexans et al., 1996).

Jannasch and Wirsen (1980) developed a retrievable injection corer specially designed for deep-sea incubations and deployed by manned submersible. This gear allows for injection of labeled substrate into a sediment core at approximately 2, 4, and 7 cm below the sediment surface and incubation *in situ* until recovery (though no *in situ* sample fixation prior to surfacing). Bacterial utilization of ^{14}C-acetate during an *in situ* incubation for 69 h at 3,580 m of depth in the North Atlantic was compared to that measured in similar controls incubated at *in situ* temperature (3°C) but under atmospheric pressure. On average, the amount of acetate metabolized by microbial populations within the three 3-cm thick sediment layers was about 2.5-fold higher in the decompressed controls than in the samples incubated *in situ*.

Cahet and Sibuet (1986) and Cahet et al. (1990) also used a specially modified corer deployed by manned submersible. This corer was fitted with special syringes to inject the labeled substrate into near-bottom water enclosed in the corer. At the end of the incubation period, samples were formalin-fixed *in situ* and the corer retrieved shipboard with the submersible. The authors measured microbial uptake of the added substrates in the surficial (0–1.5-cm thick) sediment layer after a 24-h incubation period at 2,000-m depth. They observed an enhanced *in situ* microbial metabolism relative to that measured in similar cores decompressed during recovery and incubated under atmospheric pressure conditions. These results contradict those obtained by Jannasch and Wirsen (1980), likely reflecting the different experimental protocols.

To my knowledge, no data have been published from sediment corers that enable sampling, retrieval and incubation shipboard without any change of hydrostatic pressure. A recent European project (see http://www.geotek. co.uk/hyacinth/), however, has the goal to develop of a special device that will allow for such experimentation, including the transfer and sub-sampling of sediment cores under pressure for chemical, microbiological and petrophysical studies.

3.4. SAMPLING EXTREME DEEP-SEA ENVIRONMENTS

The small-scale and spatially targeted sampling of hot deep environments at submarine hydrothermal vents has required manned submersible operations fitted with thermally tolerant sampling gear for collection of superheated geothermal fluids under deep-sea pressure (see details and review in Deming and Baross, 1993; Baross and Deming, 1995; Deming, 2001). Very recently, a new hydrothermal vent sampler (also deployed by manned submersible) was designed to capture hot fluids and incubate them *in situ* at conditions characteristic of deep-sea hydrothermal vents for periods of hours to days. This new system seems to have proved its capacity for *in situ* capture and incubation for biological and geochemical research in the harsh smoker environment (Phillips et al., 2003).

To explore the possibility of microbial activity in the extreme conditions of deep hypersaline anoxic basins (DHABs), a strategy has been developed for targeted sampling inside the small-scale and spatially constrained DHABs without manned submersible. This goal has been achieved by collecting brine samples in a 5-liter high-pressure sampler (HP-5L) that maintains all physical and chemical parameters at ambient conditions during retrieval and incubation and is fitted on the MODUS/ SCIPACK system (see http://www.geo.unimib.it/BioDeep/Project.html). This gear includes a remotely-operated vehicle (ROV) that provides for sampling at accurate depth and position in a relatively thin interface layer. On board the research vessel, brine samples collected with the HP-5L pressure-retaining sampler are transferred without decompression into 500-ml high-pressure bottles (HPBs). The HP-5L and 500-ml HP bottles were first autoclaved and sealed tightly until filling. Such processing prevents sample contamination, a possible source of artifacts when studying microbial life in extreme environments. Similarly, samples collected using HP bottles are not exposed to the atmosphere, avoiding their contamination by molecular oxygen.

3.5. HIGH-PRESSURE DEVICES FOR LABORATORY EXPERIMENTS

Isolation of the first pure culture of piezophilic bacterium was reported in 1979 by Yayanos et al. (1979). The spirillum-like bacterial strain CNPT-3 had a rapid doubling rate at 50 MPa but did not form colonies at atmospheric pressure for several weeks. In the ensuing years, many piezophilic isolates were obtained from the cold (and hot) deep sea (e.g. see reviews by Yayanos and Delong, 1987; Yayanos, 1995; Deming and Baross, 2000). Rapid progress in assessing deep-sea biodiversity and obtaining new strains of barophilic bacteria is now being made by Japanese scientists and their colleagues (e.g. Kato et al., 1998; Nogi et al., 1998; Qureshi et al., 1998; Abe and Kato, 1999; Kato and Qureshi, 1999) as a result of significant new funding for deep-sea research in microbiology and biotechnology (see http://www.jamstec.go.jp).

Important advances in laboratory pressure vessels were made by Yayanos and colleagues (Yayanos and Van Boxtel, 1982; Yayanos et al., 1984) as well shown and documented in Deming (2001). Such vessels enabled isolation and cultivation of deep-sea bacteria (Dietz and Yayanos, 1978), study of concomitant effects of high-pressure and low temperature on pure cultures of piezophilic, piezotolerant and piezosensitive bacteria (Yayanos and Delong, 1987; Yayanos, 1995; Yayanos and Chastain, 1999), adaptation of the membrane lipids of a piezophilic bacterium (Delong and Yayanos, 1985), and isolation of genes regulated by hydrostatic pressure (Bartlett et al., 1989; Bartlett, 1999; Bartlett, 2002).

The role of prokaryotes attached to particles and their relative contribution to global biogeochemical cycles are not well known; the study of attached Archaea is in its infancy. We are developing a system intended to simulate the descent of particles through the mesopelagic and bathypelagic zones. With this particle-sinking simulator

(PSS), we have studied the effects of increasing pressure on both, microbial community structure and mineralization processes (Tamburini et al., submitted). This knowledge will permit us to estimate rates of transformation of particulate organic carbon to dissolved organic carbon and of regeneration of mineral elements such as silica and carbonates, often limiting factors in primary production and the source of ballast minerals for sinking particles.

4. Pressure Effect on Microbial Life

The first reports of deep-sea microbial measurements made without pressure changes were perfomed by Jannasch and Wirsen (1973), who concluded that "elevated pressure decreases rates of growth and metabolism of natural microbial populations collected from surface waters as well as from the deep sea" (Jannasch and Wirsen, 1973). Contrary to this conclusion, all other research groups have shown that deep-sea microorganisms (depending on certain conditions discussed below) are adapted to high-pressure conditions. Hence incubation of deep-sea samples at atmospheric pressure underestimates *in situ* activity under ambient (high) hydrostatic pressure conditions (ZoBell, 1970; Tabor and Colwell, 1976; Deming et al., 1980; Tabor et al., 1981; Jannasch and Wirsen, 1982; Yayanos, 1995; Kato and Qureshi, 1999; Tholosan et al., 1999; Deming, 2001; Tamburini et al., 2002; Tamburini et al., 2003).

This discrepancy in conclusions deserves to be discussed. The early experiments of Jannasch and collaborators were performed at the water-sediment interface, the endpoint of sinking particles. This benthic interface appears to be a melting pot where autochthonous deep-sea microorganisms and allochthonous surface-derived organisms colonizing sinking particles both accumulate (Deming, 1985). The former population is adapted to ambient high-pressure conditions, while the latter one is not and thus is often inhibited by deep-sea conditions (Jannasch and Wirsen, 1973; Turley and Lochte, 1990; Turley, 1993; Bianchi and Garcin, 1994). Moreover, the early measurements of of Jannasch and colleagues used excessive concentrations of added labeled substrates (compared to the extremely low concentrations available in deep-sea environments) and incubated samples for excessive periods of time (several weeks), thus missing first-order rates early in the experiments.

Based on 106 pairs of microbial activity measurements made in the water column, without a pressure shock (V_{HP}) and with decompression (V_{DEC}), Bianchi et al., (submitted) calculated that the mean difference between the ratio of the two measured rates is positive and highly significant: $V_{HP}/ V_{DEC} = 3.1 \pm 5.3$ (mean ± SD, $n = 106$, $p < 0.0002$). The high standard deviation reflects that, even in deep water masses, microbial activity is far from constant throughout the year. This variability is not restricted to the seasonal scale; important variations in metabolic rates may be observed in the same water masses over a few days in the case, for example, of high fecal pellet fluxes (Bianchi et al., 1999b). Seasonal change in pressure effects has been detected with the shift from stratified (spring/fall) to mixed (winter) features of the water column (Bianchi and Garcin, 1994), with the stratified conditions allowing the differentiation of pressure-adapted communities in the deep layers of the Mediterranean Sea. According to these findings, the microbial response to decompression has been proposed as a suitable parameter for evaluating the presence of adapted microbial populations in deep-sea areas.

Changes in hydrostatic pressure when handling deep-sea samples affect all steps in microbially mediated mineralization: uptake and respiration of labile organic matter bacterial production, and ectoenzymatic activity (TABLE 1). The collective V_{HP}/V_{DEC} results provide strong support for the piezophilic behavior of deep-sea microbial communities.

TABLE 1. Mean (± S.D.) pressure effect (Pe) calculated from literature data on microbial processes in deep-sea waters, where Pe is the ratio of the rate measured under high pressure (V_{HP}) and the rate in decompressed samples (V_{DEC}).

Process measured	Pe = V_{HP}/V_{DEC} mean ± SD
Uptake of labeled compounds (^{14}C-amino-acids, ^{14}C-sugar, etc.) (1)	3.0 ± 6.0 (n = 69)
Bacterial production (^{3}H-leucine incorporation) (2)	3.1 ± 3.9 (n = 34)
Ectoenzymatic activity (aminopeptidase, phosphatase) (3)	2.2 ± 0.7 (n = 6)

References:
(1) Seki and Robinson, 1969; Jannasch and Wirsen, 1973; Jannasch et al., 1976; Deming et al., 1980; Tabor et al., 1981; Bianchi and Garcin, 1994; Bianchi et al., 1999; Tholosan et al., 1999; Tamburini et al., 2003
(2) Tholosan et al., 1999; Tamburini et al., 2002
(3) Tamburini et al., 2002

In Mediterranean deep-sea waters, Bianchi and collaborators showed than when specific conditions bring surface-derived bacteria to depth, i.e. during a surface-water cascading event (Bianchi and Garcin, 1994) or a bloom of vertically migrating zooplankters (Bianchi et al., 1999a), microbial communities collected at depth are not observed to be adapted to high-pressure conditions (TABLE 2). In such cases, metabolic rates are boosted by decompression. Confirming the hypothesis that microorganisms attached to particles are not adapted to elevated hydrostatic pressures, experiments simulating the descent of diatom aggregates through the water column have demonstrated that attached, microbially-mediated enzymatic hydrolysis rates were slowed by increasing pressure (Tamburini, 2002; Tamburini et al., submitted). TABLE 2 attempts to summarize and interpret, in simplified manner, the effect of decompression on prokaryotic activity measurements depending on the deep environment sampled.

TABLE 2. Interpretation of the microbial response to high-pressure conditions in diverse marine environments and estimation of the subsequent effect of sample decompression on microbial activity.

Marine environment	If sample is decompressed, *in situ* metabolic rates are	Microbial population is
Stratified water column [(1)]	underestimated	Adapted
Water column under mixed winter conditions[(2)]	overestimated	not adapted
Sinking particles [(3)]	overestimated	not adapted
Water-sediment interface[(4)]	overestimated	both adapted & not adapted
Sediments[(5)]	underestimated	Adapted

References and remarks:
[(1)] ZoBell, 1970; Tabor and Colwell, 1976; Deming et al., 1980; Tabor et al., 1981; Jannasch and Wirsen, 1982; Yayanos, 1995; Kato and Qureshi, 1999; Tholosan et al., 1999; Deming, 2001; Tamburini et al., 2002; Tamburini et al., 2003.
[(2)] Bianchi and Garcin, 1994
[(3)] Bianchi and Garcin, 1994; Tamburini, 2002; Tamburini et al., submitted
[(4)] Jannasch and Wirsen, 1973; Deming, 1985; Turley and Lochte, 1990; Turley, 1993.
[(5)] Cahet and Sibuet, 1986; Cahet et al., 1990

5. Microbial Role in Organic Matter Mineralization in Mesopelagic and Bathypelagic Waters

For several decades, the scientific community has been confronting the increasing concentration of carbon dioxide in the atmosphere. The role of the Ocean in regulating atmospheric CO_2 is intensely studied but the supporting data are derived mainly from studies of the surface ocean (euphotic zone) where about 50% of the organic carbon produced daily by photosynthetic organisms (in the temperature ocean) is processed by bacteria, either for respiration or to produce new biomass (Ducklow and Carlson, 1992). Microbial activity rates measured in the deep-water layers of the ocean appear low relative to those described in the surface layer. However, when the measured rates are integrated through the depth of each water layer, the mineralization mediated by microbial populations below 200 m appear far from negligible (Lefèvre et al., 1996; Del Giorgio et al., 1997; Bianchi et al., 1998; Tamburini et al., 2002).

Deeper in the water column, microbial communities consume between 40 and 100% of the sinking particulate organic carbon (POC) (Simon et al., 1992). Several models describe the relationships between the nutritional resources available for deep-sea microorganisms, the role of these organisms in the mineralization of peptide compounds, and consequent implications to the global carbon cycle (Amon and Benner, 1996; Kiørboe and Jackson, 2001). A primary role has been deduced from the streaming of dissolved organic matter (DOM) (originating from both biotic and abiotic processes; Long & Azam, 2001) behind sinking particles of marine snow (Figure 3). DOM streaming indicates that when particle fluxes are sufficient, a tight coupling exists between sinking particles and free-living microorganisms in the mesopelagic (Cho and Azam, 1988; Smith et al., 1992) and bathypelagic layers (Nagata et al., 2000). Tamburini et al. (2003) confirmed in the NW Mediterranean Sea (DYFAMED site) that relatively high fluxes of fresh POC enhanced the efficiency of microbial growth. However, when the flow of particles is extremely low, prokaryotes must extract carbon and energy from the semi-refractory part of the dissolved organic bulk (Figure 3).

To obtain information on the uptake of specific organic compounds, Teira et al. (2004) used the MICRO-CARD-FISH method and observed that the percentage of deep water microorganisms (with *Bacteria* and *Archaea* contributing about equally) classified as metabolically active (able to assimilate labeled L-amino acids) was similar to that of surface waters. Moreover, a significantly higher percentage of *Crenarchaeota* took up D-amino acids than either *Bacteria* or *Euryarchaeota* in the deep meso- and bathypelagic realm. All of these measurements, however, were made under surface pressure conditions. The fraction of metabolically active prokaryotes under *in situ* pressure conditions remains unknown, providing a major goal for forthcoming research.

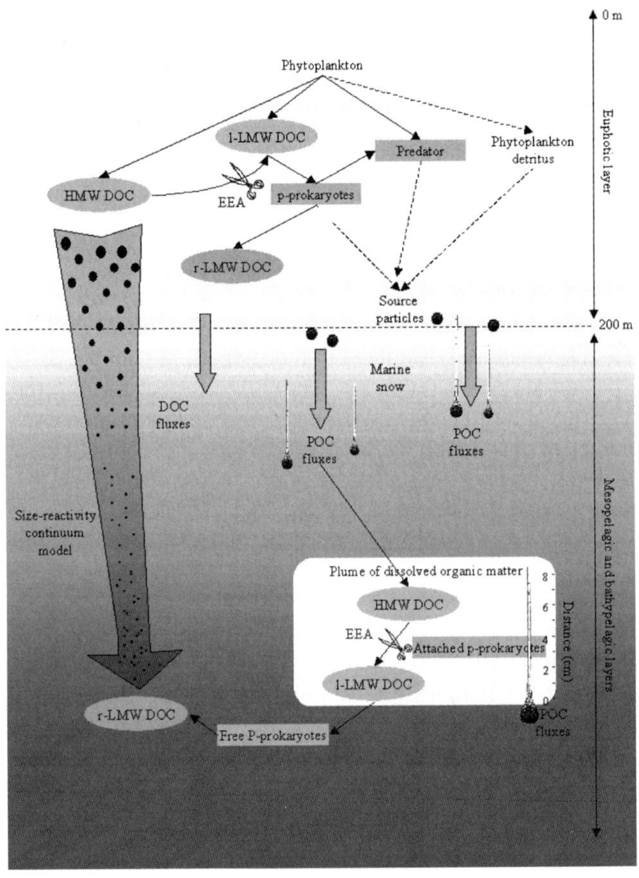

Figure 3. Microbial participation in the mineralization of organic matter, where HMW DOC refers to high molecular weight dissolved organic carbon, l-LMW DOC to labile low molecular weight DOC, r-LMW DOC to refractory-LMW DOC, P-microorganisms to Bacteria and Archaea adapted to high hydrostatic pressure, and p-microorganisms to those not adapted to high hydrostatic pressure (modified from Tamburini et al., 2003).

6. Conclusion

Specific high-pressure devices have allowed for valuable measurements of microbial activity in deep seawater. The evolution of high-pressure technology has brought us to the point of working "routinely" with deep-sea water samples without loss of pressure during retrieval, incubation and sub-sampling. Some laboratories (e.g. NIOZ Institute) are presently designing and equipping with high-pressure water sampling systems. The next challenge will be to develop readily available and affordable high-pressure systems for all interested research groups.

An immediate task for marine microbiologists is to improve the sensitivity of microbiological techniques so that the substrate can be added to mimic natural concentrations and thus measure actual metabolic rates in deep-sea waters (rather than the potential rates discussed throughout this chapter). With rates that reflect both the oligotrophic and high-pressure state of deep-sea waters, realistic calculations of the flow of matter and energy processed via microbial activity throughout the water column, including the deep sea, should be possible. By combining conventional approaches for determining metabolic activity of the bulk microbial (*Archaea* and *Bacteria*) community with recent developments on single cell approaches in the deep ocean, we can expect to better understand microbial roles in the mineralization of organic matter in the mesopelagic and bathypelagic zones. Moreover, the application of molecular techniques to deep-sea samples fixed *in situ* will eventually reveal not only the phylogenetic diversity of the twilight and deep-sea zones but also the dominant functional genes being expressed *in situ* at the time of sample fixation. These advances will bring new data and insights to microbial ecology and models of biogeochemical cycling in the oceans and potential applications in molecular environmental biotechnology.

5. References

Abe, F. and Horikoshi, K. (2001). The biotechnological potential of piezophiles. Trends in Biotechnology, **19:** 102-108.

Abe, F. and Kato, C. (1999). Barophysiology (Piezophysiology). In: K. Horikoshi and K. Tsujii (eds.) *Extremophiles in Deep-Sea Environments*, Springer-Verlarg, Tokyo. pp. 227-248.

Amon, R.M.W. and Benner, R. (1996). Bacterial utilisation of different size classes of dissolved organic matter. Limnology and Oceanography, **41:** 41-51.

Baross, J.A. and Deming, J.W. (1995). Growth at high temperatures: isolation and taxonomy, physiology and ecology. In: D. M. Karl (ed.) *The Microbiology of Deep-Sea Hydrothermal Vents*, CRC Press, New York, N.Y. pp. 169-217.

Bartlett, D., Wright, M., Yayanos, A.A. and Silverman, M. (1989). Isolation of a gene regulated by hydrostatic pressure in a deep-sea bacterium. Nature, **342:** 572-574.

Bartlett, D.H. (1999). Microbial Adaptations to the Psychrosphere/Piezosphere. Journal of Molecular Microbiology and Biotechnology, **1:** 93-100.

Bartlett, D.H. (2002). Pressure effects on in vivo microbial processes. Biochimica et Biophysica Acta (BBA) - Protein Structure and Molecular Enzymology, **1595:** 367-381.

Bianchi, A. and Garcin, J. (1993). In stratified waters the metabolic rate of deep-sea bacteria decreases with decompression. Deep-Sea Research I, **40:** 1703-1710.

Bianchi, A. and Garcin, J. (1994). Bacterial response to hydrostatic pressure in seawater samples collected in mixed-water and stratified-water conditions. Marine Ecology Progress Series, **111:** 137-141.

Bianchi, A., Garcin, J., Gorsky, G., Poulicek, M. and Tholosan, O. (1999a). Stimulation du potentiel de dégradation dans les eaux marines profondes par les bactéries barotolérantes colonisant les fèces du plancton migrateur. Comptes Rendus de l'Académie des Sciences de Paris, **322:** 1113-1120.

Bianchi, A., Garcin, J. and Tholosan, O. (1999b). A high-pressure serial sampler to measure microbial activity in the deep sea. Deep-sea Research I, **46:** 2129-2142.

Bianchi, A., Tamburini, C. and Garcin, J. (submitted). Effect of pressure conditions on microbial activity measurements in deep-sea water: a review. Aquatic Microbial Ecology.

Bianchi, A., Van Wambeke, F. and Garcin, J. (1998). Bacterial utilization of glucose in the water column from eutrophic to oligotrophic pelagic areas in the eastern North Atlantic Ocean. Journal of Marine Systems, **14:** 45-55.

Boetius, A. and Lochte, K. (1994). Regulation of microbial enzymatic degradation of organic matter in deep-sea sediments. Marine Ecology Progress Series, **104:** 299-307.

Cahet, G., Daumas, R. and Sibuet, M. (1990). *In situ* experimentation at the water/sediment interface in the deep sea: 2. Biotransformation of dissolved organic substrates by microbial communities at 2000 m depth in the Bay of Biscay. Progress in Oceanography, **24:** 169-178.

Cahet, G. and Sibuet, M. (1986). Activité biologique en domaine profond: transformations biochimiques *in situ* de composés organiques marqués au carbone-14 à l'interface eau-sédiment par 2000 m de profondeur dans le golfe de Gascogne. Marine Biology, **90:** 305-307.

Certes, A. (1884). Sur la culture, à l'abri des germes atmosphériques, des eaux et des sédiments rapportés par les expéditions du Traveilleur et du Talisman. Comptes Rendus de l'Académie des Sciences de Paris, **98:** 690-693.

Cho, B.C. and Azam, F. (1988). Major role of bacteria in biogeochemichal fluxes in the ocean's interior. Nature, **332:** 441-443.

Colquhoun, J.A., Heald, S.C., Li, L., Tamaoka, J., Kato, C., Horikoshi, K. and Bull, A.T. (1998). Taxonomy and biotransformation activities of some deep-sea actinomycetes. Extremophiles, **2:** 269-277.

Corliss, J.B., Dymond, J., Gordon, L.I., Edmond, J.L., Von Erzen, R.P., Ballard, R.D., Green, K., Williams, D., Baindridge, A., Crane, J. and Van Andel, T.H. (1979). Submarine thermal springs on the Galapagos Rift. Science, **203:** 1073-1083.

Del Giorgio, P.A., Cole, J.J. and Cimbleris, A. (1997). Respiration rates in bacteria exceed phytoplankton production in unproductive aquatic systems. Nature, **385:** 148-151.

Delong, E.F. and Yayanos, A.A. (1985). Adaptation of the membrane lipids of a deep-sea bacterium to changes in hydrostatic pressure. Science, **228:** 1101-1113.

Deming, J.W. (1985). Bacterial growth in deep-sea sediment trap and boxcore samples. Marine Ecology Progress Series, **25:** 305-312.

Deming, J.W. (1993). ^{14}C tracer method for measuring microbial activity in deep-sea sediments. In: P.F. Kemp, B.F. Sherr, E.B. Sherr and J.J. Cole (eds) *Handbooks of Methods in Aquatic Microbial Ecology*. Lewis Publishers, Boca Raton, Ann Arbor, London, Tokyo, pp. 405-414.

Deming, J.W. (2001). Unusual or extreme high-pressure marine environments. In: C.J. Hurst, G.R. Knudsen, M.J. McInerney, L.D. Stetzenbach and M.V. Walter, (eds.) *ASM Manual of Environmental Microbiology*, ASM press.Washington DC, pp. 478-490.

Deming, J.W. and Baross, J.A. (1993). Deep-sea smokers: windows to a subsurface biosphere? Geochimica et Cosmochimica Acta, **57:** 3219-3230.

Deming, J.W. and Baross, J.A. (2000). Survival, dormancy, and nonculturable cells in extreme deep-sea environments. In: R.R. Colwell and D. J. Grimes (eds.) *Nonculturable Microorganims in the Environment*, ASM Press.Washington D.C., pp. 147-180.

Deming, J.W., Tabor, P.S. and Colwell, R.R. (1980). Deep ocean microbiology. In: F. Diemer, J. Vernberg and D. Mirkes (eds.) *Advanced Concepts in Ocean Measurements for Marine Biology*, University of South Carolina Press.Columbia St, pp. 285-305.

Dietz, A.S. and Yayanos, A.A. (1978). Silica gel media for isolating and studying bacteria under hydrostatic pressure. Applied and Environmental Microbiology, **36:** 966-968.

Ducklow, H.W. and Carlson, C.A. (1992). Oceanic Bacterial Production. In: K.C. Marshall (ed.) *Advances in Microbial Ecology*, Plenum Press.New York, pp. 113-181.

Jannasch, H.W. and Taylor, C.D. (1984). Deep-sea microbiology. Annual Review of Microbiology, **38:** 487-514.

Jannasch, H.W. and Wirsen, C.O. (1973). Deep-sea microorganisms: *in situ* response to nutrient enrichment. Science, **180:** 641-643.

Jannasch, H.W. and Wirsen, C.O. (1977). Retrieval of concentrated and undecompressed microbial populations from the deep sea. Applied and Environmental Microbiology, **33:** 642-646.

Jannasch, H.W. and Wirsen, C.O. (1980). Studies on the microbial turnover of organic substrates in deep-sea seadiments. In: R. Daumas (ed.) *Biogéochimie de la Matière Organique à l'Interface Eau-Sédiment Marin - Marseille 25-27 Avril*, Editions du CNRS.Paris, pp. 285-290

Jannasch, H.W. and Wirsen, C.O. (1982). Microbial activities in undecompressed microbial populations from the deep seawater samples. Applied and Environmental Microbiology, **43:** 1116-1124.

Jannasch, H.W. and Wirsen, C.O. (1984). Variability of pressure adaptation in deep sea bacteria. Archives of Microbiology, **139:** 281-288.

Jannasch, H.W., Wirsen, C.O. and Winget, C.L. (1973). A bacteriological pressure-retaining deep-sea sampler and culture vessel. Deep-Sea Research, **20:** 661-664.

Kato, C., Li, L., Nogi, Y., Nakamura, Y., Tamaoka, J. and Horikoshi, K. (1998). Extremely barophilic bacteria isolated from the Mariana Trench, Challenger Deep, at a depth of 11,000 meters. Applied and Environmental Microbiology, **64:** 1510-1513.

Kato, C., Nakasone, K., Qureshi, M.H. and Horikoshi, K. (2000). How do deep-sea microorganisms respond to changes in environmental pressure? In: K.B. Storey and. J. Storey (eds.), *Environmental Stressors and Gene Responses*, pp. 277-291.

Kato, C. and Qureshi, M.H. (1999). Pressure response in deep-sea piezophilic bacteria. Journal of Molecular Microbiology and Biotechnology, **1:** 87-92.

Kiørboe, T. and Jackson, G.A. (2001). Marine snow, organic solute plumes, and optimal chemosensory behavior of bacteria. Limnology and Oceanography, **46:** 1309-1318.

Lefèvre, D., Denis, M., Lambert, C.E. and Miquel, J.-C. (1996). Is DOC the main source of organic matter remineralization in the ocean water column? Journal of Marine Systems, **201:** 281-291.

Meyer-Reil, L.A. (1986). Measurement of hydrolytic activity and incorporation of dissolved organic substrates by microorganisms in marine sediments. Marine Ecology Progress Series, **31:** 143-149.

Meyer-Reil, L.A. and Köster, M. (1992). Microbial life in pelagic sediments: the impact of environmental parameters on enzymatic degradation of organic material. Marine Ecology Progress Series, **81:** 65-72.

Nagata, T., Fukuda, H., Fukuda, R. and Koike, I. (2000). Bacterioplankton distribution and production in deep Pacific waters: Large-scale geographic variations and possible coupling with sinking particle fluxes. Limnology and Oceanography, **45:** 426.

Nogi, Y., Kato, C. and Horikoshi, K. (1998). *Moritella japonica* sp. nov., a novel barophilic bacterium isolated from the Japan Trench sediment. Journal of Genetic Applied Microbiology, **44:** 289-295.

Phillips, H., Wells, L.E., Ii, R.V.J., Elliott, S. and Deming, J.W. (2003). LAREDO: A new instrument for sampling and *in situ* incubation of deep-sea hydrothermal vent fluids. Deep Sea Research I, **50**.

Picon, P. 2000. Diagénèse précoce de la matière organique à l'interface eau-sédiment dans un environnement de marge continentale, Université de Droit, d'Economie et des Sciences d'Aix-Marseille, 221 pp.

Prieur, D., Erauso, G. and Jeanthon, C. (1995). Hyperthermophilic life at deep-sea hydrothermal vents. Planetary and Space Science, **43:** 115-122.

Qureshi, M.H., Kato, C. and Horikoshi, K. (1998). Purification of two pressure-regulated c-type cytochromes from a deep-sea barophilic bacterium, *Shewanella* sp. strain DB-172F. FEMS Microbiology Letters, **161:** 93-99.

Relexans, J.C., Deming J.W., Dinet A., Gaillard J.F. & Sibuet M. (1996). Sedimentary organic matter and micro-meiobenthos with relation to trophic conditions in the northeast tropical Atlantic. Deep Sea Research, **43:** 1343-1368.

Saunders, P.M. and Fofonoff, N.P. (1976). Conversion of pressure to depth in the ocean. Deep Sea Research, **23:** 109-111.

Schwarz, J.R. and Colwell, R.R. (1975). Heterotrophic activity of deep-sea sediment bacteria. Applied Microbiology, **30:** 639-649.

Seki, H. and Robinson, D.G. (1969). Effect of decompression on activity of microorganisms in seawater. Internationale Revue gesamten Hydrobiologie, **54:** 201-205.

Simon, M., Welschmeyer N.A. and Kirchman D.L. (1992). Bacterial production and the sinking flux of particulate organic matter in the subarctic Pacific. Deep Sea Research, **39:** 1997-2008.

Smith, D.C., Simon, M., Alldredge, A.L. and Azam, F. (1992). Intense hydrolytic enzyme activity on marine aggrgates and implications for rapid particle dissolution. Nature, **359:** 139-142.

Tabor, P. and Colwell, R.R. (1976). Initial investigation with a deep ocean *in situ* sampler. Proc. MTS/IEEE OCEANS'76, Washington D.C., pp. 13D-11 - 13D-14.

Tabor, P.S., Deming, J.W., Ohwada, K., Davis, H., Waxman, M. and Colwell, R.R. (1981). A pressure-retaining deep ocean sampler and transfer system for measurement of microbial activity in the deep sea. Microbial Ecology, **7:** 51-65.

Tamburini, C. 2002. La dégradation du matériel organique profond par les microflores profondes : de la mesure des vitesses potentielles au flux de CO_2 généré *in situ*, Université de la Méditerranée, 222 pp.

Tamburini, C., Garcin, J. and Bianchi, A. (2003). Role of deep-sea bacteria in organic matter mineralization and adaptation to hydrostatic pressure conditions in the NW Mediterranean Sea. Aquatic Microbial Ecology, **32:** 209-218.

Tamburini, C., Garcin, J., Grégori, G., Leblanc, K., Rimmelin, P. and Kirchman, D.L. (submitted). Pressure effects on marine prokaryotes responsible for biogenic silica dissolution during a diatom sinking experiment. Submitted to Limnology and Oceanography.

Tamburini, C., Garcin, J., Ragot, M. and Bianchi, A. (2002). Biopolymer hydrolysis and bacterial production under ambient hydrostatic pressure through a 2000 m water column in the NW Mediterranean. Deep-Sea Research II, **49:** 2109-2123.

Teira, E., Reinthaler, T., Pernthaler, A., Pernthaler, J. and Herndl, G.J. (2004). Combining catalyzed reporter deposition-fluorescence *in situ* hybridization and microautoradiography to detect substrate utilization by bacteria and Archaea in the deep ocean. Applied and Environmental Microbiology, **70:** 4411-4414.

Tholosan, O., Garcin, J. and Bianchi, A. (1999). Effects of hydrostatic pressure on microbial activity through a 2000 m deep water column in the NW Mediterranean Sea. Marine Ecology Progress Series, **183:** 49-57.

Turley, C. and Lochte, K. (1990). Microbial response to the input of fresh detritus to the deep-sea bed. Palaeogeography, Palaeoclimatology, Palaeoecology, **89:** 3-23.

Turley, C.M. (1993). The effect of pressure on leucine and thymidine incorporation by free-living bacteria and by bacteria attached to sinking oceanic particles. Deep-Sea Research I, **40:** 2193-2206.

Van Der Wielen, P.W.J.J., Bolhuis, H., Borin, S., Daffonchio, D., Corselli, C., Giuliano, L., De Lange, G.J., Huebner, A., Varnavas, S.P., Thompson, J., Tamburini, C., Marty, D., Mcgenity, T.J., Timmis, K.N. and BioDeep Scientific Party (2005). The enigma of prokaryotic life in deep hypersaline Anoxic basins. Science **(307):** 121-123.

Viollier, E., Rabouille, C., Apitz, S.E., Breuer, E., Chaillou, G., Dedieu, K., Furukawa, Y., Grenz, C., Hall, P., Janssen, F., Janssen, F., Morford, J.L., Poggiale, J.-C., Roberts, S., Shimmield, T., Taillefert, M., Tengberg, A., Wenzhöfer, F. and Witte, U. (2003). Benthic biogeochemistry: state of the art technologies and guidelines for the future of *in situ* survey. Journal of Experimental Marine Biology and Ecology, **285-286:** 5-31.

Witte, U., Wenzhofer, F., Sommer, S., Boetius, A., Heinz, P., Aberle, N., Sand, M., Cremer, A., Abraham, W.R., Jorgensen, B.B. and Pfannkuche, O. (2003). *In situ* experimental evidence of the fate of a phytodetritus pulse at the abyssal sea floor. Nature, **424:** 763-766.

Yayanos, A.A. (1995). Microbiology to 10,500 meters in the deep-sea. Annual Review of Microbiology, **49:** 777-805.

Yayanos, A.A. and Chastain, R.A. (1999). The influence of nutrition on the physiology of piezophilic bacteria. *In* Microbial Biosystels: New Frontiers. Proceedings of the 8th international Symposium on Microbial Ecology. C. R. Bell, M. Brylinsky and J.-G. P. (eds), Atlantic Canada Society for Microbial Ecology.

Yayanos, A.A. and Delong, E.F. (1987). Deep-sea bacterial fitness to environmental temperatures and pressures. In: H. W. Jannasch, R. E. Marquis and A. M. Zimmerman (eds), *Current Perspectives in High Pressure Biology*, Academic Press, London, pp. 17-32.

Yayanos, A.A. and A.S. Dietz (1982). "Thermal inactivation of a deep-sea barophilic bacterium isolate CNPT-3." Applied and Environmental Microbiology **43(6):** 1481-1489.

Yayanos, A.A., S., D.A. and Van Boxtel, R. (1979). Isolation of a deep-sea barophilic bacterium and some of its growth characteristics. Science, **205:** 808-809.

Yayanos, A.A. and Van Boxtel, R. (1982). Coupling device for quick high-pressure connections to 100 MPa. Rev. Sci. Instrum., **53:** 704-705.

Yayanos, A.A., Van Boxtel, R. and Dietz, A.S. (1984). High-pressure-temperature gradient instrument: use for determining the temperature and pressure limits of bacterial growth. Applied and Environmental Microbiology, **48:** 771-776.

ZoBell, C.E. (1970). Pressure effects on morphology and life processes of bacteria. In: H. M. Zimmerman (ed.) *High pressure effects on cellular processes*, Academic Press, New York, pp. 85-130.

ZoBell, C.E. and Johnson, F.H. (1949). The influence of hydrostatic pressure on the growth and viability of terrestrial and marine bacteria. Journal of Bacteriology, **57:** 179-189.

PART 2:
LIFE IN THE CONTEXT OF THE NEW SCIENCE OF ASTROBIOLOGY

The Origin of Life on Earth and in the Universe

Llorca
Schrader
Stano et al.

The Evolution of Life on Earth and in the Universe

Deamer, Evans
Blumberg
Carnerup et al.
Rose et al.
Bell
Lee
Saffo
Shah
Maccone

The Distribution of Life in the Universe

McKay
Trigo-Rodríguez
Ehrenfreund, Martin
Lammer
Svoboda
Lineweaver
Shostak
Krupička
Teodorani

The Density of Life in the Universe

Chela-Flores
Coyne

Biodata of **Jordi Llorca** of ***"Early precursors of life in the solar nebula"***

Dr. Jordi Llorca is currently a Research Scientist at the Department of Inorganic Chemistry of the University of Barcelona, Spain. He obtained his Ph.D. from the University of Barcelona in 1995. He has been researcher at the Materials Science Institute (CSIC, Spain), the Institute of Meteoritics (The University of New Mexico, USA), and the Catalysis Research Institute (CNRS, France). Dr. Jordi Llorca scientific interests are in the areas of heterogeneous catalysis, cosmochemistry and meteoritics. He has been awarded by the Catalan Society of Chemistry (1990) and received the Catalan Government Distinction Award for University Research (2003).

E-mail: **jordi.llorca@qi.ub.es**

Dr. Jordi Llorca

J. Seckbach (ed.), Life as We Know It, 147-162.

EARLY PRECURSORS OF LIFE IN THE SOLAR NEBULA

JORDI LLORCA
Department of Inorganic Chemistry,
University of Barcelona,
Martí i Franquès 1-11, 08028 Barcelona, Spain

1. Primordial Soups

The synthesis of biogenic molecules in the early solar system is an essential step of the chemical evolution that must have preceded the origin of life on Earth. It is widely accepted that for life to appear on Earth, it first needed a primordial soup of simple organic molecules, as suggested by the Russian biochemist Oparin (1924) and the British biologist Haldane (1929). The origin of the primordial soup was identified, back in 1953, with the famous Miller experiment (Miller, 1953). When a gaseous mixture of hydrogen, water, methane and ammonia was subjected to electrical discharges simulating the effect of atmospheric storms in what it was thought to be the primitive atmosphere of the Earth, a soup was produced containing a wide variety of organic molecules, including amino acids from which proteins are built.

In recent decades, however, the general consensus among researchers is that the primitive terrestrial atmosphere was much too oxidizing to allow the production of significant quantities of biogenic molecules (Kasting and Ackerman, 1986). So, how and where did the building blocks of the primordial soup originate?

Today, one revision of the preferred terrestrial synthesis of biogenic molecules looks at conditions in the deep ocean ridges for the reducing environment that might have produced the precursors to life on Earth. Another scenario looks to produce these organics throughout the solar nebula or in primitive bodies such as asteroids and comets, which could deliver these materials easily to the growing Earth.

As we will discuss below, comets and primitive meteorites are filled with water and organic molecules. On this basis and the fact that since their formation comets and meteorites have been colliding with our planet time after time, Oró proposed explicitly in 1961 that water and the precursors of the organic matter on Earth could have come from the comets that bombarded it. This idea was later elaborated by other authors, like Delsemme (1984), Anders (1989) and Chyba and Sagan (1992). Previously, Chamberlin suggested in 1909 a similar role for organic matter from planetesimals. The idea that a significant fraction of the carbon compounds required for the origin of life had an extraterrestrial source may be surprising, but we, each of us, have part of a star inside us. The iron in hemoglobin that allows us to breath and the calcium in bones, for example, were made in stars and distributed by the nuclear explosions that ended their life before our solar system was born.

2. Where Are You, Carbon?

Carbon is the major player in the evolutionary scheme of prebiotic molecules because of its abundance and its ability to form complex species. There are numerous theoretical frameworks that are concerned with the final stages of chemical evolution from molecular clouds to planetary systems, but their assessments depend on the experimental data that may be obtained in the laboratory. Both types of studies have been recognized as fundamental for improving our current knowledge on the physical and chemical processes experienced by interstellar material prior to and during its incorporation into planetesimals, and also for the search of links between materials present in different space environments and biogenic components (Ehrenfreund et al., 2002a).

Prinn and Fegley (1989) pioneered the study of solar nebula chemistry using thermochemical equilibrium models with adiabatic pressure-temperature profiles. They found that, if the solar nebula had been in equilibrium, C and N would have been in the form of CO and N_2 in the inner nebula, and CH_4 and NH_3 in the outer. These authors emphasized the importance of thermochemical models in identifying departures from equilibrium, such as the existence of organics in meteorites and CO in comets, and the need for chemical kinetic models to quantitatively explain them. Willacy and co-workers presented time-dependent non-equilibrium chemical models for protoplanetary disks (Willacy et al., 1998), including ionization by cosmic rays and decay of ^{26}Al, and found that the chemistry in the inner disk was dominated by neutral-neutral reactions, and also that relatively high abundances of non-equilibrium species could exist there. Kress and Tielens (2001) modeled grain-surface catalyzed CO dissociation and hydrogenation. They showed that, in the presence of iron dust grains, surface catalysis would have occurred in the nebula in the vicinity of the modern-day asteroid belt, during the epoch of the solar nebula in which the meteorites formed.

It has been recognized that grain-surface chemistry in the inner solar nebula was fundamentally different from that in the outer solar nebula (Kress, 2000). In the outer nebula, refractory grains served as a substrate on which volatile species such as H_2O, CO, CO_2, CH_4, NH_3, and N_2 could condense as ices; reactions could occur in the icy mantles, but there was little or no chemical interaction between the volatiles and the refractory cores. In the inner nebula a dynamical physical and chemical interplay existed between the gas and dust. Gas-grain chemistry there was characterized by surface-catalyzed reactions and the destruction of the grains either by evaporation or by reaction with gas-phase species. The most important example of grain-surface chemistry described in the literature which could occur in the warm inner nebula is Fischer-Tropsch catalysis. This has been considered as the only available thermally-driven route to access the carbon locked up in CO in the solar nebula (Kress, 2000), because the strength of the CO bond precludes it from participating in gas-phase reactions under nebular conditions. However, the popularity of Fischer-Tropsch synthesis as a reaction responsible for the formation of solar system organics has waxed and waned since the pioneering work by Urey and Oró in the 1950s and 1960s (for a detailed discussion see i.e. Zolotov and Shock, 2001, and references therein). While other materials have been suggested to catalyze

Fischer-Tropsch-type reactions in different astrophysical environments, the only viable catalyst material would have been metallic Fe, Ni alloy grains, which are ubiquitous in the fine-grained matrix of primitive meteorites (Nagahara, 1984). Fe, Ni grains have also been isolated from interplanetary dust particles of probable cometary origin and have been found to be embedded in and coated with carbonaceous material, strongly suggesting that heterogeneous catalysis occurred in the solar nebula (Bradley et al., 1984a).

3. Heterogeneous Catalysis in the Solar Nebula

Heterogeneous reactions between model catalysts grains and nebular gaseous mixtures under solar nebula conditions have recently become a very rich and fruitful area of study. Meteoritic and cometary data and theoretical models are providing more detail on the physical conditions during the evolution of the nebula itself. In addition, surface science and catalytic studies capable to simulate the nebular environments in the laboratory and to study the chemistry under those conditions constitute a valuable approach in order to constrain the origin of organic molecules in the solar system. Recent laboratory work supports the hypothesis that Fischer-Tropsch catalytic processes operating in heterogeneous phase may have played a crucial role in solar nebula chemistry by converting carbon monoxide into less volatile materials, which could be much more readily processed in the nebula and in parent bodies to originate more complex organics. Llorca and Casanova (1998, 2000) have obtained both saturated and unsaturated hydrocarbons from H_2+CO mixtures (H_2: CO = 250:1) at 473 K and 50 Pa as well as thiols from H_2+CO+H_2S (H_2:CO:H_2S = 250:1:0.1) over Fe,Ni grains. Ferrante et al. (2000) have obtained in a similar way hydrocarbons over Fe-doped silicates from H_2+CO.

However, heterogeneous catalysis does not explain the structural and isotopic characteristics of all the organics present, for example, in primitive meteorites and interplanetary dust particles of probable cometary origin, which are the bodies that have best preserved the record of interstellar medium and nebular processes and that we can analyze in detail in the laboratory. Organics in meteorites can exhibit extreme fractionation patterns of H, N, and C isotopes, which has been interpreted in terms of significant contribution of interstellar material and/or compounds that derive directly from interstellar molecules (Cronin and Chang, 1993). On the other hand, the composition of primitive meteorites is dominated by phyllosilicates resulting from extensive aqueous alteration that occurred in their parent bodies. At present, we can not discuss about the effect that these secondary events had over the organics because very little experimental work has been carried out. Were they destructive or allowed the development of more complex organic compounds? Understanding the relationships and interdependence of organics and solid phases is essential for comprehending the processes and mechanisms involved in the evolution of organic material in the early solar system.

4. The Meteorite Connection

It was a Sunday morning, September 28, 1969, in the town of Murchison, Australia. Suddenly, at the time that most residents were attending church, a bright, spectacular, fireball appeared, signaling the arrival of hundreds of meteorites. The specimens were scattered across a 8 km^2 area within the town, and people gathered them easily from their yards, recovering more than 500 kg. The Murchison meteorite belongs to the CM carbonaceous chondrite family. Carbonaceous chondrites are primitive stony meteorites that may contain up to 10% of their weight in water and more than 3% weight in carbon. They are classified by subtle chemical compositional differences on the basis of their similarity to several prototypical specimens, leading to the nomenclature CI, CM, CV, CO, CR, CK and CH.

The CI and CM carbonaceous chondrites are the richest in organic matter and water, and for that reason they are very friable and fragile. They must be collected quickly after their fall, before the weather has a chance to destroy them. This is one of the reasons why Murchison has been a crucial stone, hundreds of meteorites were recovered just after they fell, and there was almost no weathering and contamination. There is also another unique circumstance, Murchison fell in 1969, at the time that laboratories all around the world were settled up for the study of other planetary materials: the rocks that the Apollo program brought from the Moon. Murchison was the right meteorite that fell at the right place and at the right time. Most of the work done on meteorite organic chemistry has been carried out on Murchison. However, organic compounds in carbonaceous chondrites were first observed in 1834 by Berzelius in the Alais carbonaceous chondrite (that fell in France in 1806) and by Berthelot in 1868 with a sample of Orgueil (that also fell in France, in 1864). The qualitative and quantitative analysis of organic molecules in meteorites represents an extremely difficult task, as the quantities of some of these compounds are similar to the amounts that would be transferred to the meteorite by just a few fingerprints (Oró and Skewes, 1965). Chromatographic and mass spectrometric techniques have, therefore, been widely used with a great success. Only the accurate analysis of the interior of carbonaceous chondrites that have been recovered shortly after their fall, that is, that have not experienced extensive terrestrial weathering, yield the most valuable information. In this context, some "organized elements" described in the early sixties in the carbonaceous chondrites Orgueil and Ivuna, which were considered by some as evidence that primitive life forms were carried to the Earth by meteorites, were contamination products.

Very recently, on the morning of January 18, 2000, an asteroid weighting approximately 200 to 250 tons and five to seven meters across impacted the Earth's atmosphere. The result was that hundreds of meteorites fell in a remote, cold arctic area between British Columbia and Yukon territories (Canada) in a strewn field 16 km long and 3 km wide. The meteorite turned to be a primitive chondrite very rich in organic compounds: it contains 3.6% carbon by weight (Brown et al., 2000). This carbonaceous chondrite, named Tagish Lake, has opened a new door for the search of organic compounds in meteorites. Meteorite fragments remained frozen in ice and never touched by human hands, thus minimizing the loss of volatile organic molecules and potential contamination.

Carbonaceous chondrites are very special meteorites. They are considered to be among the most primitive solar system materials on the basis of their chemical composition. With the exception of very volatile elements, like hydrogen, helium, carbon, nitrogen and oxygen, carbonaceous chondrites have an elemental composition that corresponds to that of the Sun. Taking into account that the solar system's elemental composition is equivalent to the chemistry of the Sun, since the later contains more than 99 percent of the mass of the whole system, the close correspondence between the elemental composition of the Sun and carbonaceous chondrites indicates that this type of meteorites have not experienced processes leading to extensive chemical fractiona-tion, like those that have altered all the rocks that we find on Earth and other types of meteorites (like irons and those from Mars). For that reason, carbonaceous chondrites can be considered as a precious window that allow us to look back to the past and to infer the way the first prebiotic molecules formed in our planetary environment.

Mixtures of clays, or phyllosilicates, comprise the dominant minerals (50 to 80%) in CI and CM chondrites. These hydrous minerals resemble some terrestrial clays, and were mostly formed as the products of low-temperature aqueous alteration of pre-existing assemblages of anhydrous minerals in asteroids (Grimm and McSween, 1989). Matrix is possibly the least understood major component of carbonaceous chondrites, but the organic matter is most concentrated in those meteorites and in those portions of the meteorites that experienced the most intense aqueous alteration. At present, *in situ* textural studies addressed to determine the exact location of the organic matter are just emerging (Pearson et al., 2002), and the study and identification of organic molecules is usually done after chemical extraction from the meteorite.

The organic matter in carbonaceous chondrites occurs in multiple forms. A macromolecular material that is extremely fine in grain size, centered at about 10 nm, constitutes the major organic carbon-bearing phase in all carbonaceous chondrites, representing more than 70% of the total carbon (Gardinier et al., 2000; Cody et al., 2002; Sephton et al., 2004). This complex material is virtually insoluble in solvents and most acids, and appears to be composed of both amorphous and poorly crystalline components, the latter constituted by turbostratic carbon or highly disordered graphite. In the Allende meteorite, a wide range of polycyclic aromatic hydrocarbons (PAHs) with extensively alkylated rings have been observed to be heterogeneously distributed within the matrix. Major species include naphthalene, phenanthrene, anthracene, and their alkyl-substituted derivatives, as well as a large range of heavier PAHs (Zenobi et al., 1989). The elemental composition of the macromolecular material is about $C_{100}H_{60}N_7O_{12}S_2$ (Hayatsu and Anders, 1981). Some fractions of this insoluble material serve as hosts for a variety of distinct noble gases, including several with isotopic compositions that can only be explained in terms of interstellar processes. Fullerenes in Murchison and Allende (the fullerene molecule exhibit a hollow, cage-like structure typically made of 60 or more carbon atoms), for example, encapsulated interstellar helium prior to incorporation into the solar nebula (Becker et al., 2000).

Solvent-extractable organic molecules in Murchison are distributed among a large variety of compound families with characteristic functional groups (Llorca, 2004). Hundreds of individual organic molecules have been encountered, and there are, for sure, much more that will be identified in the future. Although these organic compounds comprise less than 0.05% of the total mass of Murchison, for example, what is really intriguing is their structural and functional diversity. Amino acids, the building blocks of proteins, and all of the organic bases in DNA, the fundamental carrier of hereditary information for life as we know it, have been found, for example, in various carbonaceous chondrites. Other functional groups are hydrocarbons, alcohols, aldehydes, ketones, carboxylic acids, amines, amides, phosphonic acids and sulphonic acids. It should be mentioned that accurate isotope analysis is required in order to identify contaminants from the terrestrial environment added to the meteorites following their fall to Earth (Sephton and Gilmour, 2001).

Most amino acids and carboxylic acids do have enantiomers. In Murchison, it appears that most of these substances do not exhibit optical activity, suggesting an abiotic mechanism for their formation. However, the search for enantiomeric excess in branched α-methyl amino acids, which are not present in terrestrial proteins, revealed that L-enantiomers predominate slightly over D-enantiomers (Cronin and Pizzarello, 1997). The origin of homochirality in biological organisms has, in fact, originated a vigorous scientific debate. During the last decade, a series of rather simple experiments have demonstrated the feasibility of producing optically active compounds from achiral materials (Avalos et al., 2000). However, there is a large gap between molecular chirality and molecular evolution. Space technology in Earth orbit is being currently used to investigate whether amino acids and peptides required for the emergence of life can be safely transported to Earth vicinity without chemical degradation and racemization (Barbier et al., 2002; Boillot et al., 2002).

The exact nature of the processes responsible for the existence of organic matter in meteorites and the locations where they occurred remain to be clearly established. Gas-grain processes requiring solid surfaces with catalytic properties and multiple gas phase reactions could have taken place both in the solar nebula and on asteroids as well as partly in the interstellar medium. Isotopic studies have revealed that large differences exist in the D/H, $^{13}C/^{12}C$, $^{15}N/^{14}N$, and $^{33}S/^{32}S$ ratios of different organic compounds in carbonaceous chondrites (Cronin and Chang, 1993), adding growing evidence that it is necessary to have more than one source region and production mechanism to explain the formation of these organics. Laboratory Fischer-Tropsch-type reactions can produce many of the organic compounds discussed above. However, none of these processes can produce D enrichments as large as those recognized in the organic acids, amino acids and macromolecular carbon of the carbonaceous chondrites. The existence of macromolecular organic material in interstellar space, similar in many respects to that present in carbonaceous chondrites, is universally accepted, since the absorption and emission of light by grains in interstellar space are entirely consistent with the presence there of complex carbon compounds (Sandford et al., 1991). It is likely that interstellar grains have silicate cores mantled by complex carbon compounds resulting from radiation processing. Carbonaceous chondrites could incorporate interstellar macromolecular organic materials or their precursors, which in turn could be thermally and chemically reprocessed in the solar nebula to varying degrees.

5. Comets and Interplanetary Dust

Of all the debris left over from the formation of the Sun and planets in the solar system, comets contain by far the largest amounts of hydrogen-, carbon-, sulfur-, and nitrogen-bearing molecules that are needed to support biogenesis. In comets, these elements are considerably more abundant than in carbonaceous chondrites. Because of the small size of comets, which implies the absence of large-scale differentiation and mixing processes, and their accretion place, far from the Sun, thus avoiding volatile molecules to be depleted, one may expect that presolar as well as nebular organic molecules are best preserved in comets.

The universal view that a comet's nucleus is an icy body that contains rocky material is by no means new. It dates back at least to Pierre-Simon Laplace in the early 19th century. However, in 1950 Fred Whipple resurrected the idea and modified it, describing the essential features of all cometary observations. Whipple's theory of "dirty snowballs" became the basis of all later work on the chemistry of comets. Water is the dominant ice in comets, and all other ices, such as carbon monoxide, carbon dioxide, methane, ammonia and methanol, may be condensed or trapped within the water ice. Physical adsorption is particularly effective on amorphous ice at low temperatures, and clathration can occur at temperatures well above the pure condensation points of the volatiles. Comets contain an average of 40-50% water (by mass), 13-16% volatile organics, 13-15% refractory organics and 22-26% silicates (Greenberg, 1998; Huebner, 2002).

In the absence of fresh cometary samples, we investigate cometary material in the laboratory through the study of interplanetary dust particles whose properties suggest a cometary origin. As the nucleus of a comet approaches the Sun, volatile and icy components sublimate, and dust as well as larger particles are emitted from the nucleus. This is viewed as a long tail, as the dust is accelerated away from the nucleus by radiation pressure. This force originates from the interaction of electromagnetic radiation (sunlight) with matter. As an example, the space probes that encountered Halley's comet in 1986 found that the comet was injecting approximately 3000 kg of dust per second into the interplanetary medium (McDonnell et al., 1986). Interplanetary dust particles have been and are still collected from a wide variety of sites using a number of different techniques. Ocean bottoms, temporary glacial melt lakes in Greenland and Antarctica, space recoverable rockets and satellites, micro-craters on mineral grains from the surfaces of lunar rocks, and the upper atmosphere represent different environments for interplanetary dust recovery (Sandford, 1987). However, collections made in the upper atmosphere are of particular importance since they provide dust particles of unquestioned interplanetary origin which have not been severely altered by capture processes. Since 1974 and after extensive work of Donald Brownlee, micrometer-sized interplanetary dust particles have been collected from the stratosphere with aircrafts flying at altitudes of about 20 km and brought back to terrestrial laboratories for detailed examination (Brownlee, 1985).

The miraculous aspect of interplanetary dust interaction with the atmosphere of the Earth, is that their deceleration is so gentle that they are slowed without severe heating and mechanical stress. Interplanetary dust enters the atmosphere at high velocities, in excess of 11 km s^{-1} (for comparison, a cannonball moves at only ~5 km s^{-1}), and slow down at altitudes above 80 km, where the air is very thin. As a result, the frictional heat builds up slowly and dust particles radiate the heat without melting. Once decelerated, the particles slowly settle out and can be collected in the stratosphere before they become mixed with terrestrial particulate material commonly found at lower altitudes. In contrast, millimeter-sized particles penetrate deeper into the atmosphere before slowing down and vaporize, originating spectacular shooting stars.

There are several lines of evidence showing that some of the collected stratospheric dust is extraterrestrial. The most convincing proofs are the presence of implanted solar wind noble gases, like He, Ne and Ar (Hudson et al., 1981), the presence of large deuterium enrichments (Zinner et al., 1983), the discovery of solar flare tracks in mineral grains within the particles (Bradley et al., 1984b), and their chondritic elemental composition. Chondritic particles, the most common type of interplanetary dust, are roughly subdivided into two major categories, chondritic porous (CP) and chondritic smooth (CS). CS dust particles are relatively compact and their mineralogy is dominated by phyllosilicates, they are similar in many respects to the matrices of CI and CM carbonaceous chondrites. In contrast, CP particles have a characteristic fragile cluster-of-grapes morphology and densities between 0.7 and 2.2 g cm^{-3}. Most of them are generally anhydrous and are constituted of a great variety of mineral grains, ranging in size from about 5 nm to a few micrometers. Examination by electron microscope techniques reveals that many grains are single minerals, while others are composed of several much smaller crystals embedded in amorphous material. Most of the amorphous material has a carbonaceous nature. Amorphous and poorly ordered low atomic number phases are abundant in CP interplanetary dust, which occur as clumped masses as well as thin coatings on some grains. Both chemical and physical features of CP interplanetary dust particles as well as their spectral properties point to a cometary origin (Bradley and Brownlee, 1986).

The average carbon content of CP interplanetary dust particles has been measured as 10 to 12% (Thomas et al., 1993), which is about 2.5 to 3 times the carbon content of the carbonaceous chondrites. Carbon is widespread throughout most IDPs. The small masses of interplanetary dust particles (about 10^{-8} g) have, thus far, precluded quantitative determination of the fraction of this carbon that is present in organic molecules. However, Clemett et al. (1993) identified the presence of individual organic molecules by detecting many polycyclic aromatic hydrocarbons (PAHs) and their alkylated derivatives in two dust particles. Also, infrared and Raman spectra of individual CP interplanetary dust are dominated by features characteristic of disordered carbonaceous materials rich in aromatic molecular units (Allamandola et al., 1987), and carbon X-ray absorption near-edge structure spectroscopy has revealed that organic matter is the "glue" that holds grains together in some anhydrous IDPs (Flynn et al., 2003).

Dust particles released from a comet are not in the same state as they were while residing inside the nucleus. They are devoid of volatile molecules which may have filled the cavities inside the particles; this is probably why these particles look like fluffy aggregates. The loose structure of cometary dust is also demonstrated by the low densities of meteors originating from comets, for which densities as low as 0.1 g cm^{-3} have been deduced (Ceplecha, 1977). Elemental abundances of Halley's comet dust obtained during flybys by the Vega and Giotto missions indicated high abundances of the elements C, H, O and N in some of the particles, the so-called CHON particles, which exhibited densities in the range 0.1-0.3 g cm^{-3}. This was interpreted in terms of organic compounds being present in the cometary dust (Jessberger et al., 1988), which appeared to be a mixture of CHON-rich and silicate-rich material present in highly variable proportions. The evidence for carbonaceous material was, in fact, one of the major results of comet Halley's exploration. Kissel and Krueger (1987) suggested through mass spectrometric measurements of Halley's dust that it contained linear and cyclic, saturated and unsaturated hydrocarbons, nitriles, amines, imines, nitrogen-bearing heterocycles, such as pyrroles, pyrimidines, pyridines, purines, etc. They suggested that the dust particles were composed of a mineral core with an organic mantle, similar in many respects to the laboratory models proposed by Greenberg (1982). Adenine, a significant organic molecule for prebiotic chemistry, could be synthesized in comets as Oró did it in 1960 from reaction of hydrogen cyanide, an important constituent of comets and also a very reactive compound, in an aqueous ammonia mixture.

Interplanetary dust particles of probable cometary origin shed some light on the intimate nature of the solid component in comets. However, our knowledge of the nature of the volatile component in cometary nucleus is based entirely on spectroscopic observations of species produced when the cometary nucleus interacts with solar radiation. As the nucleus approaches the Sun, ices from its surface sublimate. The released gases form a wrapper, named coma, that is continually lost to the interplanetary medium. Solar ultraviolet radiation dissociates and subsequently ionizes the molecules in the coma, which are, in addition, modified by fast, complex ion-molecule reactions. The ions interact with the solar wind, and are blown away into the ion tail. This complicated chain of reactions makes it difficult to reconstruct the nature and abundance of the unobserved parent molecules in the nucleus from the daughter molecular species visible in both the coma and the ion tail. Thus, our knowledge of the nature of the organic component in comets remains indirect and subject to hypotheses.

Improvements in spectroscopic techniques have made possible remote detection of volatile organic molecules outgassed from the cometary nucleus. The recent close approach to Earth of comets Hyakutake and Hale-Bopp in 1996 and 1997, respectively, allowed the first identifications of the organic species C_2H_2, CH_3CN, HC_3N, NH_2CHO, HCOOH and $HCOOCH_3$ (Bockelée-Morvan et al., 2000; Bockelée-Morvan and Crovisier, 2002), besides other molecules already reported, like CH_4, C_2H_6, CH_3OH and HCHO. The prebiotic relevance of all these organic molecules is easy to recognize.

A wide array of organic products have been produced from irradiation of cometary ice-analogs in the laboratory (see for example Strazzulla, 1997; Ehrenfreund et al., 2002b). When the resulting laboratory ices are warmed up, many of the parent species as well as the new photoproducts sublimate out of the ice and, at this point, moderately complex organic molecules, such as alcohols, amides, amines, ketones and nitriles are formed (Bernstein et al., 1999). In addition, some species with greater complexity are produced, like polyoxymethylene-related species ((-CH_2O-)$_n$) and hexamethylenetetramine ($C_6H_{12}N_4$). On Earth, many of these species are biologically important. When hexamethylenetetramine is hydrolyzed in acid, amino acids are spontaneously produced (Bernstein et al., 2002). More intriguingly, when some organic mixtures left over from warming laboratory ices to room temperature are placed in water, insoluble lipid-like droplets are formed, which show self-organizing, membrane-forming behavior (Allamandola et al., 1988; Deamer et al., 2003). One may speculate that, in the presence of liquid water, simple organic cometary compounds reacted to form more complex biochemical molecules on Earth.

6. The Infall of Organic Molecules on Earth

Just after accretion, terrestrial planets were poor in organic material because they formed in the inner, hotter regions of the solar nebula. In contrast, as the distance from the Sun increases, the amount of organic matter appears to increase. As we have seen, the asteroids begin to show signs of significant organics, and these are probably abundant in Kuiper belt objects and certainly in comets. Also, organic matter is apparently common in the moons of the giant planets (Irvine, 1998). This is one of the paradoxes of life as we know it: organic molecules thought to be possible building blocks of the first organisms are extensively found in the outer solar system, whereas the environments conducive to molecular evolution and the growth of life are found in the inner solar system. One credible suggestion for the origin of life would be to transport biogenic organic molecules from the outer solar system to the inner solar system. This is the key role that comets and meteorites may have played. Another possibility is that specific organic precursors of key importance to life which were difficult to synthesize on the Earth itself might have been supplied from cometary or meteoroidal impacts (Pizzarello, 2004).

According to some estimates, the contemporary influx of meteoritic material (meteorites, comets and interplanetary dust) into the Earth is about 10^5 kg/day (Maurette, 1998). Taken into account that collisions of comets and meteorites with planetary surfaces could result in the decomposition or alteration of their organic content to a considerable extent, interplanetary dust, which is not subjected to significant heating during atmospheric deceleration, is believed to be the best delivery vehicle for organic matter of space origin to the Earth. There are only a few interplanetary dust particles in each cubic meter of space, but the Earth sweeps up around 10^7 kg of this material each year as it travels around the Sun. About 80% of the total mass accreted by the Earth corresponds to particles weighting between 10^{-7} and 10^{-3} grams (Flynn, 1996). Concerning the survival of organic compounds, Chyba and Sagan (1992, 1997) calculated that most of the organic material in carbonaceous chondrites can survive temperatures up to 850 K for about one second, and that a typical interplanetary dust particle spends only a few seconds within 100 K of its peak temperature on atmospheric entry. Textural zoning attributable to temperature gradients observed in 50 to 100 μm diameter micrometeorites recovered from terrestrial polar ices suggests that a mechanism exists to allow the interior to remain cool.

Comets and meteorites contain secreted within them the oldest and most remote known materials as well as an impressive array of organic molecules, but their study is still difficult. A fascinating aspect of this organic matter is that some may be precursors of life. Answering questions such as which may be the product of chemical evolution of such organic materials and when and where they originated awaits the development of yet more precise analytical techniques and procedures that exist today. The ability to examine fresh cometary material in a laboratory setting will make future space missions invaluable. It could confirm and put into context previous laboratory findings that complex organic molecules could have formed in space and seeded the early Earth.

7. References

Allamandola, L.J., Sandford, S.A. and Wopenka, B. (1987) Interstellar polycyclic aromatic hydrocarbons and carbon in interplanetary dust particles and meteorites. Science **237**, 56-59.

Allamandola, L.J., Sandford, S.A. and Valero, G.J. (1988) Photochemical and thermal evolution of interstellar/precometary ice analogs. Icarus **76**, 225-252.

Anders, E. (1989) Pre-biotic organic matter from comets and asteroids. Nature **342**, 255-257.

Avalos, M., Babiano, R., Cintas, P., Jiménez, J.L. and Palacios, J.C. (2000) Chiral autocatalysis: where stereochemistry meets the origin of life. Chem. Commun. 887-892.

Barbier, B., Henin, O., Boillot, F., Chabin, A., Chaput, D. and Brack, A. (2002) Exposure of amino acids and derivatives in the Earth orbit. Planet. Space Sci. **50**, 353-359.

Becker, L., Poreda, R.J. and Bunch, T.E. (2000) Fullerenes: An extraterrestrial carbon carrier phase for noble gases. Proc. Natl. Acad. Sci. **97**, 2979-2983.

Bernstein, M.P., Sandford, S.A., Allamandola, L.J., Gillete, J.S., Clemett, S.J. and Zare, R.N. (1999) UV Irradiation of polycyclic aromatic hydrocarbons in ices: Production of alcohols, quinones and ethers. Science **283**, 1135-1138.

Bernstein, M.P., Dworkin, J.P., Sandford, S.A., Cooper, G.W. and Allamandola, L.J. (2002) Racemic amino acids from the ultraviolet photolysis of interstellar ice analogues. Nature **416**, 401-403.

Berthelot, M.P.E. (1868) Sur la matière charbonneuse des météorites. Comptes Rendus Hebd Seances Acad. Sci. **67**, 849.

Berzelius, J.J. (1834) Ueber Meteorsteine. Ann. Phys. Chem. **33**, 113-148.

Bockelée-Morvan, D., Lis, D.C., Wink, J.E., Despois, D., Crovisier, J., Bachiller, R., Benford, D.J., Biver, N., Colom, P., Davies, J.K., Gérard, E., Germain, B., Houde, M., Mehringer, D., Moreno, R., Paubert, G., Phillips, T. and Rauer, H. (2000) New molecules found in comet C/1995 O1 (Hale-Bopp). Astron. Astrophys. **353**, 1101-1114.

Bockelée-Morvan, D. and Crovisier, J. (2002) Lessons of comet Hale-Bopp for coma chemistry: Observations and theory. Earth Moon Planets **89**, 53-71.

Boillot, F., Chabin, A., Bure, C., Venet, M., Belsky, A., Bertrand-Urbaniak, M., Delmas, A., Brack, A. and Barbier, B. (2002) The Perseus Exobiology mission on MIR: Behaviour of amino acids and peptides in Earth orbit. Origins Life Evol. B. **32**, 359-385.

Bradley, J.P. and Brownlee, D.E. (1986) Cometary particles: Thin sectioning and electron beam analysis. Science **231**, 1542-1544.

Bradley, J.P., Brownlee, D.E. and Fraundorf, P. (1984a) Carbon compounds in interplanetary dust: evidence for formation by heterogeneous catalysis. Science **223**, 56-58.

Bradley, J.P., Brownlee, D.E. and Fraundorf, P. (1984b) Discovery of nuclear tracks in interplanetary dust. Science **226**, 1432-1434.

Brown, P., Hildebrand, A.R., Zolensky, M.E., Grady, M., Clayton, R.N., Mayeda, T.K., Tagliaferri, E., Spalding, R., MacRae, N.D., Hoffman, E.L., Mittlefehldt, D.W., Wacker, J.F., Bird, J.A., Campbell, M.D., Carpenter, R., Gingerich, H., Glatiotis, M., Greiner, E., Mazur, M.J., McCausland, P.JA., Plotkin, H. and Rubak Mazur, T. (2000) The fall, recovery, orbit, and composition of the Tagish Lake meteorite: A new type of carbonaceous chondrite. Science **290**, 320-325.

Brownlee, D.E. (1985) Cosmic dust: Collection and research. Ann. Rev. Earth Planet. Sci. **13**, 147-173.

Ceplecha, Z. (1977) Meteoroid populations and orbits, In: A.H. Delsemne (ed.) *Comets, Asteroids, Meteorites*. University of Toledo, Ohio, pp. 143-152.

Chamberlin, T.C. and Chamberlin, R.T. (1908) Early terrestrial conditions that may have favored organic synthesis. Science **28**, 897-911.

Chyba, C.F. and Sagan, C. (1992) Endogenous production, exogenous delivery and impact-shock synthesis of organic molecules: an inventory for the origin of life. Nature **355**, 125-130.

Chyba, C.F. and Sagan, C. (1997) Comets as a source of prebiotic organic molecules for the early Earth, In: P.J. Thomas, C.F. Chyba and C.P. McKay (eds.) *Comets and the Origin and Evolution of Life*. Springer-Verlag, New York, pp. 147-174.

Clemett, S.J., Maechling, C.R., Zare, R.N., Swan, P.D. and Walker, R.M. (1993) Identification of complex aromatic molecules in individual interplanetary dust particles. Science **262**, 721-725.

Cody, G.D., Alexander, C.M.O. and Tera, F. (2002) Solid-state (H-1 and C-13) nuclear magnetic resonance spectroscopy of insoluble organic residue in the Murchison meteorite: A self-consistent quantitative analysis. Geochim. Cosmochim. Acta **66**, 1851-1865.

Cronin, J.R. and Chang, S. (1993) Organic matter in meteorites: molecular and isotopic analyses of the Murchison meteorite, In: J.M. Greenberg, C.X. Mendoza-Gómez and V. Pirronello (eds.) *Chemistry of Life's Origins*. Kluwer Academic Pub., Dordrecht, pp. 209-258.

Cronin, J.R. and Pizzarello, S. (1997) Enantiomeric excesses in meteoritic amino acids. Science **275**, 951-955.

Deamer, D.W., Dworkin, J.P., Sandford, S.A., Bernstein, M.P. and Allamandola, L.J. (2003) The first cell membranes. Astrobiology **2**, 371-381.

Delsemme, A. (1984) The cometary connection with prebiotic chemistry. Origins Life **14**, 51-60.

Ehrenfreund, P., Irvine, W., Becker, L., Blank, J., Brucato, J.R., Colangeli, L., Derenne, S., Despois, D., Dutrey, A., Fraaije, H., Lazcano, A., Owen, T., Robert, F. and International Space Science Institute ISSI-Team (2002a) Astrophysical and astrochemical insights into the origin of life. Rep. Progr. Phys. **65**, 1427-1487.

Ehrenfreund, P., Rodgers, S.D. and Charnley, S.B. (2002b) Physico-chemistry of comets: Models and laboratory experiments. Earth Moon Planets **89**, 221-246.

Ferrante, R.F., Moore, M.H., Nuth III, J.A. and Smith, T. (2000) Formation of Carbon Compounds by Catalytic Conversion of CO on Laboratory Grain Analogs Containing Iron. Icarus **145**, 397-300.

Flynn, G.J. (1996) The delivery of organic matter from asteroids and comets to the early surface of Mars. Earth Moon Planets **72**, 469-474.

Flynn, G.J., Keller, L.P., Feser, M., Wirick, S. And Jacobsen, C. (2003) The origin of organic matter in the solar system: Evidence from the interplanetary dust particles. Geochim. Cosmochim. Acta **67**, 4791-4806.

Gardinier, A., Derenne, S., Robert, F., Behar, F., Largeau, C. and Maquet, J. (2000) Solid state CP/MAS C-13 NMR of the insoluble organic matter of the Orgueil and Murchison meteorites: quantitative study. Earth Planet. Sci. Lett. **184**, 9-21.

Greenberg, J.M. (1982) What are comets made of? A model based on interstellar dust, In: L.L. Wilkening (ed.) *Comets*. University of Arizona Press, Tucson, pp. 131-163.

Greenberg, J.M. (1998) Making a comet nucleus. Astron. Astrophys. **330**, 375-380.

Grimm, R.E. and McSween, H.Y. (1989) Water and the thermal evolution of carbonaceous chondrite parent bodies. Icarus **82**, 244-280.

Haldane, J.B.S. (1929) The Origin of Life. The Rationalism Annual. Reprinted in D. Deamer and G. Flieschaker (1994) *Origins of Life: the Central Concepts*. Jones and Barlett, Boston.

Hayatsu, R. and Anders, E. (1981) Organic compounds in meteorites and their origins. Top. Current Chemistry **99**, 1-37.

Hudson, B., Flynn, G.J., Fraundorf, P., Hohenberg, C.M. and Shirck, J. (1981) Noble gases in stratospheric dust particles: Confirmation of extraterrestrial origin. Science **211**, 383-386.

Huebner, W.F. (2002) Composition of comets: Observations and models. Earth Moon Planets **89**, 179-195.

Irvine, W.M. (1998) Extraterrestrial organic matter: A review. Origins Life Evol. B. **28**, 365-383.

Jessberger, E.K., Christoforidis, A. and Kissel, J. (1988) Aspects of the major element composition of Halley's dust. Nature **332**, 691-695.

Kasting, J. and Ackerman, T.F. (1986) Climatic consequences of very high carbon dioxide levels in the Earth's early atmosphere. Science **234**, 1383-1385.

Kissel, J. and Krueger, F.R. (1987) The organic component in dust from Comet Halley as measured by the PUMA mass spectrometer on board Vega 1. Nature **326**, 755-760.

Kress, M. (2000) New developments in inner solar nebula chemistry, In: Y.C. Minh and E. Van Dshoeck (eds.) *Astrochemistry: From Molecular Clouds to Planetary Systems*. Astronomical Society of the Pacific, San Francisco, vol. 197, pp. 537-546.

Kress, M. and Tielens, A.G.G.M. (2001) The role of Fischer-Tropsch catalysis in solar nebula chemistry. Meteorit. Planet. Sci. **36**, 75-91.

Llorca, J. (2004) Organic molecules in meteorites. Int. Microbiol. **7**, 239-248.

Llorca, J. and Casanova, I. (1998) Formation of carbides and hydrocarbons in chondritic interplanetary dust particles: A laboratory study. Meteorit. Planet. Sci. **33**, 243-251.

Llorca, J. and Casanova, I. (2000) Reaction between H_2, CO and H_2S over Fe,Ni metal in the solar nebula. Experimental evidence for the formation of sulfur-bearing organic molecules and sulfides. Meteorit. Planet. Sci. **35**, 841-848.

Maurette, M. (1998) Carbonaceous micrometeorites and the origin of life. Origins Life Evol. B. **28**, 385-412.

McDonnell, J.A.M., Alexander, W.M., Burton, W.M., Bussoletti, E., Clark, D.H., Grard, R.J.L., Grün, E., Hanner, M.S., Hughes, D.W., Igenbergs, E., Kuczera, H., Lindblad, B.A., Mandeville, J.C., Minafra, A., Schwehm, G.H., Sekanina, Z., Wallis, M., Zarnecki, J.C., Chakaveh, S.C., Evans, G.C., Evans, S.T., Firth, J.C., Littler, A.N., Massone, L., Olearczyk, R.E., Pankiewicz, G.C., Stevenson, T.J. and Turner, R.J. (1986) Dust density and mass distribution near comet Halley from Giotto observations. Nature **321**, 338-341.

Miller, S.L. (1953) Production of aminoacids under possible primitive Earth conditions. Science **117**, 528-529.

Nagahara, H. (1984) Matrices of type 3 ordinary chondrites–primitive nebular records. Geochim. Cosmochim. Acta **48**, 2581-2595.

Oparin, A.I. (1924) *Proiskhodenie Zhizni. Moscovky Rabotchii.* English translation in D. Deamer and G. Flieschaker (1994) *Origins of Life: the Central Concepts*. Jones and Barlett, Boston.

Oró, J. (1960) Synthesis of adenine from ammonium cyanide. Biochem. Biophys. Res. Commun. **2**, 407-412.

Oró, J. (1961) Comets and the formation of biochemical compounds on the primitive Earth. Nature **190**, 389-390.

Oró, J. and Skewes, H.B. (1965) Free amino acids on human fingers: the question of contamination in microanalysis. Nature **207**, 1042-1045.

Pearson, V.K., Sephton, M.A., Kearsley, A.T., Bland, B.A., Franchi, I.A. and Gilmour, I. (2002) Clay mineral-organic matter relationships in the early solar system. Meteorit. Planet. Sci. **37**, 1829-1833.

Pizzarello, S. (2004) Chemical evolution and meteorites: An update. Origins Life Evol. B. **34**, 25-34.

Prinn, R.G. and Fegley, B.Jr. (1989) Solar nebula chemistry: Origin of planetary, satellite, and cometary volatiles, In: S.K. Atreya, J.B. Pollack and M.S. Matthews (eds.) *The origin and evolution of planetary and satellite atmospheres*. Univ. Arizona Press, Tucson, pp. 78-136.

Sandford, S.A. (1987) The collection and analysis of extraterrestrial dust particles. Fund. Cosmic Phys. **12**, 1-73.

Sandford, S.A., Allamandola, L.J., Tielens, A.G.G.M., Sellgren, K., Tapia, M. and Pendleton, Y. (1991) The interstellar carbon-hydrogen stretching band near 3.4 microns: Constraints on the composition of organic material in the diffuse interstellar medium. Astrophys. J. **371**, 607-620.

Septhon, M.A. and Gilmour, I. (2001) Compound-specific isotope analysis of the organic constituents in carbonaceous chondrites. Mass Spectrom. Rev. **20**, 111-120.

Sephton, M.A., Love, G.D., Watson, J.S., Verchovsky, A.B., Wright, I.P., Snape, C.E. and Gilmour, I. (2004) Hydropyrolysis of insoluble carbonaceous matter in the Murchison meteorite: New insights into its macromolecular structure. Geochim. Cosmochim. Acta **68**, 1385-1393.

Strazzulla, G. (1997) Ion bombardment of comets, In: Y. Pendleton and A. Tielens (eds.) *From Stardust to Planetesimals*. ASP Conference Series, vol. 122, pp. 423-433.

Thomas, K.L., Blandford, G.E., Keller, L.P., Klock, W. and McKay, D.S. (1993) Carbon abundance and silicate mineralogy of anhydrous interplanetary dust particles. Geochim. Cosmochim. Acta **57**, 1551-1566.

Whipple, F.L. (1950) A comet model I. The acceleration of comet Encke. Astrophys J. **111**, 375-394.

Willacy, K., Klahr, K.K., Millar, T.J. and Henning, Th. (1998) Gas and grain chemistry in a protoplanetary disk. Astron. Astrophys. **338**, 995-1005.

Zenobi, R., Philippoz, J.M., Buseck, P.R. and Zare, R.N. (1989) Spatially resolved organic analysis of the Allende meteorite. Science **246**, 1026-1029.

Zinner, E., McKeegan, K.D. and Walker, R.M. (1983) Laboratory measurements of D/H ratios in interplanetary dust. Nature **305**, 119-121.

Zolotov, M.Y. and Shock, E.L. (2001) Stability of condensed hydrocarbons in the solar nebula. Icarus **150**, 323-337.

Biodata of **Malcolm Schrader** author of ***"Is Life on Earth Accidental? Natural Selection and the Second Law of Thermodynamics"***

Dr. Malcolm E. Schrader was born in New York City in 1925. After serving in the US Army during World War II, with award of the Purple Heart and Combat Infantry Badge, he received his PhD (1956) in Physical Chemistry from Brooklyn Polytechnic Institute. He subsequently performed fundamental and applied research for the US Navy, largely in the field of surface chemistry and nano-surface science. He pioneered in the application of radioisotopes to the study of the interface between fiber and matrix in composite materials, and was the first to use ultra-high vacuum techniques in the study of wettability. Some highlights of the achievements of his experimental research are the solving of the famous controversy on the wettability of gold by water, obtaining of a previously elusive result for water-wettability of graphite, and determination of the nano-locus of hydrolytically induced failure of the glass-resin bond in glass-reinforced plastics.

Dr. Schrader served for four years as Meeting Secretary (program manager) of the Division of Colloid and Surface Chemistry of the American Chemical Society, and as long-range coordinator of symposia on fluid-solid surface interaction of that Division. He served on the Board of Directors of the Society for Plastic Industries, and was involved with surface chemistry applications as a member of the American Vacuum Society and the American Society for Testing Materials.

On retiring as senior scientist from the U.S. Navy, he was appointed professor on the research staff of the Hebrew University of Jerusalem. In the theoretical area, he subsequently revised the thermodynamic approach to wettability, and showed that, contrary to previously accepted concepts, the adsorption of a nano-layer of its own vapor does not necessarily impede spreading of a given liquid on a solid surface. He is presently pursuing an interest in chemical evolution as Guest Scientist with the Department of Inorganic and Analytical Chemistry at the Hebrew University.

E-mail: **schrader@vms.huji.ac.il**

J. Seckbach (ed.), Life as We Know It, 163-178.

IS LIFE ON EARTH ACCIDENTAL?
Natural Selection and the Second Law of Thermodynamics

MALCOLM E. SCHRADER
Department of Inorganic and Analytical Chemistry,Institute of Chemistry, The Hebrew University of Jerusalem, 91904 Jerusalem, Israel

1. Introduction

The "Theory of Evolution", as presented by Wallace in 1855 and formulated by Charles Darwin in 1859, is in reality a combination of proposals, of which each has received a different degree of acceptance from leading biologists throughout the years. E. Mayr (1982) has resolved it into 5 separate components, or sub-theories, which he lists as follows:

1. Evolution did indeed take place.
2. Speciation occurred
3. All evolution involved descent from common ancestry.
4. Evolution occurred as a result of natural selection.
5. Evolution is gradual, rather than "saltational". This means, essentially, that a transition from one species to another does not occur as a result of one sudden large change, but rather from the accumulation of many small changes, resulting finally in a new species.

Mayr lists his five sub-theories of Darwinism according to their acceptability by a number of prominent classical biologists.

Table 1. Agreement with sub-theories of Darwinism.

	Occurrence of Evolution	Common Descent	Gradualness	Populational Speciation	Natural Selection
Lamarck	Yes	No	Yes	No	No
Darwin	Yes	Yes	Yes	Yes	Yes
Haeckel	Yes	Yes	Yes	?	In part
Neo-Lamarckians	Yes	Yes	Yes	Yes	No
Huxley	Yes	Yes	No	No	(No)
de Vries	Yes	Yes	No	No	No
Morgan	Yes	Yes	(No)	No	Unimportant

It turns out that only one of the sub-theories is accepted unanimously, and that of course, is number 1 above, the statement that evolution did indeed take place.

Furthermore, since the time of Darwin consensuses have built up supporting the other sub-theories, but not to the extent of that obtained for number 1. While each of the others are accepted, to varying degrees, as the best available, evolution itself has been accepted as fact, i.e., as a law rather than a theory.

1.1. NATURAL SELECTION

Natural selection, the particular component of Darwinism that has been regarded as the "driving force" behind the evolutionary process, itself consists of two parts, variation and selection. It is a widely observed and well-established characteristic of living things that during procreation many more individuals are born than those that survive to ultimately reproduce themselves. The term "variation", of course, refers to the fact that no two individuals are exactly the same (Barnes, 1998). The differences between the various individuals are generally considered random, either in terms of a criterion of complexity (will be elaborated on later) or of survivability to further reproduce. For example, in terms of complexity, they may be equally complex but in a different manner. Or, they may vary to lesser and greater degrees of complexity, but the mean, or average complexity would be the same as that of the parent generations. Thus, the variation component of natural selection is considered completely random, and there is no reason to expect it to lead to evolution. Any nonrandom input to natural selection, therefore, would have to come from the second component, that of selection.

The selection component is viewed as being responsible for plucking out the advantageous fluctuations in the variation component while discarding the disadvantageous. This, of course, when operating indefinitely results in a continuous increase of advantage and complexity. This proposed ability to "select" would then have to be a characteristic of the environment, which may be geographically local or widespread, constant or changing, independent or feedback controlled.

Darwin did not consider this selection to necessarily be another random process. However, in the absence (Lima de Faria, 1988) of a suggested and accepted nonrandom mechanism for selection, many have come to consider the selection component to be random as well. Thus, a perception has taken root, with considerable following, that the mechanism of the evolutionary process proposed by Darwin is based on completely random events. It is a purpose of this chapter to show that acceptance of evolution as the end result of a series of random events is in contradiction to basic principles of the physical sciences, and consequently acts to prevent the process of "reduction" of biology to laws of chemistry and physics. It is furthermore a purpose to point out that this assumption of randomness is not at all necessary, and that a very reasonable source of energy to power the process of evolution can be proposed, which is in conformity with the relevant physical law known as the second law of thermodynamics.

1.2. INITIAL AND FINAL STATES AND PATHWAYS BETWEEN THEM

The fundamental factor that allows the transition between different thermodynamic states is the free energy difference between them. The particular pathway chosen for the transition may be regarded as of secondary importance, and must always fit into the framework of the allowed thermodynamics. While the initial and final states determine the *energy difference,* the pathway determines the *rate* at which the reaction will take place. If the rate along a path is extremely slow, while that of others is reasonably fast, the latter pathways will of course be the important ones. However, they must all conform to the thermodynamic energy requirements to be used at all.

1.2.1. Mechanical Analogy

A mechanical analogy may be useful here. A ball placed halfway down a hill has a gravitational potential which provides the energy for it to roll *down to the bottom.* There may be many pathways by means of which it can roll down to the bottom. If a path is blocked, the ball will not find that path useful. If a path is well paved and suitably shaped, the ball should use it to reach the bottom rapidly. If it is a tortuous downward path, with minor obstacles that the ball can push aside, the ball may reach the bottom through it, but more slowly. There may be a large number of pathways, with a wide spectrum of difficulties, all of which may be useful to varying degrees. However, the ball cannot reach the *top of the hill* utilizing gravitational potential. All pathways going up the hill, regardless of how well paved, smooth or frictionless, are useless. The energy difference between the two states will simply not allow the transition. Thus we regard the gravitational potential as the important factor in allowing the transition, and the pathways as secondary.

Analagously, we say that natural selection is involved in choosing feasible evolutionary pathways, but is a secondary factor compared to the required energy input which is postulated to come from the sun. Without the energy input, no pathway will enable the evolutionary transition.

2. Outline

In the following sections,

(a) the first and second law of thermodynamics will be reviewed with a view to understanding the second law as a criterion for the "spontaneity", or natural occurrence, of chemical reactions.

(b) the goal of "reduction" of biology to the physical sciences (chemistry and physics), is described. Attaining that goal then requires conformance of biological evolution (from low to higher biological forms) to the second law of thermodynamics.

(c) it is shown that to attain consistency with the principle of the second law, biological evolution cannot rely on random events but rather requires an input of energy.

(d) it is proposed that the sun provides this input.

(e) it is then shown that selection of the sun for this role enables postulation of a continuous progression, energy-wise, from prebiotic chemical evolution to and through biological evolution.

3. The First Law of Thermodynamics

The first two laws of thermodynamics have been stated both mathematically and in words. In words, the first law can simply be stated as the equivalent of the law of conservation of energy. The totality of experience, since alchemy has been replaced by chemistry and physics, i.e., since the founding of thc physical sciences, has been that energy can neither be created nor destroyed. The form of energy can be changed, but if it has gone through a cycle of changes, when it comes back to its original form, its quantity will be exactly the same as that at the start. When a reliable system of units which measures amounts of energy has been established, in every change the amount disappearing from one form is equal to the amount appearing in another form. Mechanical-, electrical-, and heat-energy are examples of different forms of energy.

4. The Second Law of Thermodynamics

4.1.DEFINITIONS AND MEANING

The second law has been stated in many ways, verbally and mathematically. A very useful statement of its principle for our present purpose is that matter will not spontaneously change from a disorganized to an organized state. The spontaneous change is always from organized to disorganize. A change from disorganized to organized will not be spontaneous, and can occur only from the expenditure of energy. That is, work has to be done to create the organization.

4.1.1. Analogies

Some examples of ordered vs disordered systems are sets of toys called Erector or the more recent LEGO and Mega Bloks. These are collections of parts which when suitably put together yield a model of a structure such as, for example, a car, bridge, or house. There are various ways in which a collection of these parts can exist together. One, shake up the box so that the various pieces lie around in more or less random manner, and we have a disordered assembly. Two, place the original parts in a box in a systematic manner, for example, all bricks in one row, all beams in another, all bolts in a third row. This could be considered an ordered assembly. These 2 types of arrangement are analogous to arrangements of molecules inside inanimate materials that are considered amorphous (disordered) on the one hand, and crystalline (ordered) on the other. Both these arrangements are subject to well-defined quantitative analysis in the physicochemical area, which we will refer to later in this chapter.

There is a third way to assemble the pieces of the set, which, like the second, is also ordered. This third way would be to build a structure, such as house, car, or bridge, from the individual pieces. Clearly a house or car is a more ordered assembly than the randomly mixed pieces of type 1. However, unlike the order of the "similar-piece line-up" of type 2, the complicated structural arrangement of house, car, or bridge possesses an additional attribute to that of mere "order". That attribute may be called functionality, a characteristic which would be recognized by an intelligent, suitably informed observer. It can do things such as move (e.g., the car), or provide a framework for having things done (e.g., the house or bridge). It therefore provides a closer (although of course not exact) analogy to biological structures.

4.2. MATHEMATICAL FORMULATION

As mentioned above, the simpler type of order that we call "similar-piece-line-up" is subject to quantitative analysis by physicochemical methods. This is done through the principle of the second law via the variable called entropy. The latter is the physicochemical quantity that connects the mathematics of heat transfer with the physical concept of order/disorder. It is generally given the symbol S, with the change in entropy given the symbol ΔS. While the latter is given thermodynamically as the amount of energy absorbed, divided by the temperature of a sample at constant volume, it is also defined statistically (in treating assemblies with a large number of molecules, which, of course, constitute real, everyday matter) as proportional to the logarithm of the probability of the configuration of the molecules in the sample.

Now, the above probability is directly related to the (somewhat more nebulous) concept of organization at the molecular level. The more disorganized a given molecular configuration, at a given temperature, the greater the probability and the greater the entropy.

4.2.1. The Entropy Principle

The concept of the second law which states that matter in a state of given order can change spontaneously only to a more disordered state, is given mathematical expression through utilization of the entropy principle. The relationship may be stated

$$\Delta S_{net} > 0 \qquad (1)$$

where ΔS_{net} is the change in entropy of the system plus its environment on the irreversible absorption of heat by the system from the environment. In words, as stated by Glasstone (1947), "in any irreversible process there is a net gain in entropy of the system and its surroundings". Substituting the word "spontaneous" for its equivalent "irreversible", and "environment" for "surroundings", it is stated that for a spontaneous change, a system plus its environment must undergo a net increase in entropy.

Glasstone expands further on this theme, in a manner that turns out to be most relevant to our subject matter. He states, "Since natural, or spontaneously occurring, processes are irreversible, it must be concluded that all such processes are associated with a net increase of entropy. From some points of view, this is one of the most important consequences of the second law of thermodynamics; the law may in fact be stated in the form that all processes occurring in nature are associated with a gain of entropy of the system and its surroundings."

4.2.2. The Free Energy Principle

Now, measurement of the change in entropy of the environment is not an easy matter. In consequence, for chemical reactions, which usually occur at constant temperature and pressure (and which constitute biological changes if one accepts "reduction" of biology to chemistry and physics), a more convenient measure of spontaneity has been devised. It is called Gibbs "free energy", defined by (Gibbs, 1928; Glasstone, 1947)

$$F = H - TS \tag{2}$$

where F is the Gibbs free energy and H is the enthalpy. As stated by Sienko and Plane (1963), "The criterion for any spontaneous change is that the total entropy (system plus surroundings) must increase. However, a more common criterion for spontaneous change ignores the surroundings, in which case both entropy and energy must be considered." Utilizing this criterion, which is called free energy, wherein both entropy (S) and a form of energy (H) are considered, and consequently it refers to the system alone, it turns out that the Gibbs free energy (free energy at constant pressure) change of a spontaneous process at constant temperature and pressure must be negative, as follows,

$$\Delta F_{TP} \leq 0 \tag{3}$$

Now, at constant temperature (and pressure), we have, from (2),

$$\Delta F = \Delta H - T\Delta S \tag{4}$$

For the reaction to proceed spontaneously, ΔF must be negative. Examining equation (4), it is clear that there are two factors which determine whether or not ΔF will indeed be negative. One, whether or not the reaction gives off heat, ie, whether or not ΔH is negative, and two, whether or not the entropy of the products exceeds the entropy of the reactants, ie., whether or not ΔS is positive. If both criteria are met, ie, increased entropy and decreased enthalpy, ΔF will be negative and the reaction will proceed spontaneously. If the entropy decreases and the enthalpy increases, ΔF will be positive and the reaction will not proceed spontaneously. If the enthalpy decreases and the entropy decreases, then ΔF will be negative and it will proceed spontaneously only if the decrease in enthalpy is greater than $T\Delta S$. If the enthalpy increases and the entropy increases, then we have negative ΔF and spontaneous reaction only if $T\Delta S$ is greater than ΔH.

Now, we have pointed out that negative entropy change is associated with increased order. Also, a sufficiently large heat given off, i.e., negative ΔH, can make a negative ΔF possible even if there is a negative entropy change. Therefore, in effect, a strongly negative heat of reaction can release energy which is utilized to create measurable order in the reaction products!

5. Reduction of Biology to Chemistry and Physics

5.1. GENERAL APPROACH

Another way of stating the theory of "reduction" is to regard biology as an extension of chemistry. The history of life is regarded as a series of chemical reactions. The entire process of evolution, from inanimate matter to the present day composition of matter on Earth, is then an increasingly complicated series of chemical reactions. With time, the reactions branch out in many directions, yielding a wide variety of "reaction products", i.e., living forms.

Discussion of the phenomenon of evolution within the framework of chemistry requires definition of the important parameters. The manner and meaning of the term evolution as known to the general public, involves not only a changing world, including changes in life forms, as opposed to a static one, but also a trend of increasing complexity of the living forms, along with the change. Thus, if present day living forms include types such as humans and amoeba, and it is assumed that these forms are connected through a line of evolution, it is understood that "evolution" refers to the gradual change of simple amoeba to complicated humans, rather than the other way around. Even if there has been a degradation of complicated forms to simple amoeba types, this is not the evolution referred to. To elaborate, a diagrammatic branched tree representing evolution (species appearance vs. time), should involve trunk and branches which sometimes go up in complexity, sometimes down, and very often sideways. However, the "line" of evolution that we are interested in, and that has had great impact in the perception that the human race has of the world around it and of its place in that world, is that which has a continuous pathway starting from the very simple, and "ending" with the most complex.

5.2. THE "COMPLEXITY" VARIABLE

At this point we note that we still have not defined the word "complex". To begin with, it is not clear that an attempt at a rigid definition at this point in time is at all possible. The word has been independently adopted by many researchers, apparently for its ambiguity as well as its substantive content. We will consequently attempt to describe it by analogy and/or example. A good starting point is with respect to the building toys described above. The three types of arrangement of their components were given as random, ordered, and ordered with function. Living forms, as a type of organized matter, clearly belong to the third of these three. They differ from random arrangements by having both order and the ability to function. Functionality, especially as regards the life process, is not, at present, easily subjected to quantitative treatment. In this essay, we deal only with the "order" component of complexity. The "order" concept, in general, has been treated and evaluated according to the principles of the second law of thermodynamics. Clearly, "order" is a vital component of the organization of living things. Thus, we deal with it as a truism that "order" is a necessary, although not sufficient, ingredient of life. In addition, we assume that increased complexity of life forms contains an increased component of order, or organization, that is subject to the requirements of the second law. Thus, the process of evolution, as understood in the cultural sense and by the public in general, involves a continuous increase in order from the very lowest forms of life, e.g., single celled animal, to the highest, i.e., homo sapiens. That being the case, the principle of reduction requires that evolution, from low to high forms, be governed by the principle of the second law.

5.3. STATE TO STATE THERMODYNAMICS OF EVOLUTION

Application of thermodynamic principles to a problem, in this manner, involves the use of thermodynamic functions that are characteristic of the state of the system. Thus, a change in a thermodynamic function for a change in state of a system involves only the nature of the initial and final states, not of the particular pathway used during the transition. Examples of thermodynamic functions of this sort, already mentioned above, include energy (E), enthalpy (H), entropy (S), and free energy (F).

Therefore, analysis of the "order" component of an evolutionary change from a lower to a higher form may be performed by focusing on the states only. The states considered can be any lower and higher phenotype. For "reduction", the transition between the two states is approximated as net chemical reactions at constant temperature and pressure (the assumption of other conditions might involve different thermodynamic variables , but there is no way of circumventing the second law). As we mentioned above, the measure of spontaneity is then a negative change in Gibbs free energy, $-\Delta F$. For this change from lower to higher form to have occurred spontaneously with a concomitant increase in "order", therefore, ΔS must be negative, and the heat given off, $-\Delta H$, must exceed $-T\Delta S$, which now has a positive value. We present below what is in our opinion, a most reasonable source for this available chemical energy.

6. Prebiotic Phenomena in the Laboratory

6.1. REDUCING ATMOSPHERE AND ELECTRICAL DISCHARGES

Stanley Miller (1953) originally subjected a mixture of gases, consisting of methane, ammonia, molecular hydrogen, and water, simulating the reducing atmosphere thought to be found on earth in the early phases of the atmosphere's formation, to electrical discharges. On analysis of the gas mixture subjected to this treatment, he found compounds that are generally associated with the synthesis or breakdown of biological molecules, such as amino acids, the building blocks of proteins. The idea that the earth's early atmosphere is reducing is originally that of Urey (1951). The results of the experiments have been taken as supporting evidence for the idea that biological molecules were originally synthesized from these small molecules with the aid of energy supplied by electrical discharges in the atmosphere such as from lightning.

6.2. NEUTRAL ATMOSPHERE AND VARIOUS ENERGY INPUTS

In recent years the supposition that the early-earth atmosphere was reducing has been largely abandoned (Walker, 1977; Kasting,1993; Brasseur, 1999). It is now thought that the atmosphere was neutral or slightly reducing, containing mainly nitrogen and carbon dioxide, some water, and traces of carbon monoxide, hydrogen, methane, ammonia, hydrogen chloride, and sulfur dioxide (Brasseur, 1999). While the postulated syntheses of biological compounds has therefore been revised, there are pathways to the synthesis of biological molecules from this slightly reducing atmosphere, aided by the input of energy. The principle that an input of energy promotes the reaction of small atmospheric molecules to form precursors to larger biological molecules remains viable. In addition to the changing concept of the nature of the atmosphere, speculation as to the nature of the exciting medium has also broadened. Included now as possibilities are ultraviolet solar radiation (Sagan, 1971; Groth, 1960; Toupance, 1977; Zahnle, 1986), visible solar radiation (Woese, 1979), and thermal energy (Harada, 1964) that is present underground, in hot springs, etc.

7. Early Chemical Evolution As Steady State Phenomenon: Development Of System Of Steady State Intermediates

Description of the process of energy utilization is facilitated through use of the concept of the steady state (Eigen, 1971). A general steady state for a chemical reaction may be described as

$$X \rightarrow Y \rightarrow Z \tag{5}$$

where the rate of production of Y from X is equal to its rate of loss to yield Z. Thus the concentration of Y remains steady. A slow increase or decrease of Y over a long period of time makes it, strictly speaking, a quasi-steady state, nevertheless, for many purposes it can be usefully described simply as a steady state process. In passing, it is emphasized that the steady state system is still subject to the laws of thermodynamics, (i.e., the steady state phenomenon, as is the case for all kinetic or rate dependent phenomena, still takes place within the framework of thermodynamics).

Early prebiotic chemistry can then be described as excitation of small atmospheric molecules which then polymerize on the earth's surface, probably incorporating minerals from the surface during the process.

Now, Miller's experiments utilized electrical discharges as the excitation source for promotion of the desired reaction. However, in principle, other types of excitation can be employed as well. A few obvious possibilities are (1) thermal energy (Fox, 1964), (2) solar radiation (3) cosmic radiation. Of these "solar radiation" is assumed, in this essay, to be the prime source of the necessary energy. It is, of course, the source of energy that has been maintaining advanced forms of life for eons, to the present. That it also has been the main actor in the start of the process seems to us

quite reasonable. Photochemical excitation is an intimate and necessary part of all postulated early earth atmospheric chemistry. Thus, a continuous role for solar energy from the earliest stages of prebiotic chemistry, through evolution, on to present day maintenance of life, can be presented as a reasonable postulate. In place of electrical discharges we then have ultraviolet or visible light, depending on the requirements of the time and place.
We then write

$$A + h\nu \rightarrow A^* \tag{6}$$

$$A^* + B \rightarrow A + B^* \tag{7}$$

$$B^* + C \rightarrow D^\delta + E^\delta \tag{8}$$

$$D^\delta + E^\delta \rightarrow J^\delta + L \tag{9}$$

Here, A is an atmospheric molecule that absorbs a photon hν, and becomes A*, which designates A in a photochemical excited state. A* may now transfer its energy to a ground state molecule B, leaving it in an electronic excited state, B*. The molecule B* may then react with C, utilizing its energy to decompose both molecules, which react and form the product intermediates D^δ + Eδ. The latter two may be chemically excited species such as free radicals or ions which will react further to produce another chemical high energy molecule-fragment J^δ and the ground state product L. As the process continues, the molecules B*, D^δ, E^δ, and J^δ may continue their role as steady-state intermediates, with their rate of production equal to their rate of consumption to yield "final" products, which may or may not be of importance.

It is of importance to understand at this point that the living system under development which is described in the equations (6-9) is not L or other final products, but rather the molecules in the steady state, represented here by B*, D^δ, E^δ, and J^δ. In effect, life is defined chemically as a system of steady state intermediates, which is kept alive by the constant input of solar energy.

As time goes on, the energetic steady state intermediates polymerize to larger and more complicated components. Thus, chemical evolution is taking place. Thermodynamically, the separate steps of all the chemical actions involve decreases in free energy, made possible by the original high free energy of the active molecules caused by absorption of solar energy, directly (eq 6) or indirectly (e.g., eq 8) . The average thermodynamic free energy of the steady state intermediate system is of course higher than ground state. Within the system, examination of the thermodynamic variables will show a constant long-range decrease in average entropy of the intermediates. This corresponds to the increase in the "order" component of the intermediates as their complexity increases.

This increase of order derives ultimately from utilization of the energy input of the sun. Thus, chemical evolution has taken place as a result of conversion of a change in energy to "negentropy" (a decrease in entropy), a permissible process in conformance with the second law of thermodynamics.

7.1. CONTINUITY OF CHEMICAL AND BIOLOGICAL EVOLUTION

To summarize the role of the sun as the activator, thermodynamically as well as kinetically, of the continuous evolution of matter, from the inanimate to and through the biological stages, we have

(1) Sun-activated inorganic matter converted to organic matter in a steady-state – Analogue of *bio-evolution* called *chemical evolution.*

(2) Sun-activated *maintenance* of quasi-steady state – Analogue of *perpetuation* of species in process of biological evolution.

(3)Sun-activated *increase in complexity* of steady state material – Analogue of biological *evolution* from one stage to a higher one.

(4) Sun-activated further increase in complexity until steady-state molecules become biomolecules – chemical evolution becoming bioevolution.

(5) Sun-activated increase in complexity of biomolecules and materials – biological evolution.

Thus, it is proposed that the input of solar energy to the planet, which has the function not only of maintaining the steady state in the "chemical" stage of evolution, but of increasing the complexity and order of the material in the steady state above that of the inanimate matter, can reasonably be expected to continue to provide the energy necessary not only for the maintenance of the species of life at any stage of biological evolution, but for increasing their complexity to form new species. The latter defines the process of evolution as used in this essay.

8. Discussion

8.1. HOW ABOUT GENETICS?

The traditional approach of biological science to the question of energetics of evolution, starting as it does with inheritance, might be described as proceeding from inside to outside. It is natural that before proceeding to the question of how does one species evolve to another, the problem of how does any given species maintain itself, presumably indefinitely, would be considered. This involves first of all the problem of reproduction. A sub-problem of that of exact reproduction is that of diversity. Since no two individuals of a species are exactly the same, the reproductive process carries along with it the "seeds" of diversification, and it is natural that in seeking an explanation for reproduction, the mechanism of diversification would be considered along with it. Since diversification involves differences, and evolution involves differences, the temptation is strong to interpret the evolutionary change as another type of diversification. However, diversification is not evolution. Evolution leads to greater complexity of the organisms. Diversification — not necessarily, perhaps not at all. Thus, laws of inheritance, or genetics as they are presently known, do not at present lead to a solution to the energy problem. They, however, could and should be treated as a "mechanism" of the overall "reaction", using chemical terminology. The thermodynamics of the state to state conversion, such as, for example, from reptile phenotype to homo sapiens phenotype, determine whether the transition can occur. If it can, then it may, provided that a suitable pathway, or system of pathways (known as a mechanism), is available under prevailing conditions. In this essay we have shown that the thermodynamic requirement can be satisfied by utilization of the continuous input of solar energy. The mechanism, or genetic pathway, remains to be determined in the future.

8.2. DOES ENVIRONMENT PROVIDE POTENTIAL ENERGY FOR SELECTION?

With respect to the alleged role of selection, an argument could be made in favor of nature acting as a selector by pointing out the numerous cases of a species with a given characteristic, such as, for example, color, surviving in environments friendly to itself, while a member of the same species with a different characteristic will survive in a different environment friendly specifically to itself. However, these adaptations do not in principle lead to advancement in complexity. There could be some small advancement, along with some small deterioration, and also some status quo, with the average complexity remaining the same. In order to advance in complexity, the environment would have to possess a type of potential energy which consistently provides the force needed for advancement. Barring constant replenishment of such energy, however (we note that the idea of replenishment of potential energy through changing the entire environment has the effect of introducing an energy supply), this does not seem to be a likely mechanism for providing the energy needed for the remarkable progression of evolution for a period of a few eons (billions of years). In fact the well known and apparently widely accepted "Red Queen" concept of Van Valen (1976), states that the use of environment for advancement of evolution in one situation, ie., one particular phenotype, decreases the availability of the overall environment to act in that manner in another situation. In other words, static environment has limited capacity to advance evolution.

8.3. CONSTANT ENERGY FLOW FROM SUN

In contrast, the assumption we have made that it is the sun's radiation that advances evolution, in the form of a constant input of kinetic energy, is rather reasonable and natural. It is known that the sun is necessary to maintain life as life is at present, and as it was for eons before. Even species that have maintained themselves on, say, thermal energy, still obtain energy from the sun indirectly.

Of course, we would assume that if a species living on thermal energy directly, also undergoes evolution, it will be using some of this energy for its evolutionary advancement. But, this thermal energy may come indirectly from the sun. If not, i.e., if it is a remnant of cosmological energy trapped during the formation of the earth, then it is limited in quantity and comes under our definition of potential energy.

At any rate, the greatest advancements in evolution have taken place in our (mammalian) branch of the tree of life. With respect to our branch of the evolutionary tree, it is a most reasonable assumption that some of the same type of energy that provides our maintenance, is converted to evolutionary advancement. Or, in more basic physical terminology, the solar radiation which keeps the biomass functioning also provides the energy for increasing its complexity.

8.4. THE "ACCIDENT" QUESTION

The overall question as to whether the occurrence of life in the universe is accidental has been treated previously (Schrader, 1979). The degree of randomness involved in the coming about of life in general, anywhere, was dependent on many assumptions. However, it was pointed out that in the worst case, that is, the most random view, it was still not possible philosophically or physically to conclude that life is an "accident".

The title of this article poses a different problem, ie, given the existence of the planet earth as it is, and as it has evolved throughout its history, is the evolution of life on this planet a result of completely random events, which would seem to support the point of view that it has come about "accidentally", or is it the result of normal scientific cause and effect. We point out that if a source of energy is understood to be involved in the evolution process, evolution can result from normative physical science. The source of energy we propose is the sun.

9. Highlights

1. The doctrine of Natural Selection as a proposed explanation for the occurrence of biological evolution on the planet earth has come to rely on random events for its scientific chemical mechanism.
2. The concept that the evolution of today's biosphere, a vast network of complex, ordered structures, as a result of continuous random chemical events, runs contrary to the second law of thermodynamics.
3. The second law requires that energy be utilized to convert a state of disorder to a state of order. It is postulated here that the sun supplies this energy for evolution on the earth, through constant conversion of absorbed solar radiation to chemical energy. Absorbed solar energy, in excess of that used for maintenance of existing life, is used for evolution.
4. So-called natural selection may be involved in choosing the permitted pathways to be traveled by biological variations in going from low to higher order states, but is not the factor which causes or allows the overall transition.
5. The transitions during main-line evolution, leading to today's most complex biological structures, are driven by the ever-present high-energy molecules resulting directly or indirectly from absorbed solar radiation.
6. There is consequently no reason to regard the occurrence of life on the planet earth as an accident of nature. It is a result of the inevitable, natural, and continual transfer of energy from the sun to the planet.

10. References

Barnes, R. S. K. (ed.) (1998) *The Diversity of Living Organisms,* Blackwell Science, Oxford.

Brasseur, G. P. and Martell, E. (1999) Atmospheric Evolution and Global Perspective, In: G. P. Brasseur, J. J. Orlando and G. S. Tyndall (eds.) *Atmospheric Chemistry and Global Change,* Oxford University Press, pp. 539-550.

Darwin, C. E. (1859) *Origin of Species.*

De Vries, H. (1889) *Intracellulare Pangenesis.* Jena: Gustav Fischer. (English trans. 1910. Chicago: Open Court Publishing Co.)

Eigen, M. (1971) Molecular Self-Organization and the Early Stages of Evolution. *Quart. Rev. Biophys.* **4,** 149-210.

Gibbs, J. W. (1928) *Collected Works* **1,** 315. Yale University Press, New Haven.

Glasstone, S. (1947) *Thermodynamics for Chemists.* D. Van Nostrand, New York.

Haeckel, E. (1866) *Generelle Morphologie der Organismen: Allgemeine Grundzuge der organishen Formen-Wissenschaft, mechanisch begrundet durch die von Charles Darwin reformirte Descendenz-Theorie.* 2 vols. Berlin: Georg Reimer.

Harada, K. and Fox, S. W. (1964) Thermal Synthesis of Natural Amino-Acids from a Postulated Primitive Terrestrial Atmosphere. *Nature* **201,** 335.

Lima de Faria, A. (1988) *Evolution without Selection,* Elsevier, Amsterdam.

Ibid, p. 15.

Mayr, E. (1982) *The Growth of Biological Thought,* Harvard University Press, Cambridge.

Miller, S. L. (1953) A production of amino acids under possible primitive earth conditions. *Science* **117,** 528.

Miller, S. L. and Urey, H. C. (1959) Organic Compound Synthesis on the Primitive Earth. *Science* **130,** 245.

Morgan, T. H. (1903) *Evolution and Adaptation,* New York: Macmillan.

Sagan, C. asnd Chyba, C. (1997) The Early Faint Sun Paradox: Organic Shielding of Ultraviolet-Labile Greenhouse Gases. *Science* **276,** 1217-1221.

Sagan, C. and Khare, B. N. (1971) Long-Wavelength Ultraviolet Photoproduction of Amino Acids on the Primitive Earth. *Science* **173,** 417-420.

Schrader, M. E. (1979) Is Life an Accident? *Zygon* **14,** 323-327.

Sienko, M. J. and Plane, R. A. (1963) *Physical Inorganic Chemistry,* W. A. Benjamin, Inc. , New York. p. 2.

Toupance, G., Bossard, A. and Raulin, F. (1977) Far UV Irradiation of Model Prebiotic Atmospheres. *Origin of Life* **8,** 259-266.

Urey, H. C. (1952) On the Early Chemical History of the Earth and the Origin of Life. *Proc. Nat. Acad. Sci., U. S.* **38,** 351-363.

an Valen, L. (1976) Energy and Evolution. *Evolutionary Theory* **1,** 179-229 . This volume covers a few years, starting with 1971.

Walker, J. C. G. (1977) *Evolution of the Atmosphere.*

Wallace, A. R. (1855) On the Law Which Has Regulated the Introduction of New Species. *The Annals and Magazine of Natural History, ser. 2,* **16,** 184 (1855).

Woese, C. R. (1979) A Proposal Concerning the Origin of Life on the Planet Earth. *J. Mol. Evol.* **13,** 95-101.

Zahnle, K. J. (1986) Photochemistry of CH_4 and HCN in the Primitive Atmosphere. *Origins Of Life* **16,** 188-189.

Biodata of **Pasquale Stano, Francesca Ferri** and **Pier Luigi Luisi** authors of the chapter: ***"From the Minimal Genome to the Minimal Cell: Theoretical and Experimental Investigations"***

Prof. Pier Luigi Luisi is currently professor of Biochemistry at the Third University of Rome (IT). After he received his PhD. (1967) in chemistry at the Scuola Normale Superiore of the University of Pisa, and post-doctoral fellowships at the University of St. Petersburg (RU), Uppsala (SE), at the Macromolecular Center Strasbourg (FR), at the Institute of Molecular Biology, Eugene Oregon (US), Prof. Luisi was first "*Oberassistent*" at the Swiss Federal Swiss Institute of Technology (ETH) of Zurich, then Assistant Professor (1973), and Full Professor of Macromolecular Chemistry (since 1984). Prof. Luisi scientific interests embrace the macromolecular chemistry, enzymology, self-assembly structures as micelles, reverse micelles, liposomes and giant vesicles. Recently he focused his attention to the chemical implementation of autopoiesis, studies on the origin of cellular complexity and self-reproduction of supramolecular structures (further information on www.plluisi.org).

E-mail: **luisi@mat.ethz.ch**

Dr. Pasquale Stano is currently a research assistant of Prof. Luisi at the Third University of Rome (IT). After a physical chemistry degree on supramolecular complexes at the University of Pisa, P. Stano joined the Luisi's group (2001-2003) at the Swiss Federal Institute of Technology (ETH). His current research interests are related on the property and reactivity of vesicles and other supramolecular assemblies, dynamic light scattering, spectroscopic techniques and on the use of liposomes as drug- and gene-delivery agents.

E-mail: **stano@uniroma3.it**

Dr. Francesca Ferri is currently a Ph. D. student at the University of Bologna (IT). After a biotechnology degree with a thesis in molecular biology at the University of Bologna, F. Ferri started her Ph. D. research project, which is focused on the role of chromatin structure and DNA topology on gene transcription.

E-mail: **fraferri@katamail.com**

Pasquale Stano

Francesca Ferri

Pier Luigi Luisi

J. Seckbach (ed.), Life as We Know It, 179-198.

FROM THE MINIMAL GENOME TO THE MINIMAL CELL: *Theoretical and Experimental Investigations*

PASQUALE STANO[1,3], FRANCESCA FERRI[2] AND PIER LUIGI LUISI[3]
[1]*Centro Studi "E. Fermi"; Compendio Viminale, 00184 Roma, Italy.* [2]*Biochemistry Department, University of Bologna; Via Irnerio 48, 40126 Bologna, Italy. and* [3]*Biology Department, University of RomaTre; V.le Marconi, 446, 00146 Roma, Italy*

1. Introduction

It is known that even the simplest living cells existing on the Earth have several hundred genes, with hundreds of expressed proteins that catalyse simultaneously hundreds of reactions within the same tiny compartment – the cell – representing a maize of an enormous and fascinating complexity. On the other hand, if we think to early cells or protocells (those that played a key role in the origin of life), it is conceivable that they could display some "living" properties with a minor number of biochemical components. In particular, the present huge complexity is most likely resulted from billions of years of evolution, that developed a series of defence and security mechanisms, redundancies, metabolic loops and highly sophisticated regulatory processes.

These considerations elicit the question, whether such a complexity is really essential for life, or whether instead cellular life might be possible with a much smaller number of components. In this chapter we describe the basis of the concepts of "minimal genome" and of the "minimal cell".

The question of the minimal genome and minimal cell has been considered since many years, and one should in particular recall the work of Morowitz (1967), who, based on the enzymatic components of primary metabolism, estimated that the size of a minimal cell should be about one-tenth smaller than *Mycoplasma genitalium*. Significant earlier insights into the field are those by Jay and Gilbert (1987) as well as by Woese (1983) and Dyson (1982). More recently, the reviews by Deamer and coworkers (Pohorille and Deamer, 2002) and Luisi's group (Luisi 2002, Oberholzer and Luisi, 2002) have sharpened the question and brought it in the perspective of modern molecular tools. In fact, the last years have seen a significant revival of interest in the field of the minimal cell, as witnessed for example by two international meetings held in 2004 on the subject (Szathmáry, 2005). A more detailed version of a review on the minimal cell is in press elsewhere (Luisi et al., 2005).

2. The Notion of Minimal Cell

The minimal cell can be broadly defined as the one having the minimal and sufficient number of components to be called alive. What does "alive" means? This is a complex question, but one can choose here a quite general description, defining life at the cellular level as the concomitance of three basic properties: self-maintenance (metabolism), self-reproduction, and evolvability (Figure 1).

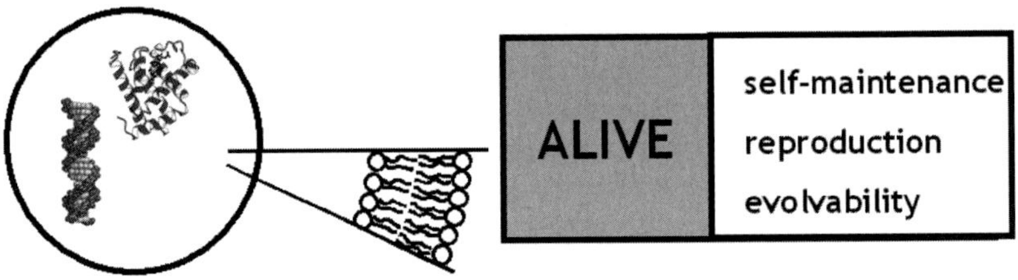

Figure 1. A cell containing the minimum and sufficient number of components to be "alive".

When all these three properties are fulfilled, there would be full-fledged cellular life. This trilogy may not be perfectly implemented, particularly in synthetic constructs, and then several kinds of approximations to cellular life can be envisaged. For example, we can have protocells capable of self-maintenance but of self-reproduction; or vice versa. Or we can have protocells in which self-reproduction is active for only a few generations; or systems that do not have the capability of evolvability. And even in a given type of minimal cell –for example one with all three attributes –one might ha ve quite different ways of implementation and sophistication. It is then clear that the term "minimal cell" depicts large families of possibilities and not simply one particular construct.

From the theoretical and experimental points of view, the aim of this research is to design and realize a cellular construct, combining the biochemistry of compartments (liposomes) with the modern molecular biology techniques. A certain number of genes, together with the transcription and translation machinery, is inserted into the aqueous core of liposomes, and the novel properties of the construct are then studied.

This kind of approach has been called "re-construction" or "semi-synthetic approach", because extant enzymes and genes are utilised in order to construct a complex entity starting from simpler components. Since the resulting cells do not exist on our biological life, the term "artificial cells" (Pohorille and Deamer, 2002) or better "semi-artificial cells" may be used.

The question of the minimal genome will be described first, focusing the attention to the smallest unicellular organisms on the Earth, in order to define the theoretical minimum number of genes capable to sustain life. Secondly, we will present a short review on the most significant molecular biology reaction carried out using liposomes as compartments, discussing the current state of the research in this field.

3. The Minimal Genome

Figure 2 compares the genome size distribution, calculated under a series of assumptions (Islas et al., 2004), of free-living prokaryotes, obligate parasites, thermophiles, endosymbionts. The values of DNA content of free-living prokaryotes can vary over a tenfold range, from the 1450 kb of *Halomonas halmophila* to the 9700 kb genome of *Azospirillum lipoferum* Sp59b. As a way of comparison, consider that *Escherichia coli* K-12 has a genome size of ca. 4640 kb.

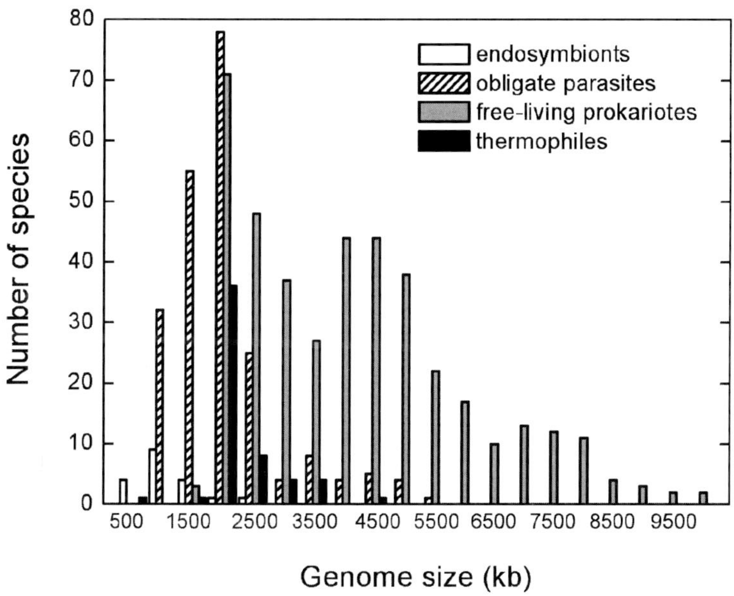

Figure 2. Prokaryotic genome size distribution (N =641). Genome sizes, complete proteomes, and the number of ORFs (open reading frames) were all retrieved from NCBI. (http://www.ncbi.nlm.nih.gov). Adapted from Islas et al., 2004.

Classification of endosymbionts as a group by themselves shows that their DNA content can be significantly smaller, the smallest sizes are then those of *Mycoplasma genitalium* and *Buchnera*, with a value that agrees well with the predictions of Shimkets (1998), according to which the minimum genome size for a living organism should approximately be around 600 kb. It is argued that these two organisms have undergone massive gene losses and that their limited encoding capacities are due to their adaptation to the highly permissive intracellular environments provided by the hosts (Islas et al., 2004). A closer analysis of the coding regions in some small genomes gives an account of the redundant genes, amounting in average to 6-20% of the all genome. Is it possible – starting from these considerations – to define the minimum number of genes of an organism?

Andres Moya and his group in Valencia have asked this question and arrived at the smaller number of 206 genes on the basis of their work with *Buchnera spp.* and other organisms (Gil et al., 2002, 2004). The results, given in Table 1, are close to that obtained by other authors on the basis of different and interesting considerations (Mushegian and Koonin, 1996; Shimkets, 1998; Mushegian, 1999; Koonin, 2000; Kolisnychenko et al., 2002).

TABLE 1. Core of a minimal bacterial gene set (Courtesy of Prof. A. Moya; Institut Cavanilles de Biodiversitat i Biologia Evolutiva; Universitat de València).

DNA metabolism		**16**
Basic replication machinery	13	
DNA repair, restriction and modification	3	
RNA metabolism		**106**
Basic transcription machinery	8	
Translation: aminoacyl-tRNA synthesis	21	
Translation: tRNA maturation and modification	6	
Translation: ribosomal proteins	50	
Translation: ribosome function, maturation and modification	7	
Translation factors	12	
RNA degradation	2	
Protein processing, folding and secretion		**15**
Protein post-translational modification	2	
Protein folding	5	
Protein translocation and secretion	5	
Protein turnover	3	
Cellular processes		**5**
Energetic and intermediary metabolism		**56**
Poorly characterized		**8**
TOTAL		**206**

We have reached the number of about 206 genes as minimal genome. This is a considerable simplification of the initial number, but it still corresponds to a formidable complexity – which again arises the question, whether and how can one go further down.

3.1. FURTHER SPECULATIONS

Obviously, only speculations can help us at this point. One way to speculate is to imagine a kind of theoretical knock down of the genome, reducing cellular complexity and at the same time part of the non-essential functionality (Luisi et al., 2002).

The first way station of this intellectual game is to imagine a cell without the enzymes (then the corresponding genes) needed to synthesize low molecular weight compounds, assuming that low molecular weight compounds, including nucleotides and amino acids, would be available in the surrounding medium and able to permeate into the cell membrane. This would be a fully permeable minimal cell. Thanks to the outside supply of substrates, such a cell should be capable of self-maintenance and of self-reproduction, including replication of the membrane's components. However it would not make low molecular weight compounds and would not have redundancies for its own defence and security (in fact, all self-repair mechanisms are missing). Also, cell division would be simply due to a physically based statistical process.

The key processes in this hypothetical minimal cell are protein and lipid biosynthesis by a modern ribosomal system, but limited to a rather restricted number of enzymes – see Table 2. This cell would have ca. 25 genes for the entire DNA/RNA synthetic machinery, ca. 120 genes for the entire protein synthesis (including RNA synthesis and the 55 ribosomal proteins), and 4 genes for the synthesis of the membrane. This would bring to a total of about 150 genes, somewhat less than Moya's figure of 206 exposed previously.

Consider now RNA and DNA polymerases. A number of data (Suttle and Ravel, 1974; Lazcano et al., 1988; Lazcano et al., 1992; Frick and Richardson, 2001) suggests that a simplified replicating enzymatic repertoire – as well as a simplified version of protein synthesis – might be possible. In particular, there is the idea that a single polymerase could play multiple roles as a DNA polymerase, a transcriptase, and a primase, is conceivable in the very early cells (Luisi et al., 2002).

The game could go on by assuming that at the time of the early cells not all "our" 20 amino acids were involved – and a lower number of amino acids would reduce the number of amino acyl-tRNA-synthetases and the number of tRNA genes.

The next victims would be ribosomal proteins. Can we take them out? There are some indications that ribosomal proteins may not be essential for protein synthesis (Zhang and Cech, 1998), and there are other suggestions about an ancient and simpler translation system (Nissen et al., 2000, Calderone and Liu, 2004). There are also some claims that the first ribosomes consisted of rRNAs associated simply with basic peptides (Weiner and Maizels, 1987).

All these considerations may help to decrease the number of genes down to a happy number of, say, 45-50 genes – see Table 2 – for a living, although certainly limping, minimal cell (Luisi et al., 2002).

Many authors would doubt that a cell with only 50 genes would be able to work. But again, the consideration goes on to the early cells, and to the consideration that the first cells could have not started with dozen of genes from the very beginning in the same compartment.

TABLE 2. A hypothetical list of gene products that defines the minimal cells according to the definitions used in this chapter, sorted by functional category (Adapted from Luisi et al., 2002).

Gene product	Number of Genes		
	Minimal DNA Cell (*a*)	"Simple-ribosome" cell	Extremely reduced cell
DNA/RNA Metabolism			
DNA polymerase III	4 (*b*)	4 (*b*)	1
DNA-dependent RNA polymerase	3 (*c*)	3 (*c*)	1
DNA primase	1	1	
DNA ligase	1	1	1
Helicases	2-3	2-3	1
DNA gyrase	2 (*d*)	2 (*d*)	1
ssDNA-binding proteins	1	1	1
Chromosomal replication initiator	1	1	
DNA Topoisomerase I and IV	1+2 (*d*)	1+2 (*d*)	1
ATP-dependent RNA helicase	1	1	
Transcript. elong. factor	1	1	
RNases (III, P)	2	2	
DNases (endo/exo)	1	1	
Ribonucleotide reductase	1	1	1
Protein biosynthesis/ translational apparatus			
Ribosomal proteins	51	0	0
Ribosomal RNAs	1 (*e*)	1 (*e*)	1 (*e*), self splicing
aa-tRNA synthetases	24	24	14 (*f*)
Protein factors required for biosynthesis and synthesis of membrane proteins	9-12 (*g*)	9-12 (*g*)	3
tRNAs	33	33	16 (*h*)
Lipid metabolism			
Acyltransferase 'plsX'	1	1	1
Acyltransferase 'plsC'	1	1	1
PG Synthase	1	1	1
Acyl carrier protein	1	1	1
TOTAL	**146-150**	**105-107**	**46**

(*a*) Based on *M. genitalium*. (*b*) Subunits a, b, y, tau. (*c*) Subunits a, b, b'. (*d*) Subunits a, b. (*e*) One operon with three functions (rRNAs). (*f*) Assuming a reduced code. (*g*) Including the possible limited potential to synthesize membrane proteins. (*h*) Assuming the third base to be irrelevant.

This last consideration permits a logical link with the notion of compartments. Suppose that a certain number of macromolecules – or their precursors – would have developed first in solution, i.e., let us forget for a moment the possibility of compartments. Then, in order to start cellular life, compartmentation should have come later on, and one would then have to assume the simultaneous entrapment of all these different genes in the same vesicle. This can be indeed regarded as highly improbably, and in fact, more reasonable is a scenario in which the complexity of cellular life evolved within the compartment from a much smaller group of components from the inside of the protocell.

4. Preliminary Considerations on Vesicles as Bioreactors

Lipid vesicles (or liposomes) are supramolecular aggregates formed by several thousand of lipid molecules. Generally they have a spherical geometry, diameters ranging from 50 nm to several microns, and represent a compartment that models and mimics the cellular membranes.

Before we proceed any further with the discussion on the use of vesicles as host for complex molecular biological reactions, there is an important field of inquiry concerning the possible analogies between vesicles and cellular membranes. In particular, the physico-chemical properties, such as stability, permeability, self-reproduction and so on, are studied to determine whether and to what extent vesicles behave similarly to cellular structures.

It has been shown that vesicles can grow and multiply themselves (Bachmann et al., 1992; Walde et al., 1994b; Luisi et al., 2004); and under certain conditions this may happens with retention of the original size distribution (the so-called matrix effect) (Bloechliger et al., 1998; Lonchin et al., 1999; Rasi et al., 2003). Vesicles are in fact able to uptake a precursor, that is spontaneously transformed in a membrane-forming compound. In this way vesicles increase their surface and eventually divide in daughter vesicles (Figure 3).

Although it is not the aim of this chapter to review all these data, it is important here to keep in mind that one of the most critical mechanism of living cells can in principle be simulated by vesicles based only on physical and chemical properties, i.e., without the use of sophisticated biochemical machineries. This consideration is relevant if one focuses on the prebiotic scenario.

Another important preliminary physico-chemical property is the membrane permeability to solutes. Several attempts have been carried out in order to overcome the limited permeability of water-soluble compound through the lipid membrane. The approaches were focused on the fine-tuning of bilayer permeability by addition of co-surfactant, or exploiting the temperature dependence membrane permeability. The use of membranes channels, on the other hand, has been recently adopted by Noireaux and Libchaber (2004), which utilized the α-hemolysin pore to supply water-soluble biochemicals into liposomes.

Another approach, still not experimentally implemented, is the use of vesicle reactivity. In fact, the problem of bring together different components might be circumvented if two or more liposomes, each containing a given substrate, could fuse together so as to produce liposomes containing all reagents. In fact, fusion of vesicle is becoming an active area of research, and interesting results have been already obtained (Stamatatos et al., 1988; Marchi-Artzner et al., 1996; Pantazatos and MacDonald, 1999; Thomas and Luisi, 2004). Fusion of compartments can be achieved also utilizing water-in-oil emulsion, and actually in this way, as we will see later on in more detail, proteins synthesis could be obtained by mixing compartments containing the various ingredients for synthesis.

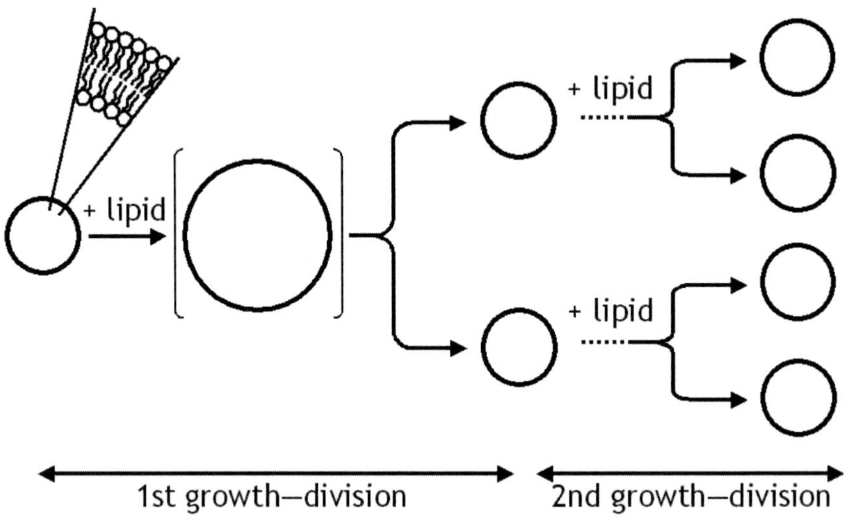

Figure 3. Schematic drawing of vesicles self-reproduction. The addition of fresh lipid molecules (or their precursors) to an initial population of vesicles can increase the vesicle number if a division process follows the initial growing process. In some cases, the size distributions of the new populations are closely related to the size distribution of the first vesicles (the so-called matrix effect). For the sake of simplicity, an ideal behaviour has been shown. These experiments model the vesicle proliferation by endogenous biosynthesis of lipids.

Concerning now the area of biochemical reactions in liposomes, there is a large amount of experimental work that has paved the way to these last developments. For example, the biosynthesis of poly(A) – a model for RNA – was reported independently from two groups (Chakrabarti et al., 1994; Walde et al., 1994a). In both cases a polynucleotide phosphorylase (PNPase) was entrapped into vesicles, and the synthesis of poly(A), which remained into the aqueous core of such vesicles, was observed. In one case (Walde et al., 1994a) the internal poly(A) synthesis proceeded simultaneously by the reproduction of the vesicle shell, due to external addition of a membranogenic precursor (oleic anhydride).

A more suggestive example was provided shortly later (Oberhozer et al., 1995b) with the use of Qβ replicase, an enzyme that replicates an RNA template. In this case too, the replication of a core component was coupled (but not synchronously) with a replication of the vesicle shell. With an excess of Qβ replicase/RNA template, the replication of RNA could proceed for a few generations. In fact, even if RNA template and vesicle shell are respectively replicating, the Qβ replicase is not continuously produced in the process, thus the system undergoes a "death by dilution". After a while, new vesicles will not contain either enzyme or template, and therefore the construct cannot reproduce itself completely.

In addition to the realization of the polymerase chain reaction into liposomes (Oberholzer et al., 1995a), Oberholzer and coworkers showed the production of poly(Phe), starting from poly(U) as mRNA, phenylalanine, ribosomes, $tRNA^{Phe}$ and elongation factors entrapped in lecithin vesicles (Oberholzer et al., 1999). The yield was 5% with respect to the experiment in water without liposomes, but the authors argued that the yield is actually surprisingly high, considering that the liposomes occupy only a very small fraction of the total volume; and that only a very few of them would contain all ingredients by the statistical entrapment.

5. Protein Expression in Liposomes

We have seen in the previous section the realization and optimisation of rather complex biochemical reactions in liposomes. In order to approach the construction of the minimal cell, one should increase the complexity of the core of the liposomes so as to reach the limits outlined in the previous session about the minimal genome. This has not been the approach used until now in the literature. Rather, researchers have first sought to insert into liposomes the conditions to express a single protein. And for reasons easy to understand, mostly from the detection facility, the green fluorescence protein (GFP) has been the target protein.

The common strategy (Figure 4) is to entrap into the aqueous core of liposomes all the ingredients for the *in vitro* protein expression; i.e., the gene for the GFP (a plasmid), an RNA polymerase, ribosomes and all the low molecular-weight components (amino acids, ATP, etc.) needed for protein expression. An overview of the work done, limited as we said to the expression of proteins in liposomes, is presented in Table 3.

Yomo and Urabe (Yu et al., 2001), for examples, have reported the expression of a mutant GFP in lecithin liposomes. Large GFP-expressing vesicles, prepared by the film hydration method, were analysed using flow cytometry as well as confocal laser microscopy.

In the procedure utilized by Oberholzer and Luisi (2002) all ingredients are added to a solution in which the vesicles are being formed by the ethanol injection method, and EGFP (enhanced GFP) production is then evidenced inside the compartments. In this case, the sample was analysed spectroscopically, monitoring the increase of the fluorescent signal of the EGFP. The disadvantage of this procedure is that entrapping efficiency is generally low, due to the small internal volume of liposomes obtained with this method. On the other hand, the observation of EGFP production inside the aqueous core of liposomes confirms that the co-entrapment of several different solutes was obtained.

Direct observation of protein expression was accomplished by the procedure utilized by Nakatani and Yoshikawa team (Nomura et al., 2003), using giant vesicles. The progress of the reaction is observed by laser-scanning microscopy and it is shown that expression of rsGFP (red shifted GFP) takes place with a very high efficiency (the concentration of rsGFP inside vesicles was greater than that in the external environment). The authors could also show that vesicles can protect gene products from external proteinase K.

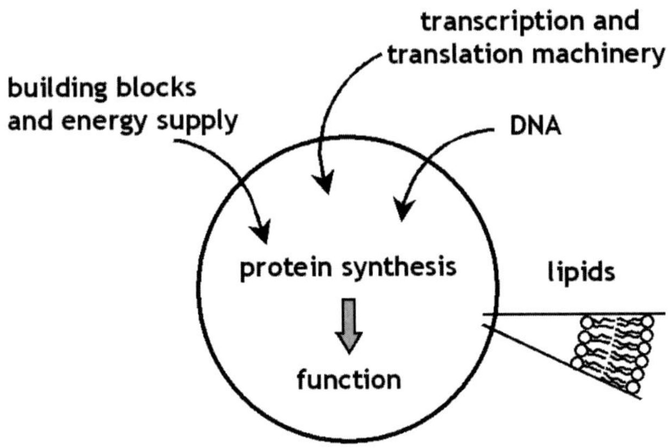

Figure 4. DNA, the whole transcription/translation machinery, building blocks and energy-rich molecules (i.e., ATP) are simultaneously entrapped into liposomes, and the resultant protein expression is studied. After the pioneering works on the GFP expression, the synthesis of a protein with a determined function could give the opportunity of realizing complex bioreactors.

More recently, based on the initial report on the expression of functional protein into liposomes (Yu et al., 2001), Yomo and coworkers were able to design and realize experimentally a two-level cascading protein expression (Ishikawa et al., 2004). A plasmid containing the T7 RNA polymerase (with SP6 promoter) and a mutant GFP (with T7 promoter) genes was constructed and entrapped into liposomes, together with an *in vitro* protein expression mixture – comprehensive of the enzyme SP6 RNA polymerase. Under these conditions, SP6 RNA polymerase drives the production of T7 RNA polymerase, which in turn induces the expression of the detectable GFP.

Of particular interest is the work by Noireaux and Libchaber (2004). Again, a plasmid encoding for two proteins was used; in particular, the authors introduced EGFP and α-hemolysin genes. In contrary with the cascading network described above, now the second protein (α-hemolysin) does not have a direct role in the protein expression, but is involved in a different task. In fact, although α-hemolysin is a water-soluble protein, it is able to self-assemble as heptamer in the bilayer, generating a pore 1.4 nm in diameter (cut-off ~ 3 kDa). In this way, it was possible to feed the inner aqueous core of the vesicles, realizing a long-lived bioreactor, where the expression of the reported EGFP was prolonged up to four days. This work represents certainly an important milestone in the road map to the minimal cell, because the α-hemolysin pore permitted the uptake of small metabolites from the external medium and thus solved the energy and material limitations typical of the impermeable liposomes.

TABLE 3. Protein Expression in Compartments.

Description of the system	**Main goal and results**	**Ref.**
Liposomes from EggPC, cholesterol, DSPE-PEG5000 used to entrap cell-free protein synthesis	Expression of a mutant GFP, determined with flow cytometric analysis	Yu et al., 2001
Expression kit entrapped during the liposome formation	GFP produced within small vesicles	Oberholzer and Luisi, 2002
Gene-expression system within cell-sized lipid vesicles	Encapsulation of a gene-expression system; high expression yield of GFP inside giant-vesicles	Nomura et al., 2003
A water-in-oil compartment system with water bubbles up to 50 μm	Expression of GFP by mixing different compartments which are able to fuse with each other	Pietrini and Luisi, 2004
A two-stage genetic network encapsulated in liposomes	A genetic network in which the protein product of the first stage (T7 RNA polymerase) is required to drive the protein synthesis of the second stage (GFP)	Ishikawa et al., 2004
E. coli cell-free expression system encapsulated in a phospholipid vesicle, which was transferred into a feeding solution containing ribonucleotides and amino acids	The expression of the α- hemolysin inside the vesicle solved the energy and material limitations; the reactor could sustain expression for up to 4 days	Noireaux and Libchaber, 2004

Finally, GFP has also been expressed in another kind of compartments, different from vesicles. These are the water cavities of a water-in-oil emulsion, where it has been shown (Pietrini and Luisi, 2004) that a functional protein can be expressed in these aqueous micrometer-sized environments, which represent a tiny volume fraction (~0.5%) of a hydrocarbon sample. Very interestingly, the desired degree of complexity, intended as the collection of all the components required for the GFP expression, was also obtained by solubilisate-exchange and/or fusion between different aqueous compartments, each one carrying a part of the whole biochemical machinery (plasmid; RNA polymerases, ribosomes, cellular extracts; amino acids).

Summarizing, in the last few years an handful of pioneering works appeared on proteins expression within liposomes, and – although this is a very new field – this approach could allow the laboratory construction of semi-synthetic minimal cells. What are the next steps to this ambitious project?

6. Perspectives

Keeping in mind the notion of minimal cell, the analysis of the data presented in this chapter makes clear what is still missing in order to proceed further in the experimental realization of a minimal cell.

Protein expression, as outlined in most salient experiments of Table 3, has been carried out without checking the number of enzymes/genes utilized in the work. We believe that it would be proper to carry out protein expression utilizing known concentrations of the single enzymes/genes – so as to know what is in the pot, and possibly have a hand on the corresponding chemistry. This operation would correspond actually to the implementation of the minimal genome inside liposomes and can pave the road to all next steps.

Another essential element that is still missing to reach the ideal case of a minimal living entity (see Figure 1) is self-reproduction of the whole construct (core and shell). In real biological systems, a cell is capable of duplicating and reproducing itself, producing a new entity containing all the elements (and properties) of the "mother" cell.

A very interesting case – currently under study – is the achievement of vesicles self-reproduction by the endogenous synthesis of vesicle lipids. In principle two strategies can be pursued: (i) incorporating the enzymes that synthesize the lipids or, (ii) starting from the corresponding genes, i.e., expressing those enzymes within the vesicles.

Early attempts have been focused on the enzymatic production of lecithin in lecithin liposomes (Schmidli et al., 1991). The metabolic pathway was the so-called salvage pathway, which converts glycerol-3-phosphate to phosphatidic acid, then diacylglycerol and finally phosphatidylcholine. The four enzymes needed to accomplish these reactions were simultaneously inserted in liposomes, and the synthesis of new phosphatidylcholine (10% yield) was followed by radioactive labelling. Liposome transformation, followed by dynamic light scattering, showed that vesicles changed their size distribution during the process.

Further studies (Luci, 2003) have been oriented to characterize the process by means of over-expression in *E. coli* and reconstitution in liposomes of the first two enzymes of the phospholipid salvage pathway, that produce phosphatidic acid, in order to obtain self-reproducing phosphatidic acid vesicles with only two enzymes.

One can imagine, for example, a vesicle that produces (starting from the corresponding genes) the enzymes needed for the lipid synthesis (Figure 5). This approach would satisfy the notion of self-generation of the boundary components, but –to reach the *status* of self-reproducing, should also duplicate the internal components (DNA, proteins, ribosomes). In other words, the core and the shell reproduction should be somehow "coupled". Therefore, a higher degree of complexity is an intrinsic feature of self-reproducing cells. Such a complexity is not compatible with the notion of minimal cell. However, one could conceive a way by which a strong coupling occurs, not because of complex biochemical regulation mechanisms, but as a consequence of a coupled chemical reaction. For example, the production of a reactive compound (in the broad sense) could elicit a membrane reaction that stimulates a division.

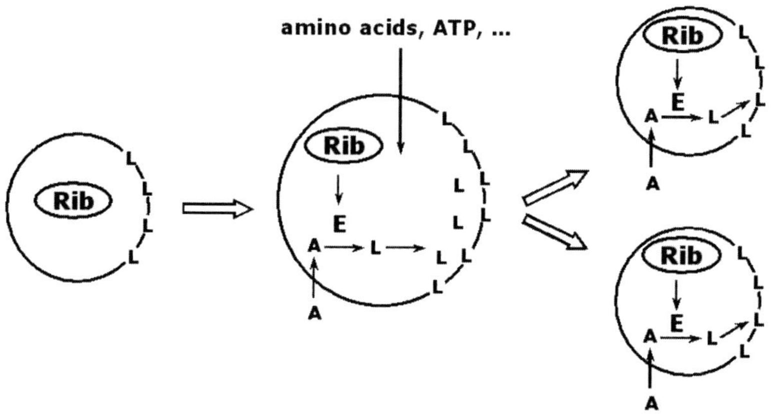

Figure 5. Schematic representation of a self-reproducing minimal cell. The cell, originally containing the whole machinery for protein expression (DNA, RNA polymerases, ribosomes, etc.), indicated as "Rib", can uptake a lipid precursor A as well as other building block and high-energy compounds (amino acids, ATP, etc.). The enzymes E, endogenously produced within the cell, transform the lipid precursor A into a membrane forming compound L. The cell growth due to the increase of its surface and eventually divide in two (or more) daughter cells. However, if a daughter cell misses at least one key component, it will undergo to "death by dilution".

In regard to the work with minimal living cell, one should also keep in mind the theory of autopoiesis, as developed by Maturana and Varela (Varela et al., 1974; Maturana and Varela, 1980; Maturana and Varela, 1998; see also Luisi, 2003). Autopoiesis gives the blue print of cellular life, emphasizing the importance of self-generation from within the boundary as the most characteristic process of life. Accordingly, a living system is a production system that generates from within its own components, which in turn self-assemble into the production system itself. It has been recently pointed out (Bitbol and Luisi, 2004) that not all autopoietic systems are living, in the sense that autopoiesis is the minimal but not sufficient conditions for life. Essential for life is also the capability of cognition, which is a specific interaction with the environment by which the bounded autopoietic system is capable of uptake substances from the environment and insert them into its own metabolism. Self-referentiality, operational closure, biological autonomy are the corollary of this situation.

In the cited work (Bitbol and Luisi, 2004), the simplest autopoietic living system is illustrated in a schematic form. Can one put this in more biological terms, so as to correspond to our notion of minimal cell?

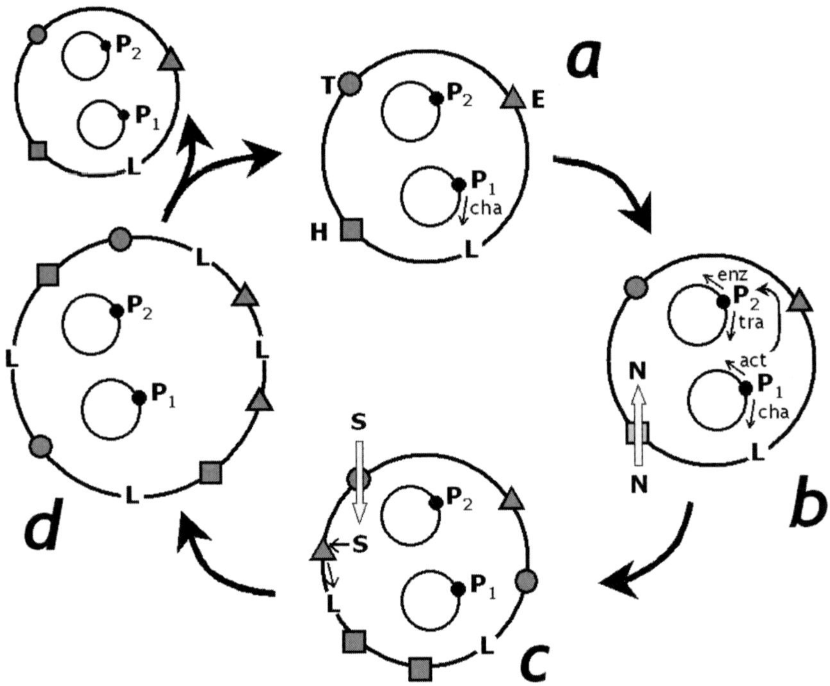

Figure 6. Cyclic process of a minimal cell considered as an autopoietic structure. Here two plasmids (inner circles) are inserted into a vesicle (large circle), together with the translation/transcription machinery for protein expression. The genes *cha* and *act*, sited on the first plasmid with a P1 promoter, codifies for a non-specific channel and for an activator of the promoter P2, respectively. The genes *tra* and *enz*, sited on the second plasmid with a P2 promoter, codifies for the specific transporter T and the enzyme E, respectively. See the text for further details.

An attempt to do so – many more can in principle be proposed – is illustrated in Figure 6. Here two plasmids (inner circles) are inserted into a vesicle (large circle), together will the translation/transcription machinery for protein expression. The gene *cha* codifies for a protein that acts as a channel for low molecular-weight compounds, such as the α-hemolysin channel utilized by Noireaux and Libchaber (2004). In addition, *act*, *tra* and *enz* are the genes for the activator of the promoter P2, the specific transporter T and the enzyme E, respectively. The cell produces the channel (e.g., α-hemolysin) constitutively (a). The cellular system is capable – thanks to the channel protein –to uptake nutrients N, and their presence triggers – through a cascade mechanism in P2 –the gene activation (b) for the expression of specific enzymes (T and E, circle and triangle in Figure 6, respectively) needed to the transport of a substrate S. This substrate is then converted into a membrane-forming compound L (c). In this way, the cell can grow and divide cyclically (d). The fact that N is selected out by the cell system, and incorporated into its own metabolism, makes the autopoietic system living (Bitbol and Luisi, 2004), and as such the system is also an adaptive one. The system has operational closure, as all information to self-generate (metabolism) and self-reproduce is contained in the internal organization.

7. Conclusions

The definition of the minimal cell as given at the beginning of this chapter appears simple and provided with its own elegance. Conversely, the experimental implementations of the minimal cells may not appear equally satisfactorily and elegant.

We have outlined the main difficulties opposing the construction of an ideal minimal cell and pointed out, for example, the several open questions about the number of genes, the permeability problems, and the synchronous core-and-shell self-reproduction.

The constructs realized in the laboratory until now represent still poor approximations of a full-fledged biological cell. Nevertheless, these forms of "limping life", represent in our opinion a very interesting part of this on-going research.

The realization in the laboratory of these partially living minimal cells may be of fundamental importance to understand the real essence of cellular life, and the historical evolutionary pathway by which this target may have been reached. Most probably, similar constructs were intermediates, which were experimented by nature to arrive at the final goal, the full-fledged biological cell.

In order to get closer to the real minimal cell, the minimal genome definition become again crucial. In all the systems of Tables 3, we are dealing with ribosomal protein biosynthesis and this implies 100-200 genes. We are still far from our ideal picture of a minimal cell and we can pose again the question on how to devise actual experiments so as to reduce this complexity.

Is this feasible experimentally? Can simple matrices be developed that are operative *in vitro* as ribosomes? Can one operate, at costs of specificity, with only a very few polymerases? Are specific tRNAs tremendously necessary? Similarly, can one conceive experiments with a limited number of amino acids?

The construction in the laboratory of semi-synthetic living cells would be a demonstration – if still needed – that *life* is indeed an emergent property. In fact in this case cellular life would be created from non-life, since single genes and or single enzymes are *per se* non-living.

All this is very challenging, and perhaps for this reason, as already mentioned, there has been an abrupt rise of interest in the minimal cell. It appears that one additional reason for this rise of interest lies in a diffuse sense of confidence, that the minimal cell is indeed an experimentally accessible target.

8. Acknowledgments

We thank the "E. Fermi" Study Center (Italy) and the European COST (D27 action) for financial support.

9. References

Bachmann, P. A., Luisi, P. L., and Lang, J. (1992). Autocatalytic self-replicating micelles as models for prebiotic structures. Nature **357**: 57-59.

Bitbol, M., and Luisi, P. L. (2004). Autopoiesis with or without cognition: defining life at its edge. J. Royal. Soc. Interface, **1**: 99-107.

Bloechliger, E., Blocher, M., Walde, P., and Luisi, P. L. (1998). Matrix Effect in the Size Distribution of Fatty Acid Vesicles. J. Phys. Chem. **102**: 10383-10390.

Calderone, C. T., and Liu, D. R. (2004). Nucleic-acid-templated synthesis as a model system for ancient translation. Curr. Opin. Chem. Biol. **8**: 645-653.

Chakrabarti, A. C., Breaker, R. R., Joyce, G. F., and Deamer, D. W. (1994). Production of RNA by a polymerase protein encapsulated within phospholipid vesicles. J. Mol. Evol. **39**: 555-559.

Dyson, F. J. (1982). A model for the origin of life. J. Mol. Evol. **18**: 344-350.

Frick, D. N., and Richardson, C. C. (2001). DNA Primases. Annu. Rev. Biochem. **70**: 39-80.

Gavrilova, L. P., Kostiashkina, O. E., Koteliansky, V. E., Rutkevitch, N. M., and Spirin, A. S. (1976). Factor-Free (Non-Enzymic) and Factor-Dependent Systems of Translation of Polyuridylic Acid by Escherichia coli Ribosomes. J. Mol. Biol. **101**: 537-552.

Gil, R., Sabater-Munoz, B., Latorre, A., Silva, F. J., and Moya, A. (2002). Extreme Genome Reduction in Buchnera spp: Toward the Minimal Genome Needed for Symbiotic Life. Proc. Natl. Acad. Sci. USA **99**: 4454-4458.

Gil, R., Silva, F. J., Peretó, J., and Moya, A. (2004). Determination of the Core of a Minimal Bacteria Gene Set. Microb. Molec. Biol. Rev. **68**: 518-537.

Ishikawa, K., Sato, K., Shima, Y., Urabe, I., and Yomo, T. (2004). Expression of a cascading genetic network within liposomes. FEBS Letters **576**: 387-390.

Islas, S., Becerra, A., Luisi, P. L., Lazcano, A. (2004). Comparative genomics and the gene complement of a minimal cell. Origins Life Evol. Biosph. **34**: 243-256.

Itaya, M. (1995). An Estimation of the Minimal Genome Size Required for Life. FEBS Letters **362**: 257-260.

Jay, D., and Gilbert, W. (1987). Basic Protein Enhances the Encapsulation of DNA into Lipid Vesicles: Model for the Formation of Primordial Cells. Proc. Natl. Acad. Sci. USA **84**: 1978-1980.

Kolisnychenko, V., Plunkett III, G., Herring, C. D., Fehér, T., Pósfai, J., Blattner, F. R., and Pósfai, G. (2002). Engineering a Reduced E. coli genome. Gen. Res. **12**: 640-647.

Koonin, E. V. (2000). How Many Genes Can Make a Cell: The Minimal-Gene-Set Concept. Annu. Rev. Genomics Human Genet. **1**: 99-116.

Lazcano, A., Guerriero, R., Margulius, L., and Oró, J. (1988). The evolutionary transition from RNA to DNA in early cells. J. Mol. Evol. **27**: 283-290.

Lazcano, A., Valverde, V., Hernandez, G., Gariglio, P., Fox, G. E., and Oró, J. (1992). On the early emergence of reverse transcription: theoretical basis and experimental evidence. J. Mol. Evol. **35**: 524-536.

Lonchin, S., Luisi, P. L., Walde, P., and Robinson, B. H. (1999). A Matrix Effect in Mixed Phospholipid/Fatty Acid Vesicle Formation. J. Phys. Chem. B **103**: 10910-10916.

Luci, P. (2003). *Gene cloning expression and purification of membrane proteins*, ETH-Z Dissertation Nr. 15108, Zurich.

Luisi, P. L. (2002). Toward the Engineering of Minimal Living Cells. Anat. Record **268**: 208-214.

Luisi, P. L. (2003). Autopoiesis: a review and a reappraisal. Naturwissenschaften **90**: 49-59.

Luisi, P. L., Ferri, F., and Stano, P. (2005). Approaches to semi-synthetic minimal cell: a review. Naturwissenschaften, *in press*.

Luisi, P. L., Oberholzer, T., and Lazcano, A. (2002). The Notion of a DNA Minimal Cell: A General Discourse and Some Guidelines for an Experimental Approach. Helv. Chim. Acta **85**: 1759-1777.
Luisi, P. L., Stano, P., Rasi, S., and Mavelli, F. (2004). A Possible Route to Prebiotic Vesicle Reproduction. Artificial Life **10**: 297-308.
Marchi-Artzner, V., Jullien, L., Belloni, L., Raison, D., Lacombe, L., and Lehn, J. M. (1996). Interaction, Lipid Exchange, and Effect of Vesicle Size in Systems of Oppositely Charged Vesicles. J. Phys. Chem. **100**: 13844-13856.
Maturana, H., and Varela, F. (1980). *Autopoiesis and Cognition: The Realization of the Living*, Reidel, Dordrecht.
Maturana, H., and Varela, F. (1998). *The Tree of Knowledge*, Shambala, Boston.
Morowitz, H. J. (1967). Biological Self-Replicating Systems. Prog. Theor. Biol. **1**: 35-58.
Mushegian, A. (1999). The minimal genome concept. Curr. Op. Gen. Develop. **9**: 709-714.
Mushegian, A., and Koonin, E. V. (1996). A Minimal Gene Set for Cellular Life Derived by Comparison of Complete Bacterial Genomes. Proc. Natl. Acad. Sci. USA **93**: 10268-10273.
Nissen, P., Hansen, J., Ban, N., Moore, P. B., and Steitz, T. A. (2000). The Structural Basis of Ribosome Activity in Peptide Bond Synthesis. Science **289**: 920-930.
Noireaux, V., and Libchaber, A. (2004). A vesicle bioreactor as a step toward an artificial cell assembly. Proc. Natl. Acad. Sci. USA **101**: 17669-17674.
Nomura, S. M., Tsumoto, K., Hamada, T., Akiyoshi, K., Nakatani, Y., and Yoshikawa, K. (2003). Gene Expression within Cell-Sized Lipid Vesicles. ChemBioChem **4**: 1172-1175.
Oberholzer, T., Albrizio, M., and Luisi, P. L. (1995a). Polymerase chain reaction in liposomes. Chemistry and Biology **2**: 677-682.
Oberholzer, T., and Luisi, P. L. (2002). The Use of Lipsomes for Constructing Cell Models. J. Biol. Phys. **28**: 733-744.
Oberholzer, T., Nierhaus, K. H., and Luisi, P. L. (1999). Protein Expression in Liposomes. Biochem. Biophys. Res. Comm. **261**: 238-241.
Oberholzer, T., Wick, R., Luisi, P. L., and Biebricher, C. K. (1995b). Enzymatic RNA replication in self-reproducing vesicles: an approach to a minimal cell. Biochem. Biophys. Res. Comm. **207**: 250-257.
Pantazatos, D. P., and MacDonald, R. C. (1999). Directly Observed Membrane Fusion Between Oppositely Charged Phospholipid Bilayers. J. Membrane Biol. **170**: 27-38.
Pietrini, A. V., and Luisi, P. L. (2004). Cell-free Protein Synthesis through Solubilisate Exchange in Water/Oil Emulsion Compartments. ChemBioChem **5**: 1055-1062.
Pohorille, A., and Deamer, D. (2002). Artificial cells: prospects for biotechnology. Trends in Biotechnology **20**: 123-128.
Rasi, S., Mavelli, F., and Luisi, P. L. (2003). Cooperative micelle binding and matrix effect in oleate vesicle formation. J. Phys. Chem. B **107**: 14068-14076.
Schmidli, P. K., Schurtenberger, P., and Luisi, P. L. (1991). Liposome-mediated enzymatic synthesis of phosphatidylcholine as an approach to self-replicating liposomes. J. Am. Chem. Soc. **113**: 8127-8130.
Shimkets, L. J. (1998). Structure and Sizes of Genomes of the Archaea and Bacteria. In: F. J. De Bruijn, J. R. Lupskin, G. M. Weinstock (eds.) *Bacterial Genomes: Physical Structure and Analysis*, Kluwer Academic Publishers, Boston, pp 5-11.
Stamatatos, L., Leventis, R., Zuckermann, M. J., and Silvius, J. R. (1988). Interactions of Cationic Lipid Vesicles with Negatively Charged Phospholipid Vesicles and Biological Membranes. Biochemistry **27**: 3917-3925.
Suttle, D. P., and Ravel, J. M. (1974). The effects of initiation factor 3 on the formation of 30S initiation complexes with synthetic and natural messengers. Biochem. Biophys. Res. Commun. **57**: 386-393.
Szathmáry, E. (2005). Life: in search of the simplest cell. Nature **433**: 469-470. DOI: 10.1038/433469a. [There was also an entire day of the 3rd COST D27 Workshop held in Crete in October 2004 (http://cost.cordis.lu/src/action_detail.cfm?action=D27), and The International School on Complexity held in Erice, Sicily, in December 2004 (http://www.ccsem.infn.it)].
Thomas, C. F., and Luisi, P. L. (2004). Novel Properties of DDAB: Matrix Effect and Interaction with Oleate. J. Phys. Chem. B **108**: 11285-11290.
Varela, F., Maturana, H. R., and Uribe, R. B. (1974). Autopoiesis: The organization of living system, its characterization and a model. Biosystems **5**: 187-196.

Walde, P., Goto, A., Monnard, P. A., Wessicken, M., and Luisi, P. L. (1994a). Oparin's Reactions Revisited: Enzymatic Synthesis of Poly(adenylic acid) in Micelles and Self- Reproducing Vesicles. J. Am. Chem. Soc. **116**: 7541-7544.

Walde, P., Wick, R., Fresta, M., Mangone, A., and Luisi, P. L. (1994b). Autopoietic Self-Reproduction of Fatty Acid Vesicles. J. Am. Chem. Soc. **116**: 11649-11654.

Weiner, A. M., and Maizels, N. (1987). tRNA-Like Structures Tag the 3' Ends of Genomic RNA Molecules for Replication: Implications for the Origin of Protein Synthesis. Proc. Natl. Acad. Sci. USA **84**: 7383-7387.

Woese, C. R. (1983). The Primary Lines of Descent and the Universal Ancestor. In: D. S. Bendall (ed.) *Evolution from Molecules to Man*, Cambridge Universiy Press, Cambridge, pp. 209-233.

Yu, W., Sato, K., Wakabayashi, M., Nakatshi, T., Ko-Mitamura, E. P., Shima, Y., Urabe, I., and Yomo, T. (2001). Synthesis of Functional Protein in Liposome. J. Bioscience Bioengineering **92**: 590-593.

Zhang, B., and Cech, T. R. (1998). Peptidyl-transferase Ribozymes: Trans Reactions, Structural Characterization and Ribosomal RNA-Like Features. Chem. Biol. **5**: 539-553.

Biodata of **David Deamer** author with co-author **John Evans** of ***"Numerical Analysis of Biocomplexity"***

Professor David Deamer is currently the acting chair of the Department of Biomolecular Engineering at the University of California, Santa Cruz. He obtained his Ph.D. from the Ohio State University in 1965 and continued his studies and research in membrane biophysics at the University of California, Davis. He moved his laboratory to Santa Cruz in 1994. Professor Deamer's scientific interests are in the areas of nanopore analysis of nucleic acids, and the role of self-assembling molecular systems in the origin of cellular life.

E-mail: **deamer@hydrogen.ucsc.edu**

John Evans graduated from Cambridge University with a degree in mathematics, then spent a number of years as a mathematics teacher and musician. In 1965 he changed to real-time software engineering, becoming an independent consultant three years later. In 1975 he became interested in the effects of electromagnetic fields on biological processes, and established a theory of vertebrate morphogenesis in which such fields play a role. He is currently testing an ELF oscillator for medical use. For the last twelve years, he has been associated with the Department of Applied Mathematics and Theoretical Physics in Cambridge and has developed the new ROSS Language for scientific computing. Further information can be found at www.ross-evans.co.uk.

E-mail: **J.Evans@damtp.cam.ac.uk**

David Deamer

John Evans

J. Seckbach (ed.), Life as We Know It, 199-212.

NUMERICAL ANALYSIS OF BIOCOMPLEXITY

DAVID DEAMER[1] AND JOHN EVANS[2]
[1]Department of Chemistry and Biochemistry University of California. Santa Cruz CA 95064. [2]Department of Applied Mathematics and Theoretical Physics. Cambridge University Cambridge UK

"When you measure what you are speaking about, and express it in numbers, you know something about it; but when you cannot measure it,. your knowledge is of a meagre and unsatisfactory kind." Lord Kelvin, 1888

1. Introduction

What do we mean when we judge that one living system is more complex than another? To answer this question, we can begin with qualitative aspects of biological complexity. For instance, it seems reasonable to believe that a bacterial cell is more complex than its growth medium, even though the cell and its growth medium are composed of exactly the same set of atoms. We can agree that a living cell is more complex than a dead cell, and that a multicellular organism is more complex than a single celled organism. We can also agree that a multicellular organism with a nervous system is more complex than an organism lacking one. These all seem like reasonable statements, but how can we answer Lord Kelvin's challenge quoted above in order to establish even the simplest semi-quantitative description of what we mean by 'more complex than'?

To narrow the question further, we can experiment with quantitative terminology and test it against our intuition. Is a bacterial cell twice as complex as its growth medium? That seems much too little. Ten times? A hundred times? Still too little, but getting closer. At the other end of the complexity scale, is a bacterium infinitely more complex? This seems too much of a leap. A bacterial cell is a finite structure, so the answer is somewhere between twice as complex and infinitely more complex, which narrows it down considerably.

2. Classifying Complexity

The exercise above suggests that it is not a hopeless task to consider biocomplexity in semi-quantitative terms. We will start by defining evolutionary complexity, phylogenetic complexity and developmental complexity as three linked yet distinct aspects of biocomplexity. Evolutionary complexity began at the time of life's origin, when biocomplexity was at some minimal value that increased dramatically when the first self-assembled unit of life appeared in the organic mixture of the prebiotic environment. We note here that our approach to complexity focuses on the unit of life, rather than the global environment. A quantitative description of complexity will therefore show an enormous jump with the appearance of a single living cell, even though one cell would be virtually invisible at the global level.

Phylogenetic complexity arose in the course of evolution, as living organisms branched to fill an ever-increasing number of niches, including those made available by the existence of other living organisms. As a result of the evolutionary process, the biosphere now contains a snapshot of evolution in which organisms range in complexity from the simplest bacteria to human beings. The groups of organisms can be understood in terms of a phylogenetic tree, and biocomplexity clearly increases as we travel up the tree from single celled prokaryotes to multicellular organisms. A bifurcation occurred when one group of organisms began to capture light energy and later evolved into multicellular plants. Photosynthetic organisms became a source of nutrients for a second group of motile organisms that evolved into animals. After achieving multicellularity and invading continental land masses, plants underwent relatively little further increments in biocomplexity, while motile animals developed nervous systems that were required for the later appearance of consciousness and intelligent behavior.

The third level of biocomplexity can be described as developmental complexity. Multicellular organisms generally begin life as a single fertilized ovum. Complexity of the organism then increases as a seed grows into a mature plant, or an embryo into an adult animal. Most multicellular organisms then undergo a second process commonly referred to as aging, which ultimately leads to death of the organism. The development of an embryo to an adult plant or animal, followed by aging and death, is a useful example for thinking about biocomplexity, because it is familiar to everyone and provides considerable insight into the complexity of biological systems.

3. Shapes of Complexity Curves Over Time

To take the first step toward a quantitative description of biocomplexity, we will first consider what an equation of developmental complexity might look like when plotted as a graph. Figure 1 illustrates the shape of a curve in which an arbitrary complexity scale is given on the Y axis. The scale clearly cannot be linear when we are dealing with growth from a single cell to millions or trillions of cells in an adult organism, thousands of genes and enzyme-catalyzed metabolic functions, and enormous numbers of receptor-mediated nervous functions. A logarithmic scale is therefore used in order to encompass large numbers of yet to be defined units. The X axis is labeled time, and if the units happened to be days we could think of this as a short-lived insect such as a fruit fly.

The rising portion of the curve is fairly obvious, taking into account increments in complexity that must occur upon fertilization, embryogenesis, development, birth and adult growth. The declining portion of the curve following death is less obvious, but contains several interesting features that help clarify what a definition of biocomplexity should take into account. For instance, the moment of death occurs when a metabolic function such as energy transduction ceases in an essential tissue like the heart or nervous system. The immediate result is an irreversible disarray of metabolic pathways within those cells, followed by a near-total loss of nervous functions that require energy. (Similar effects occur during hypoxia, cooling or anesthesia, but are reversible.) It follows that death defines the boundary between functional complexity and structural complexity, since a dead organism seconds after death is structurally identical to the same live organism. Over the next few units of time, structural complexity also begins to decrease as supramolecular function and order are lost, followed by loss of macromolecular order through denaturation. The final steps include hydrolysis of biological polymers to monomers and dispersal of monomers and other degradation products back into the environment.

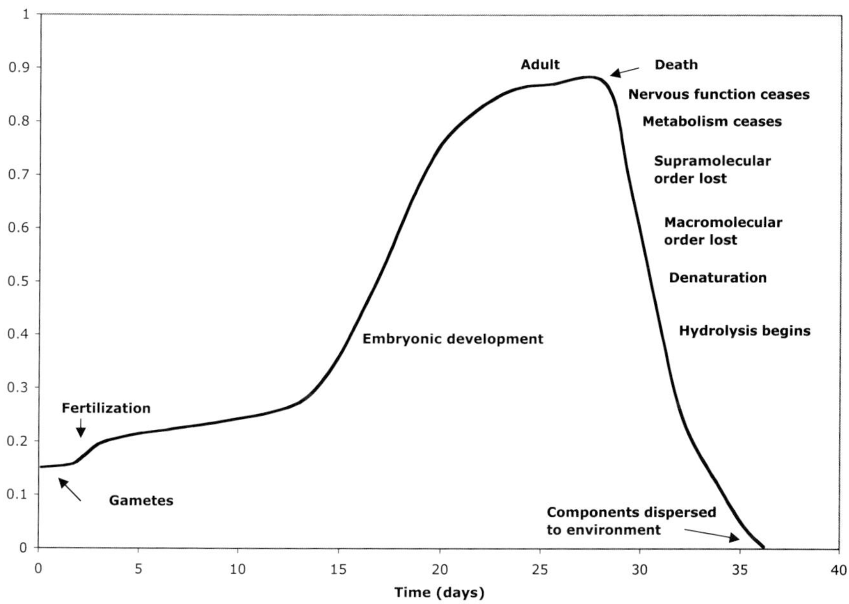

Figure 1. Developmental complexity of an individual organism.

Using some of these principles, we can next think about evolutionary complexity, which again can be considered in terms of a set of steps that each lead to an increment in complexity. A graphical illustration of the points to follow is shown in Figure 2. Arbitrary numerical values are used for the log scale of complexity, since these are simply meant to illustrate our approach.

1. At the far left of Figure 2, a sterile planetary surface such as the early Earth becomes more complex with the addition of organic compounds from endogenous or exogenous sources.

2. The solution of organic solutes becomes more complex over time as organic molecules self-assemble into molecular systems. Examples include the chemical

synthesis of random polymers from suitable monomers such as amino acids, and the assembly of membranous vesicles from amphiphilic molecules.

3. The self-assembled structures become much more complex at the origin of life when one or more of the self-assembled structures happens to have properties that allow it to use energy to accumulate simpler molecules from the environment and assemble them into reproductions of the original structure. Complexity arises from the increasing number of macromolecular structures that are organized within the unit of life, from the increasing number of catalyzed metabolic pathways incorporated, and from the increasing ability of macromolecules to transmit information from one kind of molecule to another, examples being the genetic code, replication, transcription and translation.

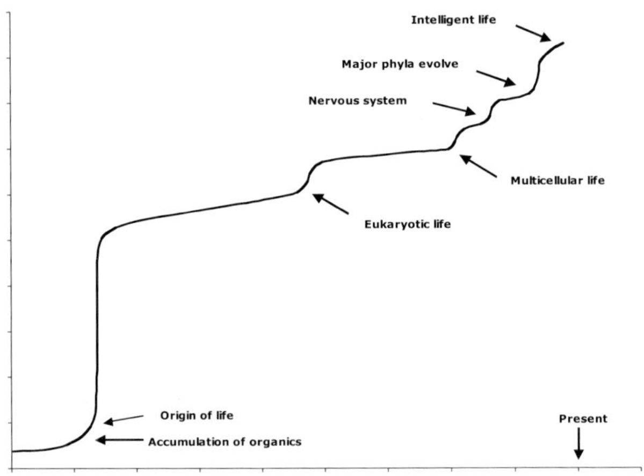

Figure 2. Increase of biocomplexity over evolutionary time.

4. After life begins, prokaryotic cellular life fills niches and evolves ever more efficient mechanisms to capture energy and nutrients. This process also represents speciation, and complexity increases with the number of species competing for the niches. However, at some point the microbial populations reach the carrying capacity of the environment, and complexity stabilizes for a period of approximately 2 billion years.

5. The microbial population becomes more complex when two or more different organisms merge to produce the first eukaryotic single-celled organisms.

6. The eukaryotic cells become more complex when two or more discover enhanced evolutionary fitness by forming the first multicellular organisms. These in turn become more complex as the cells begin to differentiate into specialized tissues, and particularly when cell-cell communication begins through junctional complexes between cells.

7. Further complexity develops when an endocrine system allows one group of cells to influence another group of cells by releasing chemical messengers that bind to cell surface receptors and activate a response. (This is not shown on the plot.)

8. Complexity increases substantially when a nervous system evolves. This allows one cell to affect not just one or a few neighboring cells but large numbers of other cells at a distance through axons that efficiently transmit a kind of digitized information. This is in contrast to endocrine information transfer, which does not have digital characteristics.

9. Complexity increases as the size and variety of multicellular organisms grows following the Cambrian explosion.

10. Complexity increases when organisms begin to interact through signals that are behaviorally mediated by sensory receptors of the nervous system. The most recent increment in biological complexity corresponds to the appearance of the conscious state and intelligence in one group of mammals.

We also note that complexity can decrease in organisms that become parasitic or live in the absence of a selective factor. We may be able to learn something about the mechanism by which complexity changes over evolutionary time from such examples, in which an obvious loss of complexity occurs when a selective factor is relaxed. In the case of cave organisms, the absent selective factor is light, with the result that eyes and pigment are gradually lost over many generations.

4. Can we Quantitatively Describe Biocomplexity?

Attempts to define biological complexity quantitatively by lumping together all of the characteristics of the organism inevitably leads to confusion. Instead we should first attempt to understand complexity by using at least three general features of a living organism. The first feature is structural complexity. We will use the concept of structural units here, a unit being the simplest bounded structure involved in a given level of complexity. Structural complexity incorporates units such as macromolecules as they become organized into molecular systems, cells and cellular systems, whole organisms and their populations. Structural complexity can be characterized by the total number of units, the kinds of units, the concentration of units in space, and organization of the units.

The second feature is functional complexity, which always includes energy and motion. At the subcellular level functional complexity involves metabolic pathways, energy transduction and biosynthesis. At the cellular level it concerns the function of cells, tissues and organs, and examples include exocytosis, hormonal secretion and heart beat. At the whole organism level functional complexity has to do with mechanisms by which the organism captures nutrients and energy from the environment, avoids predators, and reproduces. In the case of motile organisms such as animals, this includes the mechanics of motion.

The third feature is related to information transfer. Informational complexity concerns the interaction of structural units with each other in such a way that information is passed from one unit to the next. In a sense, this is a subset of functional complexity, but is characterized by information transfer rather than motion, energy transduction and metabolism. At the subcellular level it includes regulatory functions, intracellular chemical signals, and the use of genetic information to direct replication and protein synthesis. At the cellular level it includes sensory receptors that respond to chemical and physical signals in the environment. At the multicellular level it includes cell-cell communication through junctional complexes; and finally the function of a nervous system in which single specialized cells (neurons) can communicate over long distances with multiple other cells. For example, a motor neuron in the CNS controls hundreds of muscle fibers, and a pyramidal cell in the cerebral cortex interacts with thousands of other cortical cells through synaptic junctions.

5. Quantitative Biocomplexity in Nervous Systems

From the above description, it is clear that a quantitative description of biocomplexity will be made difficult by the number of parameters that must be considered along with their interactions. For this reason we choose first to attempt a mathematical model of a single cell type – neurons – and consider only their structural complexity in terms of numbers of cells and connectedness. After learning how to approach a single biological system using two parameters, we can go on to more complex multi-parameter interacting systems.

The nervous system is useful because cell numbers have been estimated, as well as the number of structural connections between nerve cells of the central nervous system. One of the first scientists to have given serious thought to this question is W. Grey Walter. In his classic book, *The Living Brain,* he asks "whether the elaboration of cerebral functions may possibly derive not so much from the number of units in the brain", as from the "*richness of their interconnection.*"

He then considers how many kinds of behavior are possible with just two cells, ***A*** and ***B***. Without assigning particular functions, he suggests that there are seven basic modes represented by

$$0,\ A,\ B,\ A{+}B,\ (A \rightarrow B),\ (B \rightarrow A),\ (A \leftrightarrow B)$$

meaning: inactive, **A** active, **B** active, **A** and **B** active independently, **A** driving **B, B** driving **A, A** and **B** driving each other.

From this, he calculates that, with 6 units, there would be a new experience every tenth of a second, assuming a time scale typical for neuron action potentials. With n units, the number of possible modes is given approximately by $M = 2^{n2 - n}$

If M is calculated for 1,000 interconnected neurons, the number of possible combinations is inconveniently large. Nonetheless, Grey Walter was encouraged by his analysis, because only a few connected units and modes of operation are required to achieve an immense variety of ways to run a system of modules, thus simulating behavioral aspects of nervous systems.

A simpler way to consider this problem is to allow just two states to each neuron, which can be thought of as 0 and 1. This has a definite physical basis, in that neurons either send an electrical pulse to other cells, or they do not. A certain minimum voltage threshold must be achieved before a nerve cell becomes active and affects other cells. Given n associated cells, the number of possible binary states at any given point of time is given by $M = 2^n$.

The active state, or action potential, lasts for just a few milliseconds. If we consider the different possible sequences of action over, say, 1 second, then we would need to multiply this figure by about 100. For 10 neurons over 1 minute, this would give about 6 million possible sequences of operations. More generally, and allowing for unlimited time, the number of possible sequences, ***S***, of ***M*** modes of behavior is given by *factorial M*: $S = M! = M(M-1)(M-2)$ **..........1.** This again leads to astronomical numbers which are not helpful in studying relative complexity. For this reason we will limit our analysis to available modes rather than near-infinite sequences.

Modern studies have shown that nerve action is in fact more flexible than the binary logic of modern computers. The dendrites of a single brain cell have thousands of synaptic connections coming from many other cells. At any point in time, the action potentials arriving at the synapses are summed both in terms of rates and spatial distribution on the cell surface. Changes in the summation can alter the activity of the neuron by inhibiting or activating its ability to produce an action potential. If we postulated z modes of behavior for each cell, then $M = z^n$.

On this basis, and using a logarithmic scale, we could define the complexity of action, **C**, of any group of associated neurons as

$$C = \log(M) = n.\ \log(z) \quad (1)$$

Although there are trillions of nerves in the human body, the ***n*** here refers to the number of interconnected neurons for performing any particular operation, and this will be only a small fraction of the total for any given action. To move a finger, for instance, let us assume that only 100 motor neurons, from brain to spinal nerve to finger muscle, are required. Using a binary basis (2 modes for each cell), then the possible modes are given by $M = 2^{100}$ which is about $\mathbf{10^{30}}$. This gives a logarithmic complexity ***C*** of ***30***. For a pianist using 10 fingers continuously, then $C = 31$. These numbers are consistent with practical experience, in that the ways of using fingers to create keyboard music are, to all intents and purposes, infinite.

For some indication of an overall estimate of modes of the nervous system, we require an expression of the form $M = p\ .\ z^n$ where ***p*** represents the number of separate units operating in the system. In the human, the 31 spinal nerves, 12 cranial nerves, and the autonomic chains on either side of the spine form the main units. These bundles of nerve fibers, each containing millions of axons from nerve cells, are relatively independent from each other, and serve particular organs and muscles. If we ignore the brain, then ***p*** is of the order of 100 for the human. However, division of activity within the cortex is much more complex. Evaluating ***C*** for the brain, we get

$$C = \log(M) = \log(p) + n.\ \log(z) \quad (2)$$

and this means that ***n***, the number of neurons in an associated bundle, remains fundamental, whatever the complexities of the subsystems in the brain.

This suggests an approach for comparing complexity between vertebrates, and particularly between the higher primates and man. For the basic functions of body, and for physical movement of muscles, both external and internal, there would seem to be no great differences between the higher vertebrates. Some primates have greater physical agility than humans, and some have less, yet all are of a comparable order. But in the use of the fingers and the voice, for instance, the vastly greater *control* complexity of parts of the cerebral system of humans becomes obvious. This is independent of rational thought: People with comparatively low intellectual ability can instinctively use their fingers and voices in remarkable ways, unlike anything to be found in the animal world. It is in the interconnections of the brain that there are significant differences.

It has long been assumed that genetic constitution defines to a large extent the range of biocomplexity. But now that we have decoded the genome of the humans and a few other vertebrates, and can determine the length of the genome in all living systems, this would seem to be far from the truth. The human genome consists of ~3 billion bases of DNA, but so does one variety of the lily. As this didn't seem to make much sense, biochemists and geneticists assumed that complexity was related to the genes that produced a protein. But the 30,000 genes of man and the higher primates are roughly the same. Even the mouse has nearly as many of the same genes as a human being. If we go back to insects with distinctly fewer genes, the genetic factor seems to have some relevance. But in the overall evolutionary picture, neither the size of the genome, nor the number of genes, are much of a guide to the final complexity of the structure and function of any organism.

As for the total number of cells, from fleas and birds to whales and elephants, we can say that, in terms of *structural* complexity, the larger creatures are more complex. But from a *functional* viewpoint - the ability to carry out difficult and unusual tasks - sometimes a surprisingly small creature can demonstrate a problem-solving capacity. A recent example is the crow that solved a food retrieval problem by bending a wire into a hook (Weir et al. 2002).

In terms of the number of brain cells of humans, there would appear to be only a modest correlation between the size of brain and mental ability. We now know that large parts of the brain are devoted just to physical control and balance, while mechanisms of memory and conceptual thought remain largely unknown in terms of cell function. The most that we can surmise is that through learning and experience, we have some innate ability to affect the interconnections of the cerebral neurons. As we have seen above, changing slightly the number of associated cells, or the modes of operation, can significantly increase the potential of any group of neurons to engage in complex functions

Lower-level factors related to cell structure and function can be brought into our formula. providing we can define meaningful parameters for each unit or function. If $\boldsymbol{a, b, c}$ are three such parameters, then $\boldsymbol{M = a \,.\, b \,.\, c \,.\, (p \,.z^n)}$ Thus on this basis,

$$C = \log(a) + \log(b) + \log(c) + \log(p) + n \,.\, \log(z) \qquad (3)$$

However, this is not really a practical formula. Comparative information about different nervous systems is sparse, and the data have not been collected in a way suitable for such a formula. Generally, all we have is some information on the total number of neurons, and perhaps some estimate of the interconnections. We can adapt the formula in the following way.

Suppose we have an estimate N for the total number of neurons, and also an estimate for the number of interconnections for each neuron. The latter gives us a value for z, the number of modes of operation for each neuron. If p is the number of subsystems - which is unknown - then the average number of neurons in each subsystem, n, is given by $n = N / p$ so that $C = log(p) + N. log(z) / p$. Now p varies between 1 and N. This means that, at the extremes, there is either one large interconnected system or N independent neurons. To get a sensible average value between these two extremes, we can integrate C for values of p between 1 and N (the area under the curve), and divide by N. Now the integral of C with respect to p is $p.log(p) - p + N.log(z).log(p)$ and from this we obtain an average value of C, given approximately by the expression $log(N)\ [1 + log(z)]$. However, this integral is based on natural logs to the base e. Converting to standard logs to the base 10, formula (4) is a more practical representation of relative complexity (the conversion factor is about 2.3).

$$C = log(N)\ [1 + 2log(z)] \tag{4}$$

Putting in real numbers for *Caenorhabditis elegans*, where N = 302 and z = 10 then $C = log(302)\ [1 + 2log(10)] \sim 7.4$. If we are off be a factor of ten in estimating z, so that z = 100, then $C = log(302)\ [1 + 2log(100)] \sim 12.4$. Equation (4) is not very parameter-sensitive, since it is based on a logarithmic scale, yet it still leads to useful estimates of relative complexity that can vary by many orders of magnitude. Table 1 shows some numerical values for the nervous systems of several organisms, ranging from *C. elegans* to *Homo sapiens*, with C values rounded to the nearest integer. Except for *C. elegans*, these are rough estimates, both for N and z, but sufficient for order-of-magnitude calculations. In our judgment, the values approximately fit our intuitive expectations for relative complexity of the range of organisms shown. It is interesting to note that the value for humans is approximately 5 orders of magnitude greater than for chimpanzees, entirely due to the greater number of cells in the human brain. This suggests the possibility that intelligence has a minimum requirement for interacting units.

Table 1. Complexity estimates for nervous systems. Complexity (C) = log(N) · [1 + 2log(z)] where N = total number of neurons and z = average number of interconnections for each neuron.

	Species	*N*	*z*	*Complexity*
1	*C. elegans*	302	10	7
2	Bee	10^6	100	30
3	Salamander	10^7	1,000	49
4	Frog	$5x10^7$	1,000	54
5	Mouse	$2x10^8$	10,000	75
6	Pigeon	10^9	10,000	81
7	Cat	$5x10^9$	10,000	87
8	Monkey	10^{10}	10,000	90
9	Chimpanzee	$3x10^{10}$	10,000	94
10	Human	10^{11}	10,000	99

6. Quantitative Biocomplexity in Molecular Systems

We will now consider whether complexity C can be described by possible quantitative relationships at the level of single cells, using the approach outlined above for the nervous system. We argue that this is not an impossible task. The number and kind of interacting structural/functional units in a living cell can now be reasonably well estimated, and Table 2 shows the actual count of subcellular units in an *Escherichia coli* cell (adapted from Neidhardt and Curtiss, 1996).

This table makes clear that we are able to count the units that contribute to complexity in a living cell. Furthermore, we can count the units in multicellular organisms by extrapolation from single cells. It should therefore be possible to apply arguments similar to that developed for the nervous system, except that the units will be molecules and molecular systems, and the connections will be the metabolic and regulatory pathways that control cell functions. It is now possible to not only count the numbers of protein species in a cell, but also to establish the number of interactions each has with other proteins. This has given rise to a new concept called the "interactome" as an extension of the concepts of genome and proteome. The interactomes are defined as the total number of protein species in a living cell, each of which is functionally linked by one of more interactions to other proteins in the cell. Interactomes have now been reported for the yeast *Saccharomyces cerevisiae*, (Uetz et al., 2000), the nematode *Caenorhabditis elegans*, (Li et al., 2004) and the fruit fly *Drosophila melanogaster*. (Giot et al., 2003).

Table 2. Number of molecules per E. coli cell.

Proteins	2.36 million
Kinds of proteins	1850
RNA as ribosomes	18,700
rRNA	18,700
tRNA	205,000
mRNA	1380
DNA	2.13 molecules (bacterial genome +plasmids)
Lipid	22 million
Lipopolysaccharide	1.2 million
Peptidoglycan	1
Glycogen	4360

We can now attempt to apply the equation C (complexity) = log(N) * [1 + 2log(z)]to one of the published interactomes. In the nervous system, we assumed that a single number Z described the number of synaptic connections between N cells. This is clearly a very broad assumption, but in the absence of actual data it is the only assumption that is presently possible, and serves to illustrate our approach to order of magnitude accuracy. However, in the interactome there are reasonably good estimates of N and Z which can be broken down into how many interactions any given protein actually has. Giot et al. (2003) analyzed the fruit fly data in terms of the number of proteins and the number of interactions for each. This ranges from ~3000 proteins having only a single interaction, to 4 proteins having 12 to 20 interactions.

A possible approach using this more detailed data is a generalization of formula (4) to :

$$\sum_{z=1}^{z\max} \log(n_z)[1+2\log(z)] \tag{5}$$

where zmax is the maximum number of interconnections, and n_z is the number of units for each z. However this is less straightforward than it first appears, requiring a normalization process for the n_z values. If applied without this, it will give absurdly high values for the simple reason that loga + logb is greater than log(a+b). It is simpler just to find the mean value of z, and apply the formula once.

$$z_{mean} = \frac{1}{N}\sum_{z=1}^{z\max} z * n_z \tag{6}$$

For *D. melanogaster*, we have the following data for N and z (from Fig. 2D, high confidence line, Giot et al., 2003). In this notation, ~3000 proteins interact with one other protein, 1000 interact with two other proteins, and so on to one protein interacting with 20 other proteins:

3000, 1; 1000, 2; 500, 3; 250, 4; 200, 5; 80, 6; 50, 7; 30, 9; 23, 10; 7, 11; 3, 15; 2, 19; 1, 12; 1, 15; 1, 18; 1, 20

Applying equation (6) to this data gives a mean value of z of 1.95, and a complexity value of 5.87. Similar data have not been presented for yeast and *C. elegans*, so we are not yet able to make comparisons. Such comparisons are probably premature in any case, since the interactomes are still incomplete.

To summarize, we have described a generalized approach by which certain kinds of biocomplexity can be analyzed and compared. The approach can be applied to systems ranging from proteins of interactomes to the neurons of nervous systems, as long as there is specific information about the number and kind of interacting units and the number of their interactions. As this type of data becomes more commonly available for other interacting biological units, it will be interesting to see whether this approach provides a useful way to estimate and compare biocomplexity between different biological systems.

7. Acknowledgements

The authors wish to thank the Esalen Institute and the participants sponsored by the Esalen Center for Evolutionary Theory for helping us to develop the concepts described in this paper.

8. References

Giot, L. et al. (2003) A protein interaction map of Drosophila melanogaster. Science **302:** 1727-36.

Li, S. et al. (2004) A map of the interactome network of the metaozan C. elegans.Science **303:** 540-3.

Neidhardt, F.C. and Curtiss, R. (1996) *Escherichia coli* and *Salmonella typhimurium* ASM Press.

Uetz, P. et al. (2000) A comprehensive analysis of protein-protein interactions in Saccharomyces cerevisiae. Nature **403:** 623-7.

Walter, W.G. (1953) *The Living Brain*. W.W. Norton & Co. (1953).

Biodata of **Baruch S. Blumberg** author of ***"Hepatitis B Virus. Conjectures on Human interactions and the Origin of Life"***

Professor Baruch S. Blumberg, M.D., Ph.D. is a Distinguished Scientist at Fox Chase Cancer Center, Philadelphia, Pennsylvania and University Professor of Medicine and Anthropology at the University of Pennsylvania in Philadelphia. From May 1999 until October 2002, he served as Director of the National Aeronautics and Space Administration (NASA) Astrobiology Institute headquartered at Ames Research Center, Moffett Field, California. From October 2000 until October 2001, he was Senior Advisor to the Administrator of NASA based in Washington D.C. From 2003-2004, he served as Principal Scientist of the NASA Division of Fundamental Space Biology. He was Master of Balliol College, Oxford University, from 1989 to 1994 and, prior to that, Associate Director for Clinical Research at Fox Chase Cancer Center from 1964. He was on the staff of the National Institutes of Health, Bethesda, Maryland, from 1957 to 1964. He earned an M.D. degree from the College of Physicians and Surgeons, Columbia University, New York, in 1951, and a Ph.D. (D. Phil.) in Biochemistry from Oxford University in 1957. He was an Intern and Resident at Bellevue Hospital and The Columbia Presbyterian Medical Center in New York City.

His research has covered many areas including clinical research, epidemiology, virology, genetics, and anthropology. He was awarded the Nobel Prize in 1976 for "discoveries concerning new mechanisms for the origin and dissemination of infectious diseases" and specifically, for the discovery of the Hepatitis B virus. In 1993, he and his co-inventor, Dr. Irving Millman, were elected to the National Inventors Hall of Fame for their invention of the hepatitis B vaccine and the diagnostic test for hepatitis B. He has taught medical anthropology at the University of Pennsylvania and elsewhere, and has been a Visiting Professor in India (Bangalore), Singapore, University of Kentucky (Lexington), Indiana University (Bloomington), the University of Otago, Dunedin, New Zealand, and Stanford University in California (department of Medicine and Program in Human Biology).

E-mail: **baruch.blumberg@fccc.edu**

J. Seckbach. J (eds.), Life as We Know It, 213-235

HEPATITIS B VIRUS:
Conjectures on human interactions and the origin of life

BARUCH S. BLUMBERG
Fox Chase Cancer Center
333 Cottman Avenue, Philadelphia, PA 19111, USA

1. Introduction

Viruses have been feared as major pathological agents, a scourge of humans and other species. However, they have a more textured relation to their hosts. This paper will focus on hepatitis B virus (HBV) - one of the most common and deadly agents that infect humans and has a powerful selective effect on survival - that also has complex and subtle interactions with its host that are not pathologic. HBV affects populations and their evolution in an imaginative manner that is a model for understanding how other microorganisms can both plague and assist their hosts at different times and in different environmental and genomic contexts. The host responses to infection with HBV are affected, in large part, by a series of polymorphic alleles at several locations on the human genome and these loci are also related to other infections. The effects of these dynamic systems on the population biology and microevolution can be profound.

I discussed these possibilities at the Croonian Lecture of the Royal College of Physicians, London in 1994 (Blumberg, 1995).

> It may be conjectured that, with massive infections and in large epidemics, HBV DNA may infect and integrate into the DNA of the host germ cells. At the time of reproduction these could enter the gene pool of the population and segregate in the same manner as ordinary host genes; it would be analogous to mutation. Although germ cell integration may be relatively rare, in areas of massive infection such as East Asia and sub-Saharan Africa the absolute number of such integrations could be large. (Sections of the human genome have homologies with HBV, but little is known about how they relate, if at all, to this conjectural model.) It would represent a form of inheritance of acquired base pair sequences, and could lead to interesting models and observations on the effect of viruses on human population genetics. HBV DNA, so integrated, could be viewed as an 'engine' of evolution.

That is, HBV, and by inference other viruses may have a significant and complex role in the evolution of human and other species. The insertion of the viral genomes is a form of mutation, but a mutation of a cassette of sequences that have already survived and adapted for many generations. The concept that viruses can contribute to the evolutionary process has been recognized for many years. Salvador Luria, the pioneering molecular biologist, stated it very succinctly (Luria, 1959).

> May we not feel that in the virus, in their merging with the cellular genome and their re-emerging from them, we observe processes which, in the course of evolution, have created the successful genetic patterns that underlie all living things?

Viruses may have been one of the earliest life forms; RNA viruses could have flourished in the RNA world thought to precede our contemporary DNA world. Viruses are sometimes said not to be "alive" because they require a host cell in order to replicate. But, in early days, viruses or their predecessors may have been able to reproduce on their own, and over the millennia evolved to the cell dependent status they now enjoy. Were HBV or its ancestor's actors in this early drama of origins? What is the connection, if any, between HBV as a possible agent of recent or contemporary human evolution and the beginning of life on Earth or elsewhere?

In this chapter, I will present the factors that bear on these issues. It includes data already published, much of it more than 30 years ago, recent data as yet unpublished (but presented at scientific meetings), and a fair amount of conjecture to bind together what is, admittedly, a heterogeneous and sometimes curious body of information. The chapter includes discussions of the following: the worldwide prevalence of HBV and the diseases that it causes, the pathological consequences of infection including data on survival, a brief discussion of molecular biology, the genetics of response to HBV infection, the vaccination programs and their profound effects on morbidity and survival, the unexpected non-pathological effects of chronic HBV infection, including alterations of sex ratio at birth and the possibility of beneficial effects of iron retention, and how all these factors may affect the co-evolution of HBV and its hosts. The chapter concludes with a discussion of how viruses may be implicated in the origins of life.

2. Prevalence of HBV and Diseases that it Causes

HBV is a partially double-stranded DNA virus belonging to the *Hepadnaviridae* (hepatitis DNA viruses) family, a category that was created when it was recognized that the newly discovered HBV (Blumberg 1967) did not fit in any previously described bin. Only a minority of those infected becomes sick. There is significant diversity in host response following infection that can be detected by serological studies. The host may asymptomatically develop antibody against the surface antigen (anti-HBs) that usually affords life-long protection against further infection with HBV; this is a very common response. Infected hosts may become carriers of the virus (HBsAg positive) often shortly after birth or at a young age. They usually remain asymptomatic carriers for years and decades, but are at high risk of, in time, developing chronic liver disease and primary cancer of the liver (hepatocellular carcinoma, HCC). There are large geographic differences in the prevalence of the carrier state. HBsAg and anti-HBs are the most common serological responses and can be considered as the polar reactions to infection.

Others, when infected will develop acute hepatitis that is usually self-limited and is followed by the development of the protective anti-HBs. In about 10% of the acute cases, the patient may remain a carrier and be susceptible to chronic disease and liver cancer. A small percentage of patients with acute hepatitis may develop fulminent hepatitis that rapidly progresses to death. The protein products of the four reading frames of the virus: HBsAg, HBcAg, HBeAg, HBxAg, and the polymerase and reverse transcriptase produced by the P reading frame may be found in blood or cells during the course of disease and recovery (see Molecular Biology below). Antibodies against several of these proteins are produced in some infected individuals: these include anti-HBc, anti-HBe, anti-HBx, and others.

The presence of any of these antigens or antibodies markers in the blood indicate past or present infection, with the exception of the hundreds of millions of people who have been vaccinated and developed anti-HBs. Based on population surveys for one or several of these markers made before the vaccination programs were in place (mid-1980s), an estimate can be made of infection prevalence in specific populations and, by extension, to national and global populations. The WHO has estimated that approximately 2 billion (2×10^9) people - one third of the world's population - have been infected with about one million deaths a year due to HBV infection. This level of engagement with a virus has major demographic effects both pathological and non - pathological that influences the future of humans.

3. HBV as a Pathologic Agent

A sizable proportion of those infected have an increased probability of serious life shortening diseases, a major selective factor in human evolution. Among the infected there are about 371 million (371×10^6) carriers who are susceptible to chronic liver disease and primary cancer of the liver; those who develop anti-HBs are far less susceptible. In 1999, H. Margolis of the Centers for Disease Control in Atlanta, GA, estimated the overall mortality among prevalent HBV carriers compared to deaths from those with prevalent HIV infection, another major viral killer (Table 1).

TABLE 1. Predicted deaths from prevalent carriers of HIV and HBV

(H. Margolis, Center for Disease Control. Paper presented at Third International Conference on Therapies for Viral Hepatitis, Maui, 12/12/99)

Virus	Number Infected (Millions)	Case Fatality Rate	*Total Deaths (Millions)*
HIV	34	100%	*34*
HBV	371	High 25%	*92.8*
		Low 15%	*55.7*

There were 34 million prevalent cases of HIV infection; of these, approximately 100% would die of this infection. (Therapy was not commonly available at that time.) There were 371 million prevalent HBV carriers; of these, there would be 56 million deaths using a low estimate of case fatality rate and 93 million using a high case fatality estimate. From this, it appears that HBV has had a profound worldwide effect on mortality that will decrease in the future as a consequence of the vaccination programs (see below).

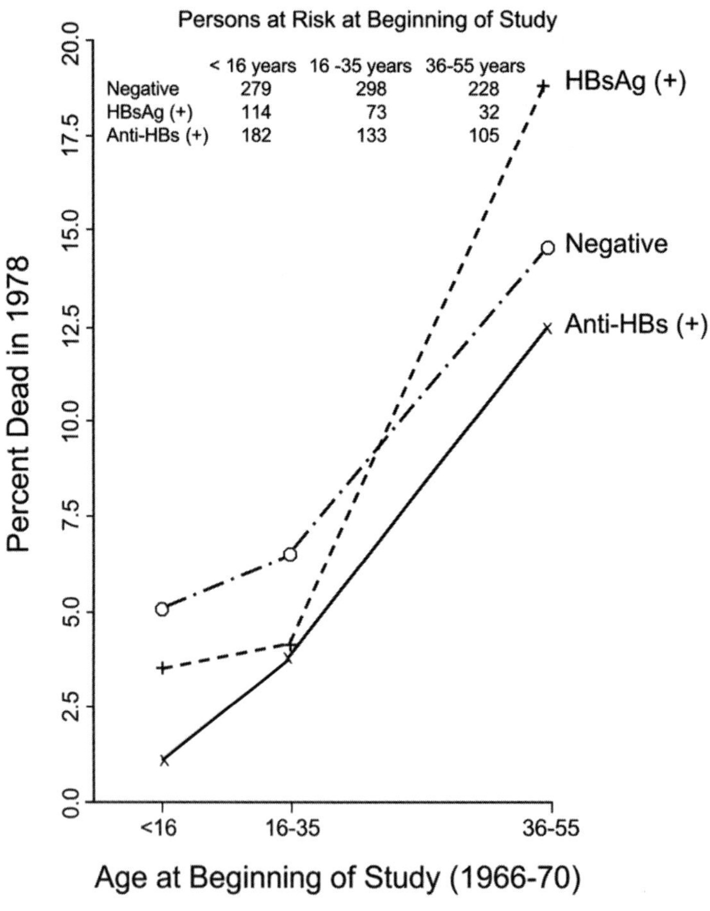

Figure 1. Responses to HBV of 1,444 individuals in the Solomon Islands in 1966-1970 compared with the mortality of these individuals in 1978. The inserted table shows the number of individuals by age and HBV status at the beginning of the study. (London 1982)

Most of the deaths associated with the HBV carrier status will occur late in life. Survival over the course of about 8 years was determined in a population in the Solomon Islands in the Central Pacific (London 1982). Sera were collected from 1966 to 1970 (before vaccination) and tested for HBsAg (the carrier state), and anti-HBs. The sites were visited again in 1978 to determine who had survived and who had not (Figure 1).

There was decreased survival of carriers compared with those who developed anti-HBs. The difference in the oldest age group (36–55) was particularly striking; nearly 19% of the carriers died compared to 12.5% of those with anti-HBs. A similar study was done in Haimen City, China (Chen, 2004). Sera were collected during 1992 and 1993 and the HBV response determined. The mortality per 100,000 was determined in 2002. As in the Solomon Islands, the mortality was much greater in carriers than non-carriers (Figure 2).

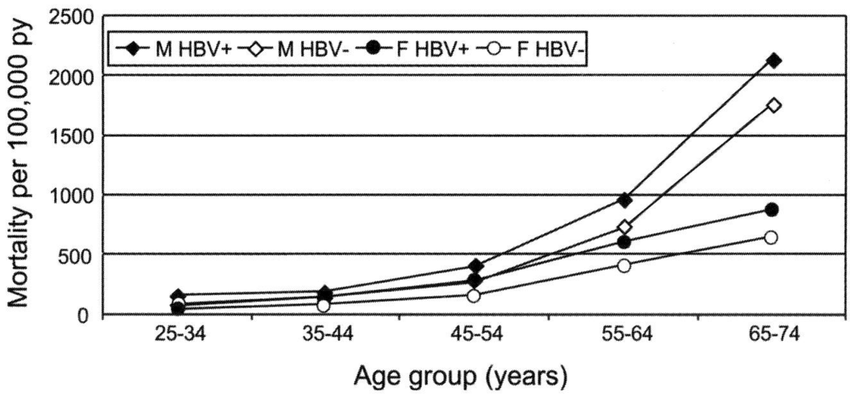

Figure 2. Mortality by age gender and HBV response in Haimen City, China. Males and females HBV positive or negative. (Chen 2004)

Males had a much higher mortality than females, but among the females, the carriers also had a significantly higher mortality. A successful vaccination program will increase the male to female ratio in the adult population because of the decrease in the differential mortality between adult males and females. Later, I will discuss the effects of the vaccination program on the male to female ratio at birth.

Causes of death were available for the Haimen City study and risk ratios for HBV carriers as compared to non-carriers were calculated for different categories of disease. As expected, the risk ratio for hepatocellular carcinoma in males (23) and females (26) was high and this was also true for chronic liver disease. But there were smaller, but significant, increases in risk for non-liver diseases in general and in non-liver cancers. If this is confirmed in subsequent studies, it suggests that HBV is implicated in diseases other than those ordinarily attributed to it and provides an approach to the study of the etiology of some of these diseases. If this is validated, the vaccination program could have an effect of survival in populations beyond that expected from the decrease only in deaths due to liver diseases.

HBV is the prototype of the *Hepadnaviridae* family examples of which are found in pre-human species. These include the woodchuck (groundhog), ducks, ground squirrel, tree squirrel, heron, and others. It would be fascinating to seek similar viruses in other species, including the phages of bacteria and archaea, to study their evolution in parallel with the evolution of their hosts. The selective effects of Hepadna viruses may also prevail in these and other species, or may have been a significant factor in the past.

4. Molecular Biology of HBV

HBV was the first human disease-causing virus that was sequenced and, despite the difficulty of growing the virus in experimental systems in vitro, its complex molecular biology has been well studied. In this section I will briefly describe aspects of the molecular structure in sufficient detail to understand how the vaccine was invented and developed. (See Web 1 in references.)

The genome of HBV is a spherical relaxed-circular, partially double-stranded DNA (rcDNA). The two strands are of unequal size; the long, full-length minus-strand is approximately 3200 nucleotides in length (one of the shortest genomes in a human virus) and has a protein (the viral polymerase) covalently bound to its 5' terminus. The shorter plus-strand can vary in length (1700–2800 nucleotides) depending on the size of the gap and the subtype of HBV. There are four open reading frames that code for seven (or more) proteins; there are in-frame start codons that permit the translation of the "extra" proteins.

The S reading frame produces the surface antigen, HBsAg made up of three envelope glycoproteins, preS1, preS2, and S. (The vaccine is made from one or more of these antigens). The core (C) reading frame produces the core antigen and the e-antigen (HBcAg, HBeAg). The P reading frame produces viral polymerase (P) that serves also as a reverse transcriptase for the integration of the viral DNA into the host DNA. The X reading frame produces the X protein (HBxAg). There are also viral promoters, enhancers and signal regions. The 5' terminus of both strands contains regions of 11 nucleotide repeats, DR1 and DR2, for priming the synthesis of the DNA strands.

The replication mechanism is reasonably well understood. It takes place mostly in the liver cells, although the virus is found in the pancreas and other tissues. As noted, it is a partially double stranded DNA virus. It is unique among DNA viruses in that it replicates its genome using an RNA intermediary controlled by a viral gene (P) that produces a reverse transcriptase/DNA polymerase. The process includes the integration of HBV viral nucleic acid sequences into the host genome that is seen in cells of HBV carriers and in patients with chronic liver disease and cancer of the liver. The replication process, viral assembly, viral entry, the immune system, the glycosylation of the surface antigens, and other processes are targets for therapies.

The whole virus is a circular particle of 42 nanometers. In addition there are small particles of about half the size and elongated particles of the same diameter but varying length that consist only of the surface antigen (HBsAg). The small particles are not infectious or pathogenic but are probably related to the defense mechanism of the virus to aloof it to evade the immune response of their host (Figure 3).

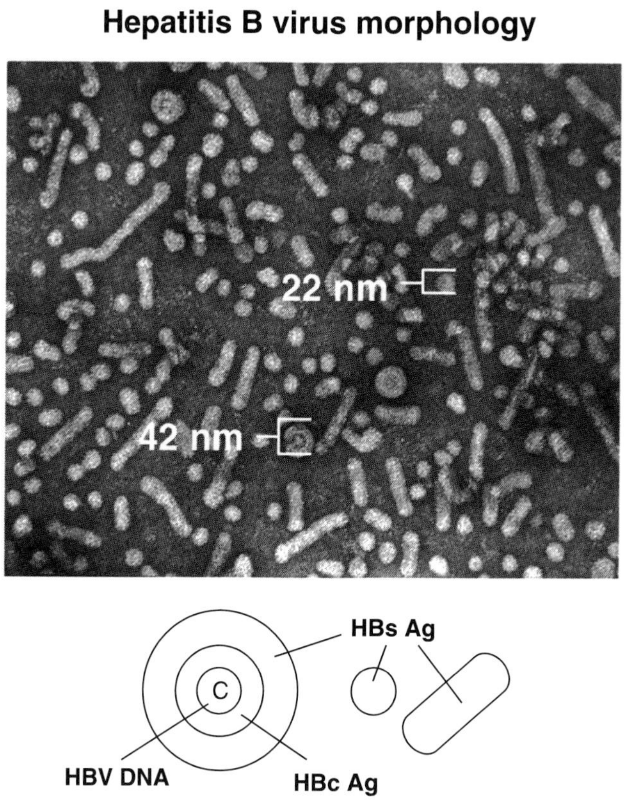

Figure 3. Electron microscope image of hepatitis B virus showing the large (42 nm) whole virus particle containing the surface (HBsAg) and core (HBcAg) antigens and the viral DNA, and the smaller particles consisting only of HBsAg. The smaller particles are either circular (22nm) or elongated rods (22nm diameter and varying length).

5. The Genetics of HBV

HBV was discovered during the course of research on serum protein polymorphisms (Blumberg 1977, 2002). Family genetic studies of the "Australia antigen", that is, the carrier state for HBV identified by its surface antigen HBsAg, were done even before we were aware that we had discovered a virus. (It was named Australia antigen because the earliest studies were done on blood from an Australian.) A summary of the family data analyses done on families in Cebu, The Philippines, and Bougainville, Solomon Islands (Blumberg1969) were consistent with the segregation of an autosomal recessive allele designated Au^{1}. When present in double dose it increased susceptibility to becoming a carrier of HBV (HBsAg positive) when the individual was exposed to the virus. Later extensive investigations identified multiple human polymorphisms in which alleles segregating at the polymorphic locus increased susceptibility to becoming a carrier of HBV (Blumberg, 1998, 2002). For a general review of inherited susceptibility see Hill (2001). Many of these identifications were based on population studies. Alleles at the polymorphic locus that were significantly more common in HBV carriers than in non-carriers were inferred to increase susceptibility to the carrier state. (There are recent additional studies on the genetics of inherited susceptibility to chronic infection with HBV (Frodsham 2005). Hill (2001) has provided a review of infectious disease susceptibility.

Often, alleles at the same susceptibility locus are related to response to several infections. I arranged these in tables using published data available in the late 1990s (Table 2).

TABLE 2. Microorganism gene affinity clusters (see text).

Locus	Allele	Agent/Disease	Chrom	Function
MHC Class II	DRB1*1302	HBV chronicity	Chrom 6	Immune response
		HBV response to interferon Malaria, cerebral		
	Pappiloma virus			
VDR	tt	HBV chronicity	Chrom 12	Vitamin D receptor Binds Vit. D 1,25D3
	tt	TB Pulmonary		
	TT	Leprosy, tuberculoid		
	Tt tt tt	Leprosy, infection Leprosy, lepromatous Bone mineral density, Osteoporosis, Prostatic		
TNF	G-308	HBV chronicity	Chrom 6	Tumor Necrosis Factor
		Malaria, cerebral Microcutaneous leish. Lepromatous leprosy Meningococcal Meningitis Trachoma Asthma		

Table 2. (cont.)

MBP	Codon 52	HBV chronicity	Chrom 10	Mannose Binding Protein
		SLE, HIV (?) Infections in childhood		
SM-1 5q31-33		Schistosomiasis	Chrom 5	Schist. Susceptibility
		? HBV chronicity		
IL-10 Promoter	Mutat. -108	HBV chronicity	Chrom 1	Immune response
	059	HBV chronicity		

For example, individuals homozygous for the *t* allele at the vitamin D receptor locus (VDR) are more likely to become carriers of HBV and they also are more susceptible to pulmonary tuberculosis. Those who are homozygous for the alternate *T* allele, when exposed to the leprosy bacillus are more likely to develop the lepromatous form, while those homozygous for *t* are more likely to develop the tuberculoid form of the disease. Another example was the Tumor Necrosis Factor (TNF) locus. TNF is a cytokine (a protein or glycoprotein involved in the regulation of cellular proliferation and function) which has many roles including the control of inflammation and the stimulation of the proliferation and the destruction of cancer cells. It also increases cachexia in cancer patients, a strange and not fully understood phenomenon which results in wasting and weight loss. There are several polymorphic sites on the TNF gene. They are related to susceptibility to HBV chronicity, cerebral malaria, lepromatous leprosy, and a form of the tropical disease caused by *Leishmania braziliensis.*

There are interesting conceptual consequences of this grouping. These infectious agents are related to each other because they are related to the same susceptibility locus or loci. The infectious agents with affinities to the same locus constitute a *Microorganism Gene Affinity Cluster* (MIGAC). (I initially used the term *Pathogen Gene Affinity Cluster* (Blumberg 1998). *Microorganism* is more appropriate since some infectious agents are not ordinarily pathogenic.) The cluster provides another method of classifying microorganisms. They can vary from population to population since the susceptibility effects of the polymorphic alleles may be dependent on their environmental and genetic context. The infectious agents and diseases they may cause are likely to co-occur and the presence of one can provide clinical aides to seek out others in the cluster. In respect to the evolutionary effects of HBV, the subject of this Chapter, an understanding of the clusters gives insights into the complex interactions of specific infectious agents and their co-evolution.

6. The Vaccine

The peripheral blood of hosts infected with HBV contains the whole virus particle with the HBsAg outer coat, core antigen and the DNA. It also contains the smaller circular and elongated particles consisting only of HBsAg (Figure 3). In 1969, my colleague Irving Millman and I devised a vaccine made from the peripheral blood of HBV carriers. The large, infectious, and pathogenic virus particles were removed and the remaining small particles consisting only of HBsAg, constituted the active component of the vaccine. This was a unique method of producing a vaccine that had not been used previously. The vaccine was tested in extensive field trials and found to be safe and highly effective; it has been given to millions of people. Subsequently, a recombinant method for production of the vaccine was made by several methods. This was the first widely used recombinant vaccine and has largely replaced the blood-derived vaccine.

7. The Vaccination Program

The current status of the worldwide childhood vaccination program was summarized by the US Center for Disease Control (Anonymous 2003). As of May 2003, 151 (79%) of 192 national members of the WHO had universal childhood vaccination programs. There are 89 member states that have been designated as having a high prevalence of HBV carriers. Sixty four (72%) have universal infant vaccination programs. It is the goal of the WHO to have vaccination programs in all countries by 2007.

The worldwide vaccination programs have resulted in striking decreases in HBV infection. Several examples will be used to illustrate these results. In the United States there was a dramatic drop in the incidence of acute hepatitis B cases. To quote (Anonymous 2004):

> During 1990-2002, a total of 13,829 cases of acute hepatitis B were reported in the United States among persons aged ≤19 years. The incidence of reported cases declined steadily during this period, from 3.03 per 100,000 population in 1990 to 0.34 in 2002, representing a decline of 89%. … Incidence declined 94% among children aged 0-4 years, 92% among children aged 5-9 years, 93% among those aged 10-14 years, and 87% among adolescents aged 15-19 years."

A universal vaccination program with a very high compliance rate was conducted among Native American populations in Alaska between 1981 and 1983. The number of acute hepatitis cases dropped from 215/100,000 before the vaccination program to 7–14 cases /100,000 in 1993 after the program was in place. By 1995, no cases were reported (McMahon 1996).

The efficacy of the program can be seen from a summary of reports from areas of high prevalence for HBV carriers. Infant vaccination programs in regions of the Peoples Republic of China have been in place since 1986. The prevalence of seropositivity for HBsAg (carrier status) decreased from 16.3% (879/5397) in historical controls to 1.43% in a cross-sectional regional study (Sun 2002). An extensive infant vaccination programs was started in 1986 in The Gambia, West Africa. By the age of four years the efficacy of the vaccine was 84% for any serological indication of infection, and 94% for the development of the carrier state. A few years later the prevalence of carriers in the young age group had dropped from 10.0% in the pre-vaccination period to 0.6% (Vivani 1996, 1999). Italy had a high frequency of HBV carriers and it was among the first of the European countries with regional and universal vaccination programs. A typical study was in the community of Afragola where the prevalence of carriers in males up to age 12 was 10.5% before the vaccination programs started in 1983. By 1993, the prevalence was 0.8% (DaVilla 1992, 1995).

The vaccination program has also decreased the incidence of primary cancer of the liver. Worldwide, infection with HBV is the major etiological agent for primary cancer of the liver (hepatocellular carcinoma, HCC) – the 3rd most common cause of death from cancer in males and the 7th most common cause in females. In a national study in Taiwan, the average annual incidence in children 6–9 years fell from 0.52/100,000 before the vaccination program to 0.13/100,000 afterwards (Chang 1997). In a three year follow-up study in Korea the unvaccinated population had 18.1 cases of HCC per 100,000 population while those who had been vaccinated or had developed anti-HBs naturally had 0.58/100.000 and 0.34/100.000 cases respectively. Both of these were significant and impressive reductions. The HBV vaccine is the first cancer preventive vaccine and, one can hope will be the exemplar of others to come.

The success, so far, of the vaccination program makes it likely that the effect of HBV as a selective factor will be decreased and, conceivably eradicated, a possibility that has been under considered since at least the 1980s (Blumberg 1990). What effect will this have on human demography and microevolution? I will come back to this question after the discussion of the non-pathological effects of HBV.

8. HBV and Gender

HBV interacts differently with males and females. The earliest population studies showed that, in general, there were more male than female carriers (Blumberg 1966a, 1966B) and this is the case in most, but not all populations that have been sampled since that time. Patients in renal dialysis units had a very high likelihood of HBV infection before the introduction of control measures. When exposed over time, males were more likely to become carriers than females while females were more likely to develop anti-HBs. The ratio of males to females is much higher in patients with chronic liver disease and with primary cancer of the liver due to HBV. These disease differentials will have an impact on sex ratio and demography in the older age groups since the morbidity and mortality of these chronic infections is greatest later in life.

But HBV also has an effect on gender at the earliest age. Because of the interest in the genetics of the carrier state (see above) we conducted many studies on the distribution of HBV carriers in families and in maternal effects on newborns. In a study in a Greek population with a very high prevalence of HBV (Table 3), we found a relation between the response of the parents to infection with HBV and the gender of their offspring (Hesser 1975, Drew 1978, Blumberg 1979).

TABLE 3. Plati, Greece. Number of male and female live births according to the responses to HBV of parents

Parent's response To HBV	Couples (No.)	Live Births		Sex ratio
		Males	Females	
Either parent HBsAg + anti-HBs –	33	60 (1.8+_0.2)	24 (0.7±0.1)	250 (161,429)*
Both parents HBsAg - : anti-HBs –	29	51 (1.8±0.2)	35 (1.2±0.2)	146 (96,230)*
Both parents HBsAg - : either parent anti-HBs +	154	24 (1.6±0.1)	22 (1.4±0.1)	109 (91,131)*

* In parentheses, the 5 percent confidence limits.
Blumberg, B.S.

If a parent was a carrier and the spouse did not have anti-HBs, then the male to female ratio at birth was significantly higher than in the families where neither parent was a carrier but one or the other had anti-HBs. The male/female ratio was intermediate in families in which the parents were not infected.

This was a very surprising finding with the implication that hepatitis B infection could influence sex ratio at birth, a very important variable for the demography and evolution of a species. The study was repeated in other populations; France (Cazal 1976), another Greek population designated Greece 2 (Livadas 1979), Papua New Guinea (Drew 1982), the Philippines (London 1982), and East Greenland (Drew 1986); the results were consistent in these very different environments Table 4.

TABLE 4. Summary of studies of HBV and sex ratio in families of six populations From Hesser (1975), Cazal (1976), Livadas (1979), Blumberg (1979), Drew (1978, 1982, 1986), Hesser (1975), London (1982), Oster (2004).

Location	*Parents carriers status*	*Sons*	*Daughters*	*Sex Ratio*
Greenland	Positive	64	60	1.07
Greenland	Negative	174	194	0.90
Kar Kar Island	Positive	63	54	1.17
Kar Kar Island	Negative	163	206	0.79
Greece 1	Positive	85	46	1.85
Greece 1	Negative	287	255	1.13
Philippines	Positive	66	41	1.61
Philippines	Negative	304	301	1.01
Greece 2	Positive	52	30	1.73
Greece 2	Negative	1006	955	1.05
France	Positive	20	12	1.66
France	Negative	149	122	1.22

In some of the studies, the increased male to female ratio was a consequence of a decreased number of female births and there was decreased fertility in the families with carrier parents. (Drew 1978, Livadas 1979, London 1982) (Figure 4).

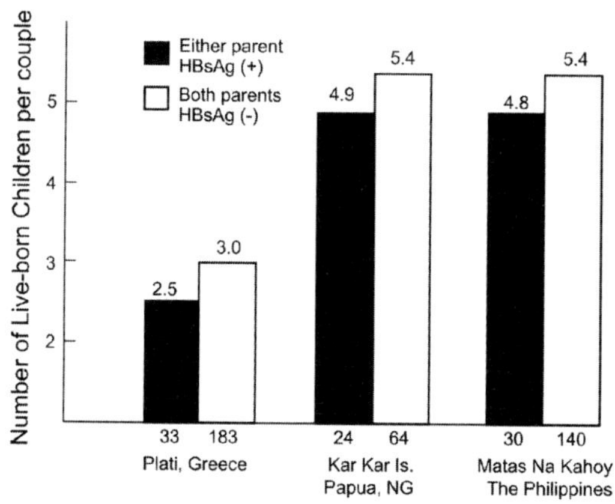

Figure 4. Relation of fertility of parents to their responses to HBV in 3 defined populations. Each bar represents the mean number of live-born children per couple according to whether either parent was HBsAg-positive or both parents were HBsAg-negative. The numbers below the bars are the numbers of couples in each category. The numbers above the bars are the mean number of children per couple. (London 1982)

There were several predictions from the family observations that could be tested in populations. In general, there should be a positive correlation between the prevalence of hepatitis B in a population and the population's sex ratio. And, if there were a causal connection, there would be a decrease in sex ratio over time in populations with effective HBV vaccination programs. Emily Oster (Oster 2004), a graduate student in Economics at Harvard University, showed that there was a strong correlation using national data for HBV prevalence and sex ratio (Figure 5).

Next, she determined if there had been a decrease in sex ratio in two regions, Alaska and Taiwan, with a documented and dramatic decrease in HBV prevalence as a consequence of the vaccination programs. Prior to the vaccination program there had been a much higher prevalence of HBV carriers (HBsAg positive) among Native Americans (Indians and Inuits (Eskimos)) than among Alaskans of European origin. Among Native Americans, the prevalence's were higher in the Inuits than the Indians. She looked at sex ratio in the communities whose predominant population belonged to one of the three groups, Native American with high HBV (Inuits), Native Americans with low HBV, and non-Native Americans (Figure 6).

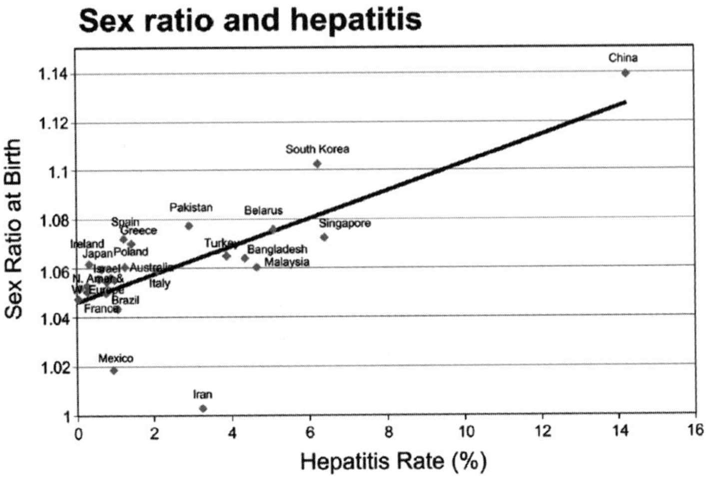

Figure 5. Sex ratio and HBV prevalence, by country. The HBV prevalence is calculated by aggregating published studies. The sex ratio is the number of boys born divided by the number of girls. The figure is taken from Oster 2004.

Notes: Sex ratio is number of boys for each girl. Only countries with more than 15,000 people used to caclulate HBV pravalence are included.

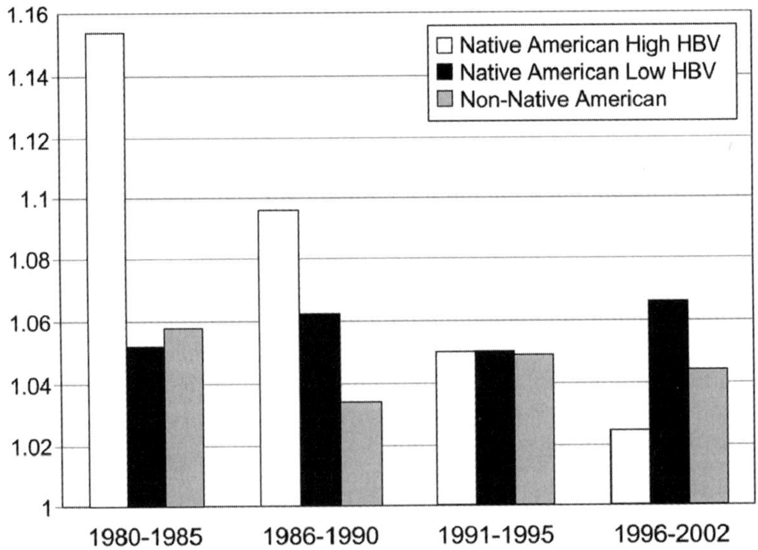

Figure 6. Changes in sex ratio in Alaska before and during vaccination program (Oster 2004).

Over the course of the vaccination program, the sex ratio had decreased in the high HBV Native Americans, but had remained the same in the other groups, as predicted.

She did a similar analysis on Taiwan an area that had a historically high prevalence of HBV that decreased dramatically during the vaccination program. In a preliminary analysis the results were similar to those in Alaska; the sex ratio had decreased during the vaccination program and presumably as a consequence of it. Similar findings in two very different populations and environments adds weight to the argument that HBV infection has (or had) a role in human gender selection.

These observations starting in the 1970s and with strong confirmation from the Oster studies encourage additional research to see if these conclusions continue to be supported. If so, they will have a profound effect on population studies and planning, particularly in Asia and Africa where HBV has been so common and is decreasing in prevalence as the vaccination programs continue. It could help to explain the "Missing Woman" problem in China (Oster 2004). It has been noted that the sex ratios in China were much higher than in Europe or the USA and suggested that this was due to bias against female births in a population where, traditionally, males are favored. Oster estimated that the HBV effect on sex ratio could explain 45% of the apparent female loss worldwide and more than 75% in China. An understanding of the effect of HBV on gender ratio could explain much of this apparent "loss" of female children.

Sex ratios have a profound effect on population size and composition with resultant evolutionary, economic, social and psychological consequences. Changes in the sex ratio, as appears to be the case in the regions with successful HBV vaccination programs, will affect these factors in a manner that is difficult to predict.

We have suggested a molecular model that is currently being tested. It postulates a region on the Y- chromosome that binds or affects the HBV genome in some manner to decrease replication. Similar regions would be less common in the female genome; the virus would be more likely to replicate and damage or kill the conceptus *in utero* and families with a carrier parent would have fewer girls, a smaller family size, and a lowered sex ratio. This could occur at an early stage of gestation and be undetected by the pregnant woman. The male conceptus of an infected mother would be more likely to survive, as would the non-replicating but retained virus, and the male would have a greater likelihood of becoming a carrier after birth. This model presumes that the virus of its protein products could have an effect *in utero*, for which there is, at present, no evidence.

These inferred biological effects on sex ratio may interact with human behavior. Consider family size planning in a community where HBV prevalence is high and where male children are preferred. If a family desires a certain number of boys then they will start having children and stop doing so when they have the requisite number of boys. If one of the parents is a carrier, then they will be more likely to have boys and procreation will be terminated relatively soon. If either of the parents is not a carrier, then they would be more likely to have girls and will continue to have children until they achieve the target number of boys. The non-carrier families will end up with more girls, a larger family, and a lower sex ratio.

The maintenance of the human sex dimorphism is a complex process. HBV infection appears to be part of this evolutionary dance; it increases the ratio of males to females at birth, but decreases the ratio at older ages when more males die from HBV infection than females. It is unlikely that HBV would be the only infectious agent that has an effect on sex ratio and it would be interesting to search for others. I briefly described the interrelations of viral and other agents that relate to the same susceptibility polymorphisms and are part of the same Microorganism Gene Affinity Cluster. This could lead to complex interactions in individuals who are simultaneously infected with several infectious agents whose intricacies can only be imagined at present. These effects are unlikely to be confined to humans; the presence of hepadna viruses similar to HBV in other species suggest that this phenomenon may occur in other species, perhaps including those deep in the ancestral chain.

9. Do Carriers have Advantages? HBV and Increased Iron Binding

An implication of polymorphisms is that there are advantages to one or more of the alternate phenotypes. Does the HBV carrier state have advantages? This is not a question that is often asked about viruses whose activities are always assumed to be unfortunate. Since our initial interest in HBV arouse from a study of polymorphism we did look for positive characteristics.

Patients with hepatitis have increased iron levels. Kolk-Vegter (1971) published a report on 11 patients in a renal dialysis unit who had liver disease without jaundice but with abnormal liver chemistries. All had a spontaneous rise in their hemoglobin level and they required fewer transfusion than did 34 other dialysis patients who did not have liver disease. We studied a group of patients with Down's Syndrome who have a high prevalence of HBV carriers. Institutionalized patients with HBsAg (carriers) were compared to patients who were not carriers, and to other patients in the same institution with diagnosis other than Downs's Syndrome, who also were not carriers (Sutnick 1974). The carriers had higher hematocrit, hemoglobin, and serum iron levels, and lower total iron binding capacity (a serum measurement that goes down when there are increased levels of iron). Several other studies confirmed these observations in different populations.

There is insufficient iron in the diet in many regions including those with a high prevalence of HBV. HBV carriers would retain larger amounts of iron than non-carriers and they would be able to produce more hemoglobin and other iron containing essential proteins. This would protect them against anemia, and the effects of some infectious agents. For example, hookworm (*Ancylostoma duodenale* and *Necator americanus*) infestation causes profound anemia that is debilitating and can be deadly. Presumably, carriers would be less affected by hookworm infestations because of their increased iron levels. There are also interactions between HBV and other agents with which the host is co-infected; there is reduced Hepatitis C virus replication when HBV also infects the host.

Advantageous genes, probably introduced by retroviruses, are known in bacteria and in humans. When the genome of *Bacillus subtilis* was completely sequenced, the investigators identified a gene transfer from a bacteriophage that helped it to resist heavy metals (Williams 1997). In humans, there is a sequence derived from viral genes that regulates the amylase gene cluster that allows the production of the enzyme in the saliva; amylase permits the digestion of digest starchy foods an important evolutionary advantage. Several other primates share this capability (Coffin 1997).

There are undoubtedly other advantages to the carrier state, but viruses are rarely studied to determine positive effects. Most of the morbidity and mortality due to HBV infection occurs late in life. In pre-modern times before the introduction of public health measures and the wide availability of health care, most carriers would have died from other causes before they developed life-shortening chronic liver disease and cancer of the liver. In earlier days, the advantages may have outbalanced the disadvantages leading to a large number of carriers. Because of frequent mother-child transmission, the high prevalence persists to this day, probably in populations where the chronic carrier state or the genes increasing susceptibility to chronicity were most advantageous. The vaccination programs appear to be reversing this evolutionary vector.

10. Co-Evolution of HBV and its Human Host

The evolutions of the infected human host and of HBV are intertwined. A "goal" of HBV – if an entity that barely makes it into the realm of a living organism and has only four coding regions can be said to have a goal – is to maintain a presence in the population in a form, primarily the carrier state, that allows for transmission from host to host and from generation to generation. Immune and other mechanisms have evolved that allow the virus to persist in the host for decades, often from nearly the moment of birth until the 5th, 6th, and later decades. Some of these mechanisms are ingenious. As noted, the virus replicating in the host cells produces large numbers of viral particles containing only the surface antigen of the virus (HBsAg). These interact with antibodies produced by the host (anti-HBs) and protect the whole virus from elimination. The C reading frame has two start codons that produce a large protein, HBcAg, that remains in the cell, and a shorter one, HBeAg, that shares many amino acids and antigenic specificities with HBcAg but can transit the cell wall and enter the blood stream to engage the host immune system. It has been conjectured that HBeAg in the circulation would make the host tolerant to HBeAg and to HBcAg, with which its shares antigenic segments. The HBcAg still present in the cell is presumably expressed on its surface. The host immune system would then, unwisely, consider the infected cell expressing HBcAg as "self" and the cell would be preserved to continue the replication and spread of the virus.

HBV has evolved to take advantage of human activities that are absolutely necessary for the perpetuation of the species; the close relation between mothers and their children during pregnancy and childbirth and later when the child is totally dependent on its mother for care and nutrition, venereal activity and sexual intercourse, sib-sib interaction and the intimacies of family and community life. One of the most common forms of HBV transmission, particularly in the regions where HBV carriers are most common, is from a carrier mother, especially from carriers with active replication to the conceptus *in utero*, at the moment of birth, or soon after when the mother and child are very close. HBV is one of the most common STDs (sexually transmitted disease) and is associated with other STDs such as gonorrhea, syphilis, and AIDS. In addition to the high risk of transmission from a carrier mother to her children, carrier sibs are highly likely to transmit the virus to uninfected sibs who often become carriers. In some societies carrier fathers are less likely to transmit the virus to their children, presumably because they interact with them less frequently than other family members. Virus transmission is also enabled by certain social practices, particularly in pre-modern societies, by group tattooing, ritual circumcision with improperly sterilized cutting instruments. There are other practices in which blood is transmitted from one person to another, for example injection of recreational or other drugs using non-sterile needles and instruments. The vaccination programs are reckoned to greatly decrease risk from all of these methods of transmission.

As noted already, males are more likely to become carriers then females and chronic HBV infection in parents favors the birth of boys rather than girls. In order to maximize the prevalence of carriers in a population, an exquisite balancing of the male/female dimorphism is required. An increased number of males increase the potential number of carriers. But, major sources of carriers are carrier mothers, and females are involved in the venereal transmission of virus. The virus uses these essential human behaviors to insure its continued transmission and maintenance in the population. It is interesting to conjecture that HBV might actually influence these behaviors as is the case with other host/parasite/saprophyte systems.

11. Viruses and the Origin of Life

In the preceding sections, I have discussed how HBV's interaction with its host could affect microevolution. If HBV affects contemporary microevolution, what role did it have in the earliest periods of the evolution of life? It is thought that life emerged on Earth between 3.85 and 4.20 billion years before present and complex microbes about 3.6 billion years ago. The date of the oldest fossil is about 3.5 billion years ago. (The Earth formed 4.56 billion years ago.) In this section, I will leap billions of years back to the time that life originated to conjecture on viruses in early times.

Edwin Kilbourne, the distinguished virologist, introduced a fascinating seminar, "Virus Paleontology, Disease, and Evolution", held at the American Philosophical Society on November 13, 2003, with a poem said to be the shortest in the English language (Kilbourne 2004).

On the Antiquity of Microbes
Adam
Had'em

Microbes, including viruses, have interacted with human hosts for a very long time, and for millions, and possibly billions, of years before that with other species that preceded humans on Earth. How far back in time did viruses exist? Can they have been one of the early forms of life or even the earliest form? Currently, viruses require a cell in order to replicate, but this may not have been so at an earlier time. Were primitive RNA viruses the format for the RNA world hypothesized to precede the current DNA life forms? Large numbers of retroviral sequences (including those of HBV) are found in the human genome. These are often the "fossil" record of viruses that were incorporated into the germ line in previous generations. They can be traced to precedent species to estimate when the infection occurred in evolutionary history. It is estimated that 8% of the modern human genome is made of retroviral sequences, a larger number than the coding sequences, and they are often highly conserved.

Astrobiologists conduct research on bacteria, archaea, viruses and other organisms that live in extreme locations that are similar to the environment of early Earth and, possibly, early Mars or other locations in space. These include locations at geothermal sites, deep under the sea and under the sea floor, in deep mines and boreholes, at low and high pH, in hyersaline liquids, in and under the ice, in very hot and very cold locations, in space, and elsewhere. The study of these viruses can provide clues as to their possible role at the time of life's origins and soon thereafter.

There is an intriguing question concerning the origin of life that could be approached by the study of viruses. If life originated once, could it have originated many times and could it be originating now? It is difficult to know how new origins could be detected but one could test the hypothesis that new extreme locations – for example, recently formed geothermal sites – might contain a different and "earlier" evolving virus and/or other organisms or pre-biotic chemicals. An argument against new origins is that existing forms would "eat" them; but evolutionary theory suggests that new forms occasionally arise that outcompete existing organisms and a balanced polymorphism could develop. Many extremophiles have gene sequences with little or no homology with previously sequenced organisms. Is this an indication of novelty?

It will be intriguing to see if HBV sequences are incorporated in these extreme viruses indicating that its "ancestors" existed at the very origins of earthly life and that descendents persist until today and continue to have a role in altering the genome.

12. Conclusions

In this chapter I have summarized the possible role of HBV as a selective factor in human populations and the effects that it may have on human evolution. There are profound pathological effects, but also non-pathological phenomenon, including the effect on sex ratio at birth. The insertion of HBV viral sequences into the genome of millions of people over many generations has directly affected the human gene pool. What effect has this had on the ongoing variation in human populations and on the selection process? The widespread application of the vaccination program will, in a few generations, alter these effects.

Viruses are relatively simple life forms. Were they present at the very origins of life? If so, were the predecessors of HBV or some portions of its current genome one of the players at this remote time?

13. References

Anonymous, (2003). Global progress toward universal childhood hepatitis B vaccination, MMWR, September 12, 2003/52**(36)**: 868-870.

Anonymous, (2004). Acute Hepatitis B among Children and Adolescents - United States, 1990-2002, *JAMA.*; **292**: 2967-2968.

Blumberg, B. S. and Melartin, L. (1966). Conjectures on inherited susceptibility to lepromatous leprosy. Internat. J. Leprosy, **34**: 60-64.

Blumberg, B. S., Melartin, L., Guinto, R. A. and Werner, B. (1966). Family studies of a human serum isoantigen system (Australia antigen). Am. J. Human Genet. **18**: 594-608.

Blumberg, B. S., Gerstley, B. J. S., Hungerford, D. A., London, W. T., and Sutnick, A. I. (1967). A serum antigen (Australia antigen) in Down's Syndrome leukemia and hepatitis. Ann. Int. Med. **66**: 924-931.

Blumberg, B. S., Friedlaender, J. S., Woodside, A., Sutnick, A. I., and London, W. T. (1969). Hepatitis and Australia antigen: Autosomal recessive inheritance of susceptibility to infection in humans. Proc. Nat. Acad. Sci. **62**: 1108-1115.

Blumberg, B. S. (1977). Australia antigen and the biology of hepatitis B. Science **197**: 17-25.

Blumberg, B. S. (1979). I. History. II. Parental responses to HBV infection and the secondary sex ratio of the offspring. Sex Differences in Response to Hepatitis B Virus. Arthritis and Rheumatism **22**: 1261-1266.

Blumberg, B. S. (1990). Proc. International Conference on Prospects for Eradication of Hepatitis B Virus. Vaccine **8** Introduction pS5; Conclusion pS139.

Blumberg, B. S. (1995). Complex interactions of hepatitis B virus with its host and environment. J. Royal College Physicians London **29**: 31-40.

Blumberg, B. S. (1998). Cancer prevention with a vaccine: hepatitis B and liver cancer, American Association for the Advancement of Science, Annual meeting, Philadelphia, PA

Blumberg, B. S. (2002). Hepatitis B. The Hunt for a Killer Virus. Princeton University Press, Princeton.

Cazal, P., Lemiare, J. M. and Robinet-Levy, M. (1976). Hepatitis B et Rapport du Masculinite. Revue franaise de transfusion et immunohmatologie **19**: 577-581.

Chang, M. H., Chen, C. J., Lai M. S., et al. (1997). Universal hepatitis B vaccination in Taiwan and the incidence of hepatocellular carcinoma in children. N. Engl. J. Med.. **336**: 1855-9.

Chen G., Lin, W., Shen, F, Iloeje, U. H., London, W. T., and Evans, A. A. (2004). Chronic hepatitis B virus infection and mortality from non-liver causes: results from the Haimen City cohort study, Inter. J. Epidemiology. In press, China.

Coffin, J. M., Hughes. S. H., and Varmus, H. E. (1997). eds., Retroviruses, Cold Spring Harbor Laboratory Press,. p 403.

Da Villa, G., Piazza, M., Iorio, R., Picciotto, L., Peluso, P., De Luca, G., and Basile, B. (1992). A pilot model of vaccination against hepatitis B virus suitable for mass vaccination campaigns in hyperendemic areas. J. Med. Virol. **36**: 274-278.

Da Villa, G., Picciottoc, L., Elia, S., Peluso, F., Montanaro, F., and Maisto, T. (1995). Hepatitis B vaccination: universal vaccination of newborn babies and children at 12 years of age versus high risk groups. A comparison in the field. Vaccine **13**: 1240-1243.

Drew, J. S., London, W. T, Lustbader, E. D., Hesser, J. E. and Blumberg, B. S. (1978). Hepatitis B virus and sex ratio of offspring. Science **201**: 687-692.

Drew, J., London, W. T., Blumberg, B. S. and Serjeantson, S. (l982). Hepatitis B virus and sex ratio on Kar Kar Island. Hum. Biol. **54**: 123-135.

Drew, J. S., Blumberg, B. S., and Robert-Lamblin, J. (1986). Hepatitis B virus and sex ratio of offspring in East Greenland. Human Biology **58**: 115-120.

Frodsham, A. J., Zhang L., Dumpis, U., Azizah, N, Taib, N. A. M., Best, S, Durham, A. , Hennig, B. J. W., Hellier, S., Knapp, S., Wright, M., Chiaramonte, M, , Bell, J. I., Graves, M., Whittle, H., Thomas, H. C., Thursz, M. R., and Hill, A. V. S., (2005) A Class II Cytokine Receptor Gene Cluster is a Major Locus for Hepatitis B Persistence, in preparation.

Hesser, J. E., Economidou, J. and Blumberg, B. S. (1975). Hepatitis B surface antigen (Australia antigen) in parents and sex ratio of offspring in a Greek population. Human Biol. **47**: 415-425.

Hill, A. V. S., (2001) The genomics and genetics of human infectious disease susceptibility, Annu. Rev. Genomics Hum. Genet, **2**: 373-400.

Kilbourne, E. D. (2004). Virus paleontology, disease, and evolution: an introduction. Proc. Amer. Philosophical Society **148**: 261 and following articles.

Kolk-Vegter, A. J., Bosch, E., van Leeuwen, A. M. (1971). Influence of serum hepatitis on haemoglobin level in patients on regular hemodialysis. Lancet **1**: 526-528.

Livadas, D., Economidou, J., Hesser, J.E., Koutras, D.A., and Hadziyannis, S. (1979) Fertility and sex ratio of offspring of female HBsAg carriers. Journal of the Royal Society of Medicine **72**: 509-512.

London, W. T., Stevens, R. G., Shofer, F. S., Drew, J. S., Brunhofer, J. E., and Blumberg, B. S. (1982). Effects of hepatitis B virus on the mortality, fertility, and sex ratio of human populations. In: W. Szmuness, H. J. Alter and J. E. Maynard (eds.) *Viral Hepatitis.*Franklin Institute Press, Philadelphia p 195-202.

Luria, S. E., (1959) in, Virus Growth and Variation, A. Isaacs and B.W. Lacey, eds., Cambridge University Press, p 1-10

McMahon, B., Mandsager, R., Wainwright, K., Wainwright, R., Williams, J., Singleton, R. (1996). The Alaska native hepatitis B control program. *Proc. IX Triennial International Symposium on Viral Hepatitis and Liver Disease,* April 1996, Rome, Italy, abstract 74.

Oster, E. (2004). Hepatitis B and the Case of the Missing Women, Seminar, Fox Chase Cancer Center, Philadelphia, Pa. USA

Sun, Z., Ming L., Zhu X., Lu, J. (2002). Prevention and control of hepatitis B in China. J. Med. Virol. **67**: 447-450.

Sutnick, A. I., Blumberg, B. S. and Lustbader, E. D. (1974). Elevated serum iron levels and persistent Australia antigen (HBsAg). Ann. Internal Med. **81**: 855-856.

Viviani, S., Jack A., Hall. A.J., (1999). Hepatitis B vaccination in infancy in the Gambia: protection against carriage at 9 years of age. Vaccine **17**: 2946-2950.

Viviani, S., Jack, A. D., Maine, N., Montesano, R., Hall, A. J., Mendy, M., and Whittle, H. (1996), The Gambia hepatitis intervention study. *Proc. IX Triennial International Symposium on Viral Hepatitis and Liver Disease,* April 1996, Rome, Italy, abstract C264.

Web 1. http://hbvrp.wits.ac.za/websitetut2.htm

Williams, N. (1997). Gram-Positive Bacterium Sequenced, Science **277**: 478.

Biodata of **Anna M. Carnerup** author (with co-authors – see below) of ***"The Record of Early Life: In search of biosignatures"***

Mrs. Anna M. Carnerup is currently a PhD student in the department of Applied Mathematics at the Australian National University in Canberra, Australia. Her dissertation topic is concerned with the abiotic inorganic precipitation of silica-carbonate aggregates, so called 'biomorphs', that display a wide range of life-like morphologies.
E-mail: **anna.carnerup@anu.edu.au**

Dr Andrew G. Christy is currently a Research Officer and Laboratory Manager in the Department of Earth and Marine Sciences, Australian National University, Canberra, Australia. He obtained his Ph.D. from the University of Cambridge in 1988 and undertook postdoctoral work in the Earth Sciences Department there, the Chemistry Department of the University of Leicester, UK, and the Synchrotron Radiation Research Division, Daresbury Laboratory, UK, before moving to Australia to take up research fellowships at the Research School of Chemistry (Disordered Materials group) of the Australian National University and then the Applied Mathematics Department of the Research School of Physical Science and Engineering, Australian National University, Canberra, Australia. Dr. Christy's principal scientific interests are in the areas of: mineralogy, petrology, crystallography, solid and aqueous geochemistry including adsorption equilibria, crystal chemistry including pressure-temperature systematics, characterisation of complex and aperiodic structures, astrobiology and geomicrobiology.
E-mail: **andyc@ems.anu.edu.au**

Professor Stephen T. Hyde is a member of the Department of Applied Mathematics, Research School of Physics and Engineering at the Australian National University in Canberra, Australia. He obtained his PhD in Physics from Monash University Melbourne, Australia, in 1986. He is interested in using concepts of modern geometry to understand morphology and structure in natural systems, from the atomic scale upward.
E-mail: **stephen.hyde@anu.edu.au**

Professor Juan Manuel García-Ruiz belongs to the Spanish Consejo Superior de Investigaciones Cientificas (CSIC) in Granada, Spain, where he founded and is director of the Laboratory for Crystallographic Studies (http://lec.ugr.es). He obtained his PhD from the University Complutense in Madrid, Spain, in 1980. His research interest focuses on pattern formation in precipitation systems under

J. Seckbach (ed.), Life as We Know It, 237-258.

diffusive environments, such as those provided by gels, microfluidics and microgravity. He applies the results of his fundamental research to problems in the frontier between geology and biology, to crystallisation techniques and industrial crystallisation, to the understanding of biomineralisation and production of biomimetic materials.

E-mail: **jmgruiz@ugr.es**

Dr. Ann-Kristin Larsson is currently at the Department of Applied Mathematics, Research School of Physics and Engineering at the Australian National University in Canberra, Australia, while on leave from a position as Associate Professor in Structural Chemistry/Electron Microscopy at Stockholm University, Sweden. She obtained her Ph.D. from the University of Lund, Sweden, in 1995 in Inorganic Chemistry/Solid State Chemistry followed by a postdoc on Electron Diffraction Solid State Chemistry at Australian National University, Canberra, Australia. Her scientific interests include electron microscopy and crystallography, in particular analyses of modulated, short-range ordered, domain and twinned structures, i.e. analysis of order and disorder at the atomic and nanoscale level.

E-mail: **akl110@rsphysse.anu.edu.au**

Anna M. Carnerup

Andrew G. Christy

Juan Manuel García-Ruiz

Stephen T.Hyde

Ann-Kristin Larsson

THE RECORD OF EARLY LIFE:
In search of biosignatures

A.M. CARNERUP[1], A.G. CHRISTY[1, 2], J.M. GARCÍA-RUIZ[3], S.T. HYDE[1] AND A.-K. LARSSON[1]
[1]Department of Applied Mathematics, Research School of Physical Sciences and Engineering, the Australian National University, 0200 Canberra, ACT, Australia. [2]Department of Earth and Marine Sciences, Australian National University, 0200 Canberra, ACT, Australia. [3]Laboratorio de Estudios Cristalograficos, Instituto Andaluz de Ciencias de la Tierra, CSIC – Universidad de Granada, E 18002 Granada Spain.

1. Introduction

The search for life's origins continues to stimulate intense debate on the earliest fossil record regarding key indicators of biogenicity. Textbooks commonly claim that microbial life was widespread 3.5 billion years ago. However, these claims are now being questioned, due to improved awareness of the complexity of structures that can be produced by abiotic geochemical processes. As a result, abiotic mechanisms for the formation of morphologies hitherto interpreted as fossils have assumed a significant role in the scientific debate. Consequently, the search for definitive evidence of Precambrian life is becoming more and more exacting. Since most Archean rocks are severely metamorphosed, the capacity to find unaltered remnants of life is limited. As these rocks have experienced high temperatures and pressures we need to be very careful in making claims about the existence of fossils within them. Here we critically review that evidence, focussing on the search for biosignatures in the geological record. Possible biosignatures can be grouped into three different categories: chemical (isotopic signatures and biomarker detection), morphological (apparent microfossils) and sedimentary structures presumed to be biologically mediated (e.g. stromatolites). We assess those signatures as applied to a few celebrated examples of ancient microfossils that reveal the intricacies and problems of distinguishing biological remnants from abiotic counterparts.

2. Chemical Evidence of Early Life

2.1. CARBON ISOTOPES

Isotopic signatures of carbon are extensively used as signatures of life. Biological carbon is enriched in the light carbon isotope ^{12}C with respect to crystallised carbonate. Carbon fractionation arises from preferential uptake of ^{12}C over ^{13}C during assimilation of CO_2 in autotrophic (in particular photosynthetic) microorganisms, and in subsequent biologically mediated reductions of the carbon. This discrimination of ^{13}C is a result of physiochemical effects during enzymatic reactions (c.f. Schidlowski, 2001). The depletion of ^{13}C is normally expressed as $\delta^{13}C$, in terms of the permil difference in the ratio $^{13}C/^{12}C$ of the sample (sa) compared to a commonly used standard (st) known as Pee Dee Belemnite (PDB):

$$\delta^{13}C = \left[\frac{\left({}^{13}C / {}^{12}C\right)_{sa}}{\left({}^{13}C / {}^{12}C\right)_{st}} - 1 \right] \times 1000 \quad (1)$$

$\delta^{13}C$ values of up to -80‰ have been reported for biogenic methane, but the aggregate of contemporary biomass carbon displays a $\delta^{13}C$ value of about –26±7‰. This is significantly lighter than atmospheric CO_2, which is about –8‰ $\delta^{13}C$ (Francey et al., 1999), while the composition of carbonate in shallow ocean water clusters around 0‰ $\delta^{13}C$ (Des Marais, 2001). The light isotopic signature of biological carbon is retained in the organic material after cell death. Subsequent degradation oxidises most of the biological matter to isotopically light CO_2. However, during diagenesis a small fraction of the carbon forms “kerogen”: an insoluble, mostly aromatic, highly condensed polymer (Durand, 1980; Schidlowski, 2001). In the geological record, isotopic carbon signatures compatible with a biological origin have been measured in samples that date back almost 4 billion years, suggesting the presence of photosynthetic microorganisms at a time when the Earth was still suffering a heavy bombardment of meteorites (Mojzsis and Harrison, 2000; Schoenberg et al., 2002; Ueno et al., 2002; Ueno et al., 2004).

Recently, ion microprobe spectrometry has been used to identify the carbon isotopic makeup of individual microfossils (House et al., 2000; Ueno et al., 2001), indicating that such signatures can play an important role in confirming a biological origin.

2.1.1. The case of Isua and Akilia

A celebrated – currently much disputed – example comes from ^{13}C-depleted graphite found in metasedimentary rocks in the Isua Supracrustal Belt and on Akilia Island in West Greenland (Mojzsis et al., 1996; Rosing, 1999; Moorbath, 2005). These rocks constitute some of the oldest known sedimentary rocks on Earth and date back to 3.7–3.8 billion years (Black et al., 1971; Moorbath et al., 1973; Nutman et al., 1996) and have been exposed to severe metamorphism (high temperature and pressure), altering the original composition and texture of the rock. The possibility of these ancient rocks hosting biological remnants has been the subject of intense debate in the past few years.

Several research groups have described findings of ^{12}C enriched carbonaceous matter in the rocks of West Greenland (Oehler and Smith, 1977; Perry and Ahmad, 1977; Mojzsis et al., 1996; Rosing, 1999; Ueno et al., 2002, van Zuilen et al., 2003). Mojzsis et al. (1996) argued that carbonaceous inclusions in minute grains of apatite (calcium phosphate, $Ca_5(PO_4)_3(F,Cl,OH)$) found in a rock interpreted to have originated as a sedimentary banded iron formation (BIF) on Akilia Island could only have been formed by biological processes, and hence that life must have emerged prior to 3.8 billion years before present. Since then, these findings have been the subject of controversy, focussing principally on whether the rocks hosting the apatite grains are in fact sedimentary. Fedo and Whitehouse (2002) argued that the rocks were instead produced by metasomatism associated with volcanic intrusions, and hence that the carbon was not necessarily biogenic. It was later suggested that the rocks analysed by Fedo and Whitehouse were different from the ones in which the apatite grains containing graphite were found (Nutman et al., 2004). Newly analysed rocks, from the very same BIF on Akilia Island, show iron and sulphur isotopic compositions indicative of a sedimentary origin (Mojzsis et al., 2003; Dauphas et al., 2004). But controversy still remains. Lepland et al. (2005) reanalysed the original rock sample in which Mojzsis et al. (1996) had found graphite inclusions in apatite, as well as 16 other rock samples from Akilia. Numerous apatite crystals were found, but none contained any graphite, casting doubt on the original study done by Mojzsis et al. (1996).

An important twist to the story of isotopically light graphite is that this feature is not necessarily exclusive to biological matter; indeed any chemical disproportionation reaction leading to carbonaceous products will lead to carbon isotopic fractionation, dependent on the reaction kinetics. Biological mediation is not essential. van Zuilen et al. (2002) showed that carbonate-rich rocks in Isua contain ^{13}C-depleted graphite associated with iron-bearing carbonates and magnetite. This association strongly suggested an abiotic origin for the graphite, by disproportionation of siderite to graphite + iron (III) oxides. Equilibrium isotope fractionation at low temperature during siderite decomposition could explain the isotopic signature of the graphite decomposition product (van Zuilen et al., 2002; van Zuilen et al., 2003). It has been shown experimentally that siderite thermally decomposes to form graphite (van Zuilen et al., 2002), non-graphitic carbonaceous matter and magnetite (McCollom, 2003) at temperatures above 300°C. Yet another explanation for the isotopically light carbon was given by Schoenberg et al. (2002) who suggested that a meteoritic origin should be considered, since debris from carbonaceous chondrites carrying insoluble carbon particles with $\delta^{13}C$ of around −18‰ could have been deposited within the sediments.

The Archean rocks of west Greenland have been interpreted as the oldest sedimentary formations, and their geological interpretation is therefore crucial to the investigation of early life. Due to their deformation and metamorphism, which has eradicated any sedimentary structures, the debate has centred on isotopic studies. The lessons of that debate are of value to all studies of early life. Perhaps the most important lesson is a simple one: carbon isotope fractionation is indeed a feature of biological activity. Nevertheless, it is also a feature of *any* physical process involving atomic transport, from chemical reactions to simple diffusion. Common geochemical processes, including siderite decomposition, serpentinization and Fischer-Tropsch processes (Holm and Charlou, 2001), can induce substantial ^{13}C depletion (Horita and Berndt, 1999). The degree of fractionation depends critically on the thermodynamic conditions of the system, including temperature and pressure. More detailed analysis of these parameters is needed to establish more firmly whether carbon isotope analysis from Isua and Akilia evidence biological remnants.

2.2. MOLECULAR BIOMARKERS

Biomolecular "fossils", or biomarkers, are chemically degraded biomolecules – typically derived from lipids and biopolymers –that are characteris tic constituents of organisms. Such molecules are found in sedimentary rocks, where in some cases they have survived for billions of years, retaining structural and isotopic compositions reminiscent of molecular species found in extant microorganisms. Derived from the phospholipid bilayer membranes of cells, these fossilised lipids can provide information on prehistoric biodiversity and environmental conditions). Since many microbes require specific habitats, discovery of corresponding biomarkers affords valuable inferences about the paleoenvironment in which the microbe once lived. During fossilisation, biolipids degrade and lose functional groups, leaving a hydrocarbon skeleton as a relic of the original molecule. Identification of biomarkers is a technically challenging art; recent advances in instrumentation make this technique feasible for surprisingly ancient samples. Biomarkers can be exceptionally specific in their origin: membrane bound lipids can act as a signature for one specific microbe. For example, it has been shown that the hydrocarbon botryococcane is exclusively produced by the alga *Botryococcus braunii* (Brocks and Summons, 2004). Such diagnostic biochemicals within the geological record can then be used to prove the presence of specific taxa at the time of sedimentation and the identification of specific "molecular fossils" offers a powerful probe of early life.

The chronological reach of biomarker identification is constrained by the degradation kinetics of biomolecular cell material. If heated or compressed under reducing conditions, as occurs during burial metamorphism, the ultimate thermodynamic fate of all organic compounds is to convert to, graphite. The persistence and exact nature of any partially decomposed biomolecular remains is strongly dependent on the pressure-temperature-time trajectory experienced by the sample. Differential degradation and recrystallisation kinetics for different compounds can result in the coexistence in a rock of graphite, "spheroidal amorphous carbon" and relatively unaltered waxy hydrocarbons, even in sillimanite-grade metasediments (Large et al., 1994).

In contrast to the recent development of the ability to detect trace quantities of biomarkers, recognition of highly degraded biochemical material in the whole rock, termed kerogen, has long been associated with fossil identification. Kerogen is intermediate in chemical character between cellular carbonaceous material and graphite. Depending on the degree of metamorphism it ranges in colour from brown (least metamorphosed) to black (most metamorphosed), as functional groups are lost and polymerisation progresses along the pathway from complex biochemicals to graphite. Kerogens are chemically ill-defined, containing complex mixtures of polycyclic aromatic and linear hydrocarbons. Spectroscopic signatures of kerogen are difficult to interpret due to the presence of broad bands. Raman spectra of kerogen display G (graphite-like) and D (disordered) bands (~1350 and ~ 600 cm^{-1} respectively). Schopf et al. (2002a) demonstrated a correlation in the ratio of D to G band intensities with the kerogen colour and preservation, proposing micro-Raman studies of kerogen in microfossils to be a useful diagnostic tool for microfossil identification. However, Raman D and G bands cannot distinguish abiotic complex organics from complex bioorganic breakdown products (Pasteris and Wopenka, 2002; Pasteris and Wopenka, 2003). For example, simple thermal curing of mixtures of formaldehyde and phenol (themselves organic but abiotic) leads to a brown-black tar whose Raman spectra are indistinguishable from kerogens in a range of purported microfossils (García-Ruiz et al., 2003).

2.2.1. The case of cyanobacteria and eukaryotes

Summons et al. (1999) showed that bitumen from the 2.5 billion-year-old Mt. McRae Shale of Hamersley basin, Western Australia contains significant amounts of 2-methylhopanes. These hydrocarbons are derivatives of 2-methylbacteriohopane-polyols, which are diagnostic for cyanobacteria (Summons et al., 1999). This strongly suggests the existence of oxygen-producing bacteria well before the generally agreed time when the terrestrial atmosphere developed considerable amounts of free oxygen. Other areas in Western Australia have shown the same biomarkers for cyanobacteria. Brocks et al. (1999) reported that steranes, characteristic of eukaryotes, are present in ~2.7 billion year-old shales of Roy Hill, Jeerinah Formation Western Australia. The possibility of contamination cannot be underestimated when examining biomarkers. The migration of younger biomarkers through cracks and pores in the rocks by circulating groundwater can provide a significant source for contamination. Younger rocks are present only outside the Hamersley Basin, at least 150 km away; migration of younger hydrocarbons over such long distances into impermeable Archean rocks is unlikely (Brocks et al., 2003). These biomarkers provide convincing evidence that eukaryotes were present in the late Archean (Brocks et al., 1999), prior to the oldest known eukaryotic microfossil evidence at 1.5-1.4 billion years from the Roper Group, Australia (Javaux et al., 2001; Javaux et al., 2003; Javaux et al., 2004). More ancient findings of allegedly eukaryotic microfossils have been reported in the 1.87 billion-year-old Negaunee iron-formation, Michigan (Han and Runnegar, 1992; Schneider et al., 2002), but the interpretation of these as fossils has been questioned, due to the lack of structural features indicative of eukaryotes (Samuelsson and Butterfield, 2001).

3. Morphological Evidence of Early Life

Perhaps the most compelling evidence of ancient life comes from structures believed to be fossilised microbes. Cellularly preserved microorganisms in Archean rocks are scarce owing to the often severe metamorphism of ancient rocks that would have obliterated any fossils once present (Barghoorn and Tyler, 1965; Schopf, 2004). Nonetheless, several microfossil occurrences have been reported from ancient but unusually well preserved sediments, of which the oldest date back almost 3.5 billion years (Schopf and Barghoorn, 1967; Awramik et al., 1983; Schopf and Walter, 1983; Walsh and Lowe, 1985; Schopf and Packer, 1987; Walsh, 1992; Schopf, 1993; Ueno et al., 2001).

Microorganisms can fossilize to preserve structural characteristics, including size, shape and texture of the outer wall, organization and colonial distribution. Following extensive scientific debates regarding biogenicity of supposed fossilized microbes (Cloud, 1973; Schopf, 1975), it is currently generally agreed that several criteria in combination afford acceptable evidence for identification of an Archean microfossil (Schopf and Walter, 1983; Buick, 1990; Schopf, 2004). To prove the antiquity of the fossil it has to be demonstrated that 1) it occurs in a sedimentary rock of Archean age (the rock must have been thoroughly characterized petrographically and geochemically, and dated); 2) it is embedded in the rock (and is not a surface contaminant introduced during collection or subsequent processing); 3) it was coeval with sedimentation (and is not an ancient microbial contaminant that invaded the rock after consolidation). In addition to demonstrating antiquity, the structure has to be shown to be of biological origin. The latter is characterised by 1) morphological similarity to contemporary microorganisms; 2) morphological evidence of cellular complexity; 3) traces of kerogen (degraded biomolecules); 4) hollow-shelled form with distinct shell and filling materials (a consequence of the fossilization process); 5) spatial association with other specimens of similar morphology, implying the existence of a viable population.

Although these criteria seem precise, establishing the biogenicity of ancient fossils has proven to be the most difficult task. This is largely due to the facts that pseudofossils can mimic convincingly the characteristics of microfossils, and that an Archean microfossil record to which objects can be compared is very rare (Schopf and Walter, 1983; Schopf, 2004).

Pseudofossils such as simple spherical objects can obviously be difficult to distinguish from their microfossil counterpart. Filamentous objects that mimic biological objects are, on the other hand according to Schopf (2004), more difficult to fabricate through inorganic processes and are therefore relatively easy to recognise as pseudofossils. True biogenic filamentous microfossils are cylindrical with uniform diameter along their length, exhibit a characteristic size, show evidence of septate internal structure, distinctive end cells and occur among others with similar morphologies. They should also display varying degrees of sinuosity correlated to filament width and length, and show examples of specimens in various states of preservation (Schopf, 2004).

All known Archean deposits containing putative microfossils are metamorphosed and hence the carbonaceous matter of the fossils is sufficiently chemically degraded to preclude definitive identification of biological molecular precursors (Schopf, 2004). The burden of proof of biogenicity for all-important ancient (Archean) samples thus rests largely with morphological criteria.

3.1. PILBARA MICROFOSSILS

The Pilbara Craton in north-western Australia contains some of the best preserved early Archean rocks in the world. It is in this region that the oldest purported microscopic traces of life were discovered. Several locations, of differing ages have been shown to contain such purported microfossils. The ~3,490 million-year-old Dresser formation of North Pole[1] and the ~3,465 million-year-old Apex Chert are two of the best known fossiliferous localities. Microfossils from both of these areas have been subject to intense debate regarding their biogenicity.

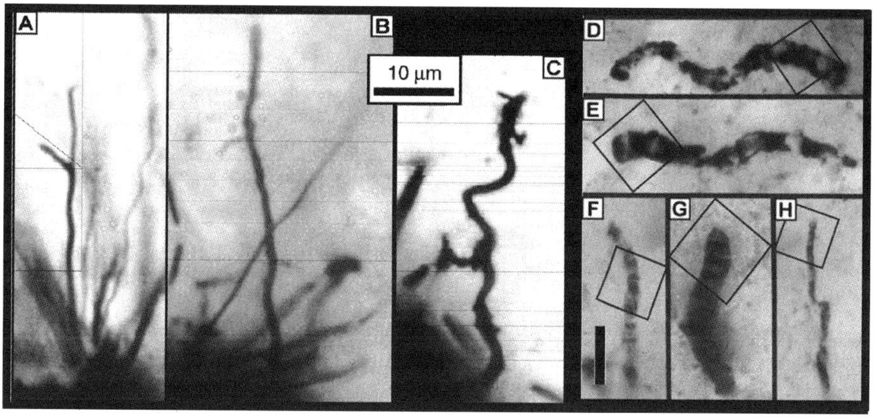

Figure 1: Carbonaceous filaments found in Archean cherts from Western Australia. A-C) Filaments found in the North Pole (Published with permission from Ueno et al., 2001). D-H) Microstructures found in the Apex Chert, scale bar in F is 10 μm and applies to D-H (Published with permission from Nature Publishing Group and Schopf et al., 2002a).

3.1.1. North Pole

The North Pole rocks contain microscopic filaments, reported to be the oldest found on Earth. Awramik et al. (1983) reported ~3.49 billion-year-old filamentous bacterial fossils from two localities (designated A and B) of laminated carbonaceous cherts, characterised as stromatolites (see section 4 below). These microscopic structures were first determined to be biological remnants because of their carbonaceous composition and their complex filamentous morphology that resemble contemporary microorganisms. These results were later challenged by Buick (1984) who suggested that the sedimentary laminae of locality B were of secondary origin, and that the filaments were therefore abiotic contaminants in secondary chert of much younger age. Awramik et al. (1988) reanalysed the fossil-bearing rock from locality A and stood by their original conclusion that the fossils are biogenic and syngenetic with the deposition of the chert and they are therefore around 3.5 billion years old. Buick (1988) argued that as the exact site of locality A is not known, the field relationships with surrounding rocks cannot be ascertained, and therefore it is impossible to draw any conclusions about the geological setting of the filaments.

---

[1] Ironically named by gold miners in the early 20th century; with daytime temperatures exceeding 40°C for much of the year!

The filaments in locality B were originally interpreted as 'possible microfossils', and are very similar to the ones discussed by Buick (1984). The sediment in which those structures were found, on the other hand, is quite dissimilar to the one described by Buick, which casts doubt on the validity of his conclusion (Awramik et al., 1988). Buick continued to maintain that the lamination in the "fossiliferous" rock was secondary and not stromatolitic (Buick, 1988; Buick, 1990). This dispute illustrates just how important a detailed and exact analysis of the geological context is for conclusions to be drawn about any microfossils. More recently, Van Kranendonk and colleagues have published detailed geological maps of this area, following a decade of mapping of the Pilbara Craton. They have established that the microfossil location lies within a chert- and barite-rich hydrothermal feeder vein system (Van Kranendonk and Pirajno, 2004; Van Kranendonk et al., 2003) rather than in the sedimentary chert (Schopf, 1993).

In 2001 Ueno et al. (2001) reported filamentous microstructures from two new localities in the 'chert-barite unit' of the North Pole (see Figure 1 A-C). About a hundred structures were found in silica veins. The black filaments are carbonaceous as determined from Raman spectroscopy, are ~1 μm wide and >10 μm long, and have spiral, thread-like or branched morphologies, comparable to modern filamentous bacteria and known microfossils. The carbonaceous matter of the filaments show an extremely light carbon isotopic composition indicative of a biological origin ($-42‰ < \delta^{13}C < -32‰$). Ueno et al. (2001) suggested that the filaments were 'probable microfossils', and it is believed that these structures together with the structures found by Awramik et al. (1983), are the oldest morphological evidence of life on Earth (Schopf, 2004).

3.1.2. The Apex Chert

In March 2002 *Nature* published two opposing articles concerning the authenticity of some of the earliest fossils of life on Earth. Schopf et al. (2002a) used laser Raman spectroscopy (cf. section 2.2) to analyse the carbonaceous residues incorporated in the microfossil structures. Various filamentous fossils from Precambrian cherts were analysed: the ~2,100 million-year-old Gunflint Formation of Ontario, Canada, ~3,375 million-year-old Kromberg Formation of South Africa and ~3,465 million-year-old Apex chert of Western Australia. All gave Raman signals indicative of carbonaceous (kerogenous) materials. These Raman spectra – together with the morphological characteristics of the structures analysed – led Schopf et al. (2002a) to conclude that the structures were indeed biogenic. Some of those materials, particularly the Apex chert specimens (see Figure 1 D-H), were classified as probable cyanobacteria (Schopf and Packer, 1987; Schopf 1993), 1 billion years older than from the microfossils identified as cyanobacteria found in the Mt. McRae Shale of the Hamersley basin (cf. section. 2.2.1); this claim has now been withdrawn (Kerr, 2002).

Brasier et al. (2002) on the other hand, dismissed the claims that the fossils from the Apex chert were of true biological origin. Morphological and Raman studies of filaments and thorough geological mapping of the fossil locality led Brasier et al., to the conclusion that the filaments were formed by thermal transformation of volcanic CO_2 into carbonaceous material, rather than as remnants of once-living microbes (Brasier et al., 2002; Brasier et al., 2004). This controversy whipped up a storm of interest in the scientific community (Dalton, 2002; Gee, 2002; Kerr 2002; Pasteris and Wopenka, 2002; Schopf et al., 2002b). Although the dust remains unsettled, it is clear that the controversy rests principally on the validity of morphological criteria.

3.2. BIOMORPHS AS PSEUDOFOSSILS

The efficacy of morphology as an indicator of life rests on the assumption of a disjunction between biotic and abiotic forms. That dichotomy is widely held to be self-evident; after all, non-living materials typically crystallise to form angular, facetted forms, while biological materials form curvilinear, rounded shapes. Closer study reveals the assumed disjunction to be a naïve one, based on consideration of specific non-living materials, such as simple plane-faced inorganic crystals. In fact, the hallmark of living form – curvature – abounds in non-living matter, provided

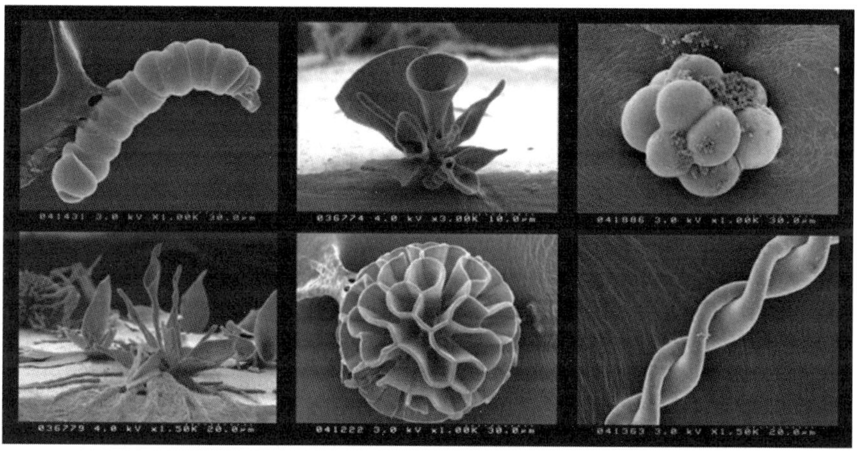

Figure 2. Scanning electron microscopy images of various silica-barium carbonate biomorphs.

one looks in the right places. Certainly, liquid crystals (once believed to be the *sine qua non* of life (Haeckel, 1917; Needham, 1968) can display identical forms at the microscopic scale to those found at the cellular level across the kingdoms of life. Smoothly rounded spherulitic growths are as commonly encountered in inorganic crystal textures, in polymeric materials, in metals, as in biological cellular material. Convoluted anticlastic (saddle-shaped) forms, characteristic of hyperbolic geometry, are exhibited at distinct length scales from the atomic, to the supermolecular, by "dead" covalent crystals, liquid crystals, polymer as well as living cell organelles (Hyde et al., 1996). Both spherulitic and hyperbolic morphologies can be found up to the micron scale in a range of single component inorganic crystals, such as chrysotile, antigorite, lizardite, "saddle" dolomite, and pyrite, as well as a variety of clays and layered silicates (Hyde, 1993; Hyde 1998 and references therein).

It is equally certain that composite assemblies of inorganic matter, though abiotic, can also adopt rounded, convoluted forms so often encountered in biology. We call such examples of abiotic materials "biomorphs". A particularly well-studied class of biomorphs are self-assembled silica-carbonate aggregates (García-Ruiz et al., 2002; García-Ruiz et al., 2003). These are routinely synthesised from an alkaline (pH 9-11) silicate solution or gel that has been enriched in barium- or strontium chloride (García-Ruiz, 1985; Hyde et al., 2004). The absorption of carbon dioxide from the air induces the precipitation of biomorphic aggregates within hours; varying the reaction conditions can produce a wide range of morphologies (see Figure 2). All structures seen in Figure 2 consist of crystalline barium carbonate nanometric crystallites embedded in continuous matrix of amorphous silica. The twisted structures may be covered by an outer skin made of colloidal silica particles. These forms include a number of shapes commonly assumed to be characteristic of life (helices, convoluted curved sheets), and display many structural features at the microscopic scale found in biological skeletal elements, such as orientational ordering of crystallites, characteristic of biominerals and both amorphous and crystalline order within a single material (Lowenstam and Weiner, 1989). Can all the morphological signatures of microfossils listed above be found in such materials? In short, yes. Filamentous biomorphs display apparent segmentation, uniform size and dimensions identical to those of simple filamentous bacterial microfossils.

The picture becomes further confused by the possibility of chemical modification of biomorphs, leaving pseudofossils with both morphological and chemical signatures common to microfossils. Weak acid leaching of twisted filamentous biomorphs results in removal of the carbonate species within the filament, leaving only the hollow silica skin as a perfect replica of the original structure (Hyde et al., 2004). Furthermore, the silica skin can act as a trap for hydrocarbons (García-Ruiz et al., 2002). Adsorption and condensation of simple hydrocarbons, such as phenol and formaldehyde, onto the filaments lead to black-brown carbonaceous biomorphs (García-Ruiz et al., 2003). These black filaments can fulfil most of the described criteria for biogenicity: 1) they have morphology similar to modern microbes; 2) they show evidence of cellular elaboration (such as segmentation); 3) they are composed of carbonaceous material (their Raman spectra resemble that of filaments from the Apex Chert, and therefore could be interpreted as kerogen); 4) they can be made hollow (the interior carbonate can be dissolved in a weak acid, retaining the overall morphology and the carbonaceous nature); 5) they occur with others of similar morphology (since they are chemical precipitates, their growth and morphology are governed by the physiochemical environment).

Although no known geological occurrence of biomorphs have been reported in the literature, the geochemical scenarios for biomorph synthesis are plausible for the Archean, suggesting that these complex carbonaceous structures could have been naturally produced on the primitive Earth (García-Ruiz, 1998; García-Ruiz, 2000; García-Ruiz et al., 2003). For instance, in the North Pole area, Western Australia, silica and barium are present in the fossiliferous 'chert-barite unit' giving the chemical requirements for biomorph synthesis, provided that pH and temperature were appropriate. Once the structures have formed, creation of the carbonaceous shell can also be explained (García-Ruiz et al., 2003). The hydrocarbons needed are small and chemically simple, and would have been readily available on the primitive Earth, whether or not life was present. This is demonstrated by the recent discovery that similar organic materials can be produced by decomposing iron-carbonate at elevated temperatures (less than 350°C) in the presence of water (McCollom, 2003). This type of reaction has been suggested as a mechanism for forming isotopically light hydrocarbons (van Zuilen et al., 2002). Iron-rich carbonates are present in the North Pole area, and given the known hydrothermal origin of some of the "microfossil"-bearing rocks, it is possible that such hydrocarbons could have been abiotically produced. Subsequent adsorption and condensation to form ^{12}C-enriched kerogen-like materials over time is equally plausible (García-Ruiz et al., 2003). Abiotic carbonaceous biomorphs, once formed, would readily be incorporated into rocks, not just those filling hydrothermal veins but also sediments downstream of the points where those veins emerged at the paleosurface as hot springs. As such, they would also fulfil the criteria for establishing the antiquity of a microfossil, since they would be proven to be indigenous and syngenetic with the sediment and not of any later type of contamination.

4. Stromatolites

In addition to microscopic evidence, micro-organisms can leave macroscopic imprints in the geological record. Biosedimentary evidence of life is visible in both ancient rocks as well as in modern, living analogues. There are several different types of biologically mediated sediments, distinguished with respect to their various fabrics, of which stromatolites are the best known (Riding, 2000). The scope of the term "stromatolite" has been widely debated, and various definitions therefore occur in the literature (Krumbein, 1983; Grotzinger and Knoll, 1999; Riding, 1999). Malcolm Walter's definition from 1976 is probably the most widely used (Altermann, 2004): *'Stromatolites are organosedimentary structures produced by sediment trapping, binding and/or precipitation as a result of the growth and metabolic activity of micro-organisms, principally cyanophytes'* (Walter, 1976). These accretionary structures have been reported from all continents (Hofmann, 1973) and they can take a wide variety of forms, ranging from planar, columnar and dome-shaped structures to complex branching systems (Hofmann, 2000). The oldest reported putative stromatolite is around 3.5 billion years old, found in the North Pole area of the Pilbara, Western Australia (Walter et al., 1980). This domical stromatolite is composed of thin laminae of dolomitic chert and has a structure comparable to younger stromatolitic examples. No microfossils have yet been described in this stromatolite. Although micro-organisms are crucial to the formation of modern stromatolites, their remnants are rarely present in ancient examples. Probably less than 1%of all stromatolites ever desc ribed contain discernible microfossils (Grotzinger and Knoll, 1999). The rapid recrystallization and the subsequent destruction of any preserved filaments are most likely the cause of the absence of preserved microfossils in stromatolites (Schopf et al., 1971).

It has been suggested that the stromatolites from Western Australia may afford at best dubious evidence for early life (Buick et al., 1981). The absence of microfossils makes it difficult to determine a biological origin of the structures. Indeed, even if fossilized microbes were to be found in a stromatolite, it is not assured that those particular microbes were the cause of the formation of the stromatolite in question (Grotzinger and Knoll, 1999). Abiotic accretions that mimic the morphology and chemistry of stromatolites are present in the geological record, further complicating assessment of biogenicity (Grotzinger and Rothman, 1996). To ensure that stromatolitic structures are biogenic in nature it is crucial to find evidence of the biological-sediment interaction that is the underlying cause of their formation. The stromatolites therefore need to be well preserved, so that the microstructures and fabrics can offer evidence of biologically mediated sedimentation. This is rarely the case for Archean stromatolites, which have undergone diagenetic alteration in which the original small-scale fabrics or textures have been lost (Cady et al., 2003). Buick et al. (1981) described 8 criteria to facilitate the recognition of true biogenic stromatolites: 1) The structures must be found in sedimentary rocks; 2) the structures have to be synsedimentary in origin; 3) they should have a majority of convex-upward structures; 4) the laminations should thicken over the crests; 5) if they are laminated, these laminations should be wavy, wrinkled and/or have several orders of curvature; 6) microfossils should be present within the structure; 7) changes in composition in microfossil assemblages should be associated with morphological changes in the stromatolite; 8) the fossils must be organized in a way indicating that the living microbes accreted the structure by trapping, binding or precipitation of sediment. Lowe (1994) has suggested that *all* putative Archean stromatolites more than 3,200 million years fail to fulfil all of these criteria. Rather, Lowe suggested that inorganic, abiological processes produced these stromatolite-like structures. Controversy persists (Buick et al., 1995; Lowe, 1995). Nevertheless, most experts agree that the North Pole structures are possibly biogenic, though conclusive evidence is still lacking.

Although stromatolites are present throughout the geological record from the early Archean right up to the present, and hence are highly significant for the reconstruction of the history of life, little is known about the microbes and processes that generate them. The most famous extant stromatolites are found in Shark Bay, Western Australia, not far from the oldest stromatolites found on Earth in the Pilbara. A recent study of these modern stromatolites has shown that they are inhabited by a vast range of prokaryotes with diverse metabolic activities (Burns et al., 2004). Future research will provide a better understanding of these ecosystems and the relationship between the various microbe types and stromatolite growth and structure.

5. Extraterrestrials?!

Science is now revisiting one of the most exciting questions to stimulate the imagination of humanity: Did life emerge somewhere else in the Universe, and if so, did it later colonise Earth by arrival in meteorites? This hypothesis arises naturally as part of the "panspermia" model (Arrhenius, 1908), in which primitive life is assumed to have travelled between planetary bodies, asteroids and meteorites, and even through interstellar space. The implications of finding life, extinct or alive, elsewhere in the Universe will be profound for our perspective of life.

Since the controversial discovery of putative microfossils and biomarkers in the martian meteorite ALH84001, discovered in the Allen Hills area of Antarctica in 1984 (McKay et al., 1996) there has been a wealth of scientific discussion and arguments for and against the conclusion that life existed (or still exists) on Mars. McKay et al. reported several lines of evidence to support their claim of extraterrestrial life, which would also be the oldest known life: the presence of indigenous microfossil-like objects resembling terrestrial fossilised microorganisms, polycyclic aromatic hydrocarbons (PAH) with restricted mass distribution, coexisting single domain magnetite (Fe_3O_4) of distinctive crystal habit, and iron sulphides associated with carbonate globules. Although, taken individually, each of these observations can be explained by abiotic processes, McKay et al. argued that the best explanation for all lines of evidence taken together is a biogenic origin (McKay et al., 1996). Since publication, all of their claims have been questioned or refuted.

The famous 'worms' identified in the meteorite were initially identified as microbial fossils on the basis of morphology. Subsequently, others pointed out that their dimensions are smaller than those of extant terrestrial bacteria and their presence in the meteorite sample was interpreted as an artefact of sample preparation (Bradley et al., 1997). As these structures are only tens of nanometers in diameter – well below generally agreed theoretical minimum cell sizes – it was concluded that they were unlikely to once have been living microbes (Morowitz, 1996). There are, however, a number of controversial reports of nanometric bacteria, whose dimensions are consistent with those of the meteoritic nanofossils. So-called "nanobacteria" (or "nanobes") have been reported in carbonate sediments (Folk, 1993) as well as in human and cow blood (Kajander and Çiftçioglu, 1998). These objects have a cell diameter of less than 200 nm and yet appear to be self-replicating in culture. The true nature of these forms remains controversial (Cisar et al., 2000). Evidently a conclusive biological origin of the martian nanofossils has not been established. Indeed, they bear a very strong resemblance to abiotic biomorphs, in both shape and size (Hyde, 1998).

The presence of PAHs in the Martian meteorite must also be treated with some scepticism, given the possibility of external contamination while the meteorite remained undisturbed in Antarctica. The PAHs present in the ALH84001 meteorite were identified as being indigenous to the rock (Clement et al., 1998) and to have a restricted mass distribution, but this does not unambiguously imply a biological origin. In fact PAHs are found in a wide range of extraterrestrial materials such as interplanetary dust particles and carbonaceous chondrites. Recall that *in situ* formation of PAH products via hydrothermal decomposition of siderite (see above) is possible; McCollom (2003) showed that a wide range of organic compounds, especially alkylated aromatic hydrocarbons, can be produced during thermal decomposition of siderite in the presence of water. This decomposition reaction is consistent with the mineralogical and organic constituents of ALH84001, suggesting an abiological origin of the PAHs found in the meteorite (McCollom, 2003). The final remaining case for biogenicity in the ALH84001 sample rested on the discovery of magnetite crystallites whose crystal habit was claimed to resemble that of biomineralised magnetites, grown in magnetotactic bacteria. It is well known that magnetotactic bacteria (such as MV-1 strain) produce intracellular single crystals of magnetite with specific morphology. Thomas-Keprta et al. (2000, 2001) described six properties of terrestrial biomineralised magnetite: single domain size with restricted width to length ratios, chemical purity, absence of crystallographic defects, unusual crystal morphology involving elongation along the [111] axis, and arrangement of crystals in chains. Five of those features were identified in elongated magnetite crystals from ALH84001. They argued that the characteristics of this martian magnetite, similar to those of magnetite grown by the MV-1 bacterium, could only be explained by biological processes since no known inorganic process can produce the same result (Thomas-Keprta et al., 2002). More recently however, it was discovered that not only are the martian magnetite crystals morphologically different from the MV-1 magnetosome magnetite crystals but they can also be abiotically reproduced by hydrothermal decomposition of Fe-rich carbonate (siderite, $FeCO_3$) (Golden et al., 2004). It may be significant that the same process can simultaneously explain the presence of PAHs in ALH84001 (described above).

It is now clear that the original claims of microfossils in the ALH84001 meteorite have not survived the intense scrutiny of scientists. As a result, no definite answer on the question of life on Mars can be given at this stage. However, the search for evidence of extra-terrestrial remnants of life remains active. Both the NASA rovers and the ESA spacecraft are currently scouring Mars and there is increasing hope of finding new pieces of evidence adding to the debate on life beyond the Earth. It has been claimed that the search for extra-terrestrial life will –if successful –shed light on the origins of life on Earth. Unless the exotic life-forms are very similar to Earth's, that hope may turn out to be misplaced: we may instead be faced with further questions.

The search for extra-terrestrial microfossils is now an active one, albeit beset by more difficulties than those impeding definitive identification of terrestrial examples. It is perhaps timely to reflect on the conclusions that could be drawn in the event that a microfossil indigenous to another planet was to be positively identified. Given that the exchange of ejecta material from meteorite impacts material within the solar system is now known to be substantial, that finding could only lend further credence to the possibility of cross-infection of planetary bodies within our solar system by bio-contaminated meteoritic fragments (Melosh, 1989). If current estimates of mass exchange between neighbouring planets in our solar system are correct, the possibility of transfer of microbes to or from the Earth and between other bodies cannot be ignored, making it very difficult to detect the original source of life in the solar system. Within this scenario, the search for life that is of definite non-terrestrial origin life will remain unconcluded until and unless a novel form of life, not found on Earth, is discovered elsewhere. That immediately begs the question of what would constitute biosignatures for life of a novel, non-terrestrial type. It is therefore incautious to rush to conclusions based on results of studies of possible extra-terrestrial life remnants using current approaches. We may be doing paleontologists a disservice in the longer term if we continue to argue that such studies can offer definitive arguments for or against the panspermia hypothesis, or the belief that life is a necessary emergent property of complex systems.

6. Conclusion

The search for ancient life – here or elsewhere –has relied on a growing series of criteria and biosignatures, from the isotopic level, to the molecular and the morphological. Unfortunately, ancient Archean signatures –necessarily indistinct and less complex than those found in more recent life remnants –can be 'forged' by abiotic materials. That complication makes fossil identification a more strenuous task than was once assumed. Nevertheless, at some point, the balance of probability will favour biotic or otherwise provenance for a putative microfossil: should the abiotic physical-chemical scenario be unlikely for the reconstructed geochemical environment during fossilisation, the likelihood of biotic formation grows, and vice versa. Probabilities, not certainties, are the best one can hope for in the near future. To refine those probabilities further, we must investigate in more detail the range of physical parameters admitting abiotic biomorphs and compare those parameters with the geochemical environment of the microfossil. In particular, the magnitude of isotope fractionation during abiotic chemical reactions must be studied. The range of dissolved salt compositions, pH, temperature and pressure that allows for biomorphic growths of a specific morphology to form also needs to be investigated; the discovery of a similar form in a geological setting may then be more convincingly judged to be an example of complex abiotic structure that is consistent with the depositional environment, rather than an example of early life.

A further, profound, issue needs to be explored before a definitive history of the emergence of life on Earth can be written. What is "Life"? The divide between earliest life and non-life is likely to be indistinct, with intermediate stages that share aspects of both. Even the comparatively accessible modern world has viruses as an example of this ambiguity. Exploration of non-living morphological analogues of simple life, such as biomorphs, self-assembled molecular containers and self-replicating molecules is essential to reconstruction of the emergence of life. Palaeontologists will continue to search the geological record for ancient microfossils. Even though that search is now known to be more difficult than once thought, it may well lead to a more thorough insight into the meeting point of abiotic and biotic chemistry: surely the key to understanding the origins of life on Earth.

7. References

Altermann, W. (2004) Precambrian stromatolites: problems in definition, classification, morphology and stratigraphy, In: P.G. Eriksson, W. Altermann, D.R. Nelson, W.U. Mueller and O. Catuneanu (eds.) *The Precambrian Earth: tempos and events,* Developments in Precambrian Geology **12**, pp. 564-574.

Arrhenius, S. (1908) *Worlds in the making.* Harper, New York, the United States of America.

Awramik, S.M., Schopf, J.W. and Walter, M.R. (1983) Filamentous fossil bacteria from the Archean of Western Australia. Precambrian Res. **20**, 357-374.

Awramik, S.M., Schopf, J.W. and Walter, M.R. (1988) Carbonaceous filaments from North Pole, Western Australia: Are they fossil bacteria in Archean stromatolites? A discussion. Precambrian Res. **39**, 303-309.

Barghoorn, E.S. and Tyler, S.A. (1965) Microorganisms from the Gunflint Chert. Science **147**, 563-577.

Black, L.P., Gale, N.H., Moorbath, S., Pankhurst, R.J. and McGregor, V.R. (1971) Isotopic dating of very early Precambrian amphibolite facies gneisses from the Godthaab district, West Greenland. Earth planet. Sci. Lett. **12**, 245-259.

Bradley, J.P., Harvey, R.P. and McSween Jr, H.Y. (1997) No 'nanofossils' on martian meteorite. Nature **390**, 454.

Brasier, M.D., Green, O.R., Jephcoat, A.P., Kleppe, A.K., Van Kranendonk, M., Lindsay, J.F., Steele, A. and Grassineau, N.V. (2002) Questioning the evidence for Earth's oldest fossils. Nature **416**, 76-81.

Brasier, M.D., Green, O.R., Lindsay, J.F. and Steele, A. (2004) Earth's oldest (~3.5 Ga) fossils and the 'Early Eden hypothesis': Questioning the evidence. Orig. Life Evol. Biosp. **34**, 257-269.

Brocks, J.J., Logan, G.A., Buick, R. and Summons, R.E. (1999) Archean molecular fossils and the early rise of eukaryotes. Science **285**, 1033-1036.

Brocks, J.J., Buick, R., Logan, G.A. and Summons, R.E. (2003) Composition and syngeneity of molecular fossils from the 2.78 to 2.45 billion-year-old Mount Bruce Supergroup, Pilbara Craton, Western Australia. Geochim. Cosmochim. Acta **67**, 4289-4319.

Brocks, J.J. and Summons, R.E. (2004) Sedimentary hydrocarbons, biomarkers for early life, In: H.D. Holland and K.K. Turekian (eds.) *Treatise on Geochemistry* **8**, pp. 63-115.

Buick, R. (1984) Carbonaceous filaments from North Pole, Western Asutralia: Are they fossil bacteria in Archaean stromatolites? Precambrian Res. **24**, 157-172.

Buick, R. (1988) Carbonaceous filaments from North Pole, Western Asutralia: Are they fossil bacteria in Archean stromatolites? A reply. Precambrian Res. **39**, 311-317.

Buick, R. (1990) Microfossil recognition in Archean rocks: An appraisal of spheroids and filaments from a 3500 M.Y old chert-barite unit at North Pole, Western Australia. Palaios **5**, 441-459.

Buick, R., Dunlop, J.S.R. and Groves, D.I. (1981) Stromatolite recognition in ancient rocks: an appraisal of irregularly laminated structures in an early Archaean chert-barite unit from North Pole, Western Australia. Alcheringa **5**, 161-181.

Buick, R., Groves, D.I. and Dunlop, J.S.R. (1995) Abiological origin of described stromatolites older than 3.2 Ga: comment and reply. Comment. Geology **23**, 191.

Burns, B.P., Goh, F., Allen, M. and Neilan, B.A. (2004) Microbial diversity of extant stromatolites in the hypersaline marine environment of Shark Bay, Australia. Env. Microbiol. **6**, 1096-1101.

Cady, S.L., Farmer, J.D., Grotzinger, J.P., Schopf, J.W. and Steele, A. (2003) Morphological biosignatures and the search for life on Mars. Astrobiology **3**, 351-368.

Cisar, J.O., Xu, D-Q Thompson, J., Swaim, W., Hu, L. and Kopecko, D.J. (2000) An alternative interpretation of nanobacteria-induced biomineralization. PNAS **97**, 11511-11515.

Clement, S.J., Chillier, X.D.F., Gillette, S., Zare, R.N., Maurette, M., Engrand, C., and Kurat, G. (1998) Observations of indigenous polycyclic aromatic hydrocarbons in "giant" carbonaceous Antarctic micrometeorites. Orig. Life Evol. Biosp. **28**, 425-448.

Cloud, P. (1973) Pseudofossils: A plea for caution. Geology **1**, 123-127.

Dalton, R. (2002) Squaring up over ancient life. Nature **417**, 782-784.

Dalton, R. (2004) Fresh study questions oldest traces of life in Akilia rock. Nature **429**, 688.

Dauphas, N., van Zuilen, M., Wadhwa, M., Davis, A.M., Marty, B. and Janney, P.E. (2004) Clues from Fe isotope variations on the origin of early Archean BIFs from Greenland. Science **306**, 2077-2080.

Des Marais, D.J. (2001) Isotopic evolution of the biogeochemical carbon cycle during the Precambrian, In: J.W. Valley and D.R. Cole (eds.) *Stable isotope geochemistry*, Reviews in Mineralogy and Geochemistry **43**, pp. 555-578.

Durand, B. (1980) *Kerogen-insoluble organic matter from sedimentary rocks*, Éditions Technip, Paris, France.
Fedo, C.M. and Whitehouse, M.J. (2002) Metasomatic origin of quartz-pyroxene rock, Akilia, Greenland, and implications for Earth's earliest life. Science **296**, 1448-1452.
Folk, R.L. (1993) SEM imaging of bacteria and nannobacteria in carbonate sediments and rocks. J. Sedim. Petrol. **63**, 990-999.
Francey, R.J., Allison, C.E., Etheridge, D.M., Trudinger, C.M., Enting, I.G., Leuenberger, M., Langenfelds, R.L., Michel, E. and Steele, L.P. (1999) A 1000-year high precision record of $\delta^{13}C$ in atmospheric CO_2. Tellus **51B**, 170-193.
Garcá-Ruiz, J.M. (1985) On the formation of induced morphology crystal aggregates. J. Cryst. Growth **73**, 251-262.
Garcá-Ruiz, J.M. (1998) Carbonate precipitation into alkaline silica-rich environments. Geology **26**, 843-846.
Garcá-Ruiz, J.M. (2000) Geochemical scenarios for th e precipitation of biomimetic inorganic carbonates, In: J.P. Grotzinger and N.P. James (eds.) *Carbonate Sedimentation and Diagenesis in the Evolving Precambrian World*, SEPM Special publication **67**, pp. 75-89.
Garcá-Ruiz, J.M., Carnerup, A., Christy, A.G., We lham, N.J. and Hyde, S.T. (2002) Morphology: An ambiguous indicator of biogenicity. Astrobiology **2**, 353-369.
Garcá-Ruiz, J.M., Hyde, S.T., Carnerup, A.M., Christy, A.G., Van Kranendonk, M.J. and Welham, N.J. (2003) Self-assembled silica-carbonate structures and detection of ancient microfossils. Science **302**, 1194-1197.
Gee, H. (2002) That's life? Nature **416**, 28.
Golden, D.C., Ming, D.W., Morris, R.V., Brearley, A.J., Lauer Jr, H.V., Treiman, A.H., Zolensky, M.E., Schwandt, C.S., Lofgren, G.E. and McKay, G.A. (2004) Evidence for exclusively inorganic formation of magnetite in Martian meteorite ALH84001. American Mineralogist **89**, 681-695.
Grotzinger, J.P. and Rothman, D.H. (1996) An abiotic model for stromatolite morphogenisis. Nature **383**, 423-425.
Grotzinger, J.P. and Knoll, A.H. (1999) Stromatolites in Precambrian carbonates: Evolutionary mileposts or environmental dipsticks? Annu. Rev. Earth Planet. Sci **27**, 313-358.
Haeckel, E. (1917) *Kristallseelen-Studien über das anorganische Leben*, Alfred Kröner Verlag, Leibzig, Germany.
Han, T.-M. and Runnegar, B. (1992) Megascopic eukaryotic algae from the 2.1-billion-year-old Negaunee iron-formation, Michigan. Science **257**, 232-235.
Hofmann, H.J. (1973) Stromatolites: characteristics and utility. Earth-Science Reviews **9**, 339-373.
Hofmann, H.J. (2000) Archean stromatolites as microbial archives, In: R.E. Riding and S.M. Awramik (eds.) *Microbial sediments*, Springer-Verlag, Berlin, pp. 315-327.
Holm, N.G. and Charlou, J.L. (2001) Initial indications of abiotic formation of hydrocarbons in the rainbow ultramafic hydrothermal system, Mid-Atlantic Ridge. Earth Planet. Sci. Let. **191**, 1-8.
Horita, J. and Berndt, M.E. (1999) Abiogenic methane formation and isotopic fractionation under hydrothermal conditions. Science **285**, 1055-1057.
House, C.H., Schopf, J.W., McKeegan, K.D., Coath, C.D., Harrison, T.M. and Stetter, K.O. (2000) Carbon isotopic compostition of individual Precambrian microfossils. Geology **28**, 707-710.
Hyde, S.T. (1993) Hyperbolic and elliptic layer warping in some sheet alumino-silicates. Phys. Chem. Minerals **20**, 190-200.
Hyde, S.T. (1998) Boxing Partula. Forma **13**, 145-178.
Hyde, S.T., Andersson S., Larsson, K., Blum, Z., Landh, T., Lidin, S., and Ninham, B.W. (1996) *The Language of Shape*, Elsevier, Amsterdam, The Neatherlands.
Hyde, S.T., Carnerup, A.M., Larsson, A.K., Christy, A.G. and Garcá-Ruiz, J M. (2004) Self-assembly of carbonate-silica colloids: between living and non-living form. Physica A **339**, 24-33.
Javaux, E.J., Knoll, A.H. and Walter, M.R. (2001) Morphological and ecological complexity in early eukaryotic ecosystems. Nature **412**, 66-69.
Javaux, E.J., Knoll, A.H. and Walter, M. (2003) Recognizing and interpreting the fossils of early eukaryotes. Orig. Life Evol. Biosp. **33**, 75-94.
Javaux, E.J., Knoll, A.H. and Walter, M.R. (2004) TEM evidence for eukaryotic diversity in mid-Proterozoic oceans. Geobiology **2**, 121-132.
Kajander, E.O. and Çiftçioglu, N. (1998) Nanobacteria: An alternative mechanism for pathogenic intra- and extracellular calcification and stone formation. PNAS **95**, 8274-8279.

Kerr, R.A. (2002) Earliest signs of life just oddly shaped crud? Science **295**, 1812-1813.
Krumbein, W.E. (1983) Stromatolites -the challenge of a term in space and time. Precambrian Res. **20**, 493-531.
Large, D.J., Christy, A.G. and Fallick, A.E. (1994) Poorly crystalline carbonaceous matter in high grade metasediments –implications for graphitisati on and metamorphic fluid compositions. Contrib. Mineral.Petrol. **116**, 108-116.
Lepland, A., van Zuilen, M.A., Arrhenius, G., Whitehouse, M.J. and Fedo, C.M. (2005) Questioning the evidence for Earth's earliest life-Akilia revisited. Geology **33**, 77-79.
Lowe, D.R. (1994) Abiological origin of described stromatolites older than 3.2 Ga. Geology **22**, 387-390.
Lowe, D.R. (1995) Abiological origin of described stromatolites older than 3.2 Ga: comment and reply. Reply. Geology **23**, 191-192.
Lowenstam, H.A. and Weiner, S. (1989) *On Biomineralization,* Oxford University Press, New York, the United States of America.
McCollom, T.M. (2003) Formation of meteorite hydrocarbons from thermal decomposition of siderite ($FeCO_3$). Geochim. Cosmochim. Acta **67**, 311-317.
McKay, D.S., Gibson, E.K., Thomas-Keprta, K.L., Vali, H., Romanek, C.S., Clemett, S.J., Chillier, X.D.F., Maechling, C.R. and Zare, R.N. (1996) Search for past life on Mars: Possible relic biogenic activity in Martian meteorite ALH84001. Science **273**, 924-930.
Melosh, J. (1989) *Impact Cratering: A Geologic Process*, Oxford University Press, New York, the United States of America.
Mojzsis, S.J., Arrhenius, G., McKeegan, K.D., Harrison, T.M., Nutman, A.P. and Friend, C.R.L. (1996) Evidence for life on Earth before 3,800 million years ago. Nature **384**, 55-59.
Mojzsis, S.J. and Harrison, T.M. (2000) Vestiges of a beginning: clues to the emergent biosphere recorded in the oldest known sedimentary rocks. GSA Today **10**, 1-6.
Mojzsis, S.J., Coath, C.D., Greenwood, J.P., Mckeegan, K.D. and Harrison, T.M. (2003) Mass-independent isotope effects in Archean (2.5 to 3.8 Ga) sedimentary sulfides determined by ion micro-probe analysis. Geochim. Cosmochim. Acta **67**, 1635-1658.
Moorbath, S. (2005) Dating earliest life. Nature **434**, 155.
Moorbath, S., O'Nions, R.K. and Pankhurst, R.J. (1973) Early Archaean age for the Isua iron formation, West Greenland. Nature **245**, 138-139.
Morowitz, H.J. (1996) Past life on Mars? Science **273**, 1639-1640.
Needham, J. (1968) *Order and life*, The MIT Press, Cambridge, Massachusetts, the United States of America.
Nutman, A.P., McGregor, V.R., Friend, C.R.L., Bennett, V.C. and Kinny, P.D. (1996) The Itsaq Gneiss Complex of southern West Greenland; the world's most extensive record of early crustal evolution (3900-3600 Ma). Precambrian Res. **78**, 1-39.
Nutman, A.P., Friend, C.R.L., Barker, S.L.L. and McGregor, V.R. (2004) Inventory and assessment of Paleoarchaean gneiss terrains and detrital zircons in southern West Greenland. Precambrian Res. **135**, 281-314.
Oehler, D.Z. and Smith, J.W. (1977) Isotopic composition of reduced and oxidized carbon in early Archaean rocks from Isua, Greenland. Precambrian Res. **5**, 221-228.
Pasteris, J.D. and Wopenka, B. (2002) Images of the Earth's earliest fossils? Nature **420**, 476-477.
Pasteris, J.D. and Wopenka, B. (2003) Necessary, but not sufficient: Raman identification of disordered carbon as a signature of ancient life. Astrobiology **3**, 727-738.
Perry, Jr., E.C. and Ahmad, S.N. (1977) Carbon isotope composition of graphite and carbonate minerals from 3.8-ae metamorphosed sediments, Isukasia, Greenland. Earth Planet. Sci. Let. **36**, 280-284.
Riding, R. (1999) The term stromatolite: towards an essential definition. Lethaia **32**, 321-330.
Riding, R. (2000) Microbial carbonates: the geological record of calcified bacterial-algal mats and biofilms. Sedimentology **47**, 179-214.
Rosing, M.T. (1999) ^{13}C-depleted carbon microparticles in 3700-Ma sea-floor sedimentary rocks from West Greenland. Science **283**, 674-676.
Samuelsson, J. and Butterfield, N.J. (2001) Neoproterozoic fossils from the Franklin Mountains, northwestern Canada: stratigraphic and palaeobiological implications. Precambrian Res. **107**, 235-251.
Schidlowski, M. (2001) Carbon isotopes as biogeochemical recorders of life over 3.8 Ga of Earth history: evolution of a concept. Precambrian Res. **106**, 117-134.

Schneider, D.A., Bickford, M.E., Cannon, W.F., Schultz, K.J. and Hamilton, M.A. (2002) Age of volcanic rocks and syndepositional iron formations, Marquette Range Supergroup: implications for the tectonic setting of Paleoproterozoic iron formations of the Lake Superior region. Can. J. Earth Sci. **39**, 999-1012.

Schoenberg, R., Kamber, B.S., Collerson, K.D. and Moorbath, S. (2002) Tungsten isotope evidence from ~3.8-Gyr metamorphosed sediments for early meteorite bombardment of the Earth. Nature **418**, 403-405.

Schopf, J.W. (1975) Precambrian paleobiology: problems and perspectives. Ann. Rev. Earth. Planet. Sci. **3**, 213-249.

Schopf, J.W. (1993) Microfossils of the early Archean Apex Chert: New evidence of the antiquity of life. Science **260**, 640-646.

Schopf, J.W. (2004) Earth's earliest biosphere: status of the hunt, In: P.G Eriksson, W. Altermann, D.R. Nelson, W.U. Mueller and O. Catuneanu (eds.) *The Precambrian Earth: tempos and events*, Developments in Precambrian Geology **12**, pp. 516-539.

Schopf, J.W. and Barghoorn, E.S. (1967) Alga-like fossils from the early Precambrian of South Africa. Science **156**, 508-512.

Schopf, J.W., Oehler, D.Z., Horodyski, R.J. and Kvennolden, K.A. (1971) Biogenicity and significance of the oldest known stromatolites. J. Paleont. **45**, 477-485.

Schopf, J.W. and Walter M.R. (1983) Archean microfossils: New evidence of ancient microbes, In: J.W. Schopf (ed.) *Earth's earliest biosphere*. Princeton University Press, Princeton, New Jersey, the United States of America, pp. 214-239.

Schopf, J.W. and Packer, B.M. (1987) Early Archean (3.3-billion to 3.5-billion-year-old) microfossils from Warrawoona Group, Australia. Science **237**, 70-73.

Schopf, J.W., Kudryavtsev, A.B., Agresti, D.G., Wdoeiak, T.J. and Czaja, A.D. (2002a) Laser-Raman imagery of Earth's earliest fossils. Nature **416**, 73-76.

Schopf, J.W., Kudryavtsev, A.B., Agresti, D.G., Wdoeiak, T.J. and Czaja, A.D. (2002b) Images of the Earth's earliest fossils? A reply. Nature **420**, 477.

Summons, R.E., Jahnke, L.L., Hope, J.M. and Logan, G.A. (1999) 2-Methylhopanoids as biomarkers for cyanobacterial oxygenic photosynthesis. Nature **400**, 554-557.

Thomas-Keprta, K.L., Bazylinski, D.A., Kirschvink, J.L., Clemett, S.J., McKay, D.S., Wentworth, S.J., Vali, H., Gibson Jr, E.K. and Romanek, C. (2000) Elongated prismatic magnetite crystals in ALH84001 carbonate globules: potential martian magnetofossils. Geochim. Cosmochim. Acta **64**, 4049-4081.

Thomas-Keprta, K.L., Clemett, S.J., Bazylinski, D.A., Kirschvink, J.L., McKay, D.S., Wentworth, S.J., Vali, H., Gibson Jr, E.K., Mckay, M.F. and Romanek, C. (2001) Truncated hexa-octahedral magnetite crystals in ALH84001: Presumptive biosignatures. PNAS **98**, 2164-2169.

Thomas-Keprta, K.L., Bazylinski, D.A., Clemett, S.J., Kirschvink, J.L., McKay, D.S., Wentworth, S.J., Vali, H., Gibson Jr, E.K. and Romanek, C. (2002) Magnetofossils from ancient Mars: a robust biosignature in the martian meteorite ALH84001. Appl. Environ. Microbiol. **68**, 3663-3672.

Ueno, Y., Isozaki, Y., Yurimoto, H. and Maruyama, S. (2001) Carbon isotopic signatures of individual Archean microfossils(?) from Western Australia. Internat. Geol. Rev. **40**, 196-212.

Ueno, Y., Yurimoto, H., Yoshioka, H., Komiya, T. and Maruyama, S. (2002) Ion micoprobe analysis of graphite from ca. 3.8 Ga metasediments, isua supracrustal belt, west Greenland: Relationship between metamorphism and carbon isotopic composition. Geochim. Cosmochim. Acta **66**, 1257-1268.

Ueno, Y., Yoshioka, H., Maruyama, S. and Isozaki, Y. (2004) Carbon isotopes and petrography of kerogens in ~3.5-Ga hydrothermal silica dikes in the North Pole area, Western Australia. Geochim. Cosmochim. Acta **68**, 573-589.

Van Kranendonk, M., Webb, G.E. and Kamber, B.S. (2003) Geological and trace element evidence for a marine sedimentary environment of deposition and biogenicity of 3.45 Ga stromatolitic carbonates in the Pilbara Craton, and support for a reducing Archaean ocean. Geobiology **1**, 91-108.

Van Kranendonk, M. and Pirajno, F. (2004) Geochemistry of metabasalt and hydrothermal alteration zones associated with c. 3.45 Ga chert and barite deposits: implications for the geological setting of the Warrawoona Group, Pilbara Craton, Australia. Geochem.: Explor. Environ., Anal. **4**, 253-278.

van Zuilen, M.A., Lepland, A. and Arrhenius, G. (2002) Reassessing the evidence for the earliest traces of life. Nature **418**, 627-630.

van Zuilen, M.A., Lepland, A., Teranes, J., Finarelli, J., Wahlen, M. and Arrhenius, G. (2003) Graphite and carbonates in the 3.8 Ga old Isua Supracrustal Belt, southern West Greenland. Precambrian Res. **126**, 331-348.

Walsh, M.M. (1992) Microfossils and possible microfossils from the Early Archean Onverwacht Group, Barberton Mountain Land, South Africa. Precambrian Res. **54**, 271-293.

Walsh, M.M. and Lowe, D.R. (1985) Filamentous microfossils from the 3,500-Myr-old Onverwacht Group, Barberton Mountain Land, South Africa. Nature **314**, 530-532.

Walter, M.R. (1976) Introduction. In: M.R. Walter (ed.) *Stromatolites*, Elsevier, Amsterdam, The Netherlands, pp. 1-3.

Walter, M.R., Buick, R. and Dunlop, J.S.R. (1980) Stromatolites 3,400-3,500 Myr old from the North Pole area, Western Australia. Nature **284**, 443-445.

Biodata of the authors of "***Ground Truth: The epistemology of searching for the Earliest Life on Earth***"

Mr. Eben Rose from Yale University works extensively in the field in both southern Africa and northwestern Australia, and has helped to map the Marble Bar 1:100,000 scale geologic map (not yet published) for the Geologic Survey of Western Australia. His previous experience includes a comprehensive review and re-examination of the Cambrian succession of the Grand Canyon.

Ms. Nicola McLoughlin is a geobiologist at Oxford University working on putative microbial fossils from the Warrawoona Group of Western Australia. She uses field mapping, microscopy and geochemical techniques to investigate the geological setting and origins of these putative fossils. Her research interests also include the role of microbial mats in Precambrian nutrient cycling, sediment dynamics and the evolution of metazoans.

J. Seckbach (ed.), Life as We Know It, 259-285.

Dr. Martin Brasier is Professor of Palaeobiology at Oxford University. He works on early biosphere evolution, integrating microfossils, biogeochemistry and chemostratigraphy from the earliest signs of life in the Archean through to the Cambrian explosion of multicellular forms. He obtained his Ph.D. degree from London University in 1973. His research interests also include the metabolism and evolution of bacterial and protistan fossil groups, especially the foraminifera.

E-mail: **martinbrasier@yahoo.co.uk**

GROUND TRUTH:
The epistemology of searching for the earliest life on earth

EBEN C. ROSE[1], NICOLA MCLOUGHLIN[2] AND MARTIN D. BRASIER[2]
[1]Department of Geology and Geophysics, Yale University P.O. Box 208109, New Haven, CT 06520-8109 USA. and [2]Department of Earth Sciences, Parks Road, Oxford, OX1 3PR, U.K.

1. Introduction

The quest for the earliest evidence for life on Earth is a 'holy grail' for astrobiology. But it leads us headlong into that infamous and dreaded morass, guarded by that monstrous riddle: "what is LIFE and how can we define it?" When a system undergoes a change in state or energy to become "life" the challenge is to match the term "life" to the referents of modern Earthly experience. This Earthly experience is teeming with diversity, complexity, and change and is packed with examples representing extremes of antagonism and competition on the one hand, and symbiosis and codependence on the other. This experience is what allows us to distinguish living systems from nonliving systems, at least for most instances of practical or historical importance. But as technology has extended our capacity to detect and identify those fundamental components which, together, make up living systems, we are confronted by a degree of complexity and interconnectedness between components within living systems, and with their non-living environment, that far outstrips our traditional capacities as interpreters of the natural world.

The search for the evidence of the earliest life on Earth has been long equated with the hunt for carbonaceous fossil cells. A generation of biologists and geologists has grown comfortable with the textbook claim (e.g. Schopf, 1999) that cellular life, especially cyanobacteria, appeared as early as the oldest well-preserved fine-grained sedimentary rocks themselves, some ~3.5 billion years ago (Ga). But iconic fossils from the Apex Chert which formed the basis for this assertion (Schopf and Packer, 1987) have proved wanting on close inspection –they appear to us to be non-biomorphic wisps and clots of degraded carbon preserved in silica that solidified in cracks within volcanic lavas, some to 100m to 1000m below the ancient land surface (Brasier *et al.*, 2002).

As might be expected, the widespread public *acceptance* of the claim for 3.5 billion-year-old life (even more so than the fossil evidence) has made its recent scientific *refutation* and abandonment a costly and difficult proposition. That is because scientific discourse across multiple disciplines has allowed this putative 'diverse fossil assemblage' in the 3.5 Ga old Apex Chert to become a keystone premise within the Earth and Life Sciences, with ramifications in the search for 'Life on Mars' and beyond.

All claims for early microbes, whether manifested as microfossil remains or as proxy signatures for microbial processes, should now be looked at with a more skeptical eye. Inevitably, new claims have arisen to seek the limelight for the oldest bona fide microbes (e.g. Furnes *et al.*, 2004, Tice and Lowe, 2004). A major challenge here, however, is the lack of universally accepted criteria that will allow the scientific community to reach consensus (*sensu* Tucker, 2003) upon evidence that is unique to or diagnostic for all known forms of microbial life.

The geological and biological communities are presently divided between those who cling to the 3.5 Ga antiquity of cellular life (the 'Early Eden' proponents sensu Brasier *et al.*, 2004a), and the growing number who are now more willing to look critically at the evidence. The aim of this chapter is to combine a discussion of the epistemology of searching for the earliest life on earth with a 'cross-examination' of the disputed fossil evidence and associated geochemical proxies. We aim to highlight the pertinent challenges to the canonical view of early life and to summarize the major lines of evidence, rather than to provide an exhaustive review of early life. To help frame this discussion in a geologic context, we first provide an overview of the rock record to portray the interpretive limitations in constraining both chronologies and environments.

2. Methods and Limitations of Geology

In the eyes of some critics, geology is at best a derivative science (Frodeman, 1995), relying upon physical and chemical principles to inject it with a degree of respectability. A more gracious characterization of geology is that it is often a narrative science rather than an experimental one. This means that our project of investigating the Earth's past seeks an ongoing refinement of its narrative, without recourse to, or at best supplemented by, tightly constrained experimentation (Cleland, 2001). Fortunately, geologists can apply to the rebus of Earth's ancient rocks a tool-kit that includes the disciplines of fluid dynamics, redox chemistry, isotope chemistry, organic chemistry, microbiology, and crystallography. Observations made upon modern and observable natural phenomena can also be applied to the ancient rock record. This approach, known as "the Principle of Uniformity", has been a guiding principle within the Earth Sciences since the time of James Hutton (1785).

Unfortunately, the "Principle of Uniformity" suffers from three fundamental weaknesses which loom ever larger as one reaches further back in the rock record: preservational bias; reconciling stratigraphic and sedimentologic time scales; and the early Earth as a temporally distant planet. The first weakness is that of preservational bias. It is well known that some environments are highly likely to be preserved and these may come to dominate the sedimentary rock record. Others lack known mechanisms for preservation and therefore remain either exceptional or unrecognized within the rock record. Representing the former are coastal marshes and dune fields, tidal flats, lowland lakes, swamps, and floodplains at or near sea level. Representing the latter are mountain environments and high-energy coasts where erosion removes the rocks; or deep ocean abyssal plains and trenches, which are rarely preserved because they are severely deformed by tectonic processes which may very rarely emplace them on to the continents. We must therefore accept that our reconstructions of ancient environments are limited to a much smaller range of physical settings than those that existed in the past.

The second weakness relates to the problem of reconciling stratigraphic and sedimentologic time scales. For example, how can short-term processes of sediment deposition in active environments that operate at time scales of days to months to tens of years, be reconciled in the rock record with geochronologies (i.e. radiometric rock ages) measured in thousands of years to tens of millions of years? Our tendency is to time-average the rocks between chronologic tie-points. This greatly understates the staccato aspect of the rock record, which may more accurately be seen as a stack of ordered snapshots rather than as a movie reel. Gaps predominate, but it is not straightforward to see how these gaps are distributed within a given geological outcrop.

The third, and perhaps most fundamental weakness of the Principle of Uniformity lies in our problem of the early Earth as a temporally distant planet. To what extent can analogues on the modern Earth serve as representatives for environments on the early Earth back to 3.5 (or more) billion years? How can we visualize physical and chemical processes from a distant time in which terrestrial vegetation did not shape surface environments, in which mat-forming microbial communities did not help stabilize substrates, or in which the cycling of carbon, phosphorus, and nitrogen was not dominated by phytoplankton? In other words, both geologists and biologists are limited in their pool of available referents for attempting to understand our early planet.

The principle of uniformity has clearly played a role in developing the investigative framework for the origins of life and early evolution of the biosphere. But these three major shortcomings, outlined above, need to be acknowledged and understood in order to claim any real knowledge about the ancient Earth.

3. Geologic and Ecologic Constraints

The contribution made by geology to the investigation of the origins of life is twofold. Firstly, it allows us to search for evidence *(patterns)* suggestive of early surface environments conducive to the evolution of biological systems. Secondly, it allows us to attempt to characterize those geological *processes* that have been altered or shaped by biological processes. In working towards these goals, geological models can be informed by studies of biomolecular architecture and function, and of metabolic substrates and their products, to help us understand the co-evolution of the physical and biological environments (e.g. Pace, 1997). But geologists and biologists are limited in their interpretive flexibility by over-representation of the modern, mature biosphere. This mature biosphere tends to dominate our lexicon of natural references and places inevitable constraints on our thinking.

The tape of history cannot be replayed and, because its oldest remaining fragments are rare and mostly stretched, squeezed, or melted beyond recognition, there are obvious limitations to the fidelity of any reconstruction we may create from them. Fragmentary and rare though it is, there is nevertheless a record of unmelted rock from the surface of the Earth going back to a time when decreasing bombardment by large meteorites marked the tapering end of our planet's initial coalescence (Kring and Cohen, 2002). Impacts continued after this time and evidence for recurrent impacts exists in the earliest rock record in the form of spherules (hardened droplets of molten rock rained down from near-ballistic altitudes) and possibly impact-related fracturing of lithified rock (Kyte *et al.*, 2003). The crater-pocked surface of the moon indicates a period of late heavy bombardment terminating between ~3.85 and ~3.82 Ga (Moorbath, 2005b). Surface-sterilizing impacts remain a possibility, if not a probability, well into the preserved rock record. Planetary resurfacing by global volcanism and sediment transport during this "Hadeo-Archean" interval (Fig. 1, Moorbath, 2005b) posed an extreme environmental constraint upon any vulnerable early biosphere. On the other hand, bolides may have delivered exotic chemicals, such as amino acids and nucleotides, "ingredients" central in many recent speculations as to the origins of life (e.g. Delsemme, 1998). It has also been suggested by some theorists that microbial life was seeded on the Earth from elsewhere during this early meteor bombardment (Mileikowski *et al.*, 2000).

It is difficult to say when the Earth became at last stable enough to sustain the biology from which today's life evolved. We should not assume that the emergence of life was inevitable after impacts became survivable. Nor should we assume that life arose only once on our planet. But whether fully native or fully exotic, life as we have come to understand it must have experienced many other extreme environmental forcings. If the end of period of late heavy bombardment is accepted to date to ~3.85 billion years, then this may serve as an oldest age limit, the first threshold, for potential life on Earth. The youngest constraint for the emergence of microbial life is a much more contentious issue that is discussed in this chapter, and equates with finding the oldest convincing and recognizable evidence for life in the rock record.

4. Epistemology and Early Life

The Gunflint Chert is a rock horizon in the Lake Superior region of North America that serves as our first milepost for assaying the antiquity of microbial life on Earth. At ~1878 million years in age (Fralick *et al.*, 2002) it pre-dates the Cambrian Explosion, the rapid radiation of multicellular forms that is at the root of most animal phyla (Knoll 2003), by some 1300 million years. The Gunflint chert contains a diversity of well-preserved microfossils (Barghoorn and Tyler, 1965; Knoll, 2003) found within algal sediments (Awramik and Semikhatov, 1978). Although discussion regarding its precise environment of deposition persist (Sommers *et al.*, 2000), and the assemblage co-exists with some abiotic mineral artifacts (e.g. Tyler and Barghoorn, 1963), the veracity of the microfossils is not in doubt. They provide a qualitative standard to which other microfossil-like morphologies and assemblages may be compared. (Putative microfossils are discussed in more detail in section 7.1 below).

The Gunflint microfossils were discovered at a time when the Precambrian time interval (i.e. rocks older than 542 million years) was thought by many to be devoid of fossils and, quite possibly, devoid of life itself (e.g., Darwin, 1859). The belated confirmation of a Precambrian fossil microflora in 1965 prompted a wave of effort to find yet older microfossils. The products of this "gold rush" for older microfossils yielded many strong claims, but few of these (the Apex "microfossils" being an exception [Brasier *et al.*, 2002, 2004a, 2004b]) have yet been subjected to close interdisciplinary and multi-person scrutiny.

Figure 1. A qualitative solution space showing the key questions in paleobiology, geology and geochemistry that need to be integrated in the search for early life. The most robust claims for early life lie in the central area, where geological (section 5), geochemical (section 6) and paleobiological (section 7) criteria are mutually reinforcing.

Attempts to plug this yawning chasm in the early fossil record between 3.5 and 1.875 Ga have, in some instances, relied upon questionable lines of logical construction. That is to say, the authors have (no doubt with good intentions) drawn attention to evidence that is "consistent with" microbial processes without *falsifying* (*sensu* Popper, 1959) other possible non-biological scenarios that may likewise be "consistent with" their evidence. Arguments of the kind: 'absence of evidence is not evidence of absence' are also highly insecure in the high-stakes search for early life, and are incompatible with evolutionary studies of all kinds. The most robust claims for early life are based upon mutually supporting lines of evidence drawn from the disciplines of geological mapping, geochemistry, and paleobiology that satisfy the criteria summarized in Fig. 1. (More specific, but by no means complete, criteria for microfossils and stromatolites are available in the literature, summarized by Brasier *et al.*, 2004b). Qualification of "biogenicity" increases towards the center of Fig. 1. Satisfaction of these criteria may be possible with reference to a large body of experimental work and a review of the literature (McKay *et al.*, 2002), but so far no single claim for early life has accomplished this and been universally accepted by the early life research community. In subsequent sections we explore how geological context (section 5), geochemistry (section 6) and paleobiology (section 7) contribute to the search for early life.

5. Geological Context

5.1. METALS, ROCK TYPES AND PLATE TECTONICS

The Archean rock record is dominated in total thickness by lavas of mafic or ultramafic composition (metal rich and silica poor), that are consistent with expectations of a higher heat flow on early Earth. The presence of ultramafic, mantle-derived lavas also shows that the Earth's crust was still maturing in some sense. There is, no evidence for modern plate tectonics prior to ~3200 Ma, and the evidence is not strong for plate tectonics between 3200 and 2800 Ma (McCall, 2003). That is not to say that plate tectonics in some form was not in operation at these times, and ~3.7-3.65 Ga orthogneisses from Green and are interpreted by some as suggestive of subduction-related precursors of continental crust. Subduction and remelting of the Earth's crust may be essential to the biosphere to prevent ecologic depletion of insoluble elements. Crucial elements that play a significant to indispensable role as enzymatic cofactors include: nickel and cobalt, which are employed in the carbon-fixing Wood-Ljungdahl (Acetyl-CoA) pathway; along with Cr, Cu, Zn, Mg, Mn, Ca, and Fe (see Williams and Frausto da Silva, 1996). Significantly, these elements occur most abundantly within mantle-derived ultramafic lavas and ashes that are uniquely abundant within crust older than 2500 Ma. Ultramafics are also a richer source than post-Archean "recycled" rocks for other biologically significant trace elements, including V, Mo, W, Se, Hf, F, I and S. It can be argued, therefore, that the highly metaliferous Archean crust was ideally suited to the synthesis and sustainment of early life (Brasier *et al.*, 2002).

In addition, there is abundant evidence discussed in this chapter for vigorous hydrothermal activity that may have given rise to Fischer-Tropsch type processes (Sherwood Lollar *et al.*, 2002; Horita and Berndt, 1999). This process produces simple carbon compounds by reacting hydrogen and carbon dioxide in the presence of iron and/or nickel catalysts, which are abundant in ultramafic rock. A schematic summary of the environmental constraints on the evolution of early life discussed here is given in Fig. 2.

The structural controls on preserved crust of Archean are dominated not by plate tectonics but by the growth of granitic diapirs (magmatic intrusions of >100 km scale), which deformed adjacent strata in a process sometimes referred to as "vertical tectonics" (Van Kranendonk *et al.*, 2004). The surrounding dense, thermally mature ultramafic rocks lost their 'buoyancy' and were pulled downward as the granite diapirs bulged upward, thereby deforming and heating all the strata squeezed between the granitic intrusions. These deformed strata then underwent mineral re-crystallization that resulted in an abundance of green minerals, such as chlorite, green amphiboles and fuchsite, and have thus been named 'greenstone belts'. The parent protolith (original rock) for metamorphosed greenstone belts is sometimes unrecognizable as anything more than altered lava or ash (and even this can be difficult at times). This means that the possibility of finding any signature of ancient life in such squeezed metamorphic rocks is very low, almost regardless of age.

Although several greenstone belts have been studied worldwide, only two (the Pilbara of Western Australia and the Barberton Mountain Land of South Africa and Swaziland) contain intact stratigraphic piles of Early Archean age (3.5-3.0 Ga). Older rocks (3.8 to 3.7 Ga) like those of Greenland and Labrador are of much higher metamorphic grade. Efforts to find the earliest life in these rocks are focused on putative meta-sedimentary rocks that are predominantly chemical in origin. Detrital meta-sediments composed of the eroded fragments of older rocks, such as sand and mudstones are much less common.

5.2. BANDED IRON FORMATIONS (BIFs) AND ATMOSPHERIC OXYGEN

A major icon in our understanding of the oxygenation history of the planet is banded iron formations (BIFs). These are laterally extensive sedimentary rocks, largely composed of alternating microcrystalline silica (chert) and iron (III) rich layers that give them a striking red or chocolate color. BIFs are particularly abundant in the Paleoproterozoic, between 2.5 and 2.0 Ga. They are commonly associated with shales, and less commonly carbonates and volcanic ash. The significance of BIFs has traditionally centered not upon the setting of large-scale silica deposition, but rather upon the global-scale redox shift that allowed soluble Fe (II) to be oxidized *en masse* to precipitate as less-labile Fe (III). The prevailing interpretation is that the Fe (II) was oxidized to Fe (III) in near-surface environments, perhaps by biologically synthesized oxygen (Konhauser 2002). However, it remains uncertain whether this oxidation was a syndepositional or post-burial (diagenetic) process.

Although BIFs are most abundant in the Paleoproterozoic, the oldest known BIF are from the ~3.8 billion year old rocks of the Isua greenstone belt. These very early BIFs are not widely regarded as evidence of free atmospheric oxygen, but rather, of very restricted local oxygen, although the argument for early oxygenation of the Earth's atmosphere has been put forward and is maintained by Ohmoto (1996,

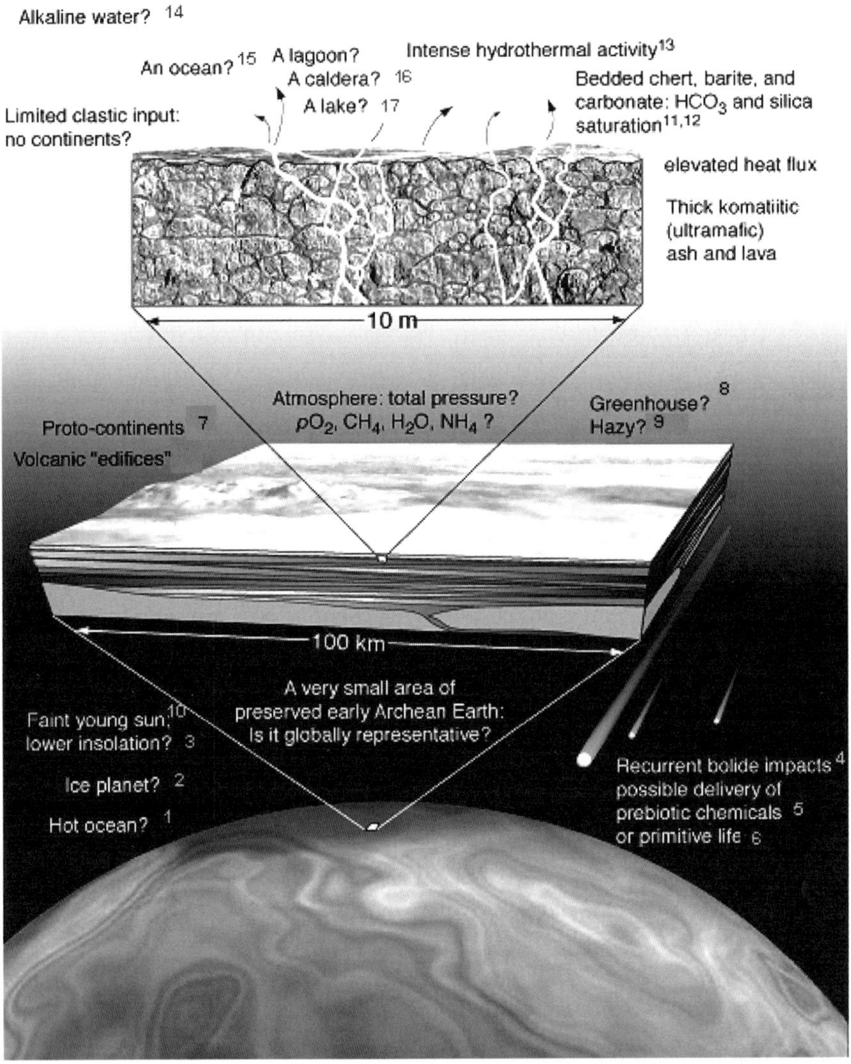

Figure 2. Schematic diagram of the Archean geological environment in which early life arose; **(base)** primordial earth subject to recurrent bolide bombardment, with perhaps a hot ocean and or periods of ice cover; **(center)** enlargement of a small part of the planetary surface, with proto-continents and volcanic edifices, enveloped by a thick atmosphere, containing unknown amounts of greenhouse gases, low oxygen levels and chemical haze; **(top)** diagram of the upper crust composed of a thick lava pile with vigorous hydrothermal circulation, overlain by an alkaline ocean with chemical sediments. **1.** Knauth and Lowe (2003). **2.** Levy and Miller (1998). **3.** Sagan and Mullen (1972). **4.** Sleep *et al.* (1989). **5.** Delsemme (1998). **6.** Mileikowsky *et al.* (1999). **7.** Bedini *et al.* (2004); McCall G.J.H. (2003). **8.** Kasting and Catling, (2003). **9.** Pavlov *et al.* (2001). **10.** Molnar and Gutowski (1995) **11.** Lowe and Worrell (1999). **12**. Siever R. (1991). **13.** Horita and Berndt (1999). **14.** Grotzinger and Kasting (1993). **15**. Groves *et al.* (1981); Nisbet and Sleep (2001); Shen and Buick (2004), **16.** Van Kranendonk *et al.* (2001); Van Kranendonk and Pirajno (2004) **17.** Buick (1992).

1997) and colleagues. The rise of atmospheric oxygen is a topic of continuing research and debate, with most participants focusing their attention on the interval spanning 2.3 and 2.1 Ma (e.g Bekker *et al.*, 2004), although biomarkers, as discussed later, may tell a different story.

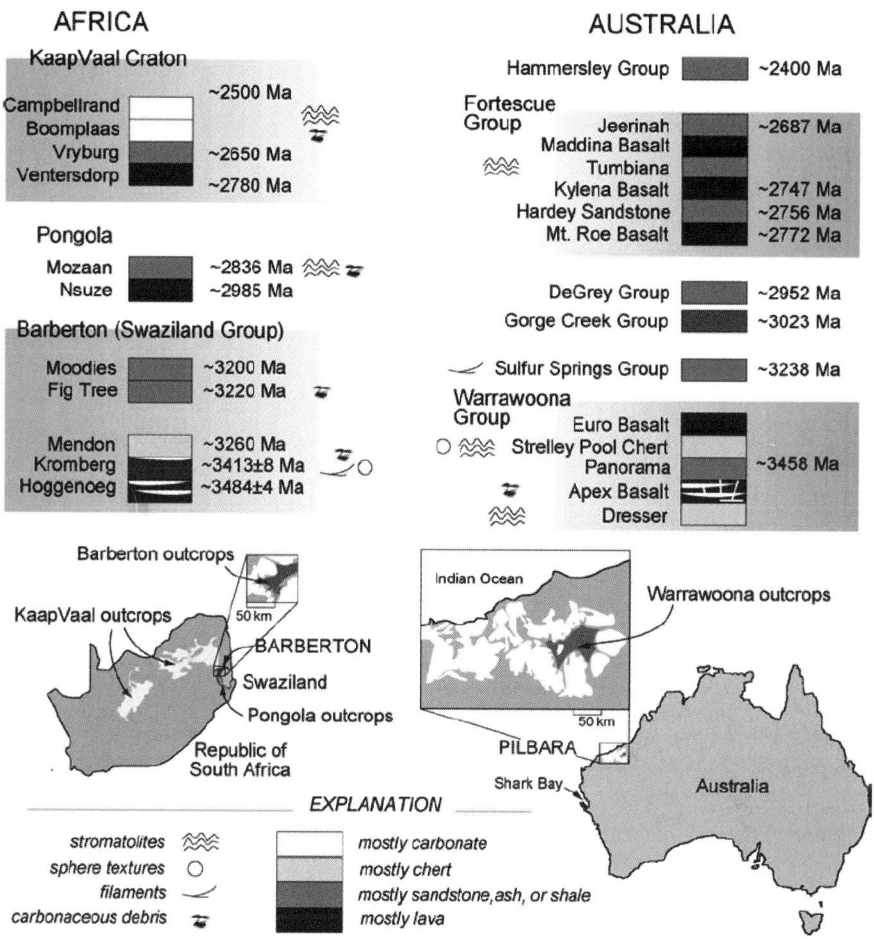

Figure 3. Simplified geological maps of the Pilbara and Barberton Cratons with a summary of the rock record, showing horizons that contain putative fossil remains: stromatolites, coccoids, filaments and carbonaceous debris.

5.3. THE PILBARA AND BARBERTON

The Pilbara and the Barberton cratons preserve rocks that span the interval ~3.5 Ga to 2.4 Ga and are shown in Fig. 3. The smaller of the two is in the Barberton Mountain Land bordering South Africa and Swaziland. The larger is in the Pilbara region of northwest Australia. (Other less well documented Archean rocks that are not discussed here, are found in the ~3.0 Ga Steep Rock Group of the Slave Craton in northwest Ontario, Canada and the ~2.9 - 2.7 Ga Belingwe Greenstone belt of Zimbabwe.) The question of whether these small windows are either truly representative of the Achaean earth surface, or exceptional for their time, is often overlooked.

Preservation of the Pilbara strata may be attributed in part, if not wholly, to the structural protection afforded them by a ring of large granite bodies (Fig. 3). The center of the ring contains strata as old as 3515 Ma (Buick *et al.*, 1995), though the most prominent and well-studied exposures comprise the slightly younger Warrawoona Group. The Warrawoona Group is dominated by thick pillow lavas that indicate subaqueous flow, between which are rare leached horizons, overlain by stratified cherts that are cross-cut by chert-filled fractures.

Preserved ancient strata of the Barberton Mountain Land are also surrounded by granitic bodies that may be the roots of supra-crustal domes (Lowe and Byerly, 1999), between which are compressional structural basins that contain the greenstone belts. In general terms, both the Pilbara and the Barberton share a broad two phase geologic history, with an early phase preceding ~3.0 Ga, where sediments and lava flows filled low-relief basins that were subsequently buckled and deformed; and a later second phase after 3.0 Ga, in which an overlying succession was deposited that remains in modern times laterally extensive and less deformed. This overlying layer in Western Australia comprises the ~2.7 Ga Fortescue Group, preserved in patches throughout the Pilbara Craton and more continuously beneath the Hamersly Basin to the south. In southern Africa it comprises the ~2.8 Ga Pongola Group in western Swaziland and its lateral equivalent, the Witwatersrand Group, which has extensive exposures to the west near Badplaas and throughout the Johannesburg region. The Pongola and Witwatersrand groups are overlain in turn by the Kaapvaal succession with extensive exposures along the Oranje River near Griquatown, South Africa, and sparsely throughout the Northern Cape and Transvaal (Fig. 3).

An important component of Archean sedimentary rock older than 3.0 Ga in both the Pilbara and the Barberton Mountain Land is chert, a silica-rich rock that was either precipitated chemically from a hydrous phase or formed as a replacement of earlier carbonate or sulfate phases. Cherts are particularly resistant to erosion and deformation, which certainly aids in their preservation. Some of these cherts contain relict textures of sedimentary grains, while others contain angular fragments that fill fractures, consistent with a hydrothermal origin. The silicification (conversion to chert) of Archean sediments is thought to be an early alteration feature. This assumption encouraged Knauth and Lowe (2003) to present oxygen isotope data (from the silica-bound oxygen atoms) that, based on temperature-dependence of oxygen fugacity at equilibrium conditions, allowed them to back-calculate the temperature of crystallization. Their claim, with all assumptions intact (including environment of deposition) was that the Archean ocean was hot, from 55° to 84°C. There is however, great uncertainty in characterizing the environment of deposition of many Archean cherts and the history of their lithification.

Following the discovery of deep sea hydrothermal vents or "black smokers" in 1979 (Spiess *et al.*), and exploration of the ecosystems that they host, models of deep-sea deposition were considered for Archean cherts, and a connection was sought between black smokers, Archean cherts, and the origin of life. This paradigm of a "hydrothermal cradle for life" has been re-invigorated by recent studies of the "Lost City" hydrothermal vent field, which is serpenitinite-hosted (Kelley *et al.*, 2005) and may provide an analogue for early-Archean ultramafic systems. Channer *et al.* (1997) and DeRonde *et al.* (1997) reported relict black smokers from cherts of the Barberton Mountain Land, preserved in modern exposure as irregular mounds or "pods" of iron oxides. However, Lowe and Byerly (2003) subsequently reported that the iron oxide goethite, which is thermally unstable above 120°C, constitutes the main component of these "pods" and has apparently not been "cooked out" of these hot "hydrothermal" rocks (which have elsewhere been heated above 120°C). For this and other reasons, the pods were reinterpreted as local accretions built up by sub-recent springs, some of which are still seeping. This case history emphasizes again, that detailed geological mapping and interpretation underpins all claims for early life.

In comparison with the Barberton exposures, the Warrawoona Group of the Pilbara craton contains more obvious dyke-filling lithologies that are identified as hydrothermal in origin (e.g. Van Kranendonk, 2000; Brasier *et al.*, 2002). For example, a well-documented barite ($BaSO_4$) dome complex in the Dresser Fm (Fig. 3) is concentrically layered with local collapse breccias and shelves, indicating that the domes formed original topographic features with meter-scale relief (Nijman *et al.*, 1998). Barite sands, variously replaced by chert, constitute a significant component of the Dresser Fm elsewhere (Van Kranendonk, 2000) and also occur in the Mapepe Fm of the Fig Tree Group in South Africa (Lowe and Nocita, 1999). Some of the Barberton dyke cherts, however, contain impact spherules that have been used to support a neptunian origin for the dyke-filling cherts, that is, by marine sedimentation from above rather than by deposition from hot hydrothermal fluids sourced below (Lowe *et al.*, 1999). But while the spherules descend to a depth of 50 meters or so, the host dykes reach at least 1km below the surface. In all other respects, these dyke systems are indistinguishable from the hydrothermal systems in Western Australia (cf. Brasier *et al.*, 2002).

6. Geochemistry

6.1. CARBON ISOTOPES

The story of the ~3.8 Ga rocks from Akilia, South west Greenland, serves to illustrate the manner in which claims for the antiquity of the biosphere have in recent years been applied to Earth's oldest supracrustal rocks. Carbon isotopes figure prominently in this debate because they have been used as a proxy chemical signature for biologic processing (Schidlowski, 2001). The following discussion is offered as a critical caveat against the use of carbon isotopes alone as evidence for early life, and serves as a preface to understanding the Akilia narrative.

We know of only one global-scale carbon cycle, and it is biologically mediated. It involves an exchange between three major reservoirs: gaseous carbon, mostly CO_2; carbonates, which are solid phases using the $(CO_3)^{2-}$ ion, and organic carbon. The stable carbon isotopes ^{13}C and ^{12}C are present in the global system in fixed amounts, but the ratio between them in any given reservoir of carbon may differ over time with conditions that favor their kinetic fractionation. Kinetic fractionation occurs because the lighter isotope ^{12}C is more easily captured and released from chemical bonds than the heavier isotope ^{13}C. Fractionation is possible whenever chemical bonds are broken and re-established at low temperatures. (At higher temperatures, e.g. above 200°C, the kinetic effects are swamped by over-excited atomic vibrations, which tend not to discriminate between isotopic mass differences.)

When fractionation takes place as a bulk process, the reservoir ratios of ^{13}C and ^{12}C will vary from some central mean. Selective fixation of ^{13}C in one reservoir will therefore lower the $^{13}C/^{12}C$ in the other. For example, the ribulose-1,5-bisphosphate carboxylase/oxygenase (Rubisco for short) is the rate-limiting enzyme in the carbon-fixing pathway (Calvin cycle) of photosynthesis and tends to favor isotopically light carbon over the heavier carbon. This results in a fractionation of ~22 per mil lower than the reservoir it draws from, as long as that reservoir remains well mixed and is large enough to approximate a relatively invariant global mean. It should be highlighted, however, that carbon isotope fractionation is by no means a uniquely biological signal. For example, abiotic processes such as Fischer Tropsch synthesis (Horita and Berndt, 1999), metamorphism and metasomatism (van Zuilen *et al.*, 2002) can produce large negative C isotope fractionations. Also meteorites (Sephton and Gilmour, 2001) and organic haze (Pavlov *et al.*, 2001) can generate a suite of organic compounds with negative carbon isotopes. In the early C isotope record there is a pronounced discontinuity in $\delta 13C_{org}$ and $\delta 13C_{carb}$ at 3.5 Ga (see Fig. 4. Schindlowski, 2001) that may largely reflect the extensive metamorphic alteration of the earlier rocks. From ~3.5 Ga onwards, however, the record is widely believed to be consistent with biological processing. But an implicit assumption in all carbon isotope studies that is sometimes overlooked is that the reservoir from which the sampled carbon was fixed is a well-mixed reservoir in equilibrium with atmospheric CO_2. But the degree of mixing in Archean reservoirs is unknown. Also, in the early rocks from the Pilbara and Barberton, it is very difficult to demonstrate the syngenicity and biogenicity of the carbon and to verify that the carbonate and organic carbon were precipitated from the same water mass. Metasomatic alteration of these rocks by CH_4/CO_2 rich fluids related to contemporaneous mafic volcanism is pervasive and has likely modified the C isotope signatures (Harnmeijer and Buick, 2005). Furthermore, graphite ("cooked" carbon) is a thermodynamically stable inorganic phase that can be formed, among other ways, by the decarbonation of iron-carbonate minerals commonly found in Archean sediments, at temperatures between 70 and 125°C, perhaps even condensing onto abiotic microfossil mimics (Garcia-Ruiz *et al.*, 2003).

The trust placed in the earliest carbon isotope record has now outrun its term, as the recent Akilia debate exemplifies. The saga begins with a strong claim for the earliest signs of life, preserved in the form of isotopically light carbon within highly metamorphosed rocks from the island of Akilia off the west coast of Greenland (Mojzsis *et al.*, 1996). At ~3800 Ma, these are some of the oldest rocks exposed on Earth. The material analyzed was graphitic carbon said to be present as inclusions within mineral grains of apatite (calcium phosphate). Using an ion microprobe, Mojzsis *et al.* (1996) reported isotopic values ranging from –20 to –50 per mil relative to PDB (Pee Dee Belemnite, an arbitrary universal standard with a value of 0 per mil). They argued that this carbon was sealed within the apatite in a low-temperature sedimentary environment and was shielded from subsequent fluid exchange during metamorphism. Thus, they argued, its isotopic signature was original (i.e. it preceded metamorphism) and could therefore be used as evidence in support of life on Earth at ~3.8 Ga.

A few years later however, a different team of geologists examined the now famous Akilia site and determined that the outcrop, which is only a few tens of square meters in total area, exhibits nothing to indicate an originally sedimentary environment. They found it to be purely metamorphic and inferred that the carbon was metamorphic as well and thus of no biological relevance (Fedo and Whitehouse, 2002). Even more concerning however, is the inability of an independent scientific team to verify the presence of carbon in the original and recollected apatite crystals (Lepland *et al.*, 2005) and it is now widely held that the Akilia rocks are "less and less plausible domiciles for biogenic tracers" (Moorbath, 2005a). A parallel debate has also occurred regarding the original biogenic interpretation of isotopically light graphite in apatite crystals from Isua (Mojzsis *et al.*, 1996). Here petrographic and geochemical studies have also rejected a biogenic origin and proposed the metamorphic decomposition of ferrous carbonate (siderite) as the source of the ^{13}C depleted organic matter (van Zuilen *et al.*, 2002). There does remain however, one locality in the Isua region where the association of graphite with meta-sedimentary rocks may still be suggestive of life (Rosing *et al.*, 2004), but this occurrence is still under debate and we therefore turn to younger rocks in the following sections. To conclude this carbon isotope discussion, we caution against using carbon isotopes alone as evidence for early life. In the early rock record carbon isotopes must be integrated with other isotope systems that purport to reflect biogenic fractionation, for example nitrogen (Pinti and Hashizume, 1999) and sulfur (Shen *et al.*, 2001), and with a complete understanding of the geological context and biogenic fractionation pathways wherever possible (Fig. 1).

6.2. BIOMARKERS

In rock horizons that do not preserve cellular microbial remains, the occurrence of hydrocarbon "biomarkers" in refractory carbon residues (kerogen) has been used as a proxy for specific biological pathways (review: Brocks and Summons, 2003). This approach has been applied to Late Archean and early Proterozoic rocks, with hopanoid and sterol-derived molecules being identified by gas chromatography on extracts from shales of the ~2.7 Ga Fortescue Group (Fig. 3) of western Australia (Brocks *et al.*, 1999; Summons *et al.*, 1999). After much careful work to demonstrate that these compounds are not contaminants and "probably syngenetic with their Archean host rock" (details in Brocks *et al.,* 2003), they have been argued as indicative of oxygen production by oxygenic photoautotrophs that was subsequently utilized in the biosynthesis of sterols (Brocks *et al.*, 1999). These claims have generated vigorous debate, as they infer production of atmospheric oxygen 500 million years before widely accepted independent lines of geological evidence that indicate oxygen's atmospheric accumulation (Rye and Holland, 1998). At the center of this debate is the uniformitarian assumption that hopanoid and sterol biosynthesis pathways, which are oxygen-requiring today, have remained unchanged over geological time. This assumuption has been found to be invalid as obligate anaerobes, such as *Geobacter sulfurreducens,* are capable of producing hopanoids under strictly anaerobic conditions (Fischer *et al.*, in press). Also Raymond and Blakenship (2005) have found that enzymes that catalyze some, but perhaps not all, of the reactions involved in steroid biosynthesis have an oxygen independent counterpart. It therefore appears that hopanoids are not safe indicators of early atmospheric oxygen, although steranes, if they are indeed sygenetic, may still be consistent with significant amounts of oxygen accumulating at least locally. This discussion reminds us of the limitations of analogous reasoning, that is, of extending modern biochemistry back to the Archean rock record.

7. Paleobiology

7.1. MICROFOSSILS AND LOOK-ALIKES

The famous Apex 'microfossils' (Fig. 4b) have been described in a series of papers (Schopf and Packer, 1987; Schopf, 1999; Schopf and Walter, 1983; Schopf *et al.*, 2002). These objects have held their key position in Archaean paleobiology because of a supposedly good state of preservation and their wide acceptance by the scientific community. This contrasts with preliminary reports of other presumed microfossils from the Warrawoona Group, dismissed as either unreliable or unreproducible (Schopf, 1993). Eleven putative species of microfossils from the Apex chert have hitherto provided the oldest accepted morphological evidence for life on Earth. Furthermore, the size range of the supposed cells (< 20μm in diameter) has been taken to suggest that oxygen-releasing cyanobacteria may have been present at least 3.45 Ga ago (Schopf, 1999). If accepted, this must imply that high levels of biological diversity were achieved at a very early stage in Earth history (Schopf, 1993), remarkably soon after the end of massive meteoritic bombardment of the inner solar system and implies an early start for the contribution of photosynthetic oxygen to the atmosphere. However, these structures are nearly a billion years older than putative cyanobacterial biomarkers (Summons *et al.*, 1999), genomic arguments for dating the first appearance of cyanobacteria (Hedges *et al.*, 2001) and evidence for an oxygenic atmosphere (Catling *et al.*, 2001).

The security of these claims is now in doubt. This is in part because major aspects of the preservation and context of this potentially important evolutionary benchmark have received little independent or detailed study, and in part because new techniques of analysis are now available. Brasier *et al.* (2002, 2004a, 2004b, sub) are taking a fresh look at Earth's olde st microfossils from the Apex Chert, using an integrated and collaborative field and laboratory program. They question both the claims for viable context and for the biological nature of these putative fossils and find that the "microfossils" are mineral reaction rims that formed in a hydrothermal dyke environment, some 100 to 1000m below the palaeosurface. Re-crystallization of silica to form spherulitic chalcedony caused the displacement of amorphous carbonaceous matter towards spherulitic margins, creating a morphological spectrum of arcuate and dendritic microstructures that include "microfossil"-like artifacts (see Brasier *et al.*, 2002).

We advance a null hypothesis for the investigation of all Archean biological claimants: that very ancient/alien microfossil-like structures (or stromatolites, or geochemical and isotopic signals) should not be accepted as being of biological origin until possibilities of their non-biological origin have been tested and can be falsified (Brasier *et al.*, 2002). This means that all Archean microfossil assemblages are currently open to question. For example, putative microfossils from the ~3.5 Ga Dresser Fm (Ueno *et al.*, 2001) are similar to the spherulitic pseudofossils described from the Apex Chert by Schopf (1993). Filaments from the ~3.24 Ga Sulphur Springs Formation (Rasmussen, 2000) look intriguing and have received some acceptance (e.g. Knoll, 2003), but their geological context (Vearncombe, 1995) and initial suggestions of biological behavior (i.e. preferred orientation dependent on substrate) need be to be further investigated. Putative microbial endoliths have also been recently claimed from subaqueous volcanic rocks, creating tubular cavities in the glassy rims of pillow lavas, from the ~3.5Ga Hoggenoeg and Kromberg Fm of the Barberton (Furnes *et al.*, 2004). But similar morphological structures can be produced by ambient inclusion trails (Tyler and Barghoorn, 1963), formed when crystal inclusion migrate under high fluid pressures through a glassy matrix or crystalline mush and this abiogenic origin has not yet been falsified (cf. Brasier *et al.*, 2004c).

Further examples of morphologically simple spheroids, ellipsoids, and filaments have been reported from the ~2.6 Ga Ghaap Subgroup of South Africa (Klein *et al.*, 1987; Lanier, 1986; Altermann and Schopf, 1995), the ~2.7 Ga Tumbiana Formation of northwest Australia (Schopf and Walter, 1983), and the ~3.41 Ga Kromberg Fm of South Africa (Walsh, 1992; and Westall *et al.*, 2001; Tice and Lowe, 2004). The perennial difficulty with interpreting all such "microfossils" is that they comprise shapes (spheres, filaments) that are difficult to distinguish from natural non-microbial mineral crystal habits that could grow under similar conditions (Fig. 4c, d). Simple abiotic experiments demonstrate the ease with which microfossil-like artifacts can be generated by geologically relevant processes (Garcia-Ruiz *et al.*, 2003; Bosak *et al.*, 2004). It may need to be demonstrated, therefore, that candidate microfossil-like structures occupy their own discrete area of morphospace, which does not overlap with that occupied by relevant abiological processes such as crystal growth. Re-examination and re-sampling of these putative microfossils in the 2.6-3.5 billion year old interval, investigating the criteria outline in Fig. 1, certainly holds exciting rewards in the search for early life.

7.2. STROMATOLITES AND WRINKLE MATS

The third major type of indicator used in the search for early life, are macroscopically layered sedimentary structures called stromatolites. These are wrinkled surfaces, layered domes, cones, and columns that were constructed by abiotic and or microbially-mediated sediment trapping, binding and precipitation. In the rock record, particularly the Archean, the challenge is to identify and separate these biotic and abiotic mechanisms of stromatolite growth. This is often made very difficult by eons of exposure to geological forces that act to destroy organic remains and microstructures. This problem lies at the root of all controversy surrounding their biogenicity.

The stromatolite debate arises for three principal reasons. Firstly, there has been an over-emphasis upon modern stromatolite analogues, particularly those from Shark Bay and the Bahamas which are forming in biological, chemical, and physical regimes that cannot be compared in a uniformitarian manner with the Archean. Secondly, there persists a reluctance to fully explore plausible abiological mechanisms, such as the growth of crystal crusts and deformation of cohesive sediments that could feasibly produce stromatolite morphologies. Thirdly, there has been limited interaction, until recently, between stromatolite paleobiologists and modelers of fluid dynamics, sedimentology and crystal growth. Fortunately, there is now a growing communication between those engaged in stromatolite field-studies and those engaged in experimentation and numerical modeling that holds great promise for understanding stromatolite morphogenesis.

The ~3.5 Ga Dresser Fm (Fig. 3) of the Warrawoona Group contains the oldest known stromatolites which are found in the North Pole Dome. These occur in both syn-depositional barite mounds and dykes (Van Kranendonk *et al.*, 2001) that form a hydrothermal complex (Nijman *et al.*, 1998;) and in the intercalated, silicified, ferruginous-carbonates (Walter *et al.*, 1980). The original stromatolites of Walter *et al.*, 1980, were reviewed by Buick *et al.*, 1981 in an informative study that attempted to define universal stromatolite biogenicity criteria, and which concluded that that the North Pole stromatolites were only "probable or possible" biogenic stromatolites. More recent studies have also described domal and stratiform stromatolites in the vents of barite dykes at the North Pole (Fig. 35 and 36, Van Kranendonk *et al.*, 2001), and have argued that these were constructed by hyperthermophilic microbes (Van Kranendonk, 2001). However, the morphology of these stromatolites is controlled by the thickness of the precipitated barite crusts and draping chert layers, and their distribution likely reflects the supply of supersaturated solutions from which they were precipitated. Robust microtextural and isotopic evidence for the involvement of microbial mats in the growth of these stromatolites has not yet been reported.

Fuel for the debate upon Archean stromatolite biogenicity (Hofmann, 2000) was provided by the discovery of a new conical stromatolite locality in the ~3.4 Ga Strelley Pool Chert, of the Warrawoona Group, Western Australia (Hofmann *et al.*, 1999). Conical stromatolites are a characteristic feature of this unit and were originally considered as biogenic in origin (Lowe, 1980), a claim that was then rescinded in favor of an abiogenic origin by evaporitic sedimentation (Lowe, 1994). The most spectacular of the Strelley Pool Chert stromatolites are displayed at the new 'Trendall locality' (Hofmann *et al.*, 1999) and are remarkable for possessing a more diverse range of conical and rare columnar morphologies (Fig. 4f), a greater variation in size, and one example of putative branching. Based largely upon morphological arguments and also rare earth element studies suggestive of a shallow marine setting, a biological origin for these structures has been re-advanced (Van Kranendonk *et al.*, 2003, Hofmann *et al.*, 1999). However, purely abiotic crystallization experiments can produce conical, crystalline bedforms (e.g. Huppert, 1990) and given that crystal fans appear to template accretion of some of the Strelley Pool Chert stromatolites (McLoughlin *et al.*, unpublished data) an abiotic origin cannot be easily falsified. An explanation is also needed for why these Trendall stromatolites are so morphologically unusual compared with the more abundant and more uniform abiogenic stromatolites (Lowe, 1994) found elsewhere in the Strelley Pool Chert.

This discussion brings into focus the question of whether stromatolite biogenicity is discernable from morphology alone. Attempts to find the "Rosetta Stone" for decoding stromatolite morphology have unearthed equations that describe fractal interface growth: the KPZ (Kardar-Parisi-Zhang) equation, which produces compacted, smoothed interfaces and the DLA (diffusion limited aggregation) equation, which yields highly ramified (i.e. branched) interfaces (review: Grotzinger and Knoll, 1999). The terms in the KPZ equation were related by Grotzinger and Rothman (1996) to purely abiotic mechanisms of stromatolite growth: chemical precipitation on a growing interface, the fallout and diffusive rearrangement of suspended sediment, and uncorrelated random noise. Other studies, however, have disputed this interpretation of the KPZ equation and used it to model light-seeking (phototrophic) biofilm growth (e.g. Batchelor *et al.*, 2004). This contradiction arises in part, because of the mathematically challenges of characterizing and separating the complex biotic and abiotic mechanisms of stromatolite growth. It appears again that stromatolite macro-morphology is an enigmatic and perhaps ambiguous indicator of biogenicity.

Figure 4. Transmitted light image of a), filamentous and coccoid microfossils from the 1.87 Ga Gunflint chert, filaments are 1-2μm across (images courtesy of A. Knoll); b), carbonaceous pseudofossils from the 3.5 Ga Apex Chert (Brasier *et al.*, 2002); scanning electron microscope images of c), twisted filamentous pseudofossils made experimentally from barium-carbonate by A. Carnerup, ~ 20μm across (Fig. 1c. Hyde *et al.*, 2004); d), rod shaped mordenite and coccoid shaped cristobalite (opal-CT) mineral crystal habits (from SEM Petrology Atlas, Welton, 1986); field photos of e), laminated barite stromatolite, that curves to the bottom left as it descends a hydrothermal dyke in the ~ 3.5 Ga Dresser Formation; f), cross section through a laminated conical stromatolite from the ~3.4 Ga Strelley Pool Chert.

A closely related phenomenon is that of “wrinkle mat” textures. Wrinkle mats are defined as micro-laminated fabrics in which the layering is created by cohesive, organic rich substrates that trap, baffle and bind sediments in a manner akin to stromatolites. Much recent attention has focused upon the significant role that microbial extracellular polymeric substances play as the adhesive binding agent. Some of the oldest wrinkle mat fabrics with putative cellular remains are from the ~2.9 Ga Pongola Supergroup (Fig. 3) of South Africa (Noffke *et al.*, 2003), and have been described as orange-peel textures on bedding surfaces with microscopic reticulated filaments of carbonaceous material among the sediment grains. The authors report carbon isotopic fractionations that are “consistent with biologic origin,” and much of the paper is devoted to probing the question, “which group of bacteria constructed the microbial mats?” Older putative wrinkle mat horizons are described from the Swaziland Supergroup (Fig. 3) of the Barberton (Fig. 3. Walsh and Lowe, 1999). These comprise carbonaceous laminae and wisps preserved in micro-crystalline silica, that include examples of plastically deformed carbonaceous fragments interpreted as microbial mat rip-up clasts, and two examples of purported filamentous microfossils (Walsh, 1992; Walsh and Lowe, 1985). These textures and morphologies are certainly very suggestive of microbial processing and arguably better preserved than anything hitherto reported from Western Australia. A prebiotic origin for this carbonaceous material needs to falsified however, given that, the range of carbonaceous textures preserved (Fig. 3-8. Walsh and Lowe) are similar to laminar and wispy textures in the hydrothermal cherts of the 3.45 Ga Apex chert (cf. Brasier *et al.*, sub) which are compactional in origin.

8. Discussion

8.1. DISTINGUISHING PREBIOTIC AND BIOTIC CARBON CYCLES

It seems reasonable to expect that the first signs of cellular life preserved in the rock record were accompanied and preceded by signs of prebiotic chemical cycling that may also be encoded in the rocks. This has been termed the transition zone, a complex, pre-life organic system in which the components of life emerged and assembled (Perry and Kolb, 2004). It is postulated that this interval saw the gradual co-evolution of self-replicating systems efficiently coupled to cycles of enzyme production and energy storage/transfer. Small incremental changes in the structure and organization of this transition zone eventually led to life. It follows however, that some of the indicators which have been utilized as early microbial signatures, including simple filamentous and coccoidal “microfossils”, fractionated carbon, carbonaceous clots, wisps and laminae, may not uniquely determine the degraded remains of modern microbial processes. Instead, these may record a continuity of organic carbon behavior from the transition zone that was inherited by later microbial processes. Envisioning such a prebiologic carbon cycle may be possible by designing imaginative prebiotic chemistry experiments (e.g. Miller and Urey, 1959; Wachtershauser, 1990). Such experimentation is our only virtual reference frame for linking the enigmatic rock record to the appearance of familiar microbial biology.

If this pre-biotic to biotic transition is preserved in Earth's rock record, how will we recognize it? Laser-raman techniques for example have been developed to investigate the origins of amorphous carbon in mid-Archean rocks but these have so far proved inconclusive (see discussion by Pasteris and Wopenka, 2003). A recent paper, claims to report "photosynthetic microbial mats in the 3,416-Myr-old ocean" (Tice and Lowe, 2004); based upon interpretations of how graphitic carbon microtextures are controlled by the paleo-environment. Those authors argue that microtextural changes in carbonaceous debris in the Buck Reef Chert (the uppermost part of the Kromberg Fm, Fig. 3) are associated with changes in paleodepth, and that this corresponds to the photic zone and hence photosynthesis. Their appraisal of paleodepth is key to their argument because the carbonaceous material itself is undiagnostic. The paleoenvironmental reconstruction employs a mix of purely descriptive (e.g. evaporite facies) and genetic (e.g. deep water facies) terms and arguably understates the very limited geological window that is preserved. Paleodepth, in particular is notoriously difficult to assess, especially in cherts that are deposited as hydrous oozes with no easily definable sediment-water interface by which to preserve sedimentary structures (e.g. ripples) formed from bottom-shear. In summary we caution that inferring paleowater depth in cherts is an insecure basis from which to argue for a light-dependent distribution of biogenic carbon. A prebiotic origin for this carbon needs also to be explored and falsified.

8.2. THE SCIENTIFIC METHOD AND SEARCHING FOR THE EARLIEST LIFE

The high-stakes field of questing for the origins of life on Earth necessitates a critical discussion of epistemological rigor and scientific method. A robust understanding of early life is only developed by an iterative process of review and criticism by a large and heterogeneous peer group that is un-coerced at both the professional and personal level (Tucker, 2003). As in all branches of science, there are a number of competitive key players and research groups who may come to influence the field. These form a stimulating focus for scientific exploration and funding (e.g. the Precambrian Paleobiology Research Group of the 80's, NASA Astrobiology Institute, and the Agouron Institute at present), but the community as a whole has to be careful to cultivate diversity and alternative opinions, and to police its own work and methodologies.

There are a number of influences on scientific practice, which if left unchecked, can impede the progress of early life research. Firstly, the academic system tends to favor certainty in phraseology, where contingency and possibility are all that may be discernable from the Archean rock record. Secondly, a career ladder that increasingly relies on citation volume and bibliometric statistics (such as 'impact factor') may create pressure to publish prematurely. Important checks and balances are provided by the cultivation of an interdisciplinary approach and plurality of perspectives to investigating early life, a portrayal of the geological context that clearly separates observation from interpretation, a willingness to explore and then falsify plausible abiotic explanations, and effective checking of grand scientific claims and duplication of spectacular results prior to their acceptance as knowledge.

9. Final Considerations

There exists no undisputed evidence for when, where, and how life began on Earth. There is a growing realization that attempts by geologists over the past few decades to ascertain the antiquity of the biosphere have been constrained by relatively over-simplistic and uniformitarian expectations of what signals early biology may have left in the rock record. An emerging cross-disciplinary synthesis holds great promise for refining the narrative of life's Earthly beginnings, but not necessarily its Earthly origins. Exciting and imaginative multidisciplinary approaches are being developed in the quest for early life, and we hope to have here persuaded our colleagues, that these new insights need to be recurrently *"ground truthed"* by the rock record.

References

Altermann, W., and Schopf, J.W. (1995). Microfossils from the Neoarchean Campbell Group, Griqualand West sequence of the Transvaal Supergroup, and their paleoenvioronmental and evolutionary implications. Precam. Res. **75**, 65-90.

Awramik, S.M., and Semikhatov, M.A. (1978). The relationship between morphology, microstructure and microbiota in three vertically intergrading stromatolites from the Gunflint Iron Formation. Can. J. Earth. Sci. **16**, 484-495.

Barghoorn, E.S., and Tyler S.A. (1965). Microorganisms from the Gunflint Chert. Science **147**, 563-577.

Batchelor, M.T., Burne R.V., Henry B.I., and Jackson M.J. (2004). A case for biotic morphogenesis of coniform stromatolites Physics A, **337**, 319-326.

Bedini, R.M., Blichert-Toft, J., Boyet, M., and Albarede, F. (2004). Isotopic constraints on the cooling of the continental lithosphere. E.P.S.L. **223**, 99-111.

Bekker, A., Holland, H.D., Wang, P.L., Rumble, D., Stein, H.J., Hannah, J.L., Ceotzee, L.L., and Beukes, N.J. (2004). Dating the rise of atmospheric oxygen. Nature **427**, 117-120.

Bosak, T., Souza E.V., Corsetti F.A., and Newman D.K. (2004). Micrometer-scale porosity as a biosignature in carbonate crusts. Geology **32**, 781-784.

Brasier, M.D., Green, O.R., Jephcoat, A.P., Kleppe, A.K., van Kranendonk, M.J., Lindsay, J. F., Steele, A. and Grassineau, N.V. (2002). Questioning the Evidence for Earth's Oldest Fossils. Nature **416**, 76-81.

Brasier, M.D., Green, O.R., Lindsay, J.F. and Steele, A., (2004a). Earth's oldest (3.5 Ga) fossils and the 'Early Eden' hypothesis: questioning the evidence. O. L. E. B. **34**, 257-269.

Brasier, M.D., Green, O.R. and McLoughlin N. (2004b). Characterization and critical testing of potential microfossils from the early Earth: the Apex 'microfossil debate' and its lessons for Mars sample return. Int. J. of Astrobio. **3**, 139-150.

Brasier, M.D., McLoughlin, N., and Perry R.S. (2004c). Testing the Biogenicity of Endolithic Microborings on Early Earth and Mars: new data from Western Australia. Geol. Soc. Am. Abs. with Prog. **36**, 88.

Brasier, M.D., Green, O.R., Lindsay, J.F., McLoughlin, N., Jephcoat, A.P., Kleppe, A.K., Steele, A. and Stoakes. C.P. *Sub.* Critical testing of Earth's oldest putative fossil assemblage from the ~3.5 Ga Apex chert, Chinaman Creek, Western Australia. Precamb. Res.

Brocks, J.J., and Summons, R.E. (2003). Sedimentary hydrocarbons, Biomarkers for early life. In: H.D. Holland and K.K. Turekian. (eds.) *Treatise on Geochemistry* **8**, 63.

Brocks, J.J., Logan, G.A., Buick, R., and Summons, R.E. (1999). Archean molecular fossils and the early rise of eukaryotes. Science **285**, 1033-1036.

Brocks, J.J., Love, G.D., Snape, C.E., Logan, G.A., Summons, R.E., and Buick, R. (2003). Release of bound aromatic hydrocarbons from late Archean and Mesoproterozoic kerogens via hydropyrolysis. G.C.A. **67**, 1521-1530.

Buick R., Dunlop, J.S.R., and Groves D.I. (1981). Stromatolite recognition in ancient rocks: An appraisal of irregularly laminated structures in an Early Archean chert-barite unit at North Pole, Western Australia. Alcheringa, **5**, 161-181.

Buick, R., (1992). The antiquity of oxygenic photosynthesis: evidence from stromatolites in sulfate-deficient Archean lakes. Science **255**, 74-77.

Buick, R., Thornett, J.R., McNaughton, N.J., Smith, J.B., Barley, M.E., and Savage, M. (1995). Record of emergent continental crust ~3.5 billion years ago in the Pilbara Craton of Australia. Nature **375**, 574-577.

Catling, D.C., Zahnle, K.J., and McKay, C.P., (2001). Biogenic Methane, Hydrogen Escape, and the Irreversible Oxidation of Early Earth. Science, **293**, 839-843.

Channer, D.M., de Ronde, C.E.J., and Spooner, E.T.C. (1997). The Cl^- Br^- I^- composition of approximately 3.23 Ga modified seawater; implications for the geological evolution of ocean halide chemistry. E.P.S.L. **150**, 325-335.

Cleland, C.E. (2001). Historical science, experimental science, and the scientific method. Geology **29**, 987-990.

Darwin, C. (1859). *Origin of Species by Means of Natural Selection*. John Murray, Albemarle Street, London, 502p.

Delsemme, A.H. (1998). Cosmic origin of the biosphere, In: A. Brock (ed.) *The Molecular Origins of Life: Assembling the Pieces of the Puzzle*. Cambridge Univ. Press, Cambridge, UK, pp. 100-118.

DeRonde, C.E.J., Channer, D.M., Fuare, K., Bray, C.J., and Spooner, E.T.C. (1997). Fluid chemistry of Archean seafloor hydrothermal vents, implications for the composition of c 3.2 Ga seawater. G.C.A. **61**, 4025-40.

Fedo, C.M., and Whitehouse, M.J. (2002). Metasomatic Origin of Quartz-Pyroxene Rock, Akilia, Greenland, and its Implications for Earth's Earliest Life, Science **296**, 1448-1452.

Fischer, W.W., Summons, R.E., Pearson, A., *in press*. Targeted genomic detection of biosynthetic pathways: anaerobic production of hopanoid biomarkers by a common sedimentary microbe. Geobiology.

Fralick, P., Davis, D.W., and Kissin, S.A. (2002). The age of the Gunflint Formation, Ontario, Canada: single zircon U-Pb age determinations from reworked volcanic ash. Can. J. Earth. Sci. **39**, 1085-1091.

Frodeman. R. (1995). Geological reasoning: geology as an interpretive and historical science: Geol. Soc. Am. Bull. **107**, 960-968.

Furnes, H., Banerjee, N.R., Muehlenbachs, K., Staudigel, H. and de Wit, M., (2004). Early Life recorded in Archean pillow lavas. Science, **304**, 578-581.

García-Ruiz, J.M., Hyde, S.T., Carnerup, A.M., Christy, A.G., Van Kranendonk, M.J., Welham, N.J. (2003). Self-Assembled Silica Carbonate Structures and Detection of Ancient Microfossils. Science **302,** 1194-1197.

Grotzinger, J.P., and Kasting J.F. (1993). New constraints on Precambrian ocean composition. J. Geology **101**, 235-243.

Grotzinger, J.P-and Knoll, A.H. (1999). Stromatolites in Precambrian carbonates; evolutionary mileposts or environmental dipsticks? Ann. Rev. E.P.S.L., **27**, 313-358.

Grotzinger, J.P., and Rothman, D.H. (1996). An Abiotic Model for Stromatolite Morphogenesis. *Nature* **383**, 423-425.

Groves, D.I., Dunlop, J.S.R., Buick, R. (1981). An early habitat of life. Sci. Am. **245**, 64-73.

Harnmeijer, J., and Buick, R. (2005). Metamorphism and Geological Survivability of Carbonaceous Biosignatures. Abstract NAI, Boulder, April 2005.

Hedges, S.B., Chen, H., Kumar, S., Wang, D., Thompson, A.S. and Watanabe, H., (2001). A genomic timescale for the origin of eukaryotes. BMC Evol. Bio., 1-4.

Hofmann, H.J. (2000). Archean Stromatolites as microbial Archives. In: R.E. Riding and S.M. Awramik (eds.) *Microbial Sediments*. Springer-Verlag 2000.

Hofmann, H.J., Grey, K., Hickman, A.H. and Thorpe, R.I. (1999). Origin of 3.45 Ga Coniform Stromatolites in the Warrawoona Group, Western Australia. Bull. Geol. Soc. Am. **111**, 1256-1262.

Horita, J., and Berndt, M.E. (1999). Abiogenic methane formation and isotopic fractionation under hydrothermal conditions. Science **285**, 1055-1057.

Huppert, H.E. (1990). The fluid mechanics of solidification. J. Fluid Mech. **212**, 209-240.

Hutton, J. (1785). Theory of the earth.

Hyde, S.T., Carnerup, A.M., Larsson, A.K. and Christy, A.G. (2004). Self-assembly of carbonate-silica colloids: between living and non-living form. *Physica*, **339**, 24-33.

Kasting, J.F., and Catling, D.C. (2003). Evolution of a habitable planet, *Ann. Rev. Astron. and Astrophy.*, **41**, 429-463.

Kelley, D.S., Karson, J.A., Früh-Green, G.L., Yoerger, D.R., Shank, T.M., Butterfield, D.A., Hayes, J.M., Schrenk, M.O., Olson, E.J., Proskurowski, G., Jakuba, M., Bradley, A., Larson, B., Ludwig, K., Glickson, D., Buckman, K., Bradley, A.S., Brazelton, W.J., Roe, K., Elend, M.J., Delacour, A., Bernasconi, S.M., Lilley, M.D., Baross, J.A., Summons, R.E., Sylva, S.P. (2005). A Serpentinite-Hosted Ecosystem: The Lost City Hydrothermal Field. Science **307**, 1428-1434.

Klein, C., Buekes, N.J., and Schopf, J.W. (1987). Filamentous microfossils in the early Proterozoic Transvaaal Supergroup: their morphology, significance, and paleoenvironmental setting. Precamb. Res. **36**, 81-94.

Knauth, L.P., and Lowe, D.R., (2003). High Archean climatic temperature inferred from oxygen isotope geochemistry of cherts in the 3.5 Ga Swaziland Supergroup, South Africa. Bull. Geol. Soc. Am., **115**, 566-580.

Knoll, A.H., (2003). Life on a young planet: the first three billion years of evolution on Earth. Princeton University Press. 277p.

Konhauser, K.O., Hamade, T., Raiswell, R., Morris, R.C., Ferris, G., Southam, G., and Canfield, D.E. (2002). Could bacteria have formed the Precambrian banded iron formations? Geology **30**, 1079-1082.

Kring, D.A., and Cohen, B.A. (2002). Cataclysmic bombardment throughout the solar system 3.9-4.0 Ga. J. Geophys. Res. **107**, 4.1-4.5.

Kyte, F.T., Shukloyokov, A., Lugmair, G.W., Lowe, D.R., and Byerly, G.R. (2003). Early Archean spherule beds 3.47-3.34 Ga- old in the Barberton greenstone belt, South Africa: A record of large meteor impacts and their influence on early crustal and biological evolution. Astrobiology **3**, 7-48.

Lanier, W.P. (1986). Approximate growth rates of early Proterozoic microstromatolites as deduced by biomass productivity. Palios **1**, 525-542.

Lepland, A., van Zuilen, M.A., Arrehnius, G., Whitehouse, M.J., Fedo, C.M., (2005). Questioning the evidence for earth's earliest life – Akilia revisited. Geology **33**, 77-79.

Levy, M., and Miller, S.L., (1998). The stability of the RNA bases: implications for the origins of life. PNAS **95**, 7933-7938.

Lowe, D.R., (1980). Stromatolites 3,400-Myr old from the Archean of Western Australia. Nature **284**, 441-443.

Lowe, D.R., (1994). Abiological origin of described stromatolites older than 3.2 Ga. Geology **22**, 387-390.

Lowe, D.R., and Byerley, G.R. (1999). Eds. *Geologic Evolution of the Barberton Greenstone Belt, South Africa.* Geol. Soc. Am. Spec. Pap. **329**, Boulder, Colorado,

Lowe, D.R., and Byerly, G.R. (2003). Ironstone pods in the Archean Barberton greenstone belt, South Africa: Earth's oldest seafloor hydrothermal vents reinterpreted as Quaternary subaerial springs. Geology 31, 909-912.

Lowe, D.R., and Nocita, B.W. (1999). Foreland basin sedimentation in the Mapepe Formation, southern-facies Fig Tree Group. In: D.R. Lowe and G.R. Byerley (eds.) *Geologic Evolution of the Barberton Greenstone Belt, South Africa*: Geol. Soc. Am. Spec. Pap. **329**, Boulder, Colorado, pp. 233-258.

Lowe, D.L., and Worrel, G.F. (1999). Sedimentology, mineralogy, and implications of silicified evaporates in the Kromberg Formation, Barberton Greenstone Belt, South Africa, In: D.R. Lowe and G.R. Byerley (eds.) *Geologic Evolution of the Barberton Greenstone Belt, South Africa*: Geol. Soc. Am. Spec. Pap. **329**, Boulder, Colorado, pp. 167-188.

McCall, G.J.H. (2003). A critique of the analogy between Archean and Phanerozoic tectonics based on regional mapping of the Mesozoic-Cenezoic plate convergent zone in the Makran, Iran. Precamb. Res. **127**, 5-17.

McKay, D.S., Thomas-Keprta, K.L., Clemett, S.J., Chillier, X.D.F. Gibson, E.K. (2002). The classification of biosignatures. Astrobiology **2**, 625-626.

Mileikowsky, C., Cucinotta, F., Wilson, J.W., Gladman, B., Horneck, G., Lindegren L., Melosh, J., Rickman H., Valtonen, M., Zheng, J.Q. (2000). Natural transfer of viable microbes in space, part 1: from Mars to Earth and Earth to Mars. Icarus **145**, 391-427.

Miller, S.L., and Urey, H.C. (1959). Organic compound synthesis on the primitive Earth. Science **130**, 245-251.

Mojzsis, S.J., Arrenhius, G., McKeegan, K.D., Harrison, T.M., Nutman, A.P., and Friend, C.R.L., (1996). Evidence for Life on Earth 3,800 Million Years Ago. Nature **384**, 55-59.
Molnar, G.I., and Gutowski, W.J. (1995). The faint young sun paradox - further exploration of the role of dynamical heat-flux feed backs in maintaining global climate stability. J.Glaciology **41**, 87-90.
Moorbath, S. (2005a). Dating earliest life, Nature **434**, 155.
Moorbath, S. (2005b). Oldest rocks, earliest life, heaviest impacts, and the Hadean–Archaean transition. Appl. Geochem. **20**, 819-824.
Nijman, W., de Bruijne, H. and Valkering, M.E., (1998). Growth Fault Control of Early Archaean Cherts, Barite Mounds and Chert-Barite Veins, North Pole Dome, Eastern Pilbara, Western Australia. Precam. Res. **88,** 25-52.
Nisbet, E.G., and Sleep, N.H., (2001). The habitat and nature of early life. Nature, **409**, 1083-1091.
Noffke, N., Hazen, R., and Nhleko, N. (2003). Earth's earliest microbial mats in a siliciclastic marine environment (2.9 Ga Mozaan Group, South Africa). Geology, **31**, 8, 673-676.
Ohmoto, H. (1997). When did the Earth's atmosphere become oxic? The Geochemical News, Fall, 1997, 12-13 and 26-27.
Ohmoto, H. (1996). Evidence in pre-2.2 Ga paleosols for the early evolution of atmospheric oxygen and terrestrial biota. Geology **24**, 1135-1138.
Pace, N.R. (1997). A molecular view of microbial diversity and the biosphere. Science **26**, 734-740.
Pasteris, J.D., and Wopenka, B., (2003). Necessary, but not sufficient: Raman identification of disordered carbon as a signature of ancient life. Astrobiology, **3**, 727-738.
Pavlov, A.A., Kasting, J.F., Eigenbrode, J.L., and Freeman K.H. (2001). Organic haze in the early atmosphere: sources of low ^{13}C Late Archean Kerogens? Geology **29**, 1003-1006.
Perry, R.S. and Kolb, V.M. (2004). On the Applicability of Darwinian Principles to Chemical Evolution that Led to Life. Int. J. Astrobiology **3**, 45.
Pinti, D.L. and Hashizume K. (2001). ^{15}N-depleted nitrogen in Early Archean kerogens: clues on ancient marine chemosynthetic-based ecosystems? A comment to Beaumont, V., Robert, F., 1999. Precam. Res. **96**, 62-82.
Popa, R. (2004). *Between Necessity and Probability: Searching for the Definition and Origin of Life*, Springer-Verlag, Berlin, 252p.
Popper, K.R. (1959). *The Logic of Scientific Discovery*. Hutchinson London 480pp.
Rasmussen, B., (2000). Filamentous microfossils in a 3,250-million-year-old volcanogenic massive sulphide deposit. Nature, **405**, 676-679.
Raymond, J., and Blakenship, R.E. (2005). Biosynthetic pathways, gene replacement and the antiquity of life. Geobiology **2**, 199-203.
Rosing, M.T., and Frei R. (2004). U-rich Archean sea-floor sediments from Greenland –indications of 3700 Ma oxygenic photosynthesis. E.P.S.L. **217**, 237-244.
Rye, R., and Holland, H.D. (1998). Paleosols and the evolution of atmospheric oxygen: a critical review. Am. J. Sci. **298**, 621-672.
Sagan, C., and Mullen, G. (1972). Earth and Mars: Evolution of Atmospheres and Surface Temperatures. Science, **177**, 52-56.
Schidlowski, M. (2001). Carbon isotopes as biogeochemical recorders of life over 3.8 Ga of Earth history: evolution of a concept. Precam. Res. **106**, 117-134.
Schopf, J.W., (1993). Microfossils of the Early Archean Apex Chert: new evidence of the antiquity of life. Science, **260**: 640-646.
Schopf, J.W., (1999). *Cradle of Life*, Princeton University Press, New Jersey, 367 pp.
Schopf, J.W., (2002). When did life begin? In: J.W. Schopf (ed.) *Life's Origin: Beginnings of Biological Evolution*. University of California Press, California pp. 158-180.
Schopf, J.W., and Packer, B.M. (1987). Early Archean (3.3 Billion to 3.5 Billion-Year-Old) Microfossils from Warrawoona Group, Australia. Science **237**, 70-73.
Schopf, J.W., and Walter, M.R. (1983). Archean Microfossils: New Evidence of Ancient Microbes. In: J.W. Schopf. (ed.), *Earth's Earliest Biosphere, its Origin and Evolution*, Princeton University Press, New Jersey, pp. 214-239.
Sephton, M.A., and Gilmour, I. (2001). Compound-specific isotope analysis of the organic constituents in carbonaceous chondrites. Mass Spectrometry Rev. **20**, 111-120.
Shen, Y., and Buick, R. (2004). The antiquity of microbial sulfate reduction. Earth-Sci. Rev. **64**, 243-272.
Shen, Y. Buick, R., and Canfield D.E. (2001). Isotopic evidence for microbial sulphate reduction in the early Archean era. Nature **410**, 77-81.

Sherwood Lollar, B., Westagate, T.D., Ward, J.A., Slater, G.F., and Lacrampe-Couloume, G. (2002). Abiogenic formation of alkanes in the Earth's crust as a minor source for global hydrocarbon reservoirs. Nature **416**, 522-524.

Siever, R. (1991). The silica cycle in the Precambrian. G. C. A. **56**, 3265-3272.

Sleep, N.H., Zahnle, K.J., Kasting, J.F., Morowitz, H.J., (1989). Annihilation of ecosystems by large asteroid impacts on early Earth. Nature **342**, 139-142.

Sommers, M.G., Awramik, S.M., and Woo K.S. (2000). Evidence for initial calcite-aragonite composition of Lower Algal Chert Member ooids and stromatolites, Paleoproterozoic Gunflint Formation, Ontario, Canada. Can. J. Earth. Sci. **37**, 1229-1243.

Spiess, F.N., Macdonald, K.C., Atwater, T., Ballard, R., Carranza, A., Cordoba, D., Cox, C., Diazgarcia, V.M., Francheteau, J., Guerrero, J., Hawkins, J., Haymon, R., Hessler, R., Juteatu, T., Kastner, M., Larson, R., Luyendyk, B., Macdougall, J.D., Miller, S., Normak, W., Orcutt, J., and Ranginn, C. (1980). East Pacific rise -Hot Springs and Geophysical experiments. Science **207**, 1421.

Summons, R.E., Jahnke, L.L., Hope, M., and Logan, G.A., (1999). 2-methylhopanoids as biomarkers for cyanobacterial oxygenic photosynthesis. Nature, **400**, 554-557.

Tice, M.M., and Lowe, D.R., (2004). Photosynthetic microbial mats in the 3,416-Myr-old ocean. Nature **431**, 549-552.

Tucker, A. (2003). The epistemic significance of concensus. Inquiry **46**, 501-521.

Tyler, S.T., and Barghoorn, E.S. (1963). Ambient Pyrite Grains in Precambrian Cherts. Am. J. of Sci. **261**, 424-432.

Ueno, Y., Maruyama, S., Isozaki, Y., and Yurimoto, H., (2001). Early Archean (ca.3.5 Ga) microfossils and ^{13}C-depleted carbonaceous matter in the North Pole area, Western Australia: Field occurrence and geochemistry, In: S. Nakashima, S. Maruyama, A. Brack and B.F. Windley (eds.) *Geochemistry and the origin of life*. Universal Academy Press Inc, Tokyo, pp. 203-236.

Van Kranendonk, M.J., (2000). *Geology of the North Shaw 1:100 000 Sheet:* Western Australia Geological Survey, 1:100 000 Geological Series Explanatory Notes, 86 p.

Van Kranendonk, M.J., (2001). Volcanic degassing, hydrothermal circulation and the flourishing of early life on earth: new evidence from the c. 3460 Ma Warrawoona Group, Pilbara Craton, Western Australia. Abstract Earth System Processes Meeting, 2001.

Van Kranendonk, M.J., and Pirajno, F. (2004). Geochemistry of metabasalts and hydrothermal alteration zones associated with c. 3.45 Ga chert and barite deposits: implications for the geological setting of the Warrawoona Group, Pilbara Craton, Australia. Geochem. Explor. Environ., Analysis **4**, 253-278.

Van Kranendonk, M.J., Webb G.E., and Kamber B.S. (2003). Geological and trace element evidence for a marine sedimentary environment of deposition and biogenicity of 3.45Ga stromatolitic carbonates in the Pilbara Craton, and support for a reducing Archean ocean. *Geobiology* **1**, 91-108.

Van Kranendonk, M.J., Hickman, A.H., Williams, I.R., and Nijman, W., (2001). *Archean Geology of the East Pilbara Granite-Greenstone Terrane Western Australia– A Field Guide*: Western Australia Geologic Survey, Record 2001/9, 134p.

Van Kranendonk, M.J., Collins, W.J., Hickman, A., and Pawley, M.J. (2004). Critical tests of vertical vs. horizontal tectonic models for the Archean East Pilbara Granite-Greenstone Terrane, Pilbara Craton, Western Australia. Precam. Res. **131**, 173-211.

van Zuilen, M.A., Lepland, A., and Arhenius, G., (2002). Reassessing the Evidence for the Earliest Traces of Life, Nature **418**, 627-630.

Vearncombe, S., Barley M.E., Groves, D.I., McNaughton N.J., Mikucki E.J., and Vearncombe J.R. (1995). 3.26Ga black smoker-type mineralisation in the Strelley Belt, Pilbara Craton, Western Australia. J. Geol. Soc. Lon. **152**, 587-590.

Wachtershauser, G. (1990). Evolution of the first metabolic cycles. Proc. Natl. Acad. Sci. **87**, 200-204.

Walter, M.R., Buick, R., and Dunlop, J.S.R., (1980). Stromatolites, 3,400-3,500 Myr old from the North Pole area, Western Australia. Nature, **284**, 443-445.

Walsh, M.M. (1992). Microfossils and possible microfossils from the early Archean Onverwacht Group, Barberton Mountain Land, South Africa. Precam. Res., **54**, 271-293.

Walsh, M.M., and Lowe, D.L. (1985). Filamentous microfossils from the 3,500 Myr-old Onverwacht Group, Barberton Mountain Land, South Africa. Nature, **314**, 530-532.

Walsh, M.M., and Lowe, D.L. (1999). Modes of accumulation of carbonaceous matter in the early Archean: a petrographic and geochemical study of the carbonaceous cherts of the Swaziland Supergroup. In: D.R. Lowe and G.R. Byerley (eds.) *Geologic Evolution of the Barberton Greenstone Belt, South Africa*: Geol. Soc. Am. Spec. Pap. **329**, Boulder, Colorado, pp. 167-188.

Welton, J.E. (1984). *SEM Petrology Atlas*: Amer. Assoc. Petrol. Geol.: Tulsa, Oklahoma, 237p.

Westall, F., de Witt, M.J., Dann, J., van der Gaast, S., de Ronde. and Gerneke, D. (2001). Early Archean fossil bacteria and biofilms in hydrothermally-influenced sediments from the Barberton greenstone belt, South Africa. Precam. Res. **106**, 93-116.

Williams, R.P.J., and Frausto da Silva, J.J.R. (1996). *The Natural Selection of the Chemical Elements.* Clarendon Press, Oxford, 646p.

Biodata of **Philip Bell** author of ***"Eukaryogenesis: the origin of the eukaryotes"***

Dr Philip J.L. Bell is a founding director of Microbiogen, a company that specializes in yeast genetics. He is also an Honorary Associate at Macquarie University. He obtained his PhD in 1999 from the University of New South Wales, Australia. His research interests are: Evolution, Genetics and Molecular Biology, and specializes in the genetics of industrial yeast strains and the development of improved yeast strains for pharmaceutical, baking, brewing and the fuel alcohol industries. He has been developing the Viral Eukaryogenesis Theory as a model for the origin of the eukaryotes since about 1992.

E-mail: **pbell@rna.bio.mq.edu.au**

J. Seckbach (ed.), Life as We Know It, 287-305.

EUKARYOGENESIS:
The origin of the eukaryotes

P. J. L. BELL
Macquarie University, New South Wales, 2109, Australia

1. Introduction

Although science can broadly state that our origins lie in the process of evolution, many questions are still unanswered. Amongst these unanswered questions are the origin of life itself, as well as the origin of the eukaryotes, the group of life to which humans belong. One way to get these questions into context is to survey the life forms that exist today and determine how they are related to each other.

If we use all the modern scientific tools available to survey the life forms on earth, three types of cellular life can be identified. These three types of life, termed domains (Woese et al., 1990), are the eukaryotes, the bacteria and the archaea. Of these, the eukaryotes are the organisms with which we are most familiar. Many of the eukaryotes, for example, plants, fungi, and animals (including ourselves), can be seen with the naked eye because they are composed of millions and in some cases billions of individual cells. Inside each of these eukaryotic cells are organelles such as nuclei, and mitochondria. The nuclei, which store the eukaryotic genome, are characteristic of these cells and responsible for giving them their name: eu, meaning well or true and karyon meaning kernel or nut. The other two domains, the bacteria and the archaea, are in general microscopic and unicellular. Due to their lack of complex internal structures, and absence of nuclei, these two domains are often considered to predate the eukaryotes and were thus collectively termed the prokaryotes ('pro' meaning before). In addition to the three cellular domains of life, there is a fourth type of nucleic acid based organism that are in some way related to life, yet not quite life as we know it. These are the viruses. Like cells, the viruses contain nucleic acids and proteins, but their replication and genetic design are quite distinct from any cellular life form.

2. Genetic Design of Nucleic Acid Based Organisms

2.1. THE VIRUSES

The simplest of nucleic acid based organisms on earth are the viruses and they can be much simpler than any living cell. For example, the polio virus genome is only 7440 bp long and encodes only one gene. This is barely one hundredth the size of the simplest bacterial genome recorded. Unlike cells, all viruses are also obligate intracellular parasites due to an inability to make their own proteins (Strauss et al., 1996). This inability to make their own proteins results from a lack of protein synthesizing machinery (the ribosomes), and obliges the viruses to enter a host to replicate. The viruses lack ribosomes, a cytoplasm or energy metabolism, and thus they specialize in invading and replicating within living cells (Figure 1). The inability of viruses to replicate outside a host cell, combined in some cases with extreme simplicity, has ensured that viruses cannot be simply classified as either alive or dead, but rather they appear to share features of the living and non-living chemical worlds.

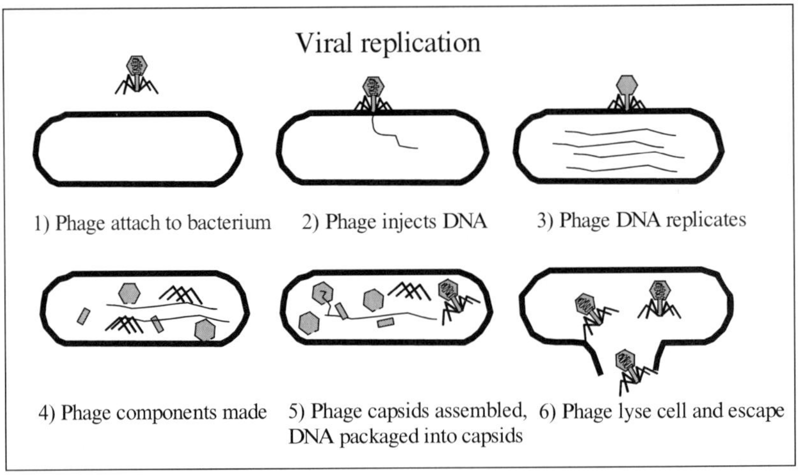

Figure 1. T4 bacteriophage replication is typical of some viral replication cycles. The viral particle does not enter the cell, but rather injects DNA directly into the cellular cytoplasm. The viral DNA is replicated and viral genes are turned on. The viral genes encode the structural proteins of the viral capsid which begins self-assembly in the cytoplasm of the host. Once the DNA is packaged and the viral particles are mature, the host is lysed and the viruses leave the cell. The viral replication cycle is dependent upon the host translational apparatus and energy metabolism since the virus lacks both these capabilities.

When viruses appeared in the evolution of life is uncertain and direct evidence for viruses at any early stage of the earth's history is unlikely to exist due to the limitations of the fossil record. However, support for an ancient origin of viruses appears to be strengthening (Benson et al., 2004). It has also been shown that viruses are consistently the most abundant biological entities (about 5-25 times the bacterial abundance) in a wide variety of marine environments (Oren et al., 1997; Fuhrman, 1999). Whilst viruses may not be living, they undoubtedly play a major role in the evolution of life due to their ability to transfer genes from one organism to another, and the selective pressure they exert on their hosts by way of infection.

Although some viruses have been studied in detail (e.g. T4 bacteriophage, HIV) studies on the viruses thriving in hot pools in Yellowstone National Park suggest that our knowledge of viral diversity is rudimentary at best (e.g. Prangishvili and Garrett, 2004). Our views of the viruses as simple non-living organisms are also likely to come under increasing challenge since the discovery and characterization of the giant Mimivirus which is as complex as many living prokaryotes (Raoult et al., 2004).

2.2. THE PROKARYOTES

The prokaryotes are the simplest living organisms. The bacterial and the archaeal domains that make up the two types of prokaryotes are remarkably self-sufficient compared to the viruses. For example, cyanobacteria (photosynthetic members of the bacterial domain) require only carbon dioxide, nitrogen gas, and water for growth, provided light and a few basic inorganic salts are available. Similarly, many archaea require only hydrogen, carbon dioxide, water and a source of nitrogen to grow. The prokaryotes achieve this autonomy because, in contrast to the viruses, their genomes have all the information required to synthesize a functional ribosome. A ribosome allows them to translate other genetic information into proteins and enzymes that ultimately allow synthesis of all cellular components including vitamins, lipids, carbohydrates and amino acids.

Despite evidence that bacterial and archaeal cells are only distantly related to each other (Woese et al., 1990), they are built and replicate along very similar lines (Figure 2). In prokaryotes a lipid membrane defines the extent of the cell and encloses the cytoplasm. The genome is composed of double stranded DNA, which is usually a single circular molecule located directly in the cell's cytoplasm. The genetic information stored in the genome is decoded by specialized enzymes (RNA polymerases), and translated into proteins by ribosomes located in the cytoplasm.

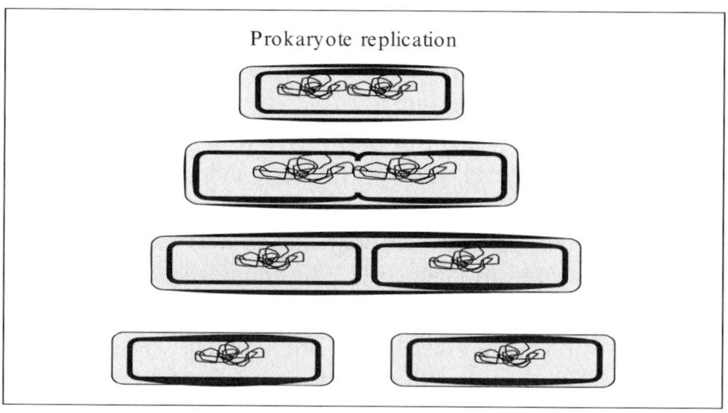

Figure 2. Prokaryotic replication and design. Typical prokaryotes consist of non compartmentalized cells, with the DNA genome located directly in the cytoplasm. In the cytoplasm with the genome are the ribosomes and all the enzymes required to convert genetic information into proteins as well as the enzymes for metabolism and energy generation. Most prokaryotes replicate via binary fission. In this process the cell begins to elongate. DNA genome is replicated and separates. Once the cell is sufficiently elongated, septa start to form and grow inwards at the centre of the cell. When the septum is fully formed, the cytoplasm is divided into two producing two daughter cells. After cell division, the septum can be separated by autolysis, leading to two independent cells.

2.3. THE EUKARYOTES

The eukaryotes share several features of cellular design with the prokaryotes. They have lipid membranes to enclose the cell's cytoplasm, they use DNA to store genetic information, they transcribe DNA into RNA, and they translate mRNA into proteins using ribosomes. However, eukaryotic cells display a much higher degree of cell to cell interaction and internal cellular complexity than seen with the prokaryotes. For example, most animals consist of hundreds of different cell types (e.g. muscle, nerve, blood), interacting in a closely co-ordinated fashion to form a functional organism.

There are also many unique features of eukaryotic cells. In particular, eukaryotic cells possess a number of organelles not found in the prokaryotes. These include the nucleus, the mitochondria, the endoplasmic reticulum, the Golgi apparatus, lysosomes, vacuoles, the cytoskeleton, and in the case of plants, chloroplasts. Each of these organelles performs highly specialized functions within the eukaryotic cell and is not found in the prokaryotes. Significantly, the nucleus stores the majority of eukaryotic genetic information, separating it from the rest of the cytoplasm by the nuclear membrane. The DNA inside the nucleus, unlike circular prokaryotic genomes, is organized into multiple linear chromosomes. Once information is transcribed from DNA into mRNA it is processed, by splicing, capping and polyadenylation, before being extruded through specialized nuclear pores into the cytoplasm. Once into the cytoplasm the information in the mRNA is read by the ribosomes and translated into proteins for use by the cell. Replication of the eukaryotic cell is also quite distinct from prokaryotic or viral replication (Figure 3). This replication cycle includes both asexual phases and sexual phases, neither of which is similar to the prokaryotic or viral replication cycles.

3. LUCA and the Origin of the Three Domains

Currently, Darwin's theory of evolution provides the best conceptual basis for understanding how the great diversity of life on the earth arose. In Darwin's theory of evolution, it is proposed that similar species are closely related to each other and descend from a common ancestor through the processes of natural selection. Given enough time and extending the principle of common decent to its logical conclusion, evolution seems capable of accounting for the development of all the diverse life forms seen today. Significant support for this concept comes from the recognizable homology observed between the ribosomal apparatus of all cellular organisms.

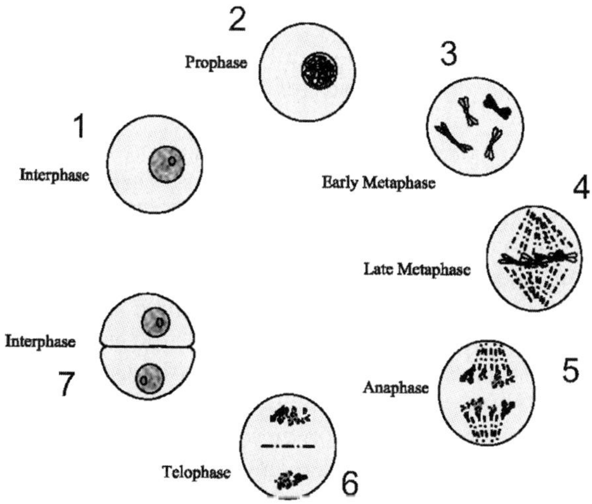

Figure 3. Eukaryotic mitosis. Eukaryotic replication requires co-ordinated replication of the nucleus and the cytoplasm. The replication of organelles such as the mitochondria are independent of the cell cycle, but must be segregated into both daughter cells if they are to be maintained. In higher eukaryotes the nucleus undergoes several steps during replication. 1) Interphase. The nucleus is active, the cell grows and the DNA divides. 2) Prophase. Each chromosome condenses and becomes a distinct thread. 3) Early metaphase. Chromosomes are highly condensed and becoming aligned, nuclear membrane breaks down. 4) Late metaphase. The chromosomes have aligned on the equatorial plate and are attached to spindle apparatus. 5) Anaphase, The replicated chromosomes are separated and moved by the spindle apparatus to the poles of the cell. 6) Telophase. Chromosomes clump at poles and begin to de-condense, the nuclear membrane reforms, and the cytoplasm starts to undergo a binary fission like process. 7) Interphase. Two daughter cells have formed. Figure 3 reprinted from Raoult *et al.*, with permission from SCIENCE 306:1344-50 (2004) published online 14 October 2004 (10.1126/science.1101485). Copyright 2004 AAAS

Although the theory of evolution provides a conceptual framework for understanding the diversity of life on earth, the theory does not explain the origin of life, nor does it explain how the three domains arose. It is elucidating these major steps in the evolution of life that has become one of the major challenges in modern biology. One particular unresolved challenge is determining the genetic design of the last universal common ancestor (LUCA) of life. At present there are several competing schools of thought on the nature of LUCA, and thus also on how the three domains arose.

In the progenote hypothesis, (Woese, 1987; Woese, 1998; Woese, 2000), LUCA was not a single organism, but rather a communal consortium of progenotes unlike modern cells. In particular their component parts had different ancestries, and the progenotes were subject to high levels of lateral gene transfer. In this hypothesis the universal ancestor was a communal, loosely knit, diverse conglomerate of cells that evolved as a unit (Woese, 1998). After the conglomerate split into several distinct communities the three primary lineages became established, and each evolved into the modern cellular lineages. Thus essentially the three types of cellular organisms, including the eukaryotes, crystallized out of a consortium of progenotes, and it was at this stage of crystallization that the three different types of cells arose. Subsequently evolution has worked to diversify each of the three domains into the variety of organisms we see today.

In an alternative hypothesis, LUCA was a eukaryote-like cell. This hypothesis was proposed originally by Reanney (1974), and has been supported using phylogenetic analysis (Brinkmann and Philippe, 1999; Forterre and Philippe, 1999; Penny and Poole, 1999). Placing the ancestor of all living organisms in the eukaryotic branch of the phylogenetic tree implies the prokaryotes evolved from the more complex eukaryote-like cells. It has been proposed that processes such as thermoreduction and r-selection selected for more efficient and faster replication, effectively streamlining the eukaryote-like design into a prokaryote design (Forterre, 1995; Poole et al., 1998; Forterre and Philippe, 1999; Penny and Poole, 1999). The theory thus rejects the concept that the prokaryotes are early primitive life forms, and replaces it with the concept that they are highly evolved, highly efficient descendents from more complex but less efficient eukaryote-like cells.

In a third current model, LUCA is located between the bacterial and the archaeal/eukaryotic lineages, and the last common ancestor was a cell of prokaryotic design. In this hypothesis, the simplicity and similarities in genetic design between the bacteria and the archaea occur because they inherited a common prokaryotic design from LUCA and the eukaryotic design is a later innovation arising from a more simple prokaryotic design. In some aspects this is similar to Whittaker's five kingdom scheme of the 1970's in which the kingdom Monera (prokaryotes) located at the base of the tree gave rise to the Protista kingdom (simple eukaryotes), which in turn gave rise to the plant, animal, and fungal kingdoms.

Several lines of evidence support a model in which LUCA was a simple prokaryote that predated the evolution of the more complex eukaryotes. Firstly, the fossil record indicates that cells of prokaryotic design arose before those of eukaryotic design. Specifically, there is evidence for cells of prokaryotic design some 3500 million years ago (Schopf, 1993), which predates both molecular fossils of eukaryotic lipids (2700 myr) and the extant eukaryotic fossil record by at least 700 million years (Brocks et al., 1999). Analysis of the bacterial, archaeal and eukaryotic genomes also suggest that the first branching of the tree of life separated the bacterial lineage from the archaeal (eukaryotic) lineage (Gogarten et al., 1989; Iwabe et al., 1989; Gribaldo and Cammarano, 1998). Further evidence for the most ancient split occurring between the bacteria and the archaea comes from the observation that sequences of the archaeal and eukaryotic ribosomal genes are more similar to each other than either is to the bacteria (Auer et al., 1989). A prokaryotic LUCA is further supported by complete genomic sequences that suggest LUCA was an organism with metabolic networks and genetic machinery similar to those of extant prokaryotic organisms (Kyrpides et al., 1999).

4. Eukaryogenesis

The tree shown in Figure 4 can be proposed as one of several models to represent the evolutionary relationship of the three domains. In this scheme LUCA was a prokaryote, and thus a non-compartmentalized simple unicellular organism. The similarities in genetic design between the bacteria and the archaea are inherited from their common ancestor, LUCA. In this model eukaryogenesis is the process by which a prokaryotic archaeon evolved into a cell with eukaryotic design.

Explaining how a eukaryotic cell could have evolved from a prokaryotic cell is difficult due to the large number of fundamental differences between them. This difficulty is made worse by the discovery that the eukaryotic genome is apparently a complex mosaic of genes from different origins. For example, genes used by the eukaryotes for replicating and processing genetic information are generally most closely related to archaeal genes, whilst metabolic genes (e.g. energy, biosynthesis) are generally most closely related to bacterial genes (Rivera et al., 1998). There are also over 300 eukaryotic genes that have no significant homology to genes in either the archaea or the bacteria (Hartman and Fedorov, 2002). Prominent amongst them are genes associated with the nucleus, the plasma membrane, the endoplasmic reticulum, the translational apparatus, and the cell cycle (Hartman and Fedorov, 2002).

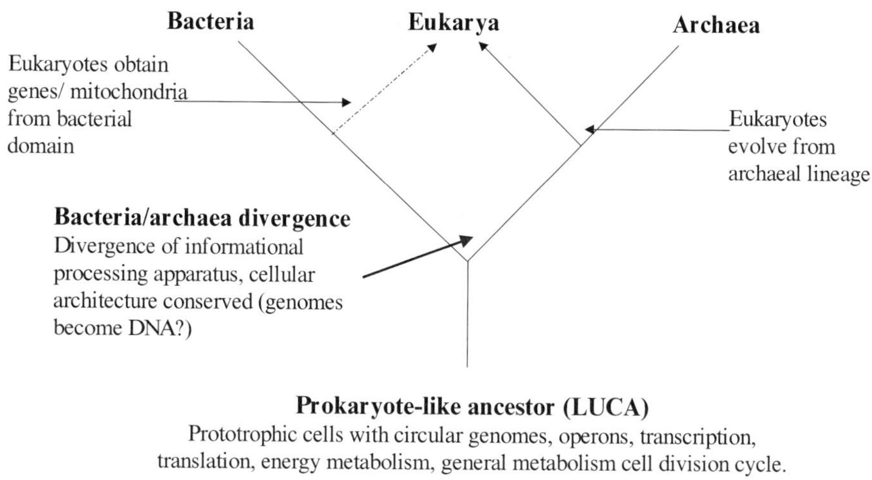

Figure 4. One model for the evolutionary relationships between the three domains. In this model, the root of the universal tree of life is placed between the two prokaryotic domains. The bacteria and the archaea are built in similar ways because they inherited this simple design from their common ancestor, LUCA. The eukaryotes emerge from the archaeal lineage, but they also obtain a significant proportion of their genes from bacteria through lateral gene transfer from bacterial endosymbionts (e.g. mitochondria, chloroplasts) and prey.

To account for these observations, several eukaryogenesis models have been proposed in which the eukaryotes arose via symbiosis, primary fusion events, or phagocytosis between the archaea and the bacteria (e.g. Cavalier-Smith, 1987; Sogin, 1991; Zillig, 1991; Lake and Rivera, 1994; Gupta and Golding, 1996; Margulis, 1996; Gupta, 1998; Martin and Müller, 1998; Moreira and López-Garcia, 1998; Ribeiro and Golding, 1998; Vellai et al., 1998; López-Garcia and Moreira, 1999; Bell, 2001; Hartman and Federov, 2002).

In the Serial Endosymbiosis Theory (SET) (Margulis, 1996), eukaryotic organelles, such as the mitochondria, chloroplasts and flagellae are derived from free living bacteria that were captured by the ancient nucleocytoplasmic (eukaryotic) ancestor. The observation that mitochondria possess their own independent genomes, and the existence of many similarities between the mitochondria and free-living alpha-proteobacteria provide are supportive of the SET theory (e.g. Gray et al., 1999). Similarly the presence of chloroplast genomes and both phylogenetic and biochemical correspondences between the chloroplasts and free living cyanobacteria provide a convincing case that the chloroplasts are derived from an endosymbiotic photo-synthetic bacterium (cyanobacterium). However, in the case of the flagella, its lack of a genome, and the lack of homology between flagella proteins and those of bacteria have not allowed a convincing case of an endosymbiotic origin to be made for this organelle. Thus some aspects of the SET theory have been confirmed, whilst other aspects have not. In particular the theory has been most widely accepted when biochemical and genetic similarities exist between the organelle and the free-living organism.

While the SET theory has successfully elucidated the ancestry of at least some of the organelles of the eukaryotes, the evolutionary origin of the eukaryotic nucleus itself is more difficult. Several different models have been proposed to address this problem.

In one model a 'chronocyte' is hypothesized which possesses an RNA genome and was host to a prokaryotic endosymbiont that gave rise to the nucleus (Hartman and Federov, 2002). Alternatively, Sogin (1991) has proposed that the protoeukaryote and protobacterial lineages derive from an original 'progenote' that embarked on different evolutionary pathways. At some stage, the protoeukaryote engulfed an archaeon leading to the formation of the nucleus. In this model, the protoeukaryotic lineage invents the cytoskeleton and has maintained and refined many RNA world characteristics (Sogin, 1991).

An endosymbiotic model for the origin of the nucleus has been proposed in the endokaryotic hypothesis (Lake and Rivera, 1994). In this model, a bacterium engulfed an archaeal cell, and the archaea eventually evolved into the nucleus. Two further models for the origin of the eukaryotes based on metabolic symbiosis have also been proposed (Martin and Müller, 1998; Moreira and López-Garcia, 1998). Both propose that eukaryotes arose through metabolic symbiosis (syntrophy) between bacteria and methanogenic archaea. In these models interspecies hydrogen transfer mediates the symbiosis. In both models, an archaeal/bacterial assemblage evolved side by side for long periods, facilitating the evolution of the eukaryotes through a process similar to that proposed in the endokaryotic pathway (Lake and Rivera, 1994).

Each of the current models provides plausible explanations for the mosaic nature of the eukaryotic genome. That is, why there is a major contribution from both a bacterial and an archaeal ancestor to the eukaryotic genome. However, none of them clearly explain why the eukaryotic nucleus evolved such novel features as mRNA capping, linear chromosomes, mRNA extrusion across a membrane, nuclear pores, separation of transcription and translation, etc. One model that addresses some of these problems is the Viral Eukaryogenesis Theory that proposes a viral ancestry for the eukaryotic nucleus (Bell, 2001; Bell, 2004).

5. The Viral Eukaryogenesis Theory

The Viral Eukaryogenesis Theory is founded on the observation that unique features of the eukaryotic nucleus are also found in some complex viruses. In particular, it appears that the eukaryotic nucleus shares several features with the poxviruses. The poxviruses possess a large complex linear DNA genome consisting of hundreds of genes. Included in the genome are genes such as the mRNA capping gene that are usually associated with eukaryotic genomes. As shown in Table 1 they also share other fundamental similarities in genetic design with nuclei that are not shared by prokaryotes. Thus characteristic features of the eukaryotic nucleus, such as linear double stranded DNA genomes, mRNA processing (capping and polyadenylation), and the absence of translation ability are all found in the poxviruses.

Table 1. Comparison between nuclei, the pox viruses and free-living prokaryotes.

	Nucleus	Pox virus	Prokaryote
Lack of translational apparatus	√	√	X
Lack of ATP production	√	√	X
Linear chromosomes	√	√	X
Cytoplasmic replication	√	√	X
Disassemble membrane during replication	√	√	X
Membrane assembled at ER	√	√	X
Transcription	√	√	X
mRNA export across membrane	√	√	X
Capping of mRNA	√	√	X
Import of proteins	√	√	X

By examining several complex DNA viruses, including the poxviruses, algal viruses, insect viruses and the African Swine Fever virus, it has been deduced that they belong to an ancient group of viruses, the NCLDV viruses (Iyer et al., 2001). At least 31 genes were likely to be present in the ancient NCLDV ancestor including the mRNA capping enzyme, the enzyme responsible for producing 'eukaryote-like' capped mRNA. The genome also includes a DNA polymerase, three RNA polymerase subunits, and several other nucleotide metabolism genes essential for genetic information processing (Iyer et al., 2001). Together these suggest that the ancestral NCLDV virus had an elaborate system for genome replication and the expression of 'eukaryotic like' capped and polyadenylated mRNA.

In the poxvirus, all the enzymes necessary for transcription are packaged in the virus, allowing the poxvirus to have an early transcription system that is independent of the nucleus (Blasco, 1995; Moss, 1996). This early transcription system includes a mechanism for the synthesis of capped mRNA that is extruded from the membrane bound viral core into the host cytoplasm for translation (Moss, 1996).

The Viral Eukaryogenesis Theory incorporates aspects of SET theory for the origin of the mitochondria, but further proposes that the eukaryotic nucleus evolved from a DNA virus. Accordingly, the eukaryotes are a consortium of three different nucleic acid based organisms: the ancestor of the eukaryotic cytoplasm and its ribosomes was an archaeon, the ancestor of the mitochondrion was an alpha-proteobacterium, and the ancestor of the nucleus was an NCLDV virus. It is further proposed that many of the other unique features of eukaryotic cellular design arose by superimposing the non-cellular design of a virus over the cellular design of a prokaryote.

Due to a number of considerations (Bell, 2001), the archaeal ancestor of the eukaryotic cytoplasm is considered to have been similar to 'Methanoplasma elizabethii' (Rose and Pirt, 1981) a member of a methane-producing consortium composed of both archaea and bacteria. Like 'M. elizabethii', it lacked a cell wall and grew in a filamentous form, presumably to maximize contact with bacterial members of the consortium. The bacterial members of the consortia metabolized complex organic materials for growth, and produced hydrogen and carbon dioxide as waste products. These waste products were used by the methanogen as raw materials for growth.

Amongst the many viruses that were likely to infect the bacterial and archaeal members of the consortium, it is proposed that an NCLDV-like virus infected the methanogen. The virus would enter the methanogen via a membrane fusion process (hence the requirement for a cell-wall less methanogen), and then replicate in a similar fashion to that observed with the pox virus today (Figure 5). As observed with the present day poxvirus, the virus would have carried its own autonomous transcription apparatus, allowing it to synthesize capped (and polyadenylated) mRNA independent of the host genome.

Two significant modifications from the poxviral replication cycle are required for the Viral Eukaryogenesis Theory. Firstly, the virus must have been able to infect an archaeal rather than a eukaryotic host. However, since the archaea and eukaryotes utilize similar information processing genes (Marsh et al., 1994; Rivera et al., 1998) and the poxvirus can replicate independently of the nucleus (Villareal et al., 1984), this is unlikely to have been a major problem. Indeed, at the time of the origin of the eukaryotes, the information processing genes of the eukaryote ancestor would have been identical to the archaeal genes since the two lineages had not yet diverged from each other. Secondly, in eukaryotes, NCLDV viruses such as the poxviruses and others are packaged within a membrane derived from collapsed vesicles of the endoplasmic reticulum (Andres et al., 1998; Tolonen et al., 2001,), an organelle absent from any known prokaryote. To account for the production of these vesicles, it is proposed that the membranes required for virus formation were derived from invaginations of the cell membrane of the methanogen. These invaginations were pinched off to make spherical vesicles which would then collapse to produce the 'viral crescents' observed in poxvirus assembly (Figure 5). The filamentous form of the methanogen and the membrane fusion proteins provided by the virus would have facilitated this process. This pinching off of the archaeal membrane represented the beginnings of the evolution of the eukaryotic endomembrane system.

The first eukaryogenesis step occurred when the virus established a persistent infection in the archaeal host. This persistent infection presumably involved the evolution of a mechanism to regulate the copy number of the virus, as well as a mechanism to ensure that the virus was segregated into each of the two daughter cells. During the persistent infection phase, the infected host would contain two sets of mRNA transcripts. The first set would be the normal archaeal mRNA produced by the host genome for the production of essential proteins and enzymes. The second set would be modified 'eukaryote like' viral mRNA (capped and polyadenylated) encoding the proteins and enzymes required for maintenance of the virus.

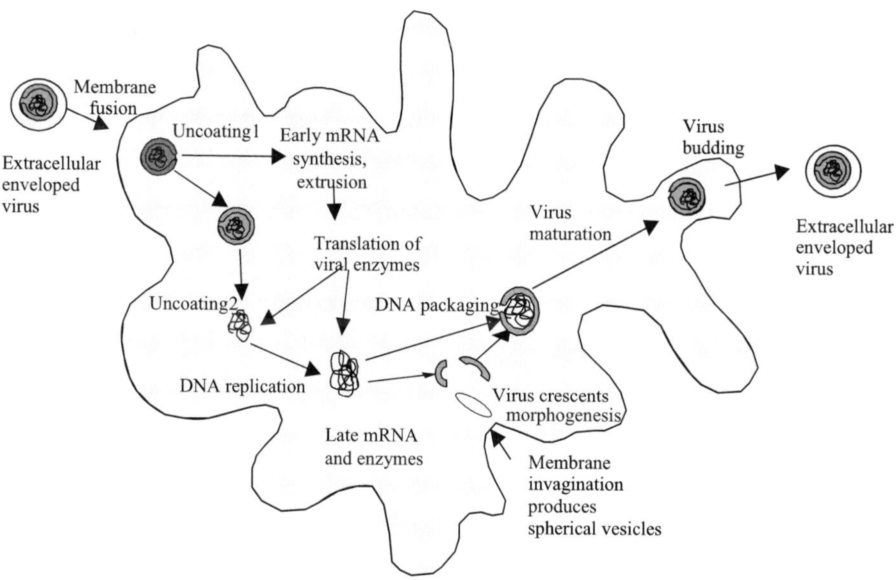

Figure 5. Proposed viral replication cycle of the viral ancestor of the nucleus. The schematic representation of the viral life cycle is based on that of a poxvirus in a mammalian cell (Moss, 1996). The extra-cellular enveloped virus fuses with the plasma membrane using viral proteins. Fusion releases virus core into the cytoplasm. Virus produces own capped mRNA which is extruded into the host cytoplasm. The capped mRNA is stable, and allows virus to take over cellular metabolism. The early mRNA also encodes proteins to uncoat the core of the virus. After uncoating, the viral DNA replicates. Recombination is used to resolve the viral genome into monomers. After DNA replication, virus directs synthesis of proteins for construction of viral particle. Internal spherical membrane vesicles are produced by invagination and pinching off of the plasma membrane. Vesicles collapse into 'viral crescents'. After assembly of the viral crescents, the DNA is transported into maturing viruses, along with essential enzymes such as RNA polymerases, DNA ligases etc. After viral assembly and maturation the viruses are disseminated by budding through the host membrane, regenerating the extra-cellular enveloped virus.

After the persistent infection was established, it is proposed that the viral membrane fusion processes, essential to viral infection, replication and morphogenesis evolved into a simple form of phagocytosis. The evolution of this primitive phagocytosis, building on the primitive endomembrane system, was facilitated by the host's filamentous form, its lack of a cell membrane, and the viral requirement to generate internal membrane bound vesicles for the construction of infectious viral particles. This primitive phagocytosis allowed alpha-proteobacteria to be captured and become endosymbionts, where they maintained their original function of providing hydrogen and carbon dioxide to the host.

It is further proposed that the close constant physical contact between the virus, the host and the endosymbionts facilitated lateral gene transfer between the three members of the consortium. Genes transferred to the viral genome were more effectively translated into proteins than genes on archaeal chromosome, due to capping and polyadenylation, thus providing a selective advantage to the transfer of genes to the viral genome. As a result, genes were inexorably transferred to the future nuclear genome from both the endosymbionts and the host archaeal genome.

Ultimately, the linear viral genome obtained the host archaeal genes for the ribosomal apparatus, allowing the virus to be become relatively independent of the host archaeal genome. In the course of evolution, the host archaeal genes were either transferred to the viral genome or were lost as the host chromosome became increasingly obsolete due to increasing dominance of genes located on the viral genome. One environmental factor contributing to the ultimate degradation of the host archaeal genome may have been a rise in oxygen levels during the period of eukaryogenesis. Since methanogenesis is a strictly anaerobic process, the increasing abundance of oxygen may have made the consortium more dependent upon obtaining energy and nutrients from prey or endosymbionts, rather than through methanogenesis. As the genes for methanogenic metabolism were gradually lost, the consortium would have become more committed to capturing bacteria for survival, and thus more like a eukaryotic predator.

The Viral Eukaryogenesis Theory thus describes an evolutionary process that leads to the appearance of an amoeba-like predator. The dependence upon the capture of bacteria for raw materials would have ensured that all the earliest eukaryotes would have possessed ancestors of the mitochondria, as well as complex membrane fusion based processes. It would also have ensured that the eukaryotes embarked on a different evolutionary trajectory to the prokaryotic domains.

6. Support for the Viral Eukaryogenesis Theory

The SET theory for an alpha-proteobacterial ancestor of the mitochondria is widely accepted because of similarities in design between the mitochondria and bacteria as well as phylogenetic support for an alpha-proteobacterial ancestor of many eukaryotic genes associated with the mitochondria. The same criteria can also be used to evaluate the Viral Eukaryogenesis Theory.

In the case of genetic design, the NCLDV viruses share a number of significant features with the eukaryotic nucleus. These include linear chromosomes, the lack of a translation apparatus, and the ability to cap and polyadenylate mRNA prior to extrusion across a membrane (Bell, 2001). Phylogenetic support for a relationship between several crucial nuclear genes and viral genes is also observed. For example, the mRNA capping enzymes are not found in any prokaryotes, but are found in all eukaryotes and in the NCLDV viruses. Phylogenetic analysis of the mRNA capping enzymes indicates that viral and eukaryote mRNA capping enzymes share a common ancestor, which existed before the eukaryotes diverged from each other (Bell, 2001). This is consistent with the hypothesis that the mRNA capping enzyme evolved in a viral ancestor prior to evolution of the eukaryotes. Studies also suggest a specific relationship between viral and eukaryotic DNA polymerases, leading to the suggestion that the DNA virus replication proteins gave rise eukaryote replication proteins (Villarreal and DeFillippis, 2000, Takemura, 2001).

The SIRV1 virus of the thermophilic archaeon *Sulfolobus* supports the existence of complex viruses like the NCLDV ancestor in the archaea. The complete sequence of the viral genome reveals that this archaeal virus shares structural features in common with the NCLDV viruses including the Poxviruses (Blum et al., 2001).

The recently described Mimivirus (Raoult et al., 2004) also provides significant support for the Viral Eukaryogenesis Theory. This member of the NCLDV viruses is the largest and most complex virus yet known, and significantly blurs the distinction between 'non-living' viruses and 'living' cells. When originally discovered infecting an amoeba, it was thought to be bacterial endosymbiont. However it is now known to be a virus that is physically as large and complex as many prokaryotes. This includes possessing a genome larger than many prokaryotes. Despite its large linear genome, the Mimivirus does not possess a functional translational apparatus, and thus like all viruses, it can only replicate inside host cells that provide access to a translation apparatus. Complete sequencing of the Mimivirus genome has revealed that it possesses seven of the 63 universally conserved genes found in all cellular organisms (Raoult et al., 2004). In excellent agreement with the Viral Eukaryogenesis Theory, phylogenetic analysis of these genes (Figure 6) shows that they branch from near the base of the eukaryotic radiation and are equidistant from the four main eukaryotic kingdoms (Protista, Animalia, Plantae and Fungi).

If the root of the universal phylogenetic tree is placed between the archaea and the bacteria, the genes in the Mimivirus appear to descend from a common ancestor to those of the eukaryotes, and this common ancestor existed before the eukaryotes diverged from each other. With its complex linear genome, and the 'eukaryotic' ability to generate capped mRNA, the Mimivirus displays many of the features to be expected from a 'missing link' between a virus and the eukaryotic nucleus.

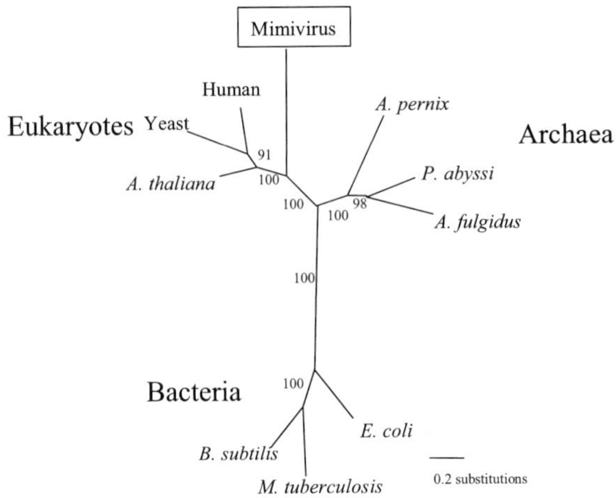

Figure 6. Phylogenetic analysis of Mimivirus genes common to all cellular life. (From Raoult et al., 2004). The position of the Mimivirus is consistent with the Viral Eukaryogenesis theory. If the root of the tree is placed between the bacteria and the archaea, the Mimivirus diverges from the eukaryotic lineage, before the eukaryotes diverge from each other. *A. thaliana*; *Arabidopsis thaliana*, Yeast; *Saccharomyces cerevisiae*, Human; *Homo sapiens*, *A. pernix*; *Aeropyrum pernix*, *P. abyssi*; *Pyrococcus abyssi*, *A. fulgidus*; *Archaeoglobus fulgidus*, *E. coli; Eschericia coli*, *M. tuberculosis*; *Mycobacterium tuberculosis*, *B. subtilis*; *Bacillus subtilis*.

7. Conclusions and Perspectives

Of the three domains of life, it is the eukaryotes that have evolved into complex organisms, with the two prokaryotic domains remaining predominantly at unicellular or colonial levels. This is perhaps surprising since on a number of levels, the bacteria and the archaea appear to be more efficient, more self reliant, and more adaptable than typical eukaryotic cells. In addition the fossil record indicates that the prokaryotes have had a significantly longer period in which to evolve complexity. These observations suggest that in the case of the prokaryotes, selection is not driving their evolution toward organismal complexity but rather to biochemical complexity and rapid replication (Forterre, 1995). Why is it the eukaryotic lineage that has evolved towards complex organisms such as ourselves? It seems likely that the answers lie in the differences in cellular design and life strategies between eukaryotes and prokaryotes.

One clue to the origin of eukaryotic complexity may lie in the evolution of the nucleus. This organelle allows separation of the functions of genetic information storage and processing away from other functions of the cell and this may have facilitated the higher levels of mRNA regulation and control observed in the eukaryotes (e.g. Mattick, 2003). Also associated with the evolution of the nucleus is the eukaryotic sexual cycle, which allows for controlled shuffling of genomes in each sexual generation. This shuffling allows for the elimination of deleterious alleles from large complex genomes, as well as facilitating the assemblage of novel combinations of genes.

Another clue to the evolution of complexity in the eukaryotes may lie in the origin of the mitochondria, an organelle descended from a free-living bacterium. Its presence, or traces of its previous presence, in all living eukaryotes (Gray et al., 1999) suggests that it was present in the last common ancestor of all eukaryotes (LECA). So, an ancestor of the mitochondria either invaded LECA, or LECA was capable of engulfing the mitochondrial ancestor. If LECA was capable of engulfing bacteria via a phagocytosis like process, then LECA was most likely a phagocytic predator. A novel predatory life style for the eukaryotes would have established a unique evolutionary trajectory for the eukaryotes compared to the bacteria and the archaea. In particular, a eukaryotic predator is likely to have evolved towards organismal complexity to facilitate the capture of prey.

Finally, it should be noted that the process of evolution of the eukaryotic cellular design from the last universal ancestor of all life is still not resolved. In the endeavor to elucidate this fundamental evolutionary process, it would be therefore unwise to discount any organisms that may have contributed to the process. In particular, although viruses are the most abundant nucleic acid based entities on earth, they are often omitted from discussions of the origin and evolution of life, presumably because they fail to meet the arbitrary definition of life. However, as the discovery of the Mimivirus has demonstrated, viruses can become as complex as some cellular life forms, and thus challenge the current definitions of life. Furthermore, it has recently been proposed that the fundamental transition from the RNA world into a DNA world was mediated by viruses (Forterre, 2001). Given the abundance of viruses, their demonstrated ability to transfer genes between organisms, and their noted effects on populations (e.g. influenza, ebola, HIV, smallpox, etc), it seems likely that viruses have played a significant role in the evolution of each of the three domains. If the Viral Eukaryogenesis Theory is correct viruses may be at the core of our own existence.

8. References

Andres G., Garcia-Escudero R., Simon-Mateo, C., Vinuela, E. (1998) African Swine Fever virus is enveloped by a two membraned collapsed cisterna derived from the Endoplasmic Reticulum. J. Virol. **72,** 8988-9001.

Auer J., Spicker, G., Böck, A. (1989) Organisation and structure of the *Methanococcus* transcriptional unit homologous to *Escherichia coli* "spectinomycin operon". Implications for the evolutionary relationship of 70 S and 80 S ribosomes. J. Mol. Evol. **209,** 21-36.

Bell P.J.L. (2001) Viral eukaryogenesis: was the ancestor of the nucleus a complex DNA virus. J. Mol. Evol. **53,** 251-256.

Bell P.J.L. (2004) The Viral Eukaryogenesis Theory, In: J. Seckbach (ed) *Origins Genesis, Evolution and Diversity of Life*. Kluwer Academic Publishers, Dordrecht, The Netherlands, pp. 347-394.

Benson S.D., Bamford, J.K., Bamford, D.H., Burnett, R.M. (2004) Does common architecture reveal a viral lineage spanning all three domains of life? Mol. Cell **16,** 673-685.

Blasco R. (1995) Evolution of poxviruses and African swine fever virus, In: A. Gibbs, C.H. Calisher, F. Garcia-Arenal (eds.) *Molecular basis of virus evolution.* Cambridge University Press, Cambridge, pp. 255-269.

Blum H., Zillig W., Mallok S., Domdey H., Prangishvili D. (2001) The genome of the archaeal virus SIRV has features in common with genomes of eukaryal viruses. Virology **281,** 6-9.

Brinkmann H. and Philippe H. (1999) Archaea sister group to Bacteria? Indications from tree construction artifacts in ancient phylogenies. Mol. Biol. Evol. **16,** 817-825.

Brocks J.J., Logan G.A., Buik R., Summons R.E. (1999) Archean molecular fossils and the early rise of the eukaryotes. Science **285,** 1033-1036.

Cavalier-Smith T. (1987) The origin of eukaryote and archaebacterial cells. Ann. N.Y. Acad. Sci. **503,** 17-54.

Forterre P. (1995) Thermoreduction, a hypothesis for the origin of prokaryotes. C.R. Acad. Sci. III **318,** 415-422.

Forterre P. (2001) Genomics and early cellular evolution. The origin of the DNA world. Life Sciences **324,** 1067-1076.

Forterre P. and Philippe H. (1999) Where is the root of the universal tree of life. Bioessays **21,** 871-879.

Fuhrman J.A. (1999) Marine viruses and their biogeochemical and ecological effects. Nature **399,** 541-548.

Gogarten J.P., Kibak H., Dittrich P., Taiz P., Bowman E.J., Manolson M.F., Poole P.J., Data T., Oshima T., Konishi J., Denca K., Yoshida M. (1989) Evolutionary relationship of archaebacteria, eubacteria and eukaryotes inferred from phylogenetic trees of duplicated genes. Proc. Natl Acad. Sci. USA **86,** 9355-9359.

Gray M.W., Burger G., Lang B.F. (1999) Mitochondrial evolution. Science **283,** 1476-1481.

Gribaldo S. and Cammarano P. (1998) The root of the universal tree of life inferred from anciently duplicated genes encoding components of the protein-targeting machinery. J. Mol. Evol. **47,** 508-516.

Gupta R.S. (1998) Protein phylogenies and signature sequences: a reappraisal of evolutionary relationships among archaeabacteria, eubacteria, and eukaryotes. Microbiol. Mol. Biol. Rev. **62,** 1435-1491.

Gupta R.S. and Golding G.B. (1996) The origin of the eukaryotic cell. Trends Biochem. Sci. **21,** 166-171.

Hartman H. and Federov A. (2002) The origin of the eukaryotic cell: a genomic investigation. Proc. Natl Acad. Sci. USA **99,** 1420-1425.

Iyer L.M., Aravind L., Koonin E.V. (2001) Common origin of four diverse families of large eukaryotic DNA viruses. J. Virol. **75,** 11720-11734.

Iwabe N., Kuma K., Hasegawa M., Osawa S., Miyata T. (1989) Evolutionary relationship of archeabacteria, eubacteria and eukaryotes inferred from phylogenetic trees of duplicated genes. Proc. Natl Acad. Sci. USA **86**, 9355-9359.

Kyrpides N., Overbeek R., Ouzounis C. (1999) Universal protein families and the functional content of the last universal ancestor. J. Mol. Evol. **49,** 413-423.

Lake J.A. and Rivera M.C. (1994) Was the nucleus the first endosymbiont? Proc. Natl Acad. Sci. USA **91,** 2880-2881.

López-Garcia P. and Moreira D. (1999) Metabolic symbiosis at the origin of eukaryotes. Trends Biochem. Sci. **24,** 88-93.

Margulis L. (1996) Archaeal-eubacterial mergers in the origin of eukarya - phylogenetic classification of life. Proc. Natl Acad. Sci. USA **93,** 1071-1076.

Marsh T.L., Reich C.I., Whitelock R.B., Olsen G.J. (1994) Transcription factor IID in the Archaea: sequences in the *Thermococcus celer* genome would encode a product closely related to the TATA-binding protein of eukaryotes. Proc. Natl Acad. Sci. USA **91,** 4180-4184.

Martin W. and Müller M. (1998) The hydrogen hypothesis for the first eukaryote. Nature **392,** 37-41.

Mattick J.S. (2003) Challenging the dogma: the hidden layer of non-protein-coding RNAs in complex organisms. Bioessays **25,** 930-939.

Moreira D. and López-Garcia P. (1998) Symbiosis between methanogenic archaea and delta-proteobacteria as the origin of eukaryotes: the syntrophic hypothesis. J. Mol. Evol. **47,** 517-530.

Moss B. (1996) Poxviridae: The viruses and their replication, In: B.N. Fields, D.M. Knipe, P.M. Howley, R.M Chanock, J.L. Melnich, T.P. Monath, B. Roizman, S.E. Strauss. (eds.) *Fundamental Virology.* Lippencott-Raven Publishing, Philadelphia, third ed., pp. 1164-1185.

Oren A., Bratbak G., Heldal M. (1997) Occurence of virus-like particles in the Dead Sea. Extremophiles **1,** 143-149.

Penny D. and Poole A. (1999) The nature of the last universal common ancestor. Curr. Opin. Gen. Dev. **9,** 672-677.

Poole A.M., Jeffares D.C., Penny D. (1998) The path from the RNA world. J. Mol. Evol. **46**, 1-17.
Prangishvili D. and Garrett R.A. (2004) Exceptionally diverse morphotypes and genomes of crenarchaeal hyperthermophilic viruses. Biochem. Soc. Trans. **32,** 204-208.
Reanney D.C. (1974) On the origin of the prokaryotes. J. Theor. Biol. **48,** 243-251.
Raoult D., Audic S., Robert C., Abergel C., Renesto P., Ogata H., La Scola B., Suzan M., Claverie J.M. (2004) The 1.2-megabase genome sequence of Mimivirus. Science **306,** 1344-1350.
Ribeiro S. and Golding G.B. (1998) The mosaic nature of the eukaryotic nucleus. Mol. Biol. Evol. **15,** 779-788.
Rivera M.C., Jain R., Moore J.E., Lake J.A. (1998) Genomic evidence for two functionally distinct gene classes. Proc. Natl Acad. Sci. USA **95,** 6239-6244.
Rose C.S. and Pirt S.J. (1981) Conversion to fatty acids and methane: Role of two mycoplasmal agents. J. Bacteriol. **147,** 248-252.
Schopf J.W. (1993) Microfossils of the early archean apex chert: New evidence of the antiquity of life. Science **260,** 640-646.
Sogin M.L. (1991) Early evolution and the origin of the eukaryotes. Curr. Opin. Genet. Dev. **1,** 457-463.
Strauss E.G., Strauss J.H., Levine A.J. (1996) Virus evolution, In: B.N. Fields, D.M. Knipe, P.M. Howley, R.M Chanock, J.L. Melnich, T.P. Monath, B. Roizman, S.E. Strauss. (eds.) *Fundamental Virology.* Lippencott-Raven Publishing, Philadelphia, third ed.
Takemura M. (2001) Poxviruses and the origin of the eukaryotic nucleus. J. Mol. Evol. **52,** 419-425.
Tolonen N., Doglio L., Schleich S., Locker J.K. (2001) Vaccinia virus DNA replication occurs in endoplasmic reticulum-enclosed cytoplasmic mini-nuclei. Mol. Biol. Cell **12,** 2031-2043.
Vellai T., Takacs K., Vida G. (1998) A new aspect to the origin and evolution of eukaryotes. J. Mol. Evol. **46,** 499-507.
Villarreal E.C., Roseman N.A., Hruby D.E. (1984) Isolation of vaccinia virus mutants capable of replicating independently of the host cell nucleus. J. Virol. **51,** 359-366.
Villarreal L.P., DeFillippis V.R. (2000) A hypothesis for DNA viruses as the origin of eukaryotic replication proteins. J. Virol. **74,** 7079-7084.
Woese C.R. (1987) Bacterial evolution. Microbiol. Rev. **51,** 221-271.
Woese C.R. (1998) The universal ancestor. Proc. Natl Acad. Sci. USA **95,** 6854-6859.
Woese C.R. (2000) Interpreting the universal phylogenetic tree. Proc. Natl Acad. Sci. USA **97,** 8392-8396.
Woese C.R., Kandler O., Wheelis M.L. (1990) Towards a natural system of organisms: proposal for the domains Archaea, Bacteria, and Eucarya. Proc. Natl Acad. Sci. USA **87,** 4576-9.
Zillig W. (1991) Comparative biochemistry of Archaea and Bacteria. Curr. Biol. 1, 544-551.

Biodata of **John J. Lee** author of ***"Symbiosis Forms of Life"***

Dr. J.J. Lee is a Distinguished Professor of Biology at City College of City University of New York and a Research Associate at the American Museum of Natural History in New York and at the IOLR National Center for Mariculture in Eilat Israel. He earned his Ph.D. at New York University in 1960. Dr. Lee and his students have published more than 80 papers and book chapters on symbiosis in "Living Sands" (larger foraminifera). He was one of the founders of the International Society of Symbiosis and served as its first President. He also has served as President of the Society of Protozoologists and was the Editor-in-Chief of the first and second editions of *The Illustrated Guide to the Protozoa.* Dr. Lee was named an Honorary member of the Society of Protozoologists in 2001 and was given the Joseph A Cushman Award by the Society for Foraminiferal Research in 1999.

E-mail: **jlee@sci.ccny.cuny.edu**

J. Seckbach (ed.), Life as We Know It, 307-324.

SYMBIOTIC:
Forms of life

JOHN J. LEE
Department of Biology, City College of City University of New York,
Convent Ave @ 138 Street, New York, NY 10031, USA

1. Introduction–What is Symbiosis?

Symbiosis is a term very loosely applied to a range of associations of two or more different species living together. Most commonly, the term is used for permanent or long-lived non-parasitic associations of organisms of different sizes in which the larger organism (the host) exploits the capabilities of one or more smaller organisms (symbionts) (Smith, 1990). This concept avoids trying to rigorously define benefits to the organisms involved. However, many of those who study symbiosis often opine on the adaptive advantages of symbiosis. Symbiosis is a phenomenon that follows the attainment of an intimate union of two or more organisms that, over time, build new structures and metabolic interactions. In this perspective, symbiosis is a mechanism for the host to survive and prosper in what might be considered marginal or extreme habitats (e.g. corals, hydrothermal vent fauna, lichens) or to exploit previously unavailable (e.g. indigestible) resources (e.g. ruminants, termites). Viewed in another way, symbiosis can be thought as a phenomenon by which two sets of genes (genomes) can be brought together to collaboratively provide abilities for adaptation and exploitation that neither partner alone could provide.

Symbiosis as a generalized concept can be envisioned as a phenomenon that involves a continuum of relationships among partners. At one end of the spectrum is the evolution of the eukaryotic cell recognizing the endosymbiotic origin of mitochondria and chloroplasts (e.g. Margulis, 2004, Kowallik, 1997; Keeling, 2004). When one contemplates the contribution of algal and higher plant photosynthesis (food and oxygen) to the biological economy of life on earth one can only conclude that these fundamental symbioses sparked the evolution of life on this planet. The loss of genes by the symbionts (e.g. proto-mitochondria and proto-chloroplasts) and the transfer of other genes to the nucleus of the host represent extreme integration of formerly independent biological species. Close to this end of the spectrum are some the bacterial-insect symbioses (*Wolbachia*; *Buchneria*; Latorre et al., 2003). At the other end of the spectrum we recognize symbioses where the relationship between the host and its symbionts is not fastidious. The same host can form stable symbiotic relationships with more than one species or variety of symbiont, such as the algal endosymbionts of corals and those of larger foraminifera. The hosts cannot be grown, even when they are fed, without their symbiotic algae (Lee and Anderson, 1991). However, the algae can be cultured in the light in relatively simple media (e.g. Lee et al., 1980, 2005, McLaughlin and Zahl, 1959).

The flexibility of symbionts in corals is viewed as critical in understanding the long-term resilience of coral reef ecosystems. The possibility of recombination of the host with different partners leads to significant shifts in symbiont community structure during environmental changes (LaJeunesse, 2004, LaJeunesse et al., 2003 Goulet and Coffroth, 2003, Baker, 2003 and Buddemeir and Fautin, 1993). In summary, at one end of the symbiotic spectrum there is evolutionary pressure to make the relationship as specific, indivisible and integrated as possible and at the other end of the spectrum there are survival advantages for hosts which have the ability to switch to alternate symbionts which are better adapted as a particular environment changes.

2. Brief Introduction to Some Model Symbiotic Systems

Symbiotic relationships abound in every habitat, particularly in the most extreme habitats. For example: in the deep seas at the hydrothermal rifts and vents, life is fueled by symbiotic associations between vent worms and clams (*Alvinella*, *Lamellibrachia*, *Parescarpia*, *Escarpia*, *Bathymodiolus*, *Solemya*) and chemolithotrophic sulfur oxidizing bacteria (Table 1). The bacteria are housed in organs called trophosomes and fix the organic carbon needed to sustain the association.

Lichens, symbiotic associations between ascomycete fungi and algae or cyanobacteria, are the only multicellular forms of life found at the boundaries of extreme cold and dryness. For example, one can find lichens above the tree line in cold and high altitude or high latitude habitats where no other photosynthetic multicellular organisms are able to grow (Table 1).

Algal symbiosis in coral reefs and living sands is viewed as an adaptation to successful growth in nutrient-poor, well-illuminated, shallow tropical and semitropical marine habitats. Global warming and bleaching events have spurred world-wide concern about losses of coral reef habitats. Recently molecular systematic approaches have shown that symbiont diversity is much wider than it was thought to be (LaJeunesse, 2001). Although symbiont diversity can be seen as a possible mechanism for hosts to adapt to environmental changes (Baker, 2003), many question whether mankind has gone too far in producing greenhouse gases and nutrient pollution for these symbiotic systems to adjust to anthropogenically introduced environmental changes.

Symbiosis apparently has driven the evolution of larger foraminifera (Lee and Hallock, 1987). Popularly known as "living sands" because they have durable calcareous shells, they live in the same habitats as corals. They are "giants" among the protists reaching mm to cm in length or diameter. Some of them are as large as small coins (Figs. 1B and 2B); others form macroscopically visible "star sands". They can be quite abundant and form the sand on the beaches of some tropical Pacific islands. The enormity of cumulative generations of living sands is impressive when one stands at the base of the mountains of their fossil remains which formed limestone breccias in the Tethys seas (Figs. 1A and 1B). Less well known, but an easily remembered fact, is that the same limestone built of foraminiferal host remains was quarried to build the massive Egyptian pyramids. Larger foraminifera are collectively the hosts for more different types of algae than any other host type (Lee and Anderson, 1991). In common with corals and many other invertebrates some larger foraminifera, the Soritinae, are the hosts of endosymbiotic dinoflagellates (*Symbiodinium* spp.). Another subfamily, Archaiasinae, is the host of green algae (*Chlamydomonas* spp.). Larger foraminifera are also the hosts for two unusual groups of endosymbiotic algae: red algae and diatoms. The Peneroplidae are the only known hosts for endosymbiotic unicellular red algae (*Porphyridium*). Perhaps more significant, and possibly unique, these symbionts are not separated from the rest of the cytoplasm by a host membrane (symbiosome membrane) (Lee 1990). Conceptually, the loss of a host membrane surrounding a symbiont is considered an important step toward integrating a symbiont into becoming an organelle. A host membrane surrounding a symbiont always has the potential to revert back to a phagocytic membrane able to dock with lysosomes, with the result that the symbiont can be digested by the host. However, unlike organelles, these red algae have retained their complete genome and can be experimentally released from their host and grown independently in relatively simple media (Hawkins and Lee, 1990).

Although diatoms are among the most abundant photosynthetic protists in the sea, they are rarely found as endosymbionts. Yet several unrelated families of larger foraminifera (Alvelolinidae, Calcarinidae, Amphisteginidae) are the hosts for a variety of endosymbiotic diatoms (e.g. Fig. 2A). The diatoms do not form frustules in their hosts (Fig. 3C), but they do form them when they are isolated from their hosts and are grown in culture giving them a morphological basis for identification (Fig. 2A) (Lee, 1994). Diatom-bearing hosts are morphologically adapted to host their symbionts (Figs. 3A-D). For example, each diatom symbiont in *Amphistegina* is held in a cup-like expansion of a poor rim (Fig. 3C.) The pores in the foraminiferan shell wall are like pipes and bring fresh seawater to each entrained symbiont (Fig. 3B).

In addition, several families of foraminifera feed on diatoms and retain their plastids for several months as temporary functioning organelles (Correia and Lee 2002). Even though these foraminifera also feed on dinoflagellates and chlorophytes, the chloroplasts of these algal groups are not retained. There are some sea slugs and a variety of planktonic ciliates which also retain chloroplasts from the algae they feed upon (Trench, 1975; Stoecker and Silver, 1987, Laval-Peuto,1990). The phenomenon is called by many names: chloroplast-symbiosis, plastid-symbiosis, chloroplast-husbandry, kleptoplasty, kleptochloroplasty. The plastids are photo-synthetically active in their hosts. The sea slugs (elysioid sacoglossian gastropods) are so green one investigator was inspired to call them "crawling leaves" (Trench, 1975). The chloroplasts are sequestered in the phagocytic digestive cells that line the highly branched digestive diverticulum (it arises from the stomach and ramifies throughout the body). The slugs (*Elysia viridis*, *Tridachia crispate*, *Tridachiella diamedea*, and *Placobranchus ianthobapsus*) retain chloroplasts from green seaweeds (Order Siphonales). Their feeding apparatus punctures the cells of the seaweeds and sucks out their contents. Experimental studies of fed, then starved, different species of slugs, have demonstrated that starved animals differ in their abilities to retain functional chloroplasts (8 days-9 months) (Rumpho et al., 2000). Survival of chloroplasts for this length of time raises many questions. Even the largest chloroplast genomes investigated contain coding for < 25% of the gene products needed for plastid function. The robustness of the chloroplasts isolated from their cells and the stability of the chloroplast

Figure 1. A. Mountain of fossilized larger foraminiferan tests in wadi Taba, Sinai, Egypt. **B.** A sample taken from the above mountain showing coin-like fossil foraminiferan algal-bearing hosts' shells (tests) which formedthe limestone.

Table 1. Brief citation of some well studied symbiotic systems.

Host	Symbionts	Major reference or review
Hydrothermal vent worms and clams- *Riftia Alvinella, Lamellibrachia, Parescarpia, Escarpia, Bathymodiolus, Solemya*	Sulfur oxidizing eubacteria (γ and ε_proteobacteria,	Felbbeck, 1990, McMullin et al., 2003.
Amoeba proteus	X-bacteria	Jeon, 1987, 1993
Ciliates-*Paramecium bursaria*	*Chlorella vulgaris*	Reisser, W., 1987
Ciliates- *Strombidium, Laboea, Tontonia*	kleptochloroplasts	Laval-Peuto,1990
Ciliates-*Paramecium* spp	*Holospora*	Görtz et al., 1993
Ciliates-diverse species (e. g. Armophoridae and Clevelandellidae)	Hydrogenosomes and methanogens	Hackstein et al., 2002
Larger foraminifera-Peneroplidae	*Porphyridium pupereum*	Lee, 2002
Larger foraminifera- Numulitidae Alveolinidae Amphisteginidae	~20 species of diatoms: e.g. *Nitzschia f. symbioticum Nitzschia panduriformis Nanofrustulum shiloi*	Lee, 2002
Larger foraminifera-Soritidae	*Symbiodinium* spp.	Lee et al, 1997; Pouchon et al., 2004
Larger foraminifera-Archaiidae	*Chlamydomonas hedleyi Chlamydomonas provasolii*	Lee, 2002
Foraminifera- *Elphidium, Haynesina, Metarotaliella*	Diatom chloroplasts	Correia and Lee 2002, Lee, 2002
Various coral species, anemones, giant clams and other marine invertebrates	Symbiodinium species	LaJeunesse, 2001
Sponges (e.g. *Chondrilla nucula Cliona veridis Dydidea herbecea*	e.g. *Aphanocapsa feldmanni, Synechocystis* spp, *Oscillatoria* spp.	Wilkinson, 1992
Didemmid ascidians	*Prochloron didemni*	Lewin and Cheng, 1989 Kühl and Larkum, 2002
Chlorohydra viridis	*Chlorella*	Muscatine et al., 1975; Rahat, 1992
White flies, fruit flies, beetles, butterflies, isopods, parasitoid wasps	*Wolbachia*	Stouthamer et al., 1999
Aphids	*Buchneria*	Douglas, 1998
Reticulotermes and other termites	Spirochetes, *Cytophaga, Flavobacterium,* α and_ δ_proteobacteria, parabasalians	Dyer, 2002
Squid (*Euprymna scolopes, Photololigo noctiluca, Sepiola robusta*) and some marine fish (*Monocentris japonica*)	*Vibrio fisheri, V. logei Photobacterium leiognathi*	Ruby, 1996 Ruby and Nealson, 1976
Mycorrhiza ~240,000 plant species including ferns, gymnosperms and angiosperms	Glomales of Zygomycota and others	Bonfante et al. 2002
Legumes-root nodules	*Rhizobium, Azorhizobium, Bradyrhizobium, Sinorhizobium*	Udvardi and Day, 1997
Lichens—Many species (>20,000) in 16 orders of Ascomycota	*Tribauxia, Trentepohlia, Sthiococcus, Myrmecia, Nostoc, Coccomyxa*	Paracer and Ahmadjian, 2000 Honegger, 1992

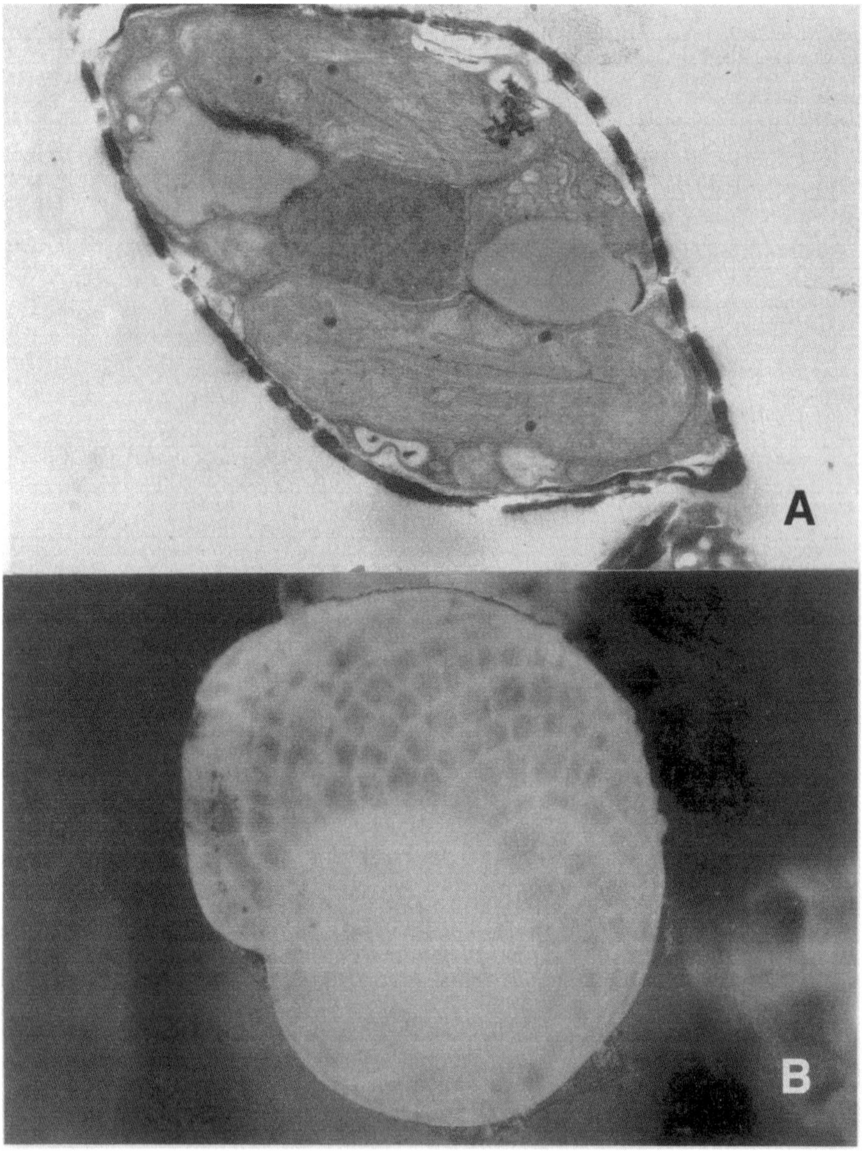

Figure 2. A. TEM of *Nitzschia frustulum symbiotica*, the most common species of endosymbiotic diatoms hosted by foraminifera. **B.** Light micrograph of *Heterostegina depressa*, a close living relative of the numulitic foraminifera found as fossils in the above limestone. Size is 3.2 mm.

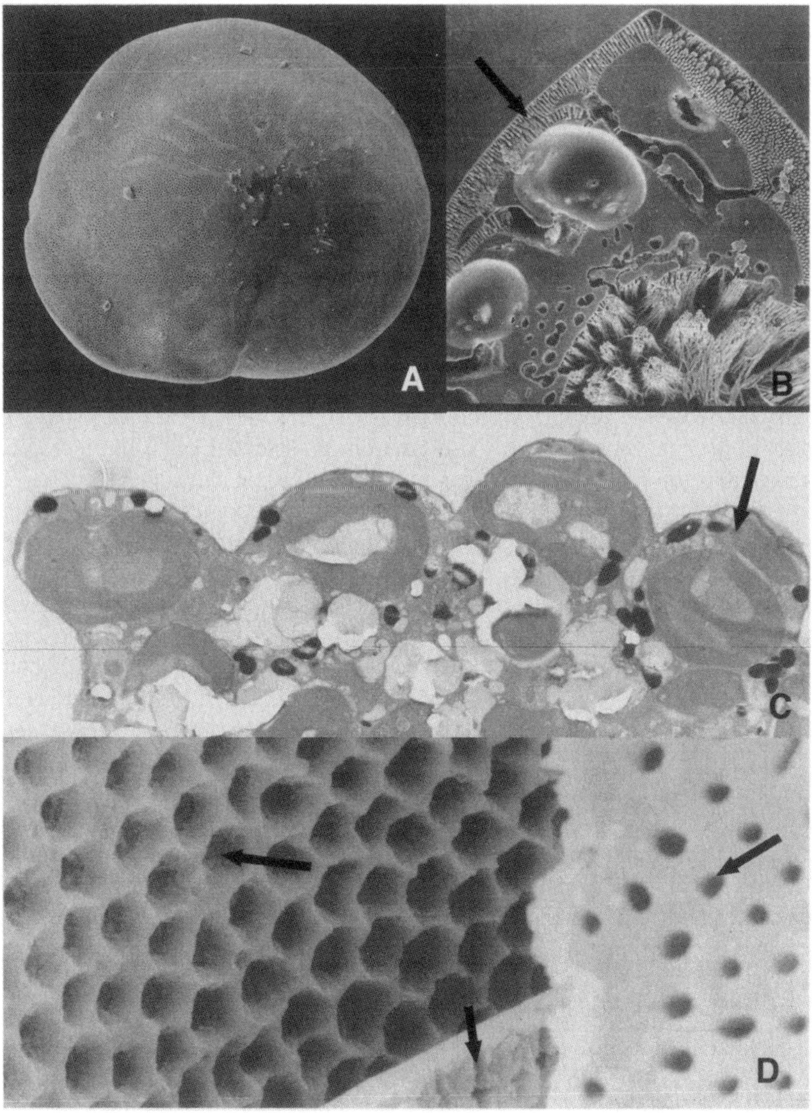

Figure 3. A. A biconvex lens-like diatom-bearing larger foraminifer,.*Amphistegina lessonii.* If you examine the surface of the test carefully you can observe the pores in it. Size is 2.3 mm. B. An epoxy cast of the foraminiferan observed in an SEM. Arrow points to the pore canals that lead from the exterior surface of the test (shell) into the interior surface of the test where the diatom symbionts are located. C. Transmission electron micrograph of a section of the foraminiferan just below the shell. Each of the diatom symbionts, *Nitzschia frustulum symbiotica*, (arrow) fits in a pore rim. D. An SEM of a broken shell of *Amphistegina* with fragments oriented to show different aspects of interest. Arrows show the inner surface with pore rims, the outer surface with pores and a cross-section of the pores passing through the shell.

proteins are key to understanding kleptoplasty. Because of the relative mass of sea slugs compared to protists involved in kleptoplasty, the former system can provide more material for molecular probing of the phenomenon. Experimental evidence points to some chloroplast protein synthesis during their retention. Transcripts of two chloroplast genes were still found in *E chlorotica* after 8 months of culture without feeding on algae (Mujer et al., 1996). Experimental evidence using radionuclide tracers also detected transcripts of nuclear encoded, plastid targeted, genes being translated after 8 months (Mujer et al., 1996; Pierce et al., 1996). Perhaps transcripts destined for the mitochondria or involved in animal gluco-neogenesis and pentose phosphate pathways may be directed to and function in the sequestered plastids.

Among the most interesting cases of symbioses are the relationships between insects and their bacterial endosymbionts. In scale insects, symbiotic bacteria strongly influence the genetic system of the insect. In some species, the males are the product of asexual reproduction and arise from unfertilized eggs. In other species, the males develop from fertilized eggs but all the chromosomes they get from their fathers are deactivated. In still other species the chromosomes from the fathers are present in some cells but not in others. Symbiotic bacteria seem to manufacture essential amino acids needed by the insects by utilizing some of the energy in the sugars taken in by the insects from the phloem sap. Transmission of the bacteria from one generation to the next is assured because bacteria from the symbiosome are deposited into the yolk of eggs. The bacteria, however, seem to kill only males (Nomark, 2004). The culprit, an α proteobacterium named *Wolbachia* that is related to the rickettsiae, has been found in 16% of insect species (including *Tribolium* and ladybird beetles, *Drosophila,* and Acreae butterflies) and a number other invertebrate hosts including mites, isopods and filarial nematodes. In all its hosts *Wolbachia* is distinguished by its ability to modify reproductive phenotypes including cytoplasmic incompatibility, parthogenesis, and feminization. Antibiotic treatment of filarial nematodes abolishes *Wolbachia* and leads to defective development and retarded growth of the filarial worms. The bacteria seem to have genes that are important for the development of the worms and code for the lipopolysaccharides which cause the disease symptoms (inflammatory reactions) suffered by those infected by the nematodes.

Aphids, close relatives to the scale insects, host bacteriocytes with *Buchnera.* Neither host nor symbiont can reproduce independently of each other. Although *Buchnera* is a close relative to *E. coli*, its genome is only 1/7 the size of the latter. While it has genes coding for the biosynthesis of essential amino acids, it cannot synthesize some non-essential amino acids and cell surface components. All this suggests that *Buchneria* is completely dependant upon its host and viable only in its limited niche, the bacteriocyte, where it may be supplied with metabolites it is unable able to synthesize (Shigenobu et al., 2000). It is speculated that the symbiotic relationship was established long ago (200-250 Mybp) and led to the co-speciation of the hosts and their symbionts.

Perhaps less conspicuous, but no less important are the many symbiotic relationships between plants and fungi and/or bacteria. Mycorrhizae are the symbiotic associations between most vascular plant roots and fungi. The association is critical in low fertility soils where the fungus mobilizes and takes up inorganic nutrients and water needed by the plant and in return receives organic carbon metabolites synthesized by the photosynthetic partner. As a result, plants with mycorrhizae are able to tolerate environmental stresses. In associations involving trees and some woody plants the fungus grows between cortical cells in the root and in the soil and this association is referred to as ectomycorrhizal (EM). The net-like structures between root cells are called the Hartig net. Often hyphal strands aggregate to form visible structures known as rhizomorphs. Over 4,000 fungal species belonging to the Basidiomycotina (mushrooms, puffballs, etc.) and some Ascomycotina are involved in ectomycorrhizal associations. In other associations (vesicular-arbuscular mycorrhiza, AM or VAM) involving woody plants and herbaceous species, mycorrhizal fungae belonging to the order Glomales, form highly branched structures called arbuscules which grow within the cortical root cells (endomycorrhizae). Ericaceous plants (rhododendron, blueberries and azaleas) which dominate the acidic highly organic heathland soils of subartic and subalpine habitats are colonized by ascomycete fungi (ericoid mycorrhiza). Arbuscules are not formed in the ericaceous plants. Some trees (e.g. Alders, willows, poplars and eucalyuptus) can have both AM and EM on the same plant roots (Paracer and Ahmadjian, 2000 Honegger, 1992).

Mycorrhizae of orchids are very interesting. The seeds of orchids are typically very small and have stored little nutrients. Mycorrhizal fungi (basidiomycetes) nurse the plant embryos after germination by providing metabolites and vitamins. In achlorophyllous species, the fungal partner feeds the orchid for its whole life aiding the digestion of materials in the substrate and the uptake of nutrients.

Some symbiotic associations involve a considerable amount of nitrogen fixation for plant growth. Recent molecular systematic studies of the soil bacteria involved in the associations (*Rhizobium, Bradyrhizobium, Sinorhizobium*) suggests that they are only distantly related to each other. For example *S. meliloti,* which causes alfalfa to nodulate, is more closely related to the plant pathogen *Agrobacterium* than it is to the rhizobia that nodulate soybeans. This suggests that the symbiotic genes may be acquired by horizontal transfer rather than by lineal descent. The root nodules of legumous plants (clover, soybeans) are best known.

Azolla, an aquatic fern and *Gunnera*, a tropical plant, cycads and many lichens form nitrogen fixing symbiotic associations with cyanobacteria. *Azolla* plays a very important role in rice production. For centuries *Azolla* and its nitrogen-fixing partner, *Anabaena*, have been used as "green manure" in China and other Asian countries to fertilize rice paddies and increase production.. Extensive propagation research is being conducted in China to produce new varieties of *Azolla* that will flourish under different climatic and seasonal conditions. According to some reports on the internet, *Azolla* can increase rice yields as much as 158% per year. They claim rice can be grown year after year, several crops a year, with little or no decline in productivity.

Another well known group of symbionts are the lichens. They are a symbiotic association of fungi and algae (and/or a cyanobacterium). The fungal filaments surround the algae to provide the physical bulk and shape of the lichen. Lichens grow almost everywhere on soils, rocks, buildings, stone walls and tree bark. The secretions of lichens growing on rocks contain enzymes and acids that speed the degradation of rocks into soil. The layered structure of lichens seems to give them an ability to survive long periods of drought. Some foliose (bush-like) lichens survive for decades and some crustose lichens are estimated to be even older (centuries). Although lichens are abundant in temperate and tropical climates, they are most conspicuous as ground cover in the tundra ("reindeer moss"), and as epibionts on rocks above the tree-line. They also survive in scorching deserts. Lichens are quite colorful and attractive and commonly are green, gray, red, orange and yellow. While some species grow almost anywhere, others seem to be accurate monitors of air quality in a given place. Lichens have been used as food (soups and salads), winter feed for animals (Eskimos and Lapps), for making dyes, in perfume manufacture and in medicines.

3. Steps, Mechanisms and Requirements for the Establishment of Symbiosis

As interesting as it is to describe the breadth of symbiotic associations, many symbiologists are focused on relating experimental studies and observations to the mechanisms by which the associations get started and are maintained. How are potential symbionts recognized by the host? In a very simple case with which I am familiar, we questioned how do the larger foraminifera, who feed on diatoms, recognize their symbiotic "diatom friends" from their "food"? (Details on this system below) Once the symbionts are recognized, how does the symbiosis become established? What factors are required to maintain the association? Partners of endosymbiotic associations exchange different types of signals and at different levels of organization. Communication between symbionts is the basis of recognition, specificity, and regulation. These features imply stable symbiotic systems and not separate entities or addition of combined partner properties.

Experiments done with the diatom-bearing larger foraminiferal host, *Amphistegina* lessoni, showed selective gradients when more than one symbiont was presented at the same time to the host. The hosts, which feed mainly on diatoms, were rendered nearly aposymbiotic by incubating them in the absence of food with DCMU and then they were transferred to flasks with fresh seawater and fed mixtures of 10 different diatoms and chlorophytes (Lee et al., 1983, 1986). Each experimental mixture contained two species of diatoms that had been isolated as endosymbionts and a free-living diatom or a chlorophyte species. The experimental flasks were incubated at either 10 or 20m depth in the Red Sea (different light levels). After a week, the foraminifera regained color. The results showed that some endosymbiotic diatom species were selected over, or were more competitive, than others. Many introduced species replaced the few surviving endosymbionts previously established within the hosts. They found that the "pecking order" of successful species was affected by the depth of incubation. In histological sections they observed that as many as 15% of the diatoms were reproducing within their hosts at the time of preparation. None of the free-living diatoms or the chlorophytes survived within the hosts

A number of investigators have done what could be described as "cafeteria" experiments. They offered the host potential symbiont species and saw which ones successfully established themselves as long-term or permanent symbionts. Studies of the green hydra, *Hydra viridissima*, first come to mind. In his review, Rahat (1992) described experiments in which aposymbiotic *H. viridissima* were offered 46 different strains and species of chlorellae. He also offered aposymbiotic *H. viridissima* 21 strains of *Symbiococcum hydrae* derived from the brown hydra, *H. magnipillata.* Similar experiments were done with the brown hydra. The experimental hydra were followed microscopically for several weeks. He obtained three types of results: 1) Some algae disappeared from the hydras within a day or two. 2) Some algae persisted in the hydra tissues for several weeks but eventually declined in numbers. 3) The algae in the third group persisted and formed a stable and inheritable symbiosis. Native zoochlorellae persisted as the sole symbionts. Free-living strains did not form stable persistant symbioses with *H. viridissima*; they were all digested.

Along the same lines, egg clutches of the sepiolid squids, *Euprymna hyllebergi* and *E. tasmanica*, were maintained aposymbiotically in artificial seawater for 18-21 days until juveniles were ready to hatch (Nishiguchi, 2002). Then various strains of *Vibrio fisheri* and *Photobacterium leignathi* were added singly, and in competition with each other, to vessels containing the aposymbiotic squid. All strains of *V. fisheri* isolated from different populations of *E. tasmanica* infected juvenile *E. tasmanica* equally well, but strains isolated from other species of *Euprymna* were not as successful. Very similar results were found using *E. hylleberg* as the experimental host. The isolate of *P. leiognathi* did not become a successful endosymbiont in *E. hylleberg.* In competition experiments, two strains of *V. fisheri* were dominant over all other inter-specific symbionts tested. The above are all examples of specificity in environmentally transmitted symbionts. If environmental transmission increases the competitive abilities of a particular symbiont to infect a particular host, then the host itself must have some mechanism(s) that allows the symbiont to increase its fitness as a benefit of the mutualism. This benefit must also outweigh the costs of continually accepting new (and possibly different) symbionts with every new generation of hosts.

Having shown that there are preferences of partners in symbiotic associations, a number of workers have wondered what are the mechanisms by which a certain host is usually associated with one or at least a limited number of partner types. Since we asked questions on this aspect of symbiosis in my own laboratory our results come easily to mind. As mentioned above, diverse families of larger foraminifera host some species of diatoms while at the same time they capture and digest diatoms as their main source of food. How do they sort symbionts from food? We found evidence to suggest that all the endosymbiotic diatoms share similar surface molecules that act as molecular signals for their recognition and retention by their hosts (Chai and Lee, 1999, 2000). A 104 kDa glycoprotein (CSSA, common symbiont surface antigen) was found in the frustules of 11 symbiotic diatom species, but not found in the frustules of 5 species that were free-living. When the CSSA was reacted with antisera in blocking experiments, the symbionts were digested. Fine structural immuno-gold labeled studies showed that there are receptors for the CSSA on the surfaces of the host pseudopodia. This suggests that the initial contact of the diatoms by the foraminifera is mediated by surface molecular interaction. Once inside the host in a symbiosome vacuole the symbiont looses its normal frustule but retains the CSSA. The CSSA is still produced by the symbiotic diatom even after it becomes mature and assumes its "normal" position in a pore cup of the host (Fig. 3C), *Amphistegina lobifera* or *A. lessonii*, in which they were studied (Chai and Lee, 2000).

The enormous economic potential has stimulated many investigators to study the molecular interactions between leguminous plants and their nitrogen-fixing bacterial symbionts. A single infected root cell of a legume can contain up to 20,000 bacteroids enclosed in peri-bacterial membrane bounded sacs. The bacterium involved, *Rhizobium*, has a life cycle that changes from a free-living gram negative rod to a pleomorphic bacterioid via a definable progression of expression of genes. The communications that occur between the plant and the rhizobia are among the most extensively studied signal transductions in plant systems. The expression of nodulation genes in the bacteria is activated by signals from the plant roots. Bacteria synthesize signals that induce a nodule meristem and enable the bacteria to enter the meristem via a plant made infection thread The legume harvests the bacteroid's nitrogenase activity via a tightly regulated system of reactions that direct synthesis of nodule-specific host proteins (nodulins; approx. 20) that compartmentalize the bacteroids. Expression of *Sym* genes in the microsymbiont direct the infection process and subsequently results in *de novo* synthesis of nodule specific *Rhizobium* proteins (bacteroidins), which are characteristic of the bacteroid. Nitrogen fixing genes (*nif*) are located on a plasmid (NGR234; 536,165-bp) of some species (*R. trifolii*, *R. leguminosarum*, *R. meliioyi*) and on the chromosome in other species (*Bradyrhizobium japonicum*) (Freiberg et al., 1997). Ammonia is secreted by the bacteroids that is converted to ureides by tropical legumes and amides by temperate-zone legumes. Abiotic factors (e.g. concentration of nitrate in the soil, pH, salinity, temperature, desiccation) have strong inhibitory effects on all stages of pre-infection.

4. Metabolic Interactions

Many researchers focus on studying the carbon and nitrogen balance of particular symbiotic systems. Lichens attract researchers because their resistance to desiccation and low temperatures adapts them to an opportunistic and episodic life-style that can be useful in designing experiments. Lichens usually grow where nutrients are in short supply and it has been observed that they do not grow when nutrients are increased. The working hypothesis is that, when nutrients are in excess, the fungus takes advantage of the excess nutrients and the alga is discriminated against. Under normal conditions, it is the algae that synthesize the organic carbon for the system. Carbohydrates excreted by the algae are rapidly transported to the fungus where they are metabolized and changed into fungal storage compounds such as mannitol. Interestingly, "lichenized" algae release carbohydrates, but when they are isolated and cultured, they fail to do so. The mechanism for the releasing of carbohydrates is unknown, however there are several good working hypotheses (Feig and Jensen, 1992): 1) The fungus may release one or more signaling molecules that enter the alga and change its metabolism. 2) The fungus may induce the synthesis of a new carbohydrate carrier. 3) The permeability properties of the cell membrane of the alga are changed by the intimate contact between the fungus and the alga. 4) The microenvironment (CO_2, O_2, moisture) of the two types of cells, when they are intimately associated, could affect the metabolism of the alga. Experiments suggest that the fungus does not need to be living, or import the carbohydrates, for the alga to be "lichenized". Usually only one of four types of carbohydrates (ribitol, erythritol, sorbitol glucose) is released by the algae or cyanobacteria in a particular lichen species. Some lichens contain filamentous cyanobacteria (in structures known as cephalodia) as well as algae. Some cells within the filamentous chain are differentiated into heterocysts, cells that are known to fix nitrogen. A number of investigators have demonstrated the presence of nitrogenase activity and the transport of the fixed nitrogen from the algae into the fungal thallus. Calculations suggest that the nitrogen fixed by the cyanobacterial heterocysts make significant contributions to the growth of lichens.

Many researchers have concentrated on the carbon and nitrogen budgets of corals. Globally this is important. It has been estimated that the global net photosynthetic production by zooxanthellae (symbiotic dinoflagellates) exceeds 4.6 X 10^8 metric tons of carbon each year (Muscatine et al., 1981). While most unicellular algae release <5% of their photosynthetically fixed carbon, zooxanthellae may release a large proportion (25-60% or more; depending on species and experimental approach used to make the determination) of their fixed carbon to their coelenterate hosts. One experimental study (Muscatine et al., 1984) estimated more than 97% of the carbon fixed by the zooxanthellae of *Styophora pistillata* was translocated to the coral. Experimentally it has been shown that a homogenate of host cells can stimulate symbiotic algae to release their carbon. Most research suggests that glycerol and lipids are the major forms of carbon trasnslocated from the algae to the corals.

In a brief chapter, one can only select a few examples to illustrate a process that has been a major theme of life on earth. While the discovery and description of new forms of symbiosis will continue to be an important aspect of future research, the focus of the field has become more oriented toward the mechanistic, adaptive and genetic aspects of the phenomenon. Some of this change has been driven by important global practical issues (e.g. Global warming/coral bleaching, need for non-leaching fertilizers/ N_2 fixation, etc.) but most of the redirection of effort has been driven by changes in intellectual curiosity toward molecular and genetic issues (e.g. Freiberg et al., 1997). This is an exciting time to be a student of symbiosis and advances in the field are eagerly awaited.

5. References

Baker, A.C. (2003) Flexibility and specificity in coral-algal symbiosis: diversity, ecology and biogeography of *Symbiodinium*. Ann. Rev. Ecol. Evol. System. **34**, 661 - 689.

Battey, J.F. (1992) Carbon metabolism in zooxanthellae-coelenterate symbioses. In: W. Reisser (ed.) *Algae and Symbioses*.. Biopress Ltd. Bristol. pp. 154 - 187.

Bonfante, P., Bianciotto, V., Ruiz-Lozano, J.M., Minerde, D. Lumini, E. and S. Perotto, S. (2002). Arbuscular mycorrhizal fungi and their endobacteria. In: J. Seckbach (ed.) *Symbiosis: mechanisms and model systems*. Kluwer Academic. Publishers, Dordrecht, The Netherlands. pp. 323 - 338.

Buddemeir, R.W. and Fautin, D.G. (1993) Coral bleaching as an adaptive mechanism. Bioscience, **43**, 320 - 326.

Chai, J. and Lee, J.J. (2000) Recognition, establishment and maintenance of diatom endosymbiosis in foraminifera. In: J.J. Lee and P.H. Muller (eds.) *Advances in the Biology of Foraminifera* (Micropaleontol. **46** (supplement 1), 182 - 196.

Chai, J. and Lee, J.J. (1999) Initial recognition of endosymbiotic diatom surface antigens by the larger foraminifer *Amphistegina lobifera. Symbiosis* **26**, 39 - 53.

Correia, M. and Lee, J. J. (2002) How long do the plastids retained by *Elphidium excavatum* last in their host? Symbiosis **32**, 27 - 38.

Douglas, A.E. (1998). Nutritional interactions in insect-microbial symbioses: aphids and their symbiotic bacteria, *Buchneria*. Ann. Rev. Entomol. **43,** 17 - 37.

Douglas, A.E. (1994). *Symbiotic interactions*. Oxford Univ. Press, N.Y.

Dyer, B.D., (2002) Termite hindgut symbionts. In: J. Seckbach (ed.) *Symbiosis: mechanisms and model systems*, Kluwer Academic. Publishers, Dordrecht, The Netherlands. pp. 703 - 713.

Feig, G.B. and Jensen, J. (1992) Basic carbon and nitrogen metabolism of lichens.. In: W. Reisser (ed.) *Algae and Symbioses*. Biopress. Bristol. pp. 277 - 300.

Felbeck, H. (1990) Symbiosis of bacteria with invertebrates in the deep sea. In: P. Nardon, V. Gianinazzi-Pearson, A.M. Grenier, L. Margulis and D.C. Smith (eds.) *Endocytobiology IV* Institut National De La Recherche Agronomique, Paris. pp. 327 - 334.

Freiberg, C., Fellay, R., Baioch, A., Rosenthal, A. and Peret, X. (1997) Molecular basis of symbiosis between *Rhizobium* and legumes. Nature **387**, 394 - 401.

Gütz, H.D., Benting, J., Freiburg, M. Wiemann, M. and Fokin. (1992) The infection of paramecia by *Holospora* bacteria. In: S. Sato, M.Ishida, and H. Ishikawa (eds.) *Endocytobiology V*. Tübingen Univ. Press, Tübingen, Germany. pp. 509 - 514.

Goulet, T.L. and Coffroth, M.A. (2003) Genetic composition of zooxanthellae between and within colonies of the octocoral *Plexaura kuna*, based on small subunit rDNA fingerprinting. Mar. Biol. **142**, 233 - 239.

Hackstein, J.H.P., van Hoek, A.H.A.M., Leunissen, J.A.M. and Huynen, M. (2002) Anaerobic ciliates and their methanogenic endosymbionts. In: J. Seckbach, (ed.) *Symbiosis: mechanisms and model systems*. Kluwer Academic. Publishers, Dordrecht, The Netherlands. pp. 453 - 464.

Hawkins, E.K. and Lee, J.J. (1990) Fine structure of the cell surface of a cultured endosymbiotic strain of *Porphyridium* (Rhodophyta). Trans. Am. Microsc. Soc. **109**, 352 - 360.

Honegger, J. (1992). Lichens: mycobiont-photobion relationships. In: W. Reisser (ed.) *Algae and Symbioses*. Biopress Ltd. Bristol. pp. 255 - 275.

Jeon, K.W. (1993) Symbiosis and endosymbiosis: advantages and disadvantages. In: S. Sato, M. Ishida, and H. Ishikawa (eds.) *Endocytobiology V* Tübingen Univ. Press, Tübingen. pp. 1 - 8.

Jeon, K.W. (1987) Change of cellular "pathogens" into required cell components. In: J.J. Lee and J.F. Fredrick (eds.) *Endocytobiology III*. Ann. NY Acad. Sci. **503**, 359 - 371.

Keeling, P. (2004) A brief history of plastids and their hosts. Protist. **155**, 3 - 7.

Kowallik, K.V. (1997) Origin and evolution of chloroplasts: current status and future perspectives. In: H.E.A, Schenk, R.G. Hermann, K.W. Jeon, N.E. Müller, W. Schwemmler. (eds.) *Eukaryotism and symbiosis*. Springer-Verlag, Berlin, pp. 3 - 23.

Kühl, M. and Larkum, A.W.D. (2002) The microenvironment and photosynthetic performance of *Procloron* sp. In: symbiosis with didemnid ascidians. In: J. Seckbach. (ed.) *Symbiosis mechanisms and model systems*. Kluwer Academic. Publishers, Dordrecht, The Netherlands., pp. 273 - 290.

LaJeunesse, T. (2004) "Species" radiations of symbiotic dinoflagellates in the Atlantic and Indo-Pacific since the Meiocene-Pliocene transition. Molec. Biol. Evol. **22**, 570 - 581.
LaJeunesse, T. (2001). Investigating the biodiversity, Ecology, and phylogeny of endosymbiotic dinoflagellates in the genus *Symbiodinium* using the ITS region: in search of a "species" level marker. J. Phycol **37**, 866 - 880.
LaJeunesse, T.C., Lob, W.K.W., van Woesik, R., Hoeg-GuldBerg, O., Schmidt, G.W. and Fitt, W.K. (2003) Low symbiont diversity in southern Great Barrier Reef corals relative to those of the Caribbean. Limnol. Oceanogr. **48**, 2046 - 2054.
Latorre, A., Gil, R., Silva F.J., Martînez-Tores, D. and Moya, A. (2003). Tempo and mode of genomic evolution in endosymbiotic bacteria of insects: the case of *Buchneria aphidicola*. Symbiosis **34**, 301 - 316.
Lee, J.J. (1990) Fine structure of the rhodophycean *Porphyridium purpureum* in situ in *Peneroplis pertusus* ((Forskå) and *P. acicularis* (Batsch) and in axenic culture. J. Foram. Res. **20**, 162 - 169.
Lee, J.J. (1994) Diatoms and their chloroplasts as Endosymbiotic partners for larger foraminifera. *Symposium on Living and fossil diatoms: IX* Mem. Calif. Acad. Sci., 21 - 36.
Lee, J.J. and Anderson (1991) Symbiosis in foraminifera. In: J.J. Lee. and O.R. Anderson (eds.) *Biology of Foraminifera.*. Academic Press, London. pp. 157 - 220.
Lee, J.J. and Hallock, P. (1987) Algal symbiosis as the driving force in the evolution of larger foraminifera. Ann. N.Y. Aca. Sci. **503**, 330 - 347.
Lee, J.J., Morales, J., Bacus, S. Diamont, A. Hallock P., Pawlowski, J. and Thorp, J. (1997). Progress in characterizing the endosymbiotic dinoflagellates of soritid foraminifera and related studies on some stages in the life cycle of *Marginopora vertebralis*. J. Foram. Res. **27**, 254 - 263.
Lee, J.J. (2002) Living sands: symbiosis between foraminifera and algae. In: J. Seckbach (ed.) *Symbiosis: mechanisms and model systems.* Kluwer Academic. Publishers, Dordrecht, The Netherlands, pp. 489 - 506.
Lee, J.J., Erez, J., McEnery, M.E., Lagziel, A. and Xenophontos, X. (1986) Experiments on persistence of endosymbiotic diatoms in the larger foraminifer, *Amphistegina lessonii*. Symbiosis. **1**, 211 - 226.
Lee, J.J., Saks, N.M., Kapiotou, F., Wilen, S.H., and Shilo, M. (1983) Experimental studies of symbiont persistence in *Amphistegina lessoni*, a diatom-bearing species of larger foraminifera from the Red Sea. In: In: H.E.A. Schenk and W. Schwemmler (eds.) *Endocytobiology*. II, Walter de Gruyter and Co., Berlin. pp. 487-514.
Leval-Peuto, M. (1990) Plastid symbiosis and myxotrophy in the protozoa. In: P. Nardon, V. Gianinazzi-Pearson, A.M. Grenier, L. Margulis and D.C. Smith s.) *Endocytobiology IV*. Institut National De La Recherche Agronomique, Paris. pp. 211 - 214.
Lewin, R.A. and Cheng, L. (1989) *Prochloron a microbial enigma.* Chapman and Hall, New York.
Margulis, L. (2004). Serial Endosymbiotic Theory (SET) and composite individuality. Microbiol. Today **31**, 172 - 174.
McLaughlin, J.J.A. and Zahl, P. (1959) Axenic zooxanthellae from various invertebrate hosts. Ann. N.Y.Acad. Sci. **77**, 55 - 72.
McMullin, E.R., Hourdez, S., Schaeffer, S.W. and Fisher C.R. (2003) Phylogeny and biogeography of Deep sea vestimentiferan tubeworms and their bacterial symbionts. Symbiosis **34**, 1 - 41.
Mujer, C.V., Andrews, D.L., Manhart, J.R., Pierce, S.K. and Rumpho, M.E. (1996) Chloroplast genes are expressed during intracellular symbiotic association of *Voucheria litorea* plastids with the sea slug *Elysia chlorotica*. Proc. Natl. Acad. Sci. U. S. A. **93**, 12333 - 12338.
Muscatine, L., Cook, C.B., Pardy, R.L. and Pool, R.R. (1975) Uptake, recognition and maintenance of symbiotic *Chlorella* by *Hydra viridis*. Symp. Soc. Exptl. Biol. **29**, 175 - 203.
Muscatine, L., McCloskey, L.R. and Marian, R.R. (1981) Estimating the daily contribution of carbon from zooxanthellae to coral animal respiration. Limnol. Oceanogr. **26**, 601 - 611.
Muscatine, L.R., Falkowski, P.G., Porter, J.W. and Dubinsky, Z. (1984) Fate of photosynthetic carbon in light and shade-adapted coloniesof the symbiotic coral, *Stylophora pistillata*. Proc. Roy. Soc. London B. **222**, 181 - 202.
Nishiguchi, M.K. (2002) Host-symbiont recognition in the environmentally transmitted sepiolid squid-*Vibrio* mutualism. Microbial Ecol. **22**, 10 - 18.
Nomark, B.B. (2004). The sex lives of scales. Nat. Hist. **113** (7), 38 - 44.
Paracer, S. and Ahmadjian, V. (2000) *Symbiosis.* Oxford University Press.

Pierce, S.K., Biron, R. and Rumpho, M. (1996) Endosymbiotic chloroplasts in molluscan cells contain proteins synthesized after plastid capture. J. Exp. Biol. **199**, 2323 - 2330.

Pouchon, X., LaJeunesse, T.C. and Pawlowski, J. (2004) Biogeographic partitioning and host specialization among foraminiferan dinoflagellate symbionts. Mar. Biol. **146**, 17 - 27.

Rahat, M. (1992) Symbioses of algae with *Hydra* spp. In: W. Reisser (ed.) *Algae and Symbioses* Biopress Ltd. Bristol. pp. 111 - 152.

Reisser, W. (1987) Naturally occurring and artificially established associations of ciliates and algae: Models for different steps in the evolution of stable endosymbioses. In: J.J. Lee and J.F. Fredrick (eds.) *Endocytobiology III* Ann N.Y. Acad. Sci. **503**, 316 – 329.

Ruby, E.G. (1996) Lessons from a cooperative, bacterial-animal association: the *Vibrio fisheri-Euprymna scolopes* light organ symbiosis. Ann. Rev. Microbiol. **50**, 591 - 624.

Ruby, E.G. Nealson, K.H. (1976) Symbiotic associations of *Photobacterium fischeri* with the marine luminous fish. *Monocentris japonica*: a model of symbiosis based on bacterial studies. Biol. Bull. **151**, 574 - 586.

Rumpho, M.E., Summer, E.J. and Manhart, J. R. (2000) Solar-powered sea slugs. Mollusc/algal chloroplast symbiosis. Plant Physiol. **123**, 29 - 38.

Shigenobu, S., Wantanabe, H., Hattori, M. Sakaki, Y. and Ishikawa, H. (2000). Genome sequence of the endocellular bacterial symbiont of aphids *Buchneria* sp. Nature **407**, 81 - 86.

Smith, D.C. (1990). Symbiosis: an overview, a definition and an examination of the aims of the colloquium. In: P. Nardon, V. Gianinazzi-Pearson, A.M. Grenier, L. Margulis and D.C. Smith, (eds.) *Endocytobiology IV*. Institut National De La Recherche Agronomique, Paris. pp. 25 - 35.

Stoecker, D.K. and Silver, M.W. (1987) Chloroplast retention by marine planktonic Ciliates. In: J.J. Lee and J.F. Fredrick (eds.). *Endocytobiology III*. Ann N.Y. Acad. Sci. **503**, 562 - 566.

Stouthamer, R., Breeuwer, J. and Hurst, G.D. (1999) *Wolbachia pipientis*: microbial manipulator of arthropod reproduction. Ann. Rev. Microbiol. **53**, 71 - 102.

Trench, R.K. (1975) Of "leaves that crawl": functional chloroplasts in animal cells. In: D.H. Jennings and D.L. Lee (eds.) *Symbiosis*. Symposia of the Soc. Exptl. Biol. **XXIX**, 229 - 266.

Udvardi, M. and Day, D.A. (1997) Ann. Rev Pl. Physiol. Pl. Mol. Biol. **48**, 493 - 523.

Wilkerson, C.R. (1992). Symbiotic interactions between marine sponges and algae. In: W. Reisser (ed.) *Algae and Symbioses*. Biopress Ltd. Bristol. pp. 41 - 62.

Biodata of **Mary Beth Saffo** author of ***"Symbiosis: The Way of All Life"***

Dr. Mary Beth Saffo is currently Adjunct Scientist at the Marine Biological Laboratory, Woods Hole and an Associate of the Museum of Comparative Zoology, Harvard University. She obtained her Ph.D. from Stanford University in 1977. From 1976 to 1978, she was a Miller Research Fellow in the Dept. of Botany at the University of California, Berkeley. Since then she has taught at Swarthmore College, the University of California at Santa Cruz, and Arizona State University West, where she was Professor of Physiology until 2002. She was a Fellow at the Radcliffe Institute for Advanced Study, Harvard University, from 2002-2003, and has been a Fellow of the American Association for the Advancement of Science since 1990. Dr. Saffo's chief scientific interests include the biology of endosymbiosis, protist phylogeny, marine biology, biomineralization, and the physiological ecology of nitrogen excretion.

E-mail: **mbsaffo@oeb.harvard.edu**

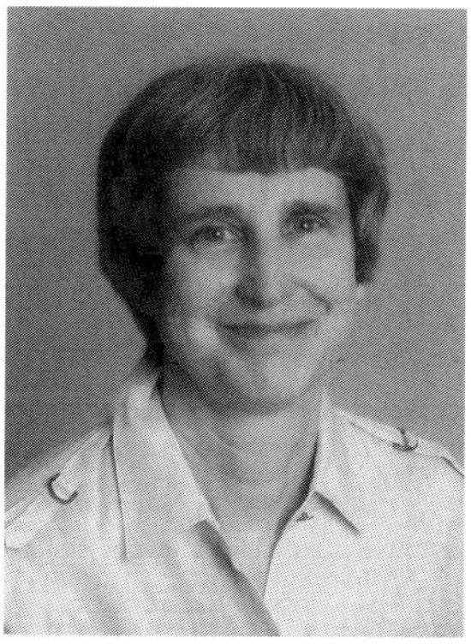

J. Seckbach (eds.), Life as We Know It, 325-339.

SYMBIOSIS
The Way of All Life

MARY BETH SAFFO
Marine Biological Laboratory (Woods Hole, MA) and Harvard University, MCZ Labs 408, Harvard University, 26 Oxford St., Cambridge, MA 02138-2902 USA

1. Introduction

When we look at an organism – a tree, a bird, an insect, a cow – we first see that organism as an individual. But in reality, that "individual" is more properly viewed as the site of a tightly-bound community, a life intimately intertwined with that of other species. This intimate interaction, or "living together", between species is called *symbiosis*. As in human social relationships (Saffo, 2001), "living together" can have many outcomes: symbiosis can be beneficial to both (or all) species partners (mutualism), or beneficial to one partner, but harmful (parasitism), or of no apparent effect (commensalism) on the other. Although multiple definitions of symbiosis continue to plague the literature, there are good historical, biological and pragmatic reasons for defining symbiosis in this broadest sense (Saffo, 1992a). Interspecies *intimacy*, not interaction outcome, is what distinguishes symbiosis from other forms of interspecies interaction. I explore here the impact of these intimate interactions on the nature of life.

2. Symbiotic Associations: Ubiquitous but Unexpected

While symbiotic associations may seem exotic, even bizarre, to us, they are found, in fact, in all ecosystems and virtually all taxonomic groups, and they have had an impact on the architecture, ecological relationships and evolutionary history of living things.

The ubiquity of symbiosis continues to surprise us, especially because it exists alongside another characteristic of life: the ubiquity of anti-"non-self" defenses (that is, defenses against foreign organisms or foreign or aberrant genomes) among organisms, ranging from restriction enzymes among bacteria to immune defense systems among animals and plants. As with symbioses, the more we look for chemical defenses among organisms, the more such defenses we find: all organisms have molecular and structural mechanisms that distinguish, often with exquisite precision, between "self" and "non-self", with the capability of barring, eliminating or destroying foreign cells and genes. Recognition of self and non-self, and protection of self from non-self, are clearly essential properties of life.

So life seems to be a contradiction, a simultaneous embodiment of both isolation and connection. In their molecular and physiological architecture, organisms are xenophobic to their very core; but organisms are equally, and simultaneously, defined by pervasive, symbiotic interactions with other species. How can intimate interspecies interactions persist in the face of ubiquitous, effective defenses against foreign organisms?

The fact that species-recognition processes play important roles in initiation and maintenance of all symbiotic interactions, while adding to this paradox, also suggests an approach to its resolution: perhaps it is more fruitful to view the forces of species isolation and species connection not so much as contradictions, but as interdependent phenomena. Is it the prevalence of symbiotic associations that necessitates self-non-self recognition and immune defenses in organisms, and do connections between species exist only when species manage to overcome those recognition and defense systems? Or could it also be only *because* of highly developed abilities of self-non-self recognition that symbiotic associations can occur?

3. Examples of Symbiotic Associations: Varieties of Interspecies Intimacy

*Endo*symbiotic interactions, in which one or more species live within the cells or tissues of another species, are the most intimate forms of symbiotic interaction. In this case, the "host" species is not merely the site of symbiotic interactions, but also the habitat for its species partners; and invasion of this habitat by endosymbionts necessitates exposure to the full armada of host immune defenses. But endosymbiosis is nevertheless very common. Many microbial, plant and animal lineages lead exclusively endosymbiotic lives, and virtually all eukaryotes (organisms with nuclei, including plants, animals, fungi, protozoa and algae), as well as – if one also considers viruses as endosymbionts – many prokaryotes (bacteria and archaea), are hosts to endosymbiotic organisms.

A plausible case can be made for the possibility that the number of endosymbiotic species on earth at least matches, and perhaps even exceeds, that of host species. First, many hosts are colonized by more than one endosymbiotic species. The rumen of a single cow or sheep, for instance, typically contains hundreds of specialized prokaryote species, as well as several species of protozoa and fungi.

Second, while some endosymbionts have broad host ranges, many others have narrow host specificities, infecting only one host species, or a cluster of closely related hosts. Among two notable examples, 52 species of intestinal helminth ("worm") parasites were isolated from 45 individuals of lesser scaup ducks; at least 8 of these parasites are specialists on this duck species (Bush and Holmes, 1986a,b). Although most ectomycorrhizal fungi seem to have broad host ranges, many others do not: about 250 species of ectomycorrhizal fungi have been identified in the Pacific northern hemisphere pine *Pseudotsuga* that are *specific* to this host genus (Molina et al., 1992; Smith and Read, 1997).

Endosymbiosis also occurs in virtually all regions of host tissues and cells. Among plants and animals, no organ or region seems immune to infection: blood, lungs, gills, hearts, kidneys, gut cavities, nervous systems, and gonadal tissues of animals, and roots, stems, leaves, flowers, seeds, xylem and phloem of plants all serve as occasional or regular habitats for symbionts While some tissues (such as animal intestinal cavities) are more heavily and commonly colonized than others, no barrier seems insuperable. Some symbiotic habitats seem absurdly inaccessible: for instance, *Nephromyces*, a protistan symbiont (itself infected with hereditarily-transmitted intracellular bacteria) of molgulid ascidian tunicates (a family of "sea squirts") must cross at least two or three epithelial layers, passing through the double gauntlets of digestive and phagocyte-laden blood, to colonize their exclusive microhabitat, a ductless organ (the "renal sac"), from which there is apparently no escape until the death of the host . Yet the symbionts make this journey readily and efficiently, taking less than 24 hours to reach the renal sac from surrounding sea water, infecting 100% of their hosts, population after population, generation after generation (Saffo and Nelson, 1983).

Intracellular symbiosis—life within the cells of other species—seems the most improbable endosymbiosis of all, yet even this most morphologically intimate form of endosymbiosis is very common. All apicomplexan parasites (a protistan group including *Toxoplasma, Plasmodium, Theileria* and other medically and agriculturally important parasites) inhabit host cells for part or most of their life cycle. Symbiotic dinoflagellates in corals and other marine invertebrates inhabit vacuoles in the gastrodermal cells of their hosts. Foraminiferan cells are colonized by dinoflagellates, green and red algae, and diatoms (Lee, 1993); nuclei and organelles of some parasitic red algae are injected into the cytoplasm of their red algal hosts (Goff and Coleman, 1984; Goff, 1991). Bacteria have been found in many eukaryotic microbes, including amoebae, ciliates and flagellate protozoans, unicellular algae and cells of arbuscular mycorrhizal fungi; some symbiotic ciliates that dwell in the intestines of surgeonfish even harbor bacteria in their nuclei (Grim, 1993). The bacterium *Wolbachia* inhabits oocytes of many terrestrial arthropods, in addition to other cells and tissues (Dobson et al., 1999); other bacteria inhabit specialized "bacteriocytes" in insects such as aphids and cockroaches. Bacteria are also found in the cells of other invertebrate animals, including the "fiber cells" of the barely-multicellular placozoa and the light-producing cells of the oceanic colonial chordate *Pyrosoma* (Saffo, 1992b). Nitrogen-fixing bacteria inhabit root cells in legumes, providing in-house "fertilizer" for peas, soybeans and other agriculturally important plants.

Defying their host's immune systems, some intracellular endosymbionts even take residence in the most heavily armored cells of their hosts, including cells which ordinarily specialize in destruction or digestion of foreign material. For instance, the apicomplexan parasite *Theileria parva* invades T lymphocytes of bovine ruminants (Shaw, 1997). Trypanosome parasites in mammals live exclusively in the lysosomes of macrophages; so also do leprosy and tuberculosis bacilli, which can persist in lysosomes for decades.

Should this extended list of examples of intracellular endosymbiosis still leave an impression of mere biological exotica, consider the fact that two key eukaryote organelles, mitochondria and chloroplasts, essential to eukaryote respiration and photosynthesis, respectively, are themselves descendants of former bacterial endosymbionts. Without intracellular endosymbiosis, eukaryotes would, simply, not exist (Margulis and Fester, 1991; Margulis, 1993).

Like many attempts at pigeonholing biological phenomena, "intimacy" between species is not really a rigid, yes-or-no category, but a matter of degree. Intimacy is easy to demonstrate in endosymbiotic associations, since, virtually by definition, protracted life within another organism necessitates intimate, prolonged biochemical and morphological interactions between hosts and symbionts. However, symbiotic interactions can also be largely external. Like endosymbioses, external symbioses (or "ectosymbioses") can show impressive complexity, specialization and interdependency between species partners. In theory, the intimacy of ectosymbiotic interactions is defined by the existence of either protracted physical contact or persistent physical proximity between symbiotic partners, and by the profundity of the influence of the interaction on the morphology, physiology or behavior of the species partners. But, in practice, drawing a line between ectosymbioses and other external species interactions is not always easy. Interspecific interactions traditionally not categorized as symbioses (for instance, specialized pollination or predator-prey interactions) are also characterized by highly specific, interdependent, coevolved interactions that are certainly at least kindred to, and, in many cases, virtually indistinguishable, from symbiosis.

In some ectosymbioses, hosts and symbionts are connected through direct, long-term physical contact at body surfaces. For instance, human body lice (*Pediculus humanus corporis*) and head lice (*P. humanus capitis*) are obligate ectoparasites, living their entire life cycle exclusively on human skin and hair (and, for body lice, clothes), with human blood as their chief nutrient. Fossil and comparative molecular and structural data indicate that the association between *Pediculus humanus* and hominids goes back at least a million years or more, perhaps predating even the appearance of *Homo sapiens*: the closest relative of *P. humanus* is *P. schaeffi*, found exclusively on chimpanzees, and there is some evidence that a distinct New World clade of head lice once may have coevolved with *Homo erectus* before taking up residence on *Homo sapiens*. Recent evidence also suggests that the evolution of body lice from head-lice ancestors may be at least generally correlated with the appearance of clothing in human culture about 100,000 years ago (Kittler et al., 2003, 2004; Mumcuoglu et al., 2003; Reed et al., 2004a,b).

Similarly, the 2000-odd species of Laboulbeniales are a distinct clade of obligately ectoparasitic ascomycete fungi that live exclusively on the integument of arthropods. So extensive are their adaptations to this lifestyle that the ascomycete affinities of Laboulbeniales were deciphered only recently. Like lice, Laboulbeniales also show high host specificity, with each species generally colonizing only one or a few species of arthropod hosts (Weir and Hammond, 1997; Henk et al., 2003).

In other ectosymbioses, physical contact is still important, but more transient. In these cases, intimacy is defined by in part by protracted physical proximity, and sometimes by host-symbiont specificity, but especially by specialized behavioral interactions between species partners. In some cases, ectosymbiotic partners may cohabit a shell, a burrow, or a nest. For instance, some hermit crabs "share" (or are forced to share) both their gastropod shells and their food with Nereid polychaete worms. Brood parasites such as cuckoos lay their eggs in the nests of other birds, who raise cuckoo young along with their own.

In other cases, such as cleaning symbioses, the habitat is shared in a looser sense, and physical contact between symbiont-host interaction is even more transient, though still crucial; in these cases, intimacy is defined most compellingly by the unusual, specific behavioral interchanges between species partners. In the sea, for instance, several fish and shrimp species "clean" ectoparasites from the surface tissues of reef fish. The behavioral "conversation" between the small cleaners and their larger, sometimes-carnivorous hosts comprises postures and responses unexpected for a large, hungry predatory fish (the hosts) and potential prey (the cleaner). Utterly unlike their normal feeding behavior, predatory fish remain motionless while cleaners pick at their surfaces, including the host's open mouth; during this interaction, the hosts (usually) resist attacking or ingesting their vulnerable, potentially edible cleaners.

Recent work suggests that the behavioral repertoire of cleaner fish serves not only as a specific species-recognition signal to hosts, but also may directly suppress predatory behavior in their hosts (Stummer et al., 2004, Grutter, 2004). In these functions, behavioral interactions in ectosymbioses parallel the processes of host infection by horizontally-transmitted (non-hereditary) endosymbionts. While the behavioral responses of a cleaner fish and its host "client" may seem utterly different from that of, for instance, a malaria parasite crossing the gut wall of a mosquito or rhizobial bacteria infecting the roots of a pea plant, in fact, endosymbiont-host infection processes show many resemblances to the largely behavioral interactions in ectosymbioses such as cleaning symbiosis. Like initiation of cleaning-symbiosis interactions, endosymbiotic infection processes involve serial biochemical, developmental and even behavioral "signal-and-response" interchanges between host and symbiont; as in ectosymbiotic interactions, these host-endosymbiont interchanges serve as mechanisms of species–partner recognition and host specificity, and they also play a role in evasion, suppression, or modulation of defensive (including immune) responses between species partners.

4. Symbiotic Associations have Consequences on the Biology of Organisms

The importance of symbiosis can be gauged not only by the numbers of taxa involved in symbiotic associations, but also by its impact on individual organisms and populations. Among endosymbiotic associations, for instance, the impact of symbiosis on host organisms can be measured in some cases by the *prevalence* of endosymbiosis (the percent of infected hosts) within host populations, and by the *numbers* or *biomass* of endosymbionts within individual hosts

In mutualistic symbioses, the prevalence (percentage of infected individuals) of endosymbiotic infection is often (though not always) close to 100% of the host population, and endosymbionts are also often (though not always) themselves obligately associated with their hosts. Among the many examples of obligate and/or high-prevalence mutualistic endosymbioses are: hereditary symbiosis between aphids and the bacterium *Buchnera*; gut symbioses in termites, ruminants and other specialist herbivores; microbial symbionts with molgulid ascidian tunicates; arbuscular mycorrhizae and ectomycorrhizae among land plants; lichens (associations of (usually) ascomycetous fungi with green algae and/or cyanobacteria; dinoflagellate algae in reef-building corals and other marine invertebrates; bacteria in ciliates and trypanosomes; and many others.

Like their mutualistic counterparts, parasitic and commensal endosymbionts are also often (though not always) obligate symbionts of their hosts. More surprising, perhaps, is the fact that the prevalence of parasitic symbionts among hosts can also be high. While the prevalence of parasitic symbioses tends to be more variable than in more benign associations – in particular host populations, parasite prevalence can be as low as 0% – it can also be surprisingly high, with infection rates of 20-90% of particular parasites not uncommon among some populations of animal hosts (Anderson, 1993, Korpimaki et al., 1993; Roberts and Janovy, 2005).

Perhaps not surprisingly in light of the benefits they receive from their microbial inhabitants, individual hosts can harbor impressive numbers of mutualisitc endosymbionts. For instance, in termites, the bacterial and protozoan gut symbionts (without which termites cannot survive), represent 40% of host biomass. Similarly, among ruminants (for whom microbial symbionts are a similarly obligate presence), 10^6 protistan cells and $10^{10}–10^{11}$ bacteria are found in a single milliliter of rumen fluid, whose total volume ranges up to 150 liters. Among ectomycorrhizal trees and shrubs, it has been estimated that the biomass of ectomycorrhizal fungi (which, among other effects, enhance nutrient uptake in their plant hosts) represents 20-40% of plant root biomass (Smith and Read, 1997).

But high-density infections are not limited to hosts with clearly beneficial symbionts. The population of permanent bacterial colonists in human intestines, comprising mostly commensal symbionts (though also some mutualists: see Hooper et al., 2001) is estimated in the trillions (10^{14} cells) per individual, comprising 40% of fecal weight, and outnumbering our own cells by a factor of approximately 10 to 1. In documenting the species diversities of helminth parasites in ducks (see section 3, above), Bush and Holmes (1986a, b) also counted numbers of parasite individuals, finding an average of 20,000 helminths per individual duck host: these are extraordinarily high numbers, especially given the fact that the parasites in question are not microbes, but multicellular animals.

5. Some Physiological, Behavioral, Ecological and Evolutionary Consequences of Endosymbiosis

In addition to the indirect evidence of impact offered by distribution and numbers of symbionts, the impact of symbiosis can also be gauged directly by its impact on the physiology, behavior, ecological interactions, and evolutionary history of many organisms. Focusing on endosymbiotic associations (since they provide a particularly rich source of data), I present here a few examples of the many ways in which symbioses have affected the architecture of organisms, communities and ecosystems, and the history of life.

5.1. EFFECTS OF SYMBIOSIS ON HOST BEHAVIOR

Many parasites of animals must pass through one or more "intermediate" host species in order to complete their life cycle in the "definitive" host (that is, the host in which sexual reproduction occurs). Transmission from the intermediate to the definitive host poses a special challenge for such a parasite, which, one would expect, could not have control over the location of either its intermediate or definitive host. However, many parasitic protozoa (apicomplexans) and parasitic "worms" (trematode, nematomorph, and acanthocephalan parasites) do, in fact, manipulate the behavior of their intermediate hosts; typically, these behavioral effects increase the chances of contact between the definitive and intermediate hosts (Berdoy et al., 2000; Moore, 1984, 2002; Williams et al., 2004).

A particularly memorable example of such effects can be found in the life cycle of the flatworm (trematode) parasite *Dicrocoelium dendriticum*, which lives, in turn, in the digestive gland and respiratory tissues of land snails, then in ants, and finally, in the bile ducts of sheep, its definitive hosts. Although strong host effects are often correlated with high prevalence and high densities of symbiotic colonists in host populations, this parasitic flatworm demonstrates dramatically that even small numbers of symbionts can have a large impact on host biology.

Getting oneself from an ant to a herbivorous sheep seems a tricky prospect, at best, and the tricks of *D. dendriticum* are impressive indeed. Infected ants typically harbor about 100 parasites in their blood; but one or two make their way to the brain of the ant. Despite their small number, these brain colonists have profound effects on the behavior of infected ants, inducing their hosts to migrate to the tops of plants at the times when sheep are most likely to be grazing (Moore, 2002; Roberts and Janovy, 2005).

5.2. EFFECTS OF INTIMATE INTERACTIONS ON NUTRIENT LIMITATION

A number of animals – oceanic, reef-building corals; deep-sea vestimentiferan worms; aphids; termites and ruminants, among many other herbivores–live in such nutrient-poor areas, or subsist on such nutrient-limited diets, that their survival in such situations would not be predicted by the usual ecological measures of nutrient availability. Virtually every such apparently nutrient-deprived animal harbors microbes which supplement their hosts' nutrient limitations by digestion of cellulose and other complex, difficult-to-digest plant carbohydrates; provision or recycling of nitrogen; synthesis of B vitamins; detoxification of (for instance) secondary plant compounds; and/or provision of organic carbon through photosynthesis or chemoautotrophy.

In symbioses between autotrophs and both animal and microbial heterotrophs (such as corals with dinoflagellates, vestimentiferans with sulfur-oxidizing bacteria, lichens, and various plant-microbial symbioses in nutrient-poor soils), there is bidirectional traffic in nutrients: typically, organic nutrients pass from autotrophs (dinoflagellates, sulfur-oxidizing bacteria, plants) to the heterotrophic species partners (animals, fungi, heterotrophic bacteria), and mineral nutrients (nitrogenous compounds, sulfur, phosphate) from heterotrophs to autotrophic partners (Smith and Douglas, 1987).

These symbiotic nutritional pathways are consequential for both hosts and symbionts, but not because the pathways are biochemically unique. Almost all the metabolic pathways in bacterial-animal and autotroph-heterotroph mutualisms are identical to the biochemical cycles that occur on a global scale. Like symbiotic bacteria in legume roots, there are also free-living nitrogen-fixing bacteria that release ammonia or nitrogenous compounds to ocean waters and to soils; like bacteria in ruminants and other herbivous animals, there are also free-living bacteria that digest cellulose and other plant carbohydrates, produce B vitamins, and metabolize plant and animal toxins; free-living animals and heterotrophic microbes still depend on the organic carbon and oxygen synthesized by free-living autotrophs; and free-living autotrophs make use of nutrients released by the metabolic activities of fungi and bacteria. In most cases, nutrient cycling in autotroph-heterotroph and animal symbioses differ from global nutrient cycles only in degrees of efficiency and scale. Rather, it is the *intimacy* of the interaction that is unique, enabling the rapid, efficiently targeted nutrient cycling that is a key explanation for the prevalence of so many such nutrient-cycling symbioses in nutrient-limited conditions.

5.3. SYMBIOTIC OUTCOMES ARE COMPLEX, AND AFFECTED BY ENVIRONMENTAL CONDITIONS

Most of the autotroph-heterotroph symbioses listed above are generally considered mutualisms, but in fact the outcomes of several of these "mutualisms" are not always beneficial: symbiotic outcomes are often too complex and too variable to be summarized as a static one-word category of "mutualistic", "commensal" or "parasitic." Many symbioses involving three or more species embody both antagonistic and mutualistic components simultaneously. For instance, some trypanosome parasites are obligately associated with bacterial endosymbionts, a mutualism which enables these trypanosomes to parasitize their animal hosts (Hollar et al., 1998; Saffo, 1991, 2001).

Outcomes of other symbioses vary over time, and with changes in both the biotic and physical environment. Extensive studies of mycorrhizal land plants offer rich, provocative examples of variation in outcomes of symbiotic interactions. Plants infected with arbuscular mycorrhizae, for example, benefit from mycorrhizal fungi in most natural conditions–that is, conditions in which soil phosphorus is very low. Mycorrhizal fungi enhance phosphorus uptake, thus enhance plant growth in these conditions. Arbuscular mycorrhizae are obligate symbionts, receiving organic carbon from their plant hosts. However, in higher-phosphorus conditions, or other conditions which favor increased growth of the fungal partner, plant growth can be depressed: that is, the fungus essentially parasitizes its plant host. In very high phosphorus conditions, symbiotic associations do not form at all. (Smith and Read, 1997; Saffo, 2001).

Among ectomycorrhizae with broad host ranges, mycorrhizae formed between identical fungal species but different host species form mycorrhizae not only of different morphologies, but also with differing selective outcomes: the same ectomycorrhizal fungus can be a pathogen or a mutualist, depending on the host species to which it is connected. In other words, the outcomes of symbiotic interactions, and the dynamics between species partners, can be shifted by changes in both biological and physical aspects of environmental conditions (Smith and Read, 1997; Saffo, 1992, 2001).

5.4. ENDOSYMBIOSES CAN AFFECT COMMUNITY INTERACTIONS AND ECOSYSTEMS

Some ectomycorrhizae connect *simultaneously* to two different species of plants. Some achlorophyllous plant parasites, for instance, are linked to their photosynthetic plant hosts by ectomycorrhizal connections, through which organic carbon flows from host plant to plant parasite (Smith and Read, 1997; Bruns et al., 2001). The existence of such links also suggests the possibility of broader-scale mycorrhizal-mediated nutrient flux even among non-parasitic plant species. If corroborated by additional experimental data (current data are still somewhat controversial), such symbiosis-mediated interplant links would have interesting ecological implications. How does one evaluate the interspecies interactions of trees which are separate from each other above ground, but possibly linked, nutritionally, below ground? Ecology

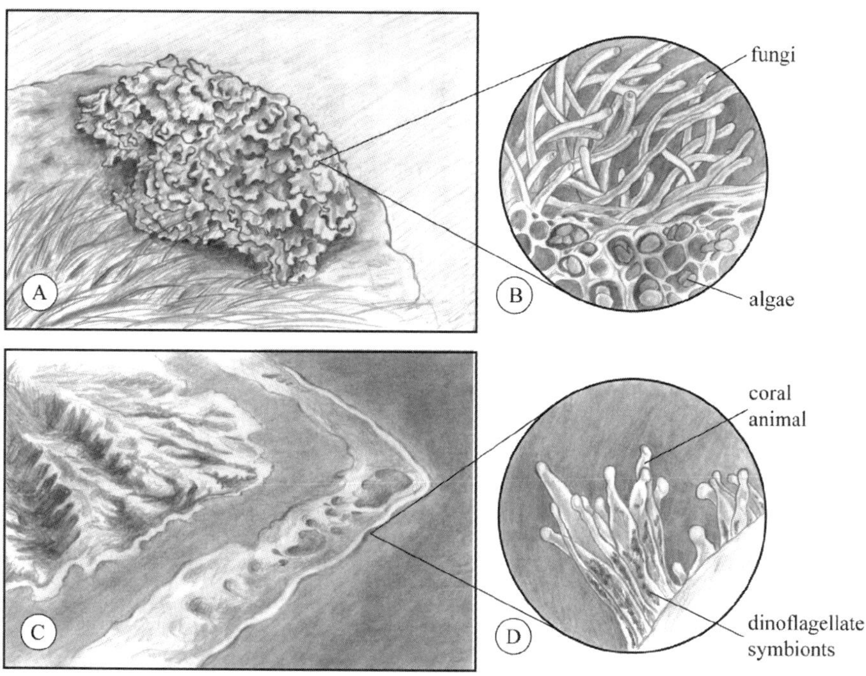

Figure 1. Lichens and reef-building corals: two ecologically important mutualistic symbioses. A. A foliose lichen in a woodland habitat. B. A close-up view of internal lichen structure, showing its algal and fungal components. C. A coral reef. D. An individual coral colony, showing internal dinoflagellate symbionts. Drawing by Tamara Clark.

has long been seen as a "tangled web" of interactions; but increased knowledge of the pervasiveness and import of symbiotic associations adds yet another dimension to what is already known of the complexities of interspecies interactions in biological communities.

Among the several ecologically important endosymbioses, several play particularly conspicous roles in the biosphere. The symbiotic interaction between corals and dinoflagellate algae enable not only the survival of these organisms in the most nutrient-poor shallow tropical seas, but also reef-building activity itself; these reefs form, literally, the backbone for reef ecosystems, oases of high productivity and high species diversity in tropical ocean "deserts" (Fig. 1). Clearly, mycorrhizal symbiosis is an integral part of the biology of most terrestrial plants: more than 80% of land plants are associated with arbuscular mycorrhizae, ectomycorrhizae, or both, and fossil evidence suggests that the evolution of land plants was bound up with mycorrhizal fungi from their very first appearance on earth. Lichens (fig. 1) are prominent members of polar ecosystems, abundant enough to serve as a major food source for reindeer in northern latitudes, and comprising 90% of terrestrial photosynthetic productivity in the Antarctic. Parasitisms have global effects, too: many animal, fungal, bacterial and protistan parasites of plants, such as species of the oomycete *Phytophthora* (the cause of potato blight and other plant diseases), have important impact on both agricultural and natural populations. Among parasites of humans, approximately 1.5-3 million human deaths are year are due to schistosomiasis and malaria alone (Roberts and Janovy, 2000).

5.5. EVOLUTIONARY CONSEQUENCES OF ENDOSYMBIOSIS

As symbiosis research has increased our appreciation for the complexities of ecological interactions, so also has it increased the complexities of interpreting evolutionary history. First, as in the Laboulbeniales, the special adaptations of many organisms for endosymbiotic life are so distinctive as to sometimes disguise the ancestry of many exclusively symbiotic lineages, and their sometimes rapid rate of evolutionary change is a reminder that not all evolutionary change occurs at uniform rates. Second, it is apparent that evolutionary history involves not just diversification of species over time, but also various degrees of fusion or combination among species, much of it through symbiotic events. Ultrastructural and genomic studies have demonstrated that many eukaryotic organisms (those organisms with nuclei, such as plants, animals, fungi, protozoa and algae) have acquired multiple endosymbionts throughout their evolutionary history, with some protists (microbial eukaryotes) showing evidence for as many as 7 genomes within a single cell. Among bacteria, horizontal transfer of genes among species is also a relatively common occurrence. Although there is still some doubt about the existence of *direct* horizontal gene transfer from prokaryotes to eukaryotes, there is no doubt that genes have been transferred from prokaryote-derived eukaryotic organelles, such as mitochondria and chloroplasts, to "host" eukaryotic nuclei within the same cell, or even between mitochondria and chloroplasts (Adams et al., 2000; Palenik, 2002; Cummings et al., 2003).

Third, symbiosis can affect short-term evolution in a number of ways, many of which transcend the usual debate about whether symbiotic infection is harmful or beneficial to hosts. The bacterium *Wolbachia*, found in a number of terrestrial invertebrates, such as parasitic (filarial) nematodes, pill bugs, and especially insects, provides a provocative and ecologically important example of the several ways in which symbionts can affect the evolution of their hosts, besides direct effects on host fitness. In some cases, *Wolbachia*'s effects can indeed be described in the usual terms of benefit or harm to hosts. In a species of parasitic wasp (*Asobara tabida*: Dedeine et al., 2001), and some mutants of the fruit fly *Drosophila melanogaster* (Starr and Cline, 2002), for instance, the presence of *Wolbachia* is required for egg production; *Wolbachia* is also required for pathogenicity of several filarial nematode parasites of humans (Saint André et al., 2002). Among the known harmful effects on its hosts, *Wolbachia* seems to suppress sexual productivity of wood ant colonies (Wenseleers et al., 2002); in other cases, *Wolbachia* infection has been shown to have measurable costs, including lethal effects, on some hosts (McGraw and O'Neill, 2004). In most cases, however, the presence of *Wolbachia* does not seem to affect the physiological health of the host *per se*: instead, it skews sex ratios among most of its arthropod hosts, typically leading to greater female/male ratios, feminization of males or even eradication of males in infected host populations. *Wolbachia* infection in species of *Drosophila* and other insects can induce mating incompatibility among host individuals of the same species, typically selecting against sexual reproduction between uninfected and infected flies, or between flies infected with different strains of *Wolbachia*. Recent evidence lends some support to hypotheses that such symbiont-engendered mating incompatibility could operate as a mechanism of sympatric speciation in *Wolbachia*-infected host populations. (Bordenstein et al., 2001; Bordenstein, 2003, Wade, 2001).

Fourth, comparative evidence suggests that persistent endosymbioses might be correlated with species diversification in some lineages. The universal presence of endosymbionts among several animal clades (see articles in Margulis and Fester, 1991) suggests not only the coevolution of such clades with their endosymbionts, but also the possibility that the endosymbiotic interaction itself might have stimulated species radiations in these clades. Evidence for large-scale co-radiation of endosymbiont and host lineages (and also, on a smaller scale, the possible effects of *Wolbachia* on *Drosophila* speciation) suggest the paradoxical possibility that morphological fusion of two species could lead to evolutionary diversification of one or both of those species partners (Bordenstein et al., 2001; Bordenstein, 2003; Wade, 2001; McGraw and O'Neill, 2004).

Overall, recent research suggests that symbiotic associations may play significant and unexpectedly diverse roles in the ecology and evolution of organisms. The increasing evidence for multiple ancestries of cell organelles and genomes also point out the inadequacies of our traditional views of the history of life. Evolution of life's diversity over time was once represented nicely as a "tree of life", its trunk representing the life's oldest, common ancestors, and branches representing the descendant species that later diverged from those ancestors. But the more we know, we have come to realize that the tree of life possesses anastomosing as well as bifurcating branches, representing an evolutionary history marked both by species fusions as well as species diversification over time.

6. Conclusion

Symbiosis is everywhere, taxonomically, histologically and geographically. The ubiquity of symbioses in nature, the high intensities and high prevalence of endosymbiosis among many groups of organisms, and the sometimes-profound effects of symbiosis on the lives of both hosts and endosymbionts, require us to consider symbiotic associations as consistent, active factors affecting the physiology, ecology and evolution of organisms.

7. References

Adams, K.L., Daley, D.O., Qiu, Y-L., Whelan, J., and Palmer, J.D. (2000). Repeated, recent and diverse transfers of a mitochondrial gene to the nucleus in flowering plants. Nature **408**, 354-357.

Anderson, R.M. (1993). Epidemiology. In: F.E.G. Cox (ed.). Modern *Parasitology* (2nd edition). Blackwell Scientific Publications, Oxford. 75-116.

Berdoy, M., Webster, J.P., and Macdonald, D.W. (2000). Fatal attraction in rats infected with *Toxoplasma gondii*. Proc. Roy. Soc. Lond. B **267**, 1591-1594.

Bordenstein, S.R. (2003). Symbiosis and the origin of species. In: K. Bourtzis and T. A. Miller (eds.). *Insect Symbiosis*. CRC Press, Boca Raton, Florida, USA. 283-304.

Bordenstein, S.R., O'Hara, F.P., and Werren, J.H. (2001). *Wolbachia*-induced incompatibility precedes other hybrid incompatibilities in *Nasonia*. Nature **409**, 707-710.

Bruns, T.D., Bidartondo, M.I., and Taylor, D.L. 2002. Host specificity in ectomycorrhizal communities: what do the exceptions tell us? Int. Comp. Biol. **42**, 352-359.

Bush, A.O., and Holmes, J.C. (1986a). Intestinal helminths of lesser scaup ducks: patterns of association. Can. J. Zool. **64**, 132-141.

Bush, A.O., and Holmes, J.C. (1986b). Intestinal helminths of lesser scaup ducks: an interactive community. Can. J. Zool. **64**, 142-152.

Cummings, M.P., Nugent, J.M., Olmstead, R.G., Palmer, J.D. (2003). Phylogenetic analysis reveals five independent transfers of the chloroplast gene *rbcL* to the mitochondrial genome in angiosperms. Curr. Genet. **43**, 131-138.

Dobson, S.L., Bourtzis, K., Braig, H.R., Jones, B.F., Zhou, W., Rousset, F., and O'Neill, S.L. (1999). *Wolbachia* infections are distributed throughout insect somatic and germ in tissues. Insect Biochem. Mol. Biol. **29**: 153-160.

Goff, L.J. (1991). Symbiosis, interspecific gene transfer, and the evolution of new species: a case study in the parasitic red algae. In: Margulis, L. and Fester, R. (eds). *Symbiosis as a Source of Evolutionary Innovation*. MIT Press, Cambridge, Mass. USA, pp 341-363.

Goff, L.J., and Coleman, A.W. (1984). Transfer of nuclei from a parasite to its host. Proc. Natl. Acad. Sci. USA **81**, 5420-5424.

Grim, N.J. (1993). Endonuclear symbionts within a symbiont: the surgeonfish intestinal symbiont, *Balantidium jocularum* (Ciliphora) is host to a Gram-positive macronuclear inhabiting bacterium. Endocyt. & Cell Res. **9**, 209-214.

Grutter, A.S. (2004). Cleaner fish use tactile dancing behavior as a preconflict management strategy. Curr. Biol. **14**, 1080-1083.

Henk, D.A., Weir, A., and Blackwell, M. (2003). *Laboulbeniopsis termitarius*, an ectoparasite of termites newly recognized as a member of the Laboulbeniomycetes. Mycologia **95**, 561-564.

Hollar L., Lukes, J., and Maslov, D.A. (1998). Monophyly of endosymbiont containing trypanosomatids: phylogeny versus taxonomy. J. Eukary.Microbiol. **45**, 293-297

Hooper, L.V., Wong, M.H., Thelin, A., Hansson, L., Falk, P.G., and Gordon, J.I. (2001). Molecular analysis of commensal host-microbial relationships in the intestine. Science **291**, 881-884.

Kittler, R., Kayser, M., and Stoneking, M. (2003). Molecular evolution of *Pediculus humanus* and the origin of clothing. Curr. Biol. **13**, 1414-1417.
Kittler, R., Kayser, M., and Stoneking, M. (2004). Erratum: Molecular evolution of *Pediculus humanus* and the origin of clothing. Curr. Biol. **14**, 2309.
Korpimaki, E., Hakkarainen, H., and Bennett, G.F. (1993). Blood parasites and reproductive success of Tengmalm's owls: deterimental effects on females but not on males? *Functional Ecology* 7: 420-426.
Lee, J. (1993). "On a piece of chalk"– updated. J. Euk. Microbiol. **40,** 395-410.
Margulis, L. (1993). *Symbiosis in Cell Evolution.* 2nd edition. W.H. Freeman, New York. USA
Margulis, L. and Fester, R. (1991). *Symbiosis as a Source of Evolutionary Innovation.* MIT Press, Cambridge, Mass. USA.
McGraw, E.A. and O'Neill, S.L. (2004). *Wolbachia pipientis*: intracellular infection and pathogenesis in *Drosophila*. Curr. Opin. Microbiol. **7**, 67-70.
Molina, R., Massicotte, H., and Trappe, J.M. (1992). Specificity phenomena in mycorrhizal symbiosis: community-ecological consequences and practical implications. In: M.F. Allen (ed.) *Mycorrhizal Functioning*. Chapman and Hall, London, pp. 357-423.
Moore, J. (1984). Altered behavioral responses in intermediate hosts – an acanthocephalan parasite strategy. Amer. Nat. **123**, 572-577.
Moore, J. (2002). *Parasites and the behavior of animals.* Oxford University Press, Oxford, UK.
Mumcuoglu, K.Y., Zias, J., Tarshis, M., Lavi, M., and Stiebel, G.D. (2003). Body louse remains found in textiles excavated at Masada, Israel. J. Med. Entomol. **40**, 585-587.
Palenik, B. (2002). The genomics of symbiosis: hosts keep the baby and the bath-water. Proc. Nat. Acad. Sci. US **99**, 11996-11997.
Reed, D.L., Smith, V.S., Hammond, S.L., Rogers, A.R., and Clayton, D.H. (2004). Genetic analysis of lice supports direct contact between modern and archaic humans. PloS Biol. **2**, 1972-1983.
Roberts, L.S. and Janovy, J., Jr. (2005). *Foundations of Parasitology*. 7th edition. McGraw Hill, Boston, USA.
Saffo, M.B. (1991). Symbiosis in Evolution. In: E.C. Dudley (ed.). *The Unity of Evolutionary Biology (Proc. 4th Intl. Congr. Syst. Evol. Biol.)*, Dioscorides Press, Portland, Oregon, USA.
Saffo, M.B. (1992a). Coming to terms with a field: words and concepts in symbiosis. Symbiosis **14**, 17-31.
Saffo, M.B. (1992b). Invertebrates in endosymbiotic associations. Amer. Zool. **32**: 557-565.
Saffo, M.B. (2001). Complexity, variety and change in symbiotic associations. Family Systems **6**, 3-19.
Saffo, M.B. and Nelson, R. (1983). The cells of *Nephromyces*: developmental stages of a single life cycle. Can. J. Bot. **61**, 3230-3239.
Saint André, A.v., Blackwell, N.M., Hall, L.R., Hoerauf, A., Brattig, N.W., Volkmann, L., Taylor, M.J., Ford, L., Hise, A.G., Lass, J.H., Diaconu, E., and Pearlman, E. (2002). The role of endoosymbiotic *Wolbachia* bacteria in the pathogenesis of river blindness. Science **295**, 1892-1895.
Shaw, M.K. (1997) The same but different: the biology of *Theileria* sporozoite entry into bovine cells. *Int. J. Parasitol.* **27**, 457-474.
Smith, D.C. and Douglas, A.E. (1987). *The Biology of Symbiosis*, Edward Arnold, London, UK.
Smith, S.E. and Read, D.J. (1997). *Mycorrhizal Symbiosis.* 2nd edition. Academic Press, San Diego, USA.
Starr, D.J. and Cline, T.W. (2002). A host-parasite interaction rescues *Drosophila* oogenesis defects. Nature **418**, 76-79.
Stummer, L.E., Weller, J.A., Johnson, M.L., and Coté, I.M. (2004). Size and stripes: how fish clients recognize cleaners. Anim. Behav. **68**, 145-150.
Weir, A., and Hammond, P.M. (1997). Laboulbeniales on beetles: host utilization patterns and species richness of the parasites. Biodiv. & Conserv. **6**, 701-719.
Wade, M. J. (2001). Infectious speciation. Nature **409**, 675-677.
Wenseleers, T., Sundstrom, L. and Billen, J. (2002). Deleterious *Wolbachia* in the ant *Formica truncorum.* Proc. R. Soc. Lond. B **269**, 623-629.
Williams, CM., Poulin, R., and Sinclair, B.J. (2004). Increased haemolymph osmolality suggests a new route for behavioural manipulation of *Talorchestia quoyanai* (Amphipoda: Talitridae) by its mermithid parasite. Funct. Ecol. **18**, 685-691.

Biodata of **Tahir Shah** author of ***"Epistemological Pluralism"*** and ***"Can there be a Limit on the Complexity of Lf"?***

Dr. Tahir Shah divides his time teaching theoretical computer science, research, and yoga. His current research interests include evolution and theories of intelligence, philosophy and foundation of mathematics. He studied physics at Karachi University, Pakistan (undergraduate 1960-64) and graduate studies in theoretical physics at Bologna University, Italy (1964-67) and at Stevens Institute of Technology, USA (1969-71). He was invited to become a reviewer for Mathematical Reviews in 1975 and took chair of mathematics at Alfateh University, Tripoli, Libya, in 1976. After designing intelligent systems for the industry in North America for several years, he returned in 1988 to pursue academic life in Trieste, at ICTP and University of Trieste, Italy. He is also member of the international editorial councils of "Problems of Theory and Methods of Education" (in Russian) and "Revista Cubana de Informática Médica."

E-mail: **shah@spin.it**

J. Seckbach (ed.), Life as We Know It, 341-357.

CAN THERE BE A LIMIT ON THE COMPLEXITY OF LIFE? *

K. TAHIR SHAH
Faculty of Engineering, University of Trieste, Trieste, Italy

1. Introduction

The question whether life will continue to evolve into increasingly complex forms has many important implications both in terrestrial and extraterrestrial contexts. But, let us first clarify the term 'complexity' in this context, which can be illustrated in the following manner. Life's adaptation due to changes in its external world can be divided into morphological and cognitive categories. Whereas morphological adaptations involve physical changes in the body, the cognitive adaptations involve changes in the way a life form interact with its external world, especially its capability to interact in a flexible manner. One of the consequences of this flexibility is that the so called 'micromanagement' of organism's interaction with its environment becomes the domain of individual judgement and control. An increase in cognitive complexity can be understood, *grosso modo*, in presently accepted theories of modular organization as an increase in the number of cognitive modules that an organism utilizes for its increasing flexible behavior. The above question can therefore be rephrased as: Will cognition continue to evolve towards greater and greater complexity as it did in 0.6 to 1.2 billion years of terrestrial history, of course, ignoring the earlier period of unicellular and multicellular organisms without nervous system? Evolution by definition is an open system since it continues whether or not there is an increase in the complexity. But, suppose the answer to above question is not in affirmative and there exists an upper limit, how can we reach such a conclusion? This is the subject matter of this chapter.

An astrobiological implication of a complexity-increasing evolutionary system hypothesis is that there can possibly be life forms more 'intelligent' than humans in the same manner as humans consider themselves more 'intelligent' than non-human primates, and non-human primate (in human judgement) more 'intelligent' than most other mammals and other terrestrial species. Perhaps there can be an endless hierarchy of intelligent beings. Of course, the underlying assumption in the above assertion is that life can originate and evolve into an ever increasing complex form anywhere in the universe. In fact, a simple linear view of the 3.8 billion years of evolution of life towards an ever increasing complexity would suggest that evolution

* Updated and adapted from (unpublished) lectures given at Cuban Neuroscience Centre, Havana, December 2001

of neuro-cognitive system is open ended. But, a careful assessment based on physical and computational constraints on the mechanisms of central nervous system would suggest that the issue is much more complex than this linear view, many other factors begin to influence natural selection. Evolution of central nervous system (CNS) is not, it seems, only a bottom-up process acting from DNA mutations to selection of species with increasing complex brains, there is also a significant contribution from top-down processes as well. Contributions to top-down processes can be of natural origins as well as of artificial origins due to life's capability to modify itself, such as the inclusion of an extra set of cognitive modules fabricated by an organism itself . Analysing both the bottom-up and top-down contributions such as nature-nurture interaction, life-environment interaction and computational and hardware constraints, we shall outline a conjecture on whether or not there can possibly be a limit on natural cognitive complexity.

2. Evolutionary and Developmental Complexity: The Evo-Devo Paradigm

A quick view of terrestrial history shows an extremely slow evolution of complex life forms initially, only unicellular life existed for some 3 billion years. The mammalian life emerged from therapsids only 230 million years ago. Although the origin of primates such as prosimians goes back to cretaceous period, higher primates, for example, *Australopithecus* (average brain volume 500 cc) appeared only 4 million years ago, *Homo habilis* (average brain volume 700 cc) about 2 million years ago, *Homo erectus* (average brain volume 1000 cc) about a million years ago and *Homo sapiens* (average brain volume 1350 cc) appeared only 160,000 years ago. The evolution of central nervous system was rapid only during the last 4 million years with the emergence of higher cognitive faculties. The full human language (i.e., including the pragmatics component), perhaps the most complex of all cognitive capabilities, seems to have been in place at least 40,000 to 50,000 years ago, around the time of the emergence of tool making and cave art (see footnote 1). Smith, and Szathmary (1999) also raised the question, why only from about 40,000 years ago, we are faced with evidence of continuing cultural innovation and not earlier. This seems most likely because of the emergence of Theory of Mind (ToM), a mental faculty prerequisite for social intelligence and the pragmatics component of human language [Shah 2004].

What this suggests is that the growth of complex life and biodiversity is not linear in time. Initially, for a long time there was very slow increase in complexity of life forms and only in the last 350 million years there was an almost exponential growth in the multicellular eukaryotic lineage in its complexity and diversity. Underlying such non-linearity we believe there lie the organization and complexity of basic mechanisms of life and developmental processes. The predominant view point today is that there is an hierarchical structure at the developmental level reflecting the evolutionary history of basic body plan [Martinez Arias and Stewart 2002, Gilbert 2003]. In the current literature this is referred to as the evo-devo paradigm.

There can possible be many measures of complexity of a species. A simple and intuitive measure of it can possibly be the proportion of the number of genes expressed during development which influences its cognitive system and thus behavioural complexity and genes expressed for all other organs. In fact, from the roughly 30,000 structural genes in the human genome, around 40 percent are uniquely expressed in the central nervous system (CNS), the remaining 60 percent are expressed both in the CNS as well as elsewhere in the body. A more detailed view of genomic complexity should take into account not only the coding regions but also much larger stretches of non-coding regions (see footnote 2).

Moreover, one should also keep in mind that both the genetic makeup and the environment determine behavior. Unfortunately, both sides of the nature-nurture debate share the same erroneous assumption that the instructions for behavior are pre-existent either in the genome or in the environment and only one or the other option is imposed on the developing organism. Instead, both genetic and environmental variables influence the process of development. The final outcome depends on their interactions with all the components of this process. As an example at the chemical level, neurotrophic signalling mechanisms seem to be one of the pre-requisites for evolution of complex nervous systems [Jaaro et al., 2003].

To analyse further the notion of complexity in more 'manageable' form we shall divide complexity of life forms into three classes: genotype complexity, neuronal complexity and cognitive complexity (normally referred to as intelligence). One may object to this division because all three levels, the genetic, neuronal and cognitive are interdependent. This is true, but the level of abstraction (or description) is independent for each case. One should not ignore this fact as evidenced by the well-known debate on mind-brain problem and its classical version, the mind-body problem – again, the mind and the brain are a differing levels of abstraction. In this division of complexity we are not asserting any philosophical viewpoint but it is simply a matter of convenience. A word of caution with respect to extrapolation of the complexity of terrestrial life to extraterrestrial life should be kept in mind given that our knowledge is restricted only to the terrestrial version of cognitive complexity and its evolution, any astrobiological speculation thus remains very much anthropomorphic and species-specific typical to *homo sapiens*.

2.1. MODULAR HIERARCHY AND NETWORKING AS BASIC STRUCTURAL DESIGN MECHANISMS OF GENOME'S COMPLEXITY

The well established scenario in evolutionary biology now is that the origin of life goes back to the RNA world with the selection of RNA over possible alternatives such as pRNA. Then, the next most important phase was the transition from the RNA world to RNA-DNA-Protein world with the appearance of the first cell, our Universal Common Ancestor. From the symbiosis of prokaryotic life, eukaryotes were born over a billion years later leading towards simple multicellularity. Genomes became more and more complex through gene copying, horizontal gene transfer, viral insertion and other mechanisms. It was not only the genome size that was responsible for increasing complexity of life forms but the internal mechanisms of genotype to phenotype conversion which became more complex and so to say, 'optimised', giving rise to mechanisms such as differential gene expression. Appearance of *Hox* and other genes gave final shape to the basic body plan. The evolution of central nervous system and its overlaying perceptual and cognitive systems were the most important events in the history of terrestrial life. All of this led to an explosion of enormous biodiversity from plants to higher mammals with their sophisticated information processing capability.

It is often said that the genome, a linear stretch of DNA, embodies the blue print of an organism. The current understanding among developmental biologists is that such an assertion is misleading one if not entirely incorrect for the following reasons. First, it is the formal linguistic structure of genes both for introns and exons that contains extra information than a simple Shannon's type information-content analysis would suggest [Searle 1992][Shah 1996]. Recall that Shannon's theory does not take into account the syntactical structure. Second, it is fairly established now that genetic programs that underlie development encode the networks of proteins, a jump to another level of abstraction, and in the end, the same molecular networks of proteins operating within and between cells executes this same program encoded in DNA. Such a multi-level hierarchical complexity structure cannot be discussed in terms of Shannon's information theory. The protein networks create cells and control basic cellular behavior associated with processes such as cell adhesion, division or movement. These are then deployed during the development of the organized cellular assemblies that eventually make up tissues, organs and ultimately the whole animals. The hierarchical structure with more than one levels of abstraction is arranged along the following line of schema. Genes encode proteins and carry information that will determine when and where these proteins are to be made. However, genes do not contain the information that determines how the proteins will assemble, or when and where they be active. This information essential in any "developmental blue print" lie in the 3-dimensional structure of the protein themselves, that is, it is encoded at chemical level of abstraction. And, from instructions embodied in this 3-dimentional structure functional and regulatory networks are assembled. The constituent proteins of these networks contribute to the composition and activity of cells, determining and modulating their environment, shape, information processing ability, and pattern division and differentiation. At the same time, the activity of the network feeds information back to genes that encode their elements, to create regulatory circuits of gene expression. In essence, whereas proteins can be said to be the creation of the genetic machinery of transcription and translation, cells are the creation of the interaction between genes and proteins. Different combinations of these elements generate different cell types which will assemble into tissues and contribute to the shape and size of an organism [see, e.g., Von Mering et al., 2003].

From the computational perspective, the molecular networks can be thought of as the "hardware" of the developmental system, a hardware made up of proteins that read and interpret different "programs" written in various DNA sequences from genes to full genomes, to produce different outputs, i.e., different animals, different organs, and supracellular aggregates. The programs are written in the regulatory regions of the genome and determine the sequence of patterns of the gene expression characteristic of each program. This also means that the hardware that reads and interprets the programs is itself determined by gene expression program, a defining characteristic of life.

2.2. UNIFIED NATURE OF DEVELOPMENTAL MECHANISMS

Developmental biology has demonstrated that all multi-cellular organisms share many of the same molecular building blocks and many of the same regulatory genetic pathways. Underneath the immense diversity of animal design lie a few unifying organisational principles and common developmental processes. This biodiversity can be attributed to a few basic processes and a general body plan with only a few basic concepts: fate maps, asymmetric division, induction, competence, positional information, determination and lateral inhibition. A concrete confirmation of this unity comes from the discovery of similar genes (e.g., homo-box) that both vertebrates and invertebrates express during the development of roughly comparable body regions. In fact, it would not be an exaggeration to say that there are not very many families of proteins of importance in developmental biology. It is now known that *Wnt*, *Hedgehog*, *TGF- beta*, and *FGF* genes are used for signalling and about a dozen classes of transcription factors, another dozen for cell surface constituents, as well as machinery for transcription, translation, transport, and cytoskeleton could possibly be sufficient to 'construct' a complex animal [Rudel and Sommer 2003]!

There are great similarities at genotype level among animals that were once considered entirely different. Currently, there is evidence for the following similarities: anteroposterior body plan controlled by *HOX* genes, dorsoventral body plan controlled by *TGF*-beta signals and their antagonists, heart development controlled by *Tinman/Nkx2.5* and *MEF2*, eye development controlled by *Pax6/eyeless/aniridia*, CNS midline regulation of axon extension, branching morphogenesis of lung and trachea, limb dorsoventral and distal-proximal patterning, and Toll signalling in insect and mammalian immunity. The picture one get is that over all structure is not changing from one animal to another except that specific gene system is committed to an specific developmental purpose [Scott 1997].

It is often stated by those biologist who accept the hypothesis of evolutionary convergence that certain morphological structures evolved independently many times over such as, tooth for more than 20 times, eyes over 40 times and so on, despite the complexity of signalling protein. What has been emerging is that this view is an overstatement of the *de novo* evolution of such morphological traits. Another view seems more likely. "The genetic program for a structure may have evolved once, and then was stashed away intact (perhaps used for another purpose?), only to become reactivated later. With this view, intermediate forms need not to have retained all the structure common to their ancestors and descendents." [Scott 1997]

2.3. ROLE OF EXTERNAL WORLD DURING DEVELOPMENT: THE ACTIVATION-DEPENDENCE OF NEURONAL CIRCUITS

The role of external environment shows up in many important observations. For instance, it is now established that there is an increase in the size of those subsystems of the nervous system which are utilized more actively. In a similar manner, inactivated systems eventually cease to exist in the course of many generations. The experimental evidence shows that frequent activation due to an specific tasks performance can influence gene expression, known as the activation-dependent gene expression. The development of the brain is strongly connected to activity-dependent modification of neural circuits. It has also been suggested [Singer 1984] that neural activity is a crucial factor in the self-organization of the developing brain. This is because external stimuli modulate neural activity during post natal life thereby influencing self-organization process. For instance, removal of one eye before birth permanently change the cellular organization and synaptic connectivity of visual centres in the primate brain. The visual cortex is known to be altered in a most notable way as: 1) ocular dominance columns fail to develop; 2) unusual synaptic connections were formed between the intact eye and the geniculate neurons that have lost their normal input (from the removed eye part); 3) the number of layers developed in the lateral geniculate nucleus were less than the normal, i.e., only two instead of five; 4) the activity-dependent connectivity rules resemble Hebb rules, i.e., strength of connections depends on pre- and post synaptic activity. In mammalian cortex, such activity- dependent modifications are restricted to a critical period in the post-natal life. There is also evidence of activity-dependent regulation of gene expression in the visual cortex of adult monkey [Jones et al., 1990]. In this case, monocular deprivation in adult monkeys influence levels of several neuro-transmitters, peptides, or protein level of at least one receptor.

3. Complexity of Central Nervous System: The Underlying Source of Cognitive Complexity

Nervous systems are highly flexible networks evolved to process information at various levels, from neuron to cortical hyper-columns, along with many binding or data integration mechanisms. There are three levels of communication mechanisms in the brain: purely electrical, electrochemical and purely chemical using hormones. The purely electrical system is used for a simple and fast response for life and death events. In the electrochemical system, the synaptic cleft mechanism allows for a large number of combinatorial variations of 'switches' with the help of two neurotransmitters and a neuropeptide along with receptors of many kinds. Within a brain, a large number of neurotransmitters type (corresponding to their respective neuron type) provides an interference-free environment for many subsystems performing different computational tasks, consequently supporting hardware modularity. On the other hand, a large number of receptor types provides precise addressing mechanism, signal delay, memory and other combinatorial mechanisms. A further subdivision into excitatory and inhibitory neurons provides further capacity for modulation of pre- and post synaptic signal inputs. Extremely large fan-in (e.g., about 80,000 for human Purkinjee cells) provides communication facility with a large number of neurons from various other subsystems. In addition to the above mechanisms, there are modulation of internal neuronal processes, such as diversity of electrical conduction mechanism, polymorphism at many levels through electrical coupling and synchrony, hierarchical organization using repetition of homologous circuits to avoid metabolic hot spots and load imbalance, and the signal integration mechanism at various levels using electrical coupling and synchronous firing. The brain also uses hormones through the hypothalamus-pituitary system to control various bodily functions. All in all, the brain is perhaps the most complex information processing machine we know of in this universe [for a general reference, see e.g., Kendal, Schwartz and Jessell 1992].

3.2. BRAIN'S COMPUTATIONAL PERSPECTIVE

Brain's characteristics from the computational perspective, like for any other networks made of processing elements (logic gates, finite automata, threshold elements etc.), is determined by two main factors:

a. The information processing capability of individual neural cells. This includes the type of computation, its speed and the grain-size of the computation.
b. The interconnection topology during one single computational task determines the output for a given set of inputs. It is called the *functional topology* in contrast to fixed interconnection static (anatomical) topology, which remains unchanged during the execution of all computational tasks. In fact, the brain's enormous computing power comes not only from the massive parallelism (e.g., over around 100-150 billion neurons in human brain), but also from immense number of possible topologies of interconnections between neurons. On the average, there are 10,000 connections per neuron and thus about one million billion interconnections in such nervous system. One may safely say that it is possible to have an extremely large number of topologies of interconnections constructed through normal anatomical interconnections and superimposed modulatory processes at various level of neural organization - from dendrites to axon to receptors [Katz 1999]. Through modulation, the brain saves processors and wiring resources for many more circuits that are made possible simply by the dynamic change of topology. A single circuit with the modulation capability is equivalent to many circuits with different interconnections [Getting 1989]. It can be said to be similar to massively parallel re-configurable machines capable of simulating more than one fixed topology machines, but their mechanism to achieve reconfiguration may be different from that of the brain. Since human genome contains only between five to six billion nucleotides, the brain's wiring is not directly coded in the genome, other extra-genetic mechanisms such as the use of neurotrophins seems to be responsible for creating such astronomical complexity of the brain's circuitry [Jaaro et al., 2002].

Other characteristics of a nervous system which are a direct consequence of (a) and (b), and those independent of its size are: its switching capability, modulation, the change of its role by a single cell and the reconfiguration capability of some networks. Some experiments demonstrated cross modality of neural circuits [Sur et al., 1990] and therefore fit into Gould's idea of exaptation where some functions have recruited cortical circuits that evolved for very different purpose. Sur and his group (1993) showed experimentally how afferents from the visual cortex can be used by the auditory modules, a good example of cross-modality use of circuits. These studies suggest that such changes in the use of a prototype circuits is far more common in nature than radically different re-routing of neural connections or creation of absolutely new circuits. Cross-modal plasticity studies by Sur and his group suggest that some basic processing modules are possible in all sensory cortices that perform stereotypical transformation on their inputs regardless of the modality.

3.3. MODULARITY AND DOMAIN-SPECIFICITY

Looking at mammalian CNS one begins to wonder why, for instance, it has so many visual areas? A fairly plausible answer lies in the fact that nature tends to select those structures which 'minimize' the use of resources and avoids *de novo* construction. Modular strategies fits into this scheme. Adding of new visual areas is a way of avoiding constraints due to modifications on existing visual areas which are required for the ongoing functional requirements [Kaas 1987]. Thus, increasing the number of visual or other cortical areas is an effective and apparently common mechanism for evolving new functionalities. Any *de novo* construction of hardware or mechanism from scratch is too costly in evolutionary terms. Such a modular approach has, at least, three advantages:

1. Increasing the number of visual areas increases the number of visual abilities.
2. Multiple areas simplify the problem of interconnecting functionally related groups of neurons.
3. Adding new visual areas overcomes constraints stated above and allows new capacities to evolve.

This clearly illustrates that some of the parallel processing issues we are facing in computer science seemed to have closely related counterparts in neuroscience. In particular, the question of how to map a task optimally onto a set of processors underlies domain-specificity and seems to be crucial for evolutionary adaptation. This problem is analyzed by Bower and Nelson (1990) and they describe the following kinds of mappings:

Continuous Maps: A computationally relevant parameter is represented in a smooth and continuous manner. These maps are optimal for computations that are characterized by predominantly local interactions in the problem space. Communication cost can be minimized by mapping neighbouring parts of the image to neighbouring processors. Load-balancing, an important processes to avoid hot metabolic spots in the brain, can be achieved in this class of mapping by increasing or decreasing appropriate regions of maps, but preserving their spatial relationship.

Scattered Maps: These are non-topographic maps which arise also in parallel computers as nearby optimal solutions for a variety of computations, where the problem space lacks any systematic structure.

Patchy Maps: In maps of this kind the spatial organization is somewhere in between patchy and continuous maps. This class of maps are characterized by a number of patches, within each patch the representation is smooth and continuous, but nearby patches may represent very different parts of the problem-space.

Each of the categories of maps described above are found in mammalian CNS. In parallel computers the optimal mapping structure reflects the load-balance and communication cost. This raises the general question of whether or not there are correlates of load-balancing and communication cost in the nervous system. We mention here two examples of neural correlates of load-balancing and communication overhead: 1) In monkeys, a prolonged increased use of a particular finger leads to an expansion of the corresponding region map in the somato-sensory cortex [Merzenich 1992]; 2) The intrinsic structure of piriform cortex which consists of an extensive network of association fibers, interconnecting all regions, confirms the non topographic nature of the olfactory cortex map [Haberly 1985].

4. Cognitive Complexity: A Measure of Complexity of Species

We used the term 'cognitive complexity' so far without clarifying its meaning. Cognitive states and processes are theoretical constructs, which provide a functional characterization of operations of the CNS and may or may not be objects of conscious awareness. They receive inputs from other states and processes (including perception) and have outputs to other cognitive states and processes and ultimately to an organism's behavior. Cognitive states are not observable. A major assumption implicit in the classical cognition approach is that the same set of cognitive processes could be implemented or instantiated in a variety of different neuro-anatomical structures or neuro-physiological processes [Pylyshyn 1984]. The following three are the most important characteristics of cognition at psychological level of abstraction.

Domain-specificity: It was suggested by Cosmides and Tooby (1994) that the origin of domain-specificity is in the evolution of functional organization of the brain resulting in species-typical psychological architecture. Evolutionary biology provides a series of reasons why it is implausible and un-parsimonious to assume that the human mind is a general purpose machine. It thus supports the evolution of domain-specificity. The roots of the idea of domain-specificity goes back to Chomsky's theory of grammar, modular approach to knowledge, constraints on induction, philosophical insights into knowledge structures, the learning, memory and problem solving, and the wisdom gained from a comparative man-animal evolutionary perspective and cross cultural studies.

Modularity of mind: Fodor (1983) and Chomsky (1988) concluded that the mind is modular meaning that thought is domain-specific and there is a large number of modules at different hierarchical levels of human cognition. The ontological commitments of modularity is different from other theories of cognition, it has evolved to deal with different levels of abstraction and functional separation. Another reason why domain-specificity and modularity of mind are essential characteristics of complex cognitive systems is because a combinatorial explosion can choke any system that is truly domain-general.

Evolutionary Continuity: Studies on the origins of great ape intelligence and comparative primate cognition increasingly show that all great apes share a grade of cognition distinct from that of other non-human primates and humans, so it offers the best available model of the cognitive platform from which human cognition evolved linking non-human primates with further lower level of complexity of cognition.

The term 'cognitive complexity' though can be equated with the notion of 'intelligence', it gives a more precise description of intelligence in evolutionary term. The term "intelligence" is one of those commonly used notions which is not only difficult to define, but there is a large number of rather context-dependent operational descriptions of what it means when this term is utilized. Quite often it is confused with individual performance or even cultural achievements. What is the characteristic of animal intelligence? First of all, we consider it as species-specific innate capacity to attain flexible behavior. Although, classical cognitive science provides us a useful hardware-independent cognitive architecture for intelligence, it completely separates the brain from cognition. The implication is that the cognitive architecture remains invariant both under radically different paradigms (e.g., symbol processing, connectionist or others) and physical processes (e.g., electrochemical, electronic, optical or others). Such a paradigm of hardware independence may have relevance in the astrobiological context because extraterrestrial planetary environments are not necessarily like ours. On the other hand, we should not forget that every cognitive system evolved to deal with properties of objects located in space-time. That is, the architecture of cognitive system's hardware was influenced by those laws of nature and principles which govern physical and biological phenomena. We should not ignore the importance and role played by hardware!

Given this premise, human intelligence or human-level cognitive complexity can be considered as a set of mental modules. The existence of the following set of cognitive capabilities or their equivalent variants is necessary for human level cognitive complexity, though it may not be sufficient. These are symbolic representation and symbol processing; higher-order intentionality or theory of mind (ToM)); recursion (unbounded generativity in linguistic and perceptual domains); granularity of mind; and meta-mind (capacity to envisage mental constructs at multiple levels of abstraction). In addition, complex linguistic capability is necessary [Premack 2004]. Of course, all of the above imply the existence of memory and perceptual apparatus. One may add more modules and sub-modules but their existence shall not influence the line of our argument.

The notion of intelligence given above is not really defined as one thing, architecture or function, but it is really a collection of cognitive modules. Each module contributes to a specific function, such as ToM, a pre-adaptation for social intelligence and is necessary for a life form to function as an active member of its society. Since a pre-requisite for the emergence of a technologically advanced civilization is communication and cooperation among its members, ToM cognitive module is necessary for such social interactions.

5. Life-Environment Interaction

So far our attention was on the characteristics of neuro-cognitive system and its complexity. We come now to an important factor not discussed widely in evolutionary studies, the bi-directional life-environment interaction. Normally, only the influence of external world is emphasized in natural selection process. But, life's modification of its external environment and the change in the direction of its own evolution as a consequence of such modification are both equally important contributions to top-down processes mentioned earlier. The so called 'oxygen revolution' due to photosynthesis is one important example which lead to metabolically more efficient life on Earth some two billion years ago [Lunine 1999]. Other kinds of substantial changes can occur when a species acquires enormous mental capabilities leading eventually to the creation of artificial life forms (e.g., androids) by this species. In addition, the species become capable of constructing and replacing its own body parts and thus changing the direction of evolution in a very significant way. Such is the case of human species as it seems now. Assuming that the same basic body plan continues since nature is not going to change drastically, one may speculate on the possible bifurcation of *Homo sapiens* lineage into a new "*Super sapiens*" species. If so, what kind of selective pressure would do it? No specific morphological changes are expected since human civilization is able to protect against minor climatic and other environmental changes. Of course, a major change either in climate or by a series of impacts on Earth by large bodies may change drastically the direction of evolution with mass extinctions, including the disappearance of human civilization. But we are not interested such extreme scenario. Since neither a prediction of long-term changes in the environment nor an observation of visibly significant effects on human species in short terms seems plausible at present, it is a difficult exercise to predict the direction of human evolution due to natural causes.

Nevertheless, a rational guess about the effect of man-made modification of its environment and artificial creation of simple lower life forms is relatively a less difficult exercise. The following recent laboratory results and construction of devices give some hints of things to come.

Neural Prosthetics: In the field of technological development of brain-computer interface, there is a clear evidence for radically innovative use of cognitive manipulation of external material devices. Such is the case of systems in which only thought manipulates a device directly. Certainly, this is not to say that some magical power is induced to humans. It is simply a technical achievement. In another example of such system, the human perceptual interface, that is, eye gaze is used to manipulate objects. In a typical application, a jet pilot uses only gaze towards an appropriate area of the instrument panel to control the flight of an aircraft. In other applications of brain-computer interface which goes deeper into neural modules, it possible to predict the intention by eavesdropping on neurons in cognitive area. But this system does not the execute any function [Kaiser 2004, Mussalm et al., 2004].

Synthetic Biology and Redesigning Life: This area of research and development includes laboratory assembly of viruses, re-wiring of bacterial cells for new functionalities and the synthesis of biological material. For instance, complete synthesis of poliviruses from mail-order material has been achieved. Genome modification and genome synthesis has been done experimentally in various laboratories [Ball 2004]. This includes synthesis of transcriptional regulators [Elowitz and Leibler 2000], construction of genetic a toggle switch in simple organisms, e.g., in E. Coli [Gardner, Cantor and Collins 2000], re-programming control of an allosteric signalling switch [Duebar et al., 2003], control of cell population (controlling evolution) and generating synthetic genome from synthetic oligonucleotides of Φx 174 bacteriophage [Smith et al., 2004]. In one the most ambitious project of redesigning genetics has its gaol of redesigning DNA to create artificial genetic system [Pennisi 2004, Benner 2004]. Synthetic life is different from what is known as artificial life which is nothing but simulation experiments on computers – material existence and simulation existence are two entirely different modes ontologically.

Stem Cells: Stem cells are unspecified cells that can self-renew indefinitely and also differentiate into more mature cells with specialized function. Their use in the regeneration of body organs and parts, cloning, repair and modification is in experimental stage and expected to become a technology in the near future.

Hybrid Intelligence systems: The hybrid cognitive capabilities refers to mixed biological intelligence and those of machines designed by biological life forms. These artificial intelligent systems may even be based on the same biological principle and material as that of the life form which has designed and constructed them. The combined use of intelligent androids and the life form leads to what has been called hybrid intelligence. This entails, for instance, a high level of human and machine cooperation in a hybrid society of human and human-made intelligent bio-androids. It also entails the existence of collective memory (storage and transmission of knowledge in time and space) [Ford et al., 1995]. A joint human-android cognitive enterprise is very likely to emerge as a form of hybrid intelligence in the near future. Such systems presently include, visual gaze activation systems, brain-wave reading-and-activation system, android-human systems, biologically generated additions to the brain (self-made mental modules) and remote controlled intelligent systems described above.

Other environmental factors which can effect the evolution of life are: diet habits, construction of sophisticated habitats to minimize environmental variations inside habitat, availability of food at all times, man-made changes in the atmosphere such as an increase in emission gasses and temperature and the gravity-free environment in future space colonies of human, plants and other animals.

6. Conclusion: Future Directions of Evolution

The objective of this chapter has not been to demonstrate the openness or closure property of biological complexity, that would be a tall order, but to outline a schema of analysis in terms of the basic mechanisms of life, including the influence of the brain's characteristics and its constraints on cognitive complexity. Having said that, a few brief comments concludes this chapter.

Let us consider some of the evolutionary changes, such as an increase in morphological complexity of neurons, the number of cells, and the diversification of neuron types. Such an increase in morphological complexity has known to accompany the brain evolution in the form of local enlargement of various areas. This occurred with the development of local circuit inter-neurons or by new long distance interconnections that were not present between two brain regions. The number of neurons also increased along with the diversification of their types in order to accommodate computational requirements for new tasks to be tackled. To support this increase and diversification of neural cells there occur an enlargement of the support system thus increasing glial-neuron ratio so that metabolic supply and other non-computational support processes remain sufficient for the need.

The past evolutionary history of an increasing neuro-cognitive complexity would suggests an open-ended ever increasing complex life. However, considerations of constraints on neural communication and computational processes point towards the existence of an upper bound, a sort of "maximal" neuro-cognitive system, beyond which further increase in complexity seems not to be feasible. It is not possible even to guess whether we have reached that limit or not. The number of factors is too large and some are still unknown. For instance, an increase in the number of cognitive modules requires further processing capability, which cannot be possible without an increase in the number of neurons, interconnections, metabolic energy, glial cells for support functions and the brain's total weight. Another important consequence of additional structure would be larger distances for neural signals to travel within the brain thus slowing down the speed of computation that need to be compatible with the survival of species requirement. Can a species with a too large, but too "slow" neurocognitive system survive in a competitive world of finite resources? Even if the present modular structure with separation of high level cognitive functions and low level fast perceptual and stimulus-response system continues in future, the metabolic energy requirement of a larger brain and of the body to carry a heavier brain should set an upper limit on the total size and complexity.

A further complication in our understanding of evolution of neuro-cognitive system comes from another perspective. Taking evolution as a series of changes over time at the genotype level that determine brain's structure and function, neural evolution poses a fundamental problem because of the intricate nature-nurture interaction during the period of neural development. This is true in particular for mammals where both the structure and function of the mature brain is a result of such interaction between the genetic apparatus and pre- and post natal environment in which a given nervous system develops. For this reason, facts about the mature brain represent both nature and nurture. How the human invented neuro-cognitive capabilities will influence the nurture part is difficult to predict, but over a long period of time many such additional modules and novel environmental conditions (e.g., zero gravity in future space colonies) are expected to influence the nature of cognition, perhaps, in a significant way. How life may continues to evolve, both due to natural as well as artificial causes (e.g., self-modification), shall remain an open question perhaps as far life continue to inhibit this universe!

7. Footnotes

1. We emphasised the great importance of non-coding regions as early as 1995 by arguing that non-coding DNA (e.g., introns, ALU sequences, satellite DNA) played a major role not only in evolution but also their regulatory action during developmental processes [Shah 1995]. For a recent update on this issue, please see [Mattick 2001].

2. The full human language (i.e., including both FLB and FLN of Hauser, Chomsky and Fitch [2002]) has four major components, phonology, syntax, semantics and pragmatics. However, for the purpose of

defining language, the phonological level should be unimportant, because sign system can be considered as full language. The data obtained from sign languages is in contrast to the hypothesis of the co-evolution of language and the neuro-anatomical mechanism of speech production [Emmorey 2002, Malle 2002]. Moreover, whereas phonology is mainly about the speech apparatus, the other three components are purely mental faculties. The unbounded generation of language using only a finite number of syntactical rules is uniquely human and has no counterpart even among other primates. The late comer on the evolutionary scene seems to be the pragmatics component containing the capacity for 'conversational implicature' of Grice, 'speech acts' of and Searle, and other such inferential-intentional aspects. Recent works on "Theory of Mind" (ToM) showed that pragmatics could have evolved only once ToM was in place. This is also the case at developmental level where ToM is known to precede language, that is, having the ability to infer others' communicative intentions is a precondition for language. *Homo sapiens* is the only species that possesses a well developed ToM, which means that only they have the ability to attribute the full range of both goal states and epistemic states to themselves and to others. One should not, therefore, equate evolution of human speech apparatus with full human language and consequently conclude that full human language originated at the time when it became possible to make complex sounds. Baron-Cohen (1999) estimated that ToM was in place some 40,000 years ago and Crow's (2000) estimate shows it be present around 50,000 years ago. Since pragmatics requires ToM, as research on Autism has clearly demonstrated [Baron-Cohen (2001), the full human language cannot be predated older than ToM.

8. References

Ball, P. (2004), Starting from scratch, Nature **431**, 624-626.

Baron-Cohen, S. (1999), The evolution of a theory of mind, in: M.C. Corballis and S.E.G. Lea (eds.) The Decent of Mind, psychological perspectives on hominid evolution, Oxford University Press, pp. 261-277.

Baron-Cohen, S. (2001), Theory of mind and autism: a review, International Review of Mental Retardation (special issue) **23**, 169.

Benner, S.A. (2004), Redesigning Genetics, Science **306**, 625-626.

Bower, J.M. and Nelson, M.E. (1990), Brain maps and parallel computers, Trends in Neurosci, **13**, 403-408.

Chomsky, N. (1988), Language and Problems of Knowledge, MIT Press, Cambridge, MA, USA.

Cosmides, L. and Tooby, J. (1994), Origin of domain-specificity: The evolution of functional organization, in: L.A. Hirschfeld and S.A. Gelman (eds.), Mapping the Mind, Domain-specificity in Cognition and Culture, Cambridge University Press.

Crow, Y.J. (2000), Schizophrenia as the price that Homo sapiens pays for language: a resolution of the central paradox in the origin of species, Brain Research Interactive **31**, 118-129.

Dueber, J.E., Yeh, B.J., Chak, K. and Lim, W.A. (2003), Reprogramming Control of an Allosteric Signalling Switch Through Modular Combination, Science **301**, 1904-1908.

Elowitz, M.B. and Leibler, S. (2000), A synthetic oscillatory network of transcriptional regulators, Nature **403**, 335-338.

Emmorey, K. (2002), Language,Cognition, and the Brain: Insight From Sign Language Research, L. Erlbaum Assoc. Publishers, Mahwah, N.J, USA.

Fodor, J. (1983), Modularity of Mind, MIT Press, Cambridge, MA, USA.

Ford, K.M., Glymore, C. and Heyes, P.J. (1995), Android Epistemology, The MIT Press, Cambridge (MA), USA.

Gardner, T.S., Cantor, C.S. and Collins, J.J. (2000), Construction of genetic toggle switch in Escherichia coli, Nature **403**, 339-340.

Getting, P.A. (1989), Emerging Principles Governing the Operation of Neural Networks, Annu. Rev. Neurosci., **12,** 185.

Gilbert, S.F. (2003), Developmental Biology, Sinauer Assoc. Inc., Sunderland (MA), USA.

Haberly, L.B. (1985), Chem. Senses, **10**, 219-238.

Hauser, M.D., Chomsky, N. and Fitch, W.T. (2002), The Faculty of Language: What Is It, Who Has It, and How Did It Evolve? Science **298**, 1569-1579.

Jaaro, H., Beck, G., Conticello S. and Fainzilber, M. (2001), Evolving better brains: a need for neurotrophins?, Trends in Neuroscience, available at http://tins.trend.com.

Jones, E.G., Benson, D.L., Hendry, S.H.C. and Isakson, P.J. (1990), Activity-dependent Regulation of Gene Expression in Adult Monkey Visual Cortex, Cold Spring harbor Symposia on Quantitative Biology, volume 55, Cold Spring Harbor Laboratory Press.

Kass, J.H., (1987), The Organization of Neocortex in Mammals: Implications for Theories of Brain Function, Ann. Rev. Psychol, **38**, 129-151.

Katz, P.S. (1999), Beyond Neurotransmission: Neuromodulation and its Importance for Information Processing, Oxford University Press.

Kaiser, J. (2004), Brain-Computer Interface Adds New Dimension, Science **306**, 1878.

Karten, H. and Lumsden, A. (1999), Evolution of the nervous system, Editorial overview, Current Opinion in Neurobiology **9**, 579-581.

Kendal, E.R., Schwartz, J.H. and Jessell, T.M. (1991), Principles of Neural Science, Elsevier.

Lunine, J.I. (1999), Earth, Evolution of A Habitable World, Cambridge University Press, Cambridge, UK.

Malle, B.F. (2002), The Relation Between Language and Theory of Mind in Development and Evolution, in: T. Givon and B.F. Malle (eds.), The evolution of language and pre-language, Benjamins, Amsterdam.

Martinez Arias, A. and Stewart, A. (2002), Molecular Principles of Animal Development, Oxford University Press.

Mattick, J.S. (2001), Non-coding RNAs: the architects of eukaryotic complexity, EMBO reports **3**, 986-991.

Merzenich, M.M. (1992), How the Brain Functionally Rewires Itself, Lectures, College on Neurophysics, ICTP, Trieste, Italy.

Mussalm, S., Corneil, M.D., Gregar, B., Scherberger, H. and Andersen, R.A. (2004), Cognitive Control Signals for Neural Prosthetics, Science **305**, 258-262.

Pennisi, E. (2004), Searching for the Genome's Second Code, Science **306**, 632-635.

Premack, D. (2004), Is Language the Key to Human Intelligence? Science **303**, 318-321.

Pylyshyn, Z.W. (1984), Computation and Cognition, The MIT Press, Cambridge (Mass), USA.

Rudel, D. and Sommer, R.J. (2003), The evolution of developmental mechanisms, Developmental Biology, **264**, 15-37.

Scott, M.P. (1997), Summary: A Common Language, Cold Spring Harbor Symposia on Quantitative Biology, **62**, 555-562.

Searls, D.B. (1992), The Linguistics of DNA, American Scientist **80**, 579-591.

Singer, W. (1984), Learning to See: mechanisms in Experience-dependent Development, in: P. Marler and H.S. Terrace (eds.), The Biology of Learning, Springer, pp. 461-477.

Shah, K.T. (1995), The Role of Information Processing in the Evolution of Complex Life Forms, in: J. Chela-Flores, M. Chada, M. Negron-Mendoza, and T. Oshima (eds.), Chemical Evolution: Self-organization of the Macromolecules of Life, A. Deepak Publishing, Hampton, VA (USA).

Shah, K.T. (1996), Information-Processing Genes, in: J. Chela-Flores, and F. Raulin (eds.), Chemical Evolution: Physics of the Origin and Evolution of Life, Kluwer Academic Publishers, Dordrecht, pp. 323-332.

Shah, K.T. (2004), Evolution of Language as Innate Mental Faculty in: J. Seckbach, J. Chela-Flores, T. Owens, and F. Raulin (eds.), Life in the Universe, Kluwer Academic Press, pp. 333-334.

Smith, J.M. and Szathmary, E. (1999), The Origin of Life, Oxford University Press, Oxford.

Smith, H.O., Hutchison III, C.A., Pfannkoch, C. and Venter, J.C. (2003), PNAS **100**, 15440-15445.

Sur, M. (1993), Cortical Specification: Microcircuits, Perceptual Identity, and Overall Perspective, Perspective on Develomental Neurobiology, **1**, 109-113.

Sur, M. et al. (1990), Cross-modal plasticity in cortical development: differentiation and specification of sensory neocortex, Trends in Neurosci., **13**, 227-233.

Von Mering, C. et al. (2003), Genome evolution reveals biochemical networks and functional modules, PNAS **100**, 15428-15433.

Biodata of **Claudio Maccone** author of ***"Evolution of Intelligence"***

Dr. Maccone obtained his PhD at the Department of Mathematics of the University of London King's College in 1980. He then joined the Space System Group of Aeritalia (now called Alenia Spazio S.p.A.) in Turin (Italy) as a technical expert for the design of artificial satellites, and got involved in the design of space missions. He also published two books "*Telecommunications, KLT and Relativity*" in 1994 and "*The Sun as a Gravitational Lens: Proposed Space Missions*" in 1998 (both at IPI Press, USA). His second book was awarded the "1999 Book Award for the Engineering Sciences" by the International Academy of Astronautics (IAA). In 2000 he was elected as Co-Vice Chair of the SETI Committee of the IAA. In 2001 the Asteroid 11264 was named "Claudiomaccone" in his honor by the International Astronomical Union (IAU). In 2002 Dr. Maccone was awarded the "Giordano Bruno Award" by the SETI League. He has published over 70 scientific and technical papers, most of them in "Acta Astronautica."

E-mail: **clmaccon@libero.it**

J. Seckbach (ed.), Life as We Know It, 359-372.

EVOLUTION OF INTELLIGENCE

CLAUDIO MACCONE
Co-Vice Chair, SETI Permanent Study Group, International Academy of Astronautics. Address: Via Martorelli, 43 - Torino (TO) 10155 – Italy. E-mail: clmaccon@libero.it

1. Introduction

Human intelligence on earth has evolved a lot during the last three thousand years. It actually increased a lot, although not at a uniform pace.

A mathematical description of how much has human intelligence increased, however, is difficult to achieve. In this chapter, we cast a mathematical model the evolution of human intelligence over the last three thousand years that seems to reflect reasonably well both what is known from past and might be extrapolated into the future.

Let us start by considering two seminal books:

1) "Cosmos" by Carl Sagan (1980) (see the references), widely known to millions of people all over the western world because of the corresponding TV series, and
2) "Interstellar Migration and the Human Experience" (by Ben Finney and Eric Jones, Editors, 1983) (see the references). Back in May 1983, a Conference was held at Los Alamos about "Interstellar Migration and the Human Experience". The relevant proceedings were published in 1985 by the University of California Press under the same title and some copies of this book still are available nowadays. The book is outstanding: it offers an unprecedented perspective of human past history from the point of view of the evolution of intelligence and of SETI. In other words, it provides answers to the question: "knowing what happened to various human civilizations in the past on the earth only, how can we extrapolate this amount of information into the future and predict what we might expect in the future from humans expanding into space and contacting ETs ?

These two books thus re-read human history in the new perspectives open up by SETI. But, although wonderfully innovative from the point of view of the ideas, these two books also seemed to this author to have one drawback: they were NOT mathematical, inasmuch as they only offer wordy descriptions of facts, but no equation was given. Real scientific progress, however, only stems out of a profound mathematical understanding of facts. So, this author sought to "cast into equations" at least a few of the lessons learned from "Cosmos" and "Interstellar Migrations and the Human Experience". And this chapter is a popular description of his mathematical results. To keep the mathematics as easy as possible, we start from a simple algebraic equation of the third degree, traditionally called "a cubic". We think that, in some sense, this cubic mirrors a trend that often happened in human history: a civilization rises from obscurity, reaches a peak, then may suffer setbacks, but finally rises again, and at such a high speed, that its own previous achievements are really dwarfed by the latest, new achievements. In other words, we assume that our mathematical model for the evolution of intelligence on Earth behaves like the cubic depicted in Figure 1.

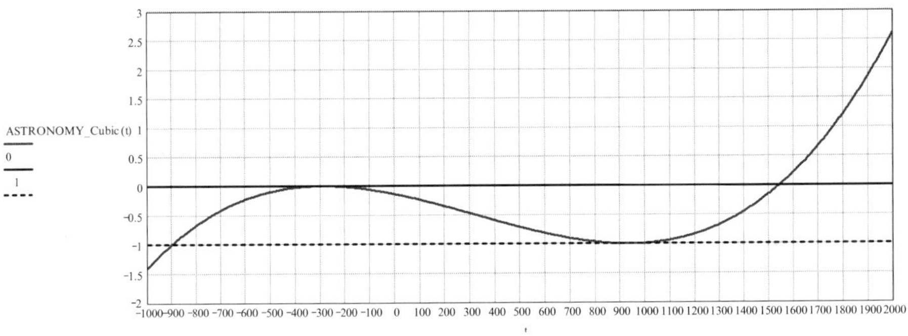

Figure 1. Evolution of Human Intelligence exemplified by the History of Astronomy as a Cubic.

2. History of Astronomy as the Proof of Evolving Human Intelligence

The first example of the cubic law we wish to investigate is provided by the 3000 years-old History of Astronomy. After Astronomy had arisen from obscurity in pre-historic times, the Ancient Greeks were the first to realize that the Earth revolves around the sun in the person of Aristarchus (floruit circa 290 B. C.) (peak of Greek Astronomy). Greek Astronomy, however, fell into decadence during the Roman Empire. Later still, the Dark Ages set on, and only in the Renaissance were the new and scientifically correct tracks of Astronomy rediscovered and first published by Copernicus (1543 A. D.). This section describes that string of events by virtue of a simple algebraic cubic curve, in the hope that it might be the starting point for deeper mathematical investigations by others future authors. Also, this section intends to be a tribute to Carl Sagan (1934-1996), whom this author had the honor and privilege to be acquainted to on the eve of the opening of the NASA SETI program, on October 11th, 1992, in Barstow, California. Carl Sagan's seminal book "Cosmos" (1980 - based on his 13-part television series, see the references) inspired to this author the mathematical model developed in this chapter. In particular, the figure on page 335 in Cosmos' Part XIII (entitled "Who Speaks for Earth?"), let this author think of "casting into some simple mathematical formula"

1) the achievements of ancient Greek Astronomy in the time of Aristarchus (290 B. C.), followed by
2) the Dark Ages of the Middle Ages (roughly 476 A. D. through 1492), and
3) finally followed again by the Copernican revolution (1543) and its aftermath, up to the Moon Landing and beyond, into our own future.

The simplest formula this author could conceive is just a cubic fit, as shown in Figure 1. The fit appears to be remarkable inasmuch as:

1) The inflection point of the cubic indeed corresponds to a turning point in history: the years (around 321 A. D.) when the Roman Emperor Constantine openly accepted the change in religion from Paganism to Christianity. Not a small change at all!
2) The bottom of the Dark Ages, as predicted by our cubic model, occurs around the year 932 A. D. This was indeed the time when the dissolution of the former Empire of Charlemagne reached its peak.
3) A few more notable coincidences are pointed out at the end of Section 3.

To a mathematician, the history of Astronomy is charming (apart from the charm of the topic itself!) as it clearly shows a peak (Aristarchus, 290 B. C.), then a fall (Dark Ages), and then a rise again (Copernicus, 1543, and beyond).

It is natural to think of some simple mathematical curve that would adapt to these three key features at best. The easiest fitting curve appears to be a cubic (i.e. an algebraic equation of the third degree) in which the time t obviously is the independent variable:

$$Cubic(t) = a_3 \cdot t^3 + a_2 \cdot t^2 + a_1 \cdot t + a_0. \tag{1}$$

This is just the most general algebraic equation of degree 3 with in t its four undetermined coefficients, a_3, a_2, a_1 and a_0.

Our task is then to find "intelligent formulae" for the coefficients a_3, a_2, a_1 and a_0. This may be done in different ways. In the first three sections of this chapter, we'll only describe a procedure that appears to us to fit the history of Astronomy at best:

1) We require that the cubic's maximum occurs at the time when Aristarchus firstly claimed (~290 B. C.) the solar system to have all planets revolving around the sun, and

2) We require that Copernicus rediscovered Aristarchus's result in the year 1543, the year of both his death and publication of his book "De Revolutionibus Orbium Coelestium".

Let us also agree about two conventions that we would like to adopt:

a) On the horizontal axis of the time, dates in the Christian Era are denoted by positive numbers, while the dates before Christ are denoted by negative numbers.

b) On the vertical axis, negative numbers are in correspondence to times when "the people regarded as true what nowadays is regarded by science as false". Then, in Astronomy, the cubic curve that we are seeking must have negative values for times smaller that –290 (times before Aristarchus) and after Aristarchus but before Copernicus (Dark Ages of Astronomy), while, after Copernicus, the cubic must be positive.

In so assuming, we have really requested our cubic to pass through two points:

1) The Aristarchus point

$$Maximum\_time \equiv Mt = -290 \tag{2}$$

2) The Copernicus point

$$Recovery\_time \equiv Rt = 1543. \tag{3}$$

Let us still look carefully at the graph shown in Figure 1. This is our way to "show" or "summarize" the history of Astronomy by virtue of the cubic (1). A glance to this graph reveals that:

1) Before Aristarchus all values of the cubic are negative, meaning that the truth, i.e. the Earth's 24-hours rotation and its revolution around the Sun in just one year, had not been understood yet. Aristarchus was the first to understand this around the year 290 B. C. The site http://www-gap.dcs.st-and.ac.uk/~history/Mathematicians/ Aristarchus.html, tells Aristarchus lived in between 310 and 230 B. C., and, according to the site http://www-gap.dcs.st-and.ac.uk/~history/Chronology/500 BC_1AD.html#290 he put forward his revolutionary theory around 290 B. C. Unfortunately, Aristarchus was not understood by his fellow countrymen. The Encyclopedia Britannica reports at the site http://www.britannica.com/eb/ article?eu=9551 that Cleanthes the Stoic declared that Aristarchus ought to be indicted for impiety. So, Aristarchus' greatest discovery hardly had any immediate influence of the development of Astronomy, and the whole world had to wait for 1833 more years for the truth to come out. Along our cubic curve, Aristarchus thus represents the "single point" with coordinates (Mt, 0). This is precisely the case where the cubic has two coinciding real roots and this root is a zero also, as we pointed out at the end of Section 2. In Jesus' time (0 – 33 A. D.), the decadence of the Greek civilization (negative values and negative derivative of the cubic) had begun already. This is historically correct, inasmuch as the Romans (since about 0 A. D. through 476 A. D.) were in essence culturally inferior to the Greeks.
2) After Aristarchus and before Copernicus (1543 A. D.) all the cubic's values are negative again. These were the "Dark Ages" of Astronomy, and, indeed, the Dark Ages of civilization as well. The bottom of the Dark Ages, our graph shows, was reached around the year 932, when Western Europe was lying in a state of virtually complete feudal anarchy. In addition, Vikings, Magyars and Muslims were freely raiding Western Europe. Some slow recovery from this havoc only began after 955 A. D., when Otto I (Holy Roman Emperor, i.e. German Emperor) defeated the Magyars at Lechfeld. The year 1000 A. D. is usually regarded as "when the rebirth of the Western civilization began".

Notice that, after the year 1000 A. D., the recovery of the West developed at a much faster pace than its own decadence had gone on. Actually, this pace was ***twice*** as much faster! This fact is faithfully mirrored by our cubic, though we cannot prove it mathematically here due to lack of space. After Copernicus, of course, all values of the cubic are positive. Giordano Bruno in 1582 was the first not only to fully accept the Copernican structure of the solar system, but also to understand that the sun is a just star as many, and that other live beings might well inhabit the planets revolving other stars (historic beginning of SETI). Bruno was burned at the stake in Rome on 17 February 1600 by the orders of the Roman Catholic Inquisition. But the truth could not be hidden any more. After Galileo, Kepler and Newton, science defeated all "religious beliefs".

3. What is The Real Meaning of the Cubic's Vertical Axis ("Intelligence") ?

So far, we have carefully avoided talking about the "meaning" of the cubic's vertical axis. Our convention b), where the vertical axis is mentioned, is not affected by the previous discussion, since the latter simply involves the sign of the positive vs. negative values of the vertical axis. So the natural question arises: "What is plot along the cubic's vertical axis?". Unfortunately (or fortunately!) this question may be given different answers according to the scientific research field that each author wants to investigate. This is a very important topic for the various possible applications of our "cubic recovery law". Some more answers are provided in Maccone 2004 (see the refencences), but, again, we cannot analyze these basic questions here due to lack of space. More future papers by this author will hopefully help by providing more applications to other fields of science like, for instance paleontology (recovery after the mass extinctions occurred on Earth 251 and 65 millions years ago, respectively).

4. Determining the Cubic in Terms of *Mt* and *Rt* Only

In this section we give, without any mathematical proof (that would take several pages), the anaytical expressions of the four coefficients a_3, a_2, a_1 and a_0 of the cubic (1) under the three hypotheses:

1) The time (or historical epoch) Mt of the maximum is assigned (i.e. is known).
2) The time (or historical epoch) Rt of the recovery is assigned (i.e. is known).
3) The cubic is normalized in units of falsity, that is, the ordinate of both Mt and Rt is zero and the ordinate of the minimum is, by assumption, -1. This normalization of all cubics to -1 is necessary in order to compare different cubics with one another, as we shall later see by comparing the cubic of Astronomy with the (different) cubic of SETI.

This author has proven mathematically that, under the three conditions above, the four coefficients of the cubic (1) are given by

$$\left[\begin{array}{l} a_0 = -\dfrac{27\,Mt^2\,Rt}{4(Rt-Mt)^3} \\ a_1 = \dfrac{27\,Mt(2Rt+Mt)}{4(Rt-Mt)^3} \\ a_2 = -\dfrac{27(Rt+2\,Mt)}{4(Rt-Mt)^3} \\ a_3 = \dfrac{27}{4(Rt-Mt)^3} \end{array}\right]. \tag{4}$$

The mathematical proof of the above equations is rather lengthy and will not be given here. It can be found in the 2004 paper by Maccone listed in the references list. As consequences of equations (4), one also finds the following results:

a) Time (or historical epoch) of the inflection point of the cubic:

$$t_{inflection} = \frac{2}{3} Mt + \frac{1}{3} Rt \tag{5}$$

with the corresponding ordinate given by –1/2 (because any cubic is and "odd" function around its own inflection point).

b) Time (or historical epoch) of the minimum of the cubic:

$$t_{minimum} = \frac{1}{3} Mt + \frac{2}{3} Rt. \tag{6}$$

with corresponding ordinate given by –1 by assumption (normalization in falsity units).

c) The decadence time span, defined as $t_{minimum} - t_{Maximum}$ is expressed in terms of *Mt* and *Rt* by

$$Decadence\_Time\_Span = \frac{2}{3} Rt - \frac{2}{3} Mt, \tag{7}$$

d) The recovery time span, defined as $Rt - t_{minimum}$ is expressed in terms of *Mt* and *Rt* by

$$Recovery\_Time\_Span = \frac{1}{3} Rt - \frac{1}{3} Mt. \tag{8}$$

It can be proven that

$$Recovery\_Time\_Span = \frac{Decadence\_Time\_Span}{2}. \tag{9}$$

e) We found useful to define the PreMaximum time, denoted $t_{preMaximum}$, as the time before the Maximum at which the *y* value is as low as the *y* value of the minimum will later be. In the language of the history of Astronomy, the preMaximum time is a time before Aristarchus at which the knowledge of Astronomy was as bad as it will later be only at the very bottom of the Dark Ages. Then, it can be proven that

$$t_{preMaximum} = \frac{4}{3} Mt - \frac{1}{3} Rt. \tag{10}$$

5. History of Astronomy Normalized in Units of the Worst Dark Ages

An immediate application of (4) is of course the cubic of Astronomy, now normalized to –1. This cubic is defined by only two conditions: the Aristarchus point and the Copernicus point, respectively located on the time axis at the two values

$$Mt = -290, \quad Rt = 1543. \tag{11}$$

No further condition is requested inasmuch as the vertical axis is now normalized in units of falsity (1 unit of falsity = the false ideas about Astronomy they had in the worst Dark Ages, i.e. around 932 A. D.). Numerical calculations then yield:

a) Transition from Paganism to Christianity (Emperor Constantine)

$$t_{inflection} = 321. \tag{12}$$

b) Worst Dark Ages, $t_{minimum} = 932.$ (13)

c) How far back in ancient Greek history one has to go in order to find as many false ideas about Astronomy as they had in the Worst Dark Ages

$$t_{preMaximum} = -901. \tag{14}$$

d) How many years it took for Astronomy to reach the worst level of misunderstanding after the truth had first been found by Aristarchus

$$Decadence\_Time\_Span = 1222. \tag{15}$$

e) How many years it took to Astronomy to rediscover the truth since the worst Dark Ages

$$Recovery\_Time\_Span = 611. \tag{16}$$

f) Actual coefficients of the normalized to –1 cubic of Astronomy:

$$\begin{bmatrix} a_0 = -0.14223 \\ a_1 = -8.88692 * 10^{-4} \\ a_2 = -1.05546 * 10^{-6} \\ a_3 = 1.09602 * 10^{-9} \end{bmatrix}. \tag{17}$$

g) We would like to complete this section by two historical remarks. The first one is summarized by the Latin sentence "Graecia capta ferum victorem coepit" i.e. "The captured Greeks indeed captured (by culture) their own ferocious conquerors (the Romans)". This said the Roman poet Horace (65 B. C. - 8 B. C.), and he was right! He meant that the Greeks always were culturally superior to the Romans, even close to Jesus' time, when Horace lived. And our cubic law just quantifies this statement: "The Greeks were 17% culturally superior to the Romans already in Jesus Christ's and Horace's time, and much more in later times still", as one can find by applying the two equations

$$\left[t_{Jesus\_death} = 33 \quad cubic\_for\_Astronomy = -0.17266 \right] \tag{18}$$

h) The second remark is that the Copernicus point might actually correspond to the year 1514, rather than to 1543. This is because, according to the University of St. Andrew's web site http://www-gap.dcs.st-and.ac.uk/~history/Mathematicians/ Copernicus.html Copernicus actually put forward his revolutionary ideas for the first time in 1514, rather than in 1543, in a handwritten (not printed) work called "Little Commentary". It would be interesting to develop a "theory of errors" in this regard, but we do not have space to do so here.

6. The History of SETI as a cubic

Let us now change the scenario completely.

In this section we claim that also the history of SETI (the modern Search for ExtraTerrestrial Intelligence) may basically be cast in a cubic shape. SETI, in fact, had its true historic beginnings back in 1582, when, in his work "De l'Infinito, Universo e Mondi" ("About Infinity, the Universe and the Worlds"), the Italian scholar Giordano Bruno (1548-1600), showed to have fully realized that:

a) The Copernican System was the right one.
b) Our Sun is just one star as many; it looks much brighter only because it is much closer to us.
c) Just as we know planets to orbit our Sun, so we might infer that other planets could possibly orbit other stars (a fact not proven scientifically until 1995, when Michel Mayor and Didier Queloz of Geneva discovered the first extrasolar planet around the star 51 Pegasi).
d) Just as our planet is inhabited by humans and animals, so the other planets might host living extraterrestrials. Because of these great conceptual achievements by Bruno, we claim that the Maximum of the Cubic of SETI is given by

1) The Giordano Bruno point: $Maximum\_time \equiv Mt = 1582$. (19)

   Bruno was the Aristarchus of SETI. He was burned at the stake by the orders of the Roman Catholic Inquisition on 17 February 1600 in Rome (in Campo dei Fiori, where his Memorial now stands). At least in part, he died at the stake because his doctrines were much of a blow to the ruling Roman Catholic establishment of his days. Ironically, the name of the Pope who had Bruno burned was Clemens VIII, and, in Latin, Clemens means "the mercyful" !

2) The Cocconi-Morrison point: $Recovery\_time \equiv Rt = 1959$. (20)

   Giuseppe (misspelled Guiseppe in many textbooks) Cocconi (1914-) is an Italian physicist of the school of Enrico Fermi, now retired director of the CERN proton syncrotron in Geneva. Phil Morrison (1915-2005) was physics professor at MIT. In 1959, Cocconi and Morrison published a seminal paper proposing the potential of microwaves in the search for interstellar communications. That paper marked the beginning of modern, experimental SETI, especially since in 1960 Frank Drake first searched two nearby stars (ε Eridani and τ Ceti) for possible "intelligent signals" around 1.420 GHz.

By virtue of the two conditions (19) and (20), the four equations (4) thus yield a new cubic that we call the Cubic of SETI and is plot in Figure 2.

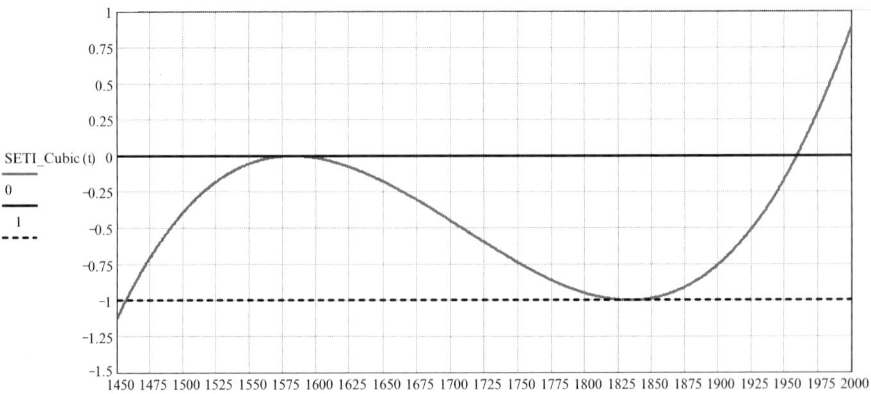

Figure 2. History of SETI as a Cubic normalized to –1, i.e. in Falsity Units.

To see how this cubic "makes sense", consider that (6) yields $t_{minimum} = 1833$. (21)

These were indeed the "Darkest Ages" of the Restoration in Europe, namely the 33 years elapsed in between the fall of Napoleon (1815) and the European Revolution of 1848. (Note: we do not regard the Parisian Revolution of July 1830 as very significant. In fact, it brought changes only to the history of France, not to the history of the whole Europe, as the subsequent Revolution of 1848 did). In other words, since Giordano Bruno's death in 1600, strong persecution by the Roman Catholic Church, as well as by most other national churches, against claims like "we are not alone in the universe" forced any "reasonable person" to keep silent about any claim the existence of any ETs at all. On the other hand, the existence of ETs could not possibly be proven scientifically until about 1950, and so the long sleep of the search for ETs lasted about 350 years.

Indeed, the decadence time span, given by (15), for SETI is, in years

$$Decadence\_Time\_Span = 251, \tag{22}$$

the SETI inflection point, by (5), falls at $t_{inflection} = 1707$ (23)

and, finally, the recovery time span, by (16), is $Recovery\_Time\_Span = 125.$ (24)

The preMaximum time falls at $t_{preMaximum} = 1456$ (25)

(Renaissance leading to a new conception of the universe), with actual coefficients of the SETI cubic

$$\begin{bmatrix} a_0 = -617.6282 \\ a_1 = 1.0961 \\ a_2 = -6.45363 * 10^{-4} \\ a_3 = 1.25974 * 10^{-7} \end{bmatrix}. \tag{26}$$

7. Extrapolating the Two Cubics up to the Warp Drive Invention (2067 A. D. ?)

So far, we only fit two cubics to the human history of past centuries. This is interpolation. Now we try to have a look at our future, and must thus do extrapolation. To this end, let us plot both the cubics of Astronomy and SETI on the same plot (Figure 3). Notice that only "the right branch" of the cubic of Astronomy appears in this plot because the time span only starts at the year 1450 A. D., just as in Figure 2. Also, the plot extends now up to the end of the present century, i.e. to 2100 A. D.

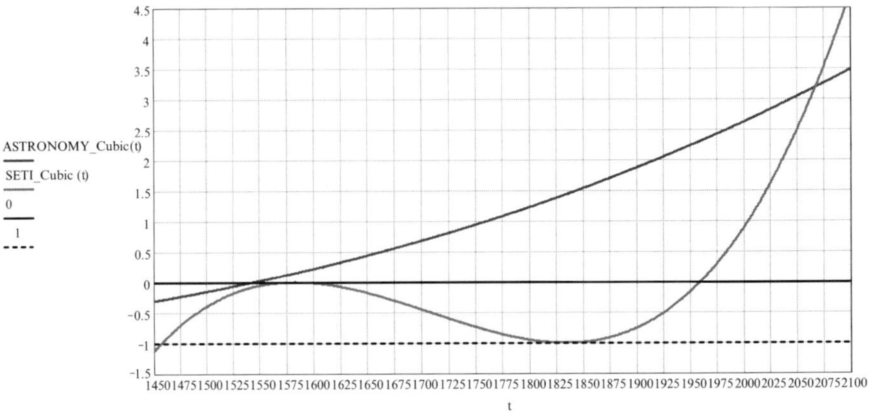

Figure 3. Combining the two Cubics of Astronomy and SETI. Finding their intercept.

One notices immediately that, after the 1959 year of Cocconi and Morrison, the SETI cubic is much "steeper" than the Astronomy cubic. In other words, SETI is developing much "faster" (in importance) than the rest of "traditional" Astronomy. The prediction is that SETI will become even more important than Astronomy itself after the year 2067 A. D., that is the abscissa of the intercept between the two cubics in the graph above.

Does this prediction of our theory make sense? Well, nobody can predict the future, of course. But… if we "are to believe" the Star Trek TV series… well the year 2063 will be extremely significant! It will be the year when Zephram Cochrane (born in 2030) invents the Warp Drive and pilots Earth's first faster-than-light space flight (see the references).

Science fiction only? May be not. In the years 1996-2002, NASA's Breakthrough Propulsion Physics Program (BPP – web site http://www.grc.nasa.gov/WWW/bpp/) was the first NASA forum to let experts in Relativity and Quantum Field Theory study the possibilities of faster-than-light (FTL) space flight. Not to mention other "photon-state teleportation" experiments that seem to be challenging Special Relativity itself!

But let us go back to Figure 3. To take a few final, bold steps of extrapolation into the future, we now assume that variable on the vertical axis represents the "distance from Earth reached by humans at various epochs". Under this assumption, the ordinate corresponding to the year 1969 must then be the Moon distance, i.e. 384,000 km. Having calibrated the two cubics in this fashion, when will humans reach the nearest star, α Centauri, at the distance of 4.29 light years? Here are some answers:

1) According to the (slow) Cubic of Astronomy, humans will reach α Centauri in 612,810 years;
2) According to the (steeper = faster) Cubic of SETI, humans will reach α Centauri in 55,803 years.

But these are rather "discouraging" predictions! More realistically, then, we may replace the Moon distance of 1969 by the Voyager 1 distance of 2004 (i.e. we may consider robotic probes instead of manned flights) and re-calibrate the vertical axis accordingly. In this case the results are more encouraging:

3) According to the (slow) Cubic of Astronomy, robotic probes will reach α Centauri in 18,220 years;
4) According to the (steeper) Cubic of SETI, robotic probes will reach α Centauri in 4,377 years.

We are now reaching the ultimate conclusion. We now claim that only the result 4) is the right one! In other words, we claim that human robotic probes will reach α Centauri in 4,377 years, or in about 4,000 years from now. The reason for our claim is that the cubic of SETI is so steep that it shows SETI will overcome general Astronomy in the public interest at least from the year 2060 onward. It is already so now (2005) to some extent: lay people "believe in UFO" (incorrectly, of course, by scientific standards), thus revealing that the "need for contact" with alien civilizations already is "in the air" and "in the lay people's mind". Also, as of the year 2005, both NASA and ESA are planning space missions intended to find ***other Earths*** outside the solar system. These two missions (robotic high-tech telescopes to be located at the Lagrangian point L2 of the Earth-Sun system) are called TPF (Terrestrial Planet Finder) and Darwin, respectively. They are scheduled for launch around 2010. When other Earths will have been found, humans will then hardly refrain from sending robotic probes towards them. Thus, we may well conclude that the evolution of human intelligence on Earth has reached an unprecedented peak, and is actually paving the way to the human expansion into Space.

8. A glance to the Future of Our Mathematical Model of the Evolution of Intelligence

We tried to model the Evolution of Intelligence on Earth by two cubics of the time. But which differential equation does a cubic fulfill? Just differentiate the cubic (1) twice:

$$\frac{d^2\, Cubic(t)}{dt^2} = 6 \cdot a_3 \cdot t + 2 \cdot a_2. \tag{27}$$

From this equation, one might naïvely infer that "The force behind progress increases linearly in time". This conclusion may be too simplistic, though. A more serious mathematical model would imply regarding the cubic (1) as the (deterministic) mean value, $Cubic(t) = \langle X(t) \rangle$, of some unknown stochastic process $X(t)$, that one would then try to determine by more advanced mathematical techniques than the simple cubics we used here. This endeavor, however, has to be deferred to a further, more profound mathematical study.

9. Conclusion

Our feeling is that we just skimmed the (mathematical) surface of some more profound theory. This theory would embrace key features in human history and sociology, as well as in the "geography" of Space, and bind them together inside a single mathematical structure. In other words, here we made just a beginning to investigate the Evolution of Human Intelligence by mathematical tools updated to the coming age of the conquest of Space.

10. Acknowledgements

The author is grateful to Dr. Steve Dick, NASA Chief Historian and Director of the NASA History Office at NASA Headquarters, for reading the preliminary version of the mathematical theory developed in this chapter, and for letting the author know about his opinions. An anonymous Referee also made constructive comments about the first version of the manuscript that were very helpful for the author's final revised version. Finally, the author would like to warmly thank Prof. Joseph Seckbach, Editor of this book, for his patience and encouragement while the author was amidst some physical and moral difficulties. Thank you, Joseph, for your magnificent editorial work!

11. References

Finney, Ben R. and Jones, Eric M., editors. "Interstellar Migration and the Human Experience". Proceedings of the Conference on Interstellar Migration, Los Alamos, May 1983. University of California Press, 1985.

Maccone, Claudio, "A Mathematical "Cubic Law of Recovery": Part 1 – Applications to History of Astronomy, SETI and Modern Europe", Frontier Perspectives (published by Temple University, Philadelphia, PA, USA), Volume 13, Number 2, Fall/Winter 2004, pages 22-33.

Okuda, Michael and Denise, "Star Trek Chronology – The History of the Future", Pocket Books, New York, 1996. See, in particular, pages 26 and 28.

Sagan, Carl, "Cosmos", Random House, New York, 1980, ISBN 0-394-50294-9 (hard cover), ISBN 0-394-71596-9 (paperback). See in particular the unnumbered figure on page 335 that inspired much of this chapter.

Biodata of **Christopher P. Mckay** author of ***"Searching other Worlds for Life as we Know It"***

Dr. Christopher P. McKay (Chris) is a planetary scientist with the Space Science Division of NASA Ames (Moffett Field, CA). He received his Ph.D. in AstroGeophysics from the University of Colorado in 1982 and has been a research scientist with the NASA Ames Research Center since that time. Chris' current research focuses on the evolution of the solar system and the origin of life. He is also actively involved in planning for future Mars missions including human settlements. Chris has been involved in research in Mars-like environments on Earth, traveling to the Antarctic dry valleys, Siberia, the Canadian Arctic, and the Atacama Desert in Chile to study life in these Mars-like environments.

E-mail: **cmckay@mail.arc.nasa.gov**

J. Seckbach (ed.), Life as We Know It, 373-381.

SEARCHING OTHER WORLDS FOR LIFE AS WE KNOW IT

CHRISTOPHER P. MCKAY
Space Science Division, NASA-Ames Research Center, Moffett Field, CA 94035-1000, USA.

1. Introduction

"Follow the water" has been the hallmark of NASA's strategy for the search for life on other worlds. But now on two worlds we have found the water and must consider what to do next. Clearly, the next step is to obtain organic material from these worlds that is likely to be remnants of past life. Here I suggest, that if we analyze this organic material in detail we may be able to distinguish between biotic from abiotic material based on the observation that biology uses a small select number of structural molecules.

The two worlds on which we have evidence for water are Mars and Europa. On Mars there is now extensive evidence from orbital data that liquid water was present and stable on the surface for geologically significant periods of time. In addition, the two MER rover missions have found geomorphological and geochemical evidence for the past presence of liquid water on the surface of Mars (Squyres et al. 2004). Europa – a moon of Jupiter – has surface features that indicate the presence, at some time, of liquid water beneath the icy surface (Kargel et al. 2000). The magnetic disturbances caused by Europa's passage through the Jovian fields indicate the presence now of a global conductor presumed to be a slightly saline ocean.

After finding the water we must search for signs of life. What is life? A concise definition still eludes us. Many definitions of life as a general phenomenon are a list of properties. Koshland (2002) lists seven features: 1) program (eg. DNA), 2) improvisation (novel responses to the environment), 3) compartmentalization, 4) energy, 5) regeneration, 6) adaptability, and 7) seclusion (chemical control and selectivity). Davies (1999) has a similar list. Perhaps the most common definition of life is a physical system that undergoes Darwinian evolution, which according to Chao (2000) is originally due to Muller (1966). Schrödinger (1945) defined life in the context of thermodynamics as "It feeds on negative entropy". Cleland and Chyba (2002) had argued that life is like water. They point out that water has a complex phenomenology but when fully understood it can be simply and precisely defined as two hydrogen atoms jointed to one oxygen atom. They propose that life is like water and when we fully understand life, a definition will be evident. However, we might instead propose that life is not like water but is like another of the Greek elementals: fire. Even with our modern understanding of statistical thermodynamics and quantum processes we cannot easily define fire. For example: Is the sun a fire? Is the element of a light bulb on fire? Is an explosion of TNT a fire? The difference between fire and water is that fire is a process – not a pure substance. It is much more likely that life is like fire and not like water. Even with a complete understanding of its underlying principles, many independent examples, and the ability to create it in the laboratory, we probably will still not be able to offer a concise definition of life.

In the search for life on other worlds it is not clear that a definition of life is required or even helpful. More useful might be clarification of what constitutes evidence of life. There are several ways to search for life. First we can be searching for life as a collective general phenomenon. However life might also be a single isolated organism. And that organism might be dead. Finally, signs of life may be fossils, artefacts or other inorganic structures. In the search for life on other worlds, any of these would be of interest.

Definitions of life typically focus on the nature of the collective phenomenon. In general, such definitions are not useful in an operational search for life on other worlds. The one exception is the proposal by Chao (2000) to modify the Viking Labelled Release (LR) experiment to allow for the detection of organisms that improve their capacity to utilize the provided nutrients. This would in principle provide a direct detection of Darwinian evolution and could unambiguously distinguish between biological metabolism and chemical reaction. Chao (2000) argues that Darwinian evolution is the fundamental property of life and other observables associated with life result from evolutionary selection. His method for searching for evolution would be practical if the right medium can be selected to promote the growth of alien microbes. Unfortunately, we now know that only a tiny fraction, $<0.1\%$, of microorganisms from an environmental sample grow in culture. This was not known at the time of the design of the Viking biology experiments, which were essentially culture experiments. The fact that most soils on Earth will grow up in a culture media is due to the vast diversity of soil microbes in these soils and not to the robustness of culturing as a way to detect organisms. We also now know that there are soils on Earth (e.g. Atacama Desert in Chile) where there are bacteria present in low numbers but nothing grows in any known culture media.

Of course growth experiments of any kind do not detect dead organisms. Yet the remains of dead organisms are potentially important evidence of life on another planet. And so are fossils. However, there is an important distinction between dead organisms and fossils. A fossil is evidence of past life but it does not reveal anything about the biochemical or genetic nature of that life. If we are searching for a second example of life then we need to be able to compare the nature of that life to Earth life. For this an organism is needed, either dead or alive, but a fossil is not sufficient.

2. Mars and Europa

The quick and the dead, shells and footprints are all signs of life and we can search for them all on Mars and Europa. However, a realistic assessment of what is possible on near term missions suggests that the organic remains of past life are the most promising target for a search for signs of alien life on either Mars or Europa.

On Mars, the search for biological remains of past life is focused on the subsurface. The surface appears to be lifeless with one or more oxidants that actively destroy organic material. Deep below the surface there may be liquid water and extant life forms (Boston et al. 1992), but this is unknown at the present time and there are no specific locations where one could search for such ecosystems. In contrast, the deep permafrost on Mars may hold remnants of past life (Smith and McKay 2005). The southern hemisphere of Mars contains heavily cratered surfaces presumably 3.8 Gyr old. In addition, poleward of about 60°, ground ice was detected in the top meter of the surface (Feldman et al. 2002). There is geological evidence (Squyres and Carr 1986) from crater morphologies that ground ice extends to much greater depths (km or more). In parts of this ancient ice-rich ground there are present strong crustal magnetic features (Acuña et al. 1999). These features indicate that the ground has not been sufficiently heated or shocked to destroy the magnetic signature. Smith and McKay (2005) identify such a location (near 76S, 195W, the target site for the ill-fated Mars Polar Lander in 1998) as a potential site for deep drilling for remnants of ancient life preserved in the 3.8 Gyr old ground ice. The organisms in the ground ice are likely to be dead from accumulated radiation dose but their organic remains could be analyzed and compared to the biochemistry of Earth life.

On Europa the near term target for a search for life is the surface. It is generally thought that below the surface ice on Europa there is an ocean (Kargel et al. 2000). Direct experimentation in that ocean is not likely in near term missions. However, it is plausible to consider examination of surface deposits in locations where the ocean water may reach the surface. The linear features on the surface of Europa are generally thought to be cracks in the ice and may be locations of ocean water reaching the surface, although it is not yet certain that this is the case. If there is life in the ocean of Europa then organic remains of that life may be present at the surface cracks. Due to the high radiation dose received from Jupiter, it is unlikely that any organisms are alive at the surface but their organic remains could persist. Eventually the high radiation flux would destroy any biological signature in the organic remains as well. The net radiation flux required to destroy the signature of life in organic material is unknown but such destruction is likely to be occurring on Europa given the high radiation flux (Cooper et al. 2001). Thus ocean material that has come onto the surface relatively recently (<100,000 yr) would be desired. This non-living material would be tested for organics that would show signs of biological origin.

I have argued previously (McKay 2004) that one way to determine if a collection of organic material is of biological origin, is to look for a selective pattern of organic molecules similar to, but not necessarily identical with, the selective pattern of biochemistry in life on Earth.

3. A Menagerie of Biochemistries: Phase Space

Life on Earth is a phenomenon of one. At the structural levels (the hardware) there is a common building kit for life. The proteins are constructed from combinations of 20 L-amino acids, the nucleic acids are composed of five nucleotide bases (A, T, C, G, and U), and the third class of large polymers in biology, the polysaccharides are composed of strings of sugars. These biopolymers together with a few other molecules are the structural molecules used in all life on Earth. At the genetic level (the software), life shares a common genetic code illustrated by the "tree of life" that shows the phylogenetic relationship of all Earth life.

Pace (2001) has argued that life everywhere will be life as we know it. He contends that the biochemical system used by life on Earth is the optimal one and therefore evolutionary pressure will cause life everywhere to adopt this same biochemical system. It is instructive to consider this argument in the context of a conceptual organic phase space. If we imagine all possible organic molecules as the dimensions of a phase space, then any possible arrangement of organic molecules is a point in that phase space. We can define biochemistries as those points in phase space that allow for life. The biochemistry of Earth life – life as we know it – represents one point in the organic phase space: we know that this one point represents a viable biochemistry. Pace's (2001) contention that biochemistry is universal is equivalent to stating that in the region of phase space of all possible biochemistries there is only one optimum biochemistry and thus any initial set of biochemical reactions comprising a system of living organisms will move toward that optimum as a result of selective pressure. If Pace (2001) is correct then the only variation between life forms that we can expect is that associated with chirality. As far as is known, the L and D forms of chiral organic molecules (such as amino acids and sugars) have no differences in their biochemical function. Life is possible that is exactly similar in all biochemical respects to life on Earth except that it has D instead of L amino acids in its proteins and L instead of D sugars in its polysaccharides.

We do not know the nature of biochemical phase space and it may well be that there is a single global optimum. However, it may also be that biochemical phase space is complex with regions that are not even connected to each other and with multiple local optima. In this case the biochemical system of a life form would depend on its initial point in the phase space. Evolution would bring it to the local optimum but this could vary between life forms. There would be a menagerie of possible biochemistries for life. The nature of the biochemical systems in these separate local optima is likely to be radically different and thus it may be beyond our current knowledge to speculate on the nature of these alternative biochemistries based only on observations of our own biochemistry and small perturbations about our own biochemistry.

The question of the number of possible biochemistries consistent with life is an empirical one and can only be answered by observations of other life forms on other worlds, or by the construction of other life forms in the laboratory. The observation or construction of even one radically alien life form would suffice to show that biochemistry as we know it is not universal.

The pattern of biochemistry of Earth life follows what I have called the "Lego principle" (McKay 2004). Life uses a small set of molecules to construct the diverse structures that it needs. This is similar to Legos in which a few different units repeated over and over again are used to construct complex structures. The biological polymers that construct life on Earth are the proteins, the nucleic acids, and the polysaccharides. These are built from repeated units of the 20 L amino acids, the 5 nucleotide bases, and the D sugars. The use of only certain basic molecules allows life to be more efficient and selective. Evolutionary selection on life anywhere is likely to result in the same selective use of a restricted set of organic molecules. As discussed above, it is premature to conclude that all life anywhere will use the same set of basic biomolecules. Thus I suggest that life will always use some basic set but it may not be the same basic set used by life on Earth. This characteristic biogenic pattern of organic molecules would persist even after the organism is dead. Given our present state of understanding of biochemistry, we are not able to propose alternative and different biochemical systems that could be the basis for life, but that may reflect a failure of our understanding and imagination rather than a restriction on the possibilities for alien life.

4. Searching for Life as We Know It

A sample from the deep permafrost in the southern hemisphere of Mars or from the crack features on the surface of Europa could be analyzed for organic material with a fairly simple detection system. If organic material was detected then it would be of interest to characterize any patterns in that organic material that would indicate a "Lego principle" pattern. Clearly one such pattern is the identical pattern of all Earth life; 20 L amino acids, the five nucleotide bases, A, T, C, G, and U. etc. However, more interesting would be a clear pattern different from the pattern known from Earth life. Figure 1 shows a schematic diagram of how a biological pattern would be different from a non-biological pattern.

Implementing this search in practical terms in near term missions will require a sophisticated ability to separate and characterize organic molecules. Currently the instrument best suited for this task is a GCMS with solvent extraction. However, new methods of fluorescence and Raman spectroscopy could provide similar information and may have a role in future mission applications.

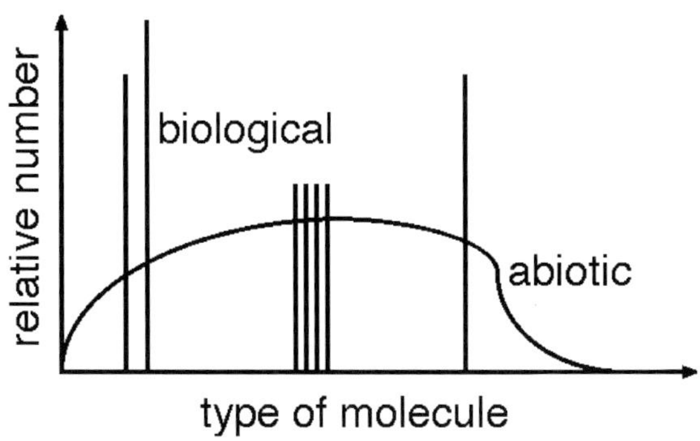

Figure 1. Schematic diagram of the distribution of organic molecules from biotic and abiotic sources. Abiotic sources results in smooth, not necessarily symmetric, distributions while biotic distributions are a series of spikes. (from McKay 2004).

5. Conclusion

The next step in the search for life on Mars and Europa is likely to be based on the detection and characterization of organic material. Natural selection has resulted in life on Earth specializing in the use of certain organic molecules in the construction of biomass. This property is likely to be common to any life form that has developed by natural selection. Thus the characterization of organics obtained from Mars and Europa could provide a general and practical way to search for life on these worlds.

6. Acknowledgements

This work was supported by subventions from the NASA Astrobiology program. I thank the reviewers for useful comments that improved the paper.

7. References

Acuña, M.H., Connerney, J.E.P., Ness, N.F., Lin, R.P., Mitchell, D., Carlson, C.W., McFadden, J., Anderson, K.A., Reme, H., Mazelle, C., Vignes, D., Wasilewski, P. and Cloutier, P. (1999) Global distribution of crustal magnetism discovered by the Mars Global Surveyor MAG/ER experiment. *Science* **284**, 790-793.

Boston, P.J., Ivanov, M. and McKay, C.P. (1992) On the possibility of chemosynthetic ecosystems in subsurface habitats on Mars. *Icarus* **95**, 300-308.

Chao, L. (2000) The meaning of life. *Bioscience* **50**, 245-250.

Cleland, C.E. and Chyba, C.F. (2002) Defining Life. *Origin Life Evol. Biosph.* **32**, 387-393.

Cooper, J.F., Johnson, R.E., Mauk, B.H., Garrett, H.B. and Gehrels, N. (2001) Energetic Ion and Electron Irradiation of the Icy Galilean Satellites. *Icarus* **149**, 133-159.

Davies, P.C.W.D. (1999) *The Fifth Miracle: The Search for the Origin and Meaning of Life.* Simon and Schuster, N.Y.

Feldman, W.C., Boynton, W.V., Tokar, R.L., Prettyman, T.H., Gasnault, O., Squyres, S.W. Elphic, C., Lawrence, D.J., Lawson, S.L., Maurice, S.G., McKinney, W., Moore, K.R. and Reedy, R.C. (2002) Global distribution of neutrons from Mars: results from Mars Odyssey. *Science* **297**, 75-78.

Kargel, J.S., Kaye, J.Z., Head, J.W., Marion, III, G.M., Sassen, R., Crowley, J.K., Ballesteros, O.P., Grant, S.A. and Hogenboom, D.L. (2000) Europa's Crust and Ocean: Origin, Composition, and the Prospects for Life, *Icarus* **148**, 226-265.

Koshland, D.E., Jr. (2002) The seven pillars of life. *Science* **295**, 2215-2216.

Levin, G.V. and. Straat, P. A. (1981) A search for a nonbiological explanation of the Viking labeled release life detection experiment *Icarus* **45**, 494-516.

McKay, C.P. (2004) What is life and how do we search for it in other worlds? *PLoS Biol.* **2**, 1260-1263.

Muller, H.J. (1966) The gene material as the initiator abnd the organizing basis of life. *American Naturalist* **100**, 493-517.

Pace, N. (2001) The universal nature of biochemistry. *Proc. Natl. Acad. Sci U.S.A.* **98**, 805-808.

Schrödinger, E. (1945) *What is Life?* Cambridge University Press, Cambridge.

Smith, H.D. and McKay, C.P. (2005) Drilling in ancient permafrost on Mars for evidence of a second genesis of life. *Planet. Space Sci.,* in press.

Squyres, S.W., Grotzinger, J.P., Arvidson, R.E., Bell, J.F., Calvin, III, W., Christensen, P.R., Clark, B.C., Crisp, J.A., Farrand, W.H., Herkenhoff, K.E., Johnson, J.R., Klingelhöfer, G., Knoll, A.H., McLennan, S.M., McSween, H.Y., Morris, .Jr., R.V., Rice, J.W., Rieder, Jr., R. and Soderblom, L.A. (2004) In Situ Evidence for an Ancient Aqueous Environment at Meridiani Planum, Mars. *Science* **306**, 1709-1714.

Squyres, S.W. and Carr, M.H. (1986) Geomorphic evidence for the distribution of ground ice on Mars. *Science* **231**, 249-252.

Biodata of **Josep M. Trigo-Rodríguez** author of the chapter ***"The role of comets and meteorites in the Origin of Life"***

Dr. Josep M. Trigo-Rodríguez is currently postdoctoral researcher of the Institute of Geophysics and Planetary Physics in the University of California Los Angeles (UCLA). He also belongs to the Astrobiology Center of UCLA. He obtained his Ph.D. from the University of Valencia (Spain) in 2002 and continued his studies and research as associated professor at the University Jaume I of Castellón, Spain. The Spanish State Secretary of Education and Universities granted him for a postdoctoral stay in UCLA. Dr. Trigo-Rodríguez scientific interests are in the areas of: evolution of the solar system, formation of planets and minor bodies: comets and asteroids, meteor and fireball interaction with the atmosphere, and origin of planetary atmospheres and life.

E-mails: **jtrigor@ucla.edu ; trigo@exp.uji.es**

J. Seckbach (ed.), Life as We Know It, 383-397.

THE ROLE OF COMETS AND METEORITES IN THE ORIGIN OF LIFE: *Survivability of organic matter during its transit through the atmosphere of the Earth.*

JOSEP M. TRIGO-RODRÍGUEZ
Astrobiology Center, Institute of Geophysics and Planetary Physics, University of California Los Angeles. Los Angeles, California, USA.

1. Introduction

Collisions of other solar system bodies (with a wide range of sizes) with the terrestrial planets have occurred since the planets first formed. The delivery and production of organic and volatile compounds during these events probably defined the course of planetary evolution. During the first 700 My of our planet's existence the impact of kilometer-sized bodies were frequent although the flux probably decreased nearly exponentially in intensity (BVSP, 1981). Such bodies are called *planetesimals* because they can be considered infinitesimal building blocks of the present planets. This intense stage known as the *heavy bombardment* remains sculpted in the surfaces of some planetary bodies, especially those with a low-density atmosphere like the Moon, Mercury or Mars. At the present time, these planetary surfaces partially preserve the ancient cratering record (Halliday, 2004; McSween, 2004; Taylor and Scott, 2004; Warren, 2004). The impact craters preserved in these bodies are a reminder of this early violent period in the evolution of the solar system (see e.g. Melosh, 1989). From this record we know that at least ten collisions producing lunar craters similar in magnitude to the basins Imbrium and Orientale occurred in the last part of this violent period (Taylor, 1982; Hörz et al., 1991).

Cometary bodies rich in volatiles and organic matter probably played an important role in the delivery of the building blocks of life to the Earth (Chamberlin and Chamberlin, 1908; Oró, 1961; Clark, 1988; Oró et al., 1990; Pierazzo and Chyba, 1999). In fact, the composition of the lunar regolith shows that about 1.5-2% is CI chondritic in composition (Anders et al., 1973; Warren et al., 1989; Haskin and Warren, 1991; Warren, 2004). This component is not observed in terrestrial rocks mainly because of the recycling of the crust due to erosion, sedimentation and plate tectonics. Although it may seem that the chemical memory of the projectiles was lost through atmospheric breakup or erosion and plate tectonics effects, studies of the abundances of noble gases in the atmosphere (Pepin, 1991), and the deuterium/ hydrogen ratio in oceanic water suggest that some clues would remain (Pavlov et al., 1999; Delsemme, 2000).

The Earth probably was derived in large part from volatile-poor meteorites that formed in the inner region of the solar nebula (see e.g. Wasson and references therein). The fast growth and likely migration of the giant planets to their present positions induced gravitational perturbations that produced an orbital scattering of comets (Ip & Fernández, 1988; Delsemme, 1992, 2000). Some comets remained as dark icy bodies stored in the Öort cloud and the Kuiper belt, but most were dispersed by energetic gravitational impulses through the outer solar system (Ip & Fernández, 1988). Probably only a small part, those dispersed inward, reached the Earth although no direct record of the earliest impacts has been preserved on the surface of our planet. It seems likely that chemical diversity in comets could explain the Earth's D/H oceanic ratio (Laufer et al., 1999; Delsemme, 2000).

This chapter is divided into four sections and the conclusions. We will focus on addressing the role carbon-rich bodies such as comets or carbonaceous chondrites had in the origin of life on the Earth. Section 2 discusses our present knowledge of the evolution of the terrestrial atmosphere and how its composition was determined by the delivery of organic matter to the Earth from exogenous sources. Section 3 deals with the study of the atmospheric interaction of meteoroids by using meteor spectroscopy. This method is a direct source of valuable information on the processes of the interaction of interplanetary matter with the terrestrial atmosphere; it also serves as a source of data on the composition of minor bodies. Typical asteroidal and cometary meteoroids larger than about 50 μm enter the terrestrial atmosphere at high velocities and are partially volatilized, thereby producing meteor phenomena. Recent advances in this field, particularly in meteor spectroscopy (Borovicka, 1993; 1994) have allowed us to obtain valuable relative chemical abundances of meteoroids from the chemical composition of the meteoric column (Borovicka et al., 1999; Trigo-Rodríguez et al., 2003). In consequence, it makes sense to use meteor spectroscopy to derive as much information as possible from known objects that may never be visited by spacecraft.

Several tentative models have been advanced to explain the mass of volatiles delivered by comets (Matsui and Abe, 1986; Ip and Fernández, 1988; Chyba et al., 1990; Chyba and Sagan, 1992; Delsemme, 1992, 1998, 2000). In general the different models bring at least one order of magnitude more water, and three orders of magnitude more gases and organics than presently occurs on Earth. This result is consistent with the present terrestrial composition because losses of volatiles are expected due to large and frequent impacts during the heavy bombardment period. Since the appearance of the work of Barak and Bar-Nun (1975), there has been a general consensus that pre-existing organic compounds would be completely dissociated in impacts although the produced radicals would later participate in the formation of new organic materials in the terrestrial environment. The reason is that most high-velocity cometary encounters produce energetic collisions where the impactor reaches a temperature high enough to destroy the bonds of volatiles and organic compounds. However, several authors have recently suggested that this problem is not clearly solved; they maintain that simulations and additional laboratory experiments are required (Blank et al., 2001; Jenniskens et al., 2004a). Finally, section 4 tries to compile evidence about the possibility that the heating processes involved in the continuous accretion of IDPs and meteoroids provide natural pathways for delivering organics and volatile compounds to the Earth.

2. Terrestrial Atmosphere during late Accretion

The composition of the early atmosphere should be studied in the context of a continuous infall of bodies decreasing in size as a function of time as seen in the lunar cratering record. In order to do this we need to increase our knowledge on the processes of accretion of minor bodies to the Earth. The physics and chemistry of collisions and catastrophic break-up in the atmosphere of comets, asteroids and meteoroids can be a key factor in understanding better the evolution of the earliest terrestrial environment. Oparin (1924) was one of the first authors to discuss the composition of the primeval atmosphere, and its implications for the origin of life. He proposed a strongly reducing atmosphere rich in hydrogen (H_2), ammonia (NH_3), methane (CH_4) and other hydrocarbons. He suggested that the reactions between these compounds would lead to increased complexity in organic compounds, which rained into the primordial ocean. Unfortunately, the available data in those days were unable to bring more clues to this topic. The first author who gave a theoretical foundation to Oparin's ideas was Urey (1952) who developed the first relatively quantitative model of the primeval atmosphere. The successful synthesis of amino acids in the laboratory realized by Miller (1953) and later by Miller & Urey (1959) supported the Oparin and Urey hypothesis that the early atmosphere was reducing. Practically at the same time, Rubey (1951, 1955) challenged this hypothesis and proposed an opposite atmospheric composition rich in oxidized components. According to this author the first atmosphere came from degassing of the Earth's interior; the volcanic gases would have existed as different species than those present today. Rubey proposed that the major species were H_2O rather than H_2, CO_2 rather than CO and N_2 rather than NH_3. In this scenario the bulk of the early atmosphere had a different oxidation state because only small amounts of reduced compounds were present (Table 1). Although both hypotheses are difficult to reconcile, Holland (1962) proposed that the primitive atmosphere passed through both stages. According to this model the early Earth would have had a reducing atmosphere because volcanic gases released prior to core formation would be rich in metallic iron. According to the model, the Earth evolved to a more neutral state but was still anoxic. In fact a common point of view in the literature is that O_2 was absent in the early atmosphere (see e.g. Holland, 1984 and references therein).

TABLE 1. Main chemical species depending on the oxidation state of the early atmosphere.

Atmosphere	H species	C species	N species	O species	Prevailing period
Reducing	H_2/H_2O	CH_4	NH_3	CO	early environment
Oxidized	H_2O	CO_2/CO	N_2	CO_2	Progressive appearance. Biogenic O_2 released ~3 Ga ago.

The composition of the earliest atmosphere of the Earth was set by continuous impacts (Zahnle et al., 1988). In consequence, endogenous and exogenous sources must be evaluated and balanced in order to explain the evolution of the earliest atmosphere of the Earth (Chyba et al., 1990; Chyba and Sagan, 1992). In this scenario considering the typical volatile species of common impactors and the chemistry of impact plumes, the steady-state atmosphere would have maintained a reducing state, although it is still an open issue. The subsequent redox state of the gas cloud generated by impacts would define the proportions of various compounds in the primeval atmosphere (see e.g. Zahnle et al., 1988; Zahnle and Sleep, 1997). As a consequence of these impacts, it has been suggested that a magma ocean was present in the early stages of the Earth. If it ever existed, its crystallisation, except near the surface, would have been dominated by higher-pressure phase equilibrium. Dense, refractory ferromagnesian constituents should have been stabilized in relatively deeper portions of the gradually thickening, largely solid mantle. In this scenario the volatile and lithophile elements concentrated toward the surface in a silicic rind surrounded by a dense primitive atmosphere (Hargraves, 1976; Shaw 1976). The upper mantle was more reduced than today, although it progressively became more oxidized as a result of the release of reduced volcanic gases and the subduction of a hydrated, oxidized seafloor (Kasting et al., 1993). Moreover, mantle redox evolution is also probably linked to the oxidation state and the early evolution of the atmosphere (Kasting & Catling, 2003).

Fortunately, some clues about the composition of the earliest terrestrial surface remain. The existence of zircons that are 4.2 Ga old suggests the presence of water in an early period of the Earth's history (Peck et al., 2001). Impacting material was, on the whole, more reduced than the crust and mantle of the differentiated Earth. The impacts tend to reduce stochastically the extent of the atmosphere and ocean. It has been proposed that as consequence of the high flux of impactors, CO rather than CO_2 could dominate the atmosphere before 4 Ga ago. A weak CO atmosphere should have been more conducive to the origin of life than CO_2 (Zahnle and Sleep, 1997). The materials excavated during impacts in the marine seafloor probably acted to remove important amounts of CO_2 from the atmosphere. However, the relative degree to which oxidation of the crust or the mantle affected the redox composition of gases emanating from the Earth is still being debated.

Fe-rich impactors to the Earth would also produce significant amounts of CH_4 and NH_3 through Fischer-Tropsch synthesis in impact plumes (Sekine et al., 2003). These conditions have also been extrapolated to other bodies containing metal particles such as chondritic asteroids or comets (Kress and McKay, 2003). The role of these processes in a transient atmosphere is directly linked with the production of hydrocarbons that settled on the Earth. In such a period, the partial pressure of N_2 was probably lower than at present because this molecule is efficiently removed by impacts (Zahnle and Sleep, 1997). In the case that the pressure and composition of the early atmosphere were significantly different than at present, it would provide changing conditions for the survivability of IDPs or meteoroids. In such a case, it would be especially important to understand the chemistry involved in the ablation of meteors. For example, under a reducing atmosphere the chemistry associated with meteors changes significantly. Ablation of the CHON component (which would be abundant in cometary meteoroids) would result in the production of CH, CN and NH radicals in meteor columns or in fireball plumes that could recombine and become a source of methane and other hydrocarbons (Kress and McKey, 2003). A $C/O>1$ ratio would promote polymerization of CH_4, whereas $C/O<1$ would lead primarily to oxidation (Sagan and Chyba, 1997).

In an anoxic atmosphere the primary reservoirs of oxygen would have been H_2O and CO_2. The abundance of water in the upper stratosphere should have been low. However, CO_2 production by volcanic emissions was probably more important in the past than at present. To maintain C/O>1, the outgassed CO_2 must have been removed from the atmosphere. Kasting (1997) and Sleep & Zahnle (2001) proposed that the rapid weathering of the ejecta from frequent large impacts could have provided a sink for atmospheric CO_2. Impact ejecta can be easily chemically attacked because much of it is more mafic than other sediments because the most frequent target is oceanic basalt. When this basic-rock-rich ejecta is subducted, along with the plates on which they are deposited, an additional sink for CO_2 is created (see e.g. Sleep & Zahnle, 2001). According to their model, methane and ammonia participated in a kind of feedback cycle, working together in two main ways: (i) methane's presence produced an organic haze due to its capacity to polymerize and (ii) this haze protected ammonia from UV photodissociation. This cycle could yield longer lifetimes for these gases, thus extending the greenhouse effect on the primeval atomsphere for longer time scales than would occur for each gas species alone. Although we have no direct data from the early evolution of the atmosphere, higher amounts of these greenhouse gases are supported by evidence for a CH_4 mixing ratio of 10^{-4} (100 ppmv) in 2.8-Ga-old paleosols (see e.g. Pavlov et al., 2000). In fact, Catling et al. (2001) suggested that the concentration of methane in the Archean atmosphere was larger than the value given by (Pavlov et al., 2000). The scenario in a CO_2-rich atomsphere changes completely. During the ablation of meteoroids all the participating CO_2 would be dissociated at 4000 K to form oxygen atoms and CO. Oxygen atoms can react with organic compounds released from the meteoroid to make carbon radicals. Experimental work on organic decay and synthesis in meteors and fireball plumes under different atmospheric compositions would be relevant in an astrobiological context (Jenniskens et al., 2004a).

3. The Intrinsic Value of the Study of Meteor Phenomena

An important part of the mass of interplanetary matter that reaches the Earth arrives as fine dust a few hundred micrometers in size. Due to the high relative velocities between this dust and the Earth (in the range of 11 to 72.6 km/s) the encounter is able to produce energetic collisions with the atoms and molecules present in the upper atmosphere of the Earth. Particles less than about 50 μm and with typical masses of 10^{-12} to 10^{-6} g have a surface-area to mass ratio high enough to radiate away the energy transferred in the collisions with the atmospheric atoms. As a consequence, such particles are efficiently decelerated, being collected later in the stratosphere as IDPs (Brownlee, 1985; Love & Brownlee, 1993; Rietmeijer, 1998, 2002a, b). The remaining particles and the bulk of the mass influx in the form of 50 μm-10 cm-sized meteoroids are ablated in the upper atmosphere producing the phenomenon called *meteors*. Particles producing meteors cannot efficiently radiate the heat produced by collisions with atmospheric components due to their lower surface-area to mass ratios. As a consequence, these particles are heated until their mineral phases reach the melting point and the meteoroids start to volatilize, losing mass as vapor, melted droplets, and small fragments. This process is called ablation and the result is the formation of a column of relatively cool plasma with a typical excitation temperature of ~4000 K (Ceplecha, 1973; Borovicka, 1993; Jenniskens et al., 2000). A recent result is that, for a study of meteoroids ranging in mass between 10^{-3} and 9 kg, and a whole range of geocentric velocities of 25 to 72 km/s, the typical temperature of the plasma remains in the range 4500 ± 1000 K (Trigo-Rodríguez et al., 2003). The lower abundance of Ca detected suggests that the volatilization of the meteoroid is not complete, and that some mineral phases survive ablation. Consequently, refractory elements would be not contributing to the emission of light. This evidence is also supported by the discovery of small cosmic spherules that are collected in the stratosphere or in different favourable environments at the terrestrial surface. These particles are spherical objects formed by melting of small particles of interplanetary dust or by meteoritic materials that leave the meteoroid forming droplets that survive complete volatilization (Taylor and Brownlee, 1991).

Meteor spectra can be taken at the same time that monitoring of meteor activity from several stations is made. From the images of the meteor from different stations, it is possible to obtain the trajectory and heliocentric orbit of the meteoroid. In these cases, it is possible to link unequivocally the meteoroid with its parent body. The chemical abundances derived from the study of meteor spectra can provide important information about minor bodies (Trigo-Rodríguez et al., 2003). The determination of chemical abundances is based on obtaining a synthetic spectrum that is directly compared with the recorded one (Borovička, 1993; Trigo-Rodríguez et al., 2003). When the triangulation of the fireball path has been done, estimates can be made of the height, temperature and relative (or absolute if the magnitude is known) abundance of chemical elements through the luminous path.

Studies of meteor spectra have shown that most of the fireball radiation is produced from a relatively small number of individual atomic lines. In the visible range of the electromagnetic spectrum the main lines are associated with the volatilization of the meteoroid. The brightest lines are usually produced by ionized Mg, Si and Ca, and neutral Na, Mg and Fe (Trigo-Rodríguez et al., 2004a). These emissions are thought to originate from warm plasma in quasi-local thermodynamic equilibrium behind the meteoroid in a region known as the wake. In the collision front of the meteoroid, a hot $T \approx 9500$ K component is produced that includes most of the ionized lines detected in meteor spectra.

Observational data also suggest that fragmentation processes in meteoroids would lead to survival of small fragments that are incompletely volatilized (Rietmeijer and Jenniskens, 2000; Rietmeijer, 2002b). Future work using IR and UV meteor spectroscopy is required in order to know the role of fragmentation and turbulence as sinks of energy. Using spectral techniques we can increase our knowledge of the processes that affected the meteoroids during ablation or in the interplanetary medium (Trigo-Rodríguez et al., 2003, 2004; Borovička et al., 1999, 2005). Survivability of organics and carbon compounds is currently debated on the basis of new meteor spectra data. For example, an additional piece of evidence for the presence of carbon-bearing species is the detection of carbon lines in a far-UV Leonid spectrum (Carbary et al., 2004). The absence of detections from the ground would suggest that the release of organics typically occurs in an early stage of ablation of the meteoroid when the luminous phase is too faint to be detected. Another possibility, especially for small meteoroids exposed during long timescales in the interplanetary medium or close perihelia (Borovička et al., 2005), is that the volatile phases are lost by thermal heating. Detailed UV meteor spectroscopy and high-sensibility CCD cameras could determine if the ablation of organics occurs at higher altitudes following the pattern of differential ablation observed in mineral phases.

Russell et al. (2000) proposed that in meteoric columns the condensation of volatiles is the main source of carbon compounds when the temperature in the meteor column is about 300 K. Additional simultaneous observations in the range of 780-840 nm can provide information on the presence of the OH Meinel band that has been recently proposed as linked to the presence of water in meteoroids (Jenniskens et al., 2004b). Despite this, a recent study of the possible increase in the column density of OH and O_2 due to the delivery from Leonid meteoroids has resulted in no clear evidence of water delivery (Trigo-Rodríguez et al., 2004c). A possible explanation is that the interfereometer sounded the atmosphere below the region of the thermosphere where this water was probably released; because Leonid meteoroids typically start ablation at ~120 km. It seems likely that most of this water is present as hydrated minerals because aqueous alteration is expected to be important in cometary meteoroids (Rietmeijer et al., 2004).

4. Delivery and Production of Organics in the Early Earth

Interplanetary Dust Particles (IDPs) have been collected in the Earth's stratosphere by NASA aircraft since the mid-1970s. Some of the first analyses made on them revealed that some samples show no evidence of heating (Brownlee, 1985). Brownlee et al. (1993) used atmospheric entry models for dust and the measured He release profiles of IDPs to determine the maximum temperature experienced by an individual IDP during entry into the atmosphere. From this approach they concluded that IDPs originate in asteroids and comets. Flynn et al. (2003) showed the presence of aliphatic hydrocarbons and of a ketone group, by detecting the CH_2, CH_3 and C=O stretching functional groups in acid-etched IDPs. Matrajt et al. (2005) studied the 3.4 μm aliphatic feature of seven IDPs and found that the aliphatic fraction is made of relatively simple compounds rather than that expected for the diffuse interstellar medium (DISM). This suggests IDP formation in environments consistent with the expected for solar system comets formed at large solar distances.

Different attempts have been made to estimate the flux of meteoroids in the early Earth. Anders (1989) estimated that the flux of IDPs would have accreted $\sim 2 \cdot 10^8$ kg·yr^{-1} of organics to the early Earth. Even under an atmosphere chemically favorable (reducing) for producing organics, such a flux of IDPs would be several orders of magnitude higher than that due to CH_4 photolysis (Chyba and Sagan, 1997). Recently,

Flynn et al. (2004) revisited these calculations and estimated a minimum flux of ~8-$15 \cdot 10^3$ $kg \cdot yr^{-1}$ of unpyrolized organic matter reaching the Earth's surface at the present time. This flux could be even higher if the typical temperatures at which organics survive are underestimated. In fact, several authors have performed experiments in the last decade in order to figure out what the temperatures are of the survivability of organics and their time scales. Glavin and Bada (1999) demonstrated that ~50% of the amino acid glycine survives a short heating pulse at 1200ºC. Also Blank et al. (2001) investigated the survival of several amino acids occurring at pressures in the range of 5.21 to 21 GPa and 412-870 K, during typical timescales of 1 to 2.7 μs. To understand heating mechanisms and their extent during the deceleration of these particles, it is useful not only to study the contribution from IDPs, but also the largest cometary meteoroids.

Blank and Miller (1998) reevaluated the problem of the survival of organic compounds during impact events. Basiuk and Douda (2001) studied the survivability of biomolecules at 500ºC, especially amino acids and poly-amino acids, during short heating events. These authors conclude that, for the short time scales (a few seconds or minutes) involved in these processes, such organic compounds would be able to survive; consequently, their flux to the Earth could be significant. In contrast to these experimental results, Glavin et al. (2004) deduced that less than 5% of micro-meteorites recovered in Antarctica that are in the 40-400 μm size contained α-aminoisobutyric acid although it is ubiquitous (~3-4 ppm) in the CM chondrite Murchison. It suggests that this amino acid is probably destroyed during atmospheric entry heating at temperatures >550ºC. They concluded that it was unlikely that large micrometeorites could have a significant role in seeding the Earth with extraterrestrial amino acids. In any case, other survival pathways are available for large meteoroids that suffer catastrophic airbursts. Chyba and Sagan (1992, 1997) suggested that these fragmentation processes could also participate in the release of energy providing ways for organic compounds to escape severe heating in the high-temperature shock front of the fireball. Cometary nuclei are expected to be mixtures of rocks, organics and ices with abundant empty spaces (porosity) and low-strength materials. The density of comets is expected to be the lowest of available solar system materials. For such objects, we can expect massive breakups during the first stages of their atmospheric entries when they are under dynamic pressures of a few MPa. Additional clues can be obtained if we look at the low-density meteorite types, i.e., CI and CM carbonaceous chondrites. With densities of ~2g/cm^3 and porosity of the surviving samples between 10 and 20% (Brit and Consolmagno, 2001) members of these meteorite classes usually disrupt during atmospheric passage, probably producing most of the type II fireballs (Ceplecha et al., 1998). The importance of their study lies in that aerial bursts are expected to be more common than crater-producing events (Wasson, 2003). The study of these processes from all possible techniques can be relevant in deciding if extraterrestrial organic matter contributed significantly to the chemical composition of the early Earth.

5. Conclusion: Comets, Carbon-Rich Asteroids, Meteors and Life

The original arguments of J. Oró (1961) on the important role of comets in the enrichment in organics of the primeval Earth have been reinforced by recent data in several research lines. The presence of living organisms on the Earth as early as ~3.5 Gy ago suggests that life likely appeared at the same time or shortly after the end of the heavy bombardment when the rate of impacts with minor bodies was still high (Schopf, 1987). However, direct impacts would not be the only source of cometary organics to the terrestrial biosphere. The contribution of smaller particles like IDPs and meteoroids that are continuously reaching the atmosphere should be taken into account and estimated by using all newly available techniques (Brownlee, 1985; Love & Brownlee, 1993; Rietmeijer and Jenniskens, 2000; Jenniskens et al., 2000; Flynn et al., 2004; Trigo-Rodríguez et al., 2004b,c,d). The ablation of large bodies in the terrestrial atmosphere would also provide natural pathways for the delivery of volatiles to the Earth, but the mechanisms are not yet well known. Additional the interaction of carbon-rich minor bodies during fragmentation processes can provide new insights on the degree of alteration of these low-strength materials and the fraction of these bodies that is able to survive. To achieve the target, observational studies of bolides are required; this involves increasing fireball monitoring from the ground. Such a program would help us clarify the role that the wide range in sizes of exogenous sources played in bringing the organic components that life needed to make its appearance on the Earth.

Were meteoroids able to deliver organic compounds efficiently to the Earth? We don't know at this point, but we cannot answer to this question by looking only at the present conditions. The recent discovery of carbon lines in a Leonid meteoroid discovered by a space-based system suggests that organic compounds should be arriving in large cometary meteoroids (Carbary et al., 2004). The small elemental abundances and the weakness of carbon lines constitute a challenge on detecting the presence of organic compounds in meteoroids from the ground. Fortunately, advances in meteor spectroscopy allow us to study the interaction of meteoroids with the Earth in extraordinary detail (Borovicka, 1993, 2005; Trigo-Rodríguez et al., 2003). The presence of OH bands or the possible overabundance of oxygen and hydrogen in meteor columns perhaps may help address this question (Jenniskens et al., 2004b). In the context of water delivery from minor bodies, the deuterium abundance of the oceans continues to be a focus of debate. Pavlov et al. (1999) suggested that the observed deuterium/hydrogen ratio could be explained as the consequence of a continuous fall of IDPs irradiated in the interplanetary medium. Other authors have suggested that there should be a large chemical diversity of comets and that the formation of these bodies in different regions of the solar system could explain the deuterium content of the oceans (Laufer et al., 1999; Delsemme, 2000). This explanation could also account for some peculiarities in the noble gas abundances measured in the terrestrial atmosphere (Owen and Bar-Nun, 1995; Notesco et al., 2003). Other authors have proposed that atmospheric Xe isotope ratios and independent estimates of the bulk chemical composition of the Earth point clearly to a CI chondrite component of the Earth (Pepin, 1991). New data on comet and low-strength meteorite compositions is required in order to distinguish the role of these two sources in the evolution of the terrestrial atmosphere.

In conclusion, the accretion of organic matter and volatile species has occurred continuously since the formation of the Earth in a wide range of masses: IDPs, meteoroids, catastrophic airbursts of medium-size meteoroids, and large impacts. The study of these different phenomena by all available techniques and with an open perspective can bring us important clues on the role of comets and carbonaceous chondrites in the origin of life on Earth.

6. Acknowledgments

The author is particularly indebted to John Wasson and Alan Rubin for advice and encouragement during preparation of this paper. The Spanish State Secretary of Education and Universities provided the funds for a postdoctoral stay in UCLA.

7. References

Anders, E. (1989) Pre-Biotic Organic Matter from Comets and Asteroids. Nature **342**: 255.

Anders, E., Ganapathy, R., Krähenbühl, U. and Morgan, J.W. (1973) Meteoritic material on the Moon. The Moon **8**: 3-24.

Barak, I. and Bar-Nun, A. (1975) The mechanisms of amino acid synthesis by high temperature shock waves. Origins of Life and Evol. of the Biosphere **6**: 483-506.

Blank, J. G., Miller, G. H., Ahrens, M. J., Winans, R. E. (2001) Experimental shock chemistry of aqueous amino acid solutions and the cometary delivery of prebiotic compounds. Origins of Life and Evolution of the Biosphere **31**: 15-51.

Blank, J. G. and Miller, G. H. (1998) The fate of organic compounds in cometary impacts. In: A. F. P. Howing, A. Paull, R. R., Boyce, P. M., Danehy, M. Hannemann, J. J., Kurtz, T. J., McIntyre, S. J., McMahon, D. J., Mee, R. J., Sandeman and Tanno H. (eds.) *Proceedings of the 21st International Symposium on Shock Waves*. Panther Press, Fyshwick, Australia, pp. 1467-1472.

Borovička, J. (1993) A fireball spectrum analysis. Astron. Astrophys. **279**: 627-645.

Borovička, J. (1994) Two components in meteor spectra. Planet. Space Sci. **42**: 145-150.

Borovička, J., Stork, R., Bocek, J. (1999) First results from video spectroscopy of 1998 Leonid meteors. Meteorit. Planet. Sci. **34**: 987-994.

Borovička, J., Koten, P., Spurný, P., Boček, J., Štork, R. (2005) A survey of meteor spectra and orbits: evidence for three populations of Na-free meteoroids. Icarus **174**: 15-30.

BVSP (1981) *Basaltic Volcanism on the terrestrial planets*. Lunar & Planetary Institute, Pergamon Press, Inc., New York.

Brit, D. T. and Consolmagno, G. J. (2001) Modeling the Structure of High Porosity Asteroids. Icarus **152**: 134-139.

Brownlee, D. E. (1985) Cosmic dust: collection and research. In: *Annual Review of Earth and Planetary Sciences*. Volume 13. Palo Alto, CA, Annual Reviews, Inc., 1985, pp. 147-173.

Brownlee, D. E., Joswiak, D. J., Love, S. G., Nier, A. O., Schlutter, D. J., Bradley, J. P. (1993) Properties of Cometary and Asteroidal IDPs Identified by He Temperature-Release Profiles. Meteoritics 28: 332.

Carbary, J. F., Morrison, D., Romick, G. J., Yee, J. H. (2004) Spectrum of a Leonid meteor from 110 to 860 nm. Adv. Space Res. **33**: 1455-1458.

Catling, D. C., Zahnle, K. J. and McKay, C. P. (2001) Biogenic methane, hydrogen escape, and the irreversible oxidation of the early Earth. Science **293**: 839-843.

Ceplecha, Z. (1973) A model of spectral radiation of bright fireballs. Bull. Astron. Inst. Czech. **24**: 232-242.
Ceplecha, Z., Borovička, J., Graham Elford, W., Revelle, D. O., Hawkes, R. L., Porubcan, V. and Simek, M. (1998) Meteor phenomena and bodies. Space Sci. Rev. **84**: 327-471.
Clark, B. C. (1988) Primeval procreative comet pound, Orig. Life Evol. Biosphere **18**: 209-238.
Chamberlin, T. C., and Chamberlin, R. T. (1908) Early extraterrestrial conditions that may have favored organic synthesis. Science **28**: 897-910.
Chyba, C. F., Thomas, P. J., Brookshaw, L., and Sagan, C. (1990) Cometary delivery of organic molecules to the early Earth, Science **249**: 366-373.
Chyba, C. F., and Sagan, C. (1992) Endogenous production, exogenous delivery and impact-shock synthesis of organic molecules: An inventory for the origins of life, Nature **355**: 125-132.
Chyba, C. F., and Sagan, C. (1997) Comets as a source of prebiotic organic molecules for the early Earth. In: P. J. Thomas, C. F. Chyba and C. P. McKay (eds.) *Comets and the origin and evolution of life*. Springer, New York, pp. 147-173.
Delsemme, A. (1992), Cometary origin of carbon, nitrogen and water on the Earth. Orig. Life **21**: 279-298.
Delsemme, A. (1998) The deuterium enrichment observed in recent comets is consistent with the cometary origin of seawater. Planet. Space Sci. **47**: 125-131.
Delsemme, A. (2000) 1999 Kuiper Prize Lecture Cometary Origin of the Biosphere. Icarus **146**: 313-325.
Flynn, G. J., Keller, L. P., Feser, M., Wirick, S., Jacobsen, C. (2003) The origin of organic matter in the solar system: evidence from the interplanetary dust particles. Geoch. et Cosmoch. Acta **67**: 4791- 4806.
Flynn, G. J., Keller, L. P., Jacobsen, C. and Wirick, S. (2004) An assessment of the amount and types of organic matter contributed to the Earth by interplanetary dust. Adv. Space Res. **33**: 57-66.
Glavin, D. P. and Bada, J. L. (1999) The Sublimation and Survival of Amino Acids and Nucleobases in the Murchison Meteorite During a Simulated Atmospheric Heating Event. 30th Ann. LPSC, Houston, abstract no. 1085.
Halliday, A. N.(2004) The Origin and Earliest History of the Earth. In: A. M. Davis (ed.) *Treatise on Geochemistry vol.1. Meteorites, Comets and planets.* Elsevier, pp. 509-557.
Hargraves, R. B. (1976) Precambrian Geologic History. Science **193**: 363-371.
Haskin, L. and Warren, P. (1991) Lunar chemistry. In: G. Heiken, D. Vaniman and B. M. French (eds.) *Lunar sourcebook: A user's guide to the Moon.* Cambridge University Press, New York, pp. 357-474.
Holland, H. D. (1962) A model for the evolution of the Earth's atmosphere. In: A. E. J. Engel, H. L. James and B. F. Leonard (eds.) *Buddington Volume*, 447-477.
Holland, H. D. (1984) *Chemical evolution of the atmosphere and oceans*. Princeton Univ. Press. New Jersey.
Hörz, F., Grieve, R., Hieken, G., Spudis, P. and Binder, A. (1991) Lunar Surface Processes. In: G. Heiken, D. Vaniman and B. M. French (eds.) *Lunar sourcebook: A user's guide to the Moon*. Cambridge University Press, New York, pp. 61-120.
Ip, W. H. and Fernández, J. A. (1988) Exchange of condensed matter among the outer and terrestrial protoplanets and the effect on surface impact and atmospheric accretion. Icarus **74**, 47-61.
Jenniskens, P., Wilson, M. A., Packan, D., Laux, C., Krüger, C. H., Boyd, I. D., Popova, O. P. and Fonda, M. (2000) Meteors: a delivery mechanism of organic matter to the Early Earth. Earth, Moon Planets **82-83**: 57-70.
Jenniskens, P., Schaller, E. L., Laux, C. O., Wilson, M. A., Schmidt, G., Rairden, R. L. (2004a) Meteors do not break exogenous organic molecules into high yields of diatomics, Astrobiology **4-1**: 67-79.
Jenniskens, P., Laux, C. O., Schaller, E. L. (2004b), Search for the OH ($X^2\Pi$) Meinel band emission in meteors as a tracer of mineral water in comets: detection of N2+ (A-X). Astrobiology **4-1**: 109-121.
Jessberger, E. K., Christoforidis, A., Kissel A. (1988) Aspects of the major element composition of Halley's dust. Nature **322**: 691-695.
Kasting, J. F. (1997) Warming early Earth and Mars. Science **276**: 1213-1215.
Kasting, J. F., Eggler; D. H. and Raeburn, S. P. (1993) Mantle redox evolution and the oxidation state of the Archean atmosphere. Journal of Geology **101**: 245-258.
Kasting, J. F., Catling; D. (2003) Evolution of a habitable planet. Annu. Rev. Astron. Astrophys. **41**: 429-463.

Kress, M. E. and McKay, C. P. (2003) Formation of methane in comet impacts: implications for Earth, Mars and Titan. *Icarus* **168**: 475-483.
Laufer, D., Notesco, G., Bar-Nun, A. and Owen, T. (1999) From the Interstellar Medium to Earth's Oceans via Comets-An Isotopic Study of HDO/H_2O. Icarus **140**: 446-450.
Love, S. G. and Brownlee, D. E. (1993) Peak atmospheric entry temperature of micrometeorites, Meteoritics **29**: 69-70.
McSween, H. Y.(2004) Mars. In: A. M. Davis (ed.) *Treatise on Geochemistry vol.1. Meteorites, Comets and planets.* Elsevier, pp. 601-621.
Matrajt, G. Muñoz Caro, G.M., Dartois, E., d'Hendecourt, L., Deboffle, D., and Borg, J. (2005) FTIR analysis of organics in IDPs: comparison with the IR spectra of the diffuse interstellar medium. Astron. Astrophys. **433**: 979-995.
Matsui, T. and Abe, Y. (1986) Evolution of an impact-induced atmosphere and magma ocean on the accreting Earth. Nature **319**: 303-305.
Melosh, H. J. (1989) *Impact Cratering: A Geological Process.* Oxford monographs on geology and geophysics #11, Oxford University Press, New York.
Miller, S. L. (1953) A production of amino acids under possible primitive earth conditions, Science **117**: 528-529.
Miller, S. L. and Urey, H. C. (1959) Organic compound synthesis on the primitive Earth, Science **130**: 245-251.
Notesco, G., Bar-Nun, A. and Owen, T. (2003) Gas trapping in water ice at very low deposition rates and implications for comets. Icarus **162**: 183-189.
Oparin, A. I. (1938) *Origin of life.* Macmillan, New York, USA.
Oró, J. (1961) Comets and the formation of biochemical compounds on the primitive Earth. Nature **190**: 389-390.
Oró, J., Miller, S. L. and Lazcano, A. (1990) The origin and early evolution of life on Earth. Annu. Rev. Earth Planet. Sci. **18**: 317-356.
Owen, T. and Bar-Nun, A. (1995) Comets, impacts and atmospheres. Icarus **116**: 30 215-226.
Pavlov, A. A., Kasting, J. F., Brown, L. L., Rages, K. A., and Freedman, R. (2000) Greenhouse warming by CH_4 in the atmosphere of early Earth. J. Geophys. Res. **105**: 11981-11990.
Pavlov, A. A., Kasting, J. F., (1999) Irradiated interplanetary dust particles as a possible solution for the deuterium/hydrogen paradox of Earth's oceans. J. Geophys. Res. **104**: E12, 30, 725-730.
Peck W., Valley, J.W., Wilde, S.A. and Graham, C.M. (2001) O isotope ratios and rare earth elements in 3.3 to 4.4 Ga zircons. Geoch.et Cosm. Acta **65**: 4215-4229.
Pepin, R. O. (1991) On the origin and early evolution of terrestrial planet atmospheres and meteoritic volatiles. Icarus **92**: 2-79.
Pierazzo, E. and Chyba, C. F. (1999) Amino acid survival in large cometary impacts, Met. Planet. Sci. **34**: 909-918.
Rietmeijer, F. J. M. (1998) Interplanetary dust particles. In: J. J. Papike (ed.) *Planetary materials*, Mineralogical Society of America, Washington D. C., pp. 1-95.
Rietmeijer, F. J. M. (2002a) The Earliest Chemical Dust Evolution in the Solar Nebula. Chemie der Erde **62-1**: 1-45.
Rietmeijer, F. J. M. (2002b) Collected extraterrestrial materials: Interplanetary Dust Particles, Micrometeorites, and Meteoric Dust. In: E. Murad and I. P. Williams (eds.) *Meteors in the Earth's atmosphere.* Cambridge University Press, pp. 215-245.
Rietmeijer, F. J. M., Jenniskens, P. (2000), Recognizing Leonid Meteoroids among the Collected Stratospheric Dust. Earth, Moon Planets **82-83**: 505-524.
Rietmeijer, F. J. M. and Nuth III, J. A. (2000) Collected extraterrestrial materials: constraints on meteor and fireball compositions. Earth, Moon and Planets **82-83**: 325-350.
Rietmeijer, F. J. M., Nuth III, J. A., Nelson, R. N. (2004) Laboratory hydration of condensed magnesiosilica smokes with implications for hydrated silicates in IDPs and comets. Meteorit. Planet. Sci. **39**: 723-746.
Rubey, W. W. (1951) Geological history of seawater: An attempt to state the problem. Geol. Soc. America Bull. **62**: 1111-1148.
Rubey, W. W. (1955) Development of the hydrosphere and atmosphere, with special reference to probable composition of the early atmosphere. In: A. Poldervaart (ed.) *Crust of the Earth.*, Geological Society of America, New York, pp. 631-650.

Russell, R. W., Rossano G. S., Chatelain M. A., Lynch D. K., Tessensohn T. K., Abendroth E., and Kim D. (2000) Mid-Infrared spectroscopy of persistent Leonid trains. Earth, Moon and Planets **82**-**83**: 439-456.

Sagan, C. and Chyba, C. (1997) The early faint Sun Paradox. Science **276**: 1217-1221.

Schopf, J. W. (1987) Early Archean (3.3-billion to 3.5-billion-year-old) microfossils from Warrawoona Group, Australia. Science **237**: 70-73.

Sekine, Y., Sugita, S., Kadono, T. and Matsui, T. (2003) Methane production by large iron impacts on early Earth. J. Geophys. Res. Planets **108(E7)**: 1-11.

Shaw, D. M. (1976) Development of the early continental crust. Part 2. Prearchean, Protoarchean and later eras. In: B. F. Windley (ed.) *The early history of the Earth*, John Wiley & Sons, New York, pp. 33-54.

Sleep, N. H. and Zahnle, K. (2001) Carbon dioxide cycling and implications for climate on ancient Earth. J. Geoph. Res. **106**: 1373-1399.

Taylor, S. and Brownleee, D. (1991) Cosmic spherules in the geologic record. Meteoritics **26**, 203-211.

Taylor, S. R. (1982) *Planetary science: a lunar perspective*. Publ. Lunar and Planetary Institute.

Taylor, G. J. and Scott, E. R. D. (2004) Mercury. In: A. M. Davis (ed.) *Treatise on Geochemistry vol. 1. Meteorites, Comets and planets*. Elsevier, pp. 477-485.

Trigo-Rodríguez, J. M. (2002) *Spectroscopic analysis of cometary and asteroidal fragments during their entry into the terrestrial atmosphere*. PhD thesis (In Spanish), University of Valencia, Spain.

Trigo-Rodríguez, J. M., Llorca, J., Borovička, J. and Fabregat, J. (2003), Chemical abundances determined from meteor spectra: I. Ratios of the main chemical elements. Meteorit. Planet. Sci. **38**: 1283-1294.

Trigo-Rodríguez, J. M., Llorca, J. and Fabregat J. (2004a) Chemical abundances determined from meteor spectra: II. Evidence for enlarged sodium abundances in meteoroids. Monthly Notices Royal Astronomical Society **348**: 802-810.

Trigo-Rodríguez, J. M., Llorca, J. and Oró J. (2004b) Chemical abundances of cometary meteoroids from meteor spectroscopy: Implications to the Earth enrichment. In: J. Seckbach, J. Chela-Flores, T. Owen and F. Raulin (eds.) *Life in the Universe: from the Miller experiment to the search for life in other worlds*. Kluwer Publishers, pp.201-204.

Trigo-Rodríguez, J. M., M. J. López González and Ortiz J. L. (2004c) OH and O_2 airglow emissions during the 1998 Leonid outburst and the 2002 Leonid storm. Earth Moon Planets 93: 191-201.

Trigo-Rodríguez, J. M., Llorca, J., Lyytinen, E., Ortiz, J. L., Sánchez Caso, A., Pineda, C., and Torrell, S. (2004d) 2002 Leonid storm fluxes and related orbital elements, Icarus 171: 219-228.

Urey, H. (1952) *The planets: Their origin and development*. Yale University Press, New Haven.

Warren, P. H., Jerde, E. A. and Kallemeyn, G. W. (1989) Lunar meteorites: siderophile element contents, and implications for the composition an origin of the Moon. Earth. Planet. Sci. Lett. **91**: 245-260.

Warren, P. H. (2004) The Moon. In: A. M. Davis (ed.) *Treatise on Geochemistry vol. 1. Meteorites, Comets and planets*. Elsevier, pp. 559-599.

Wasson, J. (1985) *Meteorites. Their Record of Early Solar-System History*. W. H. Freeman and Company. New York.

Wasson, J. (2003) Rubey Colloquium Paper. Large aerial bursts: An important class of terrestrial accretionary events. Astrobiology **3**: 163-179.

Zahnle, K. J., Kasting, J. F., Pollack, J. B. (1988) Evolution of a steam atmosphere during Earth's accretion. Icarus **74**: 62-97.

Zahnle, K. J. (1996) Dynamics and chemistry of SL9 plumes. In: K. S.Noll, H. A. Weaver and P. D. Feldman (eds.) *The collision of comet Shoemaker-Levy 9 and Jupiter*. Space Telescope Science Institute and Cambridge University Press, pp. 183-212.

Zahnle, K. J. and Sleep, N. H. (1997) Impacts and the early evolution of life. In: P. J. Thomas, C. F. Chyba and C. P. McKay (eds.) *Comets and the origin and evolution of life*. Springer, New York, pp. 175-207.

Biodata of **Pascale Ehrenfreund** and **Eduardo Martin** authors of ***"Astronomical constraints for life on terrestrial planets"***

Dr. Pascale Ehrenfreund is currently leading the Astrobiology group at the Leiden Institute of Chemistry, The Netherlands. She obtained her Ph.D. in 1990 from the University Paris VII and University of Vienna. Professor Pascale Ehrenfreund's interests are in the area of astrobiology, astrochemistry, comets, meteorites and the origin of life.

E-mail: **p.ehrenfreund@chem.leidenuniv.nl**

Dr. Eduardo Martin is currently leading the Astrobiology group at the Instituto de Astrofisica de Canarias, Spain. He is also part-time associate professor at the University of Central Florida, USA. He obtained his Ph.D. in 1994 from the University of La Laguna, Spain. Dr. Eduardo Martin's interests are in the area of astrobiology, brown dwarfs, extrasolar planets and low-mass star formation.

E-mail: **ege@iac.es**

J. Seckbach (ed.), Life as We Know It, 399-411.

ASTRONOMICAL CONSTRAINTS FOR LIFE ON TERRESTRIAL PLANETS

P. EHRENFREUND[1] AND E. L. MARTIN[2, 3]
[1]Leiden Institute of Chemistry, Astrobiology Laboratory, POBox 9502, 2300 RA Leiden, The Netherlands, [2]Instituto de Astrofisica de Canarias, La Laguna, Tenerife, E-38200, Spain, [3]Physics Department, University of Central Florida, Orlando FL-32816, USA

1. Introduction

Life as we know it is based on specific elements including H, C, O, N, S and P (Oró et al., 1990, Maciá et al., 1997). In the very early Universe, approximately 13 billion years ago, during Big Bang nucleosynthesis, only H and He and traces of a few other light elements were formed. Heavier elements, such as C, O and N, crucial for life, were produced later in the interior of stars during the course of stellar evolution. Recent cosmological data indicate that the formation of the first stars (also known as population III stars) may have occurred as early as 200 million years after the Big Bang. Those high-mass stars are born from contracting gas clouds (dominated by hydrogen and helium) in dark-matter halos under the influence of gravity (Scannapieco et al., 2003). The mass and life-time of stars strongly influences their ability to form heavy elements and terrestrial planets in their surrounding. During their evolution, stars fuse their internal H and He reservoirs to larger elements. Stars such as the Sun, with a mass of 2×10^{30} kg, have a life-time of ~10 billion years and can produce elements including C, O and N. More heavy elements can be exclusively synthesized in stars that are much larger than the Sun. All elements, which are heavier than iron, are formed during the final evolutionary period of stars which is dominated by mass loss, core degradation and stellar explosions. Those processes produce a vast number of neutrons which can be captured to form heavier nuclei. Thus, the life cycle of stars is crucial for the production and distribution of heavy elements throughout space and for the enrichment of the interstellar gas with heavy elements, astronomically also called "*metallicity*" (see Spaans 2004 for a review). Metals in the gas or incorporated into small (micron-sized) dust particles have the ability to cool down the interstellar gas more efficiently. This leads to cloud fragmentation on a smaller scale and triggers the formation of low mass stars (Scalo and Biswas, 2002).

Metallicity enhancement has thus important consequences for the formation of low-mass stars, such as the Sun, a class of stars that are characterized by a long term evolution and the ability to form terrestrial planets that eventually harbor life. The cycle of birth and death of stars constantly increases the abundance of heavy elements in the interstellar medium. Therefore it is important to consider the general distribution of metallicity in the Milky Way in order to understand the timescales and formation of terrestrial planets.

The Galaxy is composed of different stellar populations. The halo population is the oldest and has a mean metallicity about 30 times lower than that of the Sun. The thick disk population may be almost as old as the halo, and has a mean metallicity about 4 times lower than solar. The mean metallicity of the thin disk population is similar to the Sun and their ages range from 0 to 10 billion years. With increasing distance from the plane of the Milky Way, the mean metallicity decreases because the contribution of thick disk and halo stars increases (Du et al., 2004). With decreasing distance to the center of the Galaxy, the mean metallicity increases because the interstellar gas is more enriched due to a higher rate of star formation (Chen et al., 2003 and references therein). These observations are consistent with inside-out galactic star formation scenarios. Mergers do not appear to have played a major role in shaping the Galaxy. There is no clear relationship between age and metallicity for thin disk stars. It seems that the metallicity of stars in the thin disk depend more on *where* they formed than on *when* they formed.

Stars and planets are born from dust and gas in giant swirling interstellar molecular clouds (Mannings et al., 2000). Five billion years ago such a cloud collapsed under its own gravity to form our Solar System with a central condensation that became the Sun, as well as a surrounding disk, the solar nebula. In the solar nebula gas and submicron particles agglomerated to larger planetesimals and subsequently into planets. Rocky planetesimals in the inner zone of the solar nebula formed the terrestrial planets. In contrast, the outer Solar System is dominated by giant gaseous planets surrounded by numerous moons. Not all planetesimals orbiting within the solar nebula have been incorporated into planets. Interactions with the forming planets determined their variable destiny. Those objects, also known as comets and asteroids, were either ejected towards the edge of the Solar System (Kuiper Belt) or into reservoirs far beyond our Solar System (Oort Cloud). Many comets and asteroids were trapped in the asteroid belt between Mars and Jupiter or collided with the young planets and deposited substantial amounts of extraterrestrial carbon (Ip and Fernández, 1997, Fernández et al., 2002). Part of this material may have been crucial for the chemical evolution that ultimately led to the origin of life (Oró, 1961, Oró, 1983, Delsemme, 2000, Owen and Bar-Nun, 1995, Ehrenfreund et al., 2002, Llorca 2005, this volume).

There is evidence that a combination of exogenous and endogenous sources provided the first building blocks of life on the early Earth (Bada and Lazcano, 2002). Comets, asteroids and their fragments contribute extraterrestrial matter, including complex organic matter in a natural way during Solar System formation (Ehrenfreund et al., 2002). Endogenous reservoirs such as shallow basins, hydrothermal vents or porous rock space may have allowed the concentration of precursor molecules and efficient reaction pathways toward complex organics. In the following sections we discuss the interstellar raw material that builds up stars and planets (section 2), possible formation scenarios for terrestrial planets (section 3) and the relationship of planet formation with galactic metallicity (section 4). We discuss the raw material for life on terrestrial planets from the astronomical perspective (section 5) and conclude in section 6.

2. The Raw Material of Stars and Planets

The interstellar medium constitutes ~10–15% of the mass of the galaxy. H and He gas are the major components, molecules and small submicron dust particles are present in small concentration. Environments with hot gas of very low density, warm intercloud gas, and regions with denser and colder material can be distinguished. The cold, dense regions favour the synthesis of molecules both in the gas phase and on (or in) dust grains (Ehrenfreund and Charnley, 2000, Wooden et al., 2004). H_2 is by far the most abundant molecule in these clouds and the most abundant carbon-containing species is CO, with CO/H_2 of order 10^{-4}. Observations at infrared, radio, millimeter, and sub-millimeter frequencies show that a large variety of gas phase organic molecules are present in the dense interstellar medium (Charnley et al., 2003). These include organic classes such as nitriles, aldehydes, alcohols, acids, ethers, ketones, amines, and amides, as well as many long-chain hydrocarbon compounds (http://www.astrochemistry.net). Silicate and carbon-based micron-sized dust particles which are produced in the outflows of late-type stars provide a catalytic surface for a variety of reactions to occur when they are dispersed throughout the molecular cloud (Ehrenfreund and Fraser, 2003). In cold clouds such dust particles adsorb an ice mantle, comprised mostly of water, carbon dioxide, carbon monoxide and methanol, with smaller admixtures of methane, ammonia, carbonyl sulphide, formaldehyde and formic acid. An extended inventory of interstellar ice species has been established for bright high-mass star-forming regions (e.g. Ehrenfreund and Fraser, 2003). Only incomplete datasets are available for solar-type stars, though recent results from the SPITZER telescope show deep ice absorption bands towards some embedded Sun-type stars (Boogert et al., 2004).

The carbon chemistry is probably more efficient in circumstellar and diffuse interstellar clouds. In the circumstellar envelopes of carbon-rich evolved stars a complex carbon chemistry occurs that is analogous to carbon soot formation in combustion processes. An active acetylene (C_2H_2) chemistry is the starting point for the development of hexagonal aromatic rings of carbon atoms. These aromatic rings probably react further to form large aromatic networks (Frenklach and Feigelson 1989, Cherchneff et al., 1992). Apart from abundant CO gas in dense clouds, the carbon species in the Universe are predominantly in the form of aromatic molecules in the gas as well as in the solid state.

Laboratory simulations in combination with interstellar observations support the idea that the predominant fraction of carbon is incorporated into solid macromolecular carbon (e.g. Pendleton and Allamandola, 2002) or amorphous and hydrogenated amorphous carbon (Mennella et al., 1998). Interstellar gas and dust – icy or refractory in nature – provide the raw material for stars and planets. During the formation of a Solar System those ingredients are mixed, processed and partly destroyed according to their location from the star. Organic material can also be formed in the solar nebula. Detailed laboratory studies have been carried out in order to simulate the interaction between nanometer-sized metal particles and different gas mixtures under nebular-type conditions (Llorca and Casanova, 2000). Results on the reaction kinetics between kamacite grains and $H_2 + CO$ gas mixtures support the idea that gas-solid reactions in the solar nebula during CO hydrogenation represent a plausible scenario for the formation of carbides and carbonaceous materials in interplanetary dust particles, as well as for the production of hydrocarbons through Fischer-Tropsch-type reactions (Llorca and Casanova 1998, 2000). For a review on chemical processes occurring in the solar nebula the reader is referred to Llorca (2005, this volume).

3. How Are Low-Mass Stars and Terrestrial Planets Forming

The formation of stars occurs through the gravitational collapse of individual galactic molecular cloud cores within dense interstellar clouds (Mannings et al., 2000). In the protostellar phase, the central protostar, strongly embedded in its parent cloud, is accreting surrounding material. Accretion of molecular cloud material lasts about a few hundred thousand years. After about one million years the star has a mass very close to its final value, and is surrounded by a disk of dust and gas. The surrounding protoplanetary disk dissipates on timescales of ~50.000 years and invokes large scale mass and angular momentum transport. After about 50 million years the temperature and density are high enough to trigger nuclear fusion in the young stellar core. During the formation of star and planetary systems infalling interstellar matter passes through the accretion shock where significant processing can occur, see Figure 1. The heating decreases with distance of the star, therefore grains in the outer solar nebula may have escaped alteration. During the late stage of collapse phase, compression of gas in the inner nebula produces strong adiabatic heating which leads to melting of the dust and progressive variation in volatile components with increasing distance from the Sun (Chick and Cassen, 1997). Radiation chemistry involving X-rays and UV irradiation as well as turbulent motion leading to radial mixing of the products within the disk are physical processes that affect the chemistry in the solar nebula, see Figure 1 (Irvine et al., 2000; Markwick and Charnley, 2004).

Planetary formation is thought to begin during the protoplanetary disk phase. Within those disks km-sized planetesimals are formed that interact gravitationally. Those planetesimals grow by mutual collisions up to terrestrial planets if sufficient matter is available. The initially microscopically small dust grains (typical size 1 μm) collide gently due to Brownian motion, differential drift motions and gas turbulence (Weidenschilling and Cuzzi, 1993), and agglomerate due to adhesive surface forces (like, e.g., van der Waals force and hydrogen bonding). When the so-called planetesimals, kilometre-sized planetary precursors, have formed, mutual gravitational interaction leads to a rapid further growth to planetary sizes (Wetherill and Stewart, 1993). The theory of the formation of planetesimals still bears major uncertainties; details on the fragmentation physics of dust agglomerates for collisions above the fragmentation threshold are missing (Blum, 2004). The volatile-poor rocky planetesimals present in the inner Solar System accreted into a series of larger intermediate bodies which in turn formed the terrestrial planets. A short timescale for terrestrial planet formation was estimated from the hafnium-tungsten chronometry of meteorites (Yin et al., 2002). Those results are completely consistent with other evidence for rapid planetary formation, and are also in agreement with dynamic accretion models that predict a relatively short time (approximately 10 million years) for the main growth stage of terrestrial planet formation. Planetary factors may be prerequisites for life on Earth, such as being at the right distance from the star having the mass required to preserve an atmosphere, having a Jupiter-like neighbour, and the presence of a dynamic crust and oceans.

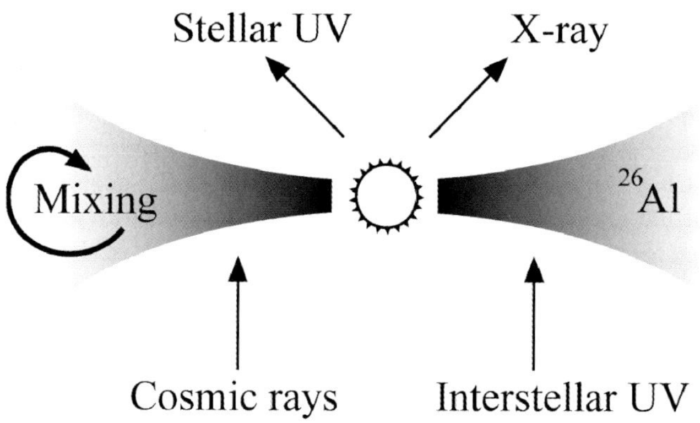

Figure 1. Schematic diagram of physical processes in the protosolar nebula. Sources of ionization such as stellar X-rays, stellar ultraviolet (UV) radiation and interstellar UV, cosmic rays and radionuclides as well as heating and transport of material affect the chemistry in those regions (adapted from Markwick and Charnley, 2004).

4. Where are Terrestrial Planets Formed?

Terrestrial extrasolar planets have been found only around the pulsar PSR B1257+12 (Konacki and Wolszczan, 2003), probably because no technique other than pulsar timing has reached the required sensitivity to find planets of masses comparable to our Earth. All the other extrasolar planets known have masses larger than 14 earths. Four of them have sizes estimated from the observations of the transits they produce when they pass in front of their host stars and occult a small fraction of the stellar disks. These diameters are larger than Jupiter's because short period planets are puffed up due to the impinging stellar radiation, and confirm that the planets are indeed gas giants.

While not much can be said on the statistics of only two planetary systems with terrestrial planets (those around the Sun and PSR B1257+12), the extrasolar giant planets discovered by radial velocity surveys and most recently by photometric transit experiments (OGLE and STARE) currently add up to 133 planets around 117 stars (http://cfa-www.harvard.edu/planets/encycl.html). A spectroscopic study of 98 stellar host stars of extrasolar planets showed that their iron abundances are on average larger by about a factor of 2 than stars of similar ages that do not have detectable planets (Santos et al., 2004). The frequency of giant planets within 4

Astronomical Units (radial velocity surveys cannot detect more distant planets because of their limited temporal baseline) of stars with iron abundances comparable to the Sun (and lower) is about 3%. However, for 25% stars with iron abundances a factor of 2 larger than the Sun extrasolar planets have been detected. The abundances of other chemical elements (C, Ca, Co, Mn, Ni, Si, V) show a similar pattern than the iron abundances (Bodaghee et al., 2003).

Does the frequency of terrestrial planets depend on metallicity? Two main scenarios have been invoked to explain the high frequency of giant exoplanets around metal-rich stars: (1) Formation scenario: giant planets within 4 AU of solar-type stars form more frequently when the metal abundance of the parental cloud is larger by about a factor of 2 than that of the solar nebula. (2) Evolutionary scenario: metal-rich stars have been contaminated by ingestion of giant planets, and a fraction of them still have giant planets that are detectable. Both scenarios have pros and cons. One of the main problems of the "evolutionary" scenario is the observational constraint that there is no connection between the metallicity of the primary stars and their masses. The depth of the convective region varies by more than a factor of 10 across the mass range of the exoplanet-host stars. For a similar amount of planetary material being tossed on the surfaces of the stars by falling planets, the more massive stars should be more contaminated because they have shallower convective regions. A possible solution to this problem has been pointed out by Vauclair (2004), who argues that the amount of metallic contamination may not depend on the mass of the star because of the action of thermohaline convection, a process that occurs in Earth's ocean. According to this model, the heavy material (rocks) deposited on the stellar surface by the planet would not be thoroughly mixed in the stellar atmosphere, but it would tend to fall to the bottom and remain deposited at the base of the convection zone.

It will take a lot more of research to determine the correct explanation for the metal-dependency of the frequency of occurrence of extrasolar giant planets among solar-type stars. But, from the Astrobiology point of view, can we infer any consequence for the probability of finding habitable terrestrial planets? We could make an extrapolation that if the frequency of extrasolar giant planets is 25% in metal-rich stars, the frequency of terrestrial planets could be even higher. The distribution of minimum masses obtained by radial velocity surveys clearly shows that lower mass planets are more common than higher mass ones (Udry et al., 2003). If metal-rich stars have many terrestrial planets, there are more chances that some of them would be habitable. The overabundance of terrestrial planets would explain the higher metallicity because some planets could become unstable due to gravitational encounters with larger planets. A fraction of those planets could fall onto the star and increase the metal content of the photosphere. Vauclair (2004) estimated that 145 planets like Earth would have to be accreted by a star of mass 1.1 solar masses in order to produce an increase in the stellar metallicity of a factor of 2 after accounting for mixing. The discovery of free-floating planets suggests that giant planets may be ejected by gravitational interactions in unstable planetary systems suggesting that planet-planet interactions may be common (Zapatero-Osorio et al., 2000). If, on the other hand, metal-rich stars have not been significantly polluted with planetary material, but they were born from metal-rich molecular clouds, the frequency of terrestrial planet could be anticorrelated with the frequency of giant planets. Metal-rich stars might not be the best targets for terrestrial planet finders. If the goal is to find Earth-twins, probably it is a good idea to search for Sun-twins first (Porto de Mello and da Silva, 1997).

The "formation" scenario has several flavors:

(1) Metallicity is probably an important factor in determining the properties of protoplanetary disks. Calculations of self-gravitating disks with different element mixtures by Wehrstedt and Gail (2003) show that low metallicity conditions result in more transparent, colder, disks. Below a metallicity about a factor of 3 lower than that of the Sun, no terrestrial planets form.
(2) Metallicity may lead to enhanced migration in protoplanetary disks. Some support for this idea has been claimed in a study that shows a possible correlation between orbital periods of extrasolar planets and the iron content of the host stars (Sozetti, 2004). Migration of giant planets across the habitable region is likely to disrupt all terrestrial planets.
(3) High metallicity and high frequency of giant planets may both be properties of stars formed in regions of very active star formation. High-mass stars may favor the formation of planets because they evaporate the gas in the disk and increase the gas/dust ratio (Throop and Bally, 2004). Moreover, gravitational interactions may be more frequent in regions with a high density of star formation, leading to planet formation by gravitational instabilities.

5. The Raw Material for Life

Most theories for the origin of life suggest that life began with an organic chemical system that somehow developed the ability to propagate itself, genetically or metabolically (Oró et al., 1990). An explanation for the source of the organic compounds thus becomes of central importance. It is generally agreed that life began relatively rapidly after conditions on the surface of the Earth allowed it, about 3.6 billion years ago, though the details regarding various rock and fossil samples are still under debate (Nisbet and Sleep, 2001). Because very little data are available regarding the atmospheric, oceanic or geological conditions on the pre-biological Earth, it is at present impossible to determine conclusively if terrestrial material (formed close to hydrothermal vents, in shallow oceanic basins, volcanic regions or in the atmosphere) or extraterrestrial molecules were most significant for the first steps in life's origin. A fundamental hurdle (among many others) remains how simple molecules (terrestrial and/or extraterrestrial) assembled into larger units under plausible geochemical conditions and later on into replicating or catalytic structures. Advances in directed evolution and membrane biophysics make the synthesis of simple living cells, if not yet foreseeable reality, an imaginable goal (Szostak et al., 2001).

Remnant planetesimals were present in high abundances in the early inner Solar System and they collided frequently with the young planets. The notion of extra-terrestrial delivery of organic matter to the early Earth has gained wide recognition because of the indisputable evidence that biomolecules, such as amino acids, have reached the Earth's surface associated with meteorites (e.g. Botta and Bada, 2002). Comets, interplanetary dust particles (IDP's) and carbonaceous meteorites were likely the most efficient sources of organic carbon in the early history of Earth. It has been estimated that the flux of organic matter reaching the Earth via comets and asteroids was significantly high during the period of heavy bombardment prior to 3.8 billion years ago (Chyba and Sagan, 1992). Comets are predominantly icy bodies, including silicates and organic refractory material (Greenberg, 1998). Many small (organic) molecules observed in cometary comae originate wholly or partially from the decomposition of much larger molecules/particles, indicating that large polymers such as polyoxymethylene (POM), polyamino-cyanomethylene (PACM) and HCN-polymers) are present in comets. The carbonaceous chondrites contain up to 2.8% carbon by weight (CI and CM type chondrites have the highest carbon content) and exhibit a range of thermal and aqueous alteration believed to have occurred on their parent bodies (Jarosewich 1990). The major part of this carbon, namely up to 90% corresponds to a macromolecular organic fraction (Gardinier et al., 2000). The large amount of extraterrestrial material ending up on young terrestrial planetary surfaces may have provided just the material needed for the assembly of the first autocatalytic molecules, that later evolved into primitive genetic material and subsequently to the first primitive cells.

TABLE 1. The cosmic timeline. (The Sun is a population I star).

Time (billion years)	Events
13.7	Big Bang
13.5	Beginning of the formation of Pop III stars
13.3	Beginning of the formation of Pop II stars
12.7	Beginning of the formation of Pop I stars
4.6	Solar System formation
4.5	Earth's formation
3.5-3.8	Traces of Life

6. Conclusion

The cycle of star-birth and late type evolution strongly influences the composition of interstellar matter. Interstellar molecules and dust within dense interstellar clouds become the building blocks for protostellar disks, from which planets, comets and asteroids form. The discovery of over a 100 of giant exoplanets has spurred a lot of activity in modeling the formation and evolution of planetary systems. It is clear now that several scenarios for planet formation are possible and that planet-planet, star-planet, and disk-planet interactions play a role in the consolidation of terrestrial-like planets. There is more than one path to planet formation, and consequently it is likely that terrestrial planets are common and diverse. Radial velocity surveys find a high frequency of extrasolar giant planets around stars that are more metal-rich than the Sun. It is still unclear what is causing this effect, and thus it is risky to extrapolate to what the frequency of habitable terrestrial planets may be as a function of metallicity. It seems premature to bias the selection of candidates for habitable stars against metal-poor stars (Turnbull and Tarter, 2003).

The fact that the same carbonaceous molecules are detected in different galactic space environments suggests that the carbon chemistry follows universal pathways throughout the Universe. There is strong evidence that aromatic material in the gas phase and in macromolecular form takes up most of the carbon in the interstellar medium, comets and meteorites and will also be the most abundant material delivered to the early planets (Ehrenfreund et al., 2002). The basic building blocks of life, at least those that are recognized on Earth, are most likely to be widespread in planetary systems in our Milky Way and other galaxies. Life on Earth is one of the outcomes of the formation and evolution of the Solar System and needed optimal conditions to develop. However, it may be extremely difficult to find any planet that is identical to Earth. The likelihood of finding extraterrestrial life depends on life's adaptability to different chemical and physical conditions throughout the Universe.

7. References

Bada, J. and Lazcano, A. (2002). Origin of life – some like it hot, but not the first biomolecules. Science **296**:1982-1983.

Blum, J. (2004) ASP Conf. Ser. 309, Grain growth and coagulation. In: A.N. Witt, G.C. Clayton, & B.T. Draine (eds.) *The Astrophysics of Dust* San Francisco: ASP, pp. 369-392.

Bodaghee, A., Santos, N., Israelian, G. and Mayor, M. (2003). Chemical abundances of planet-host stars. Results for alpha and Fe-group elements. Astronomy and Astrophysics **404**: 715-727.

Boogert, A. et al. (2004). Spitzer Space Telescope Spectroscopy of Ices toward Low-Mass Embedded Protostars. Astrophysical Journal Supplement Series **154**: 359-362.

Botta, O. and Bada, J. (2002). Extraterrestrial organic compounds in meteorites. Geophysics Survey: **23**, 411-467.

Charnley, S.B., Ehrenfreund, P. and Kuan, Y. (2003). Molecules in Space. Physics World, Institute of Physics Publishing Ltd, October 2003, 35-38.

Chen, L., Hou, J.L., Wang, J.J. (2003). On the Galactic Disk Metallicity Distribution from Open Clusters. I.New Catalogs and Abundance Gradient. The Astronomical Journal **125**: 1397-1406.

Cherchneff, I., Barker, J.R. and Tielens, A.G.G.M. (1992). Polycyclic aromatic hydrocarbon formation in carbon-rich stellar envelopes. Astrophysical Journal **401**: 269-287.

Chick, K. and Cassen, P. (1997). Thermal Processing of Interstellar Dust Grains in the Primitive Solar Environment. Astrophysical Journal **477**: 398-409.

Chyba, C.F. and Sagan, C. (1992). Endogenous production, exogenous delivery and impact-shock synthesis of organic molecules: an inventory for the origins of life. Nature **355**: 125-132.

Delsemme, A. (2000) 1999 Kuiper Prize Lecture Cometary Origin of the Biosphere, Icarus **146**: 313-325.

Du, C.H., Zhou, X., Ma, J., Shi, J.-R., Chen, A.B., Jiang, Z. and Chen, J. (2004). The metallicity distribution of F/G dwarfs derived from BATC survey data. The Astronomical Journal **128**: 2265-2273.

Ehrenfreund, P. and Charnley, S.B. (2000). Organic molecules in the ISM, comets and meteorite:A voyage from dark clouds to the early Earth. Annual Review of Astronomy and Astrophysics **38**: 427-483.

Ehrenfreund, P., Irvine, W., Becker, L., Blank, J., Brucato, J., Colangeli, L., Derenne, S., Despois, D., Dutrey, A., Fraaije, H., Lazcano, A., Owen, T. and Robert, F. (2002). Astrophysical and astrochemical insights into the origin of life. Reports on Progress in Physics **65**: 1427-1487.

Ehrenfreund, P. and Fraser, H. (2003). Ice chemistry in space. In: V. Pirronello and J. Krelowski (eds.) *Solid state Astrochemistry*. NATO ASI Series, Kluwer Academic Publishers, pp. 317-356.

Fernandez, J.A., Gallardo, T., and Brunini, A. (2002) Are There Many Inactive Jupiter-Family Comets among the Near-Earth Asteroid Population? Icarus **159**: 358-368.

Frenklach, M. and Feigelson, E.D. (1989). Formation of polycyclic aromatic hydrocarbons in circumstellar envelopes. Astrophysical Journal **341**: 372-384.

Gardinier, A., Derenne, S., Robert, F., Behar, F., Largeau, C. and Maquet, J. (2000). Solid state CP/MAS 13C NMR of the insoluble organic matter of the Orgueil and Murchison meteorites: quantitative study. Earth Planet Sci. Letters **184**: 9-21.

Greenberg, J.M. (1998). Making a comet nucleus. Astronomy and Astrophysics **330**: 375-380.

Ip, W-H and Fern?ndez, J.A. (1997). On dynamical scattering of Kuiper Belt Objects in 2:3 resonance with Neptune into short-period comets. Astronomy and Astrophysics **324**: 778-784.

Irvine, W.M., Crovisier, J., Fegley, B. and Mumma, M.J. (2000). Comets: A Link Between Interstellar and Nebular Chemistry. In: V. Mannings, A.P. Boss, S.S. Russell (eds.) *Protostars and Planets IV*, University of Arizona Press, Tucson, pp. 1159-1200.

Jarosewich, E. (1990) Chemical analyses of meteorites: a compilation of stony and iron meteorite analyses. Meteoritics **25**: 323-337.

Konacki, M. and Wolszczan, A. (2003). Masses and orbital inclinations of planets in the PSR B1257+12 system. Astrophysical Journal **591**: L147-L150.

Llorca, J. and Casanova, I. (1998) Formation of carbides and hydrocarbons in chondritic interplanetary dust particles: A laboratory study. Meteoritics & Planetary Science **33**: 243-251.

Llorca, J. and Casanova, I. (2000) Reaction between H_2, CO, and H_2S over Fe,Ni metal in the solar nebula: Experimental evidence for the formation of sulfur-bearing organic molecules and sulfides, Meteoritics & Planetary Science **35**: 841-848.

Llorca, (2005), this volume

Maciá, E., Hernández, M.V., Oró, J. (1997). Primary Sources of Phosphorus and Phosphates in Chemical Evolution, Origins of Life and Evolution of the Biosphere **27**: 459-480.

Mannings, V., Boss, A.P. and Russell, S.S. (2000). *Protostars and Planets IV*. Tucson: University of Arizona Press.

Markwick, A. and Charnley, S.B. (2004). Chemistry of Protoplanetary Disks: relation to primitive solar system material. In: P. Ehrenfreund et al. (eds.) *Astrobiology: Future perspectives*. Astrophysics and Space Science Library: Vol. **305**, pp. 33-66.

Mennella, V., Colangeli, L., Bussoletti, E., Palumbo, P. and Rotundi, A. (1998). A New Approach to the Puzzle of the Ultraviolet Interstellar Extinction Bump. Astrophysical Journal **507**: L177-180.

Nisbet, E.G. and Sleep, N.H. (2001). The habitat and nature of early life. Nature **409**: 1083-1090.

Oró, J. (1961). Comets and the formation of biochemical compounds on the primitive Earth, Nature **190**: 389-390.

Oró, J. (1983). Chemical evolution and the origin of life, Advances in Space Research **3**: 77-94.

Oró, J., Miller, S. L., Lazcano, A. (1990) The Origin and Early Evolution of Life on Earth, Annual Review of Earth and Planetary Sciences **18**: 317

Owen, T. and Bar-Nun, A. (1995). Comets, impacts and atmospheres, Icarus **116**: 215-226.

Pendleton, Y.J. and Allamandola, L.J. (2002). The Organic Refractory Material in the Diffuse Interstellar Medium : Mid-Infrared Spectroscopic Constraints. Astrophysical Journal Suppl. **138**: 75-98.

Porto de Mello, G.F. and da Silva, L. (1997). HR 6060: The Closest Ever Solar Twin?. Astrophysical Journal **482**: L89-L92.

Santos, N., Israelian, G. and Mayor, M. (2004). Spectroscopic [Fe/H] for 98 extra-solar planet-host stars Exploring the probability of planet formation: Astronomy and Astrophysics **415**: 1153-1166.

Scalo, J. and Biswas, A. (2002). Turbulent compressibility of protogalactic gas. Monthly Notices of the Royal Astronomical Society **332**: 769-776.

Scannapieco, E., Schneider, R. and Ferrara, A. (2003). The Detectability of the First Stars and Their Cluster Enrichment Signatures. Astrophysical Journal **589**: 35-52.

Sozetti, A. (2004). On the possible correlation between the orbital periods of extrasolar planets and the metallicity of the host stars. Monthly Notices of the Royal Astronomical Society **354**: 1194-1200.

Spaans, M. (2004). The synthesis of the elements and the formation of stars. In: P. Ehrenfreund et al. (eds.) *Astrobiology: Future perspectives*. Astrophysics and Space Science Library: Vol. **305**, pp. 1-16.
Szostak, J,W., Bartel, D.P. and Luisi, P,L. (2001). Synthesizing Life. Nature **409**: 387-390.
Throop, H.B. and Bally, J. (2004). Triggered formation of planetesimals by photo-evaporation, Abstract "36th DPS Meeting", Louisville (Kentucky, USA) #44.04.
Turnbull, M.C. and Tarter, J. (2003). Target selection for SETI. II. Tycho-2 dwarfs, old open clusters and the nearest 100 stars. Astrophysical Journal Supplement Series **149**: 423-436.
Udry, S., Mayor, M. and Santos, N. (2003). Statistical properties of exoplanets. I. The period distribution: Constraints for the migration scenario. Astronomy and Astrophysics **407**: 369-376.
Vauclair, S. (2004). Metallic fingers and metallicity excess in exoplanets' host stars: The accretion hypothesis revisited. The Astrophysical Journal **605**: 874-879.
Wehrstedt, M. and Gail, H.-P. (2003). Radial mixing in protoplanetary accretion disks. V. Models with different element mixtures. Astronomy and Astrophysics **410**: 917-935.
Weidenschilling, S.J. and Cuzzi, J.N. (1993). Formation of Planetesimals in the Solar Nebula. In: E.H. Levy, J.I. Lunine (eds.) *Protostars and Planets III*, Univ. of Arizona Press, Tucson, Arizona. pp. 1031-1060.
Wetherill, G.W. and Stewart, G.R. (1993). Formation of planetary embryos – effects of fragmentation, low relative velocity, and independent variation of eccentricity and inclination. Icarus: **106**: 190-209.
Wooden, D.H., Charnley, S.B. and Ehrenfreund, P. (2004). Composition and Evolution of Molecular Clouds. In: M. Festou, H.U.Keller, H. Weaver (eds.) *COMETS II*, Univ. Arizona Press, pp. 33-66.
Yin, Q., Jacobsen, S.B., Yamashita, K., Blichert-Toft, J., Telouk, P. and Albarede F. (2002). A short timescale for terrestrial planet formation from Hf-W chronometry of meteorites. Nature **418**: 949-52.
Zapatero-Osorio, M.R., Bejar, V.J.S., Martin, E.L., et al. (2000). Discovery of Young, Isolated Planetary Mass Objects in the Sigma Orionis Star Cluster. Science: **290**: 103-107.

Biodata of **H. Lammer** et al. authors of ***"Prerequisites for the Evolution of Life on Exoplanets"***

Mag. Dr. Helmut Lammer is a scientist at the *Austrian Space Research Institute*, of the *Austrian Academy of Sciences* in Graz (http://www.iwf.oeaw.ac.at) and received his Ph.D., in Geophysics from the *Karl-Franzens-University* in 1992. He is involved in various space missions and is a member of ESA's Terrestrial Exoplanet Science Advisory Team (TE-Sat). **helmut.lammer@oeaw.ac.at**

Thomas Penz is a graduate student at the *Institute of Physics*, *Karl-Franzens-University* in Graz, Austria. **thomas.penz@oeaw.ac.at**

Dr. Natalia K. Belisheva is a biologist at the *Polar-Botanical Garden Institute* of the *Kola Scientific Centre*, of the *Russian Academy of Science*s and works currently on effects due to high energetic particles on biological systems. She received her PhD in biology from in the *Institute of Cytology* of the *Russian Academy of Science* in Leningrad (St.Petersburg) in 1981. **belisheva@com.mels.ru**

Dr. Yuri N. Kulikov graduated from the *Moscow Physico-Technical Institute*'s *Department of Aerophysics and Applied Mathematics*, where he received his PhD in 1984. He was involved in several space missions of the former USSR and works at present as an atmospheric scientist at the *Polar Geophysical Institute* of the *Russian Academy of Sciences* in Murmansk. **kulikov@pgi.ru**

Prof. Dr. Helfried K. Biernat is a scientist at the *Space Research Institute* of the *Austrian Academy of Sciences* in Graz, Austria. He received his PhD in 1978 from the Institute of Theoretical Physics at the *Karl-Franzens-University* in Graz. He is an expert in magnetospheric physics and a lecturer at the *Karl-Franzens-University* in Graz, Austria. **helfried.biernat@oeaw.ac.at**

T. Penz, Yu. N. Kulikov, N. K. Belisheva, H. Lammer

H. K. Biernat

J. Seckbach (ed.), Life as We Know It, 413-429.

PREREQUISITES FOR THE EVOLUTION OF LIFE ON EXOPLANETS

HELMUT LAMMER[1], THOMAS PENZ[1, 2], NATALIA K. BELISHEVA[3], YURI N. KULIKOV[4], AND HELFRIED K. BIERNAT[1, 2]

[1]Space Research Institute, Austrian Academy of Sciences, Schmiedlstrasse 6, A-8042 Graz, Austria; [2]Institute of Physics, University of Graz, Universitätsplatz 5, A-8010 Graz, Austria, [3]Polar-Alpine Botanical Garden-Institute, Kola Scientific Centre, Russian Academy of Sciences, Fersman Str. 14, 189204, Apatity, Russian Federation; and [4]Polar Geophysical Institute, Russian Academy of Sciences, Khalturina Str. 15, Murmansk, 183010, Russian Federation

1. Introduction

Until 1995, when the first exoplanet orbiting a main-sequence star was detected (Mayor and Queloz, 1995), the question if Earth-like planets and consequently also life could exist outside our Solar System was a hypothetical one. In spring 2005 we know more than 150 giant exoplanets. Recently, the first non-Jupiter-class exoplanet with a mass of about $14M_\oplus$ orbiting a solar-like G-type star at about 0.1 AU was discovered (Santos et al., 2004). This detection represents the first discovery of a planet with a mass slightly smaller than that of Uranus. Whether this planet can be considered a migrated *ice giant* or a *super-Earth* is currently under debate in the context of the core-accretion and migration models. In fact also Earth-mass planets where detected orbiting the pulsar PSR1257+12 (Wolszcan and Frail, 1992), but it is very unlikely that life can be found in such an environment.

The discoveries of such low-mass exoplanets strengthen the major question, which has to be answered in the future, namely if life may have evolved on a habitable Earth-like exoplanet outside our Solar System. For understanding the principles that generated Earth's environment and its long-time habitable conditions and for studying similar terrestrial exoplanets, space missions like ESA's Darwin and NASA's Terrestrial Planet Finder (TPF-C, TPF-I) and their precursor photometry missions like CoRoT (CNES) and Kepler (NASA) are currently under development. The CoRoT space observatory is planned to be launched in 2006 and will be able to find exoplanets with sizes above $2R_\oplus$ by using high-precision photometry and will therefore, be the first mission in this series dedicated to discover small exoplanets at orbital distances of up to about 0.5 AU (e.g., Rouan et al., 2000; Deeg et al., 2002).

Although the necessary conditions for the emergence, survival, and evolution of life are still unknown, one requirement is widely accepted as an unavoidable necessity: liquid water. The presence of liquid water, far from being a sufficient requisite for biology, allows the identification of potential extraterrestrial habitats or, more rigorously, to exclude dry environments, where the presence of life may be ruled out. In the Solar System, and with the exception of Earth, liquid water is expected in the subsurface of Mars, on icy satellites like Europa and Callisto and in the Venusian clouds, that does not imply that life is likely to be found there, but that life may be absent elsewhere. One should also note that there are strong indications that microbial life is widespread at depth in the crust of the Earth, and life has been identified in numerous ocean vents. This life-forms do not dependent on solar energy and photosynthesis for their primary energy supply, and they are essentially independent of the surface circumstances.

This article is organized as follows: we discuss in Section 2 the galactic environments, which are suitable for the evolution of life on Earth-like exoplanets orbiting inside the circumstellar habitable zone, as outlined in Section 3. In Section 4 we study the implications of high X-ray and EUV fluxes of young and active host stars to planetary atmospheres and their water inventories. Because most stars in the Galaxy are low mass M-type stars with habitable zones in orbital distances less than 0.3 AU, we study in Section 5 the effects of coronal mass ejections on their atmospheres. In the next section we investigate the impact of high energetic particles on biological systems on the surface of the Earth as an analogue for terrestrial exoplanets. Finally, in Section 7, we discuss potential subsurface habitats on terrestrial exoplanets.

2. From Element Formation to the Galactic Habitable Zone

In the early stage of the Universe, the temperature and the particle density was suitable for nuclear fusion for about 3 minutes, the so-called Big Bang nucleosynthesis (BBN) era. During this stage, the formation of hydrogen, deuterium, the helium isotopes ^{3}He and ^{4}He, as well as ^{7}Li and ^{7}Be took place (e.g., Wagoner et al., 1967; Copi et al., 1995). The brevity of the BBN era is important because it prevented elements heavier than beryllium from forming, while at the same time allowing unburned lighter elements, such as deuterium, to exist. The elemental abundances derived from theoretical considerations are about 75% hydrogen, 25% ^{4}He, and trace amounts of deuterium, Li and Be, which is in general agreement with observations (e.g., Tytler et al., 2000; Thuan and Izotov, 2000).

All the heavier elements were formed by different following processes, which are mainly connected with fusion processes inside stars (Trimble, 1997). In main sequence stars, a number of nuclear burning sequences takes place (Pagel, 1997), starting from hydrogen burning, helium burning, and then proceed to more advanced burning stages.

Hydrogen burning is the main source of energy for all stars for more than 90% of their life times. For hydrogen, there are two main reaction chains. These are the proton-proton (pp) chain and the CNO-cycle, where the former dominates in low mass main sequence stars ($M < 1.5 M_{\mathrm{Sun}}$) like our Sun, while the latter dominates in more massive stars. When the hydrogen is exhausted in the center of a star, the star contracts, and the temperature rises and eventually helium is ignited. This usually happens when $T > 10^8$ K. Helium burning produces carbon ^{12}C by the triple-alpha process from ^{4}He and ^{16}O from ^{12}C via α- and γ-decay in the cores of all stars more massive than about 0.5 M_{Sun} (e.g., Fynbo et al., 2005). In massive stars, carbon burning, neon burning, oxygen burning, and silicon burning takes place, leading to the formation of elements up to ^{56}Fe (see Fig. 1).

Elements heavier than Fe are formed by adding neutrons and subsequent radiative decays via the so-called s-, r-, and p-processes, where the first is a slow process taking place in the Asymptotic Giant Branch (AGB) stars of the Hertzsprung-Russell diagram (Busso et al., 1999), while the latter processes appear during supernova explosions (Cowan and Sneden, 2004).

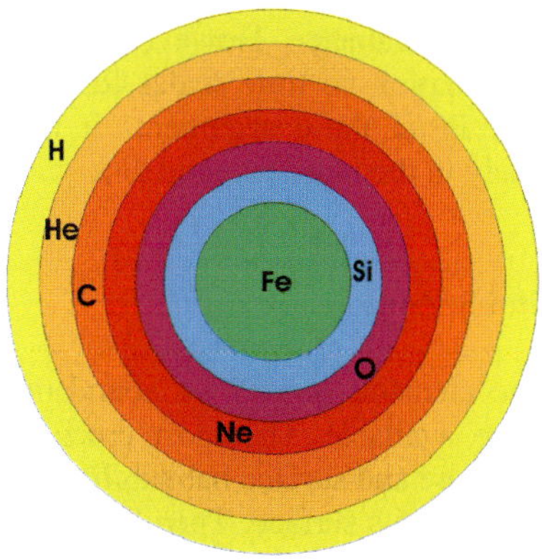

Fig. 1. Illustration of the onion-like nuclear burning structure of a massive star with $M > 8M_{\rm Sun}$.

Only hydrogen is created in abundance shortly after the Big Bang, while the other biologically important elements C, O, P, N, and S are formed during fusion processes occurring at least in massive main-sequence stars. Also the heavy elements necessary for the existence of terrestrial planets, like Si, O, Mg, Fe, and Ni originate from massive stars. Only some radioactive elements, like U and Th, which are important for the heat budget of terrestrial planets are formed in supernovae explosions. These considerations restrict the time range during which habitable terrestrial planets can evolve. During the first Gyr, the generation of all important elements took place, and only after the first generation of stars died, these heavy elements could be distributed throughout the Universe.

Probably the most crucial parameter for the formation of terrestrial exoplanets as pointed out above is the generation of heavy elements and the metallicity of the protoplanetary cloud. If the metallicty is too small, no terrestrial planets can form, because the necessary chemical elements are not available in a reasonable abundance. A low metallicity is leading to the formation of small terrestrial planets, whereby Gonzalez et al. (2001) estimated that at least half of the Sun's metallicity is needed for the formation of terrestrial planets. If the planets are too small, the heat flow and associated geophysical processes like plate tectonics and strong intrinsic magnetic moments can not be maintained over sufficient time periods, leading to efficient atmospheric loss processes.

Santos et al. (2003) showed that close-in giant exoplanets are correlated with a high metallicity of their host stars, which may gravitationally perturb the orbits of terrestrial planets in the habitable zone. Also the abundance of the radioisotopes ^{40}K, $^{235,238}U$, and ^{232}Th is important, since these elements are sources of geothermal heat, allowing a long-term climate stability via the carbon cycle.

The region in a Galaxy, which is called the Galactic Habitable Zone (GHZ) defines the favorable sites, where life most likely can evolve (Gonzalez et al., 2001). Additional constraints, like an environment free of life-extinguishing supernovae (Dar et al., 1998), were introduced in the definition by Lineweaver et al. (2004). Additionally, Lineweaver et al. (2004) introduced the time span needed for the evolution of complex life as a parameter. However, this seems to be somehow questionable, because the information on time scales for the origin of life is based only on terrestrial life-forms. From these arguments it is obvious that habitable planets can be found most likely in the inner galactic disk.

3. The Circumstellar Habitable Zone

As discussed above, biogenic elements should be abundant in our galactic neighbourhood, and heat sources are available. The most crucial question regarding the evolution of an Earth-like biosphere and surface habitability of terrestrial planets is the presence of liquid water on the planets surface. Therefore, the circumstellar habitable zone (HZ) is defined as the region, where a planet can maintain surface conditions that allow liquid water to exist.

Two important parameters, which influence the HZ, are the energy emitted by the host star and the atmospheric composition of the planet (e.g., Hart, 1979; Kasting et al., 1993; Franck et al., 2002; Kasting and Catling, 2003). The influence of the emitted energy of the host star is easy to describe. On the outer boundary of the HZ, the energy input is too small to maintain liquid water on the surface, while the inner boundary is defined as the distance, where water escapes rapidly to space. Since the energy output of the star increases over time, the location of the HZ also changes and will move outwards.

If a planet has an atmosphere, the most important atmospheric parameter for the surface temperature is the abundance of greenhouse gases, like CO_2, CH_4, and water. If there is a dense CO_2 atmosphere, the surface temperature can be raised by several 100°C, therefore the HZ can be farther away from its host star. A more conservative estimate of the outer boundary position is based on the fact that a dense CO_2 atmosphere starts to condensate in the middle atmosphere at a closer distance to the star. This can occur, if there is no atmospheric dust or another greenhouse gas, which can raise the temperature of the middle atmosphere. This condensation process will limit the amount of CO_2 present in the atmosphere, and the outer boundary of the HZ is shifted inside (Kasting, 1991). Additionally, at large orbital distances, an increase of the CO_2 pressure results in surface cooling more than heating, due to enhanced Rayleigh back-scattering of the incoming stellar radiation to space (e.g., Michelangeli et al., 1993; Forget and Pierrehumbert, 1997).

The inner boundary of the HZ is found at the distance, where a runaway greenhouse effect (Kasting, 1988) can occur, resulting in vaporization of water, that starts to increase the surface warming and thus enhance the evaporation in a positive feedback process. This results in a H_2O-rich atmosphere with surface temperatures above 100°C, in which H_2O vapor reaches the upper atmosphere, where it can be photolyzed. Within a period of the order of 10 to 100 Myr, the planet loses its hydrogen to space and becomes dry. One main parameter in the definition of the HZ is the energy input by the host star, which is related to its mass. Since stars with masses of more than $2M_{\rm Sun}$ have lifetimes of less than 2 Gyr, it is probably unlikely that complex life forms will evolve on a planets surface during this short time frame.

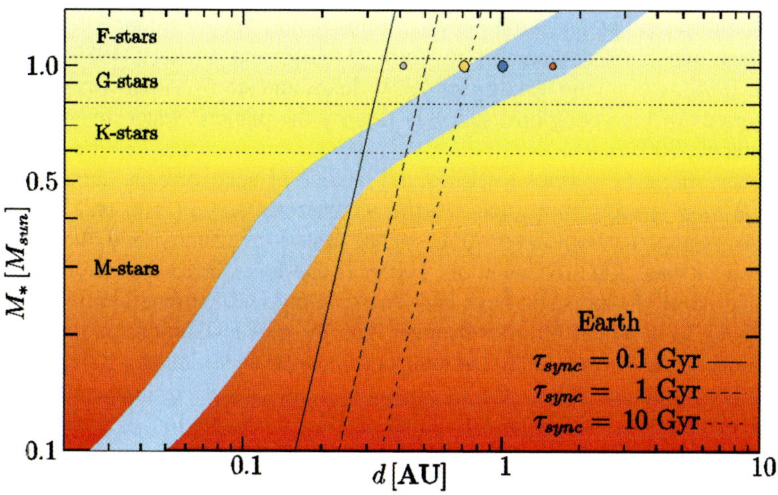

Fig. 2. Continuous circumstellar habitable zone from 0.5 Gyr after Zero Age Main Sequence to 90% of the main sequence as a function of stellar mass (Bauer and Lammer, 2004). The straight lines give the tidal-locking limit at which an Earth-mass planet become synchronized after 0.1, 1 and 10 Gyr (courtesy of F. Selsis and J.-M. Grießmeier).

Furthermore, a more massive star has a higher energy output compared to a low-mass star. Therefore, for stars with masses of $1.5M_{\rm Sun}$, the HZ is centered at about 2.5 AU, while the HZ of a star with $0.5M_{\rm Sun}$ is centered at about 0.3 AU (Fig. 2). Additionally, low mass stars evolve more slowly, thus the location of the HZ does not change significantly during several Gyr (Kasting et al., 1993).

4. Evolution of Planetary Water Inventories and Atmospheric Loss

Water can be accreted by terrestrial planets through planetesimals and planetary embryos (e.g., Levison et al., 2000; Morbidelli et al., 2000). The level of hydration of planetesimals and embryos depends on the heliocentric distance at which they are formed. A threshold distance is the so-called snowline, beyond which water condensates as ice. It is believed that, at the time of planetesimal formation in the Solar System, the snowline was at about 4 - 5 AU. Icy snowballs formed at the snow line could have drifted inwards by gas-drag and be incorporated by growing planetesimals in the inner Solar System (Cyr et al., 1998). Additionally, accretion of icy planetary embryos delivers a substantial mass in the form of water (Morbidelli et al., 2000; Raymond et al., 2004). The location, mass, number, and eccentricity of gas giants outside the HZ are important for the occurrence of impacts and the delivery of water and volatiles to terrestrial planets (Wetherill, 1994; Levison et al., 2000; Raymond et al., 2004).

Earth-like planets can evolve into a habitable world if they keep their atmospheres and water inventories during the period of heavy bombardment by asteroids and comets and during the host stars' active X-ray and extreme ultraviolet (XUV) radiation fluxes and stellar wind periods. Impacts may play a minor role for planets with a size and mass comparable to Earth (Melosh and Vickery, 1989; Brain and Jakosky, 1998), while long lasting high XUV fluxes and strong stellar winds during the active periods of the young host star may destroy the planets' water inventories (e.g., Lammer et al., 2005a).

Observations by various satellites and studies of solar proxies indicate that the early Sun was rotating more than 10 times faster its present rate and had correspondingly strong dynamo-driven high energy emissions (Zahnle and Walker, 1982; Guinan and Ribas, 2002; Ribas et al., 2005). It can be inferred that the early Sun, a representative of G-type stars, have had strong X-ray and extreme ultraviolet (XUV) emissions up to several 100 times stronger than the present Sun (Zahnle and Walker, 1982; Guinan and Ribas, 2002; Lammer et al., 2004; Ribas et al., 2005). Since the stellar XUV radiation affects the thermosphere and exosphere temperatures of the exposed planets (e.g., Bauer and Lammer, 2004), the evolution of planetary water inventories must be understood within the context of the evolving stellar energy and particle fluxes. If IR-cooling of the thermosphere can not balance the incoming stellar XUV radiation anymore, the excess thermal energy is directly converted into kinetic energy and hydrodynamic escape occurs.

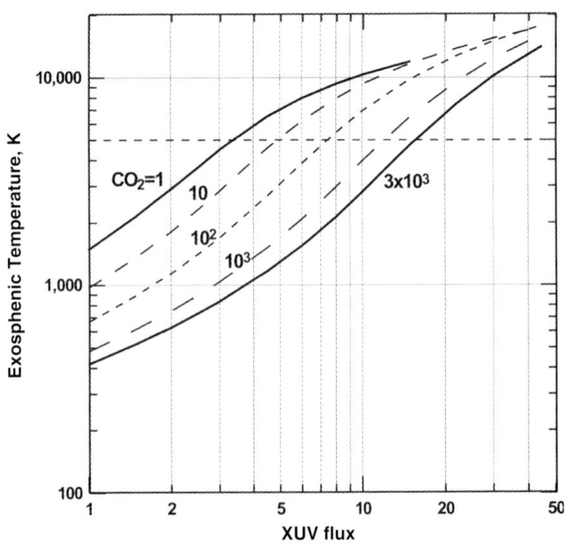

Fig. 3. Exospheric temperature at an Earth-like planet as a function of stellar XUV-radiation for various CO_2 levels expressed in PAL units (PAL = 1 corresponds to the Earth's CO_2 Present Atmospheric Level). The horizontal dashed line shows the critical *blow off* temperature for atomic hydrogen of about 5000 K.

One should also not forget that the stellar wind experiences large change over time. Wood et al. (2002) showed that solar proxies have a 100 to 1000 times denser particle environment during the early stage of their evolution.

Furthermore, the XUV radiation environments of dwarf stars seem to have saturated emission levels up to 1 Gyr and possibly longer and then decrease in an analogous way to G and K stars (Bauer and Lammer, 2004; Ribas et al., 2005; Khodachenko et al., 2005). Observations indicate that early M stars can have XUV irradiances that are about 10 – 100 times higher than solar-type G stars of the same age. Fig. 3 shows calculated exospheric temperatures, corresponding to various CO_2 levels at an Earth-like planet, by using the thermospheric model of Gordiets et al. (1982) adapted recently for high CO_2 abundance.

An interesting effect may appear at a terrestrial water-rich planet inside the HZ of a dwarf star emitting high XUV fluxes over longer time periods then a G-type star. High CO_2 levels in a planet's early atmosphere may prevent extreme hydrodynamic escape conditions like on early Earth. But after the CO_2 is removed from the atmosphere due to chemical weathering in a humid environment (Pollack et al., 1987), N_2 like on present Earth may become the dominant constituent in the atmosphere. So, the upper atmosphere is heated and large escape rates develop, which can evaporate the whole planets H_2O inventory (Bauer and Lammer, 2005a).

The conventional theory is that CO_2, in the form of H_2CO_3 dissolved in water, could be removed by reactions with exposed continental rocks to produce carbonate alteration products that were, in turn, transported to the oceans by rain and river flows for final burial in deep-sea sediments. Ultimately, the latter would be removed into the mantle by subduction of the sediment-coated oceanic crust. Plate tectonics plays an important role in the recycling of CO_2 in the atmosphere; the subducted carbonate is resupplied to the atmosphere through volcanic eruptions (Kasting and Catling, 2003). Therefore, it is very likely that geological processes, like plate tectonics and surface weathering, are essential for the long-term habitability of Earth-like exoplanets (von Bloh et al., 2003).

5. Effects of CME's on Planetary Atmospheres at Close-In Habitable Zones

Especially the surface habitability of terrestrial exoplanets orbiting around low-mass M and K stars requires a detailed investigation, because dwarf stars constitute about 75% of all main sequence stars in the Universe (e.g., Bauer and Lammer, 2004). Joshi (1997) applied a climate model to exoplanets orbiting M stars in synchronuous rotation, and showed that the climatic changes may not rule out the evolution of life, if the atmosphere is thick enough, which was confirmed by Heath et al. (1999). An additional study by Joshi (2004) showed that planets inside M star HZs may even establish a biosphere with atmospheres of about 1 bar, depending on the available land-mass and water covered surface.

The HZ of low-mass K and M stars is at orbital distances of ≤0.5 AU, where the planets become tidally locked on time scales of less than 1 Gyr (see Fig. 2). Due to slow rotation, tidally locked Earth-like planets will have weak magnetic dynamos resulting in weak intrinsic magnetic moments, which are decreased by about two orders of magnitude (Grießmeier et al., 2004; Khodachenko et al., 2005). Thus, these planets will develop only small magnetospheres, which are compressed by the stellar wind at close orbits at least during the active stellar periods. Thus, the planetary atmospheres may build a Venus-like obstacle for the stellar wind interaction, leading to an enhanced atmospheric loss (Grießmeier et al., 2004). Additionally, the continuous occurrence of Coronal Mass Ejections (CMEs) during the first 2 Gyr of the M star evolution may further enhance the atmospheric loss rates and remove planetary atmospheres of several tens of bars during short time periods (Lammer et al., 2005b).

These studies show that the atmospheres of terrestrial exoplanets are strongly affected by the incoming dense CME plasma flow at orbital distances ≤0.1 AU. CMEs can compress the magnetospheres of such planets down to atmospheric levels, where the ionized compounds of the atmosphere build an ionopause obstacle. The preliminary results based on momentum consideration models indicate that CMEs can erode an atmosphere equivalent to a pressure of about 1 bar of an Earth-like exoplanet orbiting at 0.05 AU after 40 Myr, resulting in a loss of about 25 bar after 1 Gyr.

For weakly magnetized Earth-like exoplanets at orbital distances of about 0.1 AU, an atmosphere of about 1 bar and 5 bar is eroded by CMEs after 200 Myr and 1 Gyr, respectively. On the other hand, an atmosphere equivalent to about 1 bar can be lost at weakly magnetized Earth-like exoplanets after 1 Gyr at orbital distances of about 0.15 AU. Studies with a complex test particle model (Lammer et al., 2005b), which include additional ionization of the atmospheric gas above the planetary obstacle by the high XUV fluxes of M stars, are in progress. One can expect that these model results yield even higher atmospheric erosion rates.

These results indicate that only terrestrial exoplanets in tidal-locking zones with intrinsic magnetic moments $\mathcal{M}_{\rm pl} > 10^{-1}\mathcal{M}_{\oplus}$ can develop large enough magnetospheres, so that their atmospheres can be protected against hitting CMEs and related atmospheric erosion processes. However, it should be noted that the Earth would not build up such strong magnetic moments at orbits inside close-in HZs, so the atmosphere of an Earth's twin would not be protected against CME-erosion (Khodachenko et al., 2005).

Also the stellar wind experiences large change over time. Wood et al. (2002) showed that solar proxies have a 100 to 1000 times denser particle environment during the early stage of their evolution. This has similar consequences than CMEs, especially for non- or weakly-magnetized planets, which establish a Venus-like stellar wind-atmosphere interaction, leading to erosion of the planetary atmosphere by various non-thermal loss processes.

6. High Energetic Particle Impact on Biological Systems on Planetary Surfaces

Earth like planets orbiting young or active stars are subject to large ionizing fluxes of stellar flares. During these events high energetic particles interact with the atmosphere, producing secondary energetic particles, like pions, and muons, where some of them can reach the surface causing Ground Level Enhancements (GLE) of secondary cosmic particles. This has effects on the biological activity through direct mutational enhancement or sterilization (Smith et al., 2004). Such events may especially influence Earth-like planets orbiting low-mass stars, since they are known for high flare activity (e.g., Haisch et al., 1991; Smith et al., 2004; Khodachenko et al., 2005). Furthermore, due to tidal locking, the protecting magnetospheres may be reduced (Grießmeier et al., 2005).

Smith et al. (2004) studied the transport of ionizing radiation in Earth-like exoplanetary atmospheres inside orbits of low-mass M stars and concluded that biological activity and atmospheric chemistry should be strongly influenced by the exposure to such intensely fluctuating radiation environments, although the nature and efficiency of the effects remains to be estimated.

One may expect that a large increase of solar secondary cosmic radiation on the Earth's surface due to the Great Solar Proton Events (GSPE) will also produce biological effects in living systems. The role of secondary solar cosmic radiation near Earth's surface for simple bio-systems was discovered and studied after the GSPE associated with GLEs, which occurred during 1989 (Belisheva et al., 1995; Belisheva and Gak, 2002; Belisheva and Vashenyuk, 2003; Grießmeier et al., 2005).

Experiments on three different cell cultures, growing in-vitro (Belisheva et al., 1995), were carried out during the second largest GLE (Reeves et al., 1992) in September and October 1989. It was found that during the increase of the energetic proton fluxes observed by the GOES-7 satellite (Reeves et al., 1992) in Earth's orbit, the cell fusion dynamics in all cells had a significant correlation with energetic α-particle and proton fluxes.

Fig. 4 shows examples of biological effects on cell cultures studied during and after the GSPE in 1989. Simultaneously, in diverse cell-lines, gigantic nuclei, nuclei association, chromatin fragmentation, dispersion of cell and nuclei matter, micro-cells, micro-nuclei and separate chromosomes emerged. The left panel in Fig. 4 shows intracellular nuclear associations, which are coupled with cell fusion. The right panel shows gigantic nuclei, which occurred either by cell nuclear fusion or by abortive cytokinesis usually observed after X-ray irradiation. One should note that similar distortions in diverse biological systems were found in biological experiments during spacecraft flights in Earth's orbit.

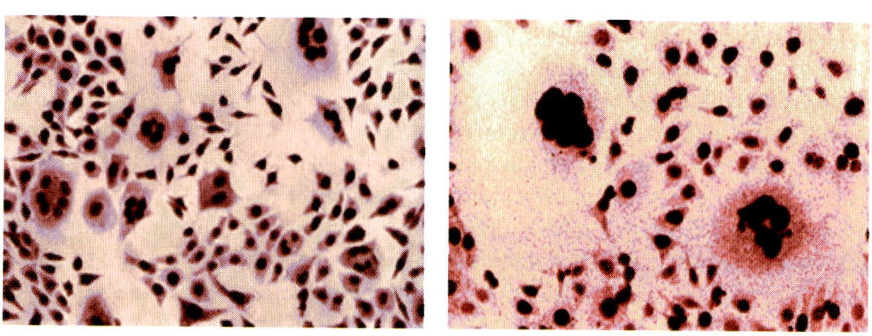

Fig. 4. Biological effects caused by secondary solar cosmic radiation on simple genetic systems observed in cell cultures in Apatity, Russia (Lat.: 67.57°; Long.: 33.40°), during the high energetic solar proton event in 1989.

Earth-like exoplanets at orbital distances of about 0.2 AU with an atmospheric pressure of about 1 bar may experience the same biological effects as observed at Earth's polar region during the GSPE in 1989 over 100% of its surface (Grießmeier et al., 2005). Thus, tidal locked Earth-like planets inside the HZ of K and M stars are less protected against high energy and secondary cosmic rays due to weaker magnetic moments and shall therefore, experience more GLEs. One can expect more mutations and biological effects on living surface organisms on terrestrial planets in orbits of close-in HZs, than at Earth-like planets at orbital distances of about 1 AU. In particular, it is unknown whether such a mutationally rich environment would enhance or suppress the rate of evolution even in simple population genetic models (Dar et al., 1998; Smith et al., 2004).

However, the effects on biological systems by high energy cosmic ray particles would not occur, if the atmosphere of a terrestrial planet were dense like on Venus (100 bar surface pressure), so that no secondary radiation can reach the surface. Contrary, the surfaces of planets with thin atmospheres like Mars (7 mbar surface pressure) have hostile surface environments due to radiation induced sterilization.

7. Unconventional Habitats in the Universe

From the discussions above one can see that the HZ defined for stars with masses $\leq 1.5 M_{\mathrm{Sun}}$ (F, G, K and M stars) is more favourable to develop a biosphere on exoplanets orbiting massive K and solar-like G and F stars. However, the definition of the HZ does not tell the whole story about planetary habitability, because it is based on the idea that only starlight is the energy source providing liquid water on a planet. Additionally, the definition in Sec. 3 proposes that liquid water must be present on the planetary surface to harbour life, and therefore atmospheric stability, of course, is a requirement. But even on Earth there exist habitats, where these restrictions are not valid. This includes ecosystems at hydrothermal vents, living without sunlight, but using chemical energy provided by chemical reactions between hot water and the surrounding rocks (e.g., Zierenberg et al., 2000). Additionally, there are life forms dwelling deep inside the continental and oceanic crust, living completely independent from Earth's surface (e.g., Gold, 1992, Krumholz et al., 1997).

Their energy supply comes from chemical sources, due to fluids that migrate upwards from deeper levels inside the Earth. In mass and volume this biosphere may be comparable with all surface life. Such microbial life may account for the presence of biological molecules in all carbonaceous materials in the outer crust, and the inference that these materials must have originated from biological deposits accumulated at the surface is therefore not necessarily valid. Subsurface life may be widespread among the planetary bodies of our solar system, since many of them have equally suitable conditions below, while having totally inhospitable surfaces. One may even speculate that such life may be widely disseminated in the universe, since planetary type bodies with similar sub-surface conditions may be common as solitary objects in space, as well as in other solar-type systems.

If there is some other energy source, like gravitational interactions or internal radioactivity, keeping water liquid, subsurface life may flourish even without a star.

Also potential chemical energy sources such as hydrogen and iron oxides are common in planetary environments.

One place in our Solar System where other heat sources are available may be the subsurface oceans of icy satellites like Europa and Callisto (Khurana et al., 1998). In this case, the heat source is the gravitational interaction between the icy satellite on one side, and the planet (Jupiter) on the other side. Due to this heat source, the ice layer is molten below 3 – 30 km, giving rise to several possible habitats, including the ice layer, the brine ocean, and the seafloor environment (Marion et al., 2003).

In other stellar systems, terrestrial planets may form far from their host stars, which might not need gravitational interactions to maintain a deep ocean. According to numerical models (e.g., Raymond et al., 2004), rocky planets could form at distances between two to four times the Earth's orbit. Since the stars heat input would evaporate less water during their formation, such distant planets would likely have deeper oceans than Earth. Internal heat sources alone can melt any ice layer deeper than 14 km below the surface on an Earth-sized planet (Vogel, 1999). Furthermore, Léger et al. (2004) have shown that massive ice-rich planets possibly form in external regions of protoplanetary disks and migrate inward. Depending on their distance to the star and properties of their atmospheres, some of them may form a surface water ocean. Such an ocean of liquid water can have a thickness of about 100 km. The results of these studies lead to the plausible conclusion that planets with deep oceans underneath ice layers may exist which provide all conditions necessary for life.

An additional scenario is proposed by Lissauer (1987), because many planetary embryos form quickly by runaway accretion during the origin of a stellar system. Some of these embryos may merge but others may be scattered into escape trajectories from the stellar system by proto-giant planets. Stevenson (1999) pointed out that such free-floating interstellar planets may also develop large oceans protected by dense atmospheres. Probably, life has much more time to evolve on such interstellar planets compared to planets in stellar systems. The expanding stars annihilate all life on planets inside the HZ after some Gyr, but an interstellar planets internal heat may keep life alive for at least 30 Gyr.

However, complex life forms require much energy to establish extended biospheres. The energy available at most of the potential habitats discussed in this section is orders of magnitudes less than at present Earth (Jakosky and Shock, 1998; Stevenson, 1999). Because, biosignatures of biological communities dwelling in a planets subsurface are hard to detect, planets with the potential to develop Earth-like extended biospheres on their surfaces are better candidates to search for life outside of our Solar System.

8. Conclusions

The evolution of biospheres and life on Earth-like exoplanets inside the region where life supporting planetary environments can exist, depends on the size and mass of the planet, greenhouse gases abundance such as H_2O, CO_2, its distance to the central star, the luminosity of the host star, etc. We show that the habitable zone is not a static or permanent region and the evolution of planetary atmospheres and water inventories is closely connected to the evolution of the X-ray and EUV activity of their host stars. Furthermore, Earth-like planets in orbits around M and K dwarfs are tidally locked after about 100 Myr, that results in weak magnetic moments, no plate tectonics, strong atmospheric erosion caused by CME's, and high energy particle exposure on the planets surface. These conditions can led to an enhancement of mutations in biological systems. Although, surface habitability and the evolution of Earth-like biospheres may not be favoured at M star planets, subsurface life may evolve. Because low-mass M dwarfs constitute about 75% of all main sequence stars in the solar neighborhood and the Galaxy, this result has important astrobiological implications, although more studies on surface habitability of Earth-like planets inside low-mass habitable zones should be carried out.

9. Acknowledgements

The authors thank the ÖAD, which supported this work by the projects I.12/04 and I.2/04. N. K. Belisheva and Yu. N. Kulikov also thank the "Russian Fund for Basic Research," which supported this study as the Russian-Austrian project RFBR No. 03-05-20003 BWTZ a, "Solar-planetary relations and space weather." Also acknowledged is the Austrian FWF, which supported this study under project P17099-N08, the support by the ÖAW, "Verwaltungsstelle für Auslandsbeziehungen", the Russian Academy of Sciences, as well as the Austrian Ministry bm:bwk for supporting the CoRoT project.

10. References

Bauer, S. J. and Lammer, H. (2004). *Planetary Aeronomy: Atmosphere Environments in Planetary Systems,* Springer, Berlin.

Belisheva N. K. and Popov, A. N. (1995). Dynamics of the morphofunctional state of cell cultures with variation in the geomagnetic field in high latitudes. *Biophysics* **40**: 737-745.

Belisheva, N. K. and Gak, E. Z. (2002). Significance of cosmic ray variations for biosystem functions. *Proceedings of the VII Int. Conf. On "Ecology and Development of North-West Russia" 2 - 7 August,* 2002, St. Petersburg, pp. 118-129.

Belisheva, N. K. and Vashenyuk, E. V. (2003). Cell systems are indicator of cosmic ray variations. *Int. Crimean Conference "Cosmos and Biosphere".* (September 28 - October 04) Crimea, Ukraine, pp. 75-76.

Brain, D. A. and Jakosky, B. M. (1998). Atmospheric loss since the onset of the Martian geologic record: Combined role of impact erosion and sputtering. *J. Geophys. Res.* **103**: 22689-22694.

Busso, M., Gallino, R. and Wasserburg, G. J. (1999). Nucleosynthesis in Asymptotic Giant Branch stars: relevance for galactic enrichment and Solar System formation. *Ann. Rev. Astron. Astrophys.* **37**: 239-309.

Chassefiere, E. (1996). Hydrodynamic escape from primitive atmospheres: Applications to the cases of Venus and Mars. *Icarus* **124**: 537-52.

Copi, C. J., Schramm, D. N. and Turner, M. S. (1995). Big-Bang Nucleosynthesis and the Baryon Density of the Universe. *Science* **267**: 192-199.

Cowan, J. J. and Sneden, C. (2004). Stellar abundances: the r-process and supernovae, In: P. Kunar, C. Wheeler and P. Hoeflich (eds.) *3-D signatures in stellar explosions.* Cambridge Univ. Press, Cambridge.

Cyr, K. E., Sears, W. D. and Lunine, J. I. (1998). Distribution and evolution of water ice in the Solar nebula: Implications for Solar System body formation. *Icarus* **135**: 537-548.

Dar, A., Laor, A. and Shaviv, N. J. (1998). Life extinctions by cosmic ray jets. *Phys. Rev. Lett.* **80**: 5813-5816.

Deeg, H. and Horne, K. (2002). Eddington's planet finding capabilities, In: B. Schürmann (ed.) *The First Eddington Workshop: Stellar Structure and Habitable Planet Finding*. ESA-SP **485**: 123-128.

Forget, F. and Pierrehumbert, R. T. (1997). Warming early Mars with carbon dioxide clouds that scatter infrared radiation. *Science* **278**: 1273-1276.

Franck, S., von Bloh, W., Bounama, C., Steffen, M., Schönberner, D. and Schnellnhuber, H.-J. (2002). Habitable zones in extrasolar planetary systems. In: G. Horneck and C. Baumstark-Kahn (eds.) *Astrobiology: The Quest for the Conditions of Life,* Springer, Heidelberg, pp. 47-56.

Fynbo, H. O. U. et al. (2005). Revised rates for the stellar triple-α process from measurements of ^{12}C nuclear resonances. *Nature* **433**: 136-139.

Gold, T. 1992. The deep, hot biosphere. *Proc. Natl. Acad. Sci.* **89**: 6045-6049.

Gonzalez, G., Brownlee, D. and Ward, P. (2001). The Galactic Habitable Zone: galactic chemical evolution. *Icarus* **152**: 185-200.

Gordiets, B. F., Kulikov, Yu., N., Markov, M. N. and Marov, M. Ya. (1982). Numerical modelling of the thermospheric heat budget. *J. Geophys. Res.* **87**: 4504-4514.

Grießmeier, J.-M., Stadelmann, A., Penz, T., Lammer, H., Selsis, F., Ribas, I., Guinan, E. F., Motschmann, U., Biernat, H. K. and Weiss, W. W. (2004). The effect of tidal locking on the magnetospheric and atmospheric evolution of "Hot Jupiters". *Astron. Astrophys.* **425**: 753-762.

Grießmeier, J.-M., Stadelmann, A., Motschmann, U., Belisheva, N. K., Lammer, H. and Biernat, H. K. (2005). Cosmic ray impact on extrasolar terrestrial planets. *Astrobiology*, submitted.

Guinan, E. F. and Ribas, I. (2002) Our changing Sun: the role of solar nuclear evolution and magnetic activity on Earth's atmosphere and climate. In: B. Montesinos, A. Giménz and E. F. Guinan (eds.) *The Evolving Sun and its Influence on Planetary Environments*. San Francisco, *ASP* **269**: 85-107.

Haisch, B., Strong, K. T. and Rodono, M. (1991). Flares on the Sun and other stars. *Ann. Rev. Astron. Astrophys* **29**: 275-324.

Hart, M. H. (1979). Habitable zones about main sequence stars. *Icarus* **37**: 251-357.

Heath, M. J., Doyle, L. R., Joshi, M. M. and Haberle, R. M. (1999). Habitability of planets around red dwarf stars. *Origins Life Evol. Biosphere* **29**: 405-424.

Jakosky, B. M. and Shock, E. L. (1998). The biological potential of Mars, the early Earth and Europa. *J. Geophys. Res.* **103**: 19359-19364.

Joshi, M. (1997). Simulations of the atmospheres of synchronously rotating terrestrial planets orbiting M dwarfs: Conditions for atmospheric collapse and the implications for habitability. *Icarus* **129**: 450-465.

Joshi, M. (2004). Climate model studies of synchronously rotating planets. *Astrobiology* **3**: 415-427.

Kasting, J. F. (1988). Runaway and moist greenhouse atmospheres and the evolution of Earth and Venus. *Icarus* **74**: 472-494.

Kasting, J. F. (1991). CO_2 condensation and the climate of early Mars. *Icarus* **94**: 1-13.

Kasting, J. F., Whitmire, D. P. and Reynolds, R. T. (1993). Habitable zones around main sequence stars. *Icarus* **101**: 108-128.

Kasting, J. F. and Catling, D. (2003). Evolution of a habitable planet. *Annu. Rev. Astron. Astrophys.* **41**: 429-463.

Khodachenko, M. L., Ribas, I., Lammer, H., Grießmeier, J.-M., Selsis, F., Leitner, M., Eirora, C., Hanslmeier, A., Biernat, H. K., Farrugia, C. J. and Rucker, H. O. (2005). CME activity of low mass M stars as an important factor for the habitability of terrestrial exoplanets I: CME impact on planetary magnetospheres. *Astrobiology*, submitted.

Khurana, K. K., Kivelson, M. G., Stevenson, D. J. , Schubert, G., Russell, C. T., Walker, R. J. and Polanskey, C. (1998). Induced magnetic fields as evidence for subsurface oceans in Europa and Callisto. *Nature* **395**: 777-780.

Krumholz, L. R., McKinley, J. P., Ulrich, G. A. and Suflita, J. M. (1997). Confined subsurface microbial communities in Cretaceous rock. *Nature* **386**: 64-66.

Lammer, H., Selsis, F., Ribas, I., Guinan, E. F., Bauer, S. J., and Weiss, W. W. (2003). Atmospheric loss of exoplanets resulting from stellar X-ray and extreme ultraviolet heating. *Astrophys. J.* **598**: L121-L124.

Lammer, H., Ribas, I., Griessmeier, J.-M., Penz, T., Hanslmeier, A. and Biernat, H. K. (2004). A brief history of the solar radiation and particle flux evolution. *Hvar Obs. Bullet.* **28**: 139-155.
Lammer, H., Kulikov, Yu. N., Penz, T., Leitner, M., Biernat, H. K. and Erkaev, N. V. (2005a). Stellar-planetary relations: atmospheric stability as a prerequisite for planetary habitability. In: R. Dvorak and S. Ferraz-Mello (eds.) *A comparison of the dynamical evolution of planetary system*, Proceedings of the 6th Alexander von Humboldt Colloquium, Kluwer, Dordrecht, in press.
Lammer, H., Lichtenegger, H. I. M., Kulikov, Yu. N., Grießmeier, J.-M., Penz, T., Selsis, F., Khodachenko, M. L., Ribas, I. and Biernat, H. K., (2005b). CME activity of low mass M stars as an important factor for the habitability of terrestrial exoplanets II: Atmospheric erosion of Earth-like exoplanets in close-in habitable zones. *Astrobiology*, submitted.
Levison, H. F. and Angor, C., (2003). The role of giant planets in terrestrial planet formation. *Astron. J.* **152**: 2692-2713.
Léger, A., Selsis, F., Sotin, C., Guillot, T., Despois, D., Mawet, D., Ollivier, M., Labèque, A., Valette, C., Brachet, F., Chazelas, B. and Lammer, H. (2004). A new family of planets? "Ocean-Planets". *Icarus* **169**: 499-504.
Lineweaver, C. H., Fenner, Y. and Gibson, B. K. (2004). The Galactic Habitable Zone and the distribution of complex life in the Milky Way. *Science* **303**: 59-62.
Lissauer, J. J. (1987). Timescales for planetary accretion and the structure of the protoplanetary disk. *Icarus* **69**: 249-265.
Marion, G. M., Fritsen, C. H., Eicken, H. and Payne, M. C. (2003). The search for life on Europa: limiting environmental factors, potential habitats, and Earth analogues. *Astrobiology* **3**: 785-811.
Mayor, M. and Queloz, D. (1995). A Jupiter-mass companion to a solar-type star. *Nature* **378**: 355-359.
Melosh, H. J. and Vickery, A. M. (1989). Impact erosion of the primordial atmosphere of Mars. *Nature* **338**: 487-489.
Michelangeli, D., Toon, O. B., Haberle, R. M. and Pollack, J. B. (1993). Numerical simulations of the formation and evolution of water-ice clouds in the Martian atmosphere. *Icarus* **100**: 261-285.
Morbidelli, A., Chambers, J., Lunine, J. I., Petit, J. M., Robert, F., Valsecchi, G. B. and Cyr, K. (2000). Source regions and timescales for the delivery of water to Earth. *Meteoritics and Planet. Science*, **35**: 1309-1320.
Pagel, B. E. J. (1997). *Nucleosynthesis and chemical evolution of galaxies*, Cambridge University Press, Cambridge.
Pollack, J. B., Kasting, J. F., Richerdson, S. M. and Poliakoff, K. (1987). The case for a wet, warm climate on early Mars. *Icarus* **71**: 203-224.
Raymond, S. N., Quinn, T. and Lunine, J. I. (2004). Making other earths: dynamical simulations of terrestrial planet formation and water delivery. *Icarus* **168**: 1-17.
Rouan, D., Baglin, A., Copet, E., Schneider, J., Barge, P., Deleuil, M., Vuillemin, A. and Léger, A. (2000). The exosolar planets program of the COROT satellite. *Earth, Moon, and Planets* **81**: 79-82.
Reeves, G. D., Cayton, T. E., Gary, S. P. and Belian, R. D. (1992). The Great Solar Energetic Particle Events of 1989 observed from geosynchronous orbit. *J. Geophys. Res.* **97**: 6219-6226.
Ribas, I., Guinan, E. F., Güdel, M. and Audard, M. (2005). Evolution of the solar activity over time and effects on planetary atmospheres: I. High-energy irradiances (1-1700 Å). *Astrophys. J.* **622**: 680-694.
Santos, N. C., Israelian, G., Mayor, M., Rebolo, R. and Udry, S. (2003). Statistical properties of exoplanets. II. Metallicity, orbital parameters, and space velocities. *Astron. Astrophys.* **398**: 363-376.
Santos, N. C., Bouchy, F., Mayor, M., Pepe, F., Queloz, D., Udry, S., Lovis, C., Bazot, M., Benz, W., Bertaux, J.-L., Curto, G. Lo., Delfosse, X., Mordasini, C., Naef, D., Sivan, J.-P. and Vauclair, S. (2004). The HARPS survey for southern extra-solar planets II. A 14 Earth-masses exoplanet around μ Arae. *Astron. Astrophys.* **426**: L19-L23.
Smith, D. S., Scalo, J. and Wheeler, J. G. (2004). Transport of ionizing radiation in terrestrial-like exoplanets atmospheres. *Icarus* **171**: 229-253.
Stevenson, D. J. (1999). Possibility of life-sustaining interstellar planets. *Nature* **400**: 32.
Thuan, T. X. and Izotov, Y. I. (2000). Big Bang nucleosynthesis and the baryonic content of the universe. *Proc. on Physics at extreme energies*, Hanoi, Vietnam, arXiv:astro-ph/0012382.
Trimble, V. (1997). Origin of the biologically important elements. *Origins Life Evol. Biosphere* **27**: 3-21.
Tytler, D., O'Meara, J. M., Suzuki, N. and Lubin, D. (2000). Review of Big Bang nucleosynthesis and primordial abundances. *Physica Scripta* **T85**: 12-31.
Vogel, G. 1999, Expanding the habitable zone. *Science* **286**: 70-71.

Von Bloh, W., Franck, S., Bounama, C. and Schnellnhuber, H.-J. (2003). Biogenic enhancement of weathering and the stability of the ecosphere. *Geomicrobio. J.* **20**: 501-511.

Wagoner, R. V., Fowler, W. A. and Hoyle, F. (1967). On the synthesis of elements at very high temperatures. *Astrophys. J* **148**: 3-49.

Wetherill, G. (1994). Possible consequences of absence of Jupiters in planetary systems. *Astrophys. Space Sci.* **212**: 23-32.

Wolszczan, A., Frail, D. A. (1992) A planetary system around the millisecond pulsar PSR1257+12. *Nature* **355**: 145-147.

Wood, B. E., Müller, H.-R., Zank, G. P. and Linsky, J. L. (2002). Measured mass loss rates of solar-like stars as a function of age and activity. *Astrophys. J.* **574**: 412-425.

Zahnle, K. J. and Walker, J. C. G. (1982). The evolution of solar ultraviolet luminosity. *Rev. Geophys*. **20**: 280-292.

Zierenberg, R. A., Adams, M. W. W. and Arp, A.-J. (2000). Life in extreme environments: Hydrothermal vents. *Proc. Natl. Acad. Sci. USA*, **97**: 12961-12962.

Biodata of **Josef Svoboda** author of chapter ***"Life as an unfolding Bicosmos"***

Josef Svoboda is a Professor Emeritus at the Department of Biology, University of Toronto at Mississauga, Canada. He studied Philosophy and Biology in Czechoslovakia but in 1949 he was arrested and sentenced by the communistic regime. He spent almost 9 years in the prisons and uranium mines' labour camps. After immigrating to Canada in 1968, he graduated from the University of Western Ontario, Canada. In 1974 he obtained his Ph.D., from the University of Alberta, Canada, for a study of productivity processes in polar semi-deserts. For more than 30 years, together with his students, he conducted research in the High Arctic focusing on plant colonization following deglaciation of polar landscapes and responses of arctic plant communities to global warming. He also published papers on evolution and the origin of life. Honours: the Northern Science Award from the Government of Canada and a Honorary doctorate from the Masaryk University, the Czech Republic.

E-mail: **jsvoboda@utm.utoronto.ca**

J. Seckbach (ed.), Life as We Know It, 431-444.

LIFE AS AN UNFOLDING BIOCOSMOS

J. SVOBODA[1]
Department of Biology, University of Toronto at Mississauga, Mississauga, Ontario, L5L 1C6, Canada.

1. Introduction

The concept of the universe has shockingly changed since Nicolas Copernicus laid the foundations of modern astronomy in the early 1500's. The absolutism of the laws of Euclidean geometry and Newtonian physics ran aground when applied to extremely small or large dimensions and speeds. A relativistic era was born. However, one aspect of the ongoing intellectual inquiry has not changed, offering some solace for the superiority of the ever-searching mind. While it takes billions of years for light to cross the immense span of the universe, the educated mind is able to encompass such a distance in an instant and view the entire cosmos "as in a nutshell" (Hawking, 2001).

The objective of this essay is to bring forth observations and ideas with respect to the *nature* of life, rather than its origins. These theoretical explorations into the realm of the living world are expansions and projections based on recent cosmological discoveries, and assumptions upheld by a majority of present cosmologists (Tyson and Goldsmith, 2004). The main premise being that our universe had at least a virtual beginning and that it has been unfolding in space and time since the Big Bang event.

Theorists about life's origin grapple with similar difficulties. The reason is the relatively sudden increase of enormous complexity, and manifestation of utterly new behavioral characteristics associated with the Life phenomenon. The event of life's emergence on Earth and possibly in the far realms of the universe remains unexplained.

2. The Inanimate Universe

2.1. ITS NATURE

Classical physics and chemistry have taught us that elements differ from each other by number of protons in their nuclei (atomic number). Their chemical properties depend on distribution of electrons around the nucleus. In other words, elements can be distinguished by their quantitative and structural characteristics, and for all practical purposes this model has worked well. However, we now learn that their corpuscles are only 'temporary entities within a wave field determined by the wave laws' and "that many processes take place *as if* these temporary entities were substantial permanent beings" (Schrödinger, 2000). Oddly or admirably, the same unknown quality has

[1] To Andrew

produced all the cosmic diversity from hypothesized sub-nuclear strings or branes (Halpern, 2004) – to galaxies and to the dazzling wings of the Monarch butterfly. The essence of the physical world, the 'ultimate reality', is a mystery.

Not too long before Hubble's discovery of the spectral red shift phenomenon of the remote galaxies in the mid-1920's (Levy, 2001), the cosmos was considered perfect, eternal and geo-centric. The Hubble telescope's red shift revealed that galaxies are not stationary but dynamically move in a centrifugal direction. This led to an a novel hypothesis, later worked out in the generally accepted theory that the universal cosmic expansion was initiated by the "Big Bang" event. According to this theory, the universe exploded into its being 13.7 Bya ago from a single point at time zero (Lemonick, 2003; Tyson and Goldsmith, 2004). It experienced an infinitesimally brief (10^{-34} sec) rapid inflation phase, during which the universe increased in volume by a factor of 10^{50} (Hawking, 1988; Guth and Steinhardt, 2000) before it began to cool down and its rate of inflation eased. However, about 8 Bya later, the cosmic expansion changed gear and began to accelerate. It is propelled by the "vacuum energy", which is stronger than the gravitational pull of the cosmic matter (Hawking, 2001; Barrow 2002).

2.2. ITS STRUCTURE AND BEHAVIOR

Compared to the concept of biological evolution, which has been around for almost 200 years, the concept of cosmic evolution is much more recent. Cosmic evolution encompasses all the physical micro- and macro-processes occurring in our universe since its earliest moments. Eventually, when a suitable stage of cosmic development has been reached this process *may* result (Dawkins 1991; Stenger 1995) or *must* result (Chaisson, 2001) in the emergence of life. As if "the properties of the expanding universe were imprinted at the time of the Big Bang" (Rees 2000). In other words, a seamless garment of evolution exists from Big Bang to Man and possibly beyond – although, as we will try to demonstrate, with a manifest change in directionality at the animate level. The idea of cosmic self-organization (Kauffman, 1995) or self-complexification (Casti, 1994) has become common and largely accepted in the current cosmological literature.

There is a discernible *asymmetry* in the cosmic macro-world owing to the fourth dimension – time. Along this cosmic "arrow of time" (Rees, 2000, Chaisson, 2001) the process of self-organization began with the first gluons being formed from the Big Bang plasma. It continued in its meso- and macrostructures until conditions were ripe for life to emerge (Davies, 1999, Kauffman, 2000). Such asymmetry is non-existent in cosmic microstructures. Atoms and their inner components are not subject to time, at least in a conventional sense. In the macro-world, however, upsetting the established equilibrium may mean progress. Asymmetry opens the door to potentialities and allows new structures to be realized.

According to the latest theories, only a small 5% fraction of the total energy (Ω) released in Big Bang emerged as ordinary matter. Of this the "visible" universe is made. The highest proportion, approximately 65% of the total, remained as "dark" or "vacuum" energy (Seife, 2003) and about 30% had "materialized" into "exotic dark matter", believed still to linger at the periphery of spiral galaxies (Rubin, 2000).

A striking feature of the universe' unfolding has been its *rapidly slowing rate of change*, inversely proportional to increasing complexity and stability of its structures. Thus "plasma condensation" was completed in the first 10^{-6}sec. The "particulate" stage, the nucleosynthesis (H, He) was accomplished within the first 3 minutes.

The first galaxies gathered in the sky within the first billion years. Since then, while the universe continuously expanded, the structural diversity has been on the rise. A few billion years later, many of the original superstars, composed of hydrogen and helium began synthesis of heavier elements up to iron, accumulating them in their core. A collapse of such a star results in giant explosions of which the supernovas are born. The debris ejected into space contains almost all the known elements. Of the supernova "ashes", stars of the second generation are born - some with planetary systems made of these heavier elements. Our Earth is such a special planet (Trefil, 2001).

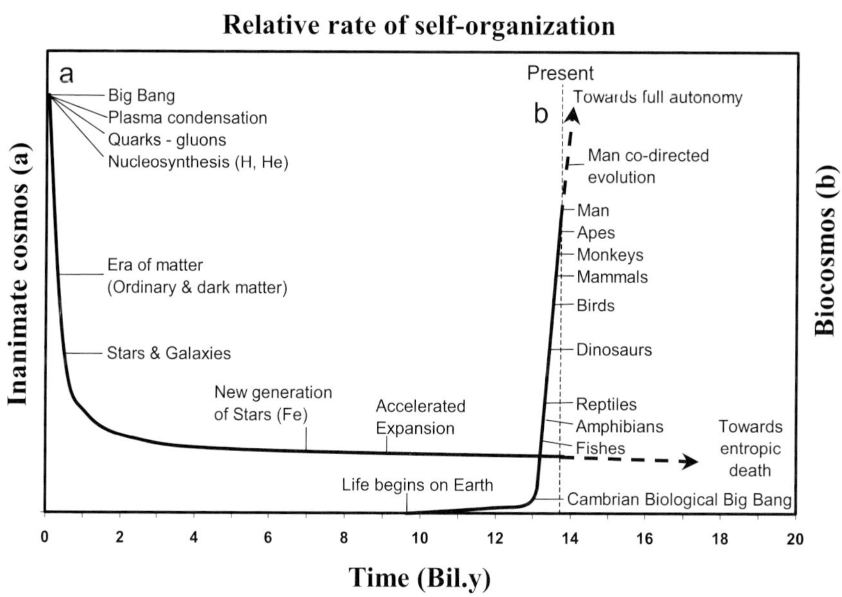

Figure 1. Contrasting "life strategies" of the inanimate and enlivened cosmos. **Figure 1a** demonstrates the decreasing efficiency in self-organization within the time expired from the Big Bang. From the explosive beginning - to the 'extended phase of peaceful expansion' (Adams, 2003) – to a passive surrender to the accelerated centrifugal momentum! Note the asymptotic approach of the curve at both its ends. Almost everything happened during the first billion years, the later stages being mere self-tuning and housekeeping. This has not excluded radical rearrangements here and there, involving local squabbles, however "cosmic" to us in proportion. **Figure 1b** shows the "hockey stick" shape of the biological self-organization (evolution). Very little happened during the first 3.2 Bya until the Cambrian explosion of life-forms (Cambrian Biological Big Bang). All remaining evolutionary progress occurred during the last 550 Mya at an accelerated pace.

When, at the end of the inflation stage (10^{-32}sec.) in the overheated kiln of the nascent universe the primordial plasma started condensing, the proto-matter, quarks and gluons (Seife, 2003) precipitated from the process. This was the first STOP in the allegorical "STOP & GO" chain of events. New realities and qualities, such as mass, speeds slower than that of light and relative stability of structures, emerged. We observe a perpetual integration – stabilization – disintegration; structuring – holding on the structure – restructuring; working for a dynamic equilibrium – reaching it – losing it, and so on. Stability is often a shaky and fragile feature of any physical structure, always temporary but it does exist. It ranges from fragments of a second to billions of years. Thus, in the material cosmos the "*STOP & GO*" *phenomenon is of cardinal importance for life, since in a favorable environment its STOP phase provides the needed, although only temporary milieu for life to emerge, sustain itself and evolve.*

The expanding visible universe may have a long life but is still temporary. Even the longest-lived atoms will eventually disintegrate and thus perish within 10^{35} years (Rees, 2000). What looks like quiet stability to an infatuated observer at night, the myriad of stars is, in fact, a gigantic, unceasing firework. And as any firework, it shall ultimately burn its fuel and fade away in a "long cold whimper" (Hawking, 2001). And a question of our ultimate concern: how long will our Sun be able to support life on this planet? Not overly long, using the cosmic yardstick, only about 1Bya. The Sun's radiation output is steadily increasing and in about 5Bya the Sun will swell and become a red giant (Chaisson, 2001). None-the-less if no other factors interfere, 1Bya would buy enough time for intelligent life to ponder and decide what to do about its future existence.

3. The Enlivened Universe – the Biocosmos

Carr and Rees (1979) brought up the idea, and Adams (2002) discusses the validity of the *Anthropic cosmological principle*. According to this premise, the universe must have anticipated arrival of an intelligent observer, "and fine-tuned its properties in such a way that intelligent life could arise". Whether one subscribes to this principle or not, achieving this stage of cosmic self-awareness has been an arduous process, on this planet lasting almost 4 Billion years. In a similar venue, Davies (1999) speaks about a *bio-friendly universe*, prone to produce and sustain life. In the waste cosmic real estate this happens rarely (Ward and Brownlee, 2000), only when a *window of opportunity* opens for life to spring up.

Life arrived belatedly yet, according to the above authors, it represents an actualization of the original blueprint of cosmic evolution. Chaisson (2001) outlines a series of progressive stages the universe passed through from time zero to the present: Particulate (origin of matter) → Galactic (origin of Milky Way) → Stellar (origin of Sun) → Planetary (origin of Earth) → Chemical-Biological (origin of Life on Earth) → Cultural (origin of humanity).

The dissimilarity in structure and function between the non-living and living things is of such a magnitude that the emergence of life can be characterized as a "*quantum leap from mechanism to organism*". As if a *wormhole* had been bored though the ceiling of the all-out deterministic inanimate world into an open and ultimately autonomous realm. A new unanticipated singularity, the living cell comes about. At that point, a brand-new process, governed by a more complex set of rules would take over and a different enlivened world of unpredictable potentialities had been born. The first protocells became new "zygotes" endowed with inconceivable *totipotency* to evolve in a complex hierarchical biosystem, as we find and study it at present. Barring for a global catastrophe, it may continue evolving beyond our capacity to imagine.

Regarding complexity, living systems are orders of magnitude superior, yet they yield themselves, as an open book, to be completely deciphered in terms of anatomy and physiology. We know pretty well how fruit flies look from inside out and how they function. This does not mean, however, that life's origin on this planet, or if at all elsewhere, is close to being determined.

The living cell is a new cosmic formation, *essentially* distinct from any non-living structures. A novel cosmic entity has come to existence based on its superior complexity. The High-school model of a cell as a little pouch, filled with fluid cytoplasm, in which mitochondria, nucleus and other organelles float freely, is far too inadequate. The present view is much more complex.

No matter how the initial protobiont arrived on Earth, it became a *seed and a starter* in a new untapped realm. Its advanced offspring, the eukaryotic cell, became the propagule of even more complex formations. In multicellular organisms, cells continue to live their own life but also participate in the life of the higher functional unit of which they are part. New world of unrealized potentialities arrived, the *BIOCOSMOS*.

Gardner (2003) considers the *entire universe,* i.e. its inanimate and enlivened components – a *Biocosm,* and in the *broader* sense the author is right. However, in this exploration we follow the trajectories of the two worlds separately, to point out the fundamental differences in their directionality and behavior. We are *narrowing down* the concept and saving the term 'Biocosmos' for the distinct, enlivened and unfolding cosmic reality.

3.1. CONDITIONS FOR LIFE EMERGENCE

Some microorganisms, the so-called extremophiles withstand, even thrive in conditions of high or low temperatures, toxic and otherwise inhospitable environments (see, Seckbach, chapter of Extremophiles in this volume). Not even the extremophiles, however, may live on the Sun or on planets too close or too far from it. In contrast, the mainstream groups of plants and animals require relatively narrow ranges of favorable conditions above freezing and below 50°C for their existence. Primarily, however, all life depends on the "properties imprinted into the universe at the time of the initial Big Bang" (Rees, 2000). The author brings to our attention six cosmic characteristics, which values define the universe in such a bio-friendly way, that life could emerge and be at least temporarily sustained. If any of the six constants and ratios would only slightly depart from what they are, there would be no cosmic evolution and no life ever possible.

While the likelihood of intelligent life's existence elsewhere is slim, the probability of its lower forms, spread throughout the universe, increases exponentially with their simplicity (Ward and Brownlee, 2000). According to Prigogine and Stengers (1984), Davies (1999) and Gardner (2003), emergence of life is a logical and radical step in the continuum of the cosmic self-organization.

3.2. ITS NATURE

Remarkably, at the threshold of life's emergence, the cosmic evolution branches. We are able to distinguish a dual strategy. While the inanimate universe, predetermined by its intrinsic properties proceeds to its apparent oblivion (Fig.1a), the disproportionately tiny allotment of the enlivened matter marches in another direction through biological and cultural evolution (Fig.1b). Noticeably, the rates of the process of self-organization between the inanimate and animate worlds have been *inversely exponential in time.*

The cosmic evolution began with a bang and accomplished its basic forms of self-structuring in a matter of minutes. Biological evolution, on the other hand, started with a "whisper" in the secrecy of yet unknown places some 3.6-3.8 Bya ago and its advancement was extremely slow. The process started to pick up pace some 550 million years ago (Cambrian explosion of life-forms) and with the arrival of modern humans, it became, not so figuratively speaking, air- and space-borne. By observing the evolutionary trajectory crossing the boundaries towards intelligence and rationality, it is tempting to suggest that evolution may lead to a "bang of its own" breaking into a new realm of existence (Svoboda, 1999b and 2001).

Life is "self-making" (Capra, 1996), ever expanding in size and complexity, at variance with the general, cosmic trend towards disorder and the second law of thermodynamics (Adams, 2002). Organisms are *open systems* driven by the external energy extracted from its surrounding (Schrödinger, 1992). They are vitally dependent on its supply and critically requiring favorable milieu to flourish. Yet, they are innovative in tapping many different sources of energy and materials from various environments. Humans, then, are the most versatile species in modifying their primeval ways of nourishment, broadening their original niche and habitats, and inventing new ones (cf. Eskimo igloos and space capsules).

Self-making and self-propagation of life are new cosmic phenomena. This new reality is endowed with special auto-kinetics, active moving around, which can be, to a smaller or greater degree, internally controlled. Life may go extinct at one place only to surface at another. It will self-propagate endlessly, gaining in sophistication in friendly environments until it reaches a state of self-awareness. The life scenario is open-ended.

Stuart Kauffman (2000) uses the term *autonomy* to characterize life in a way no other physical system could be characterized. After a long search for a specific definition of life he came to the conclusion that living organisms can be best described as "*autonomous agents, able to act on their own behalf*". How do they act on their behalf? By working, in order to survive. What makes them autonomous? A degree of behavioral indeterminism.

3.3. ITS DURATION

Many exobiologists have become comfortable with the possibility of intra-galactic travel of the primitive life-forms', seeding life in life-prone places. This *panspermia hypothesis* proposes that the inter-stellar space abounds with propagules of extremophilic microbes raining on all celestial bodies but germinating only on those with favorable conditions. However, besides the detection of important organic compounds, the potential life-building blocks in meteorites (ca 70 aminoacids; Bernstein, 2000), this hypothesis still lacks more convincing evidence.

No matter how life spreads through the cosmic space-time, whether by *neogenesis* or by being seeded externally, life requires a carrier to be attached to. This carrier must provide shelter, access to energy and be loaded with needed material supplies. Life, as we know it, requires water. This bio-friendly carrier should be relatively stable, to allow life to bud, propagate and evolve. Our planet is the only known such a carrier, however, it is plausible to conjure that there are other celestial bodies with life-friendly environment, supporting life at least at the simplest level.

We prefer to call these temporary bases carrying and supporting life, "*temporary cosmic life rafts*". Our premise is that once life emerged in the universe, it tends to last. Due to continuously increasing solar output, in time our planet will become hotter and in the future less and less hospitable. In turn, the presently cold Mars may grow warmer and thus be ameliorating at least one of the prerequisites for the life support, its climate. Hence the theoretical possibility of life moving further form the Sun. From this viewpoint, *rafting the biota* could be considered an ingenious mechanism for long-term life perpetuation. Living organisms, including the intelligent ones, ought to be considered mere temporary tenants on various celestial rafts, be they planets, moons, travelers imbedded in various cosmic debris or, in utopian terms, cruising space-time *via* sophisticated artificial devices (Crick, 1981).

On Earth, the organic life is an ephemeral entity, a fine hair on the balding global skull. Its total biomass represents only about 40 billionths of a percent of the earth's mass. No wonder that life is a vulnerable commodity, even prone to extinction, as has almost happened several times on this planet. Organisms utilize about 1/3 of the 92 natural elements; however, the bulk of their "dryweight" is made of only three elements: carbon, hydrogen and oxygen. Carbon alone accounts for less than 1% of the earth's mass (Svoboda, 1989).

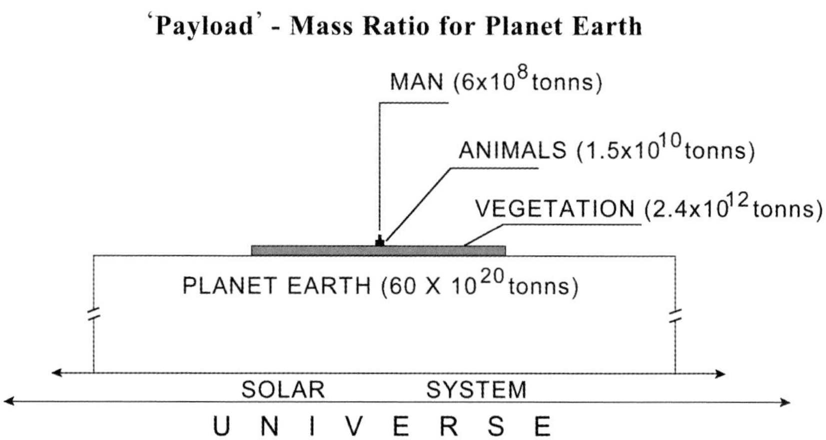

Figure 2. Eltonian pyramid of numbers for planet Earth and its biota. The Earthmass to Biomass ('Payload') Ratios for the Lithosphere : Biosphere = 25x 10^8:1, Phytosphere : Zoosphere = 160:1, and Zoosphere. Homosphere = 36:1 (Modified from Svoboda, 1989).

However, numbers still matter and could be cited at least for interest. If life emergence were the intrinsic result of cosmic evolution, then, in terms of the enlivened to inanimate mass (*payload : mass*) ratio, the exerted cosmic thrust at the big Bang event to launch the first living cell, is beyond imagination. Let's focus on the more comprehensible Geo-Bio ratios extracted from this planet. Figure 2 shows the Eltonian pyramid of numbers of the Earthmass and Biomass components for our planet and the ensuing ratios.

The disproportionality of the non-living to living mass even at the planetary level is enormous but not so surprising. More revealing is how life itself is structured. Not only biomass quantity matters but also its stratification in terms of complexity. The telescopic nature of the ecological architecture in the biosphere places plants at the base, animals in the centre and mind-bearing humans at the top of the pyramid. Or, perhaps more relevantly, in the top capsule of a three stage rocket. In other words, not life as such but the *rational life* can be seen the pinnacle of the cosmic evolution of Chaisson (2001) and many others.

3.4. ITS STRUCTURE AND BEHAVIOR

"LIFE" is an abstract concept we use to describe *the totality of enlivened matter*, which structure is manifold and hierarchical. From the aspect of their anatomy, living organisms are just physical systems with a higher level of complexity. Smart self-organization of many kinds of simple and complex compounds produced a radically new *enlivened* entity, the living cell, an amoeba. In higher organisms, due to sophisticated *quantitative* arrangements of specialized cells, a *new quality emerges*: a dew worm, fish, monkey, man.

Crick (1978) considers the discovery of general uniformity of biochemistry throughout nature (a common blueprint, DNA) as one of the most astonishing achievements, although an obvious diversity of organisms exists. One tree of life but three domains: the Archaebacteria (Archaea), the Eubacteria (Bacteria) and the Eukarya. The last domain including all the eukaryotic organisms – the protists, fungi, plants and animals (Tudge, 2000).

There is a morphological and ecological hierarchy in the living world of this planet. Individuals of the same species form *populations*, which associate in multi-specific *communities*, ultimately all forming an *ecosystem*. The biosphere is one huge ecosystem.

3.4.1. Evolution

Evolution is conservation as well as transformation (Tomlin, 1978) and Darwinian natural selection works well with Mendelian inheritance (Tallack, 2003). The inheritance is preserving what has been selected until it is replaced by a new selection.

One of the most intriguing features of our planetary evolution has been its non-linearity. Evolutionary progress has occurred in relatively short bursts after long periods of sluggishness. Eldredge and Gould (1972) and Gould (1996) named this evidently STOP & GO pattern, "punctuated equilibrium". Accelerated emergence of new species, macro-evolution, occurred after a long "stasis" when only micro-evolution had taken place. In the history of Earth, these bursts of progressive advancement were almost always triggered by a radical environmental change, often synonymous with major catastrophic events. During these occurrences (asteroid impacts, volcanic activity, even sudden climate changes), large percentages of species perished, opening emptied ecological niches to surviving groups of organisms (see chapter by Seckbach, Extremophiles, in this volume).

The last of the big catastrophes, 65 Mil y. ago, ended the Cretaceous era and brought, among others, extinction of the great reptiles. The following Tertiary era opened horizons for the proliferation of mammals. We are the end product of this fast-track post-cataclysmic evolution.

Allegorically, we may see the dynamic, competitive and in ecological terms "successional" GO phase as *"climbing"* the evolutionary pagoda to higher floors of complexity. In contrast, the sluggish "stasis" or STOP phase, with species micro-adjustment to highly specialized niches, represents *"spreading"*, forcing the occupiers to even narrower specialization.

The extremely destructive brunt of major global catastrophes on the biota followed by their phenomenal recovery, points to the importance of impacts, even their inevitability for the evolutionary furtherance (de Duve, 2002). *The macro-evolution would not and could not successfully progress without a periodic mass-devastation of the extant biota.*

3.4.2. Three crossing

Three evolutionary breakthroughs or step-like passages over the thresholds of preceding limitations can be clearly identified in the make-up of the Biocosmos (Svoboda 2000, 2001). *Emergence of life*, the breakthrough from the inanimate to the enlivened world – embodies the *first crossing. Emergence of self-awareness*, the *Mind's Big Bang* (Calvin, 2004) marks the *second crossing*. The anticipated *accelerated advancement of Mankind* may, prospectively, reach the *third crossing*. A threshold, characterized with liberation from physical limitations such as aging, gravitation, spatial confinement. New literature is replete with concepts like *Homo sapiens, sapiens, sapiens*, Ultraman, Man of the future, Metaman, Post-humans, etc., where genetic engineering, a combination of human and high-tech (cyborg) or, for some, mere spiritual development will result in a strikingly novel bio-intelligence and artificial-intelligence, or pure consciousness. For more inspiration in this respect consult Hardison (1990), Stock (1993), Darling (1996), Kurzweil (1999), Davies (1999), Calvin (2004), and others.

On this planet, its evolutionary hierarchy could be spelled out as *Geosphere – Biosphere – Homosphere – Noosphere*. The *Geosphere* has become the penultimate material base upon which life established itself.

The Biosphere is a spatial realm filled with biota interacting with the Geosphere (Lowelock, 1987). The *Homosphere* has been defined as *a Biosphere modified by Homo sapiens.* The human impact on Earth has expanded to such a magnitude that, in actual fact, the Homosphere has already supplanted the Biosphere (Svoboda, 1999a).

The Noosphere, is defined as a *realm of human minds interconnected and interacting through communication*. This 'sphere of knowing and sharing' has always existed among sentient and intelligent creatures. However, figuratively as well as literally, it has expanded beyond the boundary of this planet by us, humans. Acquisition of language, invention of writing and recent advancements of wireless communication became historical markers of the Noosphere's evolution. Its explosive expansion will continue beyond the grasp of imagination (Svoboda, 1999b).

3.5. ITS DESTINY

At the present stage, the "visible" material universe has reached its middle age, it is stabilized and predictable. No revolutionary upheavals on a large scale are expected. In contrast, creative bio-evolution is probably in progress at many galactic places. We may assume that intelligent life, wherever evolved, has attempted to liberate itself from the crib-conditions of its origins, which acted as gentle nurseries. Our planning of the Mars colonization is well in progress and future extraterrestrial habitat projects are already on the drawing board.

There are various scenarios contemplated by different authors, some optimistic, other less so. Chaisson (2001) envisages that "a mature Life Era may never fully come to pass… We now perceive the dawn of a whole new reign of cosmic development – an era of opportunity to unlock secrets of the universe… and to decipher who we are and whence we came." De Duve (2002) is not so sure about the positive outlook of the Man-directed evolution. In the past, climatic upheavals and catastrophic events exterminated up to 90% of species and there is no guaranty that such a blow won't strike again. Other, more modern threats include: AIDS, Ebola virus, and other "superbugs", upsetting the balance of the Biosphere. Stock (1993) believes that mankind will become a "superorganism". Individuals will be deprived more and more of their freedom and autonomy in favor of the centralized "plurihuman organism" Having experienced the past and present totalitarian regimes, and seen films of the Matrix-kind, Stock's argument may be quite convincing. Yet, the cultural pendulum has been swinging between enslavement and liberation for millennia. Human beings are kept going more by hope than fear.

4. Afterthought

If organisms are autonomous agents (Kauffman, 2000), could the definition of life embrace their totality and describe the Biocosmos as *the ultimate autonomous agent acting on its own behalf*? The conclusion seems to be logical, even inevitable, granted life's ability to move from place to place, to choose or abandon its carriers, and to aim for the final liberation from the entropy-stigmatized universe. In contrast to its inanimate parent, Biocosmos is open-ended and its potential is virtually limitless (Calvin, 2004).

And as for the human race? The exponential trend of bioevolution shows Man as a tip of an arrow released almost vertically towards the sky. Davies (1999) asks the penultimate question about mind's predestination. Could mind become free from physical chains and become "airborne" as a form of pure intelligence? Not even atoms live forever but life-forms of liberated intelligence? Too daring even to guess.

5. References

Adams, F. (2002). *Origins of Existence. How Life Emerged in the Universe.* Free Press. New York. 266pp.

Barrow, J. D. (2002). *The Constants of Nature. From Alpha to Omega – the Numbers that Encode the Deepest Secrets of the Universe.* Pantheon Books. New York. 352pp.

Bernstein, M. P., Sandford, S. A. and L. J. Allamandola (2000). Life's Far-Flung Raw Materials. In. D. H. Levy (ed.) *The Scientific American Book of the Cosmos.* St. Martin's Press. New York. pp. 265-271.

Calvin, W. H. (2004). *A Brief History of the Mind. From Apes to Intellect and Beyond.* Oxford University Press. New York. 219 pp.

Capra, F. (1996). *The Web of Life. A New Scientific Understanding of Living Systems.* Anchor Books. Doubleday. New York. 347 pp.

Carr, B. and Rees, M. J. (1979). The Anthropic Principle and the Structure of the Physical World. *Nature* **278**:605-612.

Castı, J. (1994). *Complexification.* Harper Collins. New York. 212pp.

Chaisson, E. J. (2001). *Cosmic Evolution. The Rise of Complexity in Nature.* Harvard, London. 274pp.

Crick, F. H. C. (1978). Developmental Biology. In: R. Duncan and M. Weston-Smith (ed.) *The Encyclopaedia of Ignorance.*. Vallaby. New York. pp. 299-303.

Crick, F. H. C. (1981). *Life Itself: Its Nature and Origin.* Simon and Schuster. New York.

Darling, D. (1996). *Zen Physics: The Science of Death, the Logic of Reincarnation.* HarperCollins. New York. 205pp.

Davies, P. (1999). *The Fifth Miracle. The Search for the Origin and Meaning of Life.* Simon & Schuster. New York. 304pp.

Dawkins, R. (1991). *The Blind Watchmaker. Why the Evidence of Evolution Reveals a Universe Without Design.* Norton. New York.

De Duve, C. (2002). *Life Evolving. Molecules, Mind and Meaning.* Oxford University Press. New York. 340pp.

Eldredge, N., and S. J. Gould (1972). Punctuated Equilibria: An alternative to phyletic gradualism. In: T. J. M. Schopf (ed.) *Models in Paleobiology,* Freeman, Cooper & Company. San Francisco. pp. 82-115.

Gardner, J. N. (2003). *Biocosm.* Inner Ocean Publ. Makawao, Maui, HI. 317pp.

Gould, S. J. (1996). *Full House. The Spread of Excellence from Plato to Darwin.* Harmony Books, Random House. New York. 244pp.

Guth, A. H. and Steinhardt, P. J. (2000). The Inflationary Universe. In: D. H. Levy (ed.). *The Scientific American Book of the Cosmos.* St. Martin's Press. New York. pp. 361-375.

Halpern, P. (2004). *The Great Beyond. Higher Dimensions, Parallel Universes and Exraordinary Search for a theory of Everything.* John Willey & Sons. New York. 336pp.

Hardison, Jr., O. B. (1990). *Disappearing through the Skylight. Culture and Technology of the Twentiest Century.* Penguine Books. New York. 389pp.

Hawking, S. (1988). *A Brief History of Time: From the Big Bang to Black Holes.* Bantam Books. London.

Hawking, S. (2001). *The Universe in a Nutshell.* Bantam Books. New York. 216 pp.

Kauffman, S. (1995). *At Home in the Universe. In Search for the Laws of Self-Organization and Complexity.* Oxford University Press. New York. 321pp.

Kauffman, S. (2000). *Investigations.* Oxford University Press. New York. 287pp.

Kurzweil, R. (1999). *The Age of Spiritual Machines.* Viking Penguin Books. New York. 352pp.

Lemonick, M. D. (2003). *Echo of the Big Bang.* Princeton University Press. 232pp.

Levy, D. H. (2001). *Cosmic Discoveries.* Prometheus Books. 232pp.

Lovelock, J. E. (1987). *Gaia. A New Look at Life on Earth.* Oxford University Press. New York. 157pp.

Prigogine, I. and Stengers, I. (1984). *Order out of Chaos. Bantam. New York.*

Rees, M. (2000). *Just Six Numbers.* Basic Books. New York. 173pp.

Rubin, V. (2000). Dark Matter in the Universe. In: D. H. Levy (ed.). *The Scientific American Book of the Cosmos.* St. Martin's Press. New York. pp. 91-96.

Seife, C. (2003). *Alpha and Omega. The Search for the Beginning and End of the Universe.* Viking. New York. 288pp.

Schrödinger, E. (1992). *What is Life.* Cambridge University Press. New York.

Schrödinger, E. (2000). What is Matter In: D. H. Levy (ed.). *The Scientific American Book of the Cosmos.* St. Martin's Press. New York. pp.25-32.

Stenger, V. J. (1995). *The Unconscious Quantum. Metaphysics in Modern Physics and Cosmology.* Prometheus Books. New York. 322pp.

Stock, G. (1993). *Metaman. The Merging of Humans and Machines into a Global Superorganism.* Simon & Schuster. New York.

Svoboda, J. (1989). The Reality of the Phytosphere and (Ultimate) values involved. *Ultimate Reality and Meaning* **12** (2):104-112.

Svoboda, J. (1999a). The Homosphere. In: D. E. Alexander and R. W. Fairbridge, (eds.): *Encyclopedia of Environmental Science* Earth Science Series. Kluwer. Boston. pp. 324-325.

Svoboda, J. (1999b). The Noosphere. In: D. E. Alexander and R. W. Fairbridge, (eds.): *Encyclopedia of Environmental Science.* Earth Science Series. Kluwer. Boston. pp. 438-439.

Svoboda, J. (2000). On the Verge of the Third Crossing. (Reflections on the Meaning of Evolution and the Future of Humanity. In: M. Huba and P. Nováček (eds.): *The World Perceived by the Heart of Europe* Palacky University Press. Olomouc. Czech republic. pp. 118-168.

Svoboda, J. (2001). Life, O Life, from where have you come? Where are you not? In: J. Elster, J. Seckbach, W. F. Vincent and O. Lhotsky (eds.) *Algae and Extreme Environments* pp.1-12. *Nova Hedwigia,* Beiheft 123. Cramer, Stuttgart. 602pp.

Tallack, P. (ed) (2003). *The Science Book.* Weidenfeld & Nicolson. London. 528pp.

Tomlin, E. W. F. (1978). Fallacies of Evolutionary Theory. *The Encyclopaedia of Ignorance.* Wallaby. New York. pp. 227-234.

Trefil, J. ed. (2001). *Encyclopedia of Science and Technology.* Routledge. New York. 552pp.

Tudge, C. (2000). *The Variety of Life. A Survey and Celebration of all the Creatures that Have Ever Lived.* Oxford University Press. New York. 684pp.

Tyson, N. D. and Goldsmith D. (2004). *Origins. Fourteen Billion Years of Cosmic Evolution.* Norton. New York. 336pp.

Ward, P. D. and Brownlee, D. (2000). *Rare Earth. Why Complex Life is Uncommon in the Universe.* Copernicus Springer Verlag. New York. 333pp.

Biodata of **Charles H. Lineweaver** author of ***"We have not detected extraterrestrial life. Or have we?"***

Dr Charles H. Lineweaver is currently a Senior Fellow at the Planetary Science Institute in the Research School of Astronomy and Astrophysics and the Research School of Earth Sciences at the Australian National University, Australia. He obtained his Ph.D., from the University of California at Berkeley in 1994 and continued his studies and research at Strasbourg Observatory and the University of New South Wales. Dr Lineweaver's scientific interests are in the areas of: planetology, cosmology and astrobiology.

E-mail: **charley@mso.anu.edu.au**

J. Seckbach (ed.), Life as We Know It, 445-457.

WE HAVE NOT DETECTED EXTRATERRESTRIAL LIFE, OR HAVE WE?

CHARLES H. LINEWEAVER
Planetary Science Institute, Australian National University, Canberra, Australia

Dorothy (to Glinda the good witch of the North): *"I think that it...that it wasn't enough just to want to see Uncle Henry and Auntie Em...and it's that if I ever go looking for my heart's desire again, I won't look any further than my own backyard; because if it isn't there, I never really lost it to begin with."*

In the movie "The Wizard of Oz" Dorothy mounts an extensive, but unsuccessful observational program to find her heart's desire. Finally, she realizes she has been misinterpreting data from her own backyard. Similarly, we have looked for extraterrestrial life and haven't found any. Maybe we haven't looked long enough or hard enough and we will eventually succeed – or maybe we need to take a lesson from Dorothy. Maybe we don't have to look any further than our own backyard. To find extraterrestrial life maybe all we need to do is redefine life by shifting our paradigm of what life is (Kuhn 1962).

In this article I challenge the assumption that we have not yet found extraterrestrial life. I argue that by redefining life in a more general way, we can legitimately conclude that we have already detected extraterrestrial life. I will also argue that this is not a semantic game – that the thermodynamic justifications for a broader definition of life are compelling and more universal than the traditional definitions of life based on DNA, "self-reproduction" and the chemical complexity of the terrestrial life most familiar to us.

1. No Evidence for Extraterrestrial Life

The canals and faces on Mars have disappeared. In 1976, results from the Viking landers were initially ambiguous, but it was concluded that life on Mars had not been observed (e.g. McKay 1997, Klein 1999). SETI searches have not found any signal. No obelisks have been found on the Moon, no Dyson Spheres have betrayed themselves in the infrared and no eyewitnesses from Roswell are talking. With all this negative evidence piling up, every reasonable scientist I know is of the opinion that no evidence for life has been found beyond Earth: "...there is no evidence from the reconnaissance of the solar system that life exists on any planet other than Earth." (McKay and Stoker 1998).

However, if I were parochial enough to define language as anything that resembles English, and then traveled to China, I would conclude that "No evidence for language has been found in China." The meaningfulness of this statement depends entirely on how meaningful my definition of language is. Similarly, the meaningfulness of the statement "No evidence for life has been found beyond earth" depends entirely on how meaningful our definition of life is. Without knowing the cosmic range of life forms, how can we determine if our terrestrial-life-based definitions are meaningful in a cosmic context?

Many hours of human thought have been dedicated to constructing definitions of life suitable for application elsewhere in the universe (Sagan 1970, Lehninger 1970, Chyba and McDonald 1995, Nealson and Conrad 1999, Cleland and Chyba 2002). Biology textbooks are full of definitions of life. These definitions have been sculpted around the flora, fauna and fungi we know on earth. They usually mention growth, self-regulation, self-reproduction and chemical complexity. Some definitions are blatantly tautological: "Life is what is common to all living beings". I am not convinced that these standard definitions of life based on terrestrial examples are broad enough to have reasonable expectations of universality. Neither was Einstein:

"Whether there are some general characteristics which would apply not only to life on this planet with its very special set of physical conditions, but to life of any kind, is an interesting but so far purely theoretical question. I once dis cussed it with Einstein, and he concluded that any generalized description of life would have to include many things that we only call life in a somewhat poetical fashion." (Bernal 1949).

How then do we construct a definition of life general enough to give us some confidence in its universality? If our intuitive ideas about what is or is not alive are too parochial, where can we hunt down some generality? Physicists and chemists are qualified to practice their trade on the planets around Alpha Centauri or anywhere in the universe, but biologists will probably have to retool. So maybe we can rely on physicists to come up with a usefully broad definition of life. Inspired by the work of Schrödinger (1944), Prigogine (1947, 1980), Bernal (1949) and Schneider and Kay (1995), I think a definition of life based on general thermodynamic principles will prove more useful in an extraterrestrial context than our current terrestrial-life based definitions.

2. Are Viruses Alive?

Viruses outnumber bacteria and all other life forms by an order of magnitude (Bamford et al. 2002). They are by far the most numerous inhabitants of our own backyard. There are 10 million viruses per milliliter in aquatic environments (Bergh et al. 1989). Their variety is staggering: circular viroids, non-pathogenic plasmids, bacteriophages, viruses infecting other viruses, satellite viruses, satellite nucleic acids, jumping genes and a whole menagerie of virus-like entities.

Despite their number, their diversity and their possibly fundamental role in the origin of traditional life, when you ask your favorite thoughtful biologist if viruses are alive, you will hear hemming and hawing. Standard biologists deny viruses the status of living beings because, it is said, viruses cannot reproduce themselves. They need help. Who doesn't? Take any standard life form, a man for example, or a woman, or both – put them into a hermetically sealed chamber in outer space and they will not self-reproduce. They will die and decay.

As far as I can tell, a single life form can do nothing by itself. Life forms are open systems deeply and obligatorily imbedded in the ecosystems in which they evolved. Because of this fundamental dependence of life forms on their environments and on the other life forms in their environments, defining life in terms of "self-reproduction" makes little sense to me. Definitions of life often use the word "self-reproduction", as if it meant something – as if life came in little isolated bags called cells that could survive by themselves. They can't. The mutual dependence of life forms runs deep into the tree of life when the earliest life forms were probably tightly knit communities of bacterial mats (some of whom produced stromatilites), not individual cells (Paerl et al. 2000).

We use the words "to reproduce" and think it different from "to be reproduced". This active/passive distinction comes intuitively to us but it is not obvious to me that nature cares about such subtleties–maybe we shouldn't use them with such confidence. For example, we say that humans 'reproduce' but laptops 'are reproduced'. However, humans can be conceived in vitro or in surrogate mothers and eventually we may mature in artificial wombs in factories, while laptops will soon control the assembly lines that manufacture laptops. With these almost inevitable developments, the distinction between "to reproduce" and "to be reproduced" will first become blurred and then become useless.

The main point I am trying to make is that our traditional definitions of life are lacking in generality. They include superficial distinctions between "to reproduce" and "to be reproduced" and they do not include viruses, possibly the most fundamental part of the biosphere (Fig. 1).

How can we measure whether our concept of life is general enough to be useful in a cosmic context? Estimates of the generality of a given property of life can be based on how fundamental or how ancient or how widespread that property is. If a property is fundamental, ancient and widespread among the life forms on Earth, it is almost certainly a property of the last common ancestor and it is a good candidate for being general enough to be present outside the Earth. Water, organic molecules, sugars, alcohols and amino acids are all over the universe – they fall from the sky in carbonaceous chondrites. These chemical building blocks are ancient, widespread and fundamental to the biochemistry of all life on Earth. Therefore, it is probably from these same standard building blocks that chemical life emerges all over the universe (if it emerges at all).

As traditional life emerged and evolved from a rather deterministic form of molecular evolution, the most important part of a life form's environment became other life forms. Selection pressures became self-referential. This type of close feedback amplified the idiosyncrasies of random mutations. Diverging life forms got quirkier with time. According to this view, we might expect bacteria or archaea on other planets because these are the earliest and simplest forms of life we know of. And we should not expect sulfur-crested cockatoos, homo sapiens or naked mole-rats on other planets. These are later, more complicated and quirkier forms of life.

The tree of life that traces the evolution of DNA-life on this planet (e.g. Pace 1997, Xue et al. 2003) tells us that there was an early divergence (3 or 4 billion years ago) that led to two separate branches: bacteria and archaea. However, the divergence between plants and animals was comparatively recent, ~1.5 billion years ago (Hedges et al. 2004). Thus, if we want to ask reasonable questions about life on other planets, it probably makes more sense to ask: Are there bacteria and archaea on the planet? (this split is labeled "A" in Fig. 1) rather than the quirkier downstream question: Are there plants and animals on the planet? (split labeled "B" in Fig. 1). I mention this chronology-of-the-tree-of-life-based criterion for reasonable questions because it is possible that viruses represent a very early form of life and thus a form of life that may be general enough to represent our best candidate for the type of life we should expect on other planets.

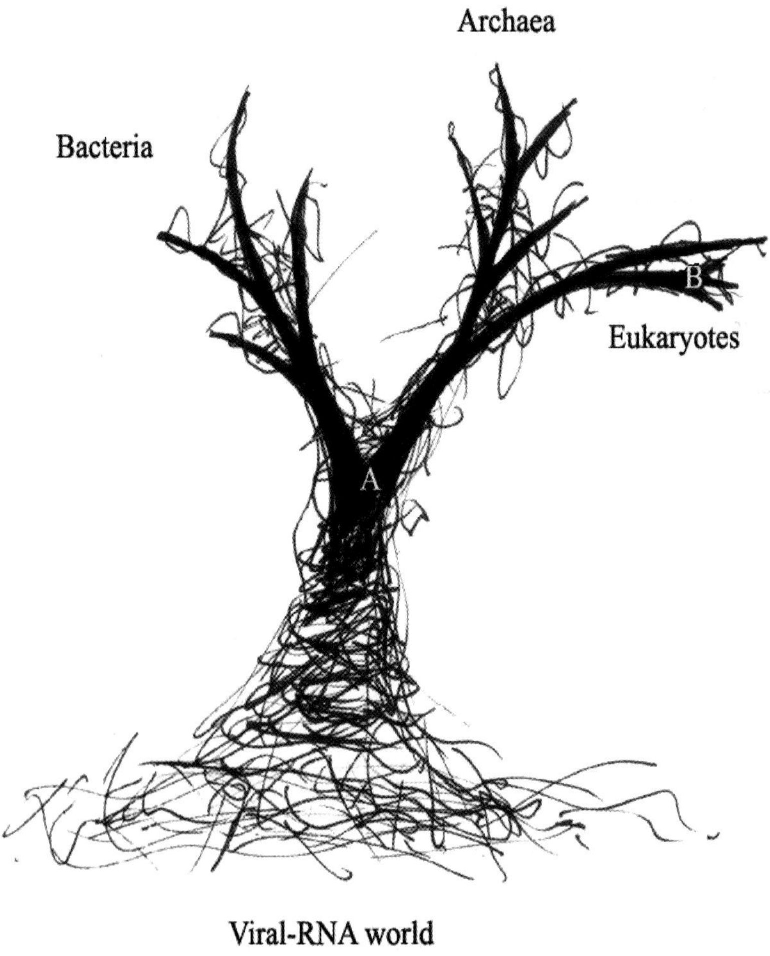

Figure 1. Tree of life emerging from roots in an RNA/Viral World. Every branch is adorned (or infected) with viruses. Viruses may well be remnants from an earlier epoch in which they were the dominant life form and no stable gene lines had yet emerged. From the viral point of view our vertically transferred genomes are frozen, hardly-evolving non-participants in the lively cut and thrust of lateral gene transfer.

Viruses are an immensely important part of the present biosphere and may have played a vital role in the origin of life. In an influential article Woese (1998) describes an RNA world of proto-organisms in which horizontal gene transfer played the dominant role. Reproductive fidelity was low but as fidelity gradually increased, "annealing" took place and organisms with gene lines emerged from a genetic chaos. Viruses still belong to such a genetic chaos and are therefore good candidates for being representative survivors from this epoch. In other words, viruses may have diverged so early from the forms of life that led to us that we have difficulty recognizing them as our ancestors. If this scenario is correct, then the most reasonable questions that one can ask about life on another planet is: Are there viruses and cellular life on the planet? Did cellular life emerge from viral life when fixed gene lines started to emerge and vertical gene transfer started to compete meaningfully with the more ancient viral horizontal gene transfer? If this scenario is even approximately correct, then one of the most important players in the origin and evolution of life on earth and one of our best guesses about the form of life elsewhere, is excluded from our traditional definition of life simply because it diverged too long ago for us to recognize it as our ancestor. This seems analogous to being able to recognize our relationship to chimpanzees but not our more distant relationship to fish. In our xenophobia, we only want ancestors who were like we are now – that is, ancestors who were preoccupied with making protein and having sex – as if the ability to outsource protein production were a bad thing!

The origin of life from non-life involves a series of transitions, and viruses may represent a stage in this series. In an evolutionary sequence, it makes little sense to break the chain at a specific link and pontificate: links before this break are not alive while links after this break are alive. Thus, to dismiss viruses as non-living because they are survivors from a very early epoch reminds me of sending everyone born before Jesus into limbo. Even if the scenario for the viral origins of life is incorrect, we need to deconstruct traditional definitions of life to make sense of the continuum between life and non-life.

3. Are Japanese Alone in the Universe?

Another way to access the generality of a particular grouping such as "life", is to consider more specific and more general groups. A fundamental question of astrobiology is: Are We Alone? The "we" is vague. Who is "we"? To remove this ambiguity a japanese website is more specific and asks: "Are humans alone in the universe?" (http://jvsc.jst.go.jp/universe/et_e/index_e.htm). But why stop at that level of specificity? We could, for example ask: "Are Japanese alone in the Universe?" There is a hierarchy of generality to these types of questions, starting from the specific and becoming more general:

IS "WHO" ALONE? FROM THE SPECIFIC TO THE GENERAL

Question	Answer
"Are Japanese alone in the Universe?"	yes
"Are Humans alone in the Universe?"	yes
"Are Mammals alone in the Universe?"	yes
"Are Vertebrates alone in the Universe?"	yes
"Are Eukaryotes alone in the Universe?"	yes
"Are DNA-based Terrestrial Life forms alone in the Universe?"	maybe
"Are Far From Equilibrium Dissipative Systems alone in the Universe"	no
"Are Terrestrial Atoms alone in the Universe"	no

On the right I have given what I believe are defensible answers. The most specific questions at the top and the most general questions at the bottom have easy answers. I think good arguments can be made that the Japanese people are alone in the universe – that is, on another planet, we will not find another Japan with Japanese people walking around speaking Japanese and playing baseball. Simpson (1964) in "The non-prevalence of humanoids" articulated the case that humans (or any given species) were a quirky product of terrestrial evolution that we should not expect to find elsewhere in any form that we could call "humanoid" (see Conway Morris 2003 for a dissenting opinion). Any given species that has evolved on the Earth will have its closest relatives here on Earth. Thus, if we consider humans to be alone on Earth, then humans are a fortiori alone in the universe.

Mammals and Vertebrates seem more generic, but these groups are monophyletic–they evolved from a single species that diverged into the range we see today. So although they are now more generic than a single species, they were a quirky single species a few hundred million years ago. We may be able to apply the same reasoning to even larger current groups, such as Eukaryotes, but we will need to go back a couple of billion years before the quirkiness of their humble and very specific beginnings are recognized (e.g. Bell 2001, Kostianovsky 2000). The next in this hierarchy of generality is DNA-based terrestrial life. Could it be that terrestrial life is as quirky as a species and therefore we should not expect it elsewhere. Opinions span the range from "our type of life is unique to Earth" to "our type of life is everywhere". Both seem reasonable to me, hence the answer "maybe" to this question.

The last two groups, Far From Equilibrium Dissipative Systems (FFEDS) and Atoms, are sufficiently general that we are sure that the answer is no; they are not alone. Detailed spectra of stars tell us that terrestrial atoms are all over the universe. Also far from equilibrium dissipative systems like stars, convection cells, atom-spheric whirlpools like Jupiter's red spot and dust devils on Mars are all over the universe.

With this hierarchy in mind, I propose that we redefine "life" by shifting it one step more general and equate life with Far from Equilibrium Dissipative Systems (Figure 2). A similar shift was necessary when we generalized our notions of life to include newly discovered microscopic life. Such a generalizing shift should already have taken place with respect to viruses.

The book Scientists on Gaia (Schneider and Boston, 1998) was dedicated to the question of whcther the biosphere as a whole is alive. It is an interesting issue that doesn't have an easy answer. Besides viruses, ant colonies, ecosystems and Gaia, there are other entities whose status as "life" forms is reluctantly bestowed. What about prions, and the artificial life we create in a computer. Is Google alive yet? Our definitions of life seem to lose their meaning for big things like ecosystems, Gaia, stars or the galaxy. They also fall apart for small things like viruses, viroids, and prions. Our traditional definition of life seems to be too narrow to classify even the range of structures we see on Earth. To divide up all the diversity found on Earth into two boxes, one called life and one called non-life may be a senseless task that we should forget about (Fig. 2, top panel). However, I believe that meaningful attempts based on more general features can be made (Fig. 2, bottom panel).

4. Thermodynamics as a Basis for a More Universal Definition of Life

Where can we hunt down some generality to improve our definition of life? I would suggest chemistry and physics and specifically thermodynamics. When a chemical system is in equilibrium, no reactions take place. The temperature is constant. There is no heat flow and no chemical gradient. Equilibrium is another name for death. When thermal gradients, chemical concentration gradients or electric potential gradients are large enough, Gibbs free energy becomes available and organized structures emerge spontaneously that act to reduce the gradient and dissipate the free energy. In a fundamental sense, these spontaneous and ubiquitous structures are nature's way of reducing gradients and obeying the second law of thermodynamics (Schneider and Kay 1994).

Convection cells in the solar photosphere are organized structures maintained by the temperature gradient between the hot subsurface and cooler surface of the Sun. Bernard cells are an example of the same phenomena in the lab. Whirlpools exploit gravitational potential energy to maintain their structure while hurricanes and dust devils run on temperature and pressure gradients. Where there is a gradient or a far from equilibrium situation, dissipative structures emerge to remove the gradient and exploit the available free energy to maintain the structure that is doing the dissipation. These gradients are all over the universe and thus, so too are the structures associated with them.

At the equator there is more carbon-based life and there are more hurricanes. This is no accident. The trees of a rain forest and hurricanes are the same sort of structure. They are far from equilibrium dissipative structures that dissipate the radiative, thermal, pressure and chemical gradients set up by the flux of the Sun. In the open ocean there is abundant sunlight but not much chemical disequilibrium and therefore not much life. Similarly, chemical equilibrium at the bottom of the ocean precludes much life there. However, on the mid-oceanic ridges, where a strong thermal gradient produces black smokers, a strong chemical gradient provides the environment for dissipative structures called life forms.

As heterotrophes we maintain our structure by dissipating the chemical gradients between our low entropy food and our high entropy waste. Traditional carbon-based

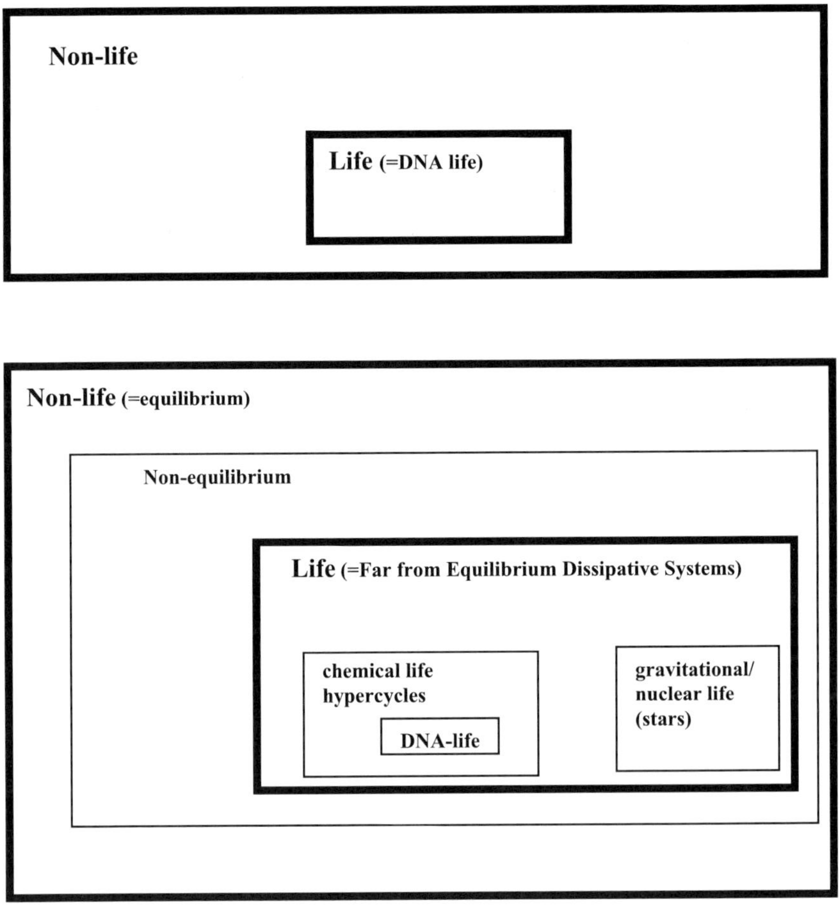

Figure 2. Redefining Life. Here are two schemes for dividing up the Universe and defining life. At the top is the traditional dichotomy based on the presence or absence of DNA. The lower panel is a scheme based on the thermodynamic concept of equilibrium. I argue that the equilibrium-based scheme more plausibly represents a classification system for extraterrestrial life and shows that DNA-life is a subset of the more general class of Far From Equilibrium Dissipative Structures. In the traditional view, we pretend that there is a discreet defensible boundary between life and non-life (thick blank line) although we are not sure in which box to put viruses, prions, ecosystems and Gaia. If we (the traditional "life forms" of Earth) belong to any generic set, it will be to the set of Far From Equilibrium life forms, along with hurricanes, red spots on Jupiter and stars (bottom panel).

life forms in general are persistent structures that reduce the electrochemical potential gradient associated with reduction/oxidation pairs – essentially electrons falling into deeper potential wells. They exploit the gradient of electron negativity in redox pairs (Nealson and Conrad 1999). Photosynthetic creatures are structures maintained by the dissipation of the energy associated with 6000 K photons in a 300 K environment. If these photons were coming from all directions and in equilibrium with a 6000 K environment, there would be no free energy available, and no far from equilibrium structure could exist: no plants and no hurricanes. With these thermodynamic considerations, far from equilibrium dissipative structures (not just the carbon-DNA-based ones) are a broad category that can be used to define life in a way that makes sense on the planets orbiting Alpha Centauri or anywhere in the Universe. It is clear that DNA-based life belongs to this group of far from equilibrium dissipative systems. It is also clear that FFEDS is a larger more generic group that includes more than just DNA-based life.

But wait! In our hunt for generality couldn't we arbitrarily say hydrogen atoms are alive – and since we've detected hydrogen everywhere in the observable universe, we could say: We have detected extraterrestrial life! That, however, would be a semantic game with no scientific content. I believe that there are many good physical, thermodynamic reasons to put traditional life into a more objective, more generic, more well-defined class of objects called far from equilibrium dissipative structures. When we get used to calling this category "life", I suspect that our previous definitions of life will seem as parochial as equating life with just flora, fauna and fungi.

If most alien civilizations are based on nuclear life on the surfaces of neutron stars (Feinberg and Shapiro 1980), what type of definitions for life would be required to call them life. A 'life = FFEDS' definition would work but a DNA-based definition would not. Far from equilibrium systems and their relation to life is discussed in Schneider and Kay (1994,1995), for an opposing view see Anderson and Stein (1986).

5. The Origin and Centralization of Information and the Prejudice against Non-DNA Life

One argument against the idea of FFEDS = life, is that hurricanes and convection cells and stars contain no information within themselves that is passed on when they reproduce. Stars for example are far from equilibrium dissipative structures metabolizing the free energy made available by a nuclear potential energy gradient. A star has no DNA inside itself and no internal structure that is passed on to the next generation of stars. However, when a massive star dies, it explodes and infuses the surrounding molecular hydrogen with heavy elements that enhance the ability of the clouds to lose energy and collapse to form more stars. The shockwave from the explosion also plays a role by kick-starting gravitational collapse. Thus, there are important links from one generation of stars to the next. However, the information controlling and sculpting the next generation of stars is in the modifications of the stellar environment, not centralized in some coded molecule.

The information content of DNA came from the environment as generation after generation of organisms were selected and filtered. Does it matter whether the information to form another system is contained centrally or distributed? In our traditional view of life, just as we may be exaggerating the importance of whether we 'reproduce' or 'are reproduced', we may be exaggerating the importance of whether the information and conditions that affect reproduction are stored internally in DNA or externally in the composition of the material that will form the new FFEDS.

Survival is the issue, not whether the gradient that led to a FFEDS has been set up artificially by centrally stored information or naturally by decentralized information.

6. The First Detection of Life as We Don't Know it

I went to a conference on artificial life where there was a serious discussion of how to build a robot – essentially a mobile computer acting on it's own, making it's own decisions about where to go etc.

We all laughed at clips of battery powered robots stumbling over little obstacles. We had all walked into the room carrying laptop computers that we opened up and fed information to. At the end of the discussion we picked up the laptops and carried them efficiently out of the room. During the discussion I mentioned that our laptops were already doing everything that we were trying to get robots to do. Our laptops have co-opted our legs to move around. I was surprised that this idea seemed absurd to all the robot-building, laptop-carrying attendees.

Similarly the detection of extraterrestrial life claimed here, based on a more general definition of life may seem absurd to the army of biochemists invested in looking to manufacture traditional life in the lab. According to the new definition, life is easy to make – just let the water out of the tub or blow smoke rings or just boil water and watch the convection cells. With a shift in focus to the most general, universal features of life – a redefinition of life – we start to see life everywhere.

The non-prevalence of terrestrial life in the universe is becoming a data set that should give us pause and motivation for a more general view of life. I have argued that the traditional definitions of life are too narrow. These definitions don't include viruses and often include a meaningless statement about 'self-reproduction'. It is possible that the universe and the life forms in it are not usefully defined by our traditional definitions of life. Therefore we should take seriously a different more universal definition: life = FFEDS.

With this more universal definition of life, we can say with conviction that we have detected extraterrestrial life and that life on Earth is not alone in the universe. We are not alone! Somebody should let the SETI people and NASA (who are "following the water") know that they've been scooped.

7. References

Anderson, P.W. and Stein, D.L. (1987) Broken Symmetry, Emergent Properties, Dissipative Structures, Life: Are They Related? In: F.E. Yates (ed.) *Self-Organizing Systems: the emergence of order*. Life Science Monographs, Plenum Press, NY, pp. 445-458.

Bamford, D.H. Burnett, R.M. Stuart, D.I. (2002) Theoretical Population Biology. **61**, 461-470.

Bell, P. (2001) Viral Eukaryogenesis: Was the Ancestor of the Nucleus a Complex DNA Virus? J. Mol. Evol. **53**, 3, 251-256.

Bergh, O. Borsheim, K.K., Bratbak, G. and Hudd, M. (1989) High abundance of viruses found in aquatic environments. Nature. **340**, 467-468.

Bernal, J.D. (1949) The physical basis of life. Proc. of the Physical Society. **62**, 10, 358B.

Chyba, C.F. and McDonald, G.D. (1995) The Origin of Life in the Solar System: current issues, Ann. Rev. Earth Planet Science. **23**, 215-49.

Cleland, C.E. and Chyba, C.F. (2002) Defining Life, Origins of Life and Evolution of the Biosphere. **32**, 387-393.

Conway-Morris, S. (2003) *Life's Solution: Alone in a Crowded Universe* Cambridge University Press.

Feinberg, G. and Shapiro, R. (1980), *Life Beyond Earth: the Intelligent Earthling's Guide to Life in the Universe*, Morrow, NY.

Hedges, S.B., Blair, J.E., Venturi, M.L., Shoe, J.L (2004) A molecular timescale of eukaryote evolution and the rise of complex multicelllular life. BioMedCentral Evolutionary Biology. **4**(2), 1-9.

Klein, H.P. (1999) Did Viking Discover Life on Mars? Orig. Life and Evol. Biosphere. **29**, 625-631.

Kostianovsky, M. (2000) Evolutionary Origin of Eukaryotic Cells, Ultrastructural Pathology. **24** (2), 59-66.

Kuhn, T. (1962) *The Structure of Scientific Revolutions*, University of Chicago Press.

Lehninger, A. L. (1970) *Biochemistry*, Worth Publishers Inc. NY.

McKay, C.P., (1997) The Search for Life on Mars, Origins of Life and Evolution of the Biosphere **27** (1-3), 263-289.

McKay, C.P. and Stoker, C.R. (1998) Gaia and life on Mars. In: S.H. Schneider and P.J. Boston (eds.) *Scientists on Gaia*, MIT press, Cambridge, MA, pp. 375-381.

Nealson, K.H. and Conrad, P.G. (1999) Life: past, present and future. Phil Trans. R. Soc. Lond B **354**, 1923-1939.

Pace, N.R. (1997) A molecular view of microbial diversity and the biosphere. Science. **276**, 734-740.

Paerl, H.W., Pinckney, J.L. and Steppel, T.F. (2000) Cyanobacterial-bacterial mat consortia: examining the functional unit of microbial survival and growth in extreme environments, Environmental Microbiology **2** (1), 11-26.

Prigogine, I. (1947) *Etude thermodynamique des phenomenes irreversibles* Dunod, Paris.

Prigogine, I. (1980) *From Being to Becoming* Freeman, NY.

Sagan, C. (1970) Life. In *Encyclopedia Britannica* 15th edition Vol. 10, New York, Macropaedia.

Schneider, S.H. and Boston, P.J. (eds.) (1991) *Scientists on Gaia*, MIT Press, Cambridge, MA.

Schneider, E.D. and Kay J.J. (1994) Life as a manifestation of the second law of thermodynamics. Mathematical and Computer Modeling. **19**, No. 6-8, 25-48.

Schneider, E.D. and Kay, J.J. (1995) Order from Disorder: The Thermodynamics of Complexity in Biology. In: M. P. Murphy and L.A.J. O'Neill (eds.), *What is Life: The Next Fifty Years. Reflections on the Future of Biology,* Cambridge University Press, pp. 161-172.

Schrödinger, E. (1944) *What is life? and other scientific essays* Double Day Anchor, Garden City, NY.

Simpson, G.G. (1964) The Non-Prevalence of Humanoids. Science. **143**, 769.

Woese, C. (1998) The Universal Ancestor. PNAS. **95**, 6854-59.

Xue, H. et al. (2003) Transfer RNA paralogs: evidence for genetic code-amino acid biosynthesis coevolution and an archaeal root of life. Gene. **310**, 59-66.

Biodata of **Seth Shostak** author of ***"The Search for Extraterrestrial Intelligence"***

Dr. Seth Shostak is a Senior Astronomer at the SETI Institute, in Mountain View, California. He has an undergraduate degree in physics from Princeton University, and a doctorate in astronomy from the California Institute of Technology. For much of his career, Dr. Shostak conducted radio astronomy research on galaxies, and has published approximately fifty papers in professional journals. He also founded and ran a company producing computer animation for TV.

Seth has written several hundred popular magazine and Web articles on various topics in astronomy, technology, film, and television. He now lectures on astronomy and other subjects at the California Academy of Sciences, and for the last five years, has been a Distinguished Speaker for the American Institute of Aeronautics and Astronautics. He has edited and contributed to a half dozen books. His first popular tome, "Sharing the Universe: Perspectives on Extraterrestrial Life" appeared in March, 1998. His most recent books are "Life in the Universe" (2002, textbook with Jeff Bennett and Bruce Jakosky) and "Cosmic Company" (2003, with Alex Barnett).

E-mail: **shostak@attglobal.net**

J. Seckbach (ed.), Life as We Know It, 459-471.

THE SEARCH FOR EXTRATERRESTRIAL INTELLIGENCE

SETH SHOSTAK
SETI Institute
515 N. Whisman Road
Mountain View, California 94043, USA

1. Introduction

It's a question that every generation has surely asked: could there be life elsewhere in the cosmos? Not just simple life – microscopic bacteria hiding in the wet rock beneath the surface of Mars, for example – but thinking, intelligent life. Could the type of aliens so often seen on television and in the movies really exist? Or is it possible that in this part of the Galaxy, or perhaps in all the universe, only humans can look at the sky above, and understand what they are seeing?

The matter of whether sentient beings are rare (or even unique to Earth) has been debated for thousands of years. In recent decades, greater understanding of biology, and discoveries in astronomy and geology, have shifted the debate from the realms of philosophy and theology to that of science. This has led to experiments that could directly affirm the existence of extraterrestrial intelligence. While earlier generations could only ask, we might be able to know.

Perhaps the discoveries that have most significantly affected the search for intelligent beings elsewhere are these:

(a) Palaeontological finds that have pushed the start of life on Earth back at least as far as 3.5 billion years. While our world is 4.6 billion years old, left-over debris from the solar system's origins pummeled Earth's surface between the time of its birth and the cessation of the heavy bombardment 3.8 billion years ago. The remarkable conclusion is that terrestrial biology got started very quickly, once our planet was no longer being pounded by planetoids. This suggests (but does not prove) that whatever mechanism is responsible for the origin of life is not improbable.

(b) The strong evidence that Europa, and quite possibly Callisto and Ganymede, might be covered with massive liquid oceans, capped by a skin of ice or rock. These moons of Jupiter have awoken us to the prospect that life could infest a wider range of habitats than once thought.

(c) The discovery of extrasolar planets. The fraction of Sun-like stars with planets is at least 10 – 20%. While the prevalence of Earth-size worlds is still unknown, the data are not inconsistent with the speculation that there could be more planets in the universe than stars, simply because planets smaller than Neptune could not, in general, be found with the experiments conducted so far.

Of course, while these advances certainly augur for the presence of life beyond Earth, they don't directly speak to the more controversial matter of *intelligent* life. Could biology be common, but sentience rare? One might hope to answer that question with investigations into how intelligence arose on Earth, and indeed that line of research is being pursued. On the other hand, experiments that might directly detect manifestations of intelligence, such as SETI (the Search for Extraterrestrial Intelligence), would short-circuit both inference and extrapolation, and tell us immediately that other thinking beings exist.

2. The History of SETI

While the possibility that simple life might have arisen on other worlds of our own solar system is still taken seriously, few scientists today would venture that any of that hypothesized biology was, or is, intelligent. Consequently, our search for thinking extraterrestrials must perforce direct its attentions to other star systems. These stellar realms are at daunting distances, unreachable by our spacecraft in less than tens of millennia, and thus robotic exploration is ruled out. So is direct contact, if the difficulties of interstellar travel are as daunting to other societies as they are to ours.

As an example of the latter, we note the case of the Pioneer and Voyager probes that NASA sent to reconnoiter the gas giant planets (Jupiter, Saturn, Uranus, and Neptune) in the 1970s. Since these probes were the first human-made, tangible artifacts destined to eventually leave the solar system, they carried "messages" in case they were ever found by extraterrestrial beings. The Pioneers were fitted with a plaque engraved with some simple graphics describing our appearance and size, and indicating Earth's location in the Galaxy. The Voyagers had a (mechanical) record carrying images, sounds, and music. While these messages to putative aliens were perhaps useful in causing us to think about the problems of intracultural communication, it seems unlikely they will ever be retrieved and read. The probes themselves would be difficult to find, and in any event will take hundreds of thousands of years to reach the distances of even the nearest stars.

Signaling, in particular with electromagnetic radiation (light and radio), is an obvious and attractive alternative to physical investigation. Karl Friedrich Gauss, the talented 19th century mathematician, devised a device – the heliotrope – that could be employed to signal with sunlight. He suggested its use for contacting inhabitants of the Moon. If communication with the lunarians could be achieved, he expected that "would be a discovery even greater than that of America."

Other 19th century scientists also suggested signaling with light, although the targets of their communication continued to be relatively nearby (the Moon and Mars). The French inventor Charles Cros proposed using lamps and mirrors to flash messages to these worlds, noting that the flashes would be more likely to get attention. In this, he foreshadowed one of modern SETI's occasional assumptions: namely, that pulsed light communications might be more easily recognized against the natural, constant glow of starlight. Cros also noted that the flashes of light from Venus that some reputable observers claimed to see might be attempts by the inhabitants of that world to get in touch with us. This is noteworthy, as it suggests a passive approach to discovering the existence of other societies, the methodology currently favored by the SETI community.

By 1959, sufficient knowledge of the true nature of both the Moon and Mars had dissuaded most scientists from expecting sophisticated inhabitants of either. But the idea of sending messages across space was quantitatively developed in an article that year in the journal *Nature* by physicists Giuseppe Cocconi and Philip Morrison. They showed that radio technology hardly more advanced than our own was adequate to send detectable transmissions across light-years of distance. Consequently, the physicists urged that a search be made for such transmissions, and added that microwave frequencies might be the most logical part of the spectrum to investigate. This was because of the natural emission line of neutral hydrogen at 1,420 MHz, the most important astrophysical line at radio frequencies, and one that any technological society would know.

Less than a year after the *Nature* article appeared, astronomer Frank Drake pioneered modern SETI experiments by turning a 26-meter dish toward the nearby Sun-like stars Tau Ceti and Epsilon Eridani in a scan for artificial signals. Although Drake failed to hear any clearly extraterrestrial emissions, the astronomical community most assuredly heard Drake. The idea of searching for radio transmissions from interstellar civilizations caught hold quickly and firmly. By the 1970s, NASA had joined the hunt, primarily because of the initiative and interest of John Billingham, working at NASA's Ames Research Center in California. In a landmark study, Billingham teamed up with engineer Barney Oliver and a slew of academics to consider the best approach for detecting sophisticated extraterrestrials. Their report, known as Project Cyclops, proposed constructing an array of large dishes, a magnificent radio telescope that would inspect vast numbers of stars in a sensitive and systematic search for microwave signals.

The Cyclops array wasn't built; the price tag was as imposing as the instrument. But ideas spawned by Drake's efforts and the Cyclops study are part and parcel of today's SETI projects.

3. The Drake Equation

A year after Drake's first search for microwave signals from other star systems, he composed a simple formula that encapsulates the basic question addressed by SETI. The Drake Equation, as it's known, estimates the number of transmitting civilizations that are in the Milky Way now. Clearly, if this number is very high, the chance that our experiments might find evidence of the aliens is good. The equation is:

$$N = R_* \, f_p \, n_e \, f_l \, f_i \, f_c \, L \qquad (1)$$

where N is the number of transmitting civilizations, R_* is the galactic rate of formation of stars suitable for hosting habitable planets (stars/year), f_p is the fraction of such stars actually having planets, n_e is the number of Earth-like worlds per planetary system, f_l is the fraction of these worlds that spawn life, f_i is the fraction of the those worlds producing intelligent life, f_c is the fraction of those that develop technical societies, and L is the average lifetime (in years) that such a society is "on the air."

The first few terms in the equation can be determined by straightforward astronomical observation. On the other hand, the fraction of habitable planets that are actually inhabited (f_l) is completely unknown, although as we've noted, the rapid appearance of life on Earth suggests that this fraction might be high. As we move farther to the right in this equation, the terms become even less well determined or, more accurately for the last two, completely speculative! Consequently, the Drake Equation appears in textbooks more for its concise organization of the factors of interest to SETI than for actually being useful for estimating N. Published values for the number of transmitting galactic civilizations range from 1 (humanity) to many millions. The values cited by SETI researchers themselves generally range from thousands (Drake's estimate) to a million or more (Carl Sagan).

Note that the Drake Equation doesn't take into account the possibility that a society might colonize other star systems. This would only increase the number of transmitting sites, of course. The equation also restricts its attention to our own galaxy, on the assumption that signals from other nebulae would be too attenuated by the intervening distance to be easily found.

4. Radio SETI

For reasons that are largely historical, most SETI experiments have been conducted at radio wavelengths.

Only a few parameters vary significantly in contemporary experiments to hunt for extraterrestrial radio signals: (a) In which directions is the search conducted; i.e., what parts of the sky are examined? (b) To what frequencies are the receivers tuned. (c) What sorts of signals are sought. (d) What is the minimum signal strength that can be detected.

We consider these parameters below:

(a) Drake conducted what is now called a "targeted search," choosing to point his antenna in the directions of Sun-like stars. This was clearly a conservative approach, as we have one example of intelligent life that's developed in this type of stellar neighborhood. Stars substantially larger than the Sun have much shorter lifetimes – for giant stars, it only takes a few tens of millions of years before they've burned through their easily accessible nuclear fuel. They then undergo drastic changes in brightness; changes that would be catastrophic for any biology on encircling worlds. On the other hand, most stars smaller than the Sun will burn with a steady intensity for many tens of *billions* of years, which is a lifetime long enough that life on any orbiting planets could evolve to intelligence.

Traditionally, smaller stars have been neglected by the SETI community. This is because the lower energy output of these dimmer objects mandates that only close-in planets would be warm enough to have liquid water on their surfaces. Close-in worlds, however, would become locked by tidal forces into a situation in which one side of the planet was always turned to its star, while the opposite hemisphere was chilled by perpetual darkness. On the other hand, recent theoretical studies have shown that an atmosphere could moderate these extremes, and there could be a habitable, circular 'ring' at the border between light and dark that was conducive to life. This is an important matter for SETI, because it increases the number of interesting star systems by about a factor of seven.

Another approach to reconnoitering the cosmos is to simply examine all of it (or at least as much of the sky as is visible to the radio telescope at hand). Such sky surveys have been conducted repeatedly in both the northern and southern hemispheres. While they have the advantage of making no assumptions as to where extraterrestrials might be, they have the disadvantage that no part of the sky is looked at for very long, and this adversely affects the sensitivity (the minimum signal strength that could be detected). The difference between targeted searches and surveys is akin to looking for birds either by aiming your binoculars persistently at a few tree tops, or by sweeping the skies.

An intermediate approach is also used in some SETI experiments: scanning not the whole sky, but selected areas such as the plane of the Milky Way, a densely packed band of stars. This exposes billions of stars to scrutiny, although on average they are tens to hundreds of times farther away than the stars examined by most targeted searches. The latter are typically a few hundred light-years distant.

(b) Cocconi and Morrison suggested that microwaves were the optimal spectral band for a radio search. One reason for this choice is that microwaves penetrate the gas and dust that fills space between the stars without significant weakening. In addition, the natural radio emission of neutral hydrogen – the largest component of the interstellar gas – which occurs at 21 cm wavelength (1,420 MHz) is an important "marker" on the radio dial of any sophisticated society. Consequently, if extra-terrestrials wish for their signal to be found, they might choose a broadcast frequency close to 1,420 MHz, as both the equipment and motivation to monitor this band would be shared by any technologically accomplished civilization. Most SETI searches examine frequencies near this cosmic "hailing channel," although other frequencies have been suggested, and occasionally tried (double the 1,420 MHz line, pi times 1,420 MHz, etc.)

(c) It's a simple engineering fact that the easiest type of signal to find in the presence of noise is one in which the transmitter power is concentrated into as narrow a range of frequencies as possible. This is analogous to whistling, where the power of your lungs is converted into an approximately monotonic (single-tone) sound, one that can be heard farther than shouted words, for instance. Radio and television broadcasts usually include a so-called "carrier" signal, which is the analog of a whistle, and would be thousands of times easier to detect than the sound or picture part of the transmission. SETI radio experiments typically search the radio spectrum for signals that are as narrow as 1 Hz wide or less, the strategy being to find a signal proving that a transmitter is on the air, and only later to address the problem of constructing the equipment necessary to record any message.

Indeed, while many people are under the impression that a SETI detection is dependent upon finding some clearly non-random signal, such as a binary representation of pi, or encoded math or music, the reality is that all radio searches simply look for monotonic signals that, at most, would vary only by slowly blinking on and off.

(d) The minimum detectable signal is set by instrumental considerations, primarily the size (total collecting area) of the antenna being used, and the amount of time spent observing at any given frequency.

As an example of contemporary practice, consider the SETI Institute's Project Phoenix, a targeted search of Sun-like star systems. Between 1995 and 2004, Phoenix observed 750 nearby (closer than ~250 light-years) targets. The range of frequencies covered was 1,200 – 3,000 MHz, and the spectral resolution was 1 Hz. Phoenix's maximum sensitivity was $\sim 10^{-25}$ watts/m^2-Hz. A signal this strong would be produced by a 100 m diameter antenna at 100 light-years distance having a transmitter power of one megawatt. If the signal was radiated in all directions equally, then the required transmitter power is substantially more, $\sim 10^{12}$ watts.

An example of a contemporary sky survey is the University of California, Berkeley's SERENDIP observing project, which has examined approximately 1/3 of the total sky multiple times across a 100 MHz frequency range centered at 1,420 MHz, and with a resolution of 0.6 Hz. The sensitivity of this survey is approximately an order-of-magnitude less than that of Project Phoenix, simply because less time is spent observing in any given direction. This project has been the source of data for processing by the popular SETI@home downloadable screen saver. More than five million home computer users have installed this free data-processing software, making SETI@home the world's largest (distributed) computing project.

The coming decade will see the implementation of equipment for radio SETI that will speed up the search by 10 – 300 times. Project SERENDIP will be able to take advantage of a new multi-horn receiver being built for the Arecibo radio telescope in Puerto Rico (the largest, single-dish antenna in the world, with a diameter of 305 m). Rather than being limited to a single patch of sky, the new receiver will simultaneously examine seven. The SETI Institute, together with the University of California at Berkeley's Radio Astronomy Laboratory, is constructing a phalanx of 350 small (6 m) antennas in northern California, called the Allen Telescope Array. This is the first radio telescope designed from the start to be used efficiently, and full-time, for SETI experiments. Within two decades, it will scrutinize more than a million star systems.

The number of radio SETI projects remains small. In addition to the SETI Institute and the University of California at Berkeley, Australia, Italy, and Argentina have experiments underway. In the past, the Soviet Union had an ambitious SETI program, both observational and theoretical.

5. Optical SETI

Despite the optimism of 19th century scientists, modern SETI practitioners were at first unenthusiastic about the idea of interstellar communication using flashing lights. It was thought that this would be far less efficient than using radio. Of course, it was known that interstellar dust limits the penetration of visible light. However, the severity of this absorption is highly dependent on exactly where you're looking, and in many galactic directions we can see stars that are thousands of light-years away. And in any case, simply switching to infrared wavelengths greatly diminishes this problem, as infrared is only slightly absorbed by the dust.

A greater concern was the fact that, because of its higher frequency, an infrared quantum of light has ~10^5 times the energy of a microwave quantum. This means that the cost per bit (in energy) of sending a message using light is orders of magnitude greater than when using radio. However, it's easier to tightly focus light. Assume, as example, that a transmitting society wants to signal another star system 100 light-years away by aiming a near-infrared (1 micron wavelength) laser in its direction. The laser is focused using a 1 m diameter mirror (or lens). While only table-top in size, this optic would concentrate the energy of the laser onto a circle about the size of Jupiter's orbit, which would presumably include all of the Earth-like planets around the target star. In contrast, a 10 m diameter radio antenna, which is approximately the same cost as the 1m mirror, would illuminate a far larger circle, roughly 2 light-years across! The optical approach clearly puts more of the transmitted photons onto the target for a comparable investment in hardware, and this roughly compensates for the greater energy "cost" of optical photons.

While there are various schemes for sending messages on a beam of light, most optical SETI experiments look for very brief pulses, typically a nanosecond (10^{-9} sec) or less. A detectable pulse might have as few as 10 photons, but even this faint flash will greatly outshine the naturally produced light of the transmitting society's host star during that nanosecond.

Typical optical SETI searches involve telescopes having ~1 m mirrors, fitted with fast photomultiplier tubes and electronics to detect nanosecond photon bursts. Usually the light from the telescope is split among two or three tubes, so that only events found by all the photomultipliers simultaneously are tallied. This eliminates false alarms caused by natural radioactivity in the tubes themselves, or strikes by incoming cosmic rays. While most optical experiments are targeted searches, similar to their radio counterparts in examining nearby, Sun-like stars (to date, several thousand have been observed), Harvard University is building an instrument that will conduct an optical sky survey. All of these experiments operate at visible wavelengths, simply because infrared is blocked by the water vapor in Earth's atmosphere. A serious infrared SETI search will require the use of space-based telescopes.

6. Searching for Artifacts

Another approach to proving the existence of other-worldly intelligence is to find, not a signal, but alien artifacts. Imagine that Hubble or some other large telescope accidentally imaged the exhaust radiation from an interstellar rocket. Or perhaps we could discover mammoth feats of astro-engineering involving the rearrangement of an alien society's entire planetary system. Another intriguing possibility is that we might find an object at one of the Lagrange points – gravitational dead spots in the Earth-moon system where an alien time capsule or probe could float in endless cold storage.

While such scenarios are easy to imagine, most are difficult to translate into meaningful experiments. Nonetheless, some searches of astronomical data for tell-tale signs of both high-powered rockets and other astro-engineering projects have been made, as well as a preliminary search of the best Lagrange point locations.

Needless to say, nothing that's clearly extraterrestrial in origin has been found so far, and this is also the case in the search for signals, both optical and radio.

7. The Fermi Paradox

Many scientists are willing to believe that the universe could be teeming with life. But what, then, to make of the fact that we still haven't found any signals from other worlds? Is that important?

Some people have claimed it is. It's a straightforward calculation to work out that the time required to colonize the Galaxy, to visit and set up a beach head on every star in the Milky Way, is rather short on a cosmological timescale: a few tens of millions of years. The physicist Enrico Fermi made this simple reckoning in 1950, and then asked "where is everybody?" If the Galaxy can be completely inhabited by an ambitious civilization in such a short time, then how can we explain our failure to notice anyone in our neighborhood? Of course, not all societies would be interested in such a colonization project, but if only *one* has found the idea appealing, there should be ubiquitous evidence throughout the Galaxy. It would be like the shores of the Mediterranean two thousand years ago, littered with Roman architecture (and Romans).

This schism between expectation and observation is called the Fermi Paradox, and the idea has been extended to self-replicating machines, a possible technical development for any advanced society. The extended argument is that even if interstellar travel is too difficult for extraterrestrials, their engineered offspring – self-replicating machines – could have spread through the Milky Way's stellar realms long ago, repairing and reproducing themselves as necessary. Their presence should be manifest, according to some.

There are many possible scenarios for resolving the conflict between the expectation of lots of cosmic intelligence, and the reality (thus far) of not finding it. Perhaps the extraterrestrials really are nearby, but we're too unsophisticated to recognize that fact. (Those who believe that some UFOs are alien craft claim that this explains the Paradox. However, most scientists are skeptical that the evidence offered so far, mostly anecdotal or based on observations at great distance, gives good reason to believe that extraterrestrials are visiting Earth.)

Frankly, there is, as yet, no convincing explanation for the Fermi Paradox. Perhaps it simply means that intelligence is extremely rare. Or that colonization is haphazard, or unappealing for some reason of which we're not aware. In any case, few SETI researchers are overly concerned about the Paradox, because it makes a very large extrapolation (the Galaxy is nearly devoid of intelligent societies) on the basis of a very local, very simple observation (we still haven't heard a signal). The amount of sky intensively examined by SETI thus far is quite limited, and this is the reason that most practitioners give for the failure, so far, to turn up a signal.

8. The Next Twenty Years

SETI offers the tantalizing possibility of answering a question of long standing and great importance. But it's been more than forty years since the first modern search, and some people will ask whether finding an hypothesized signal from aliens could be just a fool's errand. Certainly, they will argue, decades of searching without conclusive result must mean something.

This is untrue. The number of star systems observed with sensitivity adequate to find weak signals, and over a wide range of frequencies, is small, approximately 1,000. To put this in perspective, consider the usual estimates for N, the number of co-existing, transmitting galactic civilizations. As noted, the estimates by SETI practitioners typically range from thousands to millions, although these researchers would be the first to admit that their estimates are, in fact, only guesses. However, if we take these numbers at face value, and note that perhaps 10% of all the stars of the Galaxy are suitable for supporting planets with life, then that means that one in 10,000 to onc in 10,000,000 of these targets will actually shelter a broadcasting society. Clearly, if we've only carefully examined a thousand star systems, then it's hardly surprising that no conclusively extraterrestrial signals have been found. Our sample size is too small to encourage us to expect a detection.

The construction of the Allen Telescope Array, as well as other instruments, will greatly accelerate SETI's search for transmissions. Historically, the speed of these experiments has increased according to Moore's Law, an empirical rule roughly stating that the amount of compute power one can purchase per dollar doubles every 18 months. SETI, which is heavily dependent on digital technology, has also closely followed this exponential improvement with time, and will continue to do so for at least the next 15 – 20 years. If Moore's Law is applied to the ATA, then in the coming two decades, this telescope will observe approximately a million targets, or a thousand times more than were observed in SETI's first 45 years. In the near future, SETI will – for the first time – have the capability to observe meaningful samples of galactic real estate.

In addition to this enormous improvement in search technology, the launch in the coming decades of space-based telescopes designed to find, and image, Earth-size planets around other stars is likely to provide SETI with high quality targets for examination. Not only will we have a list of Earth-size worlds, but spectral analysis of the light reflected by their atmospheres could hint at the presence of life. Clearly, this foreseeable augmentation of astronomical knowledge will improve the appeal of targets examined in SETI experiments.

9. What Would Be the Consequences?

New instruments now being built to search for radio and optical signals from space will increase the number of star systems that have been carefully observed by three orders of magnitude. As an example of what this could mean, note that if ten thousand or more galactic societies are transmitting detectable signals, the new telescopes should come up with a detection in the next two decades.

However, unlike the scenarios usually depicted in films, the detection wouldn't be a dramatic event in which researchers start screaming and frantically call the media. As noted earlier, the type of signals sought by SETI experiments are narrow-band, as these are both easily found and the type of emission that only transmitters produce, as far as we know. Of course, modern society generates countless narrow-band signals, and one of SETI's greatest challenges is to sort out this man-made interference from possible extraterrestrial broadcasts. In the past, this problem encouraged the practice of siting radio telescopes in deep valleys and areas of low population density. Today, such precautions are only partially successful, as telecommunications satellites will be overhead no matter where your telescope is located.

The consequence is that narrow-band signals crop up in SETI experiments all the time. Checking their origin is done using a series of tests; for example making observations with a second telescope, referring to a large database of known terrestrial interference, or simple (but time-consuming) observing procedures that include nodding the telescope back and forth across the sky. Rather than calling the local newspapers, a SETI researcher who finds a promising candidate signal will busy himself for days verifying that it's truly extraterrestrial. One essential test will be to ask colleagues at a second, independent observatory to verify the discovery, thereby ruling out hardware and software failures, or pranks.

Since confirmation will take several days, it's unavoidable that the detection of the candidate signal will be reported by the media (there is no policy of secrecy for SETI), and the newspapers will be touting stories of an "interesting signal." Once the signal is confirmed to be truly extraterrestrial, the public will be exposed to what many in the press expect would be one of the biggest news events of all time.

There would be only a few hard facts to share, albeit of great import. Since the astronomical community would be among the first to be alerted to the signal, the patch of sky where the discovery is made will be scrutinized by every major telescope capable of doing so. This would quickly tell us which star system is home to the transmitter (assuming it *is* a star system), as well as the star's distance and brightness. Observing slow changes in the frequency of a radio signal could tell us the length of the aliens' day and year, yielding some simple parameters that would influence the climatic conditions on their world.

What follows next depends on whether we could decipher any message attached to the signal. A message seems a likely component of any broadcast, but simply being able to detect and record it might be very difficult in the case of a radio signal. This is because the message will be wide-band, with energy spread over a large swath of spectrum. To adequately receive this type of signal would typically require a radio telescope thousands of times larger than that necessary to find the narrow-band signal proving that a transmitter is switched on. (This difficulty is far less severe for optical signals.)

Presumably, the detection of an alien broadcaster would generate so much interest that whatever instrument was necessary to detect the message would be built. However, whether the content could be understood remains an open question. As can be easily seen from a quick perusal of the Drake Equation, the number of transmitting civilizations is directly proportional to L, the average lifetime of a technically competent society. If L is small, our chance of finding something is also small. So, *ex post facto*, if a signal is detected, that's an indication that many societies stay "on the air" for long periods of time – and L is thousands of years or greater. Since it's statistically improbable that we will pick up a signal from a society that will last for thousands of years, but is at the moment only a few decades beyond our own development, the most probable scenario is that a detected signal would come from beings that are technically far in advance of us.

If that's the case, then it's conceivable that we will never succeed in decoding their message, unless in some act of extraterrestrial altruism they have deliberately made it easily understandable. If so, and if they have chosen to share some of their learning, the discovery of extraterrestrial intelligence could open doors to highly advanced knowledge. This could be akin to teaching Neanderthals how to read and then offering them access to a modern library. Information from another society would obviously have great impact on our own.

But even for the alternative scenario – in which a message is found, but never understood – our civilization would be greatly affected. We would know that our knowledge, our religious beliefs and our cultures are simply one example among many. The world would have incontrovertible proof of a humbling fact: that we are just another speck of foam on a vast, cosmic sea.

10. References

General

Davies, P. (1996) *Are We Alone? Philosophical Implications of the Discovery of Extraterrestrial Life*, Basic Books, New York

Drake, F.D. and Sobel, D. (1992) *Is Anyone Out There?,* Delacorte Press New York

Goldsmith, D. and Owen, T. (2001) *The Search for Life in the Universe*, University Science Books, Sausalito, California

Shklovskii, I. and Sagan, C. (1966) *Intelligent Life in the Universe*, Holden-Day Inc., New York

Shostak, S. (1998) *Sharing the Universe: Perspectives on Extraterrestrial Life*, Berkeley Hills Books, Berkeley, California

Shostak, S. and Barnett, A. (2003) *Cosmic Company*, Cambridge Univ. Press, Cambridge

Sullivan, W. (1993) *We Are Not Alone* (revised edition), Dutton, New York

Webb, S. (2002) *If the Universe is Teeming with Aliens, Where is Everybody? Fifty Solutions to Fermi's Paradox and the Problem of Extraterrestrial Life*, Copernicus Books, New York

More Specialized

Cocconi, G. and Morrison, P. (1959) Searching for Interstellar Communications, Nature, **184**, 844

Cosmovici, C.B., Bowyer, S. and Werthimer, D. eds. (1997), *Astronomical and Biochemical Origins and the Search for Life in the Universe*, Editrice Compositori, Bologna

Ekers, R.D., Cullers, D.K., Billingham, J., and Scheffer, L.K. (2002) *SETI 2020: A Roadmap for the Search for Extraterrestrial Intelligence*, SETI Press, Mountain View, California

Harrison, A. (1997) *After Contact*, Plenum Press, New York

McConnell, B. (2001) *Beyond Contact*, O'Reilly and Associates, Sebastopol, California

Oliver, B.M., and Billingham, J. (1971) *Project Cyclops: A Design Study for a System for Detecting Extraterrestrial Intelligence*, NASA CR 114445, Washington

Ward, P. and Brownlee, D. (2000) *Rare Earth*, Copernicus Books, New York

Web Sites

SETI Institute: www.seti.org

Harvard SETI program: http://seti.harvard.edu/seti/

Berkeley SETI program: http://seti.ssl.berkeley.edu/serendip/serendip.html

Australian SETI program: http://seti.uws.edu.au

Biodata of **Jiří Krupička** author of ,***"Search for Extraterrestrial Intelligence"***.

Dr. Jiří Krupička is Professor Emeritus of the University of Alberta in Edmonton, Canada. He obtained his Dr at the Charles University in Prague in 1947. In 1950 he was sentenced to 16 years in prison for an attempt to smuggle out his manuscript entitled "*Man and Mankind*" to the free world. Half of his sentence he spent in the infamous uranium mines at Jáchymov. He was released after ten years and participated in the "Prague Spring" in 1968. Dr. Krupička took part in the organization of the International Geological Congress the same year. The Congress was broken up by the Soviet invasion. In the same year he emigrated with his family to Canada.

Dr. Jiří Krupička is the author of several books dealing with some of the mankind's acute problems. All were published in Czech, in Prague. He was awarded the Pen Club Prize for his literary work and the medal of the Czech Academy of Sciences for his geological work.

E-mail: **jkrupicka@shaw.ca**

J. Seckbach (ed.), Life as We Know It, 473-485.

SEARCH FOR EXTRATERRESTRIAL INTELLIGENCE

J. KRUPIČKA
Professor Emeritus, University of Alberta
[1701] 11200-53 Ave., Edmonton, Alberta T6H OS4, Canada

1. Introduction

In all civilizations there were curious minds that asked: are there creatures like us even elsewhere in the universe? Science has brought some modesty into the too naive formulation of the question and changed the corporeal 'creatures' into 'intelligences'. Without unduly bothering about a stricter definition of what that concept should mean.

As long as the Earth was the undeniable center of the cosmos and humans were God's unique creations such questions did not arise, at least not for the believers.

With the fall of the view regarding the Earth as the center of the material universe the question about the existence of life elsewhere in the cosmos became inevitable. First among freethinking intellectuals, later in the general public; today it is High School textbook material in many developed countries. Cosmic creatures have entered both science fiction and popular culture *en masse* and have become an integral part of much of modern entertainment.

Universe has two goals: to find proofs of the existence of extraterrestrial life even in its humblest forms, and to come across signs of *intelligence* (as defined by humans) in the depths of the cosmos. The first goal is much more modest in comparison with the second, and can use a lot of astronomical data which are being accumulated at an ever-growing pace. All of them have hitherto been of probabilistic nature; they brought information about environments, which *might* be favorable for the germination of some forms of life.

The range of opinions about the existence of intelligent extraterrestrial life may be illustrated by the following quotations:

"Technological civilizations may have developed on as many as 390 million planets in our galaxy; virtually all of them are more technologically advanced than we are. If this were so, the average distance between such civilizations is about 40 light-years" (Asimov, 1979).

"How many extraterrestrials are there? Without hesitation, I would say that millions may exist" (Heidmann, 1995).

"Many astronomers like to argue that life must exist elsewhere because there exist so many sites where it could do so. The overriding opinion of biologists, on the other hand, is that the probability is essentially nil because the number of evolutionary pathways leading to biological dead-ends is at least as large as the number of sites" (Barrow and J. Silk, 1983).

"Even if my guess is right and there is lots of life in the galaxy it is extremely unlikely that such extraterrestrial life has either the capability of or any interest in contacting humanity" (Pagels, 1988).

2. The Present Results of the Search

The effort to find traces of life in our solar system has failed hitherto. Our closest neighbor, the Moon has, in this respect, disappointed widespread expectations accompanying Man's first step into the space; all lunar material collected in the Apollo program is barren, without any trace of past life. Yet the optimism in finding them in our solar system remains high, and many scientists together with much of the educated public are convinced that the discovery is just a matter of time. The space research institutions of several nations consider the search for life an inseparable part of their programs.

Mars is now the target number one, and large sums are being invested into its exploration. This is money invested in pure research—a compliment to the breadth of the human spirit. No professional protesters have come up yet with the accusations of oil companies standing behind the screen of scientific space research. The consumer society is capable of other than only money making efforts, and it is the thirst for knowledge for knowledge's sake that may be one of its healers at times of the promised overabundance of material things. When human minds in sate bodies will be seeking for new food.

The search for planets, possible bearers of life in the outer space has, until mid 2004, discovered over 100 planet-like objects, and the discovery of new such bodies is moving forward at an ever increasing pace. The nearest is 41 light years distant from our solar system. One of the earlier discovered bodies is orbiting a star in the 55 Cancri System; it is supposed to be a large ball of ice with four times greater mass than that of Jupiter. Its orbit takes almost the same time as Jupiter's. Measured by the cosmic yardstick, the Cancri System is our very close neighbor.

All discoveries, up to this time, have been indirect deductions from small irregularities, wobbles in the movement of the suspected 'mother' star. The continuing spectacular improvements in astronomical technology can make the suspected bodies directly visible. NASA is planning to launch a very big space telescope system, the Terrestrial Planet Finder (TPF), in the year 2015. It is hoped to allow direct observation of planets in the range of our Earth's size. In the mid-future the Darwin ESA mission may be combined with TPF, if ESA and NASA agree to work together.

Radio waves between the lengths of 1 millimeter to 30 meters are the fundamental tool of the hunt for signals that might come from other cosmic civilizations. The task is not the proverbial search for a needle in a haystack but, if any such comparison is suitable at all, it resembles an attempt to find an imagined needle in the vastness of the Antarctica snows. These snows being the billions of billions of radio signals produced by the human society plus those originating from strictly physical processes in the universe. Much of the 'sifting' technology is of military origin.

Arecibo in Puerto Rico has the world's largest array of radio telescopes today. It works closely with the older Very Large Array in Socorro in New Mexico and with Jodrell Bank in England. Information is shared with some two dozen other radio telescope stations around the world.

"The search for the extraterrestrial intelligence, or SETI, is at a turning point after nearly four decades of hard work. The field has reached a high point in terms of telescope sensitivity, a top goal of the alien hunters. But it has found no extra-terrestrials so far, despite forecasts that they should have been discovered by now. Instead, it has probed the heavens with regularity and heard nothing but dead silence." (Wilford, 2001).

Forty years of disappointing results do not in the least deter the believers in the existence of intelligent extraterrestrial beings and of advanced civilizations from continuing their effort to get in touch with them. And they are supported by a steadily growing interest of the wider public in the exploration of the universe and the scientifically intriguing and ever more puzzling astronomical discoveries. Which seem to grow almost exponentially in their number.

At present, SETI focuses on our closer neighbors within the range of some 100 to 200 light years from our solar system. The fast improving technology will go on raising the range rather fast. How many 'civilized' areas might be expected to exist in our Galaxy?

Modest SETI researchers put the number at some ten thousand; genuine enthusiasts consider as quite plausible an estimate of one million in a Galaxy containing at least 100 billion, maybe as many as 400 billion stars. A civilization appearing in one star system out of one hundred thousand of them does not seem to be an exaggerated guess—if the mind accepts the inevitability of the existence of advanced forms of extraterrestrial life. This, of course, is the very big 'if'.

Both, the intellectual and emotional consequence of receiving a definitely verified *artificial* signal—just *one* such signal from the depth of the cosmos—would be enormous. Especially if the signal could be interpreted as an answer to some of the preceding messages sent out by humans from the Earth. "If extraterrestrial intelligence (ETI) exists, then positive detection of it would be the greatest scientific discovery of all time". Introductory words in Heidmann's (1995) 'Extraterrestrial Intelligence'.

Science would get the uncontroversial confirmation of its cardinal postulate of the universal validity of the natural laws as they have been grasped and defined by humans. For it would constitute a proof that other beings in the universe use the *same* laws of nature (very probably the signal would be of electromagnetic nature) as we do. And that they feel the same drive to find out if the cosmos contains other beings that resemble them at least in some respects. Among them the capacity to infer some universally valid lasting regularities from the mass of physical activity surrounding us. Yet more convincing even for diehard skeptics would be the case of an answer to an earthly message.

The psychological impact of a message from space would be the doubling of the consequences of the Copernican revolution.

When the Earth lost its central position in the universe there still remained the uniqueness of the cosmic existence of Man, of the human phenomenon. The dethroning of the Earth from its biblical and koranic status shook the faith in the literalness of the scriptural natural history but it did not undo the foundations of religious faith. Religion is still a vital factor for most people even in the present world of science and technology.

But the emergence of other intelligent forms of life where 'intelligence' does, by its very nature, include the capacity of conceiving the idea of God, would make intelligent beings Man's competitors for God's love and mercy. A multiplicity of advanced civilizations would lead to a confrontation with the very foundation of Christian faith.

Which stands and falls with Christ's role as the savior of mankind, *a one-time and unique phenomenon in the biblical cosmos*. God did love Man, his own creation, so much that he sacrificed his only son to save the sinning humanity from eternal damnation. What, in a different cosmic scenery, would be the place of the other competitors of the human race for God's grace, how often would the son have to be sacrificed?

The dethroning of the Earth as the center of the universe and of humankind as the center of God's care was a long, centuries long process in human minds. The shock was so violent that its reverberations cost some eminent men their lives. Nothing resembling the fight for the preservation of the Earth-centered view has been provoked by the idea of extraterrestrial civilizations. Although this idea undermines the very core of the Christian faith; its consequences go yet deeper than Earth's loss of its special status. The reason?

One is the utter lack of any material, any observational proof; the idea has the status of pure theoretical inference supported by firm hope of enthusiasts both among scientists and among the general public. For them the quantitative, statistical arguments based on the huge numbers of celestial bodies with which the probability arguments operate seem just too overwhelming. *The mere size of these numbers gives imagination free rein.* The second reason of the lack of religious reaction is a marked decrease of deep Scripture-based religiousness in the Christian countries, especially in Europe. In most of the literature dealing with the possibility of other than human forms of intelligence the concept itself—intelligence—is left without definition and is treated as more or less self-evident. Ferris (1997) in his discussion of ETI, handles the problem strictly practically, and chooses a definition fitting the specific topic. "To listen for a signal (from space) is an expression of faith in science and technology. It evinces the belief that 'intelligent' creatures—*here defined as those with big radio sets*—generally manage to survive…They are intelligent because they are able and willing to communicate with humans".

Attempts to find a non-circular general definition of intelligence face the same insuperable problem as do most of our abstract concepts—the impossibility of escaping the infinite regress. The search for extraterrestrial intelligence has to base its effort on the *symbol creating* capacity of any intelligence worthy of that name. For the core of the work is to recognize *the symbolic character of at least <u>one</u> signal* out of the incessant flood of billions of billions of signals crisscrossing the atmosphere and the outer space.

The universe is activity, unceasing change. Photons in the whole range from very low to very high frequencies carrying widely different amounts of energy tell of gigantic processes running in huge concentrations of matter or of very local events in the in the depths of space. They all, photons and particles just as well as their sometimes huge concentrations represent themselves and nothing else, carry no message encoded in them by another agency. It is the mind that tries to look for more, to unravel their history and to find their source; it attempts to force them to 'tell their story'. Science tries to tell the story in a collection of episodes held together by strict rules called the laws of nature. Which, of course, remains the ideal.

The task of the ETI searchers is to detect a signal carrying *more* than itself and its 'natural' story. A signal suggesting that it conceals some encoded message. It is our, human intelligence that tries to detect the existence of a message from another intelligence in the avalanche of signals reaching us from space. To interpret some uncommon feature of a common natural process as a work of beings sharing with us the capability of using material phenomena to convey more than their mere physical nature.

* * *

In the battle of arguments about the probability of the existence of extraterrestrial civilizations a question does soon arise: Why is there such a disparity of views when both [or all] sides in the dispute have—or may have if they wish—*completely the same information at their disposal*? Does the disparity of views arise from biased selection of data, from their different and often contradictory explanation, or is perhaps, at least for quite a few people, their opinion the expression of their general attitude to the world, of their emotional and ideological mind-set?

Hundred billion galaxies each with hundred billion stars in the observable universe is an awe-inspiring number. And this is the most modest estimate on the cosmological market; some go much higher. In much of the space among them 'float' vast clouds of gas and often of dust. Carbon, the element on the presence of which all life on Earth is based appears as a rather common component and so do the other bioelements.

"One inference is that the complex carbon chemistry which led to the development of life on Earth was a natural development from the only slightly less complex chemistry of the clouds of material from which the solar system formed, and that where other Earth-like *planets* exist, complex chemistry may have developed towards life in a similar way … This implies that there could be *millions of civilizations* (stressed by the author) involving life forms based on carbon chemistry in our galaxy today….

Traces of single-celled life forms … have been found in rocks more than 3.5 billion years old, which were laid down some half a billion years after the formation of the Earth. It seems to indicate that life arose almost as soon as the Earth had cooled to the point where it could exist. Which strengthens the case that the young planet was seeded with complex organic material from space, perhaps including amino acids such as glycine.

Some researchers have gone further, suggesting that the Earth was seeded with actual living cells from space. In one variation on this theme, Fred Hoyle and Chandra Wickramasinghe propose that life evolved in interstellar clouds over a very long period of time before the Solar System had even formed. Francis Crick (1981) has espoused an alternative proposal, that the Galaxy had been deliberately seeded with life by a civilization, which arose billions of years ago when the Galaxy itself was young. This idea is known as *directed panspermia*; it builds on earlier speculations … but still leaves unanswered the question of where the original civilization came from. Either proposal, though, extends the timescale available for evolution from the roughly 4.5 billion years of Earth history to three or four times as long, depending on exactly how old the universe is. The one inescapable conclusion seems to be that life did not start from scratch [that is, from simple atoms and molecules of carbon, oxygen, nitrogen and hydrogen] on the surface of the Earth….

Extending the analogy still further, Smolin suggested that the entire Universe might be regarded as alive and having evolved [in the strict Darwinian sense] from earlier generations of universes." (Gribbin, 1997).

Million civilizations according to some ETI superenthusiasts—just in our Galaxy? What hardly imaginable number of cosmic civilizations do we then get with the hundred or more billions of galaxies? Very large numbers can make the mind swirl, and if they keep growing towards the infinite *everything seems to come within the range of possibility.*

New cosmological hypotheses create not one but a virtually infinite number of universes either coexisting in time (what time?), or following one another in a never ending series (cycles) of births and deaths. Just give me a large enough number—and nothing becomes unthinkable. Either in an ever expanding Universe, suggested by the new astronomical measurements and theoretically based by the introduction of negative gravity, or in the many-worlds or multiverse versions of the cosmologists, or in Feynman's multiple, theoretically unlimited set of Universes each with its own history. *Everything seems to be possible if one lets the numbers grow without limit.*

3.The Role of Big Numbers

But in the problem of the existence of extraterrestrial *intelligence the very big numbers can speak in two contradictory languages.* Logic can use them both to support or to deny the existence of cosmic civilizations. What would be the arguments of the denial?

The numbers again. How to explain the failure to detect any signal, even the slightest trace of a signal coming from another cosmic civilization after an intensive search of many years when the numbers suggest a definitely non-negligible probability of recording such signal?

In 1960 the astronomer Francis Drake started the first scientifically sound SETI observations by training the radiotelescope at Green Bank on two stars belonging to the same category as our Sun. Since that time observations in the SETI program were intermittently carried out in quite a number of observatories. In 1992 there were already 25 large radio telescopes and their arrays in the world. Specific intensive search for extraterrestrial signals which might bear the signature of an intelligent source was conducted by the Very Large Array in Socorro in New Mexico, followed by the world's largest radio telescope array in Arecibo, Puerto Rico.

Francis Drake (1992) devised an equation giving the probability of getting in touch with an extraterrestrial civilization in our Galaxy. Some elements of the equation are observational and theoretically sound, some are conjectural. According to Drake's equation there exist a relationship between the lifetime of a galactic civilization and the probability of its communicating with other civilizations.

The length of the duration of a civilized society *as a whole* is a matter of pure guess even if one would like to call it an educated guess. In the discussion of the Drake equation the figure of ten thousand years appears quite often. If the lifetime of ten thousand years, the 'existential' period of a civilized society is taken as the probable figure, then our Galaxy ought to contain at present some ten thousand such (potentially) communicating centers of intelligent beings. Quite aside, one thing in such short estimate of a cosmic civilization's lifetime reflects a deep-seated, maybe subconscious pessimism about the fate of mankind itself. More than anything else would such attitude characterize the mindset of many of today's Western intellectuals.

Ten thousand civilizations in our Galaxy—technically capable of mutual communication? Such state of things must have existed throughout much of the Galaxy's history; there had been enough time (cosmological time) for every new civilizations to evolve, flourish and die. The fundamental premise of SETI requires the capability of communication at our present or higher level of technology in the life formations (societies) whose signals are supposed to reach the Earth.

A little more than a hundred years have passed since the first use of radio waves by the Earth's civilization. During those years—a millisecond on the cosmic calendar—a virtual knowledge revolution has taken place. Scientifically and technologically much of the human society is a different phenomenon than what it was those hundred years ago. An analogous process might be expected to have occurred in other cosmic 'societies', which have reached the stage of technology enabling them to use radio signals for communication. Reaching thereby the stage which Ferris (1997) uses for the definition of a civilization.

If this is the case then our galactic space should be, at any time, *criss-crossed in all directions by signals coming from its treasury of living and dead civilizations.* Some of which might be or might have been much more sophisticated than those used by SETI. At least some of the galactic civilizations may have enjoyed a much longer time of their development than our pitiful few thousand years and have reached technological levels that even our science fiction writers are incapable of imagining. The strength and lifetime of their signals might carry them throughout the whole Galaxy.

Such is the case, if only the modest number of the ten thousand' contemporaneous (cosmic time) civilizations in our Galaxy is considered. The superenthusiasts are thinking not of thousands but of millions of galactic civilizations. Should they be right then *our Galaxy ought to be virtually saturated with communication messages.*

It is not—at least as far as human technology can tell. We have not yet detected a *single trace of a message* coming from extraterrestrial intelligent beings, although a lot of effort has been devoted to the task in the last forty years.

The challenge does not stimulate and drive only a few relentless scientists. The Universe and its fantastic complexity have started to intrigue much of the educated public and thousands, maybe millions of people follow the cosmic news with interest and often excitement. Never in the past did the book market produce and *sell* so many works dealing both with the latest state of our general knowledge of the Universe. All with detailed studies of specific cosmic problems and, probably most of all, stimulating our curiosity of other worlds and the power of fantasy creating enigmatic and mysterious cosmic scenarios. The imagination of cosmologists using mathematics as a poet's tool has surpassed anything that could be offered by ancient mythologies.

People did always feel the need, at least at some stages of their lives, to reach out towards the Great Unknown, towards the Mysterious. Religion and mythology have been fulfilling this role since times immemorial and, at later stages of cultural development, were joined by philosophy.

Is there, in countries called developed, any intellectually intriguing antidote to the incessant operation of ugly and evil news upon human minds, a flood with which the media are saturating them? In the past, much of the spirit pervading the best literature nourished at least some elements of hope in the meaning of life, in the worth of the human community. The suffering, tragedy and death of the protagonists of their works may shock the reader but the masterpieces call forth a feeling of cleansed soul. How cleansed does one feel after having read Lolita or seen Hannibal?

Astronomers and cosmologists, in opposite to postmodern philosophers, say: we *do* know something, and our knowledge is getting ever more captivating by its strangeness. We move along huge distances of space and time, and the farther we come the more riveting and mind-stimulating things are turning out. We offer your minds, disillusioned and depressed by the goings in your small world, the knowledge that boosts their capacity to wonder in awe and think about things unthinkable in the past.

Almost all the probes of the inner planets included in their programs as a fundamental task the search for signs of present or past organic life. The disappointment caused by the failure to find any such evidence during the lunar missions was almost palpable among scientists. The effort and hopes for the next years are concentrated on Mars, where especially the recently confirmed presence of water seems to make the existence of past and maybe even present primitive organic life highly probable. No scientist or any of the interested amateurs, however, expect to find in the planetary system of our Sun advanced life forms comparable to those existing on the Earth.

The search for them must go farther afield in the cosmic space. *There is the kingdom of numbers so unimaginably large that they seem to make everything possible and can turn the most bizarre ideas into reality.* Even the existence of civilizations, which would consider us, the earthlings, living still at the level of their Stone Age past.

Staggeringly big numbers resemble, figuratively, the double nature of electrons which appear as corpuscles when observed in one way and as waves when observed in another way. *The mental game with extremely large numbers may also lead to two contradictory results.*

One either starts with the belief that the huge numbers characterizing the cosmos lead to the logical conclusion that life *must* have developed throughout the Universe and has, in some places, acquired the characteristics of intelligence. The pressure of the numbers makes the appearance of cosmic civilizations virtually inevitable. Given the 12 billion years of the evolution of the Universe, with many galaxies one or two generations older than our Galaxy, logic makes even grow the probability of the development of civilizations *surpassing* ours in the amount of accumulated knowledge and in the refinement of their technologies.

The final speed of light does fundamentally restrict the possibility of cosmic communication among the *same generations* of the civilizations to the cosmically minute region of our close galactic neighborhood. An *interchange* of information and ideas among the living could take place only with advanced life forms on the planets of the closest stars; according to our present astronomical knowledge such possibility is close to nil.

Yet relativity and the limit of the speed of light does not in the least inhibit the possibility of human contact with other cosmic civilizations. The vast majority of the signals coming from sources in the Galaxy would inform us about events that, at the moment of their reception on our Earth (cosmic time) are history for them, in the case of distant senders a history of thousands of years. And if one believes in the existence (past or present) of supercivilizations which developed technologies incomparably more sophisticated than ours capable of producing extremely strong and long-lasting signals then the history of their emission may go millions of years back.

Ten thousand civilizations ought to exist in the Galaxy *at the same time* if Drake's equation is used to estimate their number. When did the first of them start signaling their presence to whoever might be interested? The age of our Solar System is estimated at some 5 billion years; the Galaxy itself is older. Some solar planetary systems developed thus before the Solar system and life—if it developed there—had a longer time for evolution.

If the modern Cassandras are mistaken with their gloomy predictions of the inevitable rise and fall of any civilization then some intelligent societies had more time to evolve than the human society—and *may have developed quite wondrous technologies*. Capable of sending signals, in most cases electromagnetic, incomparably stronger and focussed than we have hitherto produced, signals with not only intragalactic but even intergalactic range. Our Earth might be getting *simultaneously* messages thousands and millions of years old. One would expect technologically advanced programs like SETI, running for several decades, to have detected quite a few of them. *This is what the astronomical supernumbers seem to demand when read under one set of lights.*

Yet not a single signal from intelligent life in the outer space has been detected during those decades of search although the logic of the astronomical supernumbers demands their existence. They ought to be rather common.

4. The Range of Answers

The very large cosmological numbers give thus contradictory answers to questions about the existence of extraterrestrial intelligence. All depends upon the attitude of the questioners. If they hope for a positive answer they formulate their questions accordingly and *the numbers comply.* They give a positive answer. And *the numbers comply just as well* in giving a negative answer if asked in the proper way:

1. Extraterrestrial intelligence does exist because it has had countless ways to develop in the countless turns of the cosmic evolution, say the ETI believers. The logic of the numbers makes the probability of their existence so high that it becomes indistinguishable from necessity.

2. Extraterrestrial intelligence does not exist because it has not given the slightest sign of its existence. Although the cosmic numbers are so high that they seem to guarantee that if such intelligences exist they would let us know about their existence.

The reactions to the dilemma may be summed up in the following answers:

A. There is no extraterrestrial intelligence, nothing approaching the level of mental evolution achieved by the human community.

B. The Universe abounds in different types of intelligent communities, and it is only a matter of time before we get a confirmation of their existence.

C. Extraterrestrial civilizations are a very rare phenomenon in the Universe; they occur so scattered in space and in cosmic time that the probability of establishing any mutual contacts between them is extremely low. We are incapable of even imagining the technology that would be needed to produce artificial signals strong enough to carry messages from galaxies separated by millions of light years. We will never know if we have some cosmic 'relatives'.

D. In sending out their messages the cosmic 'relatives' are using technologies completely unknown to us; we have no means to decipher them or even detect their existence.

My mind keeps telling me that the first answer (A) is closest to the cosmic reality. *The human society is a unique phenomenon in the Universe.* Just as the human brain. The answer (C) is the next possibility.

If this is the reality then the consequences ought to be decisive for our view of the world and for our feeling of *being part of a very special cosmic community.* Hitherto only few people share this feeling in more than in words.

And yet such deeply felt awareness, if it entered the minds of many people and became a firm point in their thought, might help in the desperate efforts of creating at least a semblance of a COMMUNITY out of the linguistic and racial emotional medley called humanity. This would be more effective than all the high-sounding preaching about the uniqueness and inborn dignity of every human individual. *Common origin, uniqueness and common destiny bind us more than anything else.*

A flash of cosmic self-awareness, born out of the vastness of mere blind material existence. That's what we are and what ought to be the source of our self-image as human beings. And Man will never stop asking the question of questions: Where is the beginning of the whole story? Much of the innermost life of his mind depends and will always depend on the answer he finds, or thinks he has found, to this question of all questions.

5. References

Asimov, J. (1979). *The Choice of Catastrophes*. Simon and Schuster. New York.
Barrow, J. D. and J. Silk (1983). *The Left Hand of Creation.* Basic Books. New York.
Crick, F. H. C. (1981). *Life Itself: Its Nature and Origin.* Simon and Schuster. New York.
Drake, F. D. (1992). *Is Anyone Out There?* Delacorte Press. New York.
Ferris, T. (1997). *The Whole Shebang*. Simon and Schuster. New York
Gribbin, J. (1997). *Companion to the Cosmos*. Phoenix. New York.
Heidmann, J. (1995). *Extraterrestrial Intelligence*. Cambridge University Press. New York.
Pagels, H. (1988). *The Dreams of Reason*. Simon and Schuster. New York.
Wilford, J. N. (ed.)(2001). *Cosmic Dispatches*. The New York Times Reports on Astronomy and Cosmology.

Acknowledgement. Ideas expressed in this article are also discussed in the book 'Kosmický mozek'(The Cosmic Brain) by J. Krupička, to be published in Czech by Paseka, Prague in 2005.

Biodata of **Massimo Teodorani** author of ***"An Alternative Method For The Scientific Search For Extraterrestrial Intelligence"***

Doctor Massimo Teodorani is an astrophysicist. He got his degree in Astronomy at the Bologna University (1982), where he subsequently earned is Ph.D., in Stellar Physics (1992). Since very recently he has been working as a researcher at the Radioastronomic Station of the National Institute of Astrophysics (INAF), in Medicina (BO), Italy, where he has been searching for the 22 GHz water MASER line in candidate exoplanets and in comets. In the past he has been working also at the Astronomical Observatories of Bologna and Naples, where he carried out observational and interpretative optical researches on eruptive stars of various kinds. Parallely with astrophysical researches, since about 11 years, he carries out physics researches concerning anomalous atmospheric plasma phenomena. He is a member of SETI in Italy. Since 2003 his named is cited in *Contemporary Who is Who*. He is currently and momentarily a scientific advisor and referee of a scientific publishing trade.

E-mail: **mteodorani@libero.it**

J. Seckbach (ed.), Life as We Know It, 487-503.

AN ALTERNATIVE METHOD FOR THE SCIENTIFIC SEARCH FOR EXTRATERRESTRIAL INTELLIGENT LIFE:
The "local seti"

MASSIMO TEODORANI
INAF – Istituto di Radioastronomia,
Radiotelescopi di Medicina
Via Fiorentina – 40060 Villafontana (BO) – Italy

1. The SETI Project and its Restrictions

The present SETI (Search for Extraterrestrial Intelligence) Project is devoted to the search for signals coming from technological civilizations living in other planets. Due to their capability to cross efficiently the interstellar medium (Cocconi and Morrison, 1959), radio frequencies have been chosen as the best method to receive and send communications from/to ET civilizations. This project is operational since a few decades and it is generally named Microwave Observing Project (MOP). The typically used frequency window ranges from 1 to 10 GHz. The goal of this investigation is to search for very narrow, possibly modulated radio signals, with a more or less marked polarization, which are presumably transmitted intentionally or unintentionally by emitters located on an extrasolar planet and characterized by a Doppler effect due to planet rotation and revolution around its star. The strategies which are currently adopted for this research are mainly: a) All Sky Survey (ASS); b) Targeted Search (TS); and c) Piggy-Back Mode Search (PBMS). Using the ASS mode (Tarter, 2001) all the celestial sphere is scanned until a suspect signal is found. According to the standard SETI procedures, such a signal must be reobserved with the same characteristics and with the same equatorial coordinates by the same observer and by all the other observers in the world. Using the TS mode (Tarter, 2001; Turnbull and Tarter, 2003a, 2003b), specific target stars having non-eruptive signatures and characteristics similar to the Sun are chosen for a detailed analysis. Using the PBMS mode (Montebugnoli et al., 2002), the SETI search works in parallel with standard radio astronomic observations wherever the antenna is pointed, without abstracting observation time to them. In order to reach the SETI goals adopting all the described strategies, the antenna used for this research is always connected to a Multi Channel Spectrum Analyzer (MCSA), which at the present time is able to scan simultaneously a few tenths of millions of channels, so that the radio spectra are obtained with a typical resolution ranging from 1 to 0.1 Hz or less. The probability to detect such a kind of signal, which is expected to be extremely weak, increases with the diameter and the type of the antenna, the sensitivity of the receiver, the power of the amplifier, and the effectiveness of the algorythm used to extract the signal from the noise. Since some years, the enormous computational needs requested for realtime data processing is also efficiently assisted by the SETI@home initiative (Korpela et al., 2004). In addition to the detection of radio waves, since a few years the SETI Project is operational in the optical range too (Horowitz et al., 2001; Kingsley, 2001). This variant, named Optical SETI (OSETI), is intended to search for very strong and pulsed optical beacons possibly of Laser kind coming from intelligent civilizations located in extrasolar planets, using photon-counting photometers and or high-resolution spectrographs able to reach a very high magnitude precision and temporal resolution. Alternative SETI projects involving the infrared range of the spectrum and the search for intelligent Maser signals in the microwave region, have been recently also proposed.

At the present time, no clear detection of intelligent ET signals has been obtained (Lazio et al., 2002), but many identified false alarms and or still unidentified signals which anyway didn't show any repeater, have been reported. All of the present SETI efforts require that intelligent beacons are originated from a planet orbiting around another star having equatorial coordinates which are expected to be constant in a time lapse of many years. Therefore celestial sources that are characterized by an anomalously high proper motion, involving a more or less abrupt change of its equatorial coordinates in the lapse of a few months or days, are excluded by the SETI search strategy. In such a way, what the standard SETI project will possibly have the luck to detect in the next years, will be a pure selection effect, which in substance consists of signals emitted by ET civilizations that are at a level of technological evolution comparable with ours and that still live in their home planet. Due to the restrictive research strategy and preset technological characteristics of the standard SETI detectors and processing procedures, the detection of radio, infrared and/or optical signals coming from relatively fast moving emitters is necessarily avoided. This means that using such a sophisticated but limitative search method, alien transmitters that are located on fast moving sources cannot be detected. Is there a solid scientific reason to build up and employ a sensing strategy able to record such a peculiar kind of signals? The answer is positive. Since at least 20 years several theoretical studies (Betinis, 1978; Finney, 1985; Jones, 1981; Newman and Sagan, 1981) have shown that the migration of extraterrestrial civilizations in the galaxy, whose moving transportation devices would be necessarily characterized by a high proper motion compared with the proper motion of close stars, is a possibility that can be investigated observationally. The detection of celestial sources having an anomalously high proper motion doesn't enter into the standard SETI protocols.

If galactic civilizations, which are far more advanced than us, are able to move from a star to another, they might have visited our solar system and Earth too. At present, in addition to the possible detection outside and inside our solar system of unidentified celestial sources having high-proper motion, we have the scientifical and technological capability to investigate aseptically the possibility that Earth too is being visited. Anomalous phenomena reported in our atmosphere might be a signature of such visitations. Which ones of them are due to natural phenomena and which ones are not? In the following sections it will be shown that this difficult but important goal can be reached adding a "sieve strategy" to our well-working Galilean method.

2. Strange Light Phenomena on Earth and Scientific Inquire

If we suppose that Earth is visited by alien intelligence, we should expect to see possibly transient anomalies in our atmosphere that have a technological signature and/or a non-random behavior. The difficult task here is to distinguish very carefully which ones of these anomalies are of natural origin, which ones are a product of advanced terrestrial technology, and which ones cannot be identified with the first two categories. Once the third category is possibly identified as an exogenous visitation, the next task consists in trying to understand how this category works in terms of the known law of physics. This involves both the investigation of possible propulsion systems, which might be identified from the mechanism of radiation emission in a wide range of wavelengths, and the investigation of how such devices are intelligently driven.

It is generally expected that such hypothesized intrusions in our atmosphere occur transiently and randomly on Earth, so that these occurrences cannot be predicted in order to permit researchers to be prepared with sensing instrumentation. In such a case scientific investigations would not be possible, even if several or many witnesses were reported. Witness reports are of no scientific relevance, because they are affected by a huge evaluation error, which cannot even be quantitatively estimated (Condon, 1969). In order to do science on this kind of investigation, it is necessary to acquire physical data using suitable measurement sensors, through which the signals of interest and the related measurement errors can be accurately evaluated. Fortunately, in addition to the transient occurrence of anomalous events on Earth, there is also a strong evidence that in some areas of Earth atmospheric anomalies occur with a remarkable regularity. Such locations can be suitably chosen as the best sites for scientific monitoring, in order to ascertain the origin of the phenomenon including a possible extraterrestrial origin too. It is possible to testify that in such locations anomalous phenomena are repeatable, therefore they are suitable for a systematic observational scientific investigation.

2.1. THE FIRST SCIENTIFIC INVESTIGATIONS

Anomalous atmospheric light phenomena reoccur in many locations of Earth, some of which have become a laboratory area for a rigorous instrumented study of the involved physics. At least 35 of these locations are documented with images and some scientific measurements (Teodorani, 2003). Such phenomena appear, both in the sky and close to the ground, as multicolor and large-sized (up to 30 meters) "light balls" characterized by irregular pulsation and erratic movements. The time correlation of light phenomena with oscillating magnetic fields is one of the most intriguing observational results that were obtained in some areas of Earth such as Hessdalen in Norway (Strand, 1984; Teodorani, 2004), Boulia in Australia ("Min-min" lights), Popocatepetl in Mexico, and Yakima in USA. The light phenomenon that is reported in the Hessdalen valley in central Norway is probably the most known in the world, as it is the only one under systematic scientific field study since over 20 years. Several kinds of measurement techniques, such as magnetometry, radar monitoring, radio and optical spectrometry, optical photography and video recording, have been employed so far. Therefore the Hessdalen location can be fit as a laboratory area for the study of this kind of atmospheric spatially reoccurring anomaly. In fact, a permanent automatic measurement station (Strand, 1998) has been installed in that area, from which since 1998 it is possible to acquire automatically and continuously video data and occasionally measurement data using electromagnetic instrumentation. The rich statistics that was obtained so far (Teodorani, 2004) from the data furnished by the Norwegian station, shows that the light phenomenon is spatially uniformly distributed and tends to appear more often in the winter season and in the hourly interval between 09.00 p.m. and 01.00 a.m. local time. The light events do not show any correlation with daily, monthly and yearly solar activity.

2.2. THE MOST RECENT SCIENTIFIC INVESTIGATIONS

Some Italian instrumented expeditions were carried out by several groups of physicists and engineers in the Hessdalen area in the years 2000, 2001, 2002, 2003 and 2004. The first three of these missions permitted to obtain crucial measurements that furnished a physical insight into the light phenomenon's structural characteristics and variable behavior, and into its mainly geophysical nature (Teodorani, 2004). Engineers monitored constantly the Hessdalen valley using spectrometers able to survey ELF (Extra Low Frequency) and VLF (Very Low Frequency) radio frequency ranges, and a UHF (Ultra High Frequency) pulse radar. Physicists concentrated on the acquisition of conventional, digital and telescopic photographs, videos and low-resolution spectra of the light phenomenon, and on the analysis of ELF-VLF data and collected ground samples in specific areas approached by the light phenomenon. Some portable instruments such as night scopes/IR-viewers, ultrasound and electric field detectors, a high-speed optical radiometer, and data scopes for triangulations, were used as well.

More recently, photographs, video frames and spectra of similar anomalous light phenomena that reoccur in some areas of Australia and Canada, have been analyzed after training scout observers to use diffraction gratings in order to obtain digital spectroscopic images of the light phenomenon (*SpecNet* initiative). An Italian field mission has also been carried out in the Arizona desert in 2003 in collaboration with the IEA organization (Adams and Strand, 2003). Two crucial locations of the Italian Apennine mountains have been monitored in 2003 (Teodorani, 2003) and 2004, and the study of the Italian areas of interest is currently going on.

3. Data Acquired in Hessdalen and their Physical Interpretation

In this section the main scientific results (Teodorani, 2004) that were obtained during instrumented missions in Hessdalen, Norway, are synthetically described.

3.1. THE OBSERVED PHYSICAL FEATURES

After eliminating a lot of man-made noise and well-known typical ionospheric signals, it was noticed that the ELF-VLF radio recordings showed very often unusual signals characterized by inclined lines (in a graph representing frequency vs. time) having a marked Doppler feature. The slope of the lines was almost periodically and gradually changed from negative to positive in a lapse of few seconds, showing that the source was alternately approaching and receding from the observer. From the measured frequency it was possible to determine the velocity of the emitting source, which was changing fast from 10,000 to 100,000 Km/s within several seconds. Many cycles were occurring during a time interval as long as half an hour, by starting and ending abruptly as if some transient electromagnetic event was turned on and off. In order to interpret this evidence, an ad-hoc empirical model has been proposed, according to which it is supposed that high-energy particles are accelerated and collimated by a cylindrically symmetric magnetic field whose axis is misaligned in comparison with the rotation axis of a fast rotating body. In this framework the observer is able to register periodically blue and red-shifts. Very high energies for particles—presumably electrons—and very strong magnetic fields are necessary in order to produce the observed effect. This mechanism seems to resemble a small-scale version of the synchrotron radiation (Lang, 1998), which is observed in fast spinning objects such as pulsars in astronomy. The Doppler radio phenomenology was recorded mostly when the light phenomenon was not in sight. After scanning the sky and the top of the hills with the IR-viewer it was sometimes possible to establish that a normally invisible light phenomenon was indeed detectable when a light amplification device was used.

Observers on the field were able to confirm very often the appearance of the light phenomena. The main deduced average characteristics of such phenomena are shown in Table 1. Both visually and photographically it was verified that the most common light phenomena are always preceded by very short-lasting flashes of light that appear everywhere in the valley and that emit an intrinsic power ranging from 10 to 300 W. Very often such flashes have been reported at a very short distance (about 100 m) from observers, so that the distance parameter could be approximately evaluated.

The typical three-dimensional light-distribution of the illuminated surface of most common light phenomena, which in optimal atmospheric conditions shows to be steep and rectilinear, results to be drastically different from the one – a Gaussian and exponential distribution – that is expected from a standard plasma. Luminosity shows very often a highly time-variable feature with a pulsation rate of one second or less, and a highest radiating power up to 20 kW has been measured in one specific case in which the distance (9 Km) could be determined using triangulations, radar scanning and topographic mapping. In most of the cases in which the light phenomenon is blinking, irregular or semi-regular pulsations are typically terminated after few cycles with an average event duration of 5 seconds, in other cases many cycles are continued for a period as long as several minutes. It has been possible to ascertain that the luminosity of such light phenomena increases in a drastic way because of the sudden appearance of many smaller light balls around a larger luminous core. Due to this the highest luminosity values are caused only by the dimensional increase of the total radiating surface that is formed by a cluster of light balls. Therefore the increase of the surface area is not caused by the expansion of a single light orb. Some of the secondary light balls are often ejected from the core, this also can cause a luminosity increase of the entire lighted target as seen from far away. The phenomenon produces light by maintaining a constant color-temperature, behaving like a lamp with "on" and "off" phases. The constancy of temperature is deduced both from the unchanged features of the spectra when the light phenomenon is shifting from the lowest to the highest luminosity values, and from the empirical dimension-luminosity correlation that is derived from the analysis of unsaturated video frames. In such a specific case the classical Stefan-Boltzmann law (Lang, 1998), which describes the behavior of a plasma in thermodynamic equilibrium and emitting light as an isotropic radiator, characterizes a self-sustained isothermal plasma where the radiant power in the optical range varies only when the radiating surface varies. The obtained spectrum of a cluster of three light-balls colored in white, red and blue, shows three well-distinguished peaks that are about 500 Å wide, anyone of which resembles a spectral feature that is very similar to the one produced by LED (Light Emitting Diode) lamps. The color-temperature derived from the spectrum is consistent with the colors of the light balls as they were recorded in the photograph that was obtained at the same time as the spectrum.

Approximately 10% of the light phenomena, which were reported and recorded during three of the five Italian missions, should be considered "peculiar" compared with light phenomena that were seen most often. The peculiar events were characterized by totally lighted geometric or symmetric shapes and sometimes by translucent or low-luminosity apparently structured shapes.

The light phenomenon shows often-strong radar tracks, which transiently appear and disappear, also when it is optically faint or almost invisible. In some cases in which it is visible, it shows no radar track. Velocities can reach values up to 60,000 km/h.

Some slightly radioactive powder was collected very close to a spot where it was ascertained that the light phenomenon approached the ground. A subsequent laboratory analysis that was carried out using plasma spectroscopy, X-ray diffraction and scanning electronic microscopy, showed the evidence of sphere-like iron particles of micrometric dimensions.

This is all what came out from the most important of our expeditions to Hessdalen, in 2000, 2001 and 2002. Missions carried out in 2003 and 2004 were mainly devoted to testing new instruments. It was finally verified that a natural laboratory is just out there, inside our planet. Physical science has at its disposal a lot of sites of this kind that can be scientifically monitored by using highly sophisticated sensing instrumentation (Teodorani, 2000).

3.2. TOWARDS A PHYSICAL THEORY

During the last ten years, several theories and hypotheses of natural kind have been considered to try to explain the origin and the nature of the light phenomenon and of the electromagnetic field that seems to be correlated to it. Mainly the following possible causes have been considered (Teodorani & Strand, 1998): ionosphere

TABLE 1. Main features of the Hessdalen light phenomena.

	Total Number	Clustered lights	Geometric shapes	Structured objects	Uncertain
	150	80%	5%	5%	10%
Origin		Geophysical	Unknown	Unknown	Unknown
Trigger mechanism		Piezoelectricity	Unknown	Unknown	Unknown
Characteristics		Plasma	Plasma	Unknown	Unknown
Time peak		11.30 pm	10.00 pm	Random	-
Time variability		High	High	Little	Little
Pulsation period		≤1 sec	≤1 sec	-	-
Surface variability		High	High	None	Little
Highest Luminosity		20 kW	-	-	-
Colors		W, R, B	W, Y, B, V	W, Y, G	W, B
Total duration		180 sec	240 sec	60 sec	30 sec
Velocity		60,000 km/h	-	-	-
Close to ground		70%	40%	10%	25%
Low in the sky		30%	60%	90%	75%
Radar / IR emission		Yes	Yes	Yes	Yes
VLF emission		Probably	Possibly	-	-
Reaction to Laser		-	-	-	Yes
Remnants on soil		Possibly	-	-	-

W = white, Y = yellow, R = red, B = blue, V = violet, G = grey

activity, solar activity, cosmic rays, magnetic monopoles, mini-black holes, Rydberg matter, heated nanoparticles (Abrahamson and Dinniss, 2000), piezoelectricity, quantum fluctuations of the vacuum state. For none of these possible causes, except for some aspects of piezoelectricity, it was possible to find a successful proof. Very recently one more complete theory, which is able to explain most of the recorded data, has been carefully elaborated (Teodorani, 2004). Such a theory, which was originally worked out by physical chemist David Turner (2003) in order to explain the ball lightning phenomenon (Stenhoff, 1999), has demonstrated to be very suitable to explain the Hessdalen-like phenomena too. It consists of a thermo-chemical mechanism producing and maintaining light balls whose structure and radiant characteristics seem to match at least 80% of the phenomena of which measurements were carried out in Hessdalen. In a first phase – named "Tectonic Trigger Phase" – air can be ionized by tectonic stress causing simultaneously piezoelectricity and the emission of VLF and UHF waves, so that a plasma ball can be formed from the wave-particle interaction (Zou, 1995). In the specific Hessdalen area tectonic stress can be produced by river water that penetrates into the many ground cavities present in the valley and that then freezes when temperature drops down: in such a way the ice compresses the many quartz rocks which are present in the area so that the best condition for piezoelectricity is produced. The development and maintenance of this mechanism is also assisted by the presence in the valley of a large quantity of copper, having its well-known conductive property. In a second phase (Turner, 2003) – named "Thermo-Chemical Confinement Phase" – the formed plasma can bind with water vapor and aerosols, to create a hot and sharp-edged light ball with a cool water-and-ion coat, in which electrical and thermo-chemical energy exchanges occur following the mode of a heat pump. In such a way some inward forces are able to counterbalance the external pressure and the light ball is consequently self-regulated in a sort of hydrostatic equilibrium. Surface energy re-minimization can determine both ball clustering and ball ejection effects. The typical erratic motion and kinematical characteristics can be explained by asymmetries in the layer of droplets of the light balls, which can be caused by changes in either the chemical or electrical state. In the specific Hessdalen case, a possible spontaneous production of almost mono-disperse quantum dots might come from mold spores, as the main semi-conducting elements, decomposed by the central plasma of the light ball. This could explain successfully not only the recorded LED-like spectrum but also the existence of balls of distinctly different colors. Some of the radar and infrared observed behaviors can be explained using this model as well.

Turner's thermo-chemical model is able to explain as a plasma phenomenon of geophysical nature most of the data collected in Hessdalen, but not all of the data. The residual data might be potentially considered in the light of a possible ET interpretation only if all of the other possibilities (of prosaic origin and/or of artificial human nature) can be accurately excluded.

3.3. THE PROBLEM OF PECULIAR CASES AND THE "ETV" HYPOTHESIS

Other evidence that was found in Hessdalen constitutes another anomaly inside the main anomaly (Teodorani, 2003, 2004). There are not yet confirmations that the lights are really associated with Doppler-like signals in the ELF-VLF range or with the deposition of metallic particles. Therefore a sound comparison of these findings with Turner's model is not yet possible. On the other hand, Turner's model is not able to explain the geometric shapes or structures that were recorded in a small minority of cases. It is not yet known whether these manifestations are different and rarer aspects of the same "standard" light phenomena or whether they are distinct phenomena that overlap with the standard one for some unknown reasons (see Table 1) A similar unexpected mixture of "standard earth-light phenomena" and very exotic features are reported in other locations of Earth too (Teodorani, 2003). This uncomfortable side of the anomaly constitutes valuable observational evidence and must be investigated more deeply. Also, working hypotheses different from those covering natural "earth lights" should be followed up. This research must be conducted by considering how far standard physics can take us, but also with some parallel attention to those other anomalies whose possibly spurious relevance should be investigated.

While most of light phenomena in Hessdalen and elsewhere can now be successfully explained within the framework of a natural mechanism, a residual of "locally overlapping data" remains presently unexplained and largely unexplored. To investigate them also the ETV (Extraterrestrial Visitation) working hypothesis is taken into account.

The search for ETV (SETV), which is consistent with the assumption of interstellar and galactic diffusion, demands for an extension of the Drake equation (Walters et al., 1980). This equation describes the probability of the existence of intelligent extraterrestrial civilizations in our galaxy according to a set of fixed physical and astronomical parameters that are based on stellar and planetary evolution. The assumption of interstellar diffusion means that galactic civilizations are able to migrate throughout the galaxy. This adds one more parameter to the Drake equation.

The verification of possible ET visitations that are consequent to interstellar migration, can be carried out only doing a rigorous screening of data coming originally from the study of natural and or celestial phenomena located both in the solar system in its entirety and on Earth where anomalous phenomena are often reported. Therefore the main strategy that is being adopted in studying some anomalies acquires the character of a sieve. Such a "sieve strategy" can allow scientists to distinguish "the stones from the nuggets". In the first case it is possible to expand our physical knowledge of natural anomalies that have not been studied enough so far but that, if fully understood, could be of basic help to brindle new energy sources. On the other side, the extreme carefulness with which scientific monitorings are carried out, can help physical scientists to establish some discrepant facts that might bring to the discovery of a possible extraterrestrial visitation both in the form of exotic technology and in the form of possible electromagnetic intelligent manifestations of endogenous or exogenous origin that have so far excaped scientific detection. A possible interaction between some anomalous phenomena of ascertained geophysical origin and an hypothetical "alien" intelligence is considered as well. Therefore, such a sieve strategy can help us to expand several aspects of our science simultaneously, involving fundamental physics, plasma and particle physics, physics based on some technological products of possible exogenous origin, bio physics, and some intriguing aspects of quantum mechanics.

4. The Hypothesis of Interstellar Migration and the SETV Project

The hypothesis that Earth is visited by exogenous intelligences is based on the possibility that alien civilizations are able to migrate throughout the galaxy.

By using appropriate "diffusion equations" it is possible to predict the interstellar expansion of galactic civilizations as a process that is expanding like a wave (Bainbridge, 1984; Betinis, 1978; Deardorff 1986; Finney, 1985; Jones, 1981; Newman and Sagan, 1981; Walters et al., 1980; Zuckerman, 1985). According to the most recent evaluations (Principia homepage, 2004) the wave speed comes out at $\sim 10^{-3}$ light years per year. This implies that intelligent civilization could settle the entire galaxy in only 60 million years. Compared with the age of our galaxy ($\tau = 10^{10}$ yrs), this means that galactic post-migration colonization would be completed during a time interval that is at least 150 times smaller. Earth itself may have been visited numerous times since the arrival of Homo sapiens and much before. This possibility raised, sixty years ago, the famous question by physicist Enrico Fermi, who since that time posed as a problem the well-known "Fermi Paradox", according to which the apparent absence of extraterrestrials on Earth is a proof of their non-existence (Freitas, 1983a, 1983c, 1985a; Tipler, 1980).

4.1. MOTIVATION FOR INTERSTELLAR MIGRATION

The possibility of extraterrestrial migration necessarily involves an extension of the Drake equation (Walters et al., 1980), so that this equation assumes a more dynamic character. Extraterrestrial migration is also justified by the hypothetic arrival methods and propulsion systems that extraterrestrial migrants would use. Three possible systems have been mainly hypothesized and theorized so far: a) vectored huge space stations inhabited by self-sustained biological intelligences, hibernated embryos or self-reproducing robots and or automatic probes, which after an interstellar travel lasting centuries or millennia might have been settled on energetically favourable zones of our solar system (Freitas, 1980, 1983b, 1985b; Freitas and Valdes, 1980; Freitas and Valdes, 1985; Jones 1985; Papagiannis, 1983; Rose and Wright, 2004; Valdes and Freitas, 1983); b) spacecrafts that use relativistic wormholes to jump very rapidly from a place to another of the galaxy (Crawford, 1995; Davis, 2004; Kaku, 1994; Krasnikov, 2000; Maccone, 1995, 2000; Morris et al., 1988); c) exotic transfer methods involving some concepts of quantum mechanics in the framework of a holographic universe with resonant and non-local effects (Bohm, 1980; Davis, 2004; Jahn and Dunne, 1986).

The existence of long-lived, and presumably highly technologically evolved, ET civilizations, can be indirectly predicted by the current theories of stellar evolution (Clayton, 1983; Zuckerman, 1985), especially when one considers the very long duration of a low-mass solar-type star, around which terrestrial-like planets might be orbiting. The end of the main-sequence phase (hydrogen burning) of their star and the beginning of the giant-expansion phase (stellar envelope ejection) could be one of the most logical reasons of the migration of such civilizations. In such a case the expansion of the envelope of the star towards the giant phase would encapsulate all the inner planets, where extraterrestrial civilizations would live most probably. There are good reasons to hypothesize that civilizations that are highly evolved scientifically and technologically are able to predict with high time accuracy the onset of the giant phase, in order to be able to build in time and launch large spacecrafts containing a big number of individuals that are destined to leave their stellar system, and to choose a new solar system that is most favourable for life. Otherwise, an alien civilization might decide to send automatic probes to our solar system for purely scientific purposes, or, alternatively, it might have found the way to carry out visitations directly and fast using the properties of wormholes or more exotic systems.

Therefore, together with the quantitative predictions coming from the diffusion equations, which are in their turn an application of stellar statistics merged with bio astronomy, there are sound theoretical motivations to search systematically for proofs of extraterrestrial visitation inside the solar system. Even if so far concrete evidences have not been found at all, our planet too might be a possible target of alien visitation: this possibility can be investigated as well following the "sieve strategy" that was discussed in the previous sections.

4.2. THE "SETV" PROJECT

In the general context of the SETI project a new branch named SETV (SETV homepage, 2003) was officially born at the end of the previous century and has been very recently developed (Ansbro, 2001; Cornet and Stride, 2003; Stride, 2001; Teodorani, 2000, 2001, 2003). The goal of this research, also named "Local SETI", is to study the possible evidence of visitation of "exogenous probes" inside our solar system. The SETV general strategy (Stride, 2001) is devoted to the monitoring of the entire solar system, including Earth, inside an ideal sphere having a radius of 50 astronomical units. This project requires the use in the very next future of the following measurement facilities: a) space satellites equipped with specific photon detectors such as high-sensitivity infrared CCD cameras and or sensors for the detection of ultraviolet, X-ray and Gamma-ray sources; b) ground-based observing stations equipped with high-power radars and radio telescopes and or new generation radio interferometer arrays connected with multi-channel spectrum analyzers; c) ground-based observing stations equipped with wide-field and small-aperture optical photometric and astrometric telescopes used for target search and measurement of anomalous proper motions; d) ground-based observing stations equipped with large-aperture optical photometric and spectroscopic telescopes used for the search for very low-luminosity targets and for the analysis of relatively high-luminosity targets. Most importantly, these planned monitoring operations may allow researchers to search for the possible evidence of anomalous celestial objects associated with low-luminosity probes of presumably exogenous origin that are expected to show an infrared excess (Matloff, 1994) and/or transient manifestations of high-energy propulsion (Harris, 1986). This possible detection includes huge space arks, which, if really existent, are presumably located, according to predictions, on energetically favorable zones such as the Earth-Moon libration points (Valdes and Freitas, 1983) and the asteroid belt (Papagiannis, 1983). A high-resolution monitor of the Moon (Arkhipov, 1998) and radar scanning of the lunar and terrestrial circum-planetary space is planned as well.

4.3. MONITORING EARTH

Even if it is unanimously recognized that at present no scientific proofs of extraterrestrial visitation exist (Teodorani, 2004; Tipler, 1980), the SETV project considers also the possibility to monitor some crucial areas of Earth using appropriate instrumentation (Teodorani, 2000; Stride, 2001). The presence on Earth of explorative devices of possible exogenous origin would appear necessarily like an anomaly in our atmosphere. Such an anomaly might be possibly reported in the form of luminous phenomena in the skies of some areas of Earth, both as a transient occurrence and as a spatial and temporal recurrence. If the visiting spacecrafts or automatic probes come from civilizations that are highly more advanced than ours the anomaly that they would be able to create in our atmosphere might be of a nature that cannot be predicted at all. What presumably comes from a highly evolved science, possibly possessed by a civilization that could have earned one million of years of advantage in comparison to us, might appear like "magic" even to the eyes of our present science. An exogenous probe might not be necessarily something "mechanical" as we expect from our technology, but something much more exotic. Therefore, even if we are not in a condition to extrapolate the future point of a super-civilization starting from ours, we can maybe speculate on what we could see. For instance, we cannot exclude the possibility that such a master civilization is able to instruct a "plasma ball" to acquire the functions of an "intelligent probe" based on a particle neural network that is planned to work both as a multi-sensing device and as a computer. After all, our own technology, which substantially was born only two centuries ago, is starting already now to plan a new generation of computers based both on the DNA and on quantum mechanics and it is already starting to use the very sophisticated science of nano-technology.

On the basis of the working hypothesis that ET is able to visit Earth too, in the next future we plan to employ a network of instrumented and possibly automatic sensing stations in specific areas of Earth where anomalous atmospheric events are reported very often (Teodorani, 2003). Instruments such as small automated telescopes of both photometric and spectroscopic kind connected with high quantum-efficiency CCD detectors, photometric radiometers, high-resolution optical spectrographs, sensors operating in the near infrared and near ultraviolet wavelength windows, detectors of high-energy events, radio spectrum analyzers operating both in the ultra-low frequencies (ELF-ULF) and in the microwaves (UHF), magnetometers, radars for searching and tracking any suspect target, electrostatic detectors, and gravimeters, are intended to be used as a complete instrumentation able to monitor a presumably multi-wavelength phenomenon. A pilot project is already operational (Teodorani, 2000, 2004), even if the search for ET visitation is only a corollary of a bigger research project aimed at studying the physics of anomalous light phenomena in general. In the ambit of the current investigation at present only simple and basic sensors are used. In particular, low-resolution spectrographs are employed in areas of recurrence of anomalous light phenomena. In such a way it has been already possible to exclude an alien origin in most of the investigated cases and to confirm a terrestrial origin of light phenomena both in terms of new aeronautic technology and in terms of poorly known geophysical and atmospheric phenomena (Teodorani, 2004). Projects of mathematical and cryptographic character to decode possible intentional and or unintentional intelligent signatures coming from atmospheric and or geophysical light sources of plasma-like kind that do not show any apparent technological behavior but that cannot be identified as a known natural phenomenon are being prepared too.

We are now in a condition to ascertain (Condon, 1969; Sturrock, 1999; Teodorani, 2004) that most of the anomalies on Earth's atmosphere are caused by natural phenomena similar to ball lightning, by the recent products of our aerospace technology and by the misidentification of known celestial and atmospheric natural phenomena. But some residual does remain, and we do not know yet where and how it comes from and what is the physical mechanism producing the related emission of radiation, but we do know that our present physical science and sensor technology might furnish the answer to the specific anomalies that might be crucial for the search for extraterrestrial intelligence, and maybe open new chapters of the book of physics. If our science will not take the control of the situation investigating anomalies in general and systematically, humanity will run the risk to fall into the a-critical and dogmatic style of thought that some "ufology" and pseudo science seem to diffuse everywhere right now. Therefore a clear and rational style of thought should be applied to any aspect of the physically observable reality, especially when the occurrence of some apparently anomalous reality is almost predictable due to the recurrence of strange light phenomena in very specific and recognizable laboratory areas present on our planet.

5. References

Abrahamson, J. and Dinniss, J. (2000) Ball lightning caused by oxidation of nanoparticle networks from normal lightning strikes on soil. Nature **403**, 519-521.

Adams, M. H. and Strand, E. P. (2003) Past IEA Missions. International Earthlight Alliance (IEA), http://www.earthlights.org/iea_missions.htm.

Ansbro, E. (2001) New OSETI observatory to search for interstellar probes, In: S. A. Kingsley and R. Bhathal (eds.) *The Search for Extraterrestrial Intelligence (SETI) in the Optical Spectrum III*. Proc. SPIE, Vol. 4273, pp. 246-253. USA.

Arkhipov, A. V. (1998) Earth-Moon system as a collector of alien artifacts. J. Brit. Interplanetary Soc. **51**, 181-184.

Bainbridge, W. S. (1984) Computer simulation of cultural drift: limitations on interstellar colonisation. J. Brit. Interplanetary Soc., **37**, 420-429.

Bohm, D. (1980) *Wholeness and the Implicate Order*, Routledge & Kegan eds., London (GB), Boston (USA).

Betinis, E. J. (1978) On ETI alien probe flux density. J. Brit. Interplanetary Soc. **31**, 217-221.

Clayton, D. D. (1983) *Principles of Stellar evolution and Nucleosynthesis*, The University of Chicago Press, Chicago (USA), London (GB).

Cocconi, G. and Morrison, P. (1959) Searching for interstellar communications. Nature **184**, 844-846.

Condon, E. U. (1969) *Scientific Study of Unidentified Flying Objects*, University of Colorado, Bantam Books, New York (USA), Toronto (Canada), London (GB).

Cornet, B. and Stride, S. L. (2003) Solar System SETI Using Radio Telescope Array. Contact in Context, SETI League, http://cic.setileague.org/cic/v1i2/s3eti-ata.pdf.

Crawford, I. A. (1995) Some thoughts on the implication of faster-than-light interstellar space travel. Q. J. R. Astr. Soc. **36**, 205.

Davis, E. W. (2004) Teleportation Physics Study. Special Report N. AFRL-PR-ED-TR-2003-0034 (78 pages) -Air Force Research Laboratory, Air Force Materiel Command, Edwards Air Force Base CA 93524-7048. Also In: http://www.fas.org/sgp/eprint/teleport.pdf.

Deardorff, J. W. (1986) Possible extraterrestrial strategy for Earth. Q. J. R. Astr. Soc. **27**, 94-101.

Finney, B. R. (1985) SETI and interstellar migration. J. Brit. Interplanetary Soc. **38**, 274-276.

Freitas, R. A. (1980) Interstellar probes – A new approach to SETI. J. Brit. Interplanetary Soc. **33**, 95-100.

Freitas, R. A. and Valdes F. (1980) A search for natural or artificial objects located at the Earth-Moon libration points. Icarus **42**, 442-447.

Freitas, R. A. (1983a) If they are here, where are they? Observational and search considerations. Icarus **55**, 337-343.

Freitas, R. A. (1983b) The case for interstellar probes. J. Brit. Interplanetary Soc. **36**, 490-495.

Freitas, R. A. (1983c) Extraterrestrial intelligence in the solar system: resolving the Fermi paradox. J. Brit. Interplanet. Soc. **36**, 496-500.

Freitas, R. A. and Valdes, F. (1985) The search for extraterrestrial artifacts (SETA). Acta Astronautica **12**,1027-1034.

Freitas, R. A. (1985a) There is no Fermi paradox. Icarus **62**, 518-520.

Freitas, R. A. (1985b) Observable characteristics of extraterrestrial technological civilizations. J. Brit. Interplanetary Soc. **38**, 106-112.

Harris, M. J. (1986) On the detectability of antimatter propulsion spacecraft, Ap. S.S. **123**, 297-303.

Horowitz, P., Coldwell, C. M., Howard, A. B., Latham, D. W., Stefanik, R., Wolff, J., Zajac, J. M. (2001) Targeted and all-sky search for nanosecond optical pulses at Harvard-Smithsonian, In: S. A. Kingsley and R. Bhatal (eds.) *The Search for Extraterrestrial Intelligence (SETI) in the Optical Spectrum III*. Proc. SPIE, Vol. 4273, pp. 119-127. USA.

Jahn, R. G. & Dunne, B. J. (1986) On the quantum mechanics of consciousness, with application to anomalous phenomena. Foundations of Physics **16**, 721-772.

Jones, E. M. (1981) Discrete calculations of interstellar migration and settlement. Icarus **46**, 328-336.

Jones, E. M. (1985) A manned interstellar vessel using microwave propulsion: a Dysonship. J. Brit. Interplanetary Soc. **38**, 270-273.

Kaku, M. (1994), *Hyperspace*, Oxford University Press, USA.

Kingsley, S. (2001) Optical SETI observatories in the new millennium: a review, In: S. A. Kingsley and R. Bhatal (eds.) *The Search for Extraterrestrial Intelligence (SETI) in the Optical Spectrum III*. Proc. SPIE, Vol. 4273, pp. 72-92. USA.

Korpela, E. J., Cobb, J. Fulton, S., Lebofsky, M., Heien, E., Person, E., Demorest, P., Bankay, R., Anderson, D., Werthimer, D. (2004) Three years of SETI@home: a status report, In: R. Norris and F. Stootman (eds.) *Bioastronomy 2002: Life Among the Stars*. Proceedings of IAU Symposium #213. San Francisco (USA). Astronomical Society of the Pacific, p. 419.

Krasnikov, S. (2000) Traversable wormhole. Phys. Rev. D **62**, pp. 4028-4046.

Lang, K. R. (1998) *Astrophysical Formulae: A Compendium for the Physicist and Astrophysicist*, 3rd ed., Springer Verlag, New York, USA.

Lazio, T. J. W., Tarter, J., Backus, P. R. (2002) Megachannel extraterrestrial assay candidates: no transmissions from intrinsically steady sources. Astron. J. **124**, 560-564.

Maccone, C. (1995) Interstellar travel through magnetic wormholes. J. Brit. Interplanetary Soc. **48**, 453-458.

Maccone, C. (2000) SETI via wormholes. Acta Astronautica **46**, 633-639.

Matloff, G. L. (1994) On the detectability of several varieties of low-energy starships. J. Brit. Interplanetary Soc. **47**, 17-18.

Montebugnoli, S., Monari, J., Teodorani, M., Bortolotti, C., Cattani, A., Maccaferri, A., Orlati, A., Pari, P., Poloni, M., Roma, M., Maccone, C., Cosmovici, C. B. (2002) SETI Italia: 2002 status report, In: U. Lacoste (ed.) *Second European Workshop on Exo/Astrobiology*. ESA Publications Division, The Netherlands. ESA SP-518, pp. 543-544.

Morris, M. S., Thorne, K. S., Yurtsever, U. (1988) Wormholes, time machines, and the weak energy condition. Phys. Rev. Lett. **61**, 1446-1449.

Newman, W. I. and Sagan, C. (1981) Galactic civilizations: population dynamics and interstellar diffusion. Icarus **46**, 293-327.

Papagiannis, M. D. (1983) The importance of exploring the asteroid belt. Acta Astronautica **10**, 709-712.

Principia homepage (2004) The Scientific Search for Extraterrestrial Intelligence. Chapter: "Where are they?". Section: "Astronomy", http://www.gifford.co.uk/~principia/Lectures/SETILectures/ frameset.htm.

Rose, C., Wright, G. (2004) Inscribed matter as an energy-efficient means of communication with an extraterrestrial civilization. Nature **431**, 47-49.

SETV homepage (2003) The Search for Extraterrestrial Visitation (SETV) – A Scientific Search for Visitation From Extraterrestrial Probes, http://www.setv.org/

Stenhoff, M. (ed.) (1999) *Ball lightning. An Unsolved Problem in Atmospheric Physics*. Kluwer Academic/Plenum, New York, USA.

Strand, E. P. (1984) Project Hessdalen 1984 – Final Technical Report 1984. Project Hessdalen, http://hessdalen.hiof.no/reports/hpreport84.shtml.

Strand, E. P. (1998) Automatic Measurement Station (AMS). Station Alarms. Project Hessdalen, http://hessdalen.hiof.no/station.

Stride, S. L. (2001) An instrument-based method to search for extraterrestrial interstellar robotic probes. J. Brit. Interplanetary Soc. **54**, 2-13.

Sturrock, P. (1999) *The UFO Enigma: a New Review of the Physical Evidence*, Warner Books, New York, USA.

Tarter, J. (2001) The Search for Extraterrestrial Intelligence (SETI). Ann. Rev. Astron. Astrophys. **39**, 511-548.

Teodorani, M. and Strand, E. P. (1998) Experimental methods for studying the Hessdalen phenomenon in the light of the proposed theories: a comparative overview. ØIH Rapport **1998(5)**, 1-93. Høgskolen i Østfold, Norway.

Teodorani, M. (2000) Physical data acquisition and analysis of possible flying extraterrestrial probes by using opto-electronic devices. Extraterrestrial Physical Review **1**, 32-37.

Teodorani, M. (2001) Instrumented search for exogenous robotic probes on Earth, In: P. Ehrenfreund, O. Angerer and B. Battrick (eds.) *First European Workshop on Exo/Astrobiology*. ESA Publications Division, The Netherlands. ESA SP-496, pp. 379-381.

Teodorani, M. (2003) *Fenomeni Luminosi: Investigazione Scientifica di Fenomeni Luminosi Anomali in Atmosfera*, M.I.R. Edizioni, Montespertoli (FI), Italy.

Teodorani, M. (2004) A long-term scientific survey of the Hessdalen phenomenon. J. Scientific Exploration **18**, 217-251.

Tipler, F. J. (1980) Extraterrestrial intelligent beings do not exist, Q. J. R. Astr. Soc. **21**, 267-281.

Turnbull, M. and Tarter, J. (2003a) Target selection for SETI. I. A catalog of nearby habitable stellar systems. Ap. J. Suppl. Ser. **145**, 181-198.

Turnbull, M. and Tarter, J. (2003b) Target selection for SETI. II. Tycho-2 dwarfs, old open clusters, and the nearest 100 stars. Ap. J. Suppl. Ser. **149**, 423-436.

Turner, D. J. (2003) The missing science of ball lightning. J. Scientific Exploration **17**, 435-496.

Valdes, F. and Freitas, R. A. (1983) A search for objects near the Earth-Moon lagrangian points. Icarus **53**, 453-457.

Walters, C., Hoover, R. A., Kotra, R. K. (1980) Interstellar colonization: A new parameter for the Drake equation? Icarus **41**, 193-197.

Zou, You-Suo,(1995) Some physical considerations for unusual atmospheric lights observed in Norway. Physica Scripta **52**, 726-730.

Zuckerman, B. (1985) Stellar evolution: motivation for mass interstellar migrations. Q. J. R. Astr. Soc. **26**, 56-59.

Biodata of **Julian Chela-Flores** author of "***Destinies of Life and the Universe: The final frontiers of astrobiology and cosmology***"

Professor Julian Chela-Flores was born in Caracas, Venezuela and studied in the University of London, England, where he obtained his Ph.D. (1969) in quantum mechanics. He was a researcher at the Venezuelan Institute for Scientific Research (IVIC) from 1971 till 1978 and Professor at Simon Bolivar University (USB, Caracas) until his retirement in 1990. During his USB tenure he was Dean of Research from 1979 till 1985. He is a Fellow of: The Latin American Academy of Sciences, The Third World Academy of Sciences, the Academy of Creative Endeavors (Moscow, Russia) and a Corresponding Member of the Academia de Fisica Matematicas y Ciencias Naturales (Caracas). Professor Chela-Flores current positions are: Staff Associate of the Abdus Salam International Center for Theoretical Physics (ICTP), Trieste, Research Associate, Dublin Institute for Advanced Studies (DIAS) and *Profesor Titular,* Institute of Advanced Studies (IDEA), Caracas. His particular area of expertise is astrobiology, in which he is the author of numerous publications in this new science, including some in the frontier between astrobiology and the humanities (philosophy and theology). Professor Chela-Flores has been the organizer of a series of Conferences on Chemical Evolution and the Origin of Life from 1992 till 2003 (and was its co-director from 1995 till 2003). In 1999 Professor Chela-Flores co-directed an Iberoamerican School of Astrobiology in Caracas at the IDEA Convention Center. All the proceedings of the present series of conferences and the Caracas School have been published. He is also the author of the book: *The New Science of Astrobiology From Genesis of the Living Cell to Evolution of Intelligent Behavior in the Universe,* published in 2001 by Kluwer Academic Publishers and reprinted as a paperback in 2004.

E-mail: **chelaf@ictp.trieste.it**

J. Seckbach (ed.), Life as We Know It, 505-517.

DESTINIES OF THE UNIVERSE AND LIFE:
The Final Frontiers of Cosmology and Astrobiology (*)

JULIAN CHELA-FLORES
The Abdus Salam International Centre for Theoretical Physics
Strada Costiera 11; 34136 Trieste, Italy
and Instituto de Estudios Avanzados, Caracas 1015A, Venezuela

1. The intelligibility of the universe

Up to the present the intelligibility of the universe has been a topic restricted to philosophy and especially to the philosophy of religion. Amongst the various aspects of intelligibility we wish to highlight the destinies of the related phenomena of the origin and evolution of the universe (Sec. 2) and the subsequent emergence of life (Sec. 3). The arguments presented in this chapter argue in favor of bringing these fundamental topics within the frontier of astrobiology to encourage interdisciplinary interactions (Chela-Flores, 2001). Amongst the priorities of this new science, we should consider the search for other lines of biological evolution elsewhere in the universe, a search for what has been called 'a Second Genesis' (McKay, 2001). The main conclusions of such a study are likely to be relevant not only to astrobiology, but also to have implications in ethics and philosophy. Intelligible means the capability of being understood, or comprehended. Alternatively, intelligible can signify to be apprehensible by the intellect alone. A third aspect of intelligible, closer to the significance of the term in the context of the present work, is related to something that is beyond perception.

An "intelligible universe" can be the starting point of a prolonged and systematic discussion amongst astrobiologists, as well as philosophers and theologians. The Belgian Nobel Laureate Christian de Duve, at the end of his recent review on the origin and evolution of life, asks himself the question "What does it all mean? Not only in science (de Duve, 2002), but also in philosophy (Davies, 2003) and in theology (Russell, 2001) the intelligibility of the universe raises questions that lie on the frontier between science and the humanities: often considerations on intelligibility return to the often quoted, but less frequently debated statement of Steven Weinberg, the American Physics Nobel Laureate (Weinberg, 1977):

"The more the universe seems comprehensible, the more it also seems pointless."

We begin this chapter by discussing, whether this quotation is on the same footing as the statements usually made in physics and other branches of science. In other words, we should clarify whether the quoted statement reflects rather a specific philosophical trend, or attitude, characteristic of the first half of last century.

(*) The author wishes to dedicate his chapter to the memory of John Oro, a friend, a colleague and a teacher.

Hence, if that were the case, the statement should not discourage the general dialogue at the frontier of science and religion discussed in this chapter. The common approach from both ends of the academic discourse to a Second Genesis should contribute, as we shall see below, to the progress of the philosophy of religion. Reciprocally, the questions raised in any field of the humanities arising from the discovery of a Second Genesis could, in turn, enrich the search for the place of humans in the universe.

The above quotation can be best understood in the context of a philosophical trend called Existentialism. Earlier doctrines (Rationalism, Empiricism and Idealism) arose from the increased scientific knowledge of the 17th and 18th centuries. They had maintained that the cosmos is a well determined ordered system, and hence comprehensible to all observers. In that framework there was no motivation for viewing the origin and evolution of the universe and life as being absurd, or pointless, a trend that was to arise some two hundred years later. On the other hand, the existentialists went beyond Rationalism that was represented, amongst others, by Benedict De Spinoza the Dutch-Jewish philosopher and foremost exponent of this 17th-century doctrine (Spinoza, 2002). The existentialists also went beyond Empiricism.

Another influential philosophical view stresses the central role of the ideal, or the spiritual, in the interpretation of experience. Known as Idealism, this philosophic doctrine, unlike Rationalism and Empiricism, maintains that the world, or reality, exists essentially as spirit or consciousness, that abstractions and laws are more fundamental in reality than sensory things. Rationalists, empiricists and idealists laid down solid bases on which to discuss questions related to the eventual destiny of life and of the universe. Rationalists and empiricists had argued that we could discover all natural universal laws by reason and experimentation, largely in agreement with the emergence of modern science with Copernicus, Galileo, Bruno, Digges, Newton and others.

A systematic idealist, Georg Hegel attempted to elaborate a comprehensive and systematic ontology from a logical starting point (Hegel, 1967). In other words, although differing somewhat from other idealists Hegel attempted to defend faith as being logical. This movement was influenced by the growth of science since the Enlightenment. The need was felt for a harmonious development of human culture. The bases of the philosophy of religion had to be extended to accommodate so much new revolutionary scientific knowledge. Faith in a Creator had to be seen in a new light. The Danish philosopher and religious writer Soren Kierkegaard opposed Hegel's views, particularly the concept that religious faith was logical; he explores the notion of the absurd: Abraham gets a reprieve from having to sacrifice Isaac, by virtue of the absurd (Kierkegaard, 1986). He further anticipates the still-to-come philosophical trend (Existentialism), insisting in a thesis opposite to Hegel's: humans suffer a deep anxiety (and hence need religion) because one has no certainties (Kierkegaard, 1981). In more modern terms we can paraphrase Kierkegaard and other pioneers of Existentialism by saying that life in the universe is pointless and absurd. To put it simply, according to

Kierkegaard, life in the universe is not intelligible. Consequently, the questions of our destinies could not be incorporated as a frontier between astrobiology and the humanities that would encourage a fruitful dialogue.

A third relevant aspect of Existentialism came from the implications of the philosophy of Martin Heidegger (Heidegger, 1996), as interpreted by his close follower Jean Paul Sartre (1947). Heidegger formalized Existentialism on the basis of the work of Edmund Husserl, a German philosopher who had founded Phenomenology (Husserl, 1999). This earlier method of enquiry was applied to the description and analysis of consciousness through which philosophy attempts to gain the character of a strict science. Husserl's method is an effort to resolve the opposition between the emphasis on observation that is maintained by Empiricism, and on reason that is stressed by Rationalism. Against this background, Sartre maintained that Existentialism is an attempt to live logically in a universe that is ultimately absurd. Another eloquent supporter of this doctrine was the Literature Nobel Laureate Albert Camus. In the mid-twentieth century Camus, through writings addressed the isolation of man in what he considered to be an alien universe (Camus, 1991). At the end of this long line of intellectuals that were under the influence of Existentialism, in the above quotation Weinberg reflects a view of the universe to which he was constrained by the adopted philosophic trend that influenced his generation.

In the Existentialist view of the universe there still remains some hope for the concept of a meaningful universe, namely intelligibility of the universe could be approached with the hope of the eventual emergence of a future "theory of everything". In this proposed all-embracing future theory we would hopefully discover the fundamental laws of nature in terms of a set of equations (Weinberg, 1993). Then, all phenomena should follow from these equations (the hope being that chemistry and biology could also be deduced). This is an extreme form of Reductionism, not an inevitable choice, given the many insights of current progress in science as a whole. Since the Enlightenment the ever-increasing growth of science has encouraged Reductionism. The reductionist dream has been supported by preliminary sets of successful equations that have embodied general phenomena at the most disparate scales (both microscopic and macroscopic). Today we recognize such efforts by assigning the equations the surnames of their authors: Newton, Maxwell, Einstein, Schrödinger Dirac, Salam and Weinberg. We are still at a very early stage in the comprehension of life in the universe. When the open question of the intelligibility of the universe is posed in a wider cultural context, including the earth and life sciences, Reductionism's restricted view becomes more evident. This contrast between different views on the intelligibility of the universe illustrates the relevance of the frontier of astrobiology for human culture, especially under the influence of philosophical doctrines (other than Existentialism) that will tend to encourage any future constructive dialogue between science and the humanities. In the remaining part of this chapter we shall attempt to review briefly what science has achieved in our understanding of the destinies of firstly the universe (Sec. 2). We discuss the destiny life in Sec. 3. In Sec. 4 we review approaches for getting further insights into our destinies by means of oriented efforts in space exploration. (We take into account life-detecting experiments.) Finally, in Sec. 5 we comment on the dialogue at the frontier between astrobiology and the humanities.

2. What is the likely destiny of the universe?

In Sec. 1 we have argued that the universe is comprehensible and that it is not necessary to espouse the view of pointlessness in the emergence of life. The next stage is to consider what physicists have done to formulate the problem of the evolution and destiny of the universe. The geometry of space in Friedmann's closed model is similar to that of General Relativity; however, there is a curvature to time as well as one to space. Unlike Einstein's model, where time runs eternally at each spatial point on an uninterrupted horizontal line that extends infinitely into the past and future, there is a beginning and end to time in Friedmann's model of a closed universe when material expands from, or is recompressed to infinite densities. These instants are called respectively the "Big Bang" and the "Big Crunch." The Big Bang model tells us that as time t increases, the universe cools down to a certain temperature, which at present is close to 3°K, the so called 'cosmic microwave radiation' (CMB). Since it has a typical wavelength of about 2 mm, due to the enormous red shift the CMB has suffered since the moment it was last scattered during the first moments of expansion. It may be confidently considered to be a cooled remnant from the hot early phases of the universe.-and has an 'isotropic' distribution. In other words, the CMB temperature does not vary appreciably, independent of the direction in which we are observing the celestial sphere. The isotropy is a consequence firstly of the uniformity of cosmic expansion, secondly, of its homogeneity when its age was 300,000 years and temperature of 3,000K. On the other hand, in 1992 more precise measurements of the $T = 3^{\circ}$ K radiation, began to be made by means of the satellite called the Cosmic Background Explorer (COBE): When the accuracy of the isotropy was tested with more refined measurements, it was found that there was some degree of anisotropy after all – the temperature did vary according to the direction of observation (one part in 100,000). This fact is interpreted as evidence of variations in the primordial plasma, a first step in the evolution of galaxies. The Microwave Anisotropy Probe (MAP) – an initiative of the National Aeronautics and Space Administration (NASA) – extends the precise observations of the CMB to the entire sky. The European Space Agency (ESA) will extend this work subsequently by means of the Planck spacecraft whose launching is planned for 2007.

During the very earliest times of the Big Bang (10^{-43} second – 10^{-35} second), the lowest-energy state may have corresponded in microscopic physics (quantum mechanics) to a phenomenon called a "false vacuum," This quantum state is characterized by a combination of mass density and negative pressure that results gravitationally in a large repulsive force. In Einstein's theory of General Relativity this repulsive force is called a 'cosmological constant'. The false vacuum may be thought of as producing a corresponding repulsive force that gave rise to the scale factor R of the universe to grow (or 'to inflate') extremely fast (mathematically 'exponentially fast'). This means that R may have doubled its size roughly once every 10^{-43} or 10^{-35} second. After several doublings, the temperature, which started out at over one thousand degrees K, would have dropped to values near the absolute zero. At these low temperatures the true vacuum state may have lower energy than the false vacuum state, in an analogous fashion to how solid ice has lower energy than liquid water. Such 'supercooling' of the universe may therefore have induced a rapid phase transition from the false vacuum state to the true vacuum state. The transition would have released energy (analogous to the "latent heat" released by water when it freezes). This, in turn, reheats the universe to high temperatures. In such high temperature (and the gravitational energy of expansion) the particles and antiparticles of the Big Bang cosmologies would have emerged.

There is some evidence for an accelerating expansion of the universe, a phenomenon, which is still to be understood (Riess, A. et al., 1998). We have to learn whether the constant that Einstein introduced into his equations of gravitation (the 'cosmological constant'), purely on theoretical grounds, may represent some form of gravitational repulsion, rather than attraction (Krauss, 1998; Ostriker and Steinhardt, 2001).

Only a small fraction of the matter in the universe is in the form of the familiar chemical elements found in the Periodic Table. It is assumed that a large proportion of the cosmic matter consists of 'dark matter', whose composition consists of particles that play a role in the sub-nuclear interactions, mostly foreign to our everyday experience. The term 'dark matter' is not a misnomer, for the sub-nuclear particles that contribute to it, do not interact with light. However, a remarkable aspect of cosmic matter is emerging: the sum total of the standard chemical elements and the dark matter make up a small fraction of the matter content of the universe. The remaining fraction of cosmic matter has been referred to as 'dark energy' with the astonishing property that its gravity is repulsive, rather than attractive. A possibility that has to be considered seriously in the future is that the repulsive gravity may dominate the overall evolution of the universe. This could lead to ever increasing rates of expansion. If this was to be the future of our cosmos, then future of life in the universe may hold some surprises regarding the eventual destiny of the universe.

A high-flying balloon that flew over Antarctica has given experimental support to the cosmological view of the expanding universe. It has demonstrated that the universe is "flat". In other words, the usual rules of geometry are observed. A beam of light is not bent by gravity as it propagates. The path followed is straight lines, not curves. But since Einstein's Theory of General Relativity was proposed, the possible paths followed by beams of light over cosmological distances has remained to be verified. Another result of the study is the prediction that the Universe will continue its steady expansion, which started at the Big Bang, and will not collapse into a "Big Crunch".

The new information is a map of the CMB. Small temperature variations in the CMB would allow a test of different models of the expanding universe. The map represents an image of the early Universe, about 300,000 years old. The current estimate of the age of the universe is over 12 thousand million years old (12 Gyr). The light that has been detected has traveled across the entire Universe The project to map the CMB was called Boomerang (Balloon Observations of Millimetric Extragalactic Radiation and Geophysics). The Boomerang results support a flat universe. A perfectly flat universe will keep on expanding forever, because there is not enough matter to trigger a Big Crunch. Boomerang backs the inflation theory of the universe suggesting that the whole of the cosmos expanded from the Big Bang, with the scale factor expanding exponentially fast during the first instances of the cosmic expansion.

A space mission to overcome such difficulties is the Wilkinson Microwave Anisotropy Probe (WMAP). This probe traveled to a point in space known as L2, about a million miles from the Earth (four times farther than the moon), in the direction opposite from the Sun. Anything placed at L2 orbits the Sun at just the speed needed to keep it at L2. With this probe it was possible to obtain a map of the inhomogeneities of CMB very accurately. The confrontation of these measurements with theoretical models has confirmed the emergence of fluctuations in the very early universe. It has demonstrated that the present structure of the cosmos consists of about 4 percent ordinary atoms, 23 percent matter of 'dark matter' that does not interact with radiation, and the remaining fraction, over 70 per cent, consists of a mysterious 'dark energy' having negative pressure. Last, but not least, it has given us an upper bound for the age of the universe to be about 14 billion years. To sum up, these results suggest that the universe is flat. We are entitled to entertain the hypothesis that in principle life may have an eternal abode.

3. What is the likely destiny of life?

Having reviewed our present understanding of the eventual destiny of the cosmos, our next objective, as stated in the title chosen for this chapter, is to discuss the destiny of the phenomenon of the living process that has emerged in the universe. The origin of life is not fully understood. However, the general outline of the question of the chemical evolution of the precursors of the biomolecules has greatly advanced. Progress was due to the research of Ivanovich Oparin, Stanley Miller, Cyril Ponnamperuma, Sidney Fox, John Oro and many other organic chemists. These are some of the principal scientists that have traced out the likely pathways that nature may have followed during the molecular evolution that preceded the Darwinian evolution of the living cell. The seminal work of Charles Darwin has established the basis for the second stage in the discussion of astrobiology.

Darwinian evolution is much better understood than the question of the emergence of life. In earlier works several authors have argued that evolution on Earth has taught us that evolutionary convergence is an important feature of the Earth biota. Hence, if Darwinian evolution were assumed to be a universal process (as, for instance, advocated by Dawkins, 1983), we would expect that whenever life emerges elsewhere in the universe, life would be bound largely by the same general properties that we have found on Earth.

So under this assumption (the universality of biology), we can anticipate new insights in the distribution of life in the universe. Perhaps the leading approaches for searching for life elsewhere in the cosmos are the exploration of the Solar System and the search for intelligent signals through windows of the electromagnetic spectrum. The latter is known as the search for extraterrestrial intelligence (SETI). Since the pioneering days of the 1960s, bioastronomers have followed the lead of Frank Drake by probing various windows of the electromagnetic spectrum for evidence of narrow-frequencies signals (Drake and Sobel, 1992; Ekers et al., 2002). Such output presumably would be characteristic of other civilizations, instead of being the product of natural phenomena, such as supernova explosions or regular emissions from pulsars.

The search for other intelligent civilizations in the SETI project might have some implications in the philosophy of religion: our religious traditions go back to Jewish theology. There is a sole omnipotent God who created heaven and earth, and subsequently life on earth. This view of our origins has traditionally been referred to as a 'first' Genesis. But revelation through the scriptures never raises explicitly the possibility of the plurality of inhabited worlds. In 1584 Giordano Bruno made a speculative, but significant reference to the question of ubiquitous life in the universe (Bruno, 2000). In the late 16th century Bruno's statement led to a bitter and tragic controversy in the frontier between science and religion.

However, due to the present progress both in science and religion, we are now aware that there is no evident incompatibility between religious traditions and the possibility that we may not be alone in the universe. What is exciting about the emergence of the new science of astrobiology is that we can explore in scientific terms the possibility of whether the evolution of intelligent behavior is inevitable in an evolving cosmos, as already assumed implicitly by the above-mentioned SETI project (Chela-Flores, 2001).

4. To understand further the destiny of life we should search for a Second Genesis

In the previous two sections we have looked briefly at first of all the possible destiny of the universe. Secondly, we considered the living process that has emerged through chemical and biological evolution on Earth, and possibly elsewhere. The next step in our discussion is dictated by the fact that the theory of Darwinian evolution is not a predictive theory. In order to get further insights as to what can be the eventual destiny of life in the universe that might not be evident from our current knowledge of biology, we should search for alternative manifestations of the living process. This may occur either in planets, or satellites in our own solar system, or even in other solar systems. Intelligent signals from other civilizations are in principle detectable. The fact remains though that our lives are short we would like to have further insights into our destiny.

After almost half a century of searching for intelligent life in the universe – with extraordinary technological progress in the detection equipment used in the SETI project – sadly, no intelligent signals have so far been identified. But technology has progressed especially in the exploration of the Solar System with missions planned by the main space agencies, so as to be in principle capable of detecting microscopic life.

The search for extraterrestrial life has been attempted for the first time on the surface of planet Mars. A quarter of a century ago the Viking missions were in a position to detect life, although their results were not convincing to most scientists. The search continues today with Mars being the present target of several space missions. Assuming that Darwinian evolution is a universal process (Dawkins, 1983), we, and others, have argued in previous papers in favor of the inevitability of the origin and evolution of life. We have also argued and that the role of contingency has to be seen in the restricted context of parallelism and evolutionary convergence (Akindahunsi and Chela-Flores, 2004; Conway Morris, 2003).

Convergence, however, is not restricted to biology, but it has some relevance in other realms of science. The sharp distinction between chance (contingency) and necessity (natural selection as the main driving force in evolution) is relevant for astrobiology. Independent of historical contingency, natural selection is powerful enough for organisms living in similar environments to be shaped to similar ends. For this reason, it is important to document the phenomenon of evolutionary convergence at all levels, in the ascent from stardust to brain evolution. In particular, documenting evolutionary convergence at the molecular level is the first step in this direction. Our examples militate in favor of assuming that, to a certain extent and in certain conditions, natural selection may be stronger than chance (Conway-Morris, 1998).

We, together with others (Pace, 2001), discuss the consequences of the hypothesis of the possible universality of biochemistry, one of the sciences supporting chemical evolution. We have assumed that natural selection seems to be powerful enough to shape terrestrial organisms to similar ends, independent of historical contingency. Besides, evolutionary convergence can be viewed as a 're-run of the tape of evolution', with end results that are broadly predictable; hence, if life arises again elsewhere in the cosmos, we would expect some degree of convergence with terrestrial life.

The universality of biochemistry suggests that in solar system missions, biomarkers should be selected from standard biochemistry. Given the importance of deciding whether the evolution of intelligent behavior has followed a convergent evolutionary pathway, and given the intrinsic difficulty of testing these ideas directly (by means of the SETI project), we can alternatively begin testing the lowest stages of the evolutionary pathway within the Solar System. Indeed, we are in a position to search directly for evolutionary biomarkers on Europa. We have considered that if extant microorganisms are to be encountered, a possible although not immediately feasible set of evolutionary biomarkers may be considered (Chela-Flores, 2003).

Testing evolutionary biomarkers clearly lies in the distant future, since the next mission (cf., the JIMO mission below) will not be in a position to use the submersible (a 'hydrobot') that we visualized in the late 1990s (Horwath et al., 1997). Given the length of time before we can test evolutionary biomarkers directly, a full discussion at the present time of the feasibility of carrying out a proper test is timely. In this sense the discussion of biomarkers is reasonable at the present stage, since the Galileo mission has already provided us with a wealth of information about the chemical non-water-ice elements on the icy surface (McCord et al., 1998; Carlson et al., 1999). The careful interpretation of such information might conceivably lead us to reliable bioindicators without actually penetrating Europa's icy surface. The most interesting possibility for testing for life on Europa concerns the next orbital mission. It is expected to determine specific locations where the icy surface is thin enough for an eventual submersible penetration, and for testing directly the surface for the presence of microorganisms.

The discovery of other solar systems suggests that their formation seem to be analogous to ours. This is compatible with extensive knowledge of interstellar matter (Ehrenfreund and Charnley, 2000). From the assumed universality of biology it seems inevitable that intelligent behavior will emerge in the cosmos, provided certain conditions favorable to the presence of continuous life on a given planet (or satellite) are maintained. One of these conditions is that early stages in the formation of a solar system are expected to be characterized by a heavy bombardment period for planets and their satellites. This period would end after a few hundred million years. Consequently, planetary conditions over geologic time are likely to allow the continuous presence of life, as it has already happened on our own planet, once the heavy bombardment ceased. Observational techniques continually improve (the Darwin project and ESO's large telescopes are now under construction). These new instruments will allow to test the duration of the initial heavy bombardment and subsequent quiescent period in other solar systems. This statement can gradually be put to observational test. The result should give us a more precise idea on the temporal constraints that allow continuous presence of life on a given planet. At the present time in our own solar system the most attractive site for the search for life is at almost four times as far from us as planet Mars: the Galileo mission arrived in the Jovian system in 1995 and completed its work in the year 2003. This mission has exposed an environment that can in principle support life: Europa is the second Galilean satellite with respect to its distance from Jupiter. In the 17th century Galileo discovered Europa together with three other satellites Io, Ganymede and Callisto. However, Europa remains the leading contender for being the host of an independent evolutionary line. A second evolutionary line could, in principle, be brought to our attention in the foreseeable future. Right now the next orbital mission is being planned and has been called "The Jupiter Icy Moons Orbiter" (JIMO). This is an ambitious project for a mission that is intended to orbit three planet-sized moons of Jupiter. In fact, the Galileo Mission gave us data to suggest that Callisto, Ganymede and Europa may harbor large oceans underneath their icy surfaces. The mission would be launched in 2012, or later. Not only is there strong evidence for the internal oceans in the Jovian system, but also Jupiter's large icy moons appear to have three ingredients essential for the origin and sustained evolution of life, namely, water, energy and the necessary chemical molecules. The evidence from Galileo suggests melted water on Europa has been in contact with the surface in geologically recent times and may still lie relatively close to it. Observations of Callisto and Ganymede would provide additional comparisons that would contribute towards our understanding the three moons.

The JIMO mission would support one of astrobiology's main objectives: to explore the solar system in a well-focused effort to obtain our first insights into how life is distributed in the universe, and consequently this would help us take the first steps in our understanding of the destiny of life in the universe. Even if we allow ourselves to think beyond JIMO, the eventual construction of a lander is conceivable, although our earlier hope (Horwath et al., 1997) of including even a submersible for exploring for the first time an extraterrestrial ocean is probably still premature. But for learning whether a Second Genesis has occurred, probably a lander may be sufficient, given the dynamics of the icy Europan surface: it is possible that matter from the interior may be raised to the surface itself. (The Galileo mission has led to the discovery of a phenomenon called 'lenticulae', which may be surface areas, whose origin is matter from the deep interior.)

5. Insights into our destinies from astrobiology, philosophy and theology

We have considered throughout this chapter astrobiology's frontier in common with philosophy, as well as other branches of the humanities. We have attempted to show that an interdisciplinary exchange across the border is not only possible, but also profitable for the whole of human culture. Indeed, science is not contradicted by the main monotheistic religions of the world. A conflict is not expected to arise with the potential discovery of a Second Genesis. Instead, a real conflict could emerge with a discussion of the evolution of all the attributes of man, including those that are of prime importance for theology, namely the spirit of man that may distinguish humans from the ancestors of the *Homo* line. However, if we remain within the constraints that the science of biology has imposed onto itself – namely that the life sciences are based on observation (for instance, natural selection), or on experiment (for example, molecular biology), the question of man's spirit and soul should not even enter into the biological discourse.

6. References

Akindahunsi, A. A. and Chela-Flores, J. (2004). On the question of convergent evolution in biochemistry, In: J. Seckbach, J. Chela-Flores, T. Owen, T. and F. Raulin (eds.) *Life in the Universe: From the Miller Experiment to the Search for Life on Other Worlds*. Springer, Dordrecht, The Netherlands. (In press.).

Bruno, G. (2000). De l'infinito, universo e mondi, Venice, 1584. [English translation: On the infinite universe and innumerable worlds, Cambridge, 1650]. For a more precise bibliographic reference: Giordano Bruno 1548-1600, Biblioteca di Bibliografia Italiana Vol. **164,** Roma, Leo S. Olschki Editore, pp. 105-106,

Camus, A. (1991). The myth of Sisyphus and other essays, New York, Vintage International.

Carlson, R. W., Johnson, R. E. and Anderson, M. S. (1999). Sulphuric acid on Europa and the radiolytic sulphur cycle, *Science,* Vol. **286**: 97-99.

Chela-Flores, J. (2001). The new science of astrobiology: From genesis of the living cell to evolution of intelligent behavior in the universe. Kluwer Academic Publishers, Dordrecht, The Netherlands.

Chela-Flores, J. (2003). Testing Evolutionary Convergence on Europa, *International Journal of Astrobiology* **2**: 307-312. http://www.ictp.trieste.it/~chelaf/ss13.html.

Chela-Flores, J. (2004). Astrobiology's last frontiers distribution and destiny of life in the universe, In: J. Seckbach (ed.) *Origins: Genesis, Evolution and the Biodiversity of Life*. Springer, Dordrecht, The Netherlands, pp. 667-679.

Conway Morris, S. (1998). The crucible of creation. The Burgess shale and the rise of animals. London, Oxford University Press, p. 202.

Conway Morris, S. (2003). Life's solution: Inevitable humans in a lonely universe. Cambridge, CambridgeUniversity Press.

Coyne, S. J., G. (1996). Cosmology: The universe in evolution, In: J. Chela-Flores and F. Raulin (eds.) *Chemical Evolution: Physics of the Origin and Evolution of Life*. Dordrecht, Kluwer, pp. 35-49.

Davies, P. (2003). Universal truths, The Guardian, January 23, 2003 (cf., http://www.guardian.co.uk/online/science/story/0,12450,879894,00.htm).

Dawkins, R. (1983). Universal Darwinism, In: D.S. Bendall, (ed.), *Evolution from Molecules to Men*. London, Cambridge University Press, pp. 403-425.

de Duve, C. (1995). Vital dust: life as a cosmic imperative. New York, HarperCollins, pp. 296-297.

de Duve, C. (2002). Life evolving molecules mind and meaning. New York, Oxford University Press.

Drake, F. and Sobel, D. (1992). Is there anyone out there? The scientific search for extraterrestrial intelligence. Delacorte Press: New York.

Ehrenfreund, P. and Charnley, S. B. (2000). Organic molecules in the interstellar medium, comets andmeteorites: A voyage from dark clouds to the Early Earth, Ann. Rev. *Astron. Astrophys.* **38:** 427-483.

Ekers, R. D., Kent Cullers, D., Billingham, J. and Scheffer, L. K. (eds.) (2002). *SETI 2020.* SETI Press, Mountain View CA.

Hegel, G. W. F. (1967). Philosophy of right, London, Oxford University Press.

Heidegger, M. (1996). Being and time. A Translation of Sein und Zeit (SUNY series in Contemporary Continental Philosophy), New York, State University of New York Press.

Horvath, J., Carsey, F., Cutts, J. Jones, J. Johnson, E., Landry, B., Lane, L., Lynch, G., Chela-Flores, J., Jeng, T-W. and Bradley, A. (1997). Searching for ice and ocean biogenic activity on Europa and Earth, In: R. B. Hoover (ed.) *Instruments, Methods and Missions for Investigation of Extraterrestrial Microorganisms*. The International Society for Optical Engineering, Bellingham, Washington USA. Proc. SPIE, **3111**: 490-500. http://www.ictp.trieste.it/~chelaf/searching_for_ice.html.

Husserl, E. (1999). The idea of phenomenology. Dordrecht, Kluwer Academic Publishers.

Kierkegaard, S. (1981). The concept of anxiety, Kierkegaard's Writings. Vol **8**, Princeton, University Press Princeton.

Kierkegaard, S. (1986). Fear and trembling, London, Penguin Books.

Krauss, L. M. (1998). The end of the age problem and the case for a cosmological constant revisited. *Astrophysical Journal* **501**: 461-466.

McCord, T. B., Hansen, G. B., Clark, R. N., Martin, P. D., Hibbitts, C. A., Fanale, F. P., Granahan, J. C., Segura, N. M., Matson, D. L., Johnson, T. V., Carlson, R. W., Smythe, W. D., Danielson, G. E., and the NIMS Team (1998). Non-water-ice constituents in the surface material of the icy Galilean satellites from the Galileo near-infrared mapping spectrometer investigation, *Jour. Geophys. Res.* Vol. **103**: No. E4, 8603-8626.

McKay, C. P. (2001). The search for a second genesis in our Solar System, In: J. Chela-Flores, T. Owen and F. Raulin (eds.), *The First Steps of Life in the Universe*. Dordrecht, Kluwer Academic Publishers, pp. 269-277.

Ostriker, J. and Steinhardt, P. (2001). The quintessential universe. *Scientific American,* January, pp. 37-43.

Pace, N. R. (2001). The universal nature of chemistry, *Proc. Natl. Acad. Sci.* **98**: 805-808.

Riess, A. G., Filippenko, A. V., Challis, P. (1998). Observational evidence from supernovae for an accelerating universe and a cosmological constant. *Astronomical Journal* **116**: 1009-1038.

Russell, R. (2001). "Life in the universe: Philosophical and theological issues", CTNS Bulletin The Center for Theology and the Natural Sciences 21.2 (Spring 2001). [This paper was presented at the Sixth Conference on Chemical Evolution and first appeared in Chela-Flores, J., Owen, T. and Raulin, F. (eds.) The first steps of life in the universe. Dordrecht: Kluwer Academic Publishers, pp. 365-374.

Sartre, J.P. (1947). Situations. I, Paris, Gallimard.

Spinoza, B. (2002). Ethics. London, Everyman.

Weinberg, S. (1977). The first three minutes. London, Fontana/Collins.

Weinberg, S. (1993). Dreams of a final theory. London, Vintage International.

Biodata of **George Coyne** author of ***"Destiny of Life and Religious Attitudes"***

Dr. George Coyne obtained his Ph.D., in Astronomy from Georgetown University (Washington, DC) in 1962 and the Licentiate in Theology from Woodstock College (Maryland, USA) in 1966. In addition he has received PhD degrees honoris causa from St. Peter's University (New Jersey), Loyola University (Chicago), and The University of Padua (Italy). Since 1966 he has been associated with astronomy programs at the University of Arizona (Tucson, USA) and from 1976 to 1980 he served in various capacities in the administration of the astronomical observatories at that University. Since 1978 he has been Director of the Vatican Observatory (Specola Vaticana), which has its headquarters at Castel Gandolfo (Rome, Italy) and a research branch at the University of Arizona in Tucson. His research interests have ranged from the study of the lunar surface to the birth of stars, and he pioneered a special technique, polarimetry, as a powerful tool in astronomical research. Currently he is studying cataclysmic variable stars, the interstellar dust in the Magellanic Clouds and the detection of protoplanetary disks. Parallel to his scientific research he has developed an interest in the history and philosophy of science and in the relationship between science and religion.

E-mail: **gcoyne@specola.va**

J. Seckbach (ed.), Life as We Know It, 519-535.

DESTINY OF LIFE AND RELIGIOUS ATTITUDES

GEORGE V. COYNE, S. J.
Vatican Observatory
V-00120 Vatican City State

1. Introduction

The general background to the topic I wish to address is to what extent religious thought can make a contribution to our scientific understanding of the origins and evolution of life in universe derived from astrophysics and cosmology. And, on the other hand, to what extent can what we know from science about life influence our religious attitudes. This twofold question poses the serious risk of transgressing upon the epistemological independence of the various disciplines: theology, philosophy, astrophysics and cosmology, and creating, thereby, more confusion than understanding. As the discussion proceeds we must maintain a consistent posture of preserving the integrity of each of the disciplines.

Too often discussions of the relationship between science and religion are carried out in very general terms. Such discourse can be quite unfruitful for two reasons: (1) As compared to the natural sciences religion contains a larger measure of the subjective, of human experiences not totally verifiable by objective reasons. Such subjective experiences are not, of course, limited to religion. They are present in many areas of our lives. Nor need these experiences, religious or otherwise, necessarily conflict with reason. They simply are not limited to rational explanation. They go beyond what can be rationally justified. (2) While for the natural sciences we have a rather acceptable idea of what we mean by science, the very notion of religion is ill-defined. Does it mean worship? Does it mean being a "good person"? Does it mean accepting certain moral dictates that go beyond what is commonly accepted as good and bad? Does it mean accepting those dictates out of personal conviction or out of loyalty to a certain tradition? Does it mean believing in certain doctrines? Does it mean accepting a certain authoritative and hierarchical structure, i.e. being affiliated with a certain Church? To most of us religion would imply more of an affirmative than a negative answer to all of the above. And yet the situation is further complicated by the multiplicity of religions which differ among themselves, have even warred among themselves, over the responses given to such questions as the above. Even today, if we look at some of the main religious traditions: Islam, Judaism, Christianity, Buddhism, etc., we see not only vast differences among them, but enormous divisions within any one of the traditions.

The only way, therefore, that dialogue as a rational experience can take place is that, on the part of religion, the dialogue be limited to the rational foundations for religious belief. Even then, the only way that any such dialogue could have universal significance is that we could assume that there existed common rational foundations across all religious traditions and that is simply not the case. It seems, therefore, that any fruitful dialogue requires that the rational basis for certain specific religious beliefs in certain specific religious traditions be confronted with what is known from the natural sciences. The natural sciences, in particular, have made great advances by adhering rigidly to canons of what is scientifically true. In fact, in recent years the norms for judging the scientific truth of a given theory of life's origins and evolution have been extended, it appears to me, in the direction of inviting dialogue with philosophy and theology. I would like now to discuss these epistemological methods of scientific astrophysics and cosmology with the view of applying that discussion to our knowledge of the origins of life in the universe.

Skeptics, dubious of ever being able to find a widely accepted definition of science, say that science is what scientists do. The element of truth in this statement is that science is not a univocal concept. It varies from one discipline to another, even, for instance, among the so-called hard sciences. But there is also sufficient commonality among them that the name "science" can be legitimately given to each analogically. Let us do this with astrophysics and cosmology. What are these disciplines? What do astrophysicists and cosmologists do? We begin with controlled data, that is, data which any other trained professional could independently verify. In astrophysics and cosmology these data are observations of the contents of the universe. The astronomical and astrophysical sciences in general, are unique in this regard. We observe; we do not perform controlled laboratory experiments. We can control the way we observe; but, unlike the other sciences, we cannot control what we observe. From the observed data we use mathematical analysis and physics to develop a model which best explains the data. We will later on discuss what constitutes a "best explanation." There are many assumptions involved in this process of applying mathematics and physics to developing our knowledge of life's origins in the universe. One of the principal ones is that it is valid to apply the laws of physics, which are derived from our knowledge of what happens on the earth, to the universe as a whole. At any rate the movement from observations to models is a continuously reciprocal process. We use the best model to determine what further observations must be made, we perfect the model with the new observations, etc. There is a constant going back and forth from observations to the model to the observations. It is important to note that in the very nature of this process of reciprocity we admit that we do not possess the truth. The most that we can expect is that we are continually approaching the truth. It is with this background that I would like to discuss the topic at hand.

It is arguably difficult to find a more heated topic of discussion than that concerning the origins and evolution of the universe, and especially of life and of intelligence in the universe, and whether such origins can be understand without evoking a Creator God. Responses range from the extremes of a Stephen Hawking or a Pope Pius XII to almost all conceivable intermediate positions. Hawking claims that, if his quantum cosmological theory of the origins of the universe without boundary conditions is correct, then we have no need of God. Pius XII attempted to claim that with Big Bang cosmologies scientists were coming to discover what had already been known from the Book of Genesis, namely that the universe had a beginning in God's creative action. In between we have such positions as evolutionary naturalism and episodic divine intervention. Evolutionary naturalists would claim that, although our scientific knowledge of evolution is limited, the best explanation of the universe and all that it contains is through complexification in an expanding, evolving system in which both deterministic and chance processes play out their roles in a universe abundant with opportunities, 13.7 billion years old and containing 10^{22} stars. Those who profess episodic divine intervention would claim that divine activity is required, at least in some phases of the evolutionary process and, in particular at the occurrence of human life and intelligence, because natural processes alone are not adequate to explain the end result. What is one who is both a religious believer and a scientist to make of all of this?

Our attempts to understand the universe have as much to say about ourselves as they do about the universe. In fact, in us the universe can reflect upon itself and from our reflections there grows the conviction that we are part of that upon which we are reflecting. As soon as we set out with the powerful instruments for telescopic observations, together with those of mathematics and physics, to understand the universe and our place in it, we are made aware that there appears to be a destiny to life. Is there?

Modern cosmology, as well as ancient mythologies, cosmologies and cosmogonies, bear witness to the immense power which drives us humans in our continuous search for a deeper understanding of life. They also bear witness to the insufficiency of our search for understanding, of the need for something or someone out there, beyond oneself. From time immemorial we have always sought this further understanding in a person with whom we could converse, someone who shared our capacity to love and be loved and our desire to understand and to accomplish. Are such religious inclinations, if we may call them that, acceptable in a rational discourse on life's meaning?

2. Scientific Evidence for Universal Evolution

Let us take a sweeping view of a reasonable scientific picture of things. By reasonable I mean that, while we do not have all the answers, the following picture can stand up to any scientific critique. If we look today in infrared light at the center of Orion we see boiling gas and dust. If we look even closer up we see incandescent regions buried in that gas and with the Hubble Space Telescope we see the fine separation of blue gas and red gas in the midst of a rather chaotic structure. The fact is that stars are being born in this gas. And where the hottest, most massive and, therefore, brightest stars are already born, they are irradiating the gas, and it is giving off hydrogen alpha radiation. In this way we can identify star birth regions. The region of star birth in Orion is just a little part of our Milky Way. Our Milky Way, like most other spiral galaxies, measures 100,000 light years across and it contains about a hundred billion stars. It has several beautiful spiral arms and the sun is located in one of the outer arms, about two-thirds of the distance from the nucleus of our galaxy.

How is a star born? It happens by the laws of physics. A cloud of gas and dust, containing about 100 to 1,000 times the mass of our sun, gets shocked by a supernova explosion or something similar and this causes an interplay between the magnetic and gravity field. The cloud begins to break up and chunks of the cloud begin to collapse. And as any gas collapses, it begins to heat up; as it expands, it cools down. In this case the mass is so great that the internal temperature reaches millions of degrees and thus turns on a thermonuclear furnace. A star is born. Thermonuclear energy is the source whereby a star radiates to the universe. You need a very hot piece of the universe to do this, and so you can only get this thermonuclear furnace by having a cloud collapse and raise the temperature. You can only get it, in other words, in stars, with one exception, namely, in the very hot early universe before galaxies or stars were born.

Stars also die. A star at the end of its life can no longer sustain a thermonuclear furnace and so it can no longer resist against gravity. It collapses for a final time, explodes and expels its outer atmosphere to the universe. This may happen nice and peacefully or it may happen in a violent cataclysmic explosion, called a supernova. The most famous of these is the Crab Nebula which has a pulsar at the middle as its dead star.

So stars are born and stars die. And as they die they spew left over star matter out to the universe. The birth and death of stars is very important. If it were not happening, you and I would not be here, and that is a scientific fact. In order to get the chemical elements to make the human body, we had to have three generations of stars. A succeeding generation of stars is born out of the material that is spewed out by a previous generation. But now notice that the second generation of stars is born out of material that was made in a thermonuclear furnace. The star lived by converting hydrogen to helium, helium to carbon, and if it were massive enough, carbon to oxygen, to nitrogen, all the way up to iron. As a star lives, it converts the lighter elements into the heavier elements. That is the way we get carbon and silicon and the other elements to make human hair and toe nails and all of those things. To get the chemistry to make amoebas we had to have the stars regurgitating material to the universe.

Obviously this story of star birth and death is very important for us. Out of this whole process around one star, which we call the sun, a group of planets came to be, among them the little grain of sand we call the Earth. An amazing thing happened with that little grain of sand when, in the 16th and 17th Centuries with the birth of modern science, we developed the capacity to put the universe in our heads. We do that by using mathematics and physics, and to some extent the laws of chemistry and biology. Since we have the capacity to put the universe in our heads, a further question comes to us. Where did galaxies come from? All of the material in the universe is concentrated in galaxies and their environs. Galaxies are the building blocks of the universe. Hubble Space Telescope has been able to photograph some of the most distant objects we have ever seen in the universe. They are at a distance of about ten billion light years from us. So we are seeing these objects as they were ten billion years ago. We think that Hubble is seeing proto galaxies. We see, for instance, a case of two blobs which seem to be merging and perhaps building up a galaxy. However, this is very controversial. We are uncertain about galaxy formation, whether it is bottom up with small units that build into a galaxy, or top down with a big cloud that collapses to form a galaxy, and then the stars form within it. Nevertheless, when we compare distant galaxies to nearby galaxies, we see clear differences in the stellar populations. Galaxies as they are born and age go through an evolutionary process. Galaxies are participating in the expansion of the universe. When we look at them on a large scale we see that they are not distributed homogeneously. There are large empty spaces and many dense alignments.

Let us now review what we know of the history of the expanding universe. As it aged, distances got larger in the universe. As this happened certain key events took place. Quarks combined to form elementary particles, which in turn formed atoms and then molecules. The universe became transparent and the cosmic background radiation came to be. Galaxies and stars were formed. The first microscopic life forms came to be after twelve billion years in a fifteen billion year old universe. Why did it take so long to make even an amoeba? We have already discussed one reason. We did not have the chemistry to make even an amoeba until we had had three generations of stars.

3. Origins of Intelligent Life

How did we humans come to be in this evolving universe? It is quite clear that we do not know everything about this process. But it would be scientifically absurd to deny that the human brain is a result of a process of chemical complexification in an evolving universe. After the universe became rich in certain basic chemicals, those chemicals got together in successive steps to make ever more complex molecules.

Finally in some extraordinary chemical process the human brain came to be, the most complicated machine that we know. I should make it clear that, when I speak about the human brain as a machine, I am not excluding the spiritual dimension of the human being. I am simply prescinding from it and talking about the human brain as a biological, chemical mechanism, evolving out of the universe.

Did this happen by chance or by necessity in this evolving universe? Was it destined to happen? The first thing to be said is that the problem is not formulated correctly. It is not just a question of chance or necessity because, first of all, it is both. Furthermore, there is a third element here that is very important. It is what I call "opportunity." What this means is that the universe is so prolific in offering the opportunity for the success of both chance and necessary processes that such a character of the universe must be included in the discussion. The universe is 13.7 billion years old, it contains about 100 billion galaxies each of which contains 100 billion stars of an immense variety.

We might illustrate what opportunity means in the following way. Einstein said that God does not play at dice. He was referring specifically to quantum mechanics, but it can be applied in general to his view of the universe. For him God made a universe to work according to established laws. This is referred to as a Newtonian Universe. It is like a clock that just keeps ticking away once you supply it energy. Today we might be permitted to challenge this point of view. We could claim that God does play at dice because he is certain to win. The point being made is that God made a universe that is so prolific with the possibilities for these processes to have success that we have to take the nature of the universe into consideration when we talk about how we came to be.

For 13.7 billion years the universe has been playing at the lottery. What do I mean by the lottery? When we speak about chance we mean that it is very unlikely that a certain event would happen. The "very unlikely" can be calculated in mathematical terms. Such a calculation takes into account how big the universe is, how many stars there are, how many stars would have developed planets, etc. In other words, it is not just guesswork. There is a foundation in fact for making each successive calculation.

A good example of a chance event would be two very simple molecules wandering about in the universe. They happen to meet one another and, when they do, they would love to make a more complex molecule because that is the nature of these molecules. But the temperature and pressure conditions are such that the chemical bonding to make a more complex molecule cannot happen. So they wander off, but they or identical molecules meet billions and billions of times, trillions if you wish, in this universe, and finally they meet and the temperature and pressure conditions are correct. This could happen more easily around certain types of stars than other types of stars, so we can throw in all kinds of other factors.

The point is that from a strictly mathematical analysis of this, called the mathematics of nonlinear dynamics, one can say that as this process goes on and more complex molecules develop, there is more and more direction to this process. As the complexity increases, the future complexity becomes more and more predetermined. In such wise did the human brain come to be and it is still evolving. Can we call this process "destiny?"

Let us pause for a moment to review the degree of certainty which we can place in the above scenario. We certainly do not have the scientific knowledge to say how each living creature came to be in detail. We do not know precisely how each more complex chemical system came to contribute to the process of self organization which brought about the diversity of life forms as we know them today. Most importantly, we do not know with scientific accuracy the sufficient elements in nature to have brought about the unbroken genealogical continuity in evolution that we propose actually happened. There are, in brief, epistemological gaps which prevent natural science from saying that a detailed theory of biotic evolution has been proven. What we have presented is the most adequate account conceivable at this time considering the available empirical data. And that empirical data, with respect to biotic evolution, comes from various independent scientific enterprises, including molecular biology, paleontology and comparative anatomy.

How do we know we are on the path to the truth in the scenario of life's origins just described? In other words how do we judge what is the best way to explain life's origins. In the natural sciences there are a number of criterion whereby an explanation is judged to be best. I would list the principal criteria as the following: (1) verifiability, i.e., there is, at least in principle, a way of judging whether the explanation fits the data; (2) predictability, i.e., from data on past or present events it is possible to predict future events and then observe to see that the future events actually occur; (3) simplicity or economy, i.e., the least assumptions are made to get the greatest explanatory power; (4) beauty, i.e., the explanation has an aesthetic quality about it. Although, especially for the natural sciences, this may appear to be a very subjective criterion, almost all great scientific discoveries have benefited from its application. (5) unifying explanatory power; i.e., not only are the observations at hand explained but the attempt to understand is also in harmony with all else that we know, even with that which we know outside of the natural sciences.

It is this last criterion which I would like to discuss, since it appears to me to extend the epistemological nature of the natural sciences towards the realm of other disciplines, such as religious thought. Put in very simple terms this criterion is nothing else than a call for the unification of our knowledge. One could hardly be opposed to that. The problem arises with the application of this criterion. When is the unification not truly unifying but rather an adulteration of knowledge obtained by one discipline with the presuppositions inherent in another discipline. History is full of examples of such adulterations. It is for this reason that scientists have always hesitated to make use of this criterion. And yet, if applied cautiously, it appears to me to be a most creative one for the advancement of our knowledge.

The supposition is that there is a universal basis for our understanding and, since that basis cannot be self-contradictory, the understanding we have from one discipline should complement that which we have from all other disciplines. One is most faithful to one's own discipline, be it the natural sciences, the social sciences, philosophy, literature, religious thought etc., if one accepts this universal basis. This means in practice that, while remaining faithful to the strict truth criteria of one's own discipline, we are open to accept the truth value of the conclusions of other disciplines. And this acceptance must not only be passive, in the sense that we do not deny those conclusions, but also active, in the sense that we integrate those conclusions into the conclusions derived from one's own proper discipline. This, of course, does not mean that there will be no conflict, even contradictions, between conclusions reached by various disciplines. But if one truly accepts the universal basis I have spoken of above, then those conflicts and contradictions must be seen as temporary and apparent. They themselves can serve as a spur to further knowledge, since the attempt to resolve the differences will undoubtedly bring us to a richer unified understanding.

The above discussion particularly applies when we are addressing fundamental and ultimate questions such as life's origins and meaning. Does the existence of intelligent beings in the Universe have a significance for understanding the Universe as a whole? Does our knowledge of God depend on our understanding of the Universe? In fact, a very strong piece of evidence that there is a universal basis for understanding is the very clear drive of the human being for meaning. This is seen clearly from the very dawn of human history where, with even a very primitive collection of data, our ancestors sought for the meaning of life in the physical universe, as well as in the events of their personal lives and those of society in general.

In summary, we note that the scientific and the religious approaches to the search for the meaning of life have for the most part been pursued in isolation from one another. In the past when they have met it has been mostly as antagonists. In recent times, however, there has been an increasing awareness of the need for dialogue in the quest for life's meaning. The common criteria of what is true in this search would be that the explanation be simple, aesthetically appealing, verifiable and that it have a unifying explanatory power. In practice these requirements for a good theory will not always agree among themselves and differing emphasis will be given to one or other criteria in the different disciplines.

Scientists are usually well aware of the limitations of their knowledge. Religious thinking also has its limitations. The excessively dogmatic approach which sometimes characterizes theology would do well to recognize this. I am not here referring to the faith dimension in religion. In fact, for the purposes of this discussion I am excluding that dimension which is transcendental and, if you will, a-rational (goes beyond reason) and I limit myself to a discussion of theology as a rational science. Theology must deal with the linguistic interpretation of written documents; it must interpret oral traditions; it must reconstruct history. It must establish a rational basis for accepting witnesses to historical events and it must determine when authority alone can be the source of certain truths. Above all there are the serious epistemological problems that arise from the relationship of theology to faith. Although theology is a science, a rational way of knowing in its own right, it is said to proceed from faith and to lead to an understanding of the faith (*fides quaerens intellectum*). This makes it subject to all of the false illusions that can arise from purely subjective behavior, and it must always struggle to separate those illusions from what is objectively true.

If we were to pursue the dialogue which I have outlined in this paper, we might soon come to see that a teleology and design in the universe, derived from a religious point of view, are not incompatible with cosmological models, derived from the scientific point of view. Or we would come to realize that inevitable tendency in the physical universe towards more complex structures is not incompatible with, for instance, human free will. In fact, as a deeper synthesis of the understanding of the whole unfolded through dialogue among the various disciplines it is very likely that the questions peculiar to each discipline would receive a more satisfactory answer. The important thing to realize is that in both the scientific and the religious approaches to understanding we are searching for the truth, which we do not yet possess. But it is clear that evolution is an intrinsic and proper characteristic of the universe. Neither the universe as a whole nor any of its ingredients can be understood except in terms of evolution. And evolution is a daily happening. We, for instance, are constantly exchanging atoms with the total reservoir of atoms in the universe. Each year 98% of the atoms in our bodies are renewed. Each time we breath we take in billions and billions of atoms recycled by the rest of breathing organisms during the past few weeks. Nothing in my genes was present a year ago. It is all new, regenerated from the available energy and matter in the universe. My skin is renewed each month and my liver each six weeks. In brief, human beings are among the most recycled beings in the universe.

4. The Questioning Human Brain

Once we developed the capacity to put the universe in our heads, we became passionately interested in asking all kinds of questions. As an example of our scientific quest for life's meaning I would like to ask a one among many such questions. It is one which is fundamental to the origin of life on the Earth and perhaps elsewhere. Did our planetary system come about by a miracle? Absolutely not. Although we do not know everything about how it came about, we know that it happened in conjunction with the formation of the sun. Gas and dust were left over from the birth of the sun, and this gas and dust had to form into a disk by the law of physics to conserve angular momentum. Once all of this mass is concentrated into a disk, there is a much greater chance that the particles of gas and dust will collide and, in some cases, stick together. And, just like the rolling snowball effect, planetesimals, about 100 kilometers in diameter, are built up through accretion and finally planets are accreted from the planetesimals. We do not know everything about this process, but we know enough about it to know that it did not happen by a miracle. It happened by ordinary physical and chemical processes.

So, a further question arises: Did what we have just described happen elsewhere? First of all we look at those nearby stars that we suspect may be something like the sun. We have detected thus far more than 140 planets about other stars due to the center of mass motion of the star. That is an indirect way but a very solid one of detecting planets. We detect a wobble in the star due to the fact that there is mass outside of it so that the center of mass of the system is not at the geometrical center of the star. Furthermore, with the Hubble Space Telescope we have discovered disks around very young stars. We know for certain that they are very young stars by their spectra. We call the disks proto planetary because we have indirect evidence that the first planets have begun to form in the inner regions of the disk. We are beginning to see about other stars the process that we think formed the planets about the sun.

5. Implications for Religious Belief

How are we to interpret the scientific picture of life's origins in terms of religious belief. Do we need God to explain this? Very succinctly my answer is no. In fact, to need God would be a very denial of God. God is not the response to a need. One gets the impression from certain religious believers that they fondly hope for the durability of certain gaps in our scientific knowledge of evolution, so that they can fill them with God. This is the exact opposite of what human intelligence is all about. We should be seeking for the fulness of God in creation. We should not need God; we should accept him when he comes to us.

The religious believer is tempted by science to make God "explanation." We bring God in to try to explain things that we cannot otherwise explain. "How did the universe begin?", "How did we come to be?" and all such questions. We sort of latch onto God, especially if we do not feel that we have a good and reasonable scientific explanation. He is brought in as the Great God of the Gaps. I have never come to believe in God, nor do I think anyone has come to believe in God, by proving God's existence through anything like a scientific process. God is not found as the conclusion of a rational process like that. I believe in God because God gave himself to me. That was not a miracle. It does make sense that there is a personal God who deals with me and loves me and who has given himself to me. I have never come to love God or God to love me because of any of these reasoning processes. I have come to love God because I have accepted the fact that he first made the move towards me.

The scientific picture traced above in Sec. 3 deals with the questions of origins, of how what we observe and experience today came to be? The question of creation, and therefore of a God Creator, responds to the question of why is there anything in existence. Creation is not one of the ways whereby things originated as opposed to other ways that can be thought of, including quantum cosmology and evolutionary biology. The claim that all things are created is a religious claim that all that exists depends for its existence on God. It says nothing scientifically of how things came to be, although beautiful stories are told in the Book of Genesis, to elaborate on the dependence of all things for their existence upon God.

Having opened the Pandora's box of the Bible, let us elaborate a bit upon it. The Bible is a collection of writings by various authors at various epochs using various literary genres. And so it best serves reason if one speaks of a specific book rather than of the Bible in general. It is clear that the overall intention of the authors of Genesis is to evoke religious faith, an adherence to the God of Abraham, Isaac and Jacob, and not to teach science. There is simply no scientific teaching in Genesis. In the Judaic/Christian tradition, the roots of religious belief reach to 5,000 years before Christ with the prophet Abraham. But Modern Science cannot be dated before the 16th or 17th century, roughly from the time of Galileo and then through many others to Newton, with the discovery of the universal law of gravity, the differential calculus, etc. You may even wish to go back to the beginnings of the experimental method with Roger Bacon and others in the 13th century. But, at any rate, the modern science that speaks to religion today is born much later than the religion to which it speaks. It has to be recognized that the religious tradition is historically much longer and to a certain extent has that richness of the past that modern science does not.

It is unfortunate that, at least in America, creationism has come to mean some fundamentalistic, literal, scientific interpretation of Genesis. Judaic-Christian faith is radically creationist, but in a totally different sense. It is rooted in a belief that everything depends upon God, or better, all is a gift from God. The universe is not God and it cannot exist independently of God. Neither pantheism nor naturalism is true.

But if we confront what we know of origins scientifically with religious faith in God the Creator, in the senses described above, what results? I would claim that the detailed scientific understanding of origins has no bearing whatsoever on whether God exists or not. It has a great deal to do with my knowledge of God, should I happen to believe he exists. Let me explain.

Take two rather extreme scientific views of origins: that of Stephen Gould of an episodic, totally contingent and, therefore, non-repeatable evolutionary process as contrasted to a convergent evolutionary process such as that of Christian de Duve, in which the interplay of chance, necessity and opportunity leads inevitably to life and intelligence. In either case, it is scientifically tenable to maintain an autonomy and self-sufficiency of the natural processes in a natural world, so that recourse to God to explain the origins of all that exists, is not required. It is not a question of chance in nature, excludes God; destiny in nature requires God. In neither case is God required.

If, however, I believe in God then what nature tells me about God in one case is very different from what nature tells me about God in the other. Please note that I am not calling upon faith to adjudicate between contrasting scientific viewpoints. I do think that convergent evolution is more consistent with God's revelation of himself in the Book of Scripture, so that, as Galileo was fond of stating, the Book of Scripture and the Book of Nature speak of the same God.

If we take the results of modern science seriously, it is difficult to believe that God is omnipotent and omniscient in the sense of the scholastic philosophers. Science tells us of a God who must be very different from God as seen by the medieval philosophers and theologians. Let us ask the hard question. Could, for instance, God after a billion years in a fourteen billion year old universe have predicted that human life would come to be? Let us suppose that God possessed the theory of everything, knew all the laws of physics, all the fundamental forces. Even then could God know with certainty that human life would come to be? If we truly accept the scientific view that, in addition to necessary processes and the immense opportunities offered by the universe, there are also chance processes, then it would appear that not even God could know the outcome with certainty. God cannot know what is not knowable. The theologian, of course, would have a different answer. God is transcendent, outside of space and time. All events are simultaneous to him. But I have wished to stress God's immanence in a universe where the origins of life are a challenge to our knowledge.

This stress on God's immanence is not to place a limitation upon God. Far from it. It reveals a God who made a universe that has within it a certain dynamism and thus participates in the very creativity of God. Such a view of creation can be found in early Christian writings, especially in those of St. Augustine in his comments on Genesis. If they respect the results of modern science, religious believers must move away from the notion of a dictator God, a Newtonian God who made the universe as a watch that ticks along regularly. Perhaps God should be seen more as a parent or as one who speaks encouraging and sustaining words. Scripture is very rich in these thoughts. It presents, indeed anthropomorphically, a God who gets angry, who disciplines, a God who nurtures the universe. Theologians already possess the concept of God's continuous creation. I think to explore modern science with this notion of continuous creation would be a very enriching experience for theologians and religious believers. God is working with the universe. The universe has a certain vitality of its own like a child does. It has the ability to respond to words of endearment and encouragement. You discipline a child but you try to preserve and enrich the individual character of the child and its own passion for life. A parent must allow the child to grow into adulthood, to come to make its own choices, to go on its own way in life. Words which give life are richer than mere commands or information. In such wise does God deal with the universe.

These are very weak images, but how else do we talk about God. We can only come to know God by analogy. The universe as we know it today through science is one way to derive analogical knowledge of God. For those who believe modern science does say something to us about God. It provides a challenge, an enriching challenge, to traditional beliefs about God. God in his infinite freedom continuously creates a world which reflects that freedom at all levels of the evolutionary process to greater and greater complexity. God lets the world be what it will be in its continuous evolution. He does not intervene, but rather allows, participates, loves. Is such thinking adequate to preserve the special character attributed by religious thought to the emergence not only of life but also of spirit, while avoiding a crude creationism? Only a protracted dialogue will tell.

6. An Invitation to Think of Life as Destined

A much discussed question among cosmologists over the past two decades is the one arising from the so-called anthropic principle. Many distinctions are made concerning its true meaning; they range from the so-called "weak" principle, which essentially states that, as observers in the universe, we see the universe as related to us, to the "strong" principle, which requires a certain teleology intrinsic to the universe. For our purposes it is necessary to state only the following well-established cosmological facts: (1) the existence of the human being has required a fine-tuning of the physical constants and the laws of nature which we find empirically by scientific investigation in the universe; (2) there is no general cosmological theory which explains why those constants should have the precise values they do and the laws should be as they are.

Many examples of the fine-tuning I have referred to have been discussed. The argument is essentially the following one: of the many constants of nature, e.g., the velocity of expansion of the universe, the mass and charge of the electron as compared to the proton, the gravity constant, etc., the empirically measured value is so precise that had it been only slightly different (in general, one part in one million) it would have been impossible for human beings to have emerged. Why, therefore, are the values of all the constants so precisely what they are?

Let me give just a few examples. In expanding, since its beginning in a Big Bang, the universe has cooled to the current temperature of about three degrees Kelvin (absolute zero scale). In so doing it has followed the normal, well-known law for gases: as a given volume of gas collapses it heats up; as it expands it cools down. If the current temperature of the universe were much different than it is, the Earth would not be able to dissipate its energy and it would continuously heat up. Life on the surface of the Earth would not be possible beyond a certain temperature. Why is it that the temperature of the universe is just the value that it is, after having begun at millions of degrees? Examples of this kind could be multiplied many times over. For instance, if the energy levels in helium, carbon, and nitrogen were not precisely the values they are, the thermonuclear fusion processes which have given us the heavier elements could not have taken place. Without those heavier elements we would not be here. In fact, in order to have the right proportion of elements in the universe to form the human organism, three generations of stars were required. As we have seen in Sec. 2, the only way known to scientists to manufacture the heavier elements is in the thermonuclear furnaces of stars. As a star lives out its life it converts the lighter elements (hydrogen, helium, etc.) into the heavier elements (carbon, silicon, oxygen, etc.). When it dies, it regurgitates this heavier material to the universe. The next generation of stars, born from this material, goes through the same life cycle, so that the universe is being constantly supplied with the heavier elements. To arrive at the chemical abundances required for the human organism three generations of stars had to perform in this way.

The cosmologist, of course, first seeks the answer in a general physical theory that will explain all of the values. No such theory exists. Next, we seek to explain the fine-tuning by statistics. Pure chance is ruled out because the probability that it could have happened by chance is unacceptable scientifically. The statistical argument then moves to the possibility that there are many universes, existing either simultaneously or successively. Each of these universes would have its own set of physical constants and of the laws of nature. If we have enough such universes, even an infinite number, then the probability that one such universe like ours would come to be is quite acceptable. However, none of these many universe proposals succeeds very well, either because data is lacking or they are not verifiable. Verifiability is an important and indispensable criterion of scientific validity. In the many-simultaneous-universes theory the universes are separated by distances greater than the light travel time for the total age of the universe, and, therefore, in principle non-verifiable because non-communicating. In the successive-universes hypothesis it is difficult to see how there could be any possible data which could verify the existence of a universe before the last Big Bang.

The inability to provide thus far a strictly scientific explanation to what is a strictly scientific problem, i.e., the anthropic principle, may be, according to the discussion above of the criterion of unifying explanatory power, an invitation to think that the explanation lies in a teleological consideration. It is important here to emphasis the word "invitation", so as to preserve the epistemological independence of the various disciplines. One is perfectly free to accept the invitation or not. One can stay firmly put within one's own discipline and continue to seek the answer there, uncontaminated by possible solutions arising elsewhere. But it seems to me that the invitation is a very real one and well-founded; it, therefore, also seems to me that it requires serious reasons to reject it. Those serious reasons must confront the long history of religious thought that there is a person at the source of the existence of the universe and that said person had a purpose or a design in "creating" the universe, a design which included, perhaps even centered upon, our existence.

What is being proposed, of course, is an invitation to return to an examination of the religious concept of the creation of the universe by God against the background of modern cosmologies. One of the most productive areas of research in modern cosmology is the application of quantum mechanics to an analysis of the origins and very earliest stages of the universe. It is important to note that our observational knowledge of the origins and early stages of the universe is very limited, we might say non-existent. But we can argue back quite rigorously to the physical conditions which characterized those stages by applying physics and mathematics to what we observe in the universe today. Amidst the myriads of such observational data there are three principal observations which emerge and which allow us to reconstruct the early universe: (1) from the measurements of distant galaxies and clusters of galaxies we know that the universe is expanding with very precise conditions; (2) from the measurement of the abundances of helium, lithium, deuterium and other light elements, we know that much of that material had to be created under extremely high temperature and density conditions in the early universe; (3) from a measurement of the current temperature of the universe, the so-called cosmic background radiation, we can establish the temperature conditions of the early universe. When we combine all of this and other observations we can determine the age of the universe, its approximate mass and its mean density.

This summary of the results of modern cosmology represents an amazing feat in the combination of our knowledge of elementary particle physics and observational astrophysics. But the nagging questions remain: how did it all begin? when it began were there not certain initial conditions which determined how it would evolve? Did the universe really come to be in all its specificity from quantum fluctuations at its origin. Such considerations also suffer from problems of verifiability. The question also arises as to whether they really provide ultimate explanations.

It is precisely here, I believe, that religious thought can play a role in cosmology. Many of the concepts which are essential ingredients in the cosmological models have important implications in religious thought and those implications must also enrich cosmological thinking, so that the latter may have the greatest unifying explanatory power, a criterion for its veracity. In exploring these implications, however, it is essential that the fundamental significance of the concepts in the various disciplines not be confused. On the other hand the precise thrust of interdisciplinary dialogue is that a wider perspective will be gained on the fundamental reality by inter-relating the concepts arising from the diverse disciplines.

In the Hot Big Bang cosmological models the universe had a beginning. That beginning at time equals zero is a mathematical singularity. It cannot be addressed by classical mathematics or physics. To avoid that singularity it is claimed that quantum gravity must be applied at the extreme conditions of the universe's beginning. During this quantum gravity regime, however, the concept of time is inapplicable in any simple way. Most approaches require an origin of our specific universe from quantum fluctuations of a previous state: a collapsing previous state, a region of flat space-time, a previous black hole final state, etc. Such approaches, therefore, only address relative beginnings. They still leave us wondering about the origins of the previous state upon which the quantum fluctuations played out their game. What, if anything, do these quantum gravity considerations of the origin of the universe have to do with, for instance, the religious considerations of the creation of the universe in time and from nothing (*creatio ex nihilo*)?

Any attempt to simply identify the nothing (*nihilo*) of the religious thinker with the quantum fluctuations of one of the preexisting states would, to my mind, create nothing but confusion. But the one concept may illuminate the other. The thrust of the "in time" and "from nothing" for the religious thinker is to assert the total and exclusive dependence of the universe upon God the Creator. There was no rival to God preexisting before the universe began and in its beginning and continuation it depends on God. I cannot see how the scientific concepts deny or challenge the religious ones and they may even be illuminated by them. It would be equally confusing to deny the existence of God by stating that, since no boundary conditions were required for the quantum cosmological origin of the universe, God is not required. The God of the religious person is not a boundary condition for the universe. He is the creator, whatever content that notion of creator might have.

7. Summary

It makes us dizzy to contemplate billions of years in the evolving universe and then to think that we are on a little planet orbiting a quite normal star, one of the 200 billion stars in the Milky Way. And the Milky Way is just one galaxy and not anything special among the billions of galaxies which populate the visible universe.

The search for life's meaning today is ever more human; it stimulates, provokes, questions us in ways that drive us beyond science in the search for satisfaction, while at the same time scientific data furnish the stimuli. In this context the best science to its great merit, does not pretend nor presume to have the ultimate answers. It simply suggests and urges us on, well aware that not all is within its ken. Freedom to seek understanding and not dogmatism in what is understood characterize the best of science's search for the origins and meaning of life. It is, in fact, a field where certainties lie always in the future; thus it is vital, dynamic and very demanding of those who seek to discover the secrets of life and their religious implications.

PART 3:
DEEPER ISSUES OF THE PHENOMENON OF LIFE

Philosophical and Theological Aspects of Life

Russell
Aumann
Laras
Ne'eman
Cleland
Riskin
Elitzur
Shah
Matsuno

Human Life as a Social Phenomenon

Gadoth
Gunde-Cimerman, Zalar
Beakes
Funderburke

Artificial Life

Sipper
Umiel

Biodata of **Dr. Robert John Russell** author of ***"Life in the Universe: Philosophical and Theological Issues"***

Robert John Russell is Professor of Theology and Science in Residence, The Graduate Theological Union (GTU), Berkeley, CA. He is the Founder and Director, The Center for Theology and the Natural Sciences (CTNS). Dr. Russell obtained his PhD. in physics in 1978 at the University of California, Santa Cruz. That same year he was ordained to ministry in higher education in the United Church of Christ/Congregational. He served as Assistant Professor of Physics at Carleton College, Northfield, MN, from 1978 to 1981. Dr. Russell then founded CTNS in 1981 and began teaching on the GTU faculty. He has been the General Editor of a research publications series on scientific perspectives on divine action co-sponsored by CTNS and the Vatican Observatory. This series includes: *Quantum Cosmology and the Laws of Nature (1993); Chaos and Complexity (1995); Evolutionary and Molecular Biology (1998); Neuroscience and the Person (1999)* and *Quantum Mechanics (2001).* He edited *Fifty Years in Science and Religion: Ian G. Barbour and His Legacy (2004)* and he co-edited *Physics, Philosophy and Theology: A Common Quest for Understanding (1988)* and *Resurrection: Theological and Scientific Assessments (2002)*. He is a founding co-editor of *Theology and Science.* Dr. Russell served as the Principal Investigator of the CTNS programs "Science and the Spiritual Quest," "Science and Religion Course Program", and "Outstanding Books in Theology and Science."

E-mail: **rrussell@ctns.org**

J. Seckbach (ed.), Life as We Know It, 539-551.

LIFE IN THE UNIVERSE:
Philosophical and Theological Issues

ROBERT JOHN RUSSELL
The Center for Theology and the Natural Sciences
The Graduate Theological Union
2400 Ridge Road, Berkeley, California, USA 94709

1. Introduction

Over the past four decades, the interdisciplinary field of 'theology and science' has undergone tremendous growth involving scholars from philosophy of science, philosophy of religion, the natural sciences, theology, ethics, history and related fields.[1] Topics range from comparative methodologies to the relations between theologies of creation, divine action, and redemption in light of Big Bang, inflationary and quantum cosmologies, quantum physics, evolutionary and molecular biology, the neurosciences, anthropology, sociobiology, behavioral genetics, etc. Originally the work drew primarily on Christian theology, but today representatives of many other religions are active contributors. Surprisingly underrepresented in this rapidly growing interaction, however, is a focus on the philosophical and theological issues raised by the possibility of extraterrestrial life (EL) and extraterrestrial intelligence (ETI).[2] This is particularly curious since historians of science have shown that Christian theology contributed in significant ways to the assumption that ETI does in fact exist (Dick 1982, ch. 2). It is particularly timely, then, that this Sixth Trieste Conference on Chemical Evolution includes a section on the philosophical and theological implications of extraterrestrial intelligence.

There are, in fact, a wealth of topics which could be addressed here.[3] For the purposes of this short presentation, I will focus on the following three: 1) Will the discovery that life is abundant or rare in the universe influence its value or meaning? 2) Will intelligent life be capable of both rationality and moral reasoning? 3) Will ETI experience moral failure or be entirely benign? First, two caveats: 1) I will speak from the context of Christian theology and welcome others to widen the circle of religious discussions of EL/ETI. 2) I will assume that we will share modes of sensory awareness and rationality with ETI. It may be that some extraterrestrial civilizations are millions or even billions of years older than ours; for such advanced ETI, the usual 'contact' scenarios may simply not apply (Dyson 1979, ch. 19; Dick 2000). For the present purposes, however, I will focus on the possible discovery of ETI for which 'contact' would be a reasonable hypothesis.

2. Three Issues From the Perspectives of Philosophy and Theology

2.1. WILL THE DISCOVERY THAT LIFE IS ABUNDANT OR RARE IN THE UNIVERSE INFLUENCE ITS VALUE OR MEANING?

The relative abundance of life in the universe is one of the pivotal issues of our conference. Hopefully, we will learn the answer in the reasonably near future, either by projects such as the exploration of Mars and Europa, at least regarding pre-biotic and microbial life, or by projects like SETI and the discovery of extraterrestrial intelligence. For now, attempts at estimating the abundance of life and intelligence remain highly controversial. On the one hand, Stephen Jay Gould has stressed the overwhelming role of chance in the history of life on earth (Gould 1985). Thus, if life has evolved elsewhere, we should expect radical diversities in its morphologies reflecting differing evolutionary histories shaped by strongly contingent events. Paul Davies (1998) and Christian DeDuve (1995), on the other hand, have argued that evolution is tightly governed by the laws of physics and biology. Julian Chela-Flores has taken De Duve's point one step further, arguing that once life originates, the evolution of eukaryotes will lead to procaryotes. From there the 'convergence' of life towards intelligence is assured, although there may well be significant diversities in the morphologies which underlie intelligence (Chela-Flores 1998).

Underlying these issues, however, is a more fundamental question: Is the meaning and value of life a function of its abundance in the universe, or is it essentially meaningless *regardless* of what we find through the exploration of our solar system or SETI? It is this latter question which has received considerable philosophical and theological reflection.

Some scientists have suggested that biological life *per se* has little significance whether or not we are alone in the universe. They see life as essentially meaningless, a random product of physics and chemistry of no more significance than the wetness of water or the structure of Saturn's rings. Biological processes are just what matter does when really unusual conditions occur, but the universe, "at rock-bottom", is just endless mass-energy and curving spacetime. Such 'cosmic pessimism' is of course a *philosophical* interpretation of nature; it is not science, per se, nor is it one which can be 'proved' by science, but it is one that has been widely propounded by eminent scientists such as Bertrand Russell (1903) and Jacques Monod (1972). It is certainly the impression Steven Weinberg gave in his often-quoted conclusion to *The First Three Minutes*: "(H)uman life is...just a more-or-less farcical outcome of a chain of accidents reaching back to the first three minutes....The more the universe seems comprehensible, the more it also seems pointless" (1977, 154).

Others, though, disagree with this view, arguing instead for a philosophy in which life is a clue to the meaning of the universe. In *Disturbing the Universe*, Freeman Dyson writes: "I do not feel like an alien in this universe. The more I examine the universe and study the details of its architecture, the more evidence I find that the universe in some sense must have known that we were coming" (1979, 250). And in his 1985 Gifford Lectures, *Infinite in All Directions*, Dyson explicitly rejects Weinberg's opinion, telling us instead he sees "...a universe growing without limit in richness and complexity, a universe of life surviving forever and making itself known to its neighbors across the unimaginable gulfs of space and time...Twentieth-century science provides a solid foundation for a philosophy of hope" (Dyson 1988, 117-18). Writing in a similar vein, Paul Davies has depicted life in terms of "teleology without teleology" (1999, 72-76; 1989, 1998) and William Stoeger has written about the "immanent directionality" of evolution (1998).

The difference in these views may stem in part from a further division between reductionist and non-reductionist philosophies. If one assumes that the processes and properties characteristic of living organisms can be fully explained by physics and chemistry, there may be little if any basis for attributing meaning and value to life. Non-reductionist arguments on the other hand, such as those deployed by Francisco Ayala (1974, 1985, 1988), Ernst Mayr (1985), and Charles Birch (1990, 1998), offer a basis within natural processes for attributing varying degrees of meaning and value to organisms with differing levels of complexity and organization. Non-reductionist epistemologies, in turn, play a crucial theological role in a variety of views often referred to collectively as 'theistic evolution.' This perspective includes two central themes: *creatio ex nihilo* and *creatio continua*. 1) God as transcendent creates the universe out of nothing (*creatio ex nihilo*), holding it in existence at each moment and maintaining its law-abiding character which we express scientifically as the laws of nature.[4] 2) God as immanent creates the universe continuously in time (*creatio continua*), working "in, with, under and through" the processes of nature,[5] as Arthur Peacocke nicely phrases it. Scientists and theologians have developed these themes in light of physical cosmology, quantum physics, chaos and complexity theory, evolutionary and molecular biology, anthropology, the neuro and cognitive sciences, etc.[6] Most hold that the multi-leveled complexity of living organisms points to the intrinsic value of life. Arguably the most remarkable construction in the galaxy is the primate central nervous system. The number of connections between the neurons of the human brain is greater than the number of stars in the Milky Way. This staggering complexity makes possible the almost unimaginable feat of self-consciousness, of knowing oneself as a free, rational and moral agent in the world. Thus on our planet, at least, we are privileged to discover a hint of what God's intentions might have been in creating a universe like ours, with its particular laws of physics. For when the evolutionary conditions are right as they have been on Earth, and as they may be elsewhere in our universe, God, the continuous, immanent, ongoing creator of all that is, working with and through nature, creates a species gifted with the "image of God" (the *imago dei*) including the capacities for reason, language, imagination, tool-making, social organization, and self-conscious moral choice, a species capable of entering into covenant with God and in turn with all of life.

Thus if it took the precise characteristics of this universe to allow for the possibility of the evolution of life, even if life is scarce in the universe,[7] then it is life as such that gives significance to our universe—and even if ours is only one of a countless series of universes, as some inflationary and quantum cosmologies depict.[8] In short, I see life as the enfleshing of God's intentions amidst biological evolution which, in turn, is the ongoing expression of God's purposes in creating all that is. God thus offers to nature nature's conscious experience of the God who acts within nature.

2.2. WILL INTELLIGENT LIFE BE CAPABLE OF BOTH RATIONALITY AND MORAL REASONING?

We turn here to the question of the biological origins of ethics.[9] If our human experience of moral capacity, like our capacity for rational thought and relationality, is a gift of God, the *imago dei*, rooted in our biological nature and bequeathed us by God acting in and through evolution, does this suggest that wherever evolution results in creatures capable of rationality they would also be equipped by God with a capacity for relationship and moral reasoning?

Sociobiologists and, more recently, behavior geneticists, have explored the biological basis of human social behavior in order to determine the relation between evolutionary and genetic constraints on the one hand and their cultural expression on the other. Many of them, such as E. O. Wilson, Richard Dawkins and Michael Ruse, are unabashedly reductionistic, interpreting their scientific research in strictly deterministic and functionalist accounts of human behavior. Ruse, for example, has argued extensively that both the capacity and the content of human morality are entirely the products of evolution (1979, 1982, 1985, 1994). He defines biological altruism as any cooperative behavior between organisms that increases 'evolutionary gain', while moral altruism refers to our considered choices to help others because it is seen as 'right' to do so. For Ruse, moral altruism is a product of (nonmoral) biological altruism with "... no objective foundation. It is just an illusion, fobbed off on us to promote 'altruism'" (1995, 95, 100).[10]

Geneticist Francisco J. Ayala takes a very different position. For nearly three decades he has argued against reductionism in biology (1974, 1985, 1988), disagreeing dramatically with Ruse over the evolutionary origins of human moral capacity. According to Ayala, evolution selected for intelligence in our ancestral hominid line; one of the many byproducts of intelligence is ethics. "Ethical behavior came about in evolution not because it is adaptive in itself, but as a necessary consequence of man's (sic) eminent intellectual abilities, which are an attribute directly promoted by natural selection" (1995, 118). In short, ethics is open to cultural, philosophical and religious views. We can choose to act against natural predispositions, such as selfishness, if they are judged morally unacceptable. Conversely, some moral norms, such as justice and benevolence, may be inconsistent with behaviors favored by natural selection. Similar arguments against reductionism have been developed by a wide range of scientists and philosophers, including Theodosius Dobzhansky and Ayala, Ernst Mayr, Arthur Peacocke (1976, 1979, 1986, 1993), Ian Barbour (1971, 1988, 1990, 1998), Nancey Murphy (1990, *Beyond Liberalism* 1996, "Supervenience" 1998, "Nonreductive Physicalism" 1998, 1999) and George Ellis (*Before the Beginning* 1993; "Theology of the Anthropic Principle" 1993; 1995, 359-96; Murphy and Ellis *On the Moral Nature of the Universe* 1996; 1998; 2001).

How does this bear on the question of ETI? If what we have found about life on earth can be generalized, we may expect that moral capacity will be found wherever life has evolved to the point of intelligence in the universe, and that evolution will play a relatively strong/weak role in determining its contents depending on where one stands on the reductionist/non-reductionist debate above.

What sorts of responses might Christian theologians offer to the discovery of ETI with rational and moral capacities? Physicist Paul Davies predicts it would "(shatter) completely the traditional perspective on God's relationship with man (sic)" (1983, 71). But theologian Ted Peters finds "little or no credible evidence" for Davies' view (Peters 1994, 1).[11] Instead there is rich evidence in the history of Christian thought in support of a "plurality of worlds" and extraterrestrial life in the universe.[12] Contemporary theology, too, has been genuinely open to the possibility of rational and moral ETI. Peters cites Roman Catholics, such as George van Noort, Theodore Hesburgh, Hans Küng, Karl Rahner, and Francis J. Connett, conservatives such as Billy Graham, and Protestants such as Krister Stendahl, A. Durwood Foster, and Paul Tillich (Peters 1994, 4-5).

I would find the discovery of ETI endowed with rational and moral capacities to be a wondrous exemplification of the intentions of God in creating a universe like ours, namely in order to achieve the evolution of creatures capable of bearing the *imago dei* and entering into genuine community and covenant with God and all of life.[13] I am also persuaded by the philosophical arguments of non-reductionists that ethical choice and the contents of our moral codes remains at least partially a 'free variable'. I would therefore expect that ETI will be 'accountable' for its choices in some way even as humanity is. This leads directly to the challenge of moral failure, our final issue.

2.3. WILL ET EXPERIENCE MORAL FAILURE OR BE ENTIRELY BENIGN?

This question embodies a tragic reality at the heart of human existence. Why do we act with a level of violence against our own kind and other species which far exceeds the needs of survival and the level of violence of all other forms of life on Earth? Why do we lust after unlimited power and indulge in travesties like racism, sexism and specism? Put theologically, why do we sin? Why do we fail to love and serve God, one another, and all of God's creation, and instead indulge ourselves in unbridled pride and inordinate sensuality? One form of Christian response to this foundational question has been to assert the paradox that sin is not an intrinsic part of human nature, yet we all sin inevitably (Niebuhr 1941, VI - X). Making it intrinsic would rob us of our individual and corporate responsibility on which our legal and political systems are built; failing to recognize its inevitability would lead to the false hope that we can free ourselves of it without depending on the grace of God. In sum, each of us inherits both the *imago dei*, the 'image of God', and the inevitability of sin; both are unique to our species.[14]

This traditional response conforms nicely to our contemporary scientific understanding of biological evolution, particularly with the philosophical theme of 'novelty within continuity'. Thus we as a species inherit diverse propensities from our pre-hominid past, but in *homo sapiens* something strikingly new emerges. This 'newness' is manifest both in the *imago dei*, including our capacity for relationship, abstract thought, formal language, complex technologies, art, ethics and science, and in the reality of human sin, including ruthless violence and our insatiable appetites for power and control. It is only through the grace of a loving God that our lives can be transformed into the fullness of what it truly means to be alive. Conversely, the formation of authentic human personhood requires a lifetime of genuine wrestling with tough moral choices and the repentance of moral failure.

What then about ETI and the 'domain' of moral failure: is it truly universal or is it limited to terrestrial history? I have suggested for scientific, philosophical and theological reasons that the essential characteristics of human life are a genuine clue to the nature of life in the universe and not just an evolutionary fluke of the evolutionary processes on Earth. I therefore expect that ETI will experience much the same kind of moral dilemma that characterizes human experience, though obviously differing in its 'moral morphology'—the personal and social form of ethics. Jill Tarter has argued that extraterrestrial civilizations will be far older than ours, and to achieve such longevity, they will have had to overcome the temptation to warfare.[15] I tend to agree with her observation, but it still leaves open the question of how such temptation is overcome?

My assumption is that ETI will experience an empowering for their struggle by a source which transcends ETI's natural capacities. Put into theological language, I believe—perhaps I should say I predict!—that God will be present to the struggles of life everywhere, and that God's grace will redeem and sanctify every species in which reason and moral conscience are kindled.

As a Christian theologian and scientist, this leads me to a final question: should Christians expect that a single Incarnation of Christ in the person of Jesus is sufficient for the redemption of all life in the universe, or should we expect there to be an Incarnation of Christ in each species of ETI? An adequate discussion would take us far beyond the limits of this essay, but I do want to note in closing that modest support for both options can be found among both Protestants and Roman Catholics: a single, universally efficacious Incarnation is suggested by Protestants Ted Peters and Wolfhart Pannenberg and Roman Catholics L. C. McHugh and J. Edgar Bruns, while multiple Incarnations have been considered by Protestants Paul Tillich and Lewis Ford and the Roman Catholics Karl Rahner, E. L. Mascall and Ernan McMullin.[16] All agree, however, that wherever ETI exists, it will be the creation of a loving and redeeming God.

3. Conclusion

Clearly the empowerment by God of the full flowering of authentic personhood is at the heart of the Biblical witness. Taking this to entail such flowering throughout the diverse species of ETI throughout the universe leads to a profound reformulating and creative transformation of Christian thought and action, whose consequences would affect our ongoing search for deeper religious pluralism and a fuller understanding of the relation between humanity and the plenum of species in our terrestrial environment. Thus, regardless of what lies ahead as we await 'first contact' with ETI, pursuing these kinds of questions and reflections will be immensely valuable.

4. Endnotes

[1]For a scholarly introduction, (Barbour 1990; Murphy 1990; Peacocke 1993; Polkinghorne 1994; Richardson and Wildman 1996). For a less technical introduction, (Haught 1995; Peters 1998; Southgate 1999).

[2]See Peters (1994). For a recent survey of Roman Catholic views, see Vakoch (2000). See also McMullin (2000) and Coyne (2000). For a recent survey of ethical implications see Randolph, Race and McKay (1997).

[3]In my opinion the discovery of EL/ETI would not undermine the central claims of 'established religion', particularly Christianity, as many authors assume, although it would challenge a literal and inerrant reading of scripture in *any* religion. For Christians who take scripture as normative but not as literal and inerrant, there is plenty of room for a creative interaction between faith and science, including the discovery of EL/ETI.

[4]Arguments from "t=0" in Big Bang cosmology are often used to support this view, even though the theological claim is primarily about existence *per se* and not temporal origins. For recent references, see R. Russell ("Finite Creation" 1993).

[5]Note: I am assuming a non-interventionist view of God's immanent action here, one consistent with science. For details, see the CTNS/Vatican Observatory conference publications referenced in these Notes.

[6]See references to the writings of Barbour, Birch, Ellis, Haught, Murphy, Peacocke, Peters, Polkinghorne and others.

[7]Scarcity can indicate value: The discovery of a single palm tree in a vast desert does not mean that the tree is insignificant. Instead its scarcity makes it a tremendous discovery, for a hidden spring of life-giving water lies at its roots. I feel this way about Earth whether or not life is found elsewhere in the universe. Our planet is like the palm tree in what might in fact be a vast interstellar desert. If it takes a thousand million stars to produce the conditions for the possibility of a sea urchin, if it takes a billion years of tinkering with genetic dice to produce a hummingbird, and if it takes a million years of scratching on bark and vocalizing intentions to produce a child who can reach out through human artifacts and chalkboard calculations and touch the edge of the visible universe, then life per se is surely "the pearl of great price" (Matt. 13:45-46) and a clue to the theological meaning of the universe as a whole.

[8]See the extensive discussion of "the Anthropic Principle" (AP) in the literature, including R. Russell (*Quantum Cosmology* 1993). 'Many worlds' scenarios via inflationary Big Bang and quantum cosmologies are often deployed to undermine the 'fine-tuning' argument supporting divine design in the AP.

[9]This is *not* an attempt to derive ethics from biology, the so-called 'naturalistic fallacy.'

[10]As Rolston summarizes Ruse: "Ethics is not true, though it is functional. (But) ethics cannot be functional unless it is believed to be true in an objective sense, a false belief." (Rolston 1995). Rolston is sharply critical of Ruse in *Genes, Genesis and God* (1999).

[11]I agree with Peters (see Note #2 above). As Peters points out (1994, 2, 7), it may have been the Protestant Fundamentalist reaction to UFOs in the 1970s that gave rise to the view of Christianity that Davies critiques. See also McMullin (2000, 164-67).

[12]As Peters shows, Greek atomists such as Democritus and Epicurus argued for many worlds, but Aristotle rejected their views. Thomas Aquinas sought to reconcile Aristotle with Christian tradition, defending a "one world" view for philosophical reasons. Others, such as John Buridan, Nicole Oresme, Albertus Magnus, the Jewish scholar Hasdai Crescas, and Nicholas of Cusa, favored the "many worlds" position. See Dick (1982).

[13]For God as continuous creator and a theological response to sociobiology see Peacocke (1984; 1996, 321-42; 1998, 695-712) and Hefner (1993; 1996, 401-24; 1998).

[14]It is interesting to compare the Augustinian view with contemporary cultural positions. For example, consider our two questions about sin: 1) is it in our nature? 2) is it avoidable by human effort alone? The Augustinian response is: 1), no; 2), no. A secular humanist would probably answer: 1), no; 2), yes. Interestingly, Carl Sagan seemed to take the position: 1), yes (the tripartite brain and its conflicts); 2), yes (through science).

[15]Tarter claims that ET's longevity is inconsistent with them having "organized monotheistic religions" since they are the cause of warfare on earth. Clearly sustained warfare is inconsistent with cultural longevity, but her suggestion of a causal relation between monotheism and warfare would be hard to defend critically. Of course religion is often used as an excuse for violence, but the real question is why humans are capable of such violence. It also seems odd that she speculates favorably about a "universal religion" having a "highly established code of ethics" centered on "the perpetuation of individuals" when these tenets are found in the monotheisms which she rejects. (Tarter 2000).

[16]For references, see Peters (1994, 5-6). See also Mascall (1956, 36-45) and McMullin (2000, 171-73). Coyne leaves the question open (2000). In order to pursue this systematically, one would need to consider a variety of theological issues. Though I agree with Davies that the discovery of EL/ETI will bring a profound transformation in "spirituality and religion", Christological suggestions in the literature do not seem to have what Davies caricatures as "an air of absurd theatricality" or constitute a Catholic "heresy" (Davies 2000). McMullin views the arguments by Davies as "simplistic" (2000, 172).

5. References

Ayala, F. J. (1974) Introduction, In: F. J. Ayala and T. Dobzhansky (eds.) *Studies in the Philosophy of Biology: Reduction and Related Problems*, University of California Press, Berkeley.

Ayala, F. J. (1985) Reduction in Biology: A Recent Challenge, In: D. J. Depew and B. H. Weber (eds.) *Evolution at a Crossroads,* MIT Press, Cambridge pp. 67-78.

Ayala, F. J. (1988) Can 'Progress' Be Defined as a Biological Concept?, In: M. H. Nitecki (ed.) *Evolutionary Progress*, University of Chicago Press, Chicago, pp. 75-96.

Ayala, F. J. (1995) The Difference of Being Human: Ethical Behavior as an Evolutionary Byproduct, In: H. Rolston III, (ed.) *Biology, Ethics and the Origins of Life.* Jones and Bartlett Publishers, Boston, pp. 113-135.

Barbour, I. G. (1971) *Issues in Science and Religion,* Harper & Row, New York.

Barbour, I. G. (1988) Ways of Relating Science and Theology, In: R. J. Russell, W. R. Stoeger, G. V. Coyne (eds.) *Physics, Philosophy, and Theology: A Common Quest for Understanding,* Vatican Observatory Publications, Vatican City State, pp. 21-48.

Barbour, I. G. (1990) *Religion in an Age of Science,* HarperSanFrancisco, San Francisco.

Barbour, I. G. (1998) Five Models of God and Evolution, In: R. J. Russell, W. R. Stoeger and F. J. Ayala (eds.) *Evolutionary and Molecular Biology: Scientific Perspectives on Divine Action,* Vatican Observatory Publications; CTNS, Vatican City State; Berkeley, Calif. , pp. 419-442.

Birch, C. (1990) *A Purpose for Everything,* Twenty-Third Publications, Mystic, Connecticut.

Birch, C. (1995) *Feelings,* University of New South Wales Press, Sydney, Australia.

Birch, C. (1998) Neo-Darwinism, Self-Organization, and Divine Action in Evolution, In: R. J. Russell, W. R. Stoeger and F. J. Ayala (eds.) *Evolutionary and Molecular Biology: Scientific Perspectives on Divine Action* Vatican Observatory Publications; CTNS, Vatican City State; Berkeley, Calif., pp. 225-248.

Chela-Flores, J. (1998) The Phenomenon of the Eukaryotic Cell, In: R. J. Russell, W. R. Stoeger and F. J. Ayala (eds.) *Evolutionary and Molecular Biology: Scientific Perspectives on Divine Action* Vatican Observatory Publications; CTNS, Vatican City State; Berkeley, Calif., pp. 79- 98.

Coyne, G. V., S. J. (2000) The Evolution of Intelligent Life on Earth and Possibly Elsewhere: Reflections from a Religious Tradition, In: S. Dick (ed.) *Many Worlds: The New Universe, Extraterrestrial Life, and the Theological Implications,* Templeton Foundation Press, Philadelphia.

Davies, P. (1983) *God and the New Physics,* Simon & Schuster, New York.

Davies, P. (1989) *The Cosmic Blueprint,* Touchstone, New York.

Davies, P. (1998) Teleology Without Teleology: Purpose Through Emergent Complexity, In: R. J. Russell, W. R. Stoeger, and F. J. Ayala (eds.) *Evolutionary and Molecular Biology: Scientific Perspectives on Divine Action*, Vatican Observatory Publications; CTNS, Vatican City State; Berkeley, Calif., pp. 151-162.

Davies, P. (1999) Is the Universe Absurd?, In: T. Peters (ed.) *Science and Theology: The New Consonance,* Westview, pp. 65-76.

Davies, P. (2000) Transformations in Spirituality and Religion, In: A. Tough (ed.) *When SETI Succeeds: The Impact of High-Information Contact,* Foundation for the Future, Bellevue, WA.

De Duve, C. (1995) *Vital Dust: Life as a Cosmic Imperative.* Basic Books, New York.
Dick, S. J. (1982) *Plurality of Worlds*, Cambridge University Press, Cambridge.
Dick, S. J. (2000) Extraterrestrials and Objective Knowledge, In: A. Tough (ed.) *When SETI Succeeds: The Impact of High-Information Contact,* Foundation for the Future, Bellevue, WA.
Dyson, F. (1979) *Disturbing the Universe,* Harper and Row, New York.
Dyson, F. (1988) *Infinite in All Directions,* Harper and Row, New York.
Ellis, G. F. R. (1993) *Before the Beginning: Cosmology Explained,* Boyars/Bowerdean, New York.
Ellis, G. F. R. (1993) The Theology of the Anthropic Principle, In: R. J. Russell, N. C. Murphy and C. J. Isham (eds.) *Quantum Cosmology and the Laws of Nature: Scientific Perspectives on Divine Action,* Vatican Observatory Publications; CTNS, Vatican City State; Berkeley, Calif., pp. 363-399.
Ellis, G. F. R. (1995) Ordinary and Extraordinary Divine Action: The Nexus of Interaction, In: R. J. Russell, N. C. Murphy and A. R. Peacocke (eds.) *Chaos and Complexity: Scientific Perspectives on Divine Action,* Vatican Observatory Publications; CTNS, Vatican City State; Berkeley, Calif., pp. 359-96.
Ellis, G. F. R. (1998) The Thinking Underlying the New 'Scientific' World-Views, In: R. J. Russell, W. R. Stoeger, and F. J. Ayala (eds.) *Evolutionary and Molecular Biology: Scientific Perspectives on Divine Action,* Vatican Observatory Publications; CTNS, Vatican City State; Berkeley, Calif., pp. 251-280.
Ellis, G. F. R. (2001) Quantum Theory and the Macroscopic World, In: R. J. Russell, P. Clayton, et al. (eds.) *Quantum Mechanics: Scientific Perspectives on Divine Action,* Vatican Observatory Publications; CTNS, Vatican City State; Berkeley, Calif., pp. 259-291.
Gould, S. J. (1985) *The Flamingo's Smile: Reflections in Natural History,* W. W. Norton, New York.
Haught, J. F. (1995) *Science & Religion: From Conflict to Conversion,* Paulist Press, New York.
Hefner, P. (1993) *The Human Factor,* Fortress Press, Minneapolis.
Hefner, P. (1996) Theological Perspectives on Morality and Human Evolution, In: W. M. Richardson and W. J. Wildman (eds.) *Religion and Science: History, Method, Dialogue* Routledge, New York, 401-24.
Hefner, P. (1998) Biocultural Evolution: A Clue to the Meaning of Nature, In: R. J. Russell, W. R. Stoeger and F. J. Ayala (eds.) *Evolutionary and Molecular Biology: Scientific Perspectives on Divine Action,* Vatican Observatory Publications; CTNS, Vatican City State; Berkeley, Calif., pp. 329-356.
Mascall, E. L. (1956) *Christian Theology and Natural Science,* Longmans, Green & Co., New York.
Mayr, E. (1985) How Biology Differs from the Physical Sciences, In: D. J. Depew and B. H. Weber (eds.) *Evolution at a Crossroads*, MIT Press, Cambridge, pp. 67-78.
McMullin, E. (2000) Life and Intelligence Far from Earth: Formulating Theological Issues, In: S. Dick (ed.) *Many Worlds: The New Universe, Extraterrestrial Life, and the Theological Implications*, Templeton Foundation Press, Philadelphia, 171-73.
Monod, J. (1972) *Chance and Necessity,* trans. Austryn Wainhouse, Vintage Books, New York.
Murphy, N. (1990) *Theology in the Age of Scientific Reasoning,* Cornell University Press, Ithaca.
Murphy, N. (1996) *Beyond Liberalism and Fundamentalism,* Trinity Press International, Valley Forge, Pa.
Murphy, N. (1998) Nonreductive Physicalism: Philosophical Issues, In: W. S. Brown, N. Murphy and H. N. Malony (eds.) *Whatever Happened to the Soul?: Scientific and Theological Portraits of Human Nature,* Fortress Press, Minneapolis, pp. 127-148.
Murphy, N. (1998) Supervenience and the Nonreducibility of Ethics to Biology, In: R. J. Russell, W. R. Stoeger, and F. J. Ayala (eds.) *Evolutionary and Molecular Biology: Scientific Perspectives on Divine Action,* Vatican Observatory Publications; CTNS, Vatican City State; Berkeley, Calif., pp. 463-489.
Murphy, N. (1999) Supervenience and the Downward Efficacy of the Mental: A Nonreductive Physicalist Account of Human Action, In: R. J. Russell, et al. (eds.) *Neuroscience and the Person: Scientific Perspectives on Divine Action*, Vatican Observatory Publications; CTNS, Vatican City State; Berkeley, Calif. , pp. 147-164.

Murphy, N. and Ellis, G. F. (1996) *On the Moral Nature of the Universe: Theology, Cosmology, and Ethics*, Fortress Press, Minneapolis.
Niebuhr, R. (1941 (1964)) *The Nature and Destiny of Man: I* Charles Scribner's Sons, New York.
Peacocke, A. (1976) Reductionism. Zygon **11.4**.
Peacocke, A. (1979) *Creation and the World of Science,* Clarendon Press, Oxford.
Peacocke, A. (1984) Sociobiology and Its Theological Implications. Zygon **19**, 171-84.
Peacocke, A. (1986) *God and the New Biology,* Harper and Row, San Francisco.
Peacocke, A. (1993) *Theology for a Scientific Age: Being and Becoming—Nature, Divine and Human* Fortress, Minneapolis.
Peacocke, A. (1996) The Incarnation of the Informing Self-Expressive Word of God, In: W. M. Richardson and W. J. Wildman (eds.) *Religion and Science: History, Method, Dialogue,* Routledge, New York, pp. 321-42.
Peacocke, A. (1998) Biology and a Theology of Evolution. Zygon **33.1**, 695-712.
Peters, T. (1994) Exo-Theology: Speculations on Extra-Terrestrial Life. CTNS Bulletin **14.3**, 1-9.
Peters, T. (ed.) (1998) *Science & Theology: The New Consonance,* Westview Press, Boulder, Colorado.
Polkinghorne, J. C. (1994) *The Faith of a Physicist: Reflections of a Bottom-up Thinker,* Fortress, Minneapolis.
Randolph, R. O.; Race, M. S. and McKay, C. P. (1997) Reconsidering the Theological and Ethical Implications of Extraterrestrial Life. CTNS Bulletin **17.3**, 1-8.
Richardson, W. M. and Wildman, W. (eds.) (1996) *Religion and Science: History, Method, Dialogue,* Routledge, New York.
Rolston, H. III (ed.) (1995) *Biology, Ethics and the Origins of Life,* Jones and Bartlett Publishers, Boston.
Rolston, H. III (1999) *Genes, Genesis and God,* CUP, Cambridge.
Ruse, M. (1979) *Sociobiology: Sense or Nonsense?,* Reidel, Dordrecht.
Ruse, M. (1982) *Darwinism Defended,* Addison-Wesley, Reading.
Ruse, M. (1985) *Taking Darwin Seriously,* Blackwell, Oxford.
Ruse, M. (1994) Evolutionary Theory and Christian Ethics: Are They in Harmony? Zygon **29.1**.
Ruse, M. (1995) Evolutionary Ethics, In: H. Rolston III (ed.) *Biology, Ethics and the Origins of Life*, Jones and Bartlett Publishers, Boston.
Russell, B. (1903) A Free Man's Worship, In: B. Russell, *Mysticism and Logic,* Allen & Unwin, London.
Russell, R. J. (1993) Finite Creation Without a Beginning, In: R. J. Russell, N. C. Murphy and C. J. Isham (eds.) *Quantum Cosmology and the Laws of Nature: Scientific Perspectives on Divine Action,* Vatican Observatory Publications; CTNS, Vatican City State; Berkeley, Calif., pp. 291-325.
Russell, R. J., Murphy, N. C., Isham, C. J. (eds.) (1993) *Quantum Cosmology and the Laws of Nature: Scientific Perspectives on Divine Action,* Vatican Observatory Publications; Center for Theology and the Natural Sciences, Vatican City State; Berkeley, Calif.
Southgate, C. et al. (eds.) (1999) *God, Humanity and the Cosmos: A Textbook in Science and Religion,* Trinity Press International, Harrisburg.
Stoeger, W. R. (1998) The Immanent Directionality of the Evolutionary Process, and Its Relationship to Theology, In: R. J. Russell, W. R. Stoeger and F. J. Ayala (eds.) *Evolutionary and Molecular Biology: Scientific Perspectives on Divine Action,* Vatican Observatory Publications; CTNS, Vatican City State; Berkeley, Calif., pp. 163-190.
Tarter, J. (2000) Implications of Contact with ETI Far Older Than Humankind, In: A. Tough (ed.) *When SETI Succeeds: The Impact of High-Information Contact,* Foundation for the Future, Bellevue, WA.
Vakoch, D. A. (2000) Roman Catholic Views of Extraterrestrial Intelligence: Anticipating the Future by Examining the Past, In: A. Tough (ed.) *When SETI Succeeds: The Impact of High-Information Contact*, ed. Allen Tough (Bellevue, Washington: Foundation for the Future, 2000).
Weinberg, S. (1977) *The First Three Minutes,* Basic Books, New York.

Biodata of **Robert Aumann**, author of ***"Consciousness."***

Professor Robert Aumann, born in Frankfurt in 1930, fled from Nazi Germany to New York in 1938. He studied pure mathematics at City College in New York and at MIT; after getting the Ph.D. in 1955, he became interested in Game Theory and its applications. In 1956 he moved to the Hebrew University of Jerusalem, where he has been ever since—with sabbaticals at Princeton, Berkeley, Yale, Stanford, Louvain, and Stony Brook. He is a member of the National Academy of Sciences (USA), and of the American, Israel, and British Academies; holds honorary doctorates from Chicago, Bonn, and Louvain; and has received the Israel, Harvey, EMET, Nemmers, Lanchester, and von Neumann Prizes, as well as the Nobel memorial prize in Economics. When not working, he likes to hike, ski, cook, and study the Talmud.

E-mail: **raumann@math.huji.ac.il**

J. Seckbach (ed.), Life as We Know It, 553-564.

CONSCIOUSNESS

R. J. AUMANN
Center for the Study of Rationality, and Departments of Mathematics and Economics, The Hebrew University of Jerusalem, 91904, Jerusalem, Israel

1. Introduction

Consciousness is the last great frontier of science. Sixty years ago, life was not understood; it was a mystery. With the discovery of DNA, that mystery was solved; today, we more or less understand, at least in principle, how life "works." But we do not at all understand how consciousness "works."

We start by defining our terms; already there, there are difficulties. By "consciousness," we mean, in the first instance, the ability to *experience*. To see, hear, smell, feel, taste, desire, enjoy, suffer, like, dislike, love, hate, fear, become excited by an idea, be saddened by a loss. We do *not* mean to "sense." A machine—a "sensor"—also senses. Machines read; record sounds; detect odors, touches, and flavors; win at chess. They can even be programmed to exhibit a frownie, or emit "downbeat" sounds when something isn't right. But presumably they do not *experience*.

What, exactly, does "experience" mean? Ah, that is where the difficulty lies. The word cannot be defined in technical terms that do not refer to the concept itself. "Experience" cannot be defined in terms that a machine can understand. If you yourself are not conscious—have never experienced something—then you will not understand what the term means; just as little as congenitally deaf people understand what music means. To be sure, they can understand about the vibrations of taut strings, about air waves, about the workings of musical instruments, and even about musical notation and rhythm; but they can never understand what music *is*. For that, one must *hear* it. Similarly, someone who is not conscious cannot understand experience.

Ernst Mayr (1982) distinguishes between two fundamental questions in biology: "how" and "why." "How" refers to mechanism, "why" to function. The question "how do we digest food?" is answered by describing the process, involving saliva, chewing, swallowing, processing in the stomach and intestines, absorption into the bloodstream, and so on; and, disposing of wastes. The question "why do we digest food?" is answered by saying that food must be digested in order to provide vital ingredients for the functioning of the body, particularly energy.

To these questions, we add a third, which logically comes before the other two: "what." This refers to the descriptive aspect of biology, and of science in general. The answer to the question "what is digestion?" is that it is the process whereby food is transformed to a state that the body can use. "What" questions also include observational, descriptive, and classifying matters, and also methodological or conceptual matters, like "what is a species?"

The remainder of this essay is divided into three sections: "What," "Why," and "How." In the first, we discuss what consciousness is, and how it differs radically from other scientific phenomena. In the second, we discuss a possible function of consciousness, from the evolutionary viewpoint; and the third discusses the mechanism. Unfortunately, the "How" section is particularly short: we really have nothing to say about this, other than to describe the difficulties.

2. What

We have defined consciousness as the ability to experience. This puts the phenomenon into a completely different category from other scientific phenomena, in several ways.

(i) Unlike almost every—indeed every—other scientific phenomenon, consciousness is completely *subjective*. No verifiable outside characteristics of consciousness are known. No matter how complex an organism's behavior is, a computer could conceivably be programmed to mimic that behavior. An individual can observe consciousness only in himself.

Specifically, *I* can observe consciousness in myself only; I cannot be certain that anybody else really is conscious. To be sure, since other people appear roughly similar to me, and act similarly, I may surmise that they, too, are conscious; but I'm not *certain*. Each individual can be certain of consciousness *only* in himself, where he directly observes it.

(ii) Whereas the phenomenon of consciousness is highly subjective, it is, paradoxically, the *only* phenomenon of which the observer is absolutely certain. All other phenomena and observations could conceivably be attributed to hallucinations, dreams, and/or mental illness. But also hallucinations, dreams, and the ravings of a madman are experiences; in each case, the observer is *sure* that he is experiencing—is "conscious"—and he is right (we include dreams under the heading of consciousness).

(iii) Some people express perplexity as to the nature of the problem. They do not see anything mysterious about consciousness, and do not understand in what way it is different from other neurological functions like, say, the regulation of breathing. Asked whether a computer could in principle be conscious, they answer, "why not?"

We are dumbfounded by this reaction, and can only conjecture that these people are themselves not conscious. To me, it is evident that no combination of silicon chips and wires could conceivably "experience" in the sense that I do. Consciousness involves something beyond the merely physical and mechanical.

(iv) It seems only slightly less evident that no combination of off-the-shelf chemicals could experience in the sense that I do. But here, we are entering a gray area. In an e-mail message dated March 22, 2005, the U.S. National Academy of

Sciences announced that "a worldwide ban on human reproductive cloning—a technique that attempts to produce a child—is justified." If it is being banned, it is presumably within reach. Thus, the day seems not far off when it will be possible to synthesize a human being. Will such a golem be conscious?

Each of the possible answers—"yes" and "no"—is problematic. "Yes" is problematic because a combination of chemicals is in principle no different from a combination of silicon chips and wires, which we intuitively feel cannot experience. But "no" is also problematic, because there is no reason to believe that a golem that is identical, molecule for molecule, with a live human being would not in all respects—including consciousness—be like that human being.

(v) Are animals conscious? On the face of it, there is no reason to suppose that they cannot be. But as stated above (in (i)), it is not even clear that all humans—other than me—are conscious. By analogy with me, I can surmise that other human beings *are* conscious; but the analogy is less compelling in the case of animals. The further one gets from human beings on the evolutionary scale, the less compelling the analogy. So the short answer is, "possibly; we don't know."

(vi) Consciousness may have levels. For example, dreaming is certainly an expression of consciousness, but perhaps at a lower level than waking consciousness. Newborn children, and the mentally impaired, may be conscious at a lower level. In the opposite direction, people taking certain drugs sometimes report a "heightened state of consciousness."

Here again, we are at a loss, because we cannot really imagine what it is like to be, say, a newborn child. We personally have never taken drugs, so cannot make a judgment in the opposite direction either. We are stuck in our own conceptual prison: consciousness is about subjective experience, so it is difficult to imagine levels of consciousness other than our own.

Though it may have various levels, its existence at any level already poses the conceptual problems discussed here.

(vii) Conscious experience appears to be associated with certain physiological-neurological processes in the brain: Prof. Rafael Malach of the Weizmann Institute of Science has reported that it is associated with the simultaneous firing of many neurons in a well-defined group of neurons. (Based on this observation, Malach makes the fascinating suggestion that a *group* of individuals acting in concert may also be conscious.) This, however, does not explain how consciousness works—just as little as noticing that human reproduction is associated with sexual intercourse explains how reproduction works.

(viii) Up to now we have discussed only the "input" component of consciousness: experience. There are also two other vital components. One is the "processing" component: thought, including intention. The other is the "output" component: volition—consciously choosing to do something, and doing it. True, a person could be conscious, but have no power to take any action—as when asleep, or as a result of a totally debilitating stroke. But under most normal circumstances, thought and volition are intimately associated with consciousness. Indeed, as we shall presently see, it is the combination of all three elements—experience, thought, and volition—that enables consciousness to perform its function.

3. Why

We next address the issue of "why": what is the *function* of consciousness, from, say, the evolutionary viewpoint?

The answer we propose is based on the two related notions of *decentralizing* and *decoupling*—roughly speaking, splitting a difficult or complex task into several easier or less complex tasks, often with the aid of an auxiliary *"driving force."* Here are some examples:

(i) **Tearing a Manhattan telephone book in two**—perhaps the granddaddy of all decoupled processes. Taken as is, it is very difficult. But if one first separates the book into a number of thinner parts, then one can easily tear each of these parts in two, thus accomplishing the task. Here the difficult task is tearing the whole book; the easier tasks are tearing each of the thinner parts; and the decoupler is separating the entire book into the thinner parts.

(ii) **Operating an economy.** An economy can be centrally planned, as in a Kibbutz (Israeli cooperative village). A central planner decides how much of each good will be allocated to each individual, how much—and where—each individual will work, and so on. In theory, the entire economy of a whole country could be planned in this way. That is the conception behind socialist economies, like that of the former Soviet Union.

Centrally planning an entire economy is enormously complex and difficult. To start with, the informational requirements—finding out what each person wants to consume and what he is capable of producing—seem utterly beyond reach. But even if that could somehow be achieved, the problem of getting the people to do what you want them to do, and the sheer complexity of the task, makes the efficient central planning of an entire economy practically unachievable. Indeed, socialist economies like that of the Soviet Union achieve levels of efficiency that are far below those of advanced "free" economies.

What enables free economies to work more efficiently is that they are, by and large, decentralized. Within limits, each person seeks to acquire the goods and services that he wants, and to work at the tasks that he wants and that he is able to perform. There is no problem in getting the people to do what you want them to do, as each individual makes his own choices. And, the complexity of the task is greatly reduced, as the task of operating the economy is transferred from a central planner to individuals, each planning only the segment of the economy that interests him—usually, his own consumption and production. Instead of one planner (or planning agencies) planning many billions of transactions, we have several million planners planning thousands of transactions each. The total size of the task is perhaps roughly the same, but the task is *decoupled* into millions of individual tasks—and so is much more easily accomplished.

The driving force that makes the whole process work is *individual motivation.* Each individual is motivated to seek for himself the best "deal" that he can get—the goods and the work that he likes most—and it is this that operates the entire economy. That is Adam Smith's "invisible hand."

(iii) **Money and prices.** Primitive decentralized markets work by barter: two or more people get together and exchange goods or services, to the mutual benefit of all parties to the transaction; the process may be repeated as often as desired. One may hope that the final outcome is *optimal*, in the sense that no traders could have done better by trading with each other (more precisely, that no group of traders could have improved the welfare of all its members by trading within the group only. In economic theory, such outcomes are called *core* outcomes). In practice, in reasonably large markets—or even in fairly small ones—identifying the bartering groups, and deciding on the barters to be implemented, is so complex, involved, and fraught with uncertainty that barter is unlikely to achieve an outcome that is anywhere close to optimal.

Enter the institutions of money and prices. Rather than bartering, each trader sells his goods at market prices; with the proceeds he buys the goods he desires, again at market prices. Then if the prices are "right," the market "clears": the demand for each good exactly matches the supply. Moreover, the resulting reallocation of goods is optimal (in the above sense); and in large markets, *all* optimal outcomes are achievable in this way.

Thus, the unsophisticated but highly complex, involved, and uncertain bartering process is replaced by the price mechanism, which, though a lot more sophisticated than barter, is far easier to execute. In barter, each trader needs some knowledge about the preferences of the traders with whom he trades, and also about at least some of the others, so that he will know with whom to trade. But the price mechanism requires only that each trader buy what *he* likes, given *his* budget (the proceeds from the sale, at market prices, of the goods that he brought to the market).

Here the complex task is achieving optimality; the simpler tasks are for each trader to decide what he wants, given his budget; and the decoupler is the price mechanism. And again, the driving force is *individual motivation.* Each individual is motivated to "maximize over his budget set"—sell and buy, at the given market prices, so as to be left with those goods and services that he most prefers.

An interesting aspect of this decoupling is that historically, it has emerged by itself in *every* reasonably advanced society, without being imposed by any planning entity. Not only has the price mechanism emerged by itself, but the market prices themselves also usually emerge by themselves—determined by supply and demand—without being imposed by planning entities. And when planning entities do enter the process of price determination, as in the former Soviet Union or with rent control in various cities, they often wreak havoc, causing shortages and other distortions.

Whereas this example is related to the previous one (Example ii), they are not the same. In the previous example, the point was decentralizing the economy, letting each individual see to himself. In the current example, the point is the formation of a price system. Logically, the examples are independent; a centrally planned economy can have a price system, and a decentralized economy can work on barter.

(iv) **Chess.** The ideal way to play chess is to plan the entire course of play from the beginning, taking into account anything the other player might do. In practice, this is beyond the power of any man or machine. Instead, the players "evaluate" the situation at each move, using numerical indices for the various pieces; e.g., 8 for the queen, 5 for a rook, 3 for a bishop or knight, 1 for a pawn. They also take account of the general characteristics of the position: development of the pieces, castling, passed pawns, and so on. Using such criteria, each player "looks ahead" for a few moves, trying to maximize his valuation of the position at the end of that period, and taking into account that the other player is doing likewise. Weaker players often do not look ahead more than a move or two, and even that only partially. Stronger players may look as much as five or six moves ahead, and sometimes even more; but they, too, do not examine all possibilities—all "branches of the tree"—in the process. Human players do not use a precise numerical valuation, but take a generalized view. Chess-playing computer programs basically do use a precise numerical valuation function; though even there, the "depth" of the look-ahead varies, with some branches of the tree being examined more thoroughly than others.

Here the complex task is planning the whole game beforehand; and the simpler, decoupled tasks are playing move by move, with a more or less modest look-ahead. The overall, macroscopic driving force is, of course, the desire to win; but "microscopically"—at each separate move—the driving force is to maximize the valuation several moves ahead.

In Examples (ii) and (iii) above, the decoupling is achieved by spreading the task over many individuals, each with his own motivation. In contrast, in this case the decoupling is achieved by spreading the task over time. There is a single overall motivation—winning—which is expressed at each move by looking ahead a few moves and maximizing valuation.

(v) **Solomon's judgment.** Rather than rendering his judgment (1 Kings 3, 16–28), King Solomon could have entered into a complex factual investigation of the women's claims. He "decoupled" the process by motivating them unwittingly to reveal the truth themselves. Here the complex task is determining which woman is the live baby's mother; the simpler tasks are for the women to express their preferences given the judgment; and the driving force is the women's motherly love.

(vi) **Fair division.** This may be achieved by cumbersome methods of direct measurement. Alternatively, the process may be decoupled by the method of "one cuts, the other chooses," which motivates the parties themselves to divide fairly.

(vii) **Evolution.** Suppose a Creator had wanted to create the living world as we know it. He could have designed each individual organism separately, together with the appropriate interactive adaptations. This would have been enormously complex. Alternatively, He could have decoupled the process by means of evolution—survival of the fittest—which runs by itself, automatically, with no need for "hands on" direction. Here the complex task is creating the world; the decoupled alternative tasks are for each organism—or even each gene—to adapt to its surroundings; and the driving force is survival of the fittest.

These examples should give the reader some idea of what we mean by decoupling. What we now suggest is that consciousness serves as a decoupler that allows human beings to perform tasks that otherwise would be unachievable because of their complexity. Let us illustrate.

(a) **Food.** The body needs food for energy and other vital purposes. The process of supplying food may be divided into two parts: (1) before it enters the mouth, and (2) afterwards. Part 1 consists of triggering the individual to seek food, and acquiring, preparing, and eating it. Part 2 consists of digesting it, i.e., transforming it into a state that the body can use (see Section 1, above).

Both parts are extremely complex, but Part 1 is by far the more so. We must choose the food to buy, shop for it, store it, clean it, cook it, and serve it. We must also choose and buy closets and a refrigerator to store it; sinks to wash it; stoves, ovens, gas tops, microwaves, pots, pans, cake forms, pie dishes, utensils, mixers, vegetable peelers and so on to prepare it; plates, bowls, platters, and utensils, to serve it. We must earn money to buy the food and all the auxiliary items we have mentioned, and those we have not mentioned. Earning money usually involves various skills—not the least of them social—which must be acquired.

There is another fundamental difference between the two parts. Part 2 is "hard-wired": it works "automatically," by mechanical, chemical, and electrical means, with no conscious, voluntary component. Part 1 is precisely the opposite: all conscious, all volitional. To start with, when the body runs short of the required nutrients, it must be prompted to eat. Once this is done, the food must be acquired, prepared, and eaten. Conceivably, parts of the process *could* be hard-wired: a gauge could tell the brain when the stomach is empty; the eye and brain could identify food, then send signals to the hand automatically to take it and put it in the mouth. But it seems unlikely that acquiring the food and preparing it, and earning the necessary money, could all be hard-wired.

How, then, does the process work? What drives it?

The answer is simple, even obvious: hunger. And, the other side of the same coin: enjoyment of food.

Hunger does *not* mean an empty stomach. An empty stomach by itself will not prompt us to eat. We need the pain or discomfort of hunger, and/or the pleasure of food, to make us eat. Pain and pleasure depend on consciousness. If you are not conscious, you cannot experience pain or pleasure. So consciousness is an important component of the mechanism that supplies us with nutrients. Moreover, it decouples the nutritive process into the two parts we have described: until the food reaches the mouth, and afterwards.

Indeed, it does much, much more: the entire process of acquiring, preparing, and serving food is decoupled into many small steps. Each step is conscious, with a well-defined goal; it is *motivated.* Going to market, picking the items to buy, standing in line at the check-out counter, bringing the items to your car, unloading, putting in the refrigerator, all the myriad tasks involved in cooking, all the myriad tasks involved in earning the money with which to buy the food, all the myriad tasks involved in all stages of the process—all together, and each one separately—are driven by hunger, through the medium of consciousness; they are *motivated.*

More precisely, the experiences of hunger and food enjoyment are the overall, macroscopic driving forces, like the desire to win in chess (Example iv above). The thought process, which is the second component of consciousness (see Section 2, Item viii), translates this into many small tasks—making money, baking a cake, and so on—each with its own driving force, like valuation-maximizing in chess. And then volition—the third component of consciousness—comes into play, enabling the individual actually to carry out these individual tasks.

It is important to note that while "hard-wired" processes like digestion may be highly complex, they are fairly repetitive. Digestion works on the same materials, in the same way, every day; there are few, if any, surprises. The processes of food acquisition and preparation are much less repetitive; they require a good deal of instantaneous adaptation to various different environments, environments that may be unfamiliar. Consciousness is particularly important for motivating and carrying out these non-repetitive tasks.

Finally, we remark that the mechanisms provided by nature to facilitate eating—hunger and the enjoyment of food—may sometimes "misfire." It is well known that severely undernourished people whose hunger leads them to overeat may well die as a consequence; there are documented cases of people who survived the concentration camps during the Holocaust, only tragically to die in this way. Even better known is that the enjoyment of food may cause people to overeat or to eat foods that are not nutritious. While not immediately fatal like with the concentration camps, this may nevertheless be detrimental to health; or at least, serve no useful adaptive purpose.

(b) **Sex.** Biologists identify two basic drives in living organisms: nourishment and reproduction. What we have said about food applies, *mutatis mutandis*, also to sex. It is hunger and the enjoyment of food—*not* the need for nourishment!—that makes people eat. Similarly, it is *not* the desire for offspring that makes them have sex; it is the sex drive—the enjoyment of sex. Enjoyment is a function of consciousness. You cannot enjoy if you are not conscious. Many people do consciously want offspring, but that is not why they have sex.

As with nourishment, the process of reproduction has several distinct parts: (1) before the semen enters the woman's body; (2) conception and pregnancy; and (3) childhood, when the offspring cannot fend for itself. The first two parts are decoupled by the sex drive. The second part, though highly complex, is hard-wired; once sex has taken place, there is no conscious intervention until birth. After birth, parental love takes over; like hunger and the sex drive, this depends on consciousness. As with food, the first and third parts of the process—which are not hard-wired—split up into a myriad of distinct small steps, starting with dating and earning the required money; each is conscious, each has a well-defined goal, each is motivated.

Like with food, the hard-wired part of the process is fairly repetitive, the other parts not.

Unlike food, sex is programmed to be profligate. In a single season, a single flowering tree produces many hundreds of thousands of blossoms, and billions of grains of pollen. It is doubtful if even one of these comes to fruition. In each ejaculation there are hundreds of millions of sperms, at most one of which is used. Many sexual episodes lead nowhere, many relationships lead nowhere, many dates lead nowhere. The sex drive leads to many activities that have no chance of producing offspring: sex with birth control, sex after the reproductive age, homosexuality, masturbation, oral sex, bestiality, pornography, and so on. Love and sex play dominant roles in advertising, literature, film, music, painting, photography, dance, almost all cultural activity. In the case of food, we used the term "misfire" to describe situations in which food does not provide nourishment. In the case of sex, the corresponding situations are so ubiquitous that they must be considered a part of the process.

(c) **Pain.** It has long been recognized that pain calls attention to something in the body having gone wrong, so that it can be attended to. Again, pain depends on consciousness; if you are not conscious, you cannot experience pain. Machines cannot suffer.

As before, we have here a decoupling process. Pain motivates the individual to seek medical or surgical treatment. Unlike with food and sex, though, the effect of treatment is not entirely automatic; repeated intervention may be required. Pain decouples the treatment process into many distinct small steps, motivating the individual at each step to do what is required in order to alleviate his condition. It, too, may "misfire," as when a medical or surgical treatment causes more pain than what the patient can suffer.

To summarize: consciousness enables the decoupling of highly complex, non-repetitive tasks into many simpler tasks, mainly through the element of motivation.

4. How

This is the shortest of our sections: we have little to contribute on this score, other than to say that the neurological phenomena that have been observed to be associated with consciousness do not explain how it works; the "how" remains a deep mystery.

One last remark is worth making. "How" questions are usually answered by analogy with something else, with which we are familiar. For example, the workings of the circulatory system are explained by analogy with plumbing. But consciousness is unique; there is nothing else in the world that is even remotely like it. Since there is nothing like it, what analogy can we use to explain it? And if not by analogy, how else can it be "explained"?

5. Summary

Consciousness is the last great frontier of science. This chapter discusses what it is, how it differs fundamentally from other scientific phenomena, what adaptive function it serves, and the difficulties in trying to explain how it works. The emphasis is on the adaptive function.

6. References

Ernst Mayr, E., (1982) *The Growth of Biological Thought*, Belknap Press, Cambridge, Massachusetts.

Biodata of **Giuseppe Laras** author of ***"Afterlife in Moshe Maimonides's thought"***

Rabbi Dr. Giuseppe Laras is an internationally renowned medieval philosophy scholar, lecturer, and author. He attained rabbinical ordination at Rabbinical College of Rome after studies carried on in Turin Rabbinical College and Jerusalem. He obtained degree in Law and Philosophy, than Ph.D., in Philosophy from Milan State University in 1975. Chief Rabbi in several Jewish Communities: Ancona, Leghorn (Livorno) and, for the last 25 years, Milan. He is actually Professor of History of Jewish Thought in Milan State University and President of Rabbinical Council of Italy. His outstanding contributions to Italian and European Jewry, to the support of Israel abroad and to Inter-religious Dialogue over the course of his career have made him one of the leading figures of today's Jewish world. Rabbi Laras is renowned for his studies on Medieval Philosophy, especially on Maimonides. He is the author of the following books: *Moshe Maimonides' philosophic thought* (1998); *Moshe Maimonides, the eight chapters: ethical teachings* (2001); *Science and God Providence in Maimonides and Gersonides* (1999); *Disputations among Jews and Christians in the Middle Age* (2000); *Immortality and Resurrection in Medieval Jewish thought* (2001); *The book of Qohelet* (2002); *Jewish mysticism and kabbalistic thought of Zohar* (2003); *Immortality and Resurrection in Maimonides*. He contributes to several magazines of religious and Jewish subjects.

E-mail: **ravlaras@yahoo.com**

J. Seckbach (ed.), Life as We Know It, 565-575.

AFTERLIFE IN MOSHE MAIMONIDES'S THOUGHT

RAV GIUSEPPE LARAS
President of Rabbinical Council of Italy,
Professor of History of Jewish Thought,
State University of Milan, Italy

1. The Soul and the World to Come (Olam Ha-Ba)

The soul is a spiritual substance whose origins are from another world. It enters a body, and, after death, leaves it and returns to the world from which it came. This idea of the soul, in its extremely synthetic form, is shared by Judaism and the other monotheistic religions. Nevertheless it is quite alien to many ancient thinkers – including Democritus, Epicurus, and Lucretius – because they reject the existence of an autonomous ontological reality that comes from another non-corporeal dimension and that enters a body only to abandon it at death.

From a materialistic perspective, the abovementioned idea of the soul does not exist, and when some philosophers speak about the soul, they mean the natural breath of the body that ceases to exist when the body dies.

In the scholastic thought from the Middle Ages, Jewish, Muslim, and Christian theologians adopted a conception of the soul inspired by Aristotle and his double category of *potency* and *act*. According to this concept, the soul is a mere "disposition" or is potentiality able to find fulfillment in *act* only through certain conditions that depend on an individual's behavior. In particular, the soul may become *act* insofar as it acquires knowledge of and meditates on the most important metaphysical concepts known as *Intelligibilia* (in Hebrew, *Musqaloth*): God, His attributes, His eternity, *creatio ex nihilo*, Providence, etc. When this level is achieved, the soul becomes immortal.

A similar concept of the soul is common throughout Medieval thinking, albeit with various nuances, and is found in the writings of such philosophers-theologians as Avicenna and Averroes in Islam, Albert the Great and Thomas Aquinas in Christianity, and Maimonides in Judaism.

In his introduction to the Sanhedrin, chapter 10, Maimonides examines the ideas of *Good* (*Tov* or Reward) and *Evil* (*Ra'* or Punishment), which are bestowed to human beings after they die according to their good or evil actions (if they live according to the commandments and the will of God or not). In this text, the Rambam presents five opinions held, in a more or less explicit manner, by five categories of people according to their intellectual capacity.

He identifies five opinions on this respect that are followed, in a more or less explicit way, by a corresponding number of categories of people. Each group is defined by intellectual capacities of supporters.

The first opinion maintains that *Good is represented by the Garden of Eden, the place where people eat and drink without physical labor or effort, where there are houses made from precious stone with beds draped with silk linen, where wine and perfumed oils flow freely, and other similar luxuries can be found...and Evil is Hell*[1], *a place full of fire, where corpses burn in perpetual flames, and people are tormented by all sorts of sorrows and pains so great that it is impossible to describe them.*

The people in the category that adopt this conception base their argument on expressions from the Scripture and the teachers, that are more or less literally interpreted based on need and convenience.

According to Maimonides, however, these expressions, beyond their literalness, are symbolic and were written to touch the hearts of the simplest believers, but they cannot be taken literally since their function is only to describe a condition of either great joy or pain. They must be interpreted solely in a spiritual sense, just as reality of the world to come is spiritual.

The second opinion identifies *Good* with the Messianic Era. During this time, *all humans will be like angels, alive and well for eternity.* Maimonides describes this new and extraordinary world with the image of the giants, mentioned in Genesis 6:1- 4, and with the passage from the Talmud Shabbat 30b, that portrays a land able to produce "baked bread and fine woolen clothes." *Evil* implies not being allowed to participate in this future time for those who do not deserve to live and contemplate it. Even this second opinion is supported with references to Scriptures and the Talmud; interpretation of texts is strictly literal.

The third opinion asserts that the expected *Good* consists in the *resurrection of dead. After death, humans will live again and share the company of loved ones, relatives, and friends, eating and drinking in a life without end.* On the other hand, *evil* – the feared punishment – consists in *not being resurrected after death together with those who will be resurrected.* In line with the most widespread belief of resurrection, this thesis maintains that the resurrected will die a second time and will pass *in corpore* after the resurrection to the eternal dimension of the world to come.

The forth opinion sustains that *Good* will come as a consequence of the observance of the commandments and will consist in resting the body and satisfying material desires in this world, which may include *fertility of the land, wealth, numerous progeny, longevity, the health of the body, peace and tranquility, the sovereignty of the King in Israel, and control over those who are hostile to us*. *Evil* is the opposite of the *Good* described above, *precisely like the times we are living now in exile (Galut).*

---

[1] It is cited in the text as "*Gehinnòm*" (Gheènna). This name defines the valley at the foot of Jerusalem's walls, also called Tòfet. According to an old tradition, it is the valley where resurrection will take place. This place is also named *'Emeq Refiim* (Death Valley) since it was a burial ground. In more ancient times, the valley was a garbage dump where fires were kept burning to consume the refuse and reduce the stench. From this image came the representation of flames in hell. The expression mentioned in Joel's prophecy (4:2) to define the place where judgment will take place after resurrection is, on the contrary, *'Emeq Jehoshafàt* (Valley of God who is Judging).

Biblical and Talmudic verses are again quoted in support of this fourth opinion, which seems oriented toward materialism and hedonism. In particular, the quoted texts are based on benedictions and curses described in the two passages of the Torah, Leviticus 26: 3-46 and Deuteronomy 28: 1-69, which deal with the success (and the sorrow) that will be bestowed on the people of Israel if they follow (or disobey) the commandments.

The fifth and last opinion, which attempts to put together the points described above, states that *the Messiah is coming, and He will resurrect the dead who will then enter the Garden of Eden, where they will eat, drink, and be in good health for all of eternity.*

After having analyzed these five opinions, Maimonides makes a fundamental critique: we are unable to find in any of them a deep evaluation of the true reality of the world to come; what is missing is a reflection that explores the spiritual – and not just the material – dimension.

Questions, which worry those who think about the world to come and, in particular, about resurrection, are already discussed directly in the Talmud. Will the dead be resurrected with or without clothes? Will these clothes be the same with which a person was buried, or will there only be a simple shroud? After the coming of the Messiah, will there still be rich and poor, strong and weak? (see Sanhedrin 90b).

In light of his spiritual sensitivity combined with a rigorous conceptual and methodological rationalism, Maimonides deplores the attitude of those people, both ignorant and educated alike, who are more concerned with the material aspects of the world to come and resurrection than with their spiritual realities.

Maimondies quotes Antignos of Sokho (Avot I, 3): "Do not be as servants serving the master in order to receive a reward. Rather, be as servants serving the master not in order to receive a reward. And be the fear of Heaven upon you." Then he comments: "*That pious man intended to say that we must not serve God with the expectation that He bestows benefits and mercy to us, as in this case we would be moved by expectance of a reward and we would serve Him only for that. On the contrary, we have to serve Him just as those servants who do not expect any reward or benefit for their service. I want to say with this that we have to serve Him moved by love as I have already explained in the tenth chapter of Sanhedrin*".

This image of *service moved by love* is the model of behavior also recommended by Maimonides. But even if he accepts the concept as his own, he does not take the drastic position of Antignos. In the *Laws of Repentance*, (X, 6), Maimonides asserts that humans have to study and observe Torah *for the sake of the Torah*, that is, without any other intention or goal other than fulfilling the will of God. He clearly explains that "one who serves God without other purposes" is the one who studies and observes the Torah not for fear of being punished or with the hope of gaining a reward, but only for love of God, who gave us His teaching. Nonetheless Maimonides is aware that such an attitude needs intense ethical and spiritual strength. Although he affirms this concept, he also quotes a famous Talmudic passage (Pesachim 50b): *man has to observe the Torah first for fear, since afterwards he will succeed in observing it for itself.*

The above passage conveys a profound principle of religious pedagogy: if someone starts from a lower level of observance with the purpose of receiving benefits (*service for fear)*, then it is possible for him to reach a higher level, similar to the one taught and followed by Antignos (*service for love)*.

In the same passage, Maimonides examines the theme of love for God. Just as Antignos mentions the *service for love*, the Rambam explains the meaning and modality of this love for God and how it should be manifested. In doing so, he offers an original concept, which is alien in Antignos' thought. He writes: *The love of God will not become attached within a person's heart until he becomes obsessed with it at all times as is fitting, leaving all things in the world except for this. This was implied by the command [Deuteronomy 6:5: Love God, your Lord,] with all your heart and all your soul.* And he adds: *One can only love God [as an outgrowth] of the knowledge with which he knows Him. The nature of one's love depends on the nature of one's knowledge. A small [amount of knowledge arouses] a lesser love. A greater amount of knowledge arouses a greater love.*

According to Maimonides, the aim of life is *knowing God*, and immortality can be achieved through meditation on God Himself, as mentioned above. If love for God is a kind of love that comes from one's intellect, the meaning of the expression *loving God* is the same as *knowing God*, and *knowing Him* is equivalent to *loving Him*[2].

He affirms that in order *to know God*, it is necessary to have knowledge of some preparatory sciences, such as logic, astronomy, mathematics, physics, medicine, and philosophy. This concept, together with the idea of *intellectual love*, cannot be found in Antignos of Sokho, even when this teacher shares Maimonides's ideal of obedience to God's commandments without any expectation of reward. It is difficult to find a thinker like Maimonides within Jewish thought, not only in medieval times. He is unique because his convictions are based on the pursuit of philosophical knowledge together with religious piety and an acceptance of his people's traditions.

Maimonides also describes the nature and the quality of bliss in the world to come. He makes a clear distinction between happiness that can be experienced and known in this earthly world and happiness reserved for the righteous people in the world to come. The body is unable to feel spiritual happiness just as a blind person cannot perceive colors and a deaf person cannot hear voices. Spiritual joy cannot be experienced in this material world just as fish ignore the element of fire. During this life, people can only perceive material pleasures, *vile* and *ephemeral*, and they are unable to experience spiritual joy, *eternal* and *incessant*. He concludes that there can be no relationship between the world of material pleasures and that of spiritual joy.

In the already mentioned *The Laws of Repentance* (VIII, 2), Maimonides writes: *In the world to come, there is no body or physical form, only the souls of the righteous alone, without a body, like the ministering angels*. In his commentary to a well known Talmudic passage (Berachot 17a), he affirms that *in the world to come, there is neither food nor drinks nor procreation...but righteous people are seated with crowns on their heads. They enjoy the glory of divine Presence.* The expression "*with crowns on their heads*" has a metaphysical meaning for Maimonides; it is

---

[2] This Maimonidian statement of the necessary link between knowledge of God and love for God echoes the *amor intellectualis Dei* of Spinoza. However Spinoza's God is opposite from the biblical concept of God in which Maimonides believed.

indicative of *the immortality of the soul in the eternity of the Agent Intellect – God.* As discussed above, Maimonides believes that the soul reaches immortality through intellectual activity undertaken during earthly life. The Agent Intellect is the final intelligence emanating from God, which is eternal like God Himself.

The passive intellect, given to humans at birth, becomes one with the Agent Intellect through meditation and contemplation on metaphysical truth (*Musqaloth*). As such, it is able to reach the level of *act* and is transformed into acquired intellect. The latter – acquired intellect – once unified with the Agent Intellect through the conception of eternal ideas, also becomes immortal. For the Rambam, this is the meaning of the immortality of the soul. It is precisely the activity of thinking about eternal realities that causes the soul to be eternal, thus producing a substantial identity between the thinker, the activity of thinking, and the object of the thought. This process resembles the Aristotelian god, who is conceived as the complete identity between the intellect, the intelligent, and the intelligible (*Guide for the Perplexed* I, 68).

The sentence "*righteous people enjoy the glory of Divine presence*" means that *those souls enjoy their perception and their knowledge of the true essence of God, blessed be He, in the same way the holy Chayoth* (see Ezekiel I) *and the higher angelic hierarchies do*. The highest good and the ultimate end of existence consist in being part of this supreme assembly, so sharing in its honor and dignity. In this way, the soul will continue to exist through eternal contemplation of God, from where the soul came.

For Maimonides, what was just described *is the greatest good to which no other good or joy can be compared. How can endless eternity be compared to what is ephemeral and temporary?*

If happiness and good are represented in the concepts above, suffering and evil are then the *annulment* (*Qarèth*) and annihilation of soul. Evil means that the soul will not survive or exist anymore.

Therefore punishment consists in the exclusion of the soul from the bliss of Divine contemplation, thus producing suffering and damnation. Isaiah (64:3) confirms the impossibility for corporeal senses of perceiving the world to come: "*Since the beginning of the world...neither has the eye seen this besides You, o Lord.*"[3]

2. Resurrection of the dead

Maimondises' *Treatise on Resurrection (Ma'amar Tehiyyat Hametim)*, composed in 1191, was written impulsively in order to justify his position on this fundamental principle of faith. He was informed by some his students and admirers that the Rosh

[3] In the Talmud Berershit 34b, Rabbi Yochannan refers to a tradition according to which all visions of prophets, concerning modification of the natural order of the world at the end of time, actually only refers to messianic times and not the world to come (*Olam Ha-ba*). Of the latter, it is said: "*Since the beginning of the world...neither has the eye seen this besides You, o Lord.*"

Yeshiva of Baghdad Rabbi Samuel Ben Elì was criticizing him and insinuating that he did not believe in resurrection.

Maimonides felt the urgency to calm his followers and reject allegations and suspicions of his opponents, some who were acting out of sheer malice while others did not have a profound understanding of his arguments. So he decided to compose this treatise to confute accusations. The particular mood of the author, which appears in this work, cannot be found in any other of his writings. While all of his books, treatises, commentaries, *responsa*, or letters are composed in a calm and peaceful style – almost with philosophical detachment and serenity – *Treatise on Resurrection* conveys a strong sense of irritation and clear emotional involvement, even though the it is well argued from a logical point of view.

It is evident he was disturbed by such accusations and that he had to prove his "innocence" at the end of his life precisely on a principle of faith that he had already often explored in his writings. He never failed to emphasize the religious importance of the resurrection of the dead, even though the dogmatic content of this principle was unsuitable for deep and sophisticated exegesis.

As a fundamental principle of Judaism, resurrection of the dead is presented in two different points in the *Introduction to the Tenth Chapter of Sanhedrin*: in one, it is discussed in detail, while in the second, it is only mentioned.

In the first point: *Resurrection of the dead is the base of all principles of the Torah of Moshe, our Teacher. He who does not believe in it has no religion and no relation to the People of Israel. Resurrection is reserved for righteous men as it is reported in Bereshith Rabba: Abundant rains are for both righteous and wicked men, but resurrection of the dead is only for righteous men. How can wicked men can be resurrected if they are even dead during their life?! Our Teachers explained about this: A wicked man is called dead even when alive; a righteous man is considered alive even after his death! Everyone must be conscious that man has to die and be decomposed in the elements he was made from.* The second point: *"The 13th Principle: Resurrection for which we have already spoken."*

Maimonides also mentions this principle of faith in his most well known halachic work *Hilchoth Teshuvah* (III, 6), which is part of the *Mishneh Torah*: *"The following individuals do not have a portion in the world to come. Rather, their [souls] are cut off and they are judged for their great wickedness and sins forever: the* Minim, *the* Epicursim, *those who deny the Torah, those who deny resurrection of the dead and the coming of the [Messianic] redeemer."*

Maimonides mentions this principle more than once; even if the exposition is synthetic, its meaning is so clear and unambiguous to consent no doubt, no misunderstanding, nor suspicion. Resurrection is a principle of faith that is difficult and complex. Therefore it is predictable that it could have been misunderstood and mixed up with other principles that are less difficult understand rationally, even though they still belong to the field of eschatology (*Acharìth Ha-Iamìm*/Last days). Together with criticism that was sincerely exposed in good faith by some of his opponents, other accusations also flourished in bad faith with polemical intentions. Little by little, a violent tempest gathered over Maimonides' head. When signals became more and more evident, he was obliged to compose a work expressly explaining in detail contents, meanings, analogies and characteristics of this fundamental article of faith in order to put an end to all misunderstandings, exploitations, and accusations.

It must be said that Maimonides' conception of resurrection is truly different from those traditionally accepted until his time. Before Maimonides, faith in the immortality of the soul and in the world to come was strictly linked to faith in resurrection. The world to come manifests its existence as an eternal and spiritual reality in the precise moment when those who are sleeping in the dust reawaken from death and are thus resurrected. Such resurrection or reawakening coincides with the beginning of the Messianic era when the Messiah, a descendant of King David, performs many miracles, among them the resurrection of the dead. At this very moment, the world to come will come into existence as the place where resurrected people will live forever with their bodies or their reconstituted psycho-physical unity.

Maimonides rejects such a conception, which was common during his time, and develops a new one that is based on both the tradition of some and on specific philosophical assumptions. First he makes a net separation between *life in the world to come*, which is exclusively for the souls, and *life in the world of resurrection*, which is only for the bodies. Second, he makes a clear separation between *world of resurrection* and the *Messianic era* and affirms that resurrection is not necessarily related to the coming of the Messiah or to His miracles. Resurrection does not depend on the Messiah but only on the impenetrable will of God. Only God can decide if and when resurrection will take place. Maimonides accepts Rav Shemuel's conception of the Messianic era: "*The only difference between the Messianic era and this world will be that the Jews will not live under foreign rule*" (Ber. 34b).

According to this statement, no miracle or modification in the natural order of the world will mark the coming of the Messiah. Instead the Messiah will come based on God's will. The Messiah will be filled with the desire of knowledge and fear of God, and he will "*reconstruct David's Kingdom on the earth of Israel, rebuild the Temple, and gather all of the scattered remains of the Jewish people from the four corners of the world.*"[4] Humanity will devote itself exclusively to the quest for God, while the fear of wars will disappear.

Maimonides explains that waiting for the Messianic era does not come from the desire "*to rule the world, to enslave nations, and to receive homage of peoples...but instead from the willingness to consecrate oneself to studying Torah in peace and without oppression in order to achieve eternal life in the world to come.*"[5]

Content and the exact meaning of resurrection are succinctly explained in the *Treatise*: when God wants and when His desire shows that the time has arrived, the dead will wake up and will be resurrected into a new life, just as it is foreseen by Daniel (12:2) and other teachers.

Maimonides considers the abovementioned citation of Daniel as the only reference to resurrection that must be taken literally. On the contrary, the well-known chapter 37 of Ezekiel is not so specific and determinative. Ezekiel's image of

[4] Mishneh Torah, Treatise on Kings, XI, 1.
[5] *Ibidem* XII, 4.

bones coming together, one by one, and the *spirit* descending in re-formed bodies may be a symbolic representation of the renaissance of the nation of Israel after "the apparent death" of exile and not of the allusive paradigm of resurrection.

All of the bodies – which after dying had deteriorated, scattered, and melted into the four natural elements – will be re-formed, and all of the organic components, which formed the original body, will be reassembled. Afterwards the original souls will enter their newly reconstructed bodies to revive them. Re-born individuals will have a new life.

It is clear that the process of regeneration and reconstruction induced by resurrection cannot be seen in the phenomenic world of nature, nor can it can be explained with the ordinary parameters of reason.

It is manifestly a miracle (the greatest miracle, according to Maimonides) that will happen by God's will sometime in history.

Time is not clearly defined but should probably correspond to a segment of the Messianic era, although not necessarily so.

The problem for believers, however, is how a religious fundament can be accepted and believed when it is in total and manifest conflict and irreconcilable with principles of reason.

Following the opinion of the renowned teacher Saadia Ha-Gaon[6], Maimonides observes that believers accept more easily the miracle – among the numerous miracles described in the Torah – of the *creatio ex nihilo*, although it objectively brings greater difficulties to human reason, compared to resurrection. The acceptance of the creation of *being* from *not-being* – that is, the acceptance of creation of the phenomenic reality of physics *from nothing* – seems to bring greater logical difficulty when compared to the reconstitution of bodies after the dispersion of their elements. In this last case, God would reconstitute the bodies with a material that exists, although it is scattered and transformed. On the contrary, during the creation of the world, this material must be derived from *non-existence*. On the basis of this schematic and simplifying reasoning, the Rambam concludes that it should be *less difficult* to accept the dogma of resurrection than the dogma of the creation of the world!

What is the nature and destiny of resurrected psycho-physical unities?

Maimonides considers such resurrected unities as human beings in all respects, with fully functioning bodies. They will eat, drink, procreate, and enjoy lives that are particularly happy and long "just like all of those who will be alive during the messianic era." But in the end, they will also die just like all human beings.

The Rambam firmly rejects any opposing opinions held by his contemporary opponents: that human beings, once resurrected, would never die again. He confutes this thesis with both religious and philosophical arguments. Just as God does not create anything in the world that has no utility, he cannot create organs in the resurrected bodies that are unable work.

If the opinion – that resurrected bodies can enter into the world to come *as they are*, thus unable to perform their organic functions – is accepted as truth, as his

---

[6] *Book of Doctrines and Beliefs* (Sepher ha-Emunoth ve-ha-De'oth): VII, 1.

opponents sustain, then it would mean that God created those organs as something superfluous. But this is evidently impossible. The world to come cannot receive material bodies but only souls deprived of materiality. For this reason, resurrected bodies must also die again, even after a very long life.

At this point, the misunderstanding that arose between Maimonides and his opponents becomes clear. They ignored or were unable to understand or to admit the originality and the innovation of this conception, where resurrection and the world to come are put on two different levels, therefore accusing Maimonides of not believing in resurrection.

We can take as an example the critical commentary written by Rabbi Abraham Ben David from Posquiéres (Raavàd) on Maimondes's *Laws of Repentance.* (VIII, 2). The Rambam writes: *In the world to come, there is no body or physical form, only the souls of the righteous alone, without a body, like the ministering angels.* The Raavàd comments: *I have the impression that the words of this man* (sic!) *want to affirm that the resurrection of the dead does not concern bodies but only souls. God* forbid! But Maimonides does not deny resurrection of the bodies, as most commentators usually want to highlight *in loco*, but only intends to affirm that resurrected bodies would die a second time and that only souls will enter the world to come for eternity.

In other words, resurrection is not a final aim as maintained by those who identify resurrection with the world to come. Rather, the final aim is the spiritual and eternal dimensions of the world to come, where all of the souls of righteous people will converge after losing their bodies for the second time.

3. Acknowledgement

I would like to express my appreciation to Prof. Massimo Giuliani and to Dr. Guido Borella for their generous dedication to the translation of this text into English.

Biodata of **Yuval Ne'eman** author of ***"Science and Religion: Some Suggestions for Conceptual Harmonizing Constraints"***

Professor Yuval Ne'eman was born in Tel-Aviv in 1925, a grandson of one of the city's founders. He studied mechanical and electrical engineering at the Technion (Israel Institute of Technology), Haifa, then worked as a hydrodynamical designer for a year. He joined the Hagana and when this organization gave birth to the Israel Defense Forces, he stayed on for 12 years, fighting Israel's 1948 War of Independence as a commander in the field and serving later as a Vice Chief of Operations at the High Command. As Colonel, Head of Defense Planning he laid the foundations of Israel's mobilization system and crystallized the strategic conception prevailing until 1967 and tested in the Six Days War. Ne'eman turned to science in 1958, studying at Imperial College in London while serving as Israel's Defense Attache in the U.K. In the early sixties new particles were being discovered and the number of different 'elementary' particles was nearing one hundred. Ne'eman identified the pattern and his 'SU(3)' classification (also known as the "Eightfold Way") has been compared with Mendeleyev's Periodic Chart of the Chemical Elements. It was experimentally validated (1964) when the Omega-Minus particle, predicted by this scheme, was observed at Brookhaven N. L. (USA). Ne'eman was also the first to suggest that particles experiencing the Strong Nuclear Force, such as protons or neutrons, are composite and are made of three fundamental "bricks" - later to be known as Quarks, when Gell-Mann (who had also later independently arrived at SU(3) though he never published it) and Zweig further developed the notion. Ne'eman contributed to important further advances in particle physics, cosmology, astrophysics, and epistemology.

Since 1977, he has developed an interest and a parallel line of work, generalizing the theory of Evolution and formulating a universal paradigm, then also applying it to various areas, especially Social Anthropology ad Evolutionary Epistemology.

Ne'eman was the founder and Head of the Department of Physics and Astronomy at Tel-Aviv University (1965-72), President of Tel-Aviv University (1971-75), and served as Director of the Mortimer and Raymond Sackler Institute of Advanced Studies in that institution (1979-1997). In 1969 he established the School of Engineering as founding dean (in 1997-2002 he was elected President of the Israel Association of Engineers). He also founded (1968) the Center for Particle Theory at the University of Texas (Austin). He founded the Israel Space Agency (1983) and has chaired it since. He has also served on Israel's Atomic Energy Commission (1965-84) and chaired it (1982-84) and held the position of Scientific Director in the IAEC Soreq Establishment (1961-63). Ne'eman was Israel's Chief Defense Scientist in 1974-76. He served as President of Israel's Bureau of Standards in 1972-1976 and chaired the Steering Committee to the Med-Dead Conduit Project in 1977-83. He was Israel's first Minister of Science and Development 1982-84, then again in 1990-1992, when he also served as Minister of Energy.

J. Seckbach (ed.), Life as We Know It, 577-587.

Ne'eman has published over three hundred and fifty scientific papers and twenty books. He is a Member of Israel's National Academy of Sciences, a Foreign Associate of the National Academy of Sciences of the USA, a Foreign Honorary Member of the American Academy of Arts and Sciences and a Member of the International Academy of Astronautics and several other Academies and Learned Societies. He has been awarded the Einstein Medal (Washington, 1969), the Wigner Medal (1982) the Israel Prize (1969), and Birla Science Award (Hyderabad, 1998) and honorary doctorates of universities in the U.S.A., Germany, Russia and Israel.

E-mail: **matildae@tauex.tau.ac.il**

SCIENCE AND RELIGION:
Some suggestions for Conceptual Harmonizing Constraints

YUVAL NE'EMAN[1]
Emeritus Professor of Physics
Tel-Aviv University, Tel-Aviv, Israel 69978

1. The Background - Some Historical Clashes

Clashes between Religion and Science have occurred on a number of occasions in Western cultural history. Examples range from the well-known trial of Galileo Galilei (Cane, 1959) to the less known burning of Giordano Bruno by the Catholic Inquisition for spreading Hasdai Crescas' (Ne'eman, 2003) model of an *infinite vacuum* (both concepts were taboo in the dogmas of Scholasticism, following Aristotle's views) as a background scene for the physical universe – a model that provided the necessary conceptual background for Galilei's work on inertia and Isaac Newton's First Law. More recent examples are provided by the anti-Darwinist Fundamentalist "monkey-trials" in the American "Bible-Belt" with the subsequent development of the *pseudo-scientific counter-doctrine* of "*Creationism*" (NAS, 1999) – and to very recent modern versions of the same (Berlinski, 1996 and 1998) These were attempts to reconcile the unconciliatable, namely *fundamentalism* and science, and could therefore be done in one direction only, i.e. enable fundamentalists to absorb some of the results and fruits of modern science without having to abandon the biblical view of creation, a view some elements of which reflect the status of observation and interpretation in Mesopotamia as of 2,000 BC. or thereabouts, when the Jewish faith was in its formative stage.

Clashes of a similar nature have occurred between Science and *totalitarian ideologies*. The best-known such example, also directed at Neo-Darwinism, though in the opposite (Lamarckian, or at best the neo-Lamarckian) direction which inspired Bernard Shaw's "Back to Methuselah", was the doctrine of *Lysenkoism*, then part of the USSR state-ideology; a similar confrontation occurred in the Earth-Sciences between Geophysics and *Belyusefism*, another pre-Glasnost USSR state-doctrine, which rejected Alfred Wegener's Continental Drift paradigm. There have also been clashes in Physics, e.g. the motto "Quantum Mechanics is a bourgeois mystification designed so as to hide the truth from the working class" of the Italian branch of the 1968 European students rebellion.

[1]Dedicated to granddaughter **Limor**
Who now learned the Laws of Man –
Best wishes, success and much more....
From Saba, still learning those of Nature, Amen!

Another such rejection involved the treatment of Relativity in Nazi Germany. Here the action was already at its worst within the scientific establishment itself, it did not need the state's intervention. It was lead by prominent mathematicians such as Ludwig Bieberbach and Oskar Teichmuller and physicists like Philip Lenard and Johannes Stark (Mehrtens, 1987; Chowdhury, 1995) moreover there was additional complication due to two interfering movements: the conservative group of anti-relativity scientists who mostly went along with the Nazi official rejection of the theory as "Jewish Physics" and those who accepted Relativity as physicists but invented for it another (non-Jewish) author, to protect their selection from the authorities (Brian, 1996).

Within Judaism itself, it appears that Spinoza's excommunication reflected a very special local background, namely the involvement of the Jewish leadership in Holland in the reabsorption into Judaism of the Spanish and Portuguese Marranos and that leadership's fear of the destabilizing doubts induced by some of Spinoza's ideas (Yovel, Y., 1977).

Note that as in the Nazi case or that of Giordano Bruno or of some of Lysenko's opponents, the clash involved some form of persecution – in theocracies during periods of intolerance, or when the ideology is that of a totalitarian state (one wonders whether the same might become true in the opposite situation, namely for a religious preacher in some future Plato-inspired "scientocracy"). We do not deal with these aspects here, as our interest is in the philosophical issues only. We also do not include "ideologies" in what follows. *In the rest of this study we explore the possibility of conceptual coexistence between science and religion and suggest mutually acceptable uncompromisimg constraints for the sake of intellectual coexistence between non-fundamentalist and more abstract religious views and science in its least "arrogant" moods.*

As long as "religion" mostly implied mythological polytheism ("idolatry" in later terminology) and "science" was yet another "bag of tricks", a kind of "magic", there was no case for a clash between the two. For the clash to occur, both sides had to evolve so as to imply comprehensive world-views, vying for the same minds. Note that Creationism could have conciliated the "scientist" side in the pre-world-view, when science was still in its "bag of tricks" stage.

2. How Science Became A Welt-Anschau

With *Science*, the change occurred when the term "science" came to mean *an attempt* to produce a *precise description* of *physical reality* in the form of a codex of *Laws of Nature*. This happened in Greece in the Sixth Century BC, a conceptual advance apparently induced by the adoption of Solon's Laws, i.e. of a *codex of laws* regulating *human* behavior, an idea imported from the Fertile Crescent in the Seventh Century BC (some 500 years after the borrowing of the alphabet from the same geographical source). If inter-human relations could be managed by Laws – why not try the method's applicability to nature? True, human laws were imposed by the state – but if this was the best way of achieving harmony, it might also have been used by the deities or other agents controlling existence. Previous observations in

Mesopotamia, Egypt and in Greece itself had indeed detected some *regularities*, so that it appeared that at the very least it was worth trying the method as a *descriptive* tool. The Pythagoreans thus tried to apply it to some of the fragmental notions collected by observation in the previous stage and added *mathematics* as a logical skeleton and glue, plus the requirement of *quantitative verifiability*, not just qualitative; yet another implied hypothesis was that it could be achieved by bits, a *patchwork* of surveyed sectors with gradual mergers. The first real sectorial success was Euclid's *Geometry*, which provided (and still does) a model for most sciences. Note that there was (and still is) no guarantee that this idea could be extended so as to cover *all aspects of physical existence*. It is to these aspects that Einstein refers in his statement that "the most surprising thing about the world is that it is comprehendible".

3. The "Great" Religions, Faith and Revelation

Religion – more precisely those religions for which we sometimes use the term "*great religions*" – are comprehensive world-views held together by *faith*, launched by self-appointed initiators, their self-selected roles ranging from incarnations of a declared deity or of some connecting intermediate godly agent – to human announcers of such future agents ("Messiahs", "prophets" etc). All of these are animated by *revelation*, an *internal experience of the mind* in which the individual is convinced that *an idea has emerged in his* (or *her*) *mind* (as if?) directly *inserted from the outside*.

The transition from the various forms of polytheism to the "great" religions occured in a series of local revelations ranging from the XVIIIth Century BC (Abraham), to the XIVth (Ikhnaton, Moses), the Sixth (Zoroaster, Buddha, Lao Tze, Confucius) to the First Century AD (Christ, Paul) to the Sixth (Mohamed, Ali) and to the XVh (Calvin, Luther), with smaller groups emerging every few decades, such as Babists, Bahais or Ahmedists in otherwise Islamic countries or the Mormon in the USA. Most 'great' religions follow a canonized text orignating in *revelation* and carry a core-message regulating *inter-human* and *universal* or *man-world* relations. Revelation being an *internal* experience, only *faith* may bring in the adepts and form the foundation for the "binding" systematics. Note that whatever the humanistic or universal context, *the implementation is bound to reflect the technological and social status of the times*. The Deity has to formulate its messages in both format and context so that both the prophet himself and the adepts will understand. Also, some of the rites may later lose their active role, retaining a symbolic value at best – but the *revelation-originated rite* – and much more so any of the key messages – are *immutable* because of their canonical nature In some cases, however, the system may have provided for a degree of flexibility allowing it to gradually adapt to the changed conditions.

4. The Super-Clash: Scholasticism and The Annihilation Of Greek Science

The First Age of Science lasted about a thousand years, roughly from 500 BC to 500 AD. Greek science (some ten academies spread along the Eastern Mediterranean) had made remarkable progress in Mathematics, Astronomy, Physics, etc, both as to conceptions and to quantitative determinations.

Taking Astronomy as one example – by 400 BC the Pythagoreans had determined the spherical shape of the earth and Eratosthenes (Dreyer, 1953) of Cyrene (~ 250 BC), Alexandria's greatest scientist but also a great research organizer (e.g. the Library) measured the earth's radius with an error of less than 0.5%; Heraclides of Pontus (~350 BC) explained the day/night cycle as a spinning motion of the earth, Hipparchus (~150 BC) measured the radius of the Moon and its distance from the Earth with a 10% error and Aristarchus of Samos (~270 BC) derived the Heliocentric structure of our system 1800 years before Copernicus.. In Physics, Archimedes (~250 BC) formulated the laws of moments and angular momentum, and of buoyancy in Hydrostatics, while Heron of Alexandria (~50 AD) built a steam engine and worked out the laws of mirror reflection in Optics.

The clash developed after Christianity had become the state religion of the Roman Empire. In the first phase it was just a matter of competing when trying to attract students and adepts from among the intelligentsia. Sometime quoting the writings of Alexandrine Jewish philosopher Philo Judaeus, who claimed to have proven that there was full harmony between the teachings of the Biblical prophets and those of Greek Philosophy, the intellectual leadership in 4th to 6th century Christianity gradually either took over or shut down the various academies. To justify the takeover or sharing, they also *adopted* some of the *scientific paradigms* of the moment, as formulated by Aristotle (including some of his personal prejudices). These were incorporated in the religion's dogma, "*frozen*" as they stood at that moment. One century after Constantine, the academies throughout the Empire had all either been taken over or closed, except for Athens and Alexandria.

In 415 AD, the mathematician Hypathia (Dzielska, 1995) rector of the academy in Alexandria was murdered and the Academy was erased; finally, in 550 A.D., the Athenian academy was closed by order of Justinian and scientific activity came to a standstill in the West, while a few of Aristotle's books and the pseudo-scientific addenda to the Christian teachings were kept with the dogma – the mix evolving into the doctrine of *Scholasticism*. There followed a one-thousand years *freeze* of most scientific activities, a freeze known as the "Middle Ages".

One should add at this point the transfer of nine members of the Athens Academy, with a part of their library, to Sassanid Persia at the invitation of Chosroes the Great and the establishment of a Science-oriented Academy in Persia, thereby preserving much of the achievement of Greek Science. The Arab conquest found a flourishing institution, the Abbassid Caliphs adjoined it to their palace, their Ommaiad rivals in Spain established an analogous institution in Cordoba and Science trickled back into Western Europe with the founding of the Medical School in Salerno and the universities of Bologna, Montpellier, Paris, Oxford, Cambridge, etc. Some three hundred years of a developing activity in these centers finally produced free scientists such as Gersonides in Astronomy or Crescas in Physics who could break the hold of the church-supported Aristotelian dogma and the way was open for the Renaissance. This is why it took a thousand years.

By 1000 AD, both Islam and Judaism had followed and added this dogmatic representation of Science to their own teachings. Thus there was now an automatic *veto* against any continuation of the scientific conceptual development process, and any scientific advance or even scientific self doubts could now be regarded as *heresy*. There were permanent suspicions and accusations of "idolatry", motivated in the early stages by the very open and research-oriented spirit of *neoplatonic* teachings.

Note that the incorporation of scientific notions in Judaism was somewhat more "smooth", due both to the non-existence of a central controlling and censoring authority and to the earlier established coexistence of four interpretations of Jewish scriptures, namely the Literal (i.e. fundamentalist), Allegorical (followed by Philo), Hermeneutical and Mystical (e.g. Kabbalistic or Hassidic). Aristotelian science was easiest to absorb in the Allegorical. This author's impression is that the same would hold with the suggestions of the present article.

5. The Fundamental Antinomy

A] Religion: devotion and fidelity

The Scientific Revolution which occurred in Physics in the earlier part of the XXth Century replaced Newtonian physics by General Relativity (mainly) in the large and by Quantum Mechanics (mainly) in the small, though this "quarantine" may soon become obsolete, with present advances in the construction of Quantum Gravity. In any case, the magnitude of that transition in the basic conceptual paradigms of Physical Science could easily be described as of the order of the Protestant Reformation in Christianity – except for one feature. In a religious change of dogma, a sizable fraction among the believers remains faithful to the current dogma; after all, this is faith and faithfulness is the basic requirement. As a matter of fact, most religions have had their *martyrs, who made the ultimate sacrifice rather than abjure some tenets of their faith.*

B] Science: a priori – doubt, objectivity and rejection if facts so indicate

In science the opposite principle is true and constitutes the foundation of scientific development. Suppose we deal with an issue explored in the past (at times t') and with that exploration having resulted in a fully satisfactory theory, represented by (Kuhn, T.S, 1962) a *paradigm* K(a', b', ..h') where (a, b, ..h) are the relevant variables and the primes represent values tested in experiments performed at various times t'. As the years go by, new experiments are performed, testing the theory at some virgin (previously untested) range of values (a", b") for variables a and b, while the other variables (h, etc.) stay within the previously explored range – until finally one detects a clear failure of the paradigm K. At some future time, the issue is understood and a new paradigm K* (a, b, ..h) has emerged. One *bridging condition* that has to hold is that the new paradigm should yield the same predictions as its predecessor K in the original range where K fitted so well.. Thus

$$K^*(a', b', .., h') = K(a', b',.., h')$$

The Michelson-Morley experiment (Ne'eman, 1984) provides a clear example. Take Galilean kinematics as K, expressible as the law of addition of velocities; in the simplest low-velocities example, v(1) is the velocity of a train and v(2) the velocity (with guaranteed precision verified at production) of a bullet with respect to the rifle from which it is fired; but as in the present case the rifle is fired from that train at a static target, the velocity with which that bullet will hit its target is given by

$$K: v(\text{total}) = v(1) + v(2)$$

This worked very well as long as the velocities were small relative to the velocity of light c. The Michelson-Morley experiment involved summing velocities one of which is the earth's, say of the order of its motion around the Sun, roughly v(1) = 30 km/s or 0.0001 c (the velocity of light) and the second is that of light itself v(2) = c. The experiment revealed the *invariance* of the velocity of light, whatever the velocity of the source, thereby leading to the demise of Galilean mechanics (K) and to their replacement by Special Relativity (K*). The bridging condition is fulfilled due to the factor (v/c), namely one over one-hundred-millions, for day-to-day velocities, which multiplies those terms in K* which did not appear in K.

Returning to our Science/Religion analysis, we may now state the *Invalidation* principle (Popper, K., 1959) given a paradigm K, *the experimental program should assume* that *the paradigm does not hold outside of the original range* of values of the variables for which it was established and focus on the detection of departures from the paradigm's predictions, rather than "proving" the paradigm's validity. The effort should always be aimed at disproving it, invalidating it ("falsifying it" in Popper's words, a term which is less ambigous in German, I believe).

C] The antinomy

This analysis has brought out the basic dichotomy, whenever a scientific paradigm has also been adopted by religion. A situation develops somewhat remindful of a judicial procedure in which the suspected canonized paradigm is prosecuted by science and defended by religion. This picture holds whenever, in the aftermath of an experiment, the canonized scientific paradigm has finally been invalidated. Any initiative to learn from the negative result of the experiment and modify the failed paradigm takes on a heresy profile. The lesson is clear and we list it as *Constraint R1:*

There should be no overlap, i.e. religion should never incorporate scientific material reflecting the paradigm of that day. Note that this conclusion is not directly realizable for a Fundamentalist Church, but should not be hard to follow otherwise.

6. Laws of Nature and Miracles

Science is based on the *reproducibility* of experimental results. For the believer, the successful description of Nature by a system of Laws just means that the Deity has chosen to run the Universe in an orderly manner, i.e. these are his/her Laws, and he/she operates through this instrument. Many religious establishments have come to that conclusion at some time in the past. In our quest for Harmony, we have to request that this assumption be included as a self-imposed constraint on the Religion – namely, the Laws of Nature Constraint, *God/the Gods rule and act through his/her Laws of Nature*.

There is, however, a residual problem: what about *miracles*? If they are outside of the above 'arrangement' we have a science/religion clash: (1) 'miracles' should be irreproducible, almost by definition, in any religion (in Catholic Christianity this is used as a criterion in the preparation of evidence for Beatification candidates) and (2) science cannot allow exceptions to reproducibility.

A possible solution does exist, at the price of a *constraint on science*. The requirement is that the scientific paradigm be non-deterministic, with a certain part of any scheduled action left to a randomized decision-making procedure... The nature of Probability is such that the "average" will follow the Laws of Nature while individual elements may display huge departures from that average. Miracles can then be "engineered" by exceptionally 'lucky throws' – and isn't the *control of luck* the first role any religion will attribute to its God or Gods? This solution exploits an idea (Lecomte du Nouy, 1947) due to the French biologist Lecomte du Nouy who conceived it in a different context, namely as a way in which God might have accelerated evolutionary processes, a suggestion made at a time when the geological timetable appeared too short for Evolution (the geological evidence later removed the paradox). Note that a similar difficulty had arisen in the nineteen-twenties, when the early red-shift measurements appeared to indicate a cosmological timetable too short for Nucleosynthesis and astrophysical evolution; the astronomer-cosmologist was Monseigneur the Abbe Lemaitre, but it was to Einstein's Cosmological Constant that he appealed, making it slow down the cosmological expansion. That difficulty was also a fluke that vanished with better measurements – and the most recent results indicate that the constant is accelerating the expansion rather than slowing it down.

With tongue in cheek, we can now ask "Does God then play dice, Einstein's contrary opinion notwithstanding?" With all due respect, (1) this possibly lay outside the expertise of even the greatest scientist (2) Quantum Mechanics is ruled by Uncertainty Relations, and (3) Einstein's precise wording in that letter to Born does not mention God – he writes "I believe that the Old Man hasn't yet revealed all his secrets and I don't think he plays dice" – he may have had somebody else in mind.

Luckily for our task, the Physical paradigms at the beginning of the XXI Century AD offer two open gates for luck and indeterminacy: *Quantum Mechanics* (Ne'eman, 1984) and *Chaos* (Driebe, 1999). We thus add two constraints, one on each side, namely the Miracles constraints:

For Science,
The physics of human scale systems (micron to megameter) have to include at least one way ensuring the possibility of their being affected by "luck", i.e. by random action.

And for religion,
Even Miracles might be produced within the laws of Physics.

In Jewish theology this position has been called the "Miracles in daylight" view, as against the opposite "hidden Miracles" option.

There is however, another context relating Chance to nature, this time directly at the macroscopic level. According to the present paradigm for the Generalized Theory of Evolution, as recently completed by the present author (Ne'eman, Y., 2004) evolution consists in a *population* of *systems,* each of which is controlled by a *program.* The systems exist and act in an *environment.* They undergo a vulnerabilizing routine which opens the way to *tychic intervention,* i.e. chance modifications of the program. The *mutated* program and the *mutated systems* carrying it then undergo a *stability* test. If the new system is stable it then faces an effective test of *survival* in its *environment.* Other types of tychic interventions happen in the *environment* and modify it Systems of one species may then be *destroyed* by the changed conditions while others may profit from these conditions, including *scalawag* species benefiting from the *destruction* of competing species in that same environment.

This characterization of evolutionary processes covers the entire range, from *Cosmogony*, via *Nucleosynthesis* to Neodarwinian *biological* evolution, to the evolution of human *societies*, *epistemology* and the evolution of *ideas*.

Taking biology as an example, DNA is the program, the procedure in which it has to be copied and recopied continuously makes it vulnerable and mistakes can occur. These are the "type M" mutations; "type E" mutations are the changes in the environment such as the meteor which destroyed the dinosaurs or whatever might have caused the extinction of the dodo… In Epistemology, serendipity provides the type M mutations and invalidation the type E, etc.

Returning to our discussion in section 1, rather than invent creationism (NAS, 1999) can't the deity develop its program by using evolution? One of the main features of evolution is *teleonomy.* i.e. the *impression* that from the very start there was an aim, a master plan. As far as Science is concerned it is "as if" there was such a plan, because it is not organized as an engineering design, it is more like *Tinkering*, of course, as it is produced by *chance.* But as far as religion is concerned, chance itself is at the disposal of the Deity, and the Great Engineer apparently has the necessary expertise.. In Judaism, this stand is presently that of the religious scientific establishment, as formulated by a prominent physicist, (Domb, 2000) in an editorial. This was also the position promoted in a remarkably enthusiastical treatment of Evolution by R. Avraham Hacohen Kook (1865-1935) (Kook, R.A.Y.H., 1961) an important religious leader who served as Chief Rabbi of British Mandate Palestine in the first part of the Twentieth Century. This opinion totally differs from that of another important religious leader, R. Menachem M. Schneersohn of the Chabad movement who, addressing the issue (Schneersohn, 1976), regrettably emphasizes possible sources of error in scientific theories (irrelevant to this case) rather than engaging in a serious dialogue.

7. Conclusions

For a religion or an intellectual system within a religion to spiritually coexist with the Scientific drive, it should check whether its principles fit with the following two constraints:

R1

Religion should refrain from incorporating in its dogma scientific material or positions which might just reflect the science of the day.

R2

The Laws of Nature realize Divine rule, possibly encompassing all our experience, including random sequences as in evolutionary chains and even miracles.

On the side of Science, the one necessary and sufficient constraint (which holds indeed in present paradigms) states,

S1

The physics of human scale systems (micron to megameter) have to include at least one way of ensuring the possibility of their being affected by randomized action at some lower level.

[Quantum Mechanics or Chaos or both, in present science]

8. References

Berlinski, D., The Deniable Darwin, in *Commentary* (June 1996).
Berlinski, D.,Was there a Big Bang, in *Commentary* (February 1998).
Brian, D. and *Einstein,* A. (1996) *Life.* J. Wiley and Sons pub. 990 pp., see p. 105.
Cane, P. (1959) Giants *of Science,* Pyramid Books, New York 288pp.
Chowdhury, M. R., (1995) Landau and Teichmuller, *Mathematical Intelligencer* 17 no. 2, pp. 12-14.
Domb, C. (2000) *BDD Journal of Torah and Scholarship*, #11 November, Editorial.
Dreyer, J.L.E. (1953) *A History of Astronomy from Thales to Kepler,* Dover pub., New York p. 174.
Driebe, D.J. (1999) Fully Chaotic Maps and Broken Time Symmetry, Kluwer Academic pub., 165 pp.
Dzielska, M. (1995) Hypatia of Alexamdria, Harvard U. Press, 200 pp.
Kook, Rabbi A.Y. haCohen (1963) (Hebrew), A Rising World (On Evolution) in Or*ot haQodesh ("Holy Lights")* Part II, Ch. 5, secs. 19 – 21, Rav Kook Foundation pub., Jerusalem.
Kuhn, T.S. (1962) The Structure of Scientific Revolutions, U. of Chicago Press, 350 pp.
Lecomte du Nouy, P. (1947) Human Destiny, Signet Books, 189 pp.
Mehrtens, H. (1987) "Ludwig Bieberbach & "Deutsche Mathematik", *Studies in the History of Mathematics* **26**, 195-241.
NAS *Science and Creationis* (1999) pub. by National Academy of Sciences of the USA, Washington DC, 34 pp.
Ne'eman, Y. (1984) Twentieth Century Physics (Hebrew), Israel Min. Def. Pub. (Radio University Series) 122 pp.
Ne'eman, Y. (2003) The Music of the Hadrons. In: V.N. Branbilla and G.M. Prosperi (eds.) *QuarkConfinement and the Hadron Spectrum.* World Scientific pub. pp. 236-248.
Ne'eman, Y. (2004) Paradigm Completion for Generalized Evolutionary Theory etc. In: J. Seckbach (ed.) *Origins: Genesis, Evolution and Diversity of Life,* Kluwer Academic Pub., Dordrecht, pp. 249-262.
Popper, K.R. (1965) *The Logic of Scientific Discovery,* Harper & Row pub., New York, 480 pp.
Schneersohn, Rabbi Menachem M. (1976) A letter on Science and Judaism, *Challenge – Torah Views on Science and its Problems* A. Carmeli and C.Domb eds., pub. by Ass. Orthodox Jewish Scientists and Feldheim pub., Jerusalem – New York, pp., 142-149.
Yovel, Y. (1977) The Excommunication of Spinoza, *Commentary,* Vol. 64, No. 5, pp. 46-53.

Biodata of **Carol E. Cleland** author of ***"Understanding the Nature of Life: A Matter of Definition or Theory?"***

Dr. Cleland is currently Associate Professor of Philosophy at the University of Colorado (Boulder) and a member of the NASA Astrobiology Institute (NAI). She received her B.A. degree in mathematics from the University of California (Santa Barbara) and her Ph.D. degree (1981) in philosophy from Brown University. She spent a year at Stanford University as a post-doctoral fellow at the Center for Studies in Language and Information (CSLI). Dr. Cleland's current interests are in the areas of philosophy of science and philosophy of logic (the theory of computation). In the area of philosophy of science, she has written articles on causation and change, space and time, the concept of an event, the methodology of science (e.g., differences between experimental science and historical science), and the problem of defining 'life'. She is currently working on two books, *The Quest for a Universal Biology: What is life?* and an anthology (with Mark Bedau) on classic and contemporary readings on the nature of life.

E-mail: **Cleland@colorado.edu**

J. Seckbach (ed.), Life as We Know It, 589-600.

UNDERSTANDING THE NATURE OF LIFE:
A Matter of Definition or Theory?

CAROL E. CLELAND
Philosophy Department and Center for Astrobiology, University of Colorado, Boulder, CO, USA

1. Introduction

The question "what is life?" has long been a source of philosophical discussion. In recent years, however, it has taken on increasing scientific importance. Biologists investigating the origin of life or trying to synthesize life in the laboratory from basic chemical building blocks want to know at what stage an ensemble of non-living molecules turns into a primitive living thing. What sort of chemical processes mark the transition from non-living to living? Charged with designing remote or in situ robotic experiments for detecting extraterrestrial life, astrobiologists wonder what features of terrestrial life are universal to all life. How different could life be from life as we know it? Even computer scientists find themselves mired in controversy over the nature of life as they speculate about whether their most sophisticated simulations of life are truly alive. Is life just a matter of realizing the right program; are the fundamental properties of life independent of the "stuff" that bears them? In order to resolve these issues, scientists need an ultimately general answer to the question "what is life?" They need to know what is essential to life, wherever it may be found in the universe.

The most common strategy for answering the question "what is life?" is to define 'life'. The literature on definitions of 'life' is very old, dating back at least to the ancient Greek philosopher Aristotle (De Anima 415^{a22}-415^{b2}). The contemporary scientific literature on definitions of 'life' begins, however, with Carl Sagan's classic 1970 article in *Encyclopaedia Britannica.* Canvassing the then most popular definitions of 'life', Sagan provided a counterexample to each. This caught the attention of scientists. In the ensuing years, there was a flurry of work on definitions of 'life.' Hoping to evade Sagan's counterexamples, some scientists focused on amending and refining the popular definitions. Others, less sanguine about overcoming the counterexamples, pursued new definitions. As a result of these efforts, scientists now possess an impressively large and diverse collection of definitions of 'life'; see Radu Poppa (2004, Appendix B) for an up-to-date catalogue. Yet there is still no consensus among scientists about which definition is correct.

As I have argued elsewhere (Cleland and Chyba, 2002; 2006), it is not an accident that scientists have been unable to reach a consensus on the definition of 'life'. The scientific project of defining 'life' rests upon mistaken ideas about the nature of definition and its capacity to answer questions about the natural world. In the next section of this chapter, I review these arguments, explaining why no definition of 'life' can provide a scientifically compelling answer to the question 'what is life?' The rest of the chapter is devoted to sketching a new, more promising, approach to this fundamental question.

2. Definitions: Varieties and Limitations

Definitions are concerned with language and concepts, as opposed to things in the world. This is illustrated by the following example: 'bachelor' means unmarried human male. The use of single quotation marks indicates that our concern is with a *word* ('bachelor'), as opposed to the things (certain men) to which the word applies. The definition explicates the *meaning* of 'bachelor' by dissecting the *concept* that we happen to associate with it. When a scientist seeks an answer to a question such as 'what is life?' she or he isn't interested in our present-day beliefs about life, which may be confused or mistaken. She or he wants to know about the phenomenon of life—what all terrestrial biological organisms (bacteria, fungi, fish, trees, humans, etc.) have in common that distinguishes them from non-living systems, and what any system, however different, must have in common with terrestrial organisms in order to qualify as living. Analysis of our current concept of life is unlikely to be of much help in this quest. But to fully appreciate this point, we need to look more closely at the nature of definition and its potential for settling scientific questions.

The primary function of definition is to facilitate successful communication. Different forms of definition, and there are a surprisingly large number of them, are designed to do this under special circumstances and for particular purposes. In this section, we restrict our attention to forms of definition that are relevant to the problem of defining 'life'; see Audi (1995) for a more complete list of types of definition.

Lexical definitions are familiar to everyone. Found in dictionaries, they report on the meanings of terms in a natural language such as English. At best, a lexical definition can reveal what 'life' means to a typical speaker of English. But this is not what a scientist wants to know when she or he asks "what is life?" Furthermore, lexical definitions are frequently circular, defining words in terms of their close cognates. This deprives them of utility for someone who hasn't mastered the associated concept or lacks familiarity with the particular term being used in the definition. "'Line' means linear path" provides an obvious example. But circularity can be more subtle. A salient example is defining 'cause' as 'something that produces an effect'. The terms 'cause' and 'effect' are so closely related (conceptually speaking) that it is unlikely that someone who is ignorant of the meaning of the former would know the meaning of the latter.

It is unlikely that anyone believes that the *Oxford English Dictionary* holds the answer to the question "what is life?" *Operational definitions* are another story, however. They are very popular among scientists. As an example, a number of astrobiologists, e.g., Conrad and Nealson (2001), McKay (1994), and, in an earlier work, Chyba (Chyba and McDonald, 1995), have called for the use of operational definitions in the search for extraterrestrial life.

Operational definitions are closely related to ostensive definitions, which are familiar to most people. *Ostensive definitions* by-pass concepts and directly indicate the meanings of terms via representative examples. Saying the word ‘cat’ to a small child while physically pointing to actual specimens provides a paradigmatic example. The pointing, may, however, be done verbally as when someone defines ‘play’ as “Oedipus Rex, Hamlet, Pygmalion, and The Monkey’s Paw.” The main problem with ostensive definitions is that they don’t say what the items falling under a term have in common. Figuring this out is left to the hapless recipient of the definition, who may inadvertently focus upon irrelevant similarities or be misled by unrepresentative specimens. Operational definitions face the same problem. Rather than using gestures or lists to identify specimens, they use *procedures*. A paradigmatic example is defining ‘acid’ as ‘something that turns litmus paper red,’ While the use of litmus paper allows one to determine whether a particular liquid is an acid, it doesn’t say anything about what makes a liquid an acid. For this reason it can’t provide a scientifically compelling answer to the question “what is an acid?” Analogously, despite their popularity among scientists, an operational definition of ‘life’ is unlikely to be of much help in answering the general question “what is life?” At best it may be used to identify specimens of familiar Earth life.

This brings us to philosophically ideal definitions, which are the most promising candidates for defining ‘life’. *Ideal definitions* are maximally informative. They dissect the concept associated with a word into a logical conjunction of properties. The conjunction of properties provides necessary and sufficient conditions for the application of the word. A *necessary condition* is a condition in whose absence the word will not apply and a *sufficient condition* is a condition in whose presence the word must apply. An ideal definition thus completely determines the set of things to which a term applies. The definition of ‘bachelor’, with which we began this discussion, provides a good illustration. It supplies a complete analysis of the concept of bachelor into the properties of being unmarried, human, and male. Moreover, it does this without presupposing a prior understanding of the concept of bachelor. Anything that satisfies the conjunction of properties qualifies as a bachelor and anything that is a bachelor satisfies the conjunction of properties. The upshot is that we know what all and only bachelors have in common. Definitions don’t get any better than this! Indeed, it is hard to imagine a better answer to the question “what is a bachelor?” than “an unmarried, human male.” The question is can we formulate an ideal definition for ‘life’ that will provide an equally satisfying answer to the question “what is life?”

To understand why the answer is “no,” it is useful to consider an analogous case from the history of science. Consider the plight of a thirteenth century “scientist” trying to answer the question ‘what is water?’ by defining ‘water.’ He knows nothing about molecules; molecular theory wasn’t discovered until the late eighteenth century. His knowledge of water is limited to its sensible properties–being a liquid, wet, transparent, tasteless, odorless, and a good solvent. But many of the substances that he calls “water” lack one or more of these properties. Salty water isn’t tasteless and muddy water isn’t transparent, to mention two examples. Which of the sensible properties of water should he cite in a definition of 'water'? Impressed by water's powers as a solvent, the alchemists (medieval chemists) chose solvency. As a consequence, they classified nitric acid and mixtures of hydrochloric acid as water, the former being known as "aqua fortis" (strong *water*) and the latter as "aqua regia" (royal *water*) (Roberts, 1994).

But we now know that the alchemists got it wrong. Nitric acid and mixtures of hydrochloric acid are not water. Water is distinguished from other substances by a unique molecular composition, H_2O. H_2O is what salt water, muddy water, distilled water, and even acidic solutions all have in common, despite their sensible differences. Could a thirteenth century scientist have figured this out merely by analyzing his concept of water? The answer is clearly 'no.' No amount of analysis of the thirteenth century *concept* of water could reveal that water is composed of two molecules of hydrogen and one molecule of oxygen because no one at that time knew anything about molecules, let alone the chemical elements hydrogen and oxygen. This means that no thirteenth century definition of 'water' could have provided a satisfactory answer to the scientific question 'what is water?'

But what about today? Surely the claim that water is H_2O provides an analysis of the present-day concept of water? Surprisingly, the answer is still "no". Like their thirteenth century ancestors, contemporary English speakers use the word 'water' for many things (e.g., muddy water, salt water) that are not pure H_2O. In other words, the scientific concept of water is narrower than the ordinary, everyday concept. This suggests a new strategy for the definitional approach. Perhaps the claim that water is H_2O represents a special scientific definition of 'water'?

Alas, this strategy is also doomed to failure. If the claim that water is H_2O were a definition, we couldn't make sense of the possibility of scientists discovering that they are somehow wrong about the stuff that we call "water"—that it is not H_2O after all. While it seems exceedingly unlikely that this will happen, we cannot completely rule it out. After all, this sort of thing has happened before in science. The conservation of mass was a pillar of nineteenth century physics, and yet it (along with a number of other fundamental tenets) was rejected by Einstein in the early part of the twentieth century; mass-energy (vs. mass alone) is conserved. In other words, the claim that water is H_2O represents a scientific *discovery*, and is subject to revision in light of future scientific investigations.

In stark contrast, one cannot imagine a social scientist discovering that we are wrong about the marital status of bachelors. Consider a graduate student in sociology trying to test the hypothesis that all bachelors are unmarried by interviewing men at the local shopping mall, and asking them whether (1) they are bachelors and (2) they are unmarried. The suggestion is absurd. We already know the answer in advance of any scientific investigation. This is not to say that someone couldn't stipulate that the word 'bachelor' also applies to married American men. They might even convince other people to speak in this peculiar manner. But this would not amount to discovering that we were previously wrong about the things called "bachelors"—that we erroneously thought that they were all unmarried. It represents nothing more than a decision to talk about the world in a different way.

This brings us to a crucial distinction. Philosophers distinguish between two kinds of general terms, those that designate natural kinds and those that do not.

Natural kinds are categories that are carved out by nature, and would exist even if there had been no humans; for more detail see Devitt and Sterelny (1987). Some prima facie examples of *natural kind terms* are 'water,' 'temperature,' 'lightening,' and 'bird.' Terms such as 'bachelor,' 'American,' 'hammer,' and 'chair,' in contrast, do not designate natural kinds. They designate categories that depend upon human interests and concerns. This explains why we do not need to consult the world of nature in order to decide whether bachelors are unmarried. Bachelors do not form a natural category; being a bachelor is merely a matter of human convention. Thus the concept that we associate with 'bachelor' completely exhausts its meaning. This is not the case for natural kind terms. While the concept that we associate with a natural kind term helps us to recognize putatively paradigmatic examples, it does not (as in the case of non-natural kind terms) fully determine its application. The class of things to which the term applies is ultimately determined by nature, as opposed to human convention. No amount of analysis of human concepts can tell us about a world that lies beyond them. Indeed, as the alchemist's concept of water illustrates, the concepts that we associate with a natural kind term may badly mislead us, resulting in incorrect classifications and erroneous conclusions. If our beliefs about a natural kind happen to be incorrect, a definition will serve only to reinforce these misconceptions, making it more difficult to overcome them. The upshot is that definitions cannot supply satisfactory answers to scientific queries about natural kinds. Insofar as scientists are interested in the intrinsic natures of natural kinds, it is a mistake to focus on providing definitions for the words that we use to designate them.

3. Searching for an Answer to the Question 'What is Life?'

The implications of the above considerations for the scientific project of defining 'life?' should be clear. If (as seems likely) life is a natural kind, then attempts to define 'life' are fundamentally misguided. 'Life' *means* whatever cyanobacteria, hyper-thermophilic archaeobacteria, amoeba, mushrooms, oak trees, sharks, tree frogs, lions, humans and, most importantly, *everything* else that is alive (on Earth or elsewhere) has in common. No alleged definition of 'life' can tell us what *all* of these things (known and unknown) have in common because no mere analysis of human concepts can reveal the nature of a world that lies beyond them. At best, a definition of 'life' can tell us what biologists currently *believe* about life. But this is not what someone who asks the question "what is life?" wants to know: They want to know what life *really* is. The pursuit of a scientifically compelling answer to this question requires a different strategy. As earlier, I will motivate this strategy by considering how modern chemists have successfully answered the analogous question about water.

As we all know, the question 'what is water?' has a scientifically compelling answer: H_2O. The claim that water is H_2O is an example of a *theoretical identity*; some other examples are lightning is an electrical discharge and temperature is mean kinetic energy. Rather than representing definitions, theoretical identities place natural kinds designated by terms from common language within the framework of scientific theories by picking out their most salient theoretical properties. In the case of water, this is its unique molecular composition. The saliency of H_2O has to do with its role (within the context of chemical theory) in making sense of the physical properties and behavior of water. This underscores an important feature of scientific theories. Scientific theories unify disparate and puzzling phenomena under common theoretical frameworks by providing schemes for classifying them into natural kinds. The best classification schemes tell us as much as possible about the phenomena under investigation. Modern science can explain a truly remarkable range of facts about water in terms of the chemistry of H_2O. It can explain why water is a good solvent and yet other good solvents, such as nitric acid, are not water, thus clarifying where the alchemists went wrong. It can also explain why water evaporates when heated and expands when frozen, and why it remains a liquid over an exceptionally wide range of temperatures. It can even explain the unusual behavior of water at extremely high temperatures and pressures.

Discoveries made within the context of a new scientific theory may fundamentally change our ideas about familiar natural kinds. For one cannot expect the theoretically refined categories carved out by our best scientific theories to be in perfect alignment with our prescientific categories. We may discover that what we thought was a single natural kind is actually two different natural kinds. The term 'jade,' for example, was once thought to apply to a single natural kind in virtue of the sensible properties (color, crystalline form, etc.) exhibited by certain minerals. With the advent of molecular theory, however, it became evident that 'jade' encompassed minerals with two different molecular compositions. As a consequence, mineralogists split the old category of jade into two new categories, which they named 'jadeite' and 'nephrite.' Alternatively, we may discover that what we thought were different natural kinds are actually the same natural kind. A good example is the discovery that (despite their striking sensible differences) sapphires and rubies have the same fundamental molecular composition; mineralogists now classify them both as corundum. It is important to be clear about the nature of such decisions. They do not represent mere verbal decisions to talk about the world in a different way. They represent decisions to bring language into closer correspondence with new discoveries about the world of nature.

In order to provide a scientifically compelling answer to the question 'what is life?' we need a theoretical identity analogous to water is H_2O. But as the discussion above indicates, this presupposes a theory of living systems. The requisite theory must be general. Just as modern chemical theory encompasses all forms of water (on Earth and elsewhere in the universe), we want our theory of life to encompass all forms of life, wherever it may be found in the universe. The question is do we currently have such a theory?

Many biologists believe that neo-Darwinian evolution provides us with such a theory. This explains the popularity of Darwinian definitions of 'life'. As an example, the most popular definition of 'life' within the origins-of-life community and NASA is the chemical Darwinian definition, which defines life as "a self-sustained chemical system capable of undergoing Darwinian evolution" (Joyce, 1994). Lin Chao has even proposed an operational Darwinian definition of 'life' (Chao, 2000), something that was long thought to be impossible. Although it is a mistake to try to answer the question "what is life?" with a definition of 'life,' the popularity of Darwinian definitions suggests that neo-Darwinian evolution might provide us with a sufficiently general theory of living systems.

Unfortunately, however, this is not the case. Neo-Darwinian evolution (also known as the "modern synthesis") provides a molecular and biochemical framework for understanding evolutionary processes, and the history and geographical distribution of life on Earth. It is, however, based exclusively upon our experience with familiar Earth life. And there are compelling biochemical reasons for believing that familiar Earth life does not provide us with an adequately general sample of the possibilities for life.

Terrestrial life comes in a mind-boggling array of different shapes, sizes, and structures. Put more technically, it exhibits remarkably *morphological* diversity. Some organisms consist of only a single cell. These organisms include bacteria, whose genetic structures are not enclosed by a nucleus; they are suspended in the cell's watery interior. Foraminifera (or "forams" as they are affectionately known to biologists) also consist of a single cell. But their chromosomes (genetic structures) are contained in a separate membrane-enclosed nucleus, and multi-chambered, colorful protective shells enclose them. Viewed under a microscope, they look a lot like miniature snails. In addition to microbes such as bacteria and forams, Earth contains the more familiar diversity of multicellular life. These include such marvels as jellyfish, snails, sharks, trout, sea gulls, elk, redwood trees, mushrooms, spiders, and butterflies. Indeed, the morphological diversity of life on Earth is so astonishing that one is tempted to conclude that all the possibilities for life are represented right here on Earth!

But this would be a mistake. Despite its remarkable morphological diversity, life as we know it on Earth is astonishingly similar. Bacteria, forams, jellyfish, sea gulls, deer, redwood trees, mushrooms, butterflies, and humans are all composed of cells. Moreover, these cells utilize the same basic machinery, composed of the same molecular building blocks, for growth, repair, and reproduction. The hereditary material of terrestrial life is stored on large molecules known as *nucleic acids* (DNA and RNA). These molecules are similar in ways that, from a biochemical perspective, could have been different. The DNA of terrestrial organisms utilizes the same four nucleotide bases to store its hereditary information even though there are other possibilities (Benner and Switzer, 1999). Moreover, the sugar molecules forming the sugar-phosphate backbone of the famous double helix of DNA all have the same right-handed *chirality*; like human hands, complex molecules sometimes come in right- and left-handed forms. Life could have built its DNA out of left-handed sugars. But it didn't. Similarly, life on Earth constructs its structural and (most of its) catalytic material out of large molecules known as *proteins*. Despite more than a hundred possibilities, terrestrial life constructs its proteins out of approximately twenty varieties of amino acids having the same left-handed chirality. From a biochemical perspective, life could have utilized a different suite of amino acids and or amino acids of the opposite (right-handed) chirality to construct its proteins. I could continue listing biochemical and molecular similarities: Life on Earth utilizes the same (triplet) genetic code to store its hereditary information, and the same tiny intercellular molecular machines (ribosomes) to translate the information stored on DNA into proteins that can be used for growth, repair, and reproduction. But the point should already be clear. There are a lot of fairly minor ways in which life on Earth could have been biochemically different. As biologists have recognized, the best explanation for these striking similarities is that life as we know it on Earth today has a common origin.

If all life on Earth has a common origin then we are dealing with a single example of life. One cannot safely generalize on the basis of a single example. Many of the characteristics of terrestrial life that currently strike us as fundamental may represent nothing more than chemical or physical contingencies present on the early Earth. Neo-Darwinian evolution explains the astounding morphological diversity of terrestrial life in terms of a common biochemical framework. But molecular analysis of extant organisms coupled with knowledge of the mechanisms of evolution also reveals that much of this diversity is an historical accident. Had the history of the Earth been different, life on Earth would have been different too. This brings us to the crucial question: What are the possibilities for life under different physical and chemical circumstances? Indeed, can one rule out the possibility of life forms that are so different from terrestrial life in their molecular architecture that they are able to exploit non-Darwinian mechanisms for adapting to changes in their environment? Neo-Darwinian evolution has nothing to say about this. It explains how organisms like us adapt to changes in their environment. But it does not exclude the possibility that life forms having a different physical makeup might utilize different mechanisms. One can imagine, for instance, an alien microbe whose genetic material directly and adaptively changes in response to different environmental conditions, thus realizing a Lamarckian inheritance of acquired characteristics; computer simulations of such "organisms" actually exist. Do such imaginings represent genuine possibilities for living organisms, or are they nothing more than fantasies? An adequately general theory of life would settle this and other fundamental questions about life in a decisive and compelling way. By way of analogy, chemical theory, which (like neo-Darwinian evolution) is also a synthesis of theories (e.g., molecular theory, thermodynamics) tells us not only that all water (on Earth and elsewhere) is composed predominately of H_2O but uses this information in conjunction with other information (e.g., Gibb's law, which connects the number of substances to the number of phases present in a chemical mixture) to predict and explain the behavior of water in very different physical and chemical circumstances.

The situation that we currently face in theorizing about life is analogous to trying to formulate a theory of mammals based on a single example, zebras. The most striking feature of zebras is their ubiquitous stripes. But it would be a mistake to conclude that all mammals have stripes just because all zebras have stripes. Ironically, one of the features that are most pertinent to the mammalian nature of zebras is their mammary glands, which are had only by the females. The saliency of the mammary glands of some (but not all) zebras can only be understood in the context of a theory of mammals based upon experience with other mammals and organisms that are not mammals. Similarly formulating an adequately general theory of life presupposes experience with forms of life that differ from life as we know it here on Earth. Insofar as we lack such experience, we are in no position to provide a scientifically compelling answer to the question 'what is life?'.

This is admittedly a disappointing conclusion. But it is also highly instructive, pointing us in a new direction. Rather than endlessly debating definitions of 'life,' we need to deliberately search for physical systems that challenge our current concept of life. This is tantamount to searching for systems that are difficult to classify as living or nonliving in light of our Earth-centric beliefs about life. Such systems will both resemble familiar life and differ from it in provocative and inexplicable ways. Thomas Kuhn used the word "anomaly" to designate ambiguous cases of this sort. According to Kuhn (1970, Ch. VI), anomalies are the driving force behind the formulation of new and better scientific theories. They alone stand out against the backdrop of our preconceptions, starkly revealing their inadequacies while at the same time providing crucial grist for the theoretical mill.

But how does one search for anomalies? This is not an easy question, and it is beyond the scope of this essay to pursue it in much detail. But a few things are clear. The first step involves treating the features that we use to recognize familiar forms of life as *tentative criteria*. Insofar as they are tentative, these criteria cannot be viewed as supplying operational definitions of 'life'. Almost paradoxically, their purpose is to identify physical systems that defy easy categorization as living or non-living. Once a genuinely anomalous physical system is discovered, it is important to explore it scientifically, keeping an open mind about its status as living or nonliving.

Genuinely anomalous physical systems will both resemble and differ from life as we know it. Thus it is important to use a diversity of independent characteristics of familiar life, rather than a small number of related characteristics, in our search for anomalies. Besides, we don't know in advance which features of terrestrial life have the greatest saliency for a universal theory of biology. So a diversity of independent criteria is essential. But terrestrial life exhibits an enormous variety of features, and it is clear that we can't use them all. Which ones shall we choose? This is a question for scientists to decide, and it is likely to vary depending upon the circumstances. Faced with the daunting prospect of encountering truly strange extraterrestrial life, an astrobiologist might want to begin by deploying a range of geological, biological and chemical criteria; despite their talk about definitions, this is consistent with suggestions made by Conrad and Nealson (2001). A biochemist trying to create a minimal cell in the laboratory, on the other hand, might elect to explore molecular variations on the structural or hereditary material of familiar life. It is even possible that microbiologists utilizing tentative criteria may someday encounter alien forms of terrestrial microbial life. We cannot rule out the possibility that life arose more than once on Earth and that remnants of some of these alternative origins may still survive today, heretofore undetected because they don't closely enough resemble familiar life. But this is only speculation. The important point for our purposes is the basic approach, which is to develop a strategy based upon life as we know it for searching for life as we don't know it. For it is only in the context of physical systems that challenge our prevailing beliefs about life that we will be in a position to formulate a credible theory of sufficient generality to provide a scientifically compelling answer to the question "What is life?"

4. References

Aristotle. De Anima, In: R. McKeon (ed.), 1970, *The Basic Works of Aristotle*. Random House, New York, pp. 535-603.

Audi, R., (ed.) (1995). *The Cambridge Dictionary of Philosophy*. Cambridge University Press, Cambridged.

Benner, S.A. and Switzer, C.Y. (1999). Chance and Necessity in Biomolecular Chemistry, In: H. Frauenfelder, J. Deisenhofer and P.G. Wolynes (eds.), *Simplicity and Complexity in Proteins and Nucleic Acids*. Dahlem University Press, Berlin, pp. 339-363.

Chao, L. (2000). The Meaning of Life. *Bioscience* **50**, 245-250.

Chyba, C.F. and McDonald, G.D. (1995). The origin of life in the solar system: current issues. *Annnu. Rev. Earth Planet. Sci.*, **23**, 215-249.

Cleland, C.E. and Chyba, C.F. (2002). Defining 'life.' *Orig. of Life and Evol. of the Biosph.* **32**, 387-393.

Cleland, C.E. and Chyba, C.F. (2006). Does 'life' have a definition?, In: W. Sullivan and J. Baross (eds.), *Planets and Life: The Emerging Science of Astrobiology*, Ch. 7. Cambridge: Cambridge University Press.

Conrad, P. and Nealson, K.H. (2001). A non-Earth-centric approach to life detection. *Astrobiology*, **1**, 15-24.

Devitt, M. and Sterelny, K. (1987). *Language and Reality*. MIT Press, Cambridge, pp. 67-79.

Joyce, G.E. (1994). Forward. In: D.W. Deamer and G.R. Fleischaker. (eds.) *Origins of Life: The Central Concepts*. Jones & Bartlett: Boston, pp. xi-xii.

Kuhn, T.S. (1970). *The Structure of Scientific Revolutions*. University of Chicago Press, Chicago.

McKay, C.P. (1994). Origins of Life. In: J.H. Shirley and R.W. Fairbridge (eds.) *Van Nostrand Reinhold Encyclopedia of Planetary Sciences and Astrogeology*. Van Nostrand, New York.

Poppa, R. (2004). *Between Necessity and Probability: Searching for the Definition and Origin of Life*. Springer, Berlin.

Roberts, G. (1994). *The Mirror of Alchemy*. University of Toronto Press, Toronto.

Sagan, C. (1970). Life. In: *Encyclopaedia Britannica*. Reprinted in *Encyclopaedia Britannica*, 1998, vol. **22**, pp. 964-981.

Biodata of **Shlomo Riskin** author of ***"And You Shall Choose Life"***

Rabbi Dr. Shlomo Riskin is an internationally renowned educator, speaker, and author. He attained rabbinical ordination at Yeshiva University from his mentor, Rabbi Joseph B. Soloveitchik, and obtained his Ph.D. in Near Eastern Languages from New York University in 1982. His outstanding contributions to Israel and to world Jewry over the course of his career have made him one of the leading voices of today's Jewish world. Rabbi Riskin is especially renowned for his innovative educational and social action programs, which are based upon his unique vision of an authentic Judaism sensitive to every human being and responsive to all universal concerns. On the cornerstone of this philosophy, Rabbi Riskin founded and serves as Chancellor of Ohr Torah Stone Colleges and Graduate Programs, a network of groundbreaking institutions including a rabbinical and communal leadership college for men; a college for advanced Jewish studies; the first and only program in the world training women advocates for the rabbinical courts; a women's "hesder" program combining Torah study with service in the Israeli Defense Forces; and a Legal Aid Center for *agunot* – women who are literally chained to dead or abusive marriages by recalcitrant husbands who refuse to grant them a Jewish divorce. He is also an innovative interpreter of Judaism, having authored the following books: *The Rebellious Wife: Women and Jewish Divorce* (1989); *Yad L'isha*, (in Hebrew, 2004); The *Passover Haggadah* (2000); *Around the Family Table* (2005); *Tora Lights – Genesis Confronts Life, Love and Family* (2005). The Brooklyn-born rabbi also serves as the founding Chief Rabbi of Efrat, Israel.

E-mail: **rabbiriskin@ots.org.il**

J. Seckbach (ed.), Life as We Know It, 601-606.

"AND YOU SHALL CHOOSE LIFE"

SHLOMO RISKIN
P.O.B. 1037, Efrat 90435, Israel

1. Introduction

Towards the end of the Book of Deuteronomy, the last of the five books of the Pentateuch, and towards the end of the farewell speech of Moses during the last moments of his life, the great leader - liberator of the Israelite nation gives what is his final charge: "Behold, I have given before you this day life and goodness, death and evil. I cause the heavens and the earth to bear witness before you this day: life and death have I given before you, blessing and curse; You shall choose life, in order that you and your seed may live" (Deuteronomy 30:15, 19). What does our supreme law-giver mean by commanding us "choose life"? Does not everyone wish to live? Is it then in our power to choose to live if our allotted number of years and days and hours are up? Is not G-d the final arbiter over life and death?

Apparently the Book of Books sees "life" as something beyond one's life-span and much more akin to one's life-elan, life-spirit, life-direction. Indeed, the very next Biblical verse links choosing life to loving G-d and identifying with Him: "To love the Lord your G-d, to listen to His voice and to cleave to Him, for He is your life and the length of your days." (Deuteronomy 30:20). I believe that an exploration of the Biblical concept which links life with G-d will teach us volumes about how Judaism views G-d, humanity and the manner in which we ought lead our lives.

2. Existence vs. Life

The official Aramaic translation of the Bible - written almost 2,000 years ago by a proselyte named Onkelos, accepted by the legendary Rabbi Akiva, oft-cited by Rashi and prescribed by the Code of Jewish Law to be studied alongside of the Biblical portion each week - translates (and interprets) the famous Biblical verse, "...Not by bread alone does the human being live but by that which comes forth from G-d's mouth does the human being live" (Deuteronomy 8:3) in the Aramaic "equivalent," "Not by bread alone does the human being exist (mitkayam), but by that which comes forth from the words of G-d does that human being live" (*bayei*- albeit not in all MSS versions, but I believe in the most authoritative).

Targum contrasts life with existence; to exist is merely to maintain one's physical being, to retain the necessary vital signs of breathing, ingesting nourishment, excreting wastes, communicating; in short, keeping the human organism alive. To live on the other hand, requires an awareness of and a desire to link's one's very being with the Divine and the Divine will.

3. What are G-D's Ways?

In order to attempt to understand what this really means, we must turn to a most significant Biblical passage in which Moses, the prophet of prophets, asks of G-d the question of questions: "Reveal to me Your glory (*kavod*) teach me the understanding of Your ways in directing the world (Exodus 33:18, as interpreted by Maimonides in his Guide for the Perplexed (Section 1:54); in effect, who are You, G-d-in-world?

The Almighty responds that no individual can see G-d and live, that G-d can only be revealed partially (His "back" or by means of His "footprints" in past historical occurrences), but not totally, or frontally. Nevertheless, G-d places Moses on the cleft of a rock and passes before him an aspect of this Divine being, His goodness: "Hashem Hashem (Loving, Loving - both before and after one's transgression). Power of Compassion and freely giving Grace, Long-Suffering (Patient, Tolerant), Full of Lovingkindness and Truth..." (Exodus 34:6-7). These are known as the thirteen attributes of G-d, and they are all expressions of mind and mind-set, spirit and soul, personality and character; they are not at all corporeal or physical. And the Sages of the Talmud (Sotah 14a)- as well as Maimonides (Sefer Ha'Mitzvot No. 8) - have established them as the ultimate guide to human behavior: Just as G-d is compassionate, so must you human beings (created in His image) be compassionate, ... just as G-d is full of lovingkindness, so must you human beings be full of loving-kindness.... In effect, the most significant ramification of this "G-d definition" lies in the realm of anthropology rather than theology!

4. G-D Worship vs. Idol Worship

And when we remember that ancient idolaters worshipped objects and animals, the golden calf, the powerful bull, the nourishing cow, and that the Greek pantheon at Mount Olympus idolized - and idealized - physical characteristics, such as the strength of Zeus, the speed of Mercury, the beauty of Aphrodite, then the fundamental difference between G-d worship and idolatry becomes sharply delineated. Our concept of G-d (or the gods) will go a long way in defining our definition of the human being: is the human being created in the image of a spiritual, eternal, transcendent G-d whom he can and must emulate, or are the gods formed in the physical image of finite, frail and fettered creatures who populate the earth?

5. The Human Image

Indeed, one of the most difficult verses in the Bible, "let us create the human being in our image and after our likeness" (Genesis 1:26), is interpreted by Nachmanides (Ramban dates in his Biblical commentary ad loc) in a manner which directly and inextricably binds our attitude towards the human being and his/her potential capabilities with our concept of the Divine. The obvious question engendered by the Biblical verse relates to whom G-d is addressing in the pronoun "us" and "our". Nachmanides suggests that the Almighty is speaking to the animals and beasts He has just created on that very sixth day of creation: "Let us create the human being in our image;" says G-d. On the one hand, the human being will be like you brute creatures, who are limited in time and strength, who are subject to a finite existence into which they are born, they develop, they degenerate and they die, and who require nutrition, excretion, sleep, and sexual reproduction in order to live; so shall be human beings. But at the same time the human being shall be created in My image, because of which he/she shall be free to choose between good and evil, be able to create, to love, to transcend the physical and to continue to live beyond this existence in am eternal dimension of the spirit.

So does Nachmanides view the human being, part beast and part G-d. So says the Psalmist, sweet singer of Israel: "What is man that You are mindful of him, the son of man that you take him into account? He is but a little lower than G-d, and he is crowned with honor and glory" (Psalms 8:5-6). An so intones our Neilah prayer just before the end of Yom Kippur, our great Day of Forgiveness: "The difference between man and beast is nothing, for all passes away as the vapor of a breath; But you separated the human being (from the animals) from the beginning of Creation, and You recognize him as he stands before You."

On the ceiling of the Sistine Chapel in Rome, Michelango has magnificently painted scenes from the Bible, the central painting being the Divine creation of the human being. Michelango depicts G-d as having created man with His Divine finger, finger of G-d to finger of man. But this is not the Biblical portrayal. The Bible describes: "And the Lord G-d took dust from the earth, and breathed into it the breath of life, and the human being became a living soul." (Genesis 2:7-8).

Michelangelo was an artist, a sculptor and a painter. Apparently, since his personal creativity was vested in his fingers, be transferred this power to the Divine and - different from the Biblical depiction - portrayed the creation of humanity via Divine touch. The Sacred Zohar comments on the Biblical verse cited above: "Anyone who exhales, exhales from within himself, from the inner recesses of his own being." Hence, the philosophic work of Rav Shneur Zalman of Ladi, founder of Habad Hassidut, the Tanya, explains that "the human being is a part of the Divine mammesh", in actuality, even within the physical aspects of his being. Man/women are blessed with a Divine soul, which has the power to suffuse the entire persona and enable the human being to become a "G-d-in-miniature."

6. Mechanists vs. Vitalists

There are two schools of philosophical thought as to the nature of the human being, the mechanists and the vitalists. The mechanists, as in Skinner's Walden Two, maintain that a human being is merely a complex animal who is devoid of freedom of choice and has neither the ability to change himself or change the world; the vitalists - in a long tradition extending from Plato to Henri Bergsson - maintain that a spirit of eternity, a breath of the Divine, informs every human being and endows each individual with the capacity to transcend many physical limitations, to change him/herself as well as society and ultimately to become a full partner with G-d to perfect the world. The Bible insists that one who chooses G-d is in effect choosing a life of action and not merely reaction, a significant life rather than a subservient existence, a mastery over oneself and one's destiny rather than an acceptance of genetics and environment, participation in eternity rather than merely "getting through" the moment. "Life and death have I given before you; choose life."

GLOSSARY TO THE NAMES IN THE ABOVE CHAPTER

ONKELOS THE CONVERT
(ca. 35 C.E. – ca. 120 C.E.)
Onkelos was a convert to Judaism who wrote the most popular translation of the Torah into Aramaic. He lived during the first generation after the destruction of the Second Temple. Onkelos studied Torah under the greatest Sages of the Jewish People. His translation into Aramaic is written on the side of the page in many editions of the Torah printed books.

RAMBAM
1135-1204
Rabbi Moshe ben Maimon was a rabbinic authority, codifier, philosopher and court physician to the ruler of Egypt. He wrote a Commentary on the Mishnah, Sefer HaMitzvot, the philosophical work: Guide to the Perplexed, and the 14 volume work on Jewish Law called the Mishneh Torah.

RAMBAN
1194-1270
Rabbi Moshe ben Nachman was the leading Bible and Talmud scholar in the generation following Ramabam and was also a renowned philosopher, poet and physicain. He wrote a Commentary on the Torah, and published works on Talmud and Halachah.

RASHI
1040-1105
Rabbi Shomo ben Yitzchak was the leading commentator on the Bible and the Talmud. He was born in Troyes, France in 1040. His commentaries on the Bible and Talmud have given greater understanding to the text. His commentaries are written on the same page as the text of the Bible and Talmud.

REB SHNEUR ZALMAN
(1745-1813)
Reb Shneur Zalman was the founder of Chabad Chassidus, and learned about Hasidism from Rabbi Dov Baer HaMaggid, leader of the Hasidic movement. Under The Maggid, Reb Shneur Zalman wrote a profound commentary on the Shulchan Aruch called Shulchan AruchHaRav Reb Shneur Zalman later published the **Tanya**, which was accepted as the written philosophy of Chabad Chassidus. His ability to explain even the most complex issues of Torah made his writings popular with Torah scholars everywhere.

Biodata of **Avshalom C. Elitzur** author of ***"When Form Outlasts Its Medium"***

Dr. Avshalom Elitzur is a senior lecturer at the Unit for Interdisciplinary Studies in Bar-Ilan University, Israel. He obtained his PhD from Tel-Aviv University in 1999. His publications deal with quantum measurement and non-locality, time's arrow and application of thermodynamics and information theory to the study of evolution.

E-mail: **avshalom.elitzur@weizmann.ac.il**

J. Seckbach (ed.), Life as We Know It, 607-620.

WHEN FORM OUTLASTS ITS MEDIUM:
A Definition of Life Integrating Platonism and Thermodynamics

AVSHALOM C. ELITZUR
Unit for Interdisciplinary Studies,
Bar-Ilan University, 52900 Ramat-Gan, Israel

1. Foreword

"Life," mused Søren Kierkegaard, "can only be understood backwards; but it must be lived forwards." We all know life first hand, but formulating this knowledge in scientific terms proves to be an elusive task (Fry, 2000; Korzeniewski, 2001; Lazcano, 1994; Luigi Luisi, 1998; Morowitz, 1992; Palyi et al., 2000; Palyi et al., 2002; Rizzotti, 1996; Schrödinger, 1945).

My own attempt to meet this challenge has taken a long path. I began with writing down all the characteristics of the known living organisms that seemed to be fundamental, shifting viewpoints between biology, physics, and philosophy. Naturally, the resulting list was long, way far from the categorical "Life is ..." statements that most theorists are after. But as time went by and I revised the list time and again, something began crystallizing, perhaps not unlike the way a mathematician's many-sheets scribbles eventually converge into a single elegant equation. Finally, a new "Life is..." has emerged, proposed here on Section 10.

Because some novel insights have been gained during the various stages of this search, I present in this article the entire way of thinking that has lead to the proposed definition.

2. A "Wet Biology" Prologue

When seeking to abstract life, to refine a single unifying principle that underlies its countless manifestations, perhaps there is no better starting point than an encounter with a living thing in the flesh – moving, moist, unexpected, even offensive – like the one that I had not long ago. I was crossing a garden late at night when, out of the carpet of dry leaves, pinecones and pebbles that covered the ground, a toad suddenly leaped high and scared me for a moment. It was this sudden movement of the little creature, which had been lying motionless like all the other lifeless things surrounding it, which made me acutely aware of its uniqueness amongst these inanimate objects. Spontaneity, the capability of being the *agent* rather than the *subject* of motion and change, is the first characteristic of life that meets the eye.

Surely, however, the toad was equally alive before leaping, while sitting still. It was also alive during the summer months when it laid dormant underground. Moreover, life dwells, though in different ways, also in the grass and moss, and even in the seeds within the dry pinecones scattered over the garden's soil. The dividing line between animate and inanimate is therefore not always as clear as in the case of a running, screaming and biting animal, but it is surely there. What, then, is that state that makes something "alive"?

3. Dry Abstractions: Form Outweighs Intensity

Clearly, it was the *spontaneity* of the toad's leap that has made it so distinctly alive. Newton's law relates the change in a body's state to the force acting on it: $F = ma$, or $a = F/m$, where a is the acceleration, namely, the change in the object's motion (or rest), F the force exerted on it amd m its mass. Acceleration, then, is proportionate to the force divided by the body's mass. Hence, for a pebble to leap as high the toad did, it would have to be given an appropriate kick. The toad, in contrast, leaped by itself, the only external "force" acting on it being a few photons that impinged on its retina. The acceleration in this case has no proportion to the physical magnitude of the external force which has initiated it.

Of course, the conflict is only superficial: The organism, using energy resources within it, exerts additional force on itself in response to the minor force from outside. But here a genuine hallmark of life emerges: It is the *form* of the external action, rather than its *magnitude*, that determines the magnitude of the organism's reaction. The light falling on the toad's eye from the garden's lamppost, for example, was much stronger, yet it did not elicit the frightened response as the weaker light coming from the approaching human. So, if viewing the organism as one whole – ignoring for a moment its being composed of numerous different parts – we can point out a crucial difference between an inanimate and an animate object. The former reacts to the sheer *magnitude* of forces, while the latter reacts to their *configuration*.

Notice, moreover, that the configuration is largely independent of its medium. In the case of the toad, another form of energy (sound, scent, etc.) may elicit the same response if it carries the same warning.

4. Generalizing: The Living Form Outlasts its Medium

Much as the above encounter confronts us with life's fleshy side, it also highlights life's abstract aspects. For more than two decades I have listened to toads every autumn croaking from the garden's pond, occasionally even seeing a few. Surely there were individual differences between their bodies and croaks, but I failed to notice them. Numerous toads, which to human perception are indistinguishable from the one I saw, inhabited that garden for many years, begetting one another, till the one that I encountered. It is the *form* which has prevailed, while individual animals came and perished.

This ability of form to outlast its substance appears even ontogenetically, during the life of the individual organism. The toad which I saw has probably exchanged all its molecules within the last few years, but it has retained its form. To be sure, it has been an embryo and then a tadpole, and later grew up till reaching its adult size, but even during those intermediate stages the toad's form outlasted (at least partially) the matter of which it was made.

Note that "form" equally refers to the toad's croaks and leaps. It therefore denotes not only spatial structures but temporal patterns as well.

These observations cannot fail to remind us of Plato's philosophy, where the concept of form, and that of ideas in general, featured so prominently. For Plato, pure form had greater reality than the concrete objects molded in that form. Similarly, ideas existed prior to their material realization. Abstract forms and ideas, not material objects, are the foundation of reality, which is why an abstract discipline such as mathematics is so powerful for mastering the physical world.

Aristotle took exception with this view, reasonably arguing that forms cannot exist without some matter of which they ought to be made. Yet Aristotle was an avid biologist, and it is very likely that his lasting preoccupation with taxonomy and with the notion of a species reflected his attempts to struggle with his mentor's challenge. This is even more evident in his speculations about the formation of the embryo, where the interaction of "form" and "matter" (roughly equated to "male" and "female," respectively) somewhat echoes Plato's thinking.

We need not go deeply into this ancient debate when seeking for a definition of life. Yet it is very striking that Plato's argument has a close affinity to two of life's most prominent hallmarks, namely, reproduction and metabolism. In both cases the same wonder occurs: *Matter assembles and disassembles with the inflow and outflow of matter into the living body, and with the births and deaths of countless individual bodies, but the form prevails.*

5. The Thermodynamic Aspect: The Living Form Feeds on Information

Plato's ideas referred to a world which is, by definition, ideal. As our world is far from that, we must complement Platonism with the branch of physics that studies just what makes our world so far from ideal. Thermodynamics studies, *inter alia*, the increase of entropy, which degrades all forms into chaos. Oddly, however, it is the thermodynamic laws that enable life not only to preserve forms against degradation but even to refine and improve them. Thermodynamics shows how, alongside with matter end energy, the living organism processes a third vital currency, namely, information (Elitzur, 1994, 1995; Lahav et al., 2001). A brief introduction to information's role in life would therefore be in order.

For a diver's watch to be waterproof, its designer must use a great deal of information about the conditions undersea. What, then, about a fish? The question makes us realize that "adaptation" means that a great deal of information has been incorporated in the adapting species' genome. Information processing, therefore, underlies evolution, an insight that we owe to the famous "Maxwell's demon" paradox (Leff and Rex, 1990). Consider a closed box, full with gas in equilibrium and divided by a partition into two halves. A tiny demon within the box directs the gas molecules' motions by opening and closing a microscopic door in the partition. Eventually, hot gas forms on one half of the box and cold gas on the other. Entropy has been reduced. But such a reduction normally requires a proportionate energy investment that would increase entropy elsewhere, in compliance with the Second Law. For the demon, however, the energy required for sorting molecules is negligible. A violation of the Second Law seems to ensue.

The paradox was resolved once it was pointed out that the demon needs *information* in order to perform its task. The acquisition of this information (or, more precisely, the repeated erasure of old bits for receiving new ones [Leff and Rex, 1990]) has its cost in energy, which increases entropy more than the entropy reduced by the demon's sorting. From this principle, which assigns an energetic price to information, a complementary principle follows: The use of information can *save* energy. For example, when we open a lock with a key we utilize the information embedded in the key, which makes the enormous force needed for breaking the lock unnecessary (someone else, of course, has already paid the energy cost of engraving the information on the key).

Biological adaptation, I suggest, abounds with such uses of information for saving energy. A tiger, for example, exerts enormous mechanical force to kill its prey. But in the same jungle dwells a cobra that can kill the same prey by merely spitting into its eye. What is striking in the latter case is the apparent disproportion between the negligible force exerted by the predator and the fatal result suffered by the prey. The secret lies in the snake's choice of the appropriate neurotoxin (in this case, cobrotoxin) that precisely matches the acetylcholine receptors at the ends of the prey's muscles. Similar precision is manifested by the choice of the vulnerable point in the prey's body (once the venom has penetrated the prey's eye, its own vascular system carries it from there over the entire body!). In other words, the cobra makes spectacular use of information about its prey's physiology and neurochemistry, thereby saving the energy that the tiger would have to invest for the same purpose of bringing the prey down. Notice that the tiger is taken here merely as an arbitrary baseline for assessing the cobra's efficiency; the tiger's is also a successful organism and its onslaught is also aided by a great deal of information. Still, the force exerted by snake's venom, in comparison, is literally infinitesimal – of molecular scale. It suffices because it is exerted, thanks to the utilization of information, with enormous precision. Of course, the acquisition of this information was paid dearly by the snake's ancestors during the species' evolution, enabling their fortunate descendant to save energy nowadays.

We can therefore formulate the utility of information thus: *With the aid of information, it is possible to perform a given work with much less energy than in the absence of information, as this little energy is invested at the right place and/or at the right time*. Living organisms, then, are lawful Maxwell demons: They save energy by using information, the energetic price of which being already paid by earlier generations during the harsh struggle for survival.

6. Maxwell's Demon in Action: Life Operates at the Micro-Level for Macro-Effects

With the above principles in mind, the toad's sudden spontaneous leap, which appeared to be such a unique characteristic of life, can be illuminated in a new light. If the living organism is a lawful Maxwell demon, then, with the aid of the information, it can act at the microscopic level, producing numerous microscopic effects which then converge into one large macroscopic effect. It is on this small-scale level, unknown to our ancestors, in which one of the most important features of life lies. Medieval thinkers argued that an organism is alive because a non-material soul dwells in it. Their rationale was that there is no material difference between a living toad and one that had just died. It seems to be the same object in both states, hence the only difference could be the soul that has perished or left the body.

The modern answer to this challenge of dualism is based on what we have learned about life's microscopic level since then. What seems to be a homogenous tissue of a leaf, a bone, etc., is in reality a myriad of enormously complex cells, resembling one another in numerous molecular details, eventually succumbing to death but promptly being replaced by new, almost identical ones. Death of the organism occurs when *minute* changes, too small and too many to be reversed or even noticed, occur together in *numerous* cells. It is this microscopic process, occurring immediately after death, which, unknown to our ancestors, gave them the impression that a non-material agent was moving life. Living forms, then, are maintained due to the great precision orchestration of their myriad microscopic mechanisms (Dolev and Elitzur, 1998).

The uniqueness of the biological motion is now illuminated in a new light. Compare the toad's leap from the ground with the opposite occurrence of a pebble falling on the ground. In the latter case, a highly ordered motion of the macroscopic object degrades into a myriad of disordered motions (i.e., heat) of the ground atoms, in compliance with the Second Law of Thermodynamics. In the toad's leap, however, something extraordinary occurs: *Numerous microscopic interactions between actin and myosin molecules within the toad's muscles converge into one macroscopic movement!*

Another hallmark of life thus emerges. The macroscopic manifestations of life are always a mere tip of a microscopic iceberg. This holds even for micro-organisms, as their collective actions lead to macroscopic phenomena, such as a lake turning green due to algae or a human succumbing to influenza. *Life gives rise to an extremely precise cooperation between numerous microscopic motions, separated in space and/or time, orchestrated so as to converge into the same large-scale outcome.*

7. Platonism Again: Forms Transcend Space and Time

The living organism, then, is not a "thing" in the ordinary sense. Whereas a rock is a rock and an iceberg is an iceberg only as long as their matter does not disintegrate, a toad remains a toad even though its matter *keeps* disintegrating; it is its form that prevails. And whereas the rock's or iceberg's interactions with other objects are determined mainly by their gross physical characteristics, such as mass and velocity, the toad's interactions are more determined by the minute atomic details of its DNA, the poison molecules in its paratoid glands, or the neurotransmitter secretions in its synapses. The living form thus assumes a causal role in itself, just like mass, momentum and charge. In fact, life enables the organism's form to largely override its more fundamental physical parameters.

This realization should pervade our use of biological notions. "Self preservation," for example, means that the organism preserves its form and not its matter. Similarly "survival of the fittest" favors not the fittest individuals but the fittest makeups. And a "selfish gene" in the form of a particular DNA segment often leads to its own destruction in favor of copies of it elsewhere. In short, *metabolism and reproduction, life's two most prominent features, are two aspects of form's supremacy over its medium.*

Once a form is considered to be a thing, just like a concrete chunk of matter, an important distinction emerges between the two kinds of things. A material object can reside only in one place at a time. Not so with a form: If there are many objects with exactly the same form, then there is one form that exists in many places at the same time!

This formulation might sound like a mere play of words, but in the next section I will show that it is this illocality of form that enables life to increasingly transcend space and time limitations. Notice, first, that illocality is already implicit in our biological parlance. When we say that "the gene BRCA1 is responsible for breast cancer" or "the fox is common in the British Isles," we do not refer a certain DNA segment within some individual cell, nor to a particular animal, but to one form whose copies abound in numerous places and times. True, the biological form is a far cry from Plato's ideal forms that reside out of space and time, and yet it is able to gradually transcend space and time limitations.

8. Platonism and Thermodynamics Merging: Evolution Involves an Increase of Informational Invariance

The Platonic and the thermodynamic aspects of life now begin to converge. Since the organism is not an ordinary chunk of matter but a form that survives its matter, and since form can reside in numerous places at the same time, this illocality of form enables the information accumulated within the organism's genome to become increasingly more *valuable*.

But how can one quantify the information value of a certain DNA sequence? This is a highly disputed issue, which we can avoid by addressing one special aspect of information, namely, the scope of its relevance. The Cobra's venom is so powerful because it matches the nervous systems of *all* vertebrates. Consider next plants' geotropism, namely, the mechanism that enables the organism to sense the direction of the gravitational force. Many plants develop individual forms in adaptation to the local conditions, e.g., the ground's slope or sunlight's direction. Yet a few trees, such as the fir and the cypress, have a uniform shape which is largely independent of the local conditions. Interestingly, these uniform trees always grow straight upwards, as they rely mainly on gravity, which is the same everywhere. In fact, the evolution of geotropism means that "knowledge" of gravity was long ago obtained by the plant's genome, for, had Newton's G been other than 6.67×10^{-11} N m^2/kg^2, the statoliths within the plant's cell would fail to properly sediment in the ambient fluid. Here again, the organism gains information about a feature of the environment that prevails in *all* locations, on Earth, in *all* times.

In some cases the information accumulated in an organism's genome is so subtle, hence so abstract, that it can surprise even the human mathematician. Take, for example, the 13-years cicada and its relative, the 17-years cicada. The nymphs of these species develop below ground and after 13 or 17 years they emerge and molt to the adult stage. Their massive brood emergence is believed to overwhelm predators, which are mostly birds.

Is this just another example of sibling species? No, for there are *three different cicada species, each having a 13- and a 17-years subspecies* (Grant, 2005; Williams and Simon, 1995). Clearly, these species could not have simply evolved from one another, because each species could have evolved from either the 13- or the 17-years subspecies of the other species, leaving the emergence of the complementary 17- or 13-years subspecies unexplained. It is evolution, then, that has come up, three times, with the same pair of primary numbers in its search for numbers that do not divide into smaller numbers, mainly in order to prevent convergence of prey's and predator's life-cycles (Dawkins, 1987; Gould, 1977).

These examples demonstrate an essential feature of biological information: This information concerns *invariant* features of the environment (Elitzur, 1997).The information encoded in the cobra's venom does not concern only an individual animal, nor a particular species, but something common to *all* vertebrates. The information encoded in the plant's geotropic mechanism is valuable *everywhere*. The information used by the cicada concerns the life cycles of *all* its predators. The latter case exhibits an even more invariant kind of information: Numbers are the subtlest aspect of physical reality! In fact, evolution gives a very profound clue as to how such a level of abstraction has been reached: The cicada has obtained information not merely about its environment but about the information obtained by other information-processing systems – higher-order information, so to speak. First, the cicada's predators have "learned to count the years" so as to establish their specific life cycles, in order for all males and females of the same species to reach sexual maturity at the same time. Then, after these numbers were taken, the cicadas had to "find" the numbers that do not divide with any of these cycles. This sequence is not much different from the way human arithmetic developed over the centuries: The ancient discovery of numbers has later enabled the recognition of special numbers such as primary numbers, whose properties were derived from those of the ordinary numbers. Plato would probably be pleased to learn that numbers have played a role in the lives of insects long before the appearance of human arithmetic!

Evolution, to reiterate, is a process by which information about the environment is accumulated in the species' genomes, and this information's relevance becomes increasingly broader. We are now in a position to prove a more specific hypothesis: *Evolution is a very efficient mechanism for extracting environmental information out of the environmental noise.* A simple analogy would serve as a useful introduction.

Consider the light coming from a distant star. Its information content is poor. The reason for this is not, as one might think, the light's weakness; light can easily be amplified. Rather, it is the fluctuations (atmospheric and optic), that pollute the information carried by the star's dim light. In other words, the information coming from the star is inflicted with random noise. For the small human pupil, this signal-to-noise ratio is too great to resolve. The Newtonian telescope overcomes this difficulty with the aid of a large concave mirror, up to a few meters, that collects the light signals over a large area and concentrates them on the telescope's lens. Here an efficient resolution of signal from noise takes place: The constant signals (coming from the star) are additive, while the random fluctuations (caused by other factors) are much less so. Let s and Δs denote signal and noise, respectively. Then, joining the radiation coming from n points,

$$\frac{S}{\Delta S} \rightarrow \frac{nS}{\sqrt{n}\Delta S} = \frac{\sqrt{n}S}{\Delta S}.$$

Noise thus "cancels our" in comparison to the strengthening of the signal. The same principle is utilized by all antennae.

The biological analogy is clear. Consider a single, only slightly advantageous mutation. At the individual level, this mutation can hardly affect survival; an organism possessing it might happen to fall victim to an accidental calamity while an organism lacking it might survive by sheer luck. If, however, the mutation has managed to be replicated in large numbers within a certain population, the above dynamics takes over. With n being large (many organisms over many generations), even the slightest advantage will eventually gain dominance over the population. In information theory terms, the weak environmental feature that gives a slight advantage to the mutation is amplified by evolution. Elsewhere (Elitzur, 1994) I have pointed out several other such "proto-cognitive" capabilities of evolution.

Let us summarize: *Alongside with matter and energy, organisms constantly process information.* It is common knowledge that the genome contains information as to how the organism should be assembled. But as Maynard-Smith (1999) has pointed out, this information is useful only by virtue of its reference to the particular *environment* in which this organism must live. The central concept of evolutionary theory, namely, adaptation, thus gains a novel meaning. Adapting to a certain environment necessitates, first and foremost, reliable quantitative information about it.

9. Inventorying the Hallmarks of a Living Organism

Based on the above discussions, we are ready to prepare a tentative inventory of the physical attributes of the living organism, leaving the definition of life itself for the next stage. Which of life's hallmarks should come first? Rather than trying to assign them any order of importance, let us adopt a more pragmatic order: First let those properties that immediately meet the eye be pointed out. Next shall come the properties that appear over longer observation periods and those that are revealed when going down to the smaller scale level. Still, as the recurrent cross-references below indicate, life itself defies any attempt to describe it in a linear order.

A Living Organism, then, is a system in which **Life** is embodied by exhibiting the following properties:

1. SELFHOOD: The organism's ingredients – matter and energy – form a distinct entity, as
   - 1.1. INTEGRITY: the organism's parts, which, in the absence of life, would have been dissociated, remain connected.
   - 1.2. ANIMATION: Yet the organism is not static: Its parts, which, in the absence of life, could have been still, are often in change and motion.
   - 1.3. ORGANIZATION: Both the organism's structure (1.1) and dynamics (1.2) maintain strict coordination in space and time. Nonrandom patterns such as regularity, uniformity, rhythm and symmetry make the organism's constituents causally related to one another more than to external objects.
2. AGENCY: Although the organism is subject to external forces, it is also the agent of spontaneous actions, initiated and brought about by its own energy resources, released in response to inner events.
3. COGNIZANCE: The organism's reaction to external force does not follow the straightforward $F = ma$ relation, as its own resources of potential energy release additional force during the reaction. Its reaction, therefore, depends not only on the external force's *magnitude* but on its *configuration* as well.
4. PURPOSE: The organism's spontaneous actions (2) and its reactions to external events (2), while causally following past events, turn out to be directed by future goals as well: Out of several, equally-possible actions, the organism takes only that one which is likely to bring about a particular outcome in the future
5. SELF-ACTION: on the organism itself. In the long run, *all* the organism's actions eventually affect itself,
6. PERPETUATION: preserving its selfhood (1) through
7. METABOLISM: incessant replacement of parts of the organism that leave it, such that even when the organism's form changes, it maintains its integrity (1.1), animation (1.2) and organization (1.3); as well as
8. REPLICATION: replacement of the entire organism: Even by the time the entire organism ceased to be alive, (an)other organism(s), carrying part or all of its essential characteristics, may have been produced by it.
9. NOVELTY: New elements always appear in the organism's dynamics, due mainly to entropy but
10. PROGRESS: systematically selected such that the organism's properties become more and more prominent.
11. INFORMATION PROCESSING: Together with matter and energy, the living organism incorporates, stores, exchanges and processes information. This information is accumulated during phylogeny (evolution) and /or ontogeny (learning) and takes part in all biological processes. It is information exchanged between the organism's inner constituents that enables the organism to maintain itself (1); it is information about surrounding events that enables the organism to discern (2) events that bear on its survival, to anticipate (4) them and to adequately respond (2) to them; it is the information constituting the organism's

blueprint that enables it to replicate (8); and it is information about a particular environment that underlies adaptation to that environment. Biological information increases the efficiency of any work carried out by the organism, by the principle "less and less energy, but more and more precisely at the right place and/or the right time." Biological information eventually attains higher value as
11.1. REFLEXIVITY: the living organism processes information not only about its environment but also about other information-processing systems like itself, thereby gaining higher-order information; and as
11.2. ABSTRACTION: the processing of environmental information reveals ever more subtle regularities in the environment, i.e., patterns that are ever more invariant in space and time, like natural law itself.

10. The Definition

Can an inventory of characteristics – assuming that they are indeed essential – converge into a concise definition? The main insight inspired by the encounter with the toad, and later reflected in the inventory, is that life manifests a striking degree of supremacy of form over its medium. Each organism's form is embedded, of course, in matter, but it *outlasts* the individual chunk of matter in which it is embodied. Both metabolism and replication are manifestations of this principle.

Based on this aspect of life, here is the first attempt at definition:
"**Life is** a process by which a form outlasts the medium in which it is embedded."

Immediate objections are expected. The living form does not remain unchanged, as organisms grow and species evolve. But then, growth and evolution make the organism even more capable of outlasting its medium, and of overriding the physical constraints to which an inanimate object of the same mass and chemical composition would be subject. A better proposal would therefore be
"**Life is** a process by which forms become increasingly independent of the medium in which they are embedded."

By referring to "forms" in plural our definition encompasses also the organism's change of form. It is, Platonically speaking, not just "a form" but "form" in general – the *idea* of form – that outlasts its medium, thanks to life. The adverb "increasingly" means that evolution makes forms *more and more* liberated from the physical constraints of their medium.

Still, the definition is not exclusive enough. A tornado is a very unique form, made of air and debris which are constantly replaced by new air and debris, such that only the tornado's form remains and its power even increases. Yet a tornado is certainly not alive. But we now recall that the supremacy of form over the medium holds not only for the organism's body but also for what this body reacts to, namely, environmental information (see sections 5-8). Information, by definition, is characterized by its *configuration* rather than by the type of matter or energy in which it is stored or carried. As we have observed, the organism reacts to the *information itself* rather than to its medium. This property is much more typical of living organisms, and together with the above characterization gives a sharper definition:

"**Life is** a process by which forms become increasingly independent of the medium in which they are embedded, by interacting not only with their environment's matter and energy but also by interacting with the forms, *as* forms, which are embedded in the environment."

Information, then, is a special kind of form, which abounds in the organism's environment, and the living organism discerns and utilizes these subtle forms in order to enable its own form to outlast its medium.

Information, however, rarely appears in Nature in its ideal form. This is why, alongside with Plato's lofty ideas, we summoned also the mundane science of thermodynamics which obliges form and information to constantly erode. In section 8 we pointed out the way an evolving population overcomes this informational erosion and extracts environmental signals from the environmental noise: Reproduction utilizes the statistical advantages of large numbers in order to increase the signal-to-noise ratio. We can therefore give our Platonic definition a last, information-theoretical twist:

"Life is a process by which forms become increasingly liberated of the constraints of their material medium, thereby appearing in a multitude of places and times, thereby interacting not only with the local, random aspects of the environment, but, increasingly, with the invariant spatio-temporal regularities underlying Nature, namely, physical laws themselves."

Would this definition prove satisfactory next time one encounters a living organism and wonders what makes it alive? The answer must be left to you, the reader. The next animal or plant that you will see has a unique form, assembled of numerous sub-forms, patterns, rhythms and regularities in space and time. These forms last despite the fact that the matter and energy of which they are made are constantly lost, only to be replaced by other matter and energy, over and over again, yet the form outlasts them all. And this form, which constantly interacts with other things, interacts not only with the matter and energy of which these other things are made, but mainly with their forms. Form gradually became liberated from the physical constraints of its medium, though never fully independent of it as in Plato's world of ideas. So when human thoughts long survive their originators, in language, in letters and on the Internet, this is the most natural consequence of what life is all about.

11. Requiem to the Toad: Life's Essence and Value

The living thing which I encountered was a rather unassuming species, grey-brownish, moist and coarse. But some other members of the order *Anura*, to which toads and frogs belong, are among the most beautiful creatures on Earth. Their striking colors signal that they are highly poisonous, which means that highly valuable medical information about us, their most menacing enemies, is concealed in their skins. Unfortunately, most of this knowledge is being rapidly lost to humanity. *Anura* and all amphibians are going extinct all over the world (Houlahan et al.,

2000). The devastation caused by humans to the biosphere are changes to which these little creatures turn out to be most vulnerable. Ecologists keep warning us that these species are only early indicators of a global catastrophe that threatens all life on Earth, us as well.

Much as the definition of life's *essence* is elusive, the search for its essential physical characteristics seems to enforce on our mind also life's *value*. Life is unique, amazing, precious – and sacred.

12. Acknowledgements

It is a pleasure to thank five anonymous referees and to the chief editor Joseph Professor Seckbach for many valuable comments. Thanks are due to Harald Atmanspacher and the Institut für Grenzgebiete der Psychologie und Psychohygiene at Freiburg, where the last version of this article has been completed. I dedicate this work to the memory of Shneior Lifson (1914-2001), my sorely missed mentor and friend.

13. References

Dawkins, R. (1987). *The Blind Watchmaker.* Norton, New York.

Dolev, S. and Elitzur, A.C. (1998). Biology and thermodynamics: Seemingly-opposite phenomena in search of a unified paradigm. Einstein Quarterly: J. Biol. Med. **15**: 24-33.

Elitzur, A.C. (1994). Let there be life: Thermodynamic reflections on biogenesis and evolution. J. Theor. Biol. **168**: 429-459.

Elitzur, A.C. (1995). Life and mind, past and future: Schrödinger's vision fifty years later. Perspect. Biol. Med. **38**: 433-458.

Elitzur, A.C. (1997). Constancy, uniformity and symmetry of living systems: The computational function of morphological invariance. Biosystems **43**: 41-53.

Fry, I. (2000). *The Emergence of Life on Earth: A Historical and Scientific Overview.* Rutgers University Press, New Brunswick, NJ.

Gould, S.J. (1977). *Ever since Darwin: Reflections on Natural History.* Penguin, Harmondsworth.

Grant, P.R. (2005). The priming of periodical cicada life cycles. Trends Ecol. Evol. **20**: 169-174.

Houlahan, J., Findlay, C. and Schmidt, B. (2000). Quantitative evidence for global amphibian population declines. Nature **404**: 752-755.

Korzeniewski, B. (2001). Cybernetic formulation of the definition of life. J. Theor. Biol. **209**: 275-286.

Lahav, N., Nir, S. and Elitzur, A.C. (2001). A comprehensive origin of life scenario. Prog. Biophys. Molec. Biol. **75**: 75-120.

Lazcano, A. (1994). The transition from non-living to living. In: S. Bengtson (ed.) *Early Life on Earth.* Columbia University Press, New York, pp. 60-69.

Leff, S.H. and Rex, A.F. Eds. (1990). *Maxwell's Demon: Entropy, Information, Computing.* Princeton University Press, Princeton.

Luizi, P.L. (1998). About various definitions of life. Orig. Life Evol. Biosphere **28**: 613-622.

Maynard Smith, J. (1999). The idea of information in biology. Quart. Rev. Biol. **74**: 395-400.

Morowitz, H. (1992). *Beginnings of Cellular Life.* Yale University Press, New Haven.

Pályi, G., Zucchi, C. and Hajdu, C. (2000). Theories on the origin(s) of life. Acc. Naz. Sci. Lett. Arti Modena **Ser VIII v. II** : 389-415.

Palyi, G., Zucchi, C., Caglioti, L., Eds. (2002). Fundamentals of Life, Elsevier Life Sci., Paris.

Rizzotti, M. (1996). *Defining Life.* University of Padova, Padova.

Schrödinger, E. (1945). *What is Life? The Physical Aspect of the Living Cell.* Cambridge University Press, Cambridge.

Williams, K. and Simon, C. (1995). The ecology, behavior, and evolution of periodical cicadas. *Ann. Rev. Entomol.* **40**: 269-295.

Biodata of **Tahir Shah** author of ***"Epistemological Pluralism"*** and ***"Can there be a Limit on the Complexity of Life"?***

Dr. Tahir Shah divides his time teaching theoretical computer science, research, and yoga. His current research interests include evolution and theories of intelligence, philosophy and foundation of mathematics. He studied physics at Karachi University, Pakistan (undergraduate 1960-64) and graduate studies in theoretical physics at Bologna University, Italy (1964-67) and at Stevens Institute of Technology, USA (1969-71). He was invited to become a reviewer for Mathematical Reviews in 1975 and took chair of mathematics at Alfateh University, Tripoli, Libya, in 1976. After designing intelligent systems for the industry in North America for several years, he returned in 1988 to pursue academic life in Trieste, at ICTP and University of Trieste, Italy. He is also member of the international editorial councils of "Problems of Theory and Methods of Education" (in Russian) and "Revista Cubana de Informática Médica."

E-mail: **shah@spin.it**

J. Seckbach (ed.), Life as We Know It, 621-639.

EPISTEMOLOGICAL PLURALISM:
Realities as Species-specific Interpretations of the Material World

K. TAHIR SHAH
Faculty of Engineering, University of Trieste, Trieste, Italy

"Are we not projecting on nature our own way of grasping order, because we cannot think otherwise"

R. Riedl

1. Introduction

Epistemology, the nature of knowledge about the external world, has been discussed since the dawn of civilization. Realism has been with us for more than two millennia with the view that the external reality has *a priori* existence; it does neither depend on the perceptual-cognitive apparatus of an observer nor on the physical processes that are used for observation. However, the basic ontological problem – how to determine the *a priori* existence of an entity – remained unsolved. Is it the empirical knowledge or purely inductive knowledge that determines its 'existence'? What is the relationship between ontology and epistemology? Is epistemology species-specific even if one take for granted ontological unity? Our present scientific understanding of life and the world around us strongly indicates that the notion of *a priori* reality is a purely neuro-mental construct which originated from mental extrapolation of our ordinary perceptual experience. The purpose of this chapter is to explore these philosophical questions from the perspective of evolution, neuro-ethology and neural and cognitive sciences.

What do we mean by epistemological pluralism? In the broader context of all possible terrestrial life forms and life that may exist elsewhere in our universe, our view is that it is the empirical knowledge available to humans and to any life form through their respective perceptual-cognitive system that determines the nature of reality. Since each and every perceptual-cognitive system is species-specific, all empirical knowledge and any neuro-mental construct derived from it is again species-specific. Such a point of view about the nature of knowledge shall be referred to as species-specific epistemology. If we take into account all terrestrial and extraterrestrial species (there is a concrete possibility of finding all kinds of life forms in the universe), then we find ourselves looking at an immense number of epistemologies. We shall call this viewpoint *epistemological pluralism.* Although this term has been used in social, economic and biological sciences in a slightly different manner [Rose 2002, Ziman 2004], it represents the notion of plurality of knowledge. In essence, we are dealing with the Platonic issue of distinction between 'appearance' and 'reality', but from a different perspective – *life as we know it.* Our thesis is that both the observed knowledge of our external world and the concept of *a priori* 'reality' are species-specific neuro-mental constructs, but each one is of different kind. The observed knowledge requires perceptual apparatus, while the notion of *a priori* reality requires higher cognitive faculties to extrapolate perceptual experience into an entity that is neither knowable directly nor is its existence necessarily verifiable empirically.

2. Historical Perspective: *From Naiyayikas to Android Epistemology*

All classical and modern schools of philosophy discuss only human epistemology and never consider perceptual knowledge of any other species. This began to change with ethological studies of J. von Uexküll and Konrad Lorenz (1973). Later, works of Piaget (1954), Riedl (1984) and the school of radical constructivism of von Glasersfeld (1995) departed from the traditional approaches to epistemology. A recent addition to this trend is android epistemology [Ford et al., 1995]. We outline in this section a few selected views on the nature of knowledge.

2.1. INDIAN SCHOOLS OF EPISTEMOLOGY

There are various schools of Indian epistemology, each with its own "right means" of acquiring knowledge of the external world as a combination of perception, inference, description and comparison as outlined in the following table (for details, see e.g., Coward 1983, Raddhakrishnan 1964).

School	*Right means of acquiring knowledge*
Cãrvakas	Perception
	Verbal testimony (linguistic description)
Mimãnsas	
Vaisesiks	Perception and Inference
	Perception, description and Inference
Samkhyas	
	Perception, description, Inference and Comparison
aiyayikas	
Prãbhãkrs	All four above plus presumption
Bhãttas,	All five above and non-existence
Vedantins	
Pauraňiks	All five of the above and probabilistic methods

In separating knowledge by description, inference and comparison the Nyãyayikas (or Nyãya) school (see foot note 1 and Bochenski 1961 for argument by comparison) came very close to modern thoughts. Most physical models today use comparison and analogies – describing fine structure of the material world through analogy with perceptual objects. Bohr's atomic model of early 20th century and the string theory of particle physics are two such examples. In Nyãya, the knowledge which arises from the contact of a perceptual system with the external world is deterministic (excluding perceptual illusions) and the name of the object does not bear direct relationship to its perceptual knowledge. The method of comparison is based on similarity and uses instantiation mostly within, for instance, natural categories. That is, one perceives an object and recognizes it as an instance of some category. The temporal domain of perception is the present; that of inference is the past, present, and future; while comparison is considered to be an instrument of perception, enabling one to know an object designated by a name (categorization-instantiation) [Bernard 1981, Shaw 1988, Shaw 1991].

Despite this reliance on observational mechanism, however, all schools of Indian epistemology accepted reality as *a priori*. Nevertheless, not being able to resolve the ontological issue, they referred to the external world as *maja* (illusion or appearance). They called the observed reality illusion realizing perhaps that absolute *a priori* reality and truth have simply the status of a 'belief' and not the complete and true knowledge. The term 'belief' is used here in the sense as defined in theories of artificial intelligence where it refers to statements which are considered valid (provable) until shown to be otherwise. A logical system, where a theorem may not remain (valid) provable when extra axioms are added, is called a non-monotonic system. Belief systems are non-monotonic reasoning systems. In Buddhist schools, however, reality is not something that is static, but a constantly changing dynamic reality rendered thus through a clear distinction between the reality of an object now and sometime in future.

2.2 DESCARTES, HUME, KANT AND RUSSELL

"Cogito ergo sum" – "I think, therefore, I am", though written by René Descartes in 1637, still stands as perhaps the most widely cited philosophical dictum. The Cartesian separation of brain and mind implied by this dictum contradicts what we empirically know about the brain and the mind from contemporary researches in neurobiology, psychology and computational studies of intelligence. Our existence as a neuro-cognitively complex species is not simply because we think, a purely conscious mental act, but because we can remember, a strict brain capability, and manipulate what we have learned about our world [Squire and Kendel 1999, Domasio 1995]. It is our memory, a neural capacity to record and store our experiences, the architecture of the brain and complex cognitive capabilities arising from it, all of which forces us to think and perceive the world in a particular way (see footnote 2). When memory is lost or some other parts of the brain are damaged, we are neither able to recreate our past nor to experience the external world consciously, as blind-sight and hemi-neglect disorders demonstrate. This is only one aspect of our central nervous system its storage of all our experience. But there are other equally important functions such as visual processing by the visual cortex, language comprehension and production by the Wernicke and Broca areas etc.

The central problem of Kant's *Critique of Pure Reason* is, also, the nature and limits of human knowledge. Kant's interest in this problem is perhaps due to Hume's challenge to the generally accepted practice that we are justified in using universal quantification "for all", but base it only on few instances. Hume asked for evidence that would allow us to make assertions about things we have not actually experienced. He came very close to what is known as constructivism in mathematics and philosophy by implicitly asking for a clear distinction between belief, default knowledge and (complete) knowledge. Kant believed that it is not possible to find such evidence so long as we continue to think of the mind and object perceived as separate things. He held that the mind is *actively involved* in the entity it perceives by organizing its experience into definite patterns and only through construction of these can one be certain to obtain knowledge of what may not yet been experienced. But, the implication of such a thesis is to abandon any claim to know things as they are in themselves – the reality in which the mind is not involved. In essence, Kant advocated the hypothesis of extrapolation and induction from observed reality and, for example, generalization from such observed instances into a category with all elements having the same observed attributes. A similar position was expressed earlier by Bishop Berkeley (1685-1753) where immediate objects of our senses do not exist independent of us.

Russell (1919) makes a distinction between knowledge acquired through sense perception and knowledge by description, a view very similar to Nyãya philosophy. Linguistic naming of perceptually acquired experience fixes its meaning, which is then propagated by writing or teaching – the so called 'verbal testimony' in Indian epistemologies. His epistemological view began from the assumption that our beliefs about the world must be expressible by certain primitives, he called atomic propositions, each of which involves what he called a 'name' that has its meanings in the object for which it stands. The fundamental principle in the analysis of proposition-containing description is this: Every proposition which we can understand must be composed wholly of constituents which are perceptual objects. The importance of knowledge by description is derived from the fact that we only know truths which are wholly composed of perceptually acquired terms, yet we can have knowledge of things which we have never experienced. Although Russell's solution to Hume's problem uses language-like compositionality, semantics cannot always be determined only by the total sum of meanings of respective constituents, one needs also to consider their interrelationship – an extra-dimensional understanding as it is now known in textual analysis and textual cohesion [Halliday and Hasan 1976]. One may also conclude from this doctrine of fundamental names that what can be said fundamentally also exists. We thus arrive at a view of reality that is somehow founded on our private experience and constructed using a language-like compositionality. But on the other hand, Russell too pointed out in his later writings, for example, "*The world of senses, if supposed to exist outside us, is largely an illusion.*" rejecting both dualism and realism very sharply [Russell 1968, p. 29].

2.3. LANGUAGE AND REALITY

Many philosophers from Plato to Wittgenstein have held the view that language is a reflection of external 'reality' and that the basic structure of language, the *subject-predicate* division reflects *particular-universal*, one of the basic structures of reality. The Kantian variation of this view is that language is not a reflection of reality (which is inaccessible to human cognitive system), but a reflection of our thoughts about reality. The linguistic paradigm of reality can be considered only partially useful because not all thoughts are expressible in linguistic forms, and pure perception may not always be linguistically expressible in a sufficiently comprehensive way. Today, since it is sufficiently established that human language is an innate mental faculty, therefore, even this notion of reality can be taken as strictly belonging to our species. The notion of absolute *a priori* existing reality is therefore not only a neuro-mental construct obtained through idealization and extrapolation of ordinary experience by humans, but it is as much species-specific as what classical philosophers called 'appearance'.

Piaget's constructive theory of knowledge [Piaget 1974] is based on his empirical studies of learning. It represents a major shift from the traditional focus on ontology to the world of experiences of our species. In the spirit of Kant, knowledge of the external world is necessarily determined by the knower's perceptual-cognitive system. Piaget's "re-presentation" of external reality is always a re-construction from experience stored in memory, but this ability to "re-present" objects to oneself is linked to language acquisition and thus to the structure of human language. Human language is sequential and thus any thought, however complex it may be, has to be arranged sequentially. This seems compatible with the sequential nature of our conscious processing system.

The Russellian descriptive knowledge of the external world is dependent on language, which in its turn, is dependent on such mental faculties as intentionality (or theory of mind), unbounded generativity [Hauser-Chomsky-Fitch 2003] and granularity [Shah 2003]. Such a description is a complex process and one comes across numerous problems, many of them still unresolved. For example, the use of linear linguistic expressions and natural categories results in what has been known as qualification and frame problems (see, footnote 3). There are other reasons why one should consider only approximate connection between, for instance, a visually perceived reality and its linguistic description [Edelman 2002]. Describing an object or scene does not really give full perceptual information resulting in a varied interpretation of each descriptive word by each person. One reason is that each of such descriptive words refers to a category of objects and not its specific instance. But, again, language being a species-specific innate mental faculty, knowledge by description is also species-specific.

Both Piaget's epistemology and Wittgenstein's remarks [Wittgenstein 1922] on the nature of reality influenced the new doctrine of *Radical constructivism* [von Glaserfeld 1995]. It is an approach to the problem of knowledge which takes into account both linguistic and cultural differences within humans. According to this doctrine, no matter how one defines knowledge, any kind of experience is essentially subjective. Reality is very much an individual reality and nothing else. An individual has no way of knowing whether another individual of the same species has the same reality or not. We shall deal with this issue later.

2.4. EVOLUTIONARY AND ANDROID EPISTEMOLOGIES

R. Riedl (1984) examined mental faculties from a comparative phylogenetic study of all cognitive processes with the aim to discover under what functional preconditions those mechanisms were evolved and how they differentiated later in evolutionary history. He called this approach *evolutionary epistemology* and claimed that it is not only empirically testable, but "encompass the old problems of the *a priori*, of inductive generalizations of realities …". He further stressed that his "method shows that any cognitive mechanism works correctly only in the environment for which it was selected; beyond these limits, it quite misleads us in our search for new knowledge". In evolutionary epistemology, there is a chain of cognitive systems of increasing complexity in conformity with the idea of evolution's continuity.

Recent studies in artificial intelligence about knowledge, beliefs, attitudes and other such notions has led to what has been called *android epistemology* [Ford et al.,1995]. It is the exploration of how computing machines can deal with epistemological issues of knowledge, beliefs etc. with their 'mental' states. In this domain too, we come across problems such as mind-machine dualism, similar to mind-brain dualism in human epistemological theories.

3. Mind-Brain Dualism: Is it Separation of Abstraction Levels?

In classical cognitive theories [Fodor and Pylyshyn 1988], the brain's architecture and its material implementation and cognitive architecture are considered independent of each other. Perhaps one of the reasons is that we do not know how many other levels of abstraction exist in between neural and cognitive descriptions. Mind-brain dualism is a consequence of a strict separation between two abstraction levels of the neuro-mental phenomenon: the cognitive architecture and its specific hardware implementation. Nevertheless, in real life situations, the speed of computation and other hardware characteristics are factors as important as the cognitive architecture. There are two distinct aspects of hardware characteristics: the nature of the material and physical processes involved in information processing and communication (e.g., electronic, optical or electrochemical), and its organizational architecture. Both of these aspects of material implementation not only give certain flexibility, but also place some crucial constraints on the ability of a cognitive system to deal with the real world survival problems in real-time.

Let us look at dualism from another perspective. What is the relationship between an information processing system and its material implementation, the physical system that does the task of information processing? Shall we search for an answer among the laws of physics or the laws of information? It is almost trivial to conclude that for mind-brain relationship, the laws governing the mind should be sought among the laws governing information and computation. On the other hand, the laws governing its embodiment should be sought in physics. There is perhaps an explanation why material implementation of a computing system and the algorithm that runs on it are considered independent. What is required, is to carefully distinguish information and computation from the medium that carries this information and performs computation by changing its physical states. The key point is to make a distinction between the laws of physics governing the hardware and the laws governing the organization-architecture of its components. Yes, the two are independent as neodualists assert, because physical laws and the laws of information-computation exist at different levels of abstraction and do not influence each other directly. Nevertheless, the laws of physics influence, for example, how a signal should be sent and at what speed, and whether it is electrical, chemical or optical. On the other hand, laws of information processing influence, for example, what can or cannot be computed and whether the computation will ever halt, or if it will halt in exponential time. Thus, while some factors in computational dynamics are no doubt affected by physical laws at the microscopic level, on the other hand, they remain depended only on the organization of its primitive components looked at a different level of abstraction. Since a relationship between two physical objects or systems is non-physical (e.g., probabilistic, deterministic, symbolic or otherwise) and that computation is a sequence of states, the mind or a program can be considered non-physical only in this sense. An algorithm is non-physical only in this sense and so is the mind. Such concepts are meaningless and of no practical use unless they can be implemented and executed on some hardware. The mind is therefore simply a set of time-ordered brain states representing some specific computation.

Our discussion on epistemological pluralism will take place within the computational paradigm with the assumption that cognitive complexity can eventually be reduced into neural complexity, thus connecting the mind-brain separation as argued above (see also, e.g., Churchland 1988). This justifies our use of the term 'neuro-cognitive' system by which we mean the cognitive architecture and the organization of its material implementation (see footnote 4), that is, the nervous system and other instruments, if there are, external to its implementation. The term "observer" means a neuro-cognitive system in the above sense that could be natural, artificial or hybrid. This definition includes a large spectrum of cognitive systems of animals, humans, extraterrestrials, android and so on and of varying sizes and complexities. Furthermore, we shall argue within the well-established theories of domain-specific innate mental modules and sub-modules, sometimes referred as adaptations and preadaptations in the evolutionary context.

4. Nature of Reality Construction: *Why we see "what we see"*

Not only are there a large number of steps as to how knowledge of the external world can be obtained, but a wide variety of processes and computational systems are also possible. It begins as a series of physical processes of information extraction, its coding and transmission, and ultimately coded data reception by a neuro-cognitive system. The reality construction ends with a final interpretation of the data received by the cognitive system. The notion of *a priori* reality is inferred from this final interpretation of perceptual data, but only by a sufficiently complex system with inferential capability.

4.1. INFORMATION AND REALITY

Suppose we believe that out there something really exists and we want to find out empirically through experiment and observation whether there is really something and what is its nature. How do we do this? To investigate, we need some other physical entity that is capable of interacting both with the presumed object and with our sensorial interface. Technically, what we need is to perform a scattering experiment either within the domain of one of the known fundamental physical interactions (e.g., photons or electromagnetic waves) or use other forms of exploration techniques such as chemical interaction (olfaction) or tactile interaction. The reason we need to perform either a scattering experiment or any other form of exploration is to get information about that something out there. This is possible only through some intermediate object which can extract and carry information about the object of investigation and deliver it to the sensorial interface. One may ask further as to how this intermediate physical entity interacts with a sensorial interface, an explanation certainly at a lower level of abstraction. But we leave out this further reduction in our present discussion in order not to end up with infinite regression of finer and finer theories. Terrestrial life forms use various kinds of interactions: chemical, electrical, magnetic, electromagnetic, air and solid vibrations or simply tactile interaction to explore their environments. An obser vational experiment provides information of some specific grain-size depending upon the type of interaction used and specific value of some relevant parameter such as frequency of electromagnetic waves.

What this implies is that we are always forced to use another physical entity, an intermediate entity like a photon or a molecule, to extract information about what is out there. If there is no interaction, we may even erroneously conclude that there is no object out there. In essence, we have no way to find out whether there is really something out there when we have exhausted all scattering experiments. How shall we distinguish between "does not exists" and "not knowable", i.e., between absence and failure to detect. Sometimes it is practical to accept the "*negation as failure*" hypothesis (see footnote 5). However, a realist's statement that "there is something but we cannot find out about it" has the mere status of a belief, a default assumption derived from the extrapolation of our ordinary everyday experience. It does not imply existence or non-existence of an object in any other sense. On the other hand, in our exploration using many different interactions if we detect the presence of something with each type of interaction, then of course we have no way of telling whether what we found was many different "versions" of a "single" object out there, or many distinct objects. One may even interpret this data as different objects, different realities. The result of such experiments will remain inconclusive, unless to begin with *a priori* knowledge of the object and its characteristics are taken for granted. But, how does a realist know *a priori* what to expect?

Information about how the intermediate objects (e.g., photons) are scattered tell us something about the mass distribution (shape) of a physical object that is being explored. Such an experiment may give us information at different levels of description depending upon how the experiment is performed and what electromagnetic frequencies are used. For the sake of argument, lets take a cognitive system that uses x-rays at its sensorial interface instead of ordinary light. It would perhaps interpret an object in a different manner than another cognitive system using different wave lengths of electromagnetic waves. Again, there is a conflict about the real nature of the object. The point is that there is no "real" nature of this object. This is well illustrated when one takes into account other terrestrial species which use more than one type of interaction independently to explore their environment resulting independent multiple realities. A case of multiple realities of how a snake interprets its world is discussed by Sjolander (1997).

4.2. THE ROLE OF INTEGRATIVE MECHANISM IN REALITY CONSTRUCTION

All living beings receive and process information in their central nervous systems of varying complexity. Both the complexity (i.e., its form, grain-size and resolution) of the information received and the capability of the central nervous system of a species determine its perception of the environment where it lives. Plants seem green or flowers colored to normal humans because they receive a certain kind of signal. A color-blind or completely blind person does not interpret information about colors. Their perception of the world is different. For a blind person the world consists of sounds, tactile information, smells and tastes, all fairly coarse-grained as compared to visual information. In a similar way, each species, each life form, perceives the world depending upon the type of signal it can receive through its perceptual interface and its information processing capacity. Some mental modules seems to be uniquely human. For instance, temporal event ordering capability is a trait possessed only by higher primates among all other mammals. It appears that most mammals, including some non-human primates, consider only *here and now* reality. To express temporal variants human language became more complex in terms of past-present-future. Further extrapolation of reality, from actual to possible worlds (say, counter-factual variants of reality), adds further complexity to human language and makes the human version of reality more complex than other terrestrial life forms.

In the classical version of epistemology, to know an object is to be aware of this object. Can we know something without being aware of it? Human subjects with blindsight disorder are not aware of objects in some section of their visual field, but they are able to avoid hitting them if such objects are in their way. Blindsight can be described as disassociation of visual detection and awareness. What kind of visual epistemology can be attributed to a person with this disorder? Does the object out there exist for them or not? There is no way they can find out about the presence of such objects using their visual awareness system.

To be aware of some specific object, the neuro-cognitive system must be able to separate information relating to that object from all others in its visual field. Such information is collected by neurons scattered around in many visual areas of the nervous system. Now, how do widely spaced neural regions manage to coordinate their part of the information into an image without forming hallucinatory feature integration? This process is known as *visual binding*. A problem still not fully solved is how the visual system correctly binds the features of a single object rather than those of different objects. What is the mechanism that prevents the formation of illusory conjunctions between features of different objects? In a series of papers, Triesman and her collaborators (1988) proposed a theory of visual feature integration, where a wide range of different features are represented in separate retinotopically organized maps; these features include color, motion, as well as orientation and various other elements of form.

Attention plays a crucial role in awareness, perception and cognitive performance. There appear to be three major functions of attention: maintenance of the state of alert, signal detection for conscious processing and orienting to relevant sensory events. We know with certainty that human perception has the ability to focus attention from one grain-size to another. The grain-size at which we choose to focus our attention affects not only what we can distinguish from its background, but also what becomes indistinguishable. This permits the human perceptual system to ignore details that are not essential to achieve a given objective. How such a shift in grain-size is achieved by the neuro-cognitive system is fundamental to our understanding of many interrelated processes such as consciousness and ultimately the nature of reality.

Binding is a broad and all pervading phenomenon that needs an explanation. We should distinguish different levels of description at which binding and integration is achieved in the neuro-cognitive system. One may even say that binding is the brain's ability to produce an integrated representation of the world at the level of awareness, although the information received may come from many different modalities and in a rather fragmented fashion. In the end, it all depends on the circuits that a species' brain has in order to integrate and interpret.

4.3. NEURAL CORRELATES OF BINDING

The construction of our visual reality depends on various mechanisms of integration in the nervous system. For instance, recent evidence suggests that our visual reality is constructed through synchronizing bi-directional connections between various cortical areas, also referred to as visual processing modules. In mammals such cortico-cortical connections develop mainly after birth, and they attain their final specificity through an activity-dependent selection process. The experimental evidence for inter-modular, bi-directional connections seems to be valid for most mammals except in their variations in the number and type of modules (visual, olfactory etc.). The visual binding problem requires a neuronal mechanism which allows to compare responses of spatially distributed feature detectors and to distinguish the responses to pattern elements that have certain features in common.

The discovery of synchronous collective neural activity in the visual and olfactory cortices has helped us to explain the nature of perception and consciousness. Results were obtained by Singer and his collaborators in the cat visual cortex [Singer 1995, Singer and Gray 1995, Engel and Singer 2001]. While recording electrical signals from widely spaced neurons in the brains of cats, they discovered that neurons tend to fire synchronous electrical impulses when responding to stimuli that appear to come from the same object. Some neurons respond to orientation while others respond to color and so on. Further work showed that neurons exhibited oscillatory responses to moving light stimuli and oscillations occurring in a frequency range 40 – 60 Hz. Neurons with overlapping receptive fields, when activated by a single stimulus, synchronize their oscillatory responses on the average with no phase difference, referred to as the phase-locked oscillation. This synchronization was found to occur, with high probability, for all combinations of orientation preferences and the degree of synchronization between the oscillatory responses reflects the coherence of the stimulus as it is perceived by a human observer. It also plays a role in the use-dependent long-term modifications of synaptic efficacy, i.e., memory.

It was proposed by Crick and Koch [1990] that visual awareness is the result of neural computations on an attended object expressed by phase-locked 40 Hz oscillations. These phase-locked oscillations are neural correlates of vivid visual awareness and activate the working memory. *Such synchronous oscillations are possible only if relevant neurons are connected to each other.* The construction of visual reality, it is clear, depends on these connections. A removal of these bi-directional connections will cause improper reality construction and may leave an animal partially or fully object-blind. Hemineglect, blindsight, and color-blindness are classical example of deficit in reality construction due to brain damage in specific areas. This area of research is still in its infancy and only future experiments will be able to shed further light on neural correlates of integrative mechanism.

An important issue in species-specific epistemology is the extent to which all brains belonging to the same species work in the same manner. Quine (1960) expressed extreme pessimism about the indeterminate nature of other's epistemology. This is because there is no real evidence of what somebody else's perception of reality is, I do not have access to your perceived world even though we belong to the same species. Despite this, I and you do communicate with each other. The question then is whether reality is a strictly individual experience or a collective reality of a species. Can this issue be settled empirically? A recent result of Hasson et al. (2004) showed experimentally that members of the same species (human in their studies) see the external world in the same way. In their experiment, they looked at brain activity of five subjects using functional imaging while the subjects were shown a movie under free natural conditions. Natural conditions differ from controlled laboratory conditions in that there is more flexibility of behavior and stimuli are not shown in isolation. The similarity of their response measured in this experiment implies that the human cortex is stereotypically responsive to audio-visual stimuli under natural conditions. Their result is of great empirical significance in order to dispel Quine's pessimism. Moreover, to a large extent, it confirms the species-specificity of sensory perception.

5. Epistemological Pluralism Due to Granularity

The ability to construct reality at different levels of abstraction is another fundamental component of the human neuro-cognitive system. Let us call this faculty the "granularity of mind". Any life form or machine with a higher level of intelligence must have granularity of mind in its set of mental faculties. Let us look at an example to clarify how granularity can lead to epistemological pluralism, but in a restricted sense. In sciences, the description of nature is given at different levels of description, from physics to chemistry to biology. Within physical theories, atoms, elementary particles, quarks and strings, all describe the same physical world, but each at a different level of abstraction. Many researchers [Rose 2002] interpret this as an example of epistemological pluralism since the knowledge of the world is constructed at many independent levels of abstraction.

Another example of restricted epistemological pluralism within a species comes from the ability to focus attention at various levels of details and shift focus from one level to another. Focusing at a certain grain-size helps a neuro-cognitive system to ignore many confusing details and enables it to map the complexities of external world into simple model-of-the-world that requires less computational resources than detailed models. For instance, our descriptive knowledge consists of a global model together with a large number of relatively simple, idealized and coarse-grain, local models [Hobbs 1985]. Two widely used granularity processes are *abstraction*, which corresponds to shifts in focus from general to specific (or vice-versa) and *aggregation,* which corresponds to shifts in focus through part-whole relationship.

In granularity theory, encapsulation of information at each level of abstraction is an essential aspect. For objects under observation, the observer does not need to know how information is internally "stored" and "processed". This is sometimes called encapsulation, which not only hides the internal structure of the defined objects, but also prevents improper manipulation of these objects. Nature seems to use *encapsulation principle* at many levels to protect implementation details against access from the outside to enhance the stability of the object, be it an electron, a nervous system or a cognitive system. There seems to be a step-wise refinement and a step-wise encapsulation of structures in our physical and cognitive world. Objects at each grain-size are encapsulated and reflect a specific level of reality. Thus, any cognitive system with granularity of mind should be able to detect a multifold of realities. The epistemological pluralism discussed by Rose (2002) and Ziman (2003) is of this type. In our discussion we shall refer to it as epistemological pluralism in the restricted sense to avoid confusion with our earlier definition.

6. Epistemological Pluralism Due to Diversity of Neuro-Cognitive Systems

Evolutionary foundation of epistemological pluralism has its roots both in comparative ethology and comparative psychology. Darwin in his 'Origin of Species' stated clearly that psychology will be based on "a new foundation, that of the necessary acquirement of each mental power and capacity by gradation". In fact, recent research on a large number of lower and higher animals such as great apes, earth worms, antelopes, dogs, spiders, dolphins, bees, fish, rats and others [Bekoff, Allen and Burghardt 2002] confirms a remarkable variation and species-specific behavior with domain expertise. Animals are all specialists! They are experts in domain-specific tasks. Consequently, their reality is most likely domain-specific and often single modality as compared to multi-modal generalist mammals, such as higher primates.

To illustrate the differences between the subjective, species-specific world of animals, ethologist Von Uexküll was perhaps the first to discuss the selection of sensory stimuli from the environment by different species. This subjective world is made up of the stimuli selected for attention which he called the *Umwelt.* Basically, it is both the alteration in the stimulus filtering aspect of the perceptual mechanism and the basic architecture of the sensory system from species to species that results in the variation of input data to be processed by their respective cognitive system and consequently variation in perceived object. Each terrestrial species has its own perceptual mechanism and thus has its own *Umwelt*, which could be radically different from those of other species. Saying that each species has its own *Umwelt* is equivalent to saying that reality is species-specific. Since each animal species has his its own *Umwelt,* which is different from ours and from other species of animals, its thought process, or its awareness, may be radically different from ours.

Epistemological pluralism avoids anthropomorphism or, at least, tries to pose the problem of reality of how other species see the world. In the broader astrobiological context, we should also give due consideration to extraterrestrial life forms and how they see the world keeping in mind that an organism's reaction to the external world is not based on *a priori* reality but only on its own representation of the world.

7. Inter-species Communication Problem

There are a number of suggestions to use numbers (e.g., prime numbers) and fundamental physical constants to communicate with extraterrestrial civilizations (Conway et al., 2003). The problem with this approach is that recent neurobiological data suggests arithmetic to be an innate human faculty (Dehaene 1997). We do not know if aliens can have the same or even similar number sense as ours. We do not have the experience of systematic communication even with other terrestrial species except our pets and animals in experimental laboratories. There is evidence for a certain degree of communication and varying numerical capabilities such as relative numerousness, subitizing, estimation, counting judgments, or proto-counting in many species of mammals [Davis and Pe'russe 1988], but many other important cognitive and morphological capabilities are missing in these animals. For instance, non-human primates may be capable of planning or tactical deception [Cheney and Seyfarth 1992] but they certainly lack a sophisticated theory of mind and unbounded language generation capability. To communicate with other terrestrial species, some degree of similarity between our respective neuro-cognitive system is necessary. This is supplied by common phylogenetic heritage of all higher terrestrial life forms – from mouse to humans. Any empirical evidence of communication among terrestrial species is not due to *a priori* existence of external reality but similarities of nervous systems, despite their different level of complexities.

There are perhaps more serious problems of definition and of comparability in the context of communication between terrestrials and extraterrestrial intelligent species. We may come across radically different neuro-cognitive capabilities which we cannot even imagine at present. One can only conjecture some kind of empirical verification process by terrestrials and extraterrestrials (or even by androids) in order to have a common translation protocol even though their languages and brains may be radically different. Such a translation protocol can possibly give a coherent way of comparing alien mathematics and science with terrestrial mathematics and science. Of course, we are not assuming that two kinds of mathematics are the same. It is difficult for us to imagine that alien mathematics can be different from ours because our cognitive system does not allow us to imagine anything radically different, as different as, for instance, imagining how a dog thinks about our way of looking at the world. Nevertheless, we are able to communicate with our pets on concrete objects such as food and other objects in the environment. Our common protocol (of translation) is represented by empirically verifiable world objects that both of us are capable of perceiving in however different way. Such a common protocol with our pets seems possible because the mammalian perceptual system uses the same kind of architecture for visual feature integration with only some variation in the number of modules or specific cortical areas dedicated to visual processing. We do not know whether this will make any sense for an alien species! (see comments by Luisi 2003).

8. Conclusion

From the above discussion one may conclude that all objects carry information about them in ways we do not understand at present. This statement does not contradict our thesis that reality is really an interpretation by a given cognitive system and thus species-specific. But, it leaves us with a deeper paradox. Why do objects behave the way they do? Why does an electron, for instance, always behave as an electron and not like another entity? The so called 'stability of object' is a delicate matter to understand. It nevertheless can be understood if we are careful in defining what we mean by an object and separate its perceptual component from the extrapolative component obtained through inference. We emphasize again that the notion of *a priori* reality is a purely neuro-mental construct which originated from the mental extrapolation of our ordinary perceptual experience.

One may even doubt that there are objects *per se*, since it is the information extraction, transmission and its interpretation that determines what, if there is any object, where it is located and how it "looks like". An object is really a bundle of information about how to interact with other objects of the universe. In this sense, then, an object is really a multi-fold of objects, or perhaps an infinite number of objects, depending upon how it is discovered and by which perceptual-cognitive system it is perceived and interpreted. How this information is encoded and processed we do not know. We can extract only a part of that information of a certain grain size depending upon our instruments. That is all. The information is utilized not only to control the behavior of an object, but the object itself is essentially defined in terms of the information it carries. *Reality is thus intricately intermingled with the complex nature of information and the ways in which it is extracted, perceived and interpreted.* Nevertheless, our thesis remains that both the perception-based knowledge and the concept of *a priori* 'reality' are species-specific neuro-mental constructs, but each one is of different kind. The observed knowledge requires a perceptual apparatus, while the notion of *a priori* reality requires a higher cognitive faculty of induction to extrapolate a perceptual entity into something that cannot be proven empirically to exist.

The perceptual knowledge of the external world is not only dependent on physical processes of extraction, encoding and transmission of information, but also on the structure and consequently capabilities of a neuro-cognitive system that process this information. What is knowable is always knowable only through some neuro-cognitive system and thus such knowledge is constrained by its perceptual and computational mechanisms. Of course we humans can extend our perceptual capabilities with the help of instruments and obtain a relatively fine-grained view of the external world. But that is all we can do. Since perceptual systems are species-specific, the perceptual knowledge of reality is thus strictly species-specific. Each life form with its unique way of constructing and representing the external world has its unique epistemology. In a universe of immense biodiversity there is an equally large number of ways the world external to an organism can be represented. Recent theories in physics suggest multiple universes (10^{500}) with multiple laws of nature thus resulting in an almost infinite number of epistemologies!

9. End Notes

1) Nyaya is a Sanskrit word which means analytical investigation through logical reasoning. The founder of Old Nyaya was Gautama, a contemporary of Buddah, who lived around 550 BC in North-eastern India. He merely formulated the generally accepted principle of time into a precise logical language. The New Nyaya was developed later, in the book *Nyaya-sutra*, edited in the 2nd century AD. Today, this logical system is classified as an intensional system in contrast to extenstional system invented by Greeks. Nyaya uses logical comparison in its proofs along side syllogistic reasoning. Comparison is the knowledge of a thing through its similarity to another previously known thing as follows. 1) M occurs in S, 2) M occurs in XP, 3) M does not occur in X-not-P, 4) therefore, P occurs in S

2) We shall keep in mind distinction between externally derived memories that originate from perceptions and internally derived memories that originate from imagination. External memories are based directly on sensory information, and represent events that really occurred in physical spec-time, objects that were really perceived, actions that were performed etc. Internal memories are pure mental constructs, not dependent on real-time input of direct perceptual data and represent events that have only been imagined, dreamed etc. The capability to distinguish between the two is called 'reality monitoring', which breaks down in many types of dementia and other mental disorder. For reality monitoring, see e.g., The Blackwell Dictionary of Cognitive Psychology, ed., M.W. Eysenck, 1990. In this sense, some of the mathematical constructs, such as actual infinity of Cantor and real numbers for which there are no corresponding objects in the world of externally derived memories, i.e., pure mental constructs.

3) Qualification and Frame Problems: Qualification problem arise in order to fully represent the condition for the successful performance of an action, an impractical and implausible number of qualification would have to be included in the sentence expressing them. Possible solutions are Logic by Default, Circumscription and other non-monotonic logical systems, formal systems where provability changes when additional facts (axioms) become available. For details see, J. McCarthy and P. Hayes, Some philosophical problems from the standpoint of artificial intelligence, in Machine Learning 4, eds. B. Meltzer and D. Michie, Edinburgh University Press 1969. The frame problem is that of specifying what doesn't change when an event occur – describing in a computationally reasonable manner what properties persist and what properties change as action is performed (see e.g., The Frame Problem in Artificial Intelligence, Ed. F.M. Brown, Morgan Kaufman Publ. Inc. 1987).

4) This is in the sense of functionalism, that is, the computational architecture (causal relations among its component) is relevant to psychology. Our view point regarding hardware-software or brain-mind distinction coincides with functionalism as stated by Ramsey [1989] in William M. Ramsey, Parallelism and Functionalism, *Cognitive Science* 13, 139-144 (1989). This paper was a reply to Thagard [1986], Paul Thagard, Parallel Computation and the Mind-Body Problem, *Cognitive Science* **10**, 301-318 (1986).

5) Negation as Failure Hypothesis: Although the notion of negation is fundamental to any common sense and formal system, it is not clear what it means under most circumstances. For instances, in Nyãya logic sixteen different types of double negations are discussed. In Horn clause logic programming, the notion of negation is equivalent to "negation as failure". that is, if A is a ground atom, then, the goal $\neg$A succeeds if A fails or, the goal - A fails if A succeeds. Using "negation as failure", A succeeds, but A is not a logical consequence of axioms/program P. Reiter's Closed World Assumption (CWA) falls in this category of reasoning - if something is not listed in a database, it is concluded that it does not exist.

[6] What is an object? Although this is a deep philosophical question, we shall propose a definition compatible with the arguments given in this chapter. An object is what a neuro-cognitive system is able to integrate external sensory stimuli into a unified whole, using one or more sensory modalities available at that instant. This is a perceptual object at a point in space-time. However, this is not sufficient to organize knowledge. Further structure can be inducted by using the principle of "cognitive economy", by giving it intension (a set of attributes that defines what it is to be a member of the concept) and extension (the set of entities that are members of the concept). If we now extrapolate this concept of an object using induction, deduction etc., we enter into the domain of non-perceptual objects, which we concluded to be existent only as a purely mental construct.

10. References

Bekoff, M., Allen, C. and Burghardt, G.M. (eds.) (2002) *The Cogniti-ve Animal*, The MIT Press, Cambridge, MA (USA).

Bernard, T. (1981) *Hindu Philosophy*, Motilal Banarasidass, New Delhi, India.

Bochenski, M. (1961) *A History of Formal Logic*, University of Notre Dame Press, pp. 416-447.

Cheney, D.L. and Seyfarth, R.M. (1992) Précis of how monkeys see the world. *Behav. and Brain Sci.*, **15**, 135.

Churchland, P. (1988) *Neurophilosophy*, The MIT Press, Cambridge, MA, USA.

Conway, A. et al. (2003) *An Introduction to Astrobiology*, Cambridge University Press, Cambridge, UK.

Crick, F. and Koch, C. (1990) Some Reflections on Visual Awareness. *Cold Spring Harbor Symp.*, **55**, 953-962.

Coward, H.G. (1983) *Studies in Indian Thought*, South Asia Books, India.

Davis, H. and Pe'russe, R. (1988) Numerical competence in animals. *Behav. and Brain Sci.*, **11**, 561-615.

Dehaene, S. (1997) *The Number Sense*, Oxford University Press, Oxford, UK.

Edelman, S. (2002) Constraining the neural representation of the visual world. *Trends in Cog. Sci.* **6**, 125-131.

Engel, A.K. and Singer, W. (2001) Temporal binding and the neural correlates of sensory awareness. *Trends in Cognitive Sci.* **5**, 16-25.

Fodor, J.A. and Pylyshyn, Z.W. (1988) Connectionism and Cognitive Architecture. *Cognition,* **28**, 3-71.

Ford, K.M., Glymour, C. and Heyes, P.J. (1995) *Android Epistemology*, The MIT Press, Cambridge.

Halliday, M.A.K. and Hasan, R. (1976) *Cohesion in English*, Longman, London, UK.

Hasson, U., et al. (2004) Intersubjective Synchronization of Cortical Activity During Natural Vision. *Science* **303**, 1634-1640.

Hauser, M.D., Chomsky, N. and Fitch, W.T. (2002) The Faculty of Language: What Is It, Who Has It, and How Did It Evolve? *Science* **298**, 1569-1579.

Hobbs, J.R. (1985) Granularity. *Proc. of the IJCAI-85*, Los Angeles, CA. pp. 432-435.

Lorenz, K. (1973) *Studies in Animal and Human Behaviour*, vol. I and vol. II, Harvard University Press, Cambridge, MA, USA.

Luisi, P.L. (2003) Contingency and determinism. *Phil. Trans. Roy. Soc. Lond.* A **361**, 1141-1147.

McCarthy, J. (1985) Epistemological Problems of AI, In: R.J. Brachman and H.J. Levesque (eds.) *Readings in Knowledge Representation*, Morgan Kaufman, CA, pp. 23-30.

Quine, W. van O. (1960) *Word and Object*, The MIT Press, Cambridge, MA, USA.

Radhakrishnan, (1964) *An Idealist View of Life*, Unwin Books, London.

Ramachandran, V.S. and Cobb, S. (1995) Visual Attention Modulates Meta-contrast Masking. *Nature* **373**, 66-68.

Riedl, R. (1984) *Biology of Knowledge*, John Wiley & Sons, New York, USA.

Rose, S.P.R. (2002) The Biology of the Future and the Future of Biology. *J. Mol. Biol.* **319**, 877-884.

Russell, B. (1919) *Introduction to Mathematical Philosophy*, Allen and Unwin Publishers, UK.

Russell, B. (1968) *The Art of Philosophising and Other Essays*, Philosophical Library, New York, USA.

Shah, K.T. (1998) Cognitive Universals: Abstract Psychology of Terrestrial and Extraterrestrial Intelligence, In: J. Chela-Flores and F. Raulin (eds.) *Exobiology: Matter, Energy and Information in the Origin and Evolution of Life in the Universe*, Kluwer Academic Press, pp. 161-164.

Shaw, J.L. (1988) The Nyaya on Double Negation. *Notre Dame Jour. of Logic* **29**, 139-154.

Shaw, J.L. (1991) Universal Sentences: Russell, Wittgenstein, Prior and the Nyaya. *J. Indian Philosophy* **19**, 103-119.

Singer, W. and Gray, C. (1995) Visual feature Integration and the Temporal Correlation Hypothesis. *Annu. Rev. Neurosci.* **18**, 555-586.

Singer, W. (1995) Synchronization of Neuronal Responses as a Putative Mechanism, In: M.A. Arbib (ed.) *The Handbook of Brain Theory and Neural Networks*, The MIT Press, Cambridge, MA (USA), pp. 960-964.

Sjolander, S. (1997) On the Evolution of Reality – Some Biological Prerequisites and Evolutionary Stages. *J. theor. Biol.* **187**, 595-600.

Squire, L.R., and Kendel, E.R. (1999) *Memory From Mind to Molecules*, Scientific American Book, NY.

Treisman, A. (1996) The binding problem. *Current Opinion in Neurobiology*, **6**, 171-178.

Von Glaserfeld, E. (1995) *Radical Constructivism*, The Falmer Press, London.

Wittgenstein, L. (1922) *Tractus Logico-philosophicus*, Routledge, London.

Ziman, J. (2003) Emerging out of nature into history: the plurality of the science. *Phil. Trans. Roy. Soc. Lond.* A**361**, 1617-1633.

Biodata of **Koichiro Matsuno** author of ***"Temperature Gradients as a Major Impetus: Driving Prebiotic Evolution"***

Koichiro Matsuno is currently Professor of biophysics in the Nagaoka University of Technology in Japan. He obtained his Ph.D. in physics from the Massachusetts Institute of Technology in 1971. Dr. Matsuno's research interest includes chemical evolution, cell motility and evolutionary processes. He is the author of *Protobiology: Physical Basis of Biology* (CRC Press, Boca Raton FL, 1989) and *What is Internal Measurement* (Seido-sha, Tokyo, 2000) and the co-author of *Molecular Evolution and Protobiology* with K. Dose, K. Harada, and D. L. Rohlfing (Plenum Press, New York, 1984), *The Origin and Evolution of the Cell* with H. Hartman (World Scientific, Singapore, 1992), and *Uroboros: Biology Between Mythology and Philosophy* with W. Lugowiski (Arboretum, Wroclaw Poland, 1998).

E-mail: **CXQ02365@nifty.com**

J. Seckbach (ed.), Life as We Know It, 641-653.

TEMPERATURE GRADIENTS AS A MAJOR IMPETUS:
Driving Prebiotic Evolution

KOICHIRO MATSUNO
Department of BioEngineering, Nagaoka University of Technology
Nagaoka 940-2188, Japan

1. Introduction

Prebiotic evolution leading eventually to the emergence of protobiological beings addresses two important issues. One is how monomers and oligomers of prebiotic significance could have been synthesized in the primitive environments, and the other is how the robust reaction network processing their synthesis and degradation could emerge and evolve as being subject to various disturbances from the environments. Experimental endeavors for synthesizing prebiotic monomers and oligomers, though massive and impressive (Lahav et al., 2001; Trevors, 2003), do not directly answer the question of how the chosen experimental boundary conditions could be naturalized in the real world. The laboratory experiments on prebiotic evolution suffer from a methodological limitation such that the imposed experimental conditions are not evolutionary on their own. Once one pays due attention to the fact that what could be evolving is not only monomers and oligomers but also the reaction network itself, the issue of evolving reaction network will come to the surface.

Evolution of reaction network differs from evolution of the reaction products in that what could be evolving is the context of the whole connected and related reactions going on, instead of the products as the contextual elements. Evolution of the context is already selective in that no contextual element can belong to two different contexts at the same time. The occurrence of such a contextual selection is not visible if only the contextual elements are focused. We shall examine some aspects of the contextual dynamics of evolutionary implication.

2. Contextual Dynamics: An Overview

Everything in this empirical world occurs in relation to everything else also residing there. If we want to address motions or changes of the constituent material bodies whether physical, chemical or biological, the interrelated conditions in which something of interest occurs will become a subject matter of prime concern. A context is just a brief term referring to the interrelated conditions. This perspective comes to remind us that such dynamics cannot be context-free, compared to the case of classical mechanics.

The original formulation of Newtonian classical mechanics, on the other hand, has been context-free. The first law of mechanics on inertia, the second law on force and the third law on action and reaction, all of them have been stated as the rules that linguistically describe changes in a string without reference to elements outside of the string. It goes without saying that the context-free formulation of classical mechanics has been tested extremely powerful in addressing motions and changes in the sense that it can work wherever and in whatever contexts. Nonetheless, classical mechanics does not dispense with the notion of context. It simply would not work unless the three laws are further supplemented by the context known as boundary conditions. At this point arises a new agenda of how the boundary conditions could be implemented and identified in classical mechanics. This will be a major concern for the enterprise of contextual dynamics.

One decisive factor focused on the contrast between contextual and context-free dynamics is causation. Causation is more than just a matter of philosophical dispute. In particular, mechanistic causation specifying how each individual could be related to itself and all of the others at the preceding stage remains unproblematic. Each individual in movement acts upon other individuals and vice versa. However, we also have much stronger form of mechanistic causality amounting to asserting that the movement of each individual is uniquely determined in coordination with all of the others from moment to moment. This observation suggests to us to examine how mechanistic causality could function in reality. The present problem cannot be marginalized simply by declaring it as a minor one especially in view of the fact that mechanistic causality in the sense of being acted upon by others is ubiquitous in any material systems we meet in the empirical world. We shall address ourselves to the issue of how causality could function in natural systems with a special focus upon mechanistic causality.

We usually take mechanistic causality to be of a one-to-one temporal mapping connecting an arbitrary predecessor to its successor expressed uniquely in the present tense. Newtonian equation of motion is a well-known representative case of mechanistic causality of unique specification. The mechanistic equation of motion can certainly specify how each individual constituting the equation develops in time. At the same time, the mechanistic equation of motion is subject to a constraint coming from the context denoted as the initial conditions. It can be of no use unless it is supplemented by the causation from the context. The present interplay between causation toward each individual and causation from the context as a vehicle of modeling natural systems comes to impart to the mechanistic equation of motion a unique property. Extreme sensitivity of the solution to the initial conditions sometimes reveals a pathological dependence of the mechanistic equation on the context. A slightest deviation in the context may bring about an enormous difference in the behaviors that the equation of motion would exhibit. This pathological sensitivity to the context is not something to be expected of natural systems, since in the latter a significant capacity of homeostasis resisting variations originating in the context, that is, the environment, is usually guaranteed. Newtonian equation of motion may not be a suitable model describing the behavior of natural systems because of its occasional pathological sensitivity to the context.

The pathological sensitivity to the context will more significantly be enhanced if Maxwell's equation of electromagnetic field in three-dimensional space is the case, since the context is specified by its boundary conditions in space and time. The context toward Maxwell's equation is materialized in essence in two forms of potential; one is the retarded potential, and the other is the advanced one. Although it may be relatively easier to prepare a boundary condition corresponding to the retarded potential giving rise to an expanding electromagnetic wave because of the relative smallness of the number of degrees of freedom to be controlled, the case for the advanced potential yielding a contracting wave in three dimensional space is practically impossible because of the presence of an immense number of degrees of freedom to be coordinated initially. Difficulty in fabricating the context applied to the electromagnetic field especially in terms of the advanced potential may render the notion of boundary conditions even irrelevant.

In contrast, the wavefunction of the Schrödinger's equation of motion in quantum mechanics makes the notion of the context amenable to physical processes taking place there. Coexistence of the wavefunctions propagating in opposite directions in fact generates a standing wavefunction corresponding to the occurrence of an eigen-wave-function that remains non-propagative. Empirical stability of a quantum state denoted as the standing wavefunction certainly manifests a likelihood of physical conditions giving rise to such a context allowing the wavefunctions to propagate in opposite directions equally. The existence of a standing wavefunction is due to the interference between two types of wavefunction; one is propagating in the forward direction in time and the other in the backward. The context being responsible for the genesis of a standing wavefunction actually admits two types of causation; one is forward in time and the other is backward. As a matter of fact, the stability of a quantum state against its context is due to juxtaposition of both the forward and the backward causations. The present stabilization of dynamics eliminating pathological sensitivity to the context is actually confirmed by conceiving an arbitrary dynamics carrying causations both in the forward and backward directions in time (Dubois, 1996). Causation in the backward is in essence seen as causation from the context since it can be regarded as a reflection of the preceding forward causation at a certain boundary forming the context. Reflected forward causation thus carries with itself the capacity of changing the boundary conditions or the context to be experienced by each constituent individual.

Causation propagating in the backward in time is however metaphorical at best, since every dynamic is actualized in the process of transferring the present progressive tense to the present perfect one. Causation from the context towards each individual inside is always of a retarded character. Individualization associated with causation toward each specified individual would be mechanistic insofar as both the causes originating in the preceding context and the complete specification of the individual were secured by whatever means and remain intact. In contrast, if complete specification of each elementary individual is unavailable for whatever reasons, there could arise such a possibility that some individuals may adjust their contents by themselves so as to fit into the context then available. Individualization and contextualization remain inseparable when complete specification of each individual residing within the context is not feasible (Salthe, 1993; Rosen, 2004). In fact, each individual subject to causation from the preceding context in turn comes to constitute the subsequent context. Such a contextualization is to come with causation towards the context.

Contextual dynamics addressing the context of interrelated conditions among the participating elements is unique in exhibiting the capacity of contextual selection. No contextual element can belong to two mutually incommensurable contexts at the same time, since the context is about an organization of the interrelated conditions as a coherent unity. Otherwise, the contextual unity would be jeopardized. Classical mechanics thus assumes the intervention of the external agency for its contextualization, to which all of the capacity of exercising the contextual selection and specification is relegated. Classical mechanics has in fact been complete in deciphering the mechanistic makeup of whatever material bodies in motion, while it still remains incomplete in uncovering the nature of the contextual agency. Then, thermodynamics comes to the fore.

3. Contextual Dynamics in Thermodynamics

Thermodynamics has historically been attempted as an endeavor toward an incomplete contextualization supplemented by an incomplete mechanistic underpinning. Something called heat was first introduced without recourse to explicating its mechanistic or atomic makeup. Despite this obvious drawback, however, thermodynamics has been concrete enough to introduce some quantitative figures addressing what the context of material bodies of interest would be all about, such as the amount of heat energy flowing through between two different bodies at different temperatures in contact. Take, for instance, temperature conceived in thermodynamics at least in an operational sense. Once the temperature of a material body, whatever it may be, is identified by whatever means, no contextual element of that body can participate in the contexts at different temperatures. This is just another way of saying that temperature is an intensive quantity about a unified context as a whole. Above all, an empirical law known as the first law of thermodynamics on energy transformation has anticipated the upcoming of contextual dynamics as admitting that the energy carried by the context can be preserved as a quantity even if the context is being transformed from within. In thermodynamics, as a matter of fact, a portion of heat energy can be transformed into mechanical energy without losing any amount of the energy involved.

The first law of thermodynamics on energy conservation through its transformation is concrete enough to specify the quantity of the energy to be transformed despite the fact that the mechanistic deciphering of the transformation mechanism remains yet to be seen. Then, a serious question arises with regard to what should be responsible for specifying the quantity inherent to the context. Asking the external agency for the concrete specification is of course one solution as has been practiced in classical mechanics. However, classical mechanics begs the further sturdy question of who in the world could that external agency be. In contrast, thermodynamics has anticipated that capacity of concrete specification arising from within. Since the capacity of concrete specification rests upon the act of measurement, thermodynamics comes to terms with measurement internal to thermodynamic bodies. Internal measurement is concrete and specific enough to point to and to be pointed out by whatever material bodies internally (Matsuno, 1989).

What is unique to thermodynamics is that it introduces macroscopic variables such as volume, pressure, temperature and entropy without detailing their atomistic makeup at the outset. These macroscopic variables are about the context in which the underlying microscopic elements, whatever they may be, are eventually situated. The contextual dynamics specifying the values of the macroscopic variables is constantly operative there. Even if the fundamental dynamics of microscopic elements is left unspecified, the contextual specification is to proceed. A molecule in the gas is subject to the temperature of the gas while at the same time the molecule is part of the gas substantiating the same temperature. Thus, any contextual element constituting the context comes to materialize and share the same contextual specification. Thermodynamics is unique in emphasizing the priority of contextual specification over elementary specification of each constituent element. Although mechanics is a theoretical enterprise equating elementary specification of an imposed character literally to contextual one in a crisp manner, thermodynamics is quite different in allowing an under-complete elementary specification whether or not it is of an imposed character. The present contextual specification now provides the interplay between the two of the contextual and the elementary dynamics with a possibility of influencing each other in both ways, namely, from the elementary to the contextual and vice versa.

One attempt for relating the elementary to the contextual dynamics is through a statistics of mechanics over an ensemble of elementary specifications. Statistical mechanics is grounded upon the premise that an ensemble of elementary specifications could be a substitute for the interplay between the two specifications, the contextual and the elementary ones. A justification of the ensemble of elementary specifications came from Boltzmann's Stosszahl Ansatz or hypothesis of a molecular chaos stating that molecules in the gas lose their memory of the past collisions with the others except for the latest ones. Those molecules in the gas thus come to have almost no correlation with the others or to move almost randomly with each other. This is equivalent to saying that the context which Boltzmann introduced is the one under which every contextual element moves almost randomly with each other (Matsuno, 2001). More specifically, one particular quantitative figure characterizing the Boltzmann's context is called temperature. To be sure, Boltzmann's context is found ubiquitous in physics. Nonetheless, it is no more than a heuristic candidate for fulfilling the role of contextual specification operating in thermodynamics. There certainly is another candidate for meeting the similar requirement of contextual specification. That is quantum mechanics.

4. Quantum Mechanics as a Contextual Dynamics

Atoms and molecules constituting a biological organism are placed within the material context of an extremely specific configuration. Such a specificity of the material context makes biological organization unique compared to nonliving physical organization of atoms and molecules. Needless to say, physics has its own rich history on explicating the nature of whatever material contexts available there. One example of an extreme significance is the material context discovered by Max Planck (Matsuno, 2003).

First of all, context as a limiting modifier of the contextual elements, if empirically available, must remain robust to a reasonable extent. Otherwise there would be no possibility for denoting it as such in the empirical domain. Empirical confirmation of the occurrence of such a robust context including a set of macroscopic variables in thermodynamics comes from examining the empirical record of the events of interest. The record is about the events already registered in the present perfect tense, while the ones right in the making are in the present progressive tense. The robust record rests upon the transference of events in the present progressive tense to those in the present perfect one. In fact, quantum mechanics grounds itself upon the existence of such a robust record.

When Planck introduced the notion of a quantum for the first time, the relevant empirical fact referred to was that a light wave emission from and absorption to a black body in thermal equilibrium with its surroundings are punctuated in a discrete manner. The discreteness is associated with the empirical observation that light wave emission in progress comes to shortly be punctuated by the emission completed, and light wave absorption in progress similarly comes to be punctuated by the absorption done. There is no indefinite prolongation of light wave emission and absorption over to an infinite duration in a continuous manner. The punctuated light wave referred to as a light quantum or a photon carries with itself the context within which continuous light wave is encapsulated in a coherent manner. In particular, Schrödinger identified that the coherent nature of the encapsulation is due to the occurrence of a standing wave as a coherent superposition of both the retarded and the advanced waves of material origin.

The context discovered by Planck, or Planck's context, is thus the one for those contextual elements moving almost coherently with each other. Planck's context is just a polar opposite to Boltzmann's, in the latter of which the contextual elements are taken to move almost randomly in an incoherent fashion with each other. However, the relationship between Planck's context and Boltzmann's is not mutually exclusive. Planck's context is more fundamental and more inclusive in that any material element of whatever sort is a quantum after Planck. In contrast, Boltzmann's context is subject to Planck's contexts embedded in it. At the same time, Planck's context is also subject to influences coming from the outside, because it always presumes the action of making a sharp distinction between the present progressive and present perfect tenses by both itself and others external to the context itself. What is responsible for generating the context is the robust interplay between the inside and the outside. The present interplay can now furnish a Boltzmann's context as a source matrix of the measuring agencies toward each Planck's context residing in its inside, with the capacity of modifying the latter context in time internally. Occurrence of a Boltzmann's context characterized by a quantitative figure called temperature is in fact an empirical testimony to the observation that the constituent quanta or Planck's contexts are measuring each other internally, that is to say, involved in internal measurement altogether.

In particular, the contrast between Planck's context and Boltzmann's will become more conspicuous once the nature of internal measurement involved is focused. Although Boltzmann's context rests upon the stipulation that each quantum loses the memory of the past measurements of the others shortly, Planck's context is about the persistent memory of the measurement internal to each quantum while distinguishing the movement in the present progressive mode from the one in the present perfect. Planck's context is for long memory of internal measurement, while Boltzmann's is for short memory. This interplay between Planck's context and Boltzmann's will impart to temperature dynamics a unique evolutionary implication of the contextual character.

5. Temperature Gradients as an Evolutionary Contextual Driver

The contextual character of temperature dynamics will become most conspicuous if two bodies at different temperatures come into contact each other. That is the issue for the time-honored law of Fourier's heat transfer. Once Fourier's law of heat transfer receives due attention it deserves, one basic issue will come to be focused. Fourier's law assumes two fundamental quantities. One is temperature, and the other is heat flow. Between these two, temperature as a quantitative figure is qualitatively different from heat flow as another quantitative figure. Temperature is a figure quantifying the context whose constituent elements, whatever they may be, exhibit a random movement, while heat flow is a quantitative figure being additive even infinitesimally. In short, temperature is intensive, while heat flow is extensive. That implies that Fourier's law is about a relationship between the context and the contextual elements. As a matter of fact, Fourier's law addressing heat flow through the interface connecting two different regions at different temperatures is about a dynamic behavior occurring when two different contexts at different temperatures come into contact with each other.

Furthermore, Fourier's law as a contextual law is already unique in admitting in itself a certain selective capacity. No contextual element can belong to two different contexts at the same time. Once a material element is taken to be a constituent of a certain context, it cannot be a constituent of another different context at the same time. When a tissue of an organism is at a certain temperature, the temperature of the organs constituting the tissue cannot be different from that of the tissue. When an organ of a tissue is at a certain temperature, the temperature of the constituent organelles cannot be different from that of the organ.

Selective capacity latent in Fourier's law will be more vividly focused when one pays attention directly to two different contexts at different temperatures coming into contact with each other, instead of to the amount of heat flowing through the interface. When the temperatures of the two contexts start varying because of the intervening heat flow through the interface, each context would modify itself or adjust its temperature as fast as possible. When a small hot body comes to contact a huge cold body, the actual temperature drop of the small body would be the one proceeding at the possible fastest rate since there is no chance left for the latecomers. This is simply because of the intrinsic affinity between the context and the contextual elements. When the context as a construct from the contextual elements in a bottom-up manner is allowed to change, the realizable change could only be the one proceeding at the fastest rate. When there happens to be a temperature gradient to be mitigated, the actual process of mitigation would have to be the one that could proceed at the possible fastest rate. There would be no room for the latecomers to take part in since the constituent elements cannot belong to different contexts at the same time.

Fourier's law is contextual and accordingly is intrinsically selective in its operation with regard to the contextual elements. Of course, Fourier's law is only one particular case demonstrating the significance of contextual laws in material dynamics. One more case of significance is found in quantum mechanics.

An energy quantum after Planck is about the nonlocal unit of material dynamics in space and time. The nonlocal material unit as a quantum is the context within which the material wave after de Broglie as the contextual element propagates back and forth as a standing wave. The context as the container of the material wave as the contained is characterized by its nearly perfect coherence or quantum coherence. Planck's context is thus unique in its coherence with regard to the constituent elements. This exhibits a marked contrast to the dynamic behavior addressed by Fourier's law of heat transfer, in the latter of which the constituent elements giving rise to a contextual temperature are supposed to be almost completely incoherent or random with each other after Boltzmann.

Material dynamics thus admits at least two different classes of contextual dynamics, namely, Boltzmann's and Planck's. Participation of Planck's context into Fourier's law of heat transfer can now provide a possibility of enriching the armory of the selective capacity latent in Fourier's law in itself. One likely thread connecting Boltzmann's context and Planck's is the first law of thermodynamics on energy transformation.

An energy quantum as a coherent enclosure of the standing material wave remains robust, but its enclosure is not literally complete. Otherwise, there would be no possibility of energy quanta interacting with each other. This comes to imply that when one applies Fourier's law of heat transfer to an energy quantum, there may happen to be a case of transforming the quantum with the aid of the underlying dynamics of varying the participating contexts. Prerequisite to such a transformation is the occurrence of heat flow through the interface between different contexts at different temperatures. When a small energy quantum at a certain temperature is suddenly made contact with the environment of almost an infinite heat capacity at a different temperature, Fourier's law is to implement the first law of thermodynamics through the contextual transformation or transforming the participating quantum. The contact of the quantum with the envi-ronment transforms that quantum so that its temperature may approach the environment's at the possible fastest rate. Fourier's law materializes in the form of the first law of thermodynamics as being accompanied by its intrinsic selective capacity of actualizing only the fastest contextual change (Matsuno and Swenson, 1999).

An energy quantum conceived within the framework of Fourier's law of heat transfer now suggests such likelihood that an energy quantum may function as a heat engine processing both the incoming and outgoing heat flow there. Heat engine as a quantum can in turn furnish itself with some transformational or evolutionary capacity depending upon the nature of temperature gradients it experiences.

6. Evolutionary Relevance of Temperature Gradients

Physical likelihood for the occurrence of temperature gradients may be either spatial or temporal, or both. The reaction network of chemical reactants on the primitive earth could have been put under the temperature gradients of various physical origins. Since the sources of heat flow toward the surface of the earth are quite sparse and localized in space as the light from the sun or the heat from the magmas located at the core mantle of the earth, the reactants serving as the heat acceptors on the surface are put under the huge gradients of decreasing their temperatures. Deep outer space for the sun light is at extremely low temperatures compared to the sun, and the vast ocean on the earth is quite cold compared to hot springs from hydrothermal vents.

The reaction network of those chemical reactants recycling through hot vents or experiencing close encounter with them repeatedly in the ocean is subject to a unique form of temperature gradients. The chemical reactants there come to undergo at least two possibilities, either making bigger molecules through the synthesis with others as having recourse to the activation energy available from the heat source or disintegrating them into smaller molecules as suffering thermal degradation before reaching their quenched stabilization. The interplay between the synthesis and the degradation, however, differs from the similar interplay conceivable in thermal equilibrium, in the latter of which every synthetic reaction is taken to counterbalance the reversed disintegration reaction by definition. In contrast, the synthetic reaction proceeding as utilizing the activation energy available from some specific heat sources can miss the chance of counterbalancing with the reversed disintegration reaction on the spot if the synthesized products are rapidly transferred to and quenched in cold surroundings. There may arise the likelihood for another chance of postponing the counterbalancing until several or many intermediate reaction steps are completed. Furthermore, if there happens to appear in the downstream of the reactions a specific reactant similar to the one appeared somewhere in the upstream, the whole reaction scheme would come to contain a reaction loop and turn out network-catalytic (Matsuno, 1982; Kauffman, 1986; Orgel, 1986; Morowitz, et al., 2000; Cody et al., 2001) even in the absence of template-catalytic reactions.

An experimental demonstration of network-catalytic reactions is surely available from the synthesis of oligopeptides from monomeric amino acids in the flow reactor simulating hydrothermal circulation of seawater through hot vents in the ocean (Matsuno, 1997; Imai et al., 1999). Time development of the synthesis of tetraglycine as starting from glycine as the initial reactant in the flow reactor, for example, demonstrated an exponential growth of the yields at least over an initial limited interval through the flow reactor operation. This observation provides evidence that a certain network-catalytic reaction has been in operation.

One evolutionary implication of the occurrence of temperature gradients on the primitive earth must have been to enhance the likelihood of generating network-catalytic reactions even prior to the emergence of template-catalytic reactions of biological origin.

7. Concluding Remarks

Biochemical reactions proceeding in all extant biological organisms incorporate in themselves template-catalytic reactions in one form or another. This does not, however, directly imply that the evolutionary emergence of template-catalytic reactions would have required no catalytic reactions. Quite the contrary, the emergence of catalytic templates of organic origin would have greatly been benefited if catalysts of inorganic origin were available beforehand. Metal surfaces as inorganic templates could certainly have helped the emergence of those of organic origin (Wächtershäuser, 1990), but this would not be the only possibility. What we have observed is another possibility of helping the emergence of catalytic templates of organic origin through the prior emergence of network-catalytic reactions empowered by the occurrence of temperature gradients. Even if the abiotic synthesis of catalytic templates of biological origin remains yet to be seen, the emergence of network-catalytic reactions is already available in the laboratory experiments.

What is significant to the occurrence of network-catalytic reactions is their self-focusing capability. Since the total molecular resources available are limited, the growth of certain molecular species through network-catalytic reactions would unfailingly suppress the populations of other molecular species as a consequence. Enhancement of a certain network-catalytic reaction may proceed at the expense of other chemical reactions that could have been conceivable in thermal equilibrium. This suppression may be just another side of the same observation that network-catalytic reactions leading to the emergence of protobiological beings could be accompanied by the enhancement of the specificities both in the reactants and in the reactions proceeding there. In addition, the network-catalytic reactions may also incorporate in themselves inorganic templates of metal origin if the incorporation would help enhance the capacity of the reaction network for resource acquisitions.

8. References

Cody, G. D., Boctor, N. Z., Hazen, R. M., Branders, J. A., Morowitz, H. J., and Yonder, H. S. Jr. (2001) Geochemical roots of autotrophic carbon fixation: hydrothermal experiments in the system citric acid, H_2O-(±FeS)-(±NiS). *Geochim. Cosmochim. Acta* **65**, 3557-3576.

Dubois, D. M. (1996) Introduction to the Aristotle's final causation in CAST: concept and method of incursion and hyperincursion. In: F. Pitcher. R. Moreno Diaz and R. Albrecht (eds.) *Computer Aided Systems Theory EUROCAST `95.* Lecture Notes in Computer Science, 1030. Springer-Verlag, Berlin, pp. 477-493.

Imai, E., Honda, H., Hatori, K., Brack, A., and Matsuno, K. (1999) Elongation of oligopeptides in a simulated hydrothermal environment. *Science* **283**, 831-833.

Kauffman, S. A. (1986) Autocatalytic sets of proteins. *J. Theor. Biol* **119**, 1-24..

Lahav, N., Nir, S., and Elitzur, A. C. (2001) The emergence of life on earth. *Progr. Biophys. Mol. Biol.* **75**, 75-120.

Matsuno, K. (1982) Natural self-organization of polynucleotides and polypeptides in protobiogenesis: appearance of a protohypercycle. *BioSystems* **15**, 1-11.
Matsuno, K. (1989) *Protobiology: Physical Basis of Biology*. CRC Press, Boca Raton Florida.
Matsuno, K. (1997) A design principle of a flow reactor simulating prebiotic evolution *Viva Origino* **25**, 191-204.
Matsuno, K. (2001) Thermodynamics as an interface agency tailoring quantum mechanics for the beginning of biology" In: H. H. Diebner, T. Druckrey, and P. Weibel (eds.) *Sciences of the Interface.* Genista Verlag, Tübingen Germany, pp. 143-154.
Matsuno, K. (2003) Quantum mechanics in first, second, and third person description. *BioSystems* **68**, 107-118.
Matsuno, K., and Swenson, R. (1999) Thermodynamics in the present progressive mode and its role in the context of the origin of life. *BioSystems* **51**, 53-61.
Morowitz, H. J., Kostelnik, J. D., Yang, J. and Cody, G. D. (2000) The origin of intermediary metabolism. *Proc. Nat'l. Acad. Sci* **97**, 7704-7708.
Orgel, L. E. (1986) RNA catalysis and the origins of life. *J. Theor. Biol.* **123**, 127-149.
Rosen, S. M. (2004) *Dimension of Apeiron: A Topological Phenomenology of Space, Time and Individuation.* Rodopi, Amsterdam.
Salthe, S. N. (1993) *Development and Evolution: Complexity and Change in Biology.* Bradford/MIT Press, Cambridge, Massachusetts.
Trevors, J. T. (2003) Early assembly of cellular life. *Progr. Biophys. Mol. Biol.* **81**, 201-217.
Wächtershäuser, G. (1990) Evolution of the first metabolic cycle. *Proc. Nat'l. Acad. Sci. USA* **87**, 200-204.

Biodata of **Natan Gadoth** author of **"Life during clinical death"**

Professor Natan Gadoth in currently the chairman of the department of Neurology at the Meir General Hospital, Kfar Saba (Israel) and a professor of Neurology, at the Sackler Faculty of Medicine, Tel-Aviv University. Tel-Aviv, Israel. He obtained his MD degree from the Hadassah Medical School, The Hebrew University, Jerusalem in 1969, and trained in Neurology at the Johns Hopkins University Hospital in Baltimore, USA (1972-1975). Professor Gadoth's scientific interests are in the areas of neurogenetics, clinical neurophysiology, neuromuscular disorders, sleep, and epilepsy both in children and adults.

E-mail: **gadoth@post.tau.ac.il**

J. Seckbach (ed.), Life as We Know It, 655-661.

LIFE DURING CLINICAL DEATH:
The "Near Death Experience" Phenomenon

NATAN GADOTH
Professor and Chairman, Department of Neurology, Meir General Hospital, Kfar – Saba and the Sackler Faculty of Medicine, Tel-Aviv University, Israel.

1. Introduction

The "near death experience" (NDE) applies to the subjective reports of people who were close to death and later recalled vivid encounters with bright lights and heavenly beings. Indeed, similar experiences can be found in Plato's republic, in a 15th century painting, in the accounts of survivors of Swiss mountaineering accidents and in survivors of Chinese earthquake of 1976. However, only with the introduction of universally accepted biological criteria for the establishment of the state of "death", it was possible to define scientifically a "near death state". Following that, a sizeable population of "near death" survivors is now available due to modern cardio-pulmonary successful resuscitation. This population is the source of our knowledge of the NDE.

One may assume that this outstanding phenomenon may be the result of one or more of the following process: a specific physiologic or pharmacologic brain activity which accompanies the process of death; a psychological reaction to the threat of impending death; a transcendental experience (Parnia et al., 2001).

The following is a review of what is known today about the biology of this not yet fully explained phenomenon.

2. The Subjective Description of the Experience

People of all ages who underwent successful resuscitation report almost identical experience which they recall during the state of "death". This experience consists of a distinctive state of consciousness in which their physical existence seems to them as outside their body and not bound to earthy environ (Greyson, 2001). The complex, but surprisingly consistent descriptions of this experience may be defined as "transcendental experiences precipitated by confrontation with death which do not seem to be accounted for by current medical understanding of the dying process".

It was Moody who in 1975 systematically compiled reports from 150 survivors and sorted out 7 main descriptive characteristics which were common to people of different ages, genders and cultural background (Moody, 1975):

1. An overwhelming feeling of peace and well-being, including freedom of pain.
2. A feeling of being located outside one's physical body.

3. A sensation of floating or drifting through darkness, sometimes described as a tunnel.
4. Awareness of a golden light.
5. Encountering and even communication with a "presence "(Moody's "being of light").
6. A rapid succession of visual images of one's life.
7. The experience of another world of beauty, and perhaps meeting there "familiar" spirits of deceased relatives and acquaintances with whom one might communicate.

Considering the fact that the detection and diagnosis of NDE is based on the subject's recall of the event, a validation scaling system which is supposed to delineate "true" NDE was developed by several authors (Greyson, 1983, Lange et al., 2004).

In the Greyson scale there are 16 questions which deal with descriptive terms pooled from accounts given by survivors. Those are related to the perception of the time domain, speed of thought, experiencing scenes from one's past, feelings that one can do everything, the feeing of eternal peace joy or pleasure, peace and harmony with the universe, seeing or feeling of surrounding bright light, feeling the senses more vivid, experiencing scenes from the future, feeling separated from one's body, entering another unearthly world, encountering the presence or the sound made by a mysterious being and or a deceased or religious spirit, and the feeling of getting to a point of no return. It seems that there is also a hierarchical order in the appearance of the items listed above.

The experience of peace, joy and harmony are the first to perceive followed by insight and mystical/religious experiences. The last to appear are the awareness of things occurring out of place or time.

3. How Frequent is the Phenomenon of NDE

Based on data obtained from surveys of resuscitated people, nation wide American poll of the general population and a survey of resuscitated patients from the Netherlands, about a third of survivors report of NDE, which amounts to 5% of the adult US population (Greyson, 1993).

4. Is it possible for the Brain to be the Source of NDE During Clinical Death?

Two mechanisms are now widely accepted to explain NDE i.e. a chemical change in the brain during the dying process and a psychological response to the perceived treat of impending death (Blackmore, 1999; Owens et al., 1990).

The chemical mediators include drugs especially various anesthetic drugs with hallucinatory potential, endogenic secretion of endorphins, lack of brain oxygen, increased carbon-dioxide brain concentration, neuronal excitatory receptors, brain serotonin, seizures and excitation of the limbic system which may be associated with feelings such as calmness and pleasure.

As most of the reported NDE was given by survivors of cardiac arrest, one should be familiar with the physiological and biochemical changes known to occur in cardiac arrest which are manifested in the brain.

Immediately when the cardiac efficient pumping actability is arrested there is a steep reduction in blood pressure to immeasurable levels. Even if immediate and proper chest compression is given, it is not sufficient to rise the diastolic blood pressure (this is the pressure which reflects the refilling ability of the heart and hence the size of the blood volume which will be distributed to the body and brain on the next contraction of the left ventricle), to restore sufficient brain circulation. Even with concomitant administration of potent drugs which increase blood pressure it seems impossible to reach sufficient levels of diastolic pressure for adequate brain perfusion.

Brain electrical activity recorded by traditional scalp electrodes (Electroencephalography – EEG) during the arrest shows initial slowing and complete cessation of activity within 10-20 seconds. It remains flat until the reinstitution of pulse and adequate cardiac output. However, in survivors of a prolonged arrest it may remain flat for many hours despite adequate output of blood from the resuscitated heart. Similar patterns of activity were recorded in experimental animals with indwelling depth electrodes from deep brain structures. The lack of electrical brain activity correlates well with loss of consciousness during the arrest. From studies of cerebral blood flow during and following experimental cardiac arrest in animals it became evident that in spite of the reinstitution of adequate peripheral blood supply to other body parts with successful resuscitation, the blood supply to the brain tissue is insufficient despite the reinstitution of adequate blood flow to the brain.

This "paradox" is due to the fact the superficial small blood vessels which are responsible for cerebral circulation and nourish the neural tissue are undergoing severe spasm due to over production of molecules which constrict blood vessels. This is believed to be caused by imbalance in the local production of chemicals which constrict (vasoconstrictors) and dilate (vasodilators) those vessels, largely in favor of the vasoconstrictors. This phase of cerebral vasoconstriction is expressed by flat EEG activity and lasts in the experimental animal as long as 24 hours after resuscitation. The abolished brain activity during this period of time may affect the whole brain or only small parts of it. It was also found that not all brain regions are similarly affected i.e. while the frontal lobes show a flow reduction of 85-90%, the cerebellum shows only 10-30% reduction. It seems logical to assume that the lack of blood and oxygen to the brain at the onset of the arrest is responsible for the above mentioned phenomena. When initiating resuscitation one needs to insert a tube into the trachea (windpipe) for artificial ventilation. This procedure (intubation) is usually met by resistance of the vocal cords derived by intact brain stem reflexes, however, in arrested patients the tube is inserted easily in-between the relaxed vocal cords indicating loss of brain stem reflexes. The depressed brain stem with its reticular formation, a structure which is responsible for the arousal of the cortical neuron, may be an additional explanation for the state of unconsciousness during cardiac arrest.

With this in mind, one should wonder how can the severely depressed brain perceive lucid, well structured thoughts, images or memories of self, and specific events during the arrest which can be later recalled.

It was suggested that NDE are hallucinations caused by chemical changes in the brain. However, true hallucinations caused by metabolic or physiological alterations are sensations experienced when brain is fully functioning. Moreover, it was clearly shown that thoughts or experiences during lack of oxygen with accompanying high levels of carbon-dioxide, other metabolic changes and under the effect of drugs or seizures, can never be lucid and well organized as those reported as NDE. It has recently been shown that even a minor reduction of brain perfusion is associated with impaired attention, whereas, NDE is not confusional, but is the result of heightened awareness, attention and memory formation.

The time when NDE was acquired may be just before consciousness was lost or immediately when regained. This view can be challenged by the fact that any significant brain insult is followed by loss of memory (amnesia), which its duration is related to the severity of the insult. There is no doubt that cardiac arrest is a severe brain insult and as such should cause profound amnesia both backward from the beginning of the arrest (retrograde amnesia) and forward from the arrest (anterograde amnesia). Thus events accruing at the beginning of the arrest or at the time of regained consciousness could not be recalled by the survivors.

However it might be argued that experiences such as seeing light or a tunnel may be perceived during the recovery period. However accounts such as "seeing" and recalling in detail items and their location as well as other specific details from the resuscitation period which were verified by the medical staff present cannot be explained, because such memories require some form of consciousness during the arrest.

5. Can NDE Shed some Light on the Nature of Human Consciousness?

The precise brain networks and mechanisms responsible for our thoughts and consciousness have not been yet elucidated. It is traditionally accepted that they are the product of interaction between large groups of neurons and their corresponding network and their impairment by head injury for example will affect thoughts, memory and consciousness. Indeed, specific brain areas are activated by thoughts and feelings as shown by modern functional imaging as MRI (Magnetic Resonance Imaging), SPECT (Single Photon Emission Computed Tomography) and PET (Positron Emission Tomography) (Frackowiak et al., 2003). Although this form of network, using various mediating chemicals such as neurotransmitter generates action potentials across the cell membrane which propagates to various brain regions, it is unknown how are those electrical currents and related biochemical changes translated to thoughts and feelings.

The opportunity to study in depth a large cohort of victims of cardiac arrest during and after resuscitation, comparing those with and without NDE, and utilizing modern electrophysiological and functional imaging techniques may shed light on our understanding of consciousness and other high cognitive functions.

6. References

Blackmore, S.J. (1996) Near death experience. *J. of the Royal Soc. Med.* **89**, 73-76.

Greyson, B. (1983) The Near Death Experience Scale: Construction, reliability, and validity. *J. Nervous and Mental Diseases* **171**, 369-375.

Greyson, B. (1993) Varieties of near-death experience. *Psychiatry* **56**, 390-399.

Greyson, B. (2000) Near-death experiences. In: E. Cardena, S J. Lynn, and Krippner, S.C. (eds.) *Varieties of anomalous experiences: Examining the scientific evidence. American* Psychological Association, Washington DC, pp. 315-325.

Lang, R. Greyson, B. and Houran, J. (2004) A Rasch scaling validation of a 'core' near-death. Experience. *Brit. J. Psychol.* **95**, 161-177.

Moody, R A. (1975) *Life after life*. Convington, GA; Mockingbird.

Owens, J.F., Cook, E.W..and Stevenson, I. (1990) Features of 'near death experience' in relation to whether or not patients were near death. *Lancet* **336**, 1175-1177.

Parnia, S., Waller, D.G., Yeates, R. and Fenwick P.A. (2001) Qualitative and quantitative study of incidence, features and aetiology of near death experience in cardiac arrest survivors. *Resuscitation* **48**, 149-156.

Biodata of **Nina Gunde-Cimerman** and **Polona Zalar**, authors of **"Beauty in Nature"**

Nina Gunde-Cimerman is currently Associate Professor of Microbiology at the University of Ljubljana, Biotechnical Faculty, Biology Department and principal investigator of the research microbiology group. She obtained her Ph.D. from the University of Ljubljana in 1993. Besides, she is in charge of the MZKI Fungal Culture Collection of the National Institute of Chemistry. Her scientific interests are in the areas of the microbiology, ecology, taxonomy and physiology of extremophilic fungi (primarily halophilic, xerophilic and psychrophilic) inhabiting solar salterns and hypersaline lakes as well as polar glacial and sub-glacial environments.

E-mail: **nina.gunde-cimerman@bf.uni-lj.si**

Polona Zalar is currently teaching assistant of Microbiology, at the University of Ljubljana, Biotechnical Faculty, Biology Department and researcher at the microbiology group. She obtained her M.Sc. from the University of Ljubljana in 1999. Her scientific interests are taxonomy and ecology of halotolerant and halophilic fungi.

E-mail: **polona.zalar@bf.uni-lj.si**

J. Seckbach (ed.), Life as We Know It, 663-683.

BEAUTY IN NATURE

NINA GUNDE-CIMERMAN[1,2] and POLONA ZALAR[1]
[1]*University of Ljubljana, Biotechnical Faculty, Department Biology, Večna pot 111, 1000 Ljubljana, Slovenia* [2] *National Institute of Chemistry, Laboratory for Biotechnology, Hajdrihova 19, 1000 Ljubljana, Slovenia*

"To see a World in a grain of Sand,
And a Heaven in a Wild Flower,
Hold Infinity in the palm of your hand,
And Eternity in an hour."
—**William Blake**

1. Introduction

Accepting an offer to write about something so obvious and at the same time so intangible and undefined as beauty in nature, at least from a scientist's perspective, brings one to first look into a dictionary, to see what the definition of beauty might be. According to Webster's Encyclopedic Unabridged Dictionary of the English language (1996), "beauty is the quality present in a thing or a person that gives intense pleasure or deep satisfaction to the mind, whether arising from sensory manifestation (as shape, color, sound), a meaningful design or pattern, or something else." Therefore a beautiful thing can be a work of art or a building, or something that is beautiful in nature. But undoubtedly something that is individually pleasing or attractive, whether due to the perfection of form, color, or even due to a personality in which high spiritual qualities are manifest.

As a scientist, one remains deeply troubled by the task on the horizon, at that moment not embellished by a deep orange-red hue of the setting sun, but a foggy cloud covering its contours. Since according to the definition, beauty is subjective and cannot be described in words, although it should follow some general, undefined rules, the best next step is to start thinking where to find beauty on your very personal, individual scale. What makes you halt and look twice, before going on, what fills your mind's lungs with a fresh breath, your eyes with harmony and an additional sparkle? Things that most of us share: works of art, nature in its many shapes, smiles and music. What could I, as an individual and a scientists offer additionally to this long list, crossing nations, races and continents, with a few local variations here and there? Perhaps the beauty I see in the world of microscopic organisms on which my research team has focused attention for quite some years by now: microfungi, beautifully varied in shape (Fig. 1, 10, 11), color, patterns and solutions to life's problems. These are not just any fungi, but a very special kind that are trying to live in some of the harshest environments on our planet – extremely saline waters of man-made salterns and salt lakes (Gunde-Cimerman et al., 2000; 2005). Some of them are even hidden in Arctic glaciers covered by thousands of meters of ice (Gunde-Cimerman et al., 2003).

Another aesthetic aspect in our work is the beauty we see in the ascetic extreme environments these microfungi inhabit. They are characterized by strong prevailing physicochemical forces that exclude all non-adapted organisms (Oren, 2002; Ventosa, 2004; Gunde-Cimerman et al., 2005). These environments, characterized by many hues of only few colors, have a completely different beauty than for example a tropical habitat with its luscious baroque display of life's overwhelmingly numerous facets. Hypersaline lakes show a silent, slow beauty that appears as if unchanged for thousands or even millions of years. Salterns, their man-made equivalents, repeat on a smaller, dynamic and cyclic time-scale the long-term history of salt lakes. In both cases the extremely low water activity and high ion concentrations toxic to most organisms, selectively shape adaptations of those organisms surviving within. With that I can perhaps rely on my instincts and knowledge and present our own, scientific version of beauty.

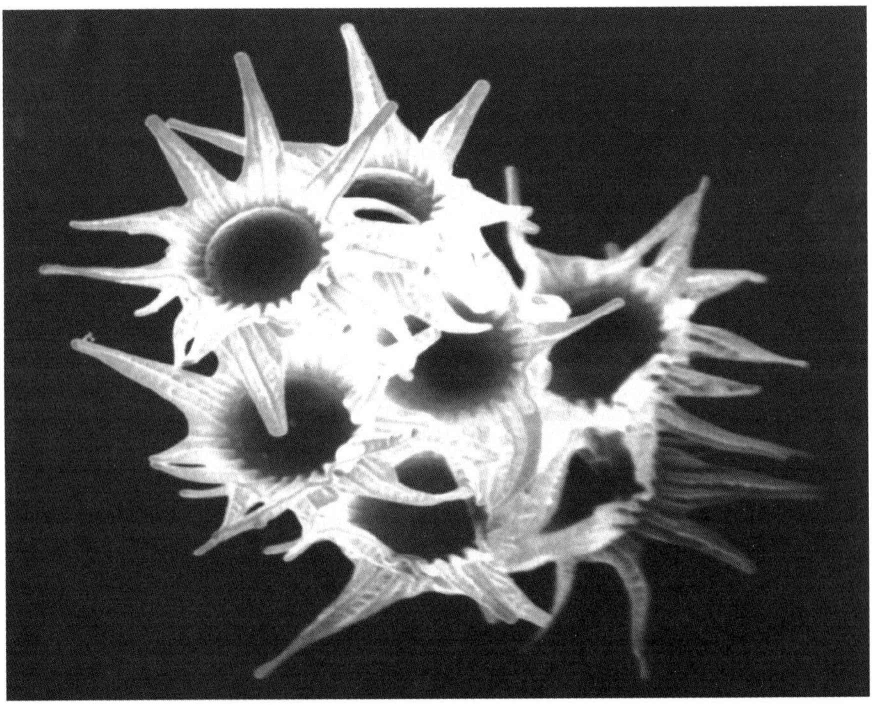

Fig. 1. *Emericella stellamaris* – ascospores.

2. Hypersaline Environments – Beauty in Simplicity

Henri Poincaré (1952) wrote: "The scientist does not study nature because it is useful; he studies it because he delights in it, and he delights in it because it is beautiful. If nature were not beautiful, it would not be worth knowing, and if nature were not worth knowing, life would not be worth living. Of course I do not here speak of that beauty that strikes the senses, the beauty of qualities and appearances; not that I undervalue such beauty, far from it, but it has nothing to do with science; I mean that profounder beauty which comes from the harmonious order of the parts, and which a pure intelligence can grasp."

This quotation represents an appropriate introduction to the beauty of the extreme environment on which our research efforts have been concentrated over the last many years: salterns (Fig. 2) and salt lakes worldwide. Their beauty does not lie only in the shapes and colors, but in the conditions that prevent saturation of senses with the diversity of life forms. Beauty of an almost empty, minimalist landscape, devoid of disturbances, where life is revealed only at the second sight – after looking through the microscope.

Fig. 2. Salterns Sečovlje, Adriatic coast.

Hypersaline environments are richer in salt than seawater, sometimes they are even saturated with salt. Such salt-rich environments are not at all abundant on our planet. They can be represented by coastal lagoons, formed by tidal waters on shallow coastlines. These areas differ in size, from little ponds on rocky shores to extensive salt marshes. Water evaporation causes elevated salt concentrations and deposition of glittering salt crystals as bitter crusts around them. More important than these small and often ephemeral phenomena are often quite extensive inland hypersaline lakes. The largest and the best known are the Dead Sea located on the border between Israel and Jordan and the Great Salt Lake in Utah, USA (Kushner, 1978; Oren, 2002). The Dead Sea is probably the saltiest place on Earth. Its viscous waters contain a total salt concentration of over 340 grams per liter. However, less than one-third of the salts is common salt; the remainder mainly consists of bitter salts magnesium chloride and calcium chloride (Oren 1993; 1999). Its shores are embellished with mind-inspiring sculptures of salt deposits.

Salt can also be found as large underground deposits of solid mineral halite. These salt deposits were formed millions of years ago when large areas covered with seawater became detached from the ocean (Oren, 2002). Many of these underground caves with reddish crystal salt deposits are nowadays known tourist attractions. Some have been mined for thousands of years to obtain salt and various objects, made of "rock salt" beautify homes of nature lovers.

Beauty is related to an intriguing blend of regularity and irregularity. When things are too regular, we usually find them to be boring because they do not challenge our mind or senses. Highly irregular phenomena, on the other hand, may be uninteresting because they lack a challenging element of predictability.

Man-made hypersaline environments – the salterns in which salt is produced from seawater by gradual evaporation – contain this regular/irregular type of beauty. Knowledge from nature is translated into a smaller spatial and temporal scale, with all the local unpredictable variations and large scale predictability. The principle to the process of salt crystallization in the salterns is beautiful in its simplicity: when seawater is pumped into shallow basins and left to dry out, the salt concentration gradually increases, from sea water to the concentration at which the first of the major salts present (calcite) in sea water precipitates, followed by the next salt in the line, gypsum. The gradual removal of unwanted salts in different ponds along the way to the final product, halite, enables the production of salt of a greater purity. The bitter magnesium and calcium salts remain in the remaining solution, appropriately named "bittern" and in this way the bitter taste is eliminated from the final product, halite (Javor, 1989).

This apparently simple procedure is in reality a quite complex multi-step chemical operation. John Keats expressed his admiration for works of art by saying: "The excellence of every art is its intensity, capable of making all disagreeables evaporate, from their being in close relationship with beauty and truth." Salterns, in their natural simplicity, could represent such "evaporating" works of art, particularly the small-scale ones, exploited with technology that has not changed for hundreds of years. But disregarding their size and operating technology, all solar salterns have a distinct beauty about them. These man-made extreme environments, usually located close to human settlements, represent an oasis of calmness and a place where time has a different dimension. It flows with the seasons and follows the changing sun.

If dynamics is succession of changes of a state during the course of time, salterns can be described as dynamical systems. Such dynamical systems can be loosely defined as anything that is in motion, such as swinging pendulums, the reactions of a chemical process or, as in this case, gradually evaporating sea-water.

In the case of the pendulum, both position and velocity change over time, while in a chemical reaction, the "motion" is represented by the ratio of reactants to reagents (Stewart, 2001).

Salterns contain different elements of motion – water currents through the system, seasonal changes of temperature, increased viscosity due to increased salt concentrations. A collection of rules determines changes of such a dynamical system over time. In periodic cycles the behavior can be predicted. A lone planet orbiting a star in an elliptical orbit is an example of a limited, periodic cycle movement, which repeats itself over and over, like vibrations of an instrument, producing a musical tone. There exist more or less periodic cycles, such as crashing of waves on the coastline, traveling of the sun across the horizon, seasonal weather.... Salterns represent a quasiperiodic system, since they never quite repeat themselves, but nevertheless at some future point return to the previous state. In salterns that operate throughout the year, life cycles evolve within individual ponds, while in less hot climates, cycles are repeated seasonally. Life in salterns dances according to the rhythm of the Sun, directing the biochemical clock of halophilic microorganisms. It runs in parallel with the cycle of day and night, salinity, temperature and oxygen.

Up until the last thirty years or so, it was believed that everything in the universe fell into either fixed point, periodic, or quasiperiodic behavior. The belief in a clockwork universe, held that, in principle, if one had an accurate measure of the state of the universe and knew all of the laws that govern the motion of everything, then one would be able to predict the future with near perfect accuracy. Recognition of chaotic systems brought a change in thinking, since we now know that they are not exceptions to the norm but are in fact, more prevalent than anyone could previously imagine. Phenomena that were once thought to be purely random are now known to be chaotic (Gleick, 1987; Hall, 1991). Chaos is everywhere: in the turbulence of water and air, in the wobble of planets as they follow complicated orbits, and in global weather patterns. In all these cases the complicated motion produced by chaos prohibits predicting the future in the long term, but they nevertheless admit prediction in the short term. Philosophers and writers have expressed these observations quite similarly. Immanuel Kant said: "Nature itself, even in chaos, cannot proceed except in an orderly and regular manner," while George Santayana concluded that "Repetition is the only form of permanence that Nature can achieve." Regularity and permanence, statistic probability and repetition of phenomena is the part where science comes in. A scientist thinks, as Henry Brooks Adams expressed it, that "Chaos often breeds life when order breeds habit", and would expect a regular diversity of life forms and life cycles within waters of a saltern. A neutral observer would only see the glittering, pinkish, almost immovable waters reflecting the sun. Crystals of salt collected in heaps would remind both of centuries-old traditions and in this case, friendly association of man and nature.

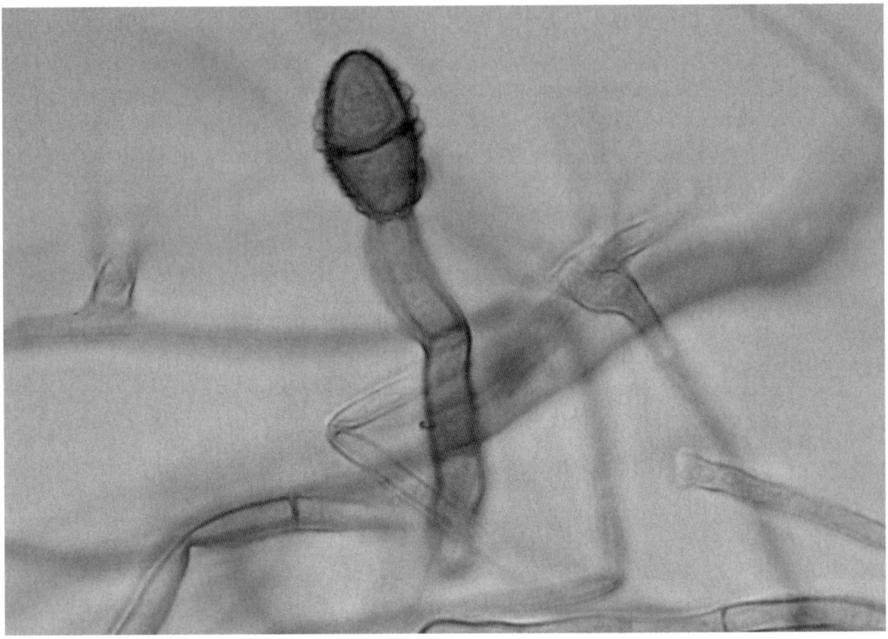

Fig. 3. *Alternaria tenuissima* – conidiophore with a single conidium.

3. The Variety of Microbial Life Forms in the Salterns – From Seawater to Saturated Brines

Ben Johnson wrote a perfect definition for the otherwise invisible beauty of microorganisms: "In small proportions we just beauty see, and in short measures life may perfect be". The diverse beauty of halotolerant and halophilic microorganisms inhabiting the salterns is revealed only after a drop of brine is observed through the lenses of a microscope.

Each salty pond – from seawater to halite crystallization and beyond, represents a small saline world of its own. Within the gradient of increasing salinity, a wide range of microorganisms, each adapted to life at a certain salt concentration range, can find the conditions favored for their growth. Halotolerant and halophilic microorganisms evolved in some cases to the extent that they can not survive if the concentrations of salt are not high.

Microbial communities that develop in the saltern ponds are not only highly diverse, but often very dense as well. Archaea alone can be present in the brines at numbers of up to a hundred million per milliliter and higher (Oren, 2002).

In the first evaporation ponds the salt concentration increases up to values around 100 grams of NaCl per liter. The diversity of microorganisms living in these ponds is often not greatly different from that found in the sea: diatoms, protozoa, marine fungi, yeasts, different types of microscopic green algae and others marine life forms (Kohlmeyer and Kohlmeyer, 1979; Post, 1983; Oren, 2002; Ventosa 2004; Gunde-Cimerman et al., 2005a; Butinar et al., 2005b). Microorganisms do not only swim or are suspended in the brine, but they are also found in the bottom sediments of the evaporation ponds. Such microbial mats mainly consist of filamentous cyanobacteria, often shaped as long bands or spirals, photosynthetic purple bacteria (Caumette, 1994; Clavero et al., 1994; Oren, 2000; 2002). But there are many others. This rich microbial life found within the benthic microbial mats and the organisms living in the saline water often attract large masses of birds. In some places sea gulls come, while in some others flamingoes embellish the shores of the ponds, calmly feeding on the biota.

When the salt concentration increases to values above 100-150 gram per liter, most of the diversity encountered before disappears, particularly the characteristic marine macro- and microorganisms. Prokaryotic microorganisms now dominate. Only a small number of more specialized and thus less adaptable eukaryotes can sustain life in this environment: some diatoms, a few protozoa, and surprisingly, many diverse specialized fungi (http://www.icbm.de/pmbio/mikrobiologischer-garten). At this salinity they are mainly represented by the melanized black yeasts such as *Hortaea werneckii* (Zalar et al., 1999; Gunde-Cimerman et al., 2000; 2005a,b) and selected species from the genera *Alternaria* (Gunde- Cimerman et al., 2005a), *Cladosporium* (Zalar et al., 2001), *Eurotium* (Butinar et al., 2005c) and *Wallemia* (Zalar et al., 2005). Abundant microbial life may be present also on the stratified bottom of the ponds.

The crystallizer ponds, in which the final stage of seawater evaporation takes place, are completely different. Thick water, almost oily in appearance, has ten to twelve times the salinity of seawater. The waters are so viscous that they are almost motionless. They are not only different in appearance from the earlier evaporation ponds, but also from a biological point of view. No profuse biofilms develop on the bottom of the ponds, within the precipitating mass of halite. Instead, most microbial life is found within the purple-reddish brine. The colors are mainly due to the presence of the unicellular green alga *Dunaliella salina* and halophilic Archaea (Antón et al., 1999; Oren, 2002; Ventosa, 2004; Gunde-Cimerman et al., 2005b). Next to these two dominating microorganisms, few fungi have found their niche as well.

4. Colors

Natural salt lakes and saltern ponds and other highly salty environments often display conspicuous colors: shades of red and orange, sometimes even purple and violet ones. Already in the Bible, in 2 Kings 3: 22-23, we can find the description of red waters:

"And when they rose early in the morning, and the sun shone upon the water, the Moabites saw the water opposite them as red as blood. And they said: "This is blood ...""

This description may well refer to a red bloom of microorganisms in the Dead Sea – the only major body of water in the area of the biblical Moab. These colors are caused by almost unimaginably dense communities of halophilic microbes, particularly Archaea and *Dunaliella salina* (Fig. 4), both containing pigments chemically related to carotene. Such colored waters are not only beautiful, but also beneficial for the operation of the salterns. The red-orange pigments of the cells absorb sunlight energy, causing a rise in temperature of the brines, which in its turn increases the rate of evaporation and the rate at which halite precipitates (Davis, 1974; Oren, 2002). This phenomenon clearly shows how organisms can have a recurrent relationship with their environment. Not only can two species mutually adapt to one another in such a way as to have a synergistic relationship, but coevolution also determines how species can coevolve with their environment. Beauty with a purpose!

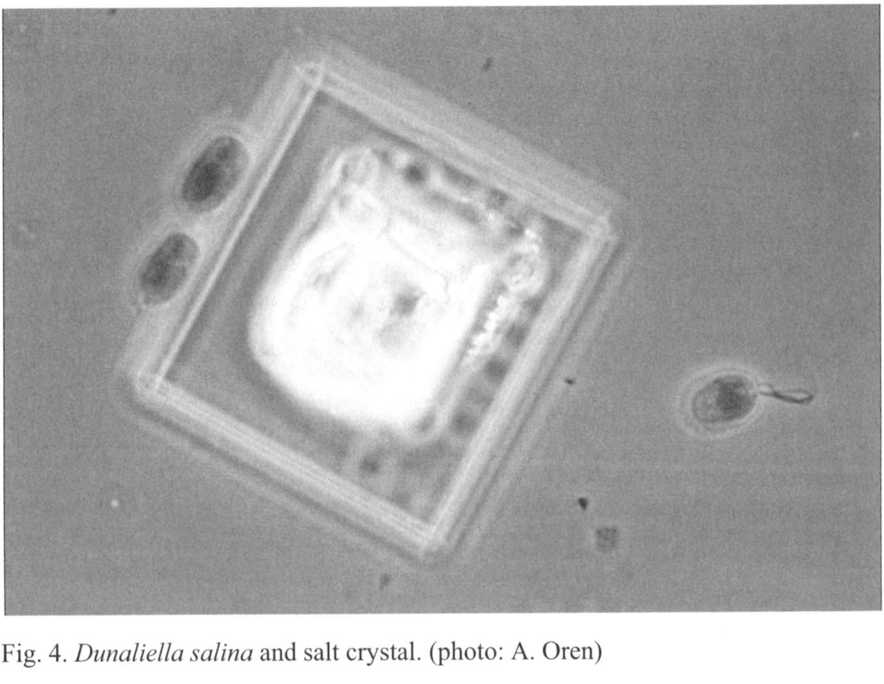

Fig. 4. *Dunaliella salina* and salt crystal. (photo: A. Oren)

Brine contains also copious amounts of spores (Fig. 8) formed by halotolerant and halophilic fungi and dispersed by air or water currents. Their thick cell walls are often impregnated with protective pigments. Diverse carotenoids or black melanins are used (Kirk et al., 2001; Kogej et al., 2004), coloring fungi (Fig. 9) in beautiful shades of black, green, blue or violet and contributing their share to the colors of the brine as well as to the green and brown hues, common in the shallow sediments.

Particularly spectacular displays of colored microorganisms are often found among the accumulating mass of gypsum crystals. Orange-brown colored cyanobacteria, dark-green cyanobacteria, and purple sulfur bacteria may form distinct layers within the gypsum crust (Canfield et al., 2004). Their position in this complex assembly is determined again by the sun. This time it is the intensity of the light that penetrates to different depths within the crust which is important. These microorganisms have pigments that absorb light of different intensity and color and therefore they have to establish in the community at the right depth.

5. Shapes and Patterns – from Crystals to Microorganisms

The most elementary fact of the beauty of shapes is symmetry, which is characteristic for patterns. Symmetry on itself is too stiff to explain all regularities in nature, but in combination with chaos and complexity it can give rise to an incredibly long list of natural patterns. Patterns are everywhere – rainbows, feathers, shells, grains of sand, salt crystals. But everywhere are also non-patterns – irregular, unpredictable things such as weather, waterfalls and cats. The result is an unusual mixture of regularity and infinite irregularity, intertwinement of flexibility and stiff mathematical rules. Thus nature works with a limited set of patterns, but they appear often in unusual combinations, creating nearly unlimited forms at different scales. Some patterns appear more often, because mathematical laws apply equally for atoms as they do for galaxies. Whether seen from space or under a microscope, these regular/irregular patterns evoke powerful responses (Flake, 1998; Stewart, 2001).

Humans like complex and repetitive patterns, full of symmetrical elegance of proportions. We encircle ourselves with such images. Wall papers, curtains, carpets, even architecture. Music is to a lesser extent composed of a rhythmic repetition of tones. Perhaps we like them because the brain has evolved in a world where the ability to recognize patterns increased the survival rate. Hypersaline environments

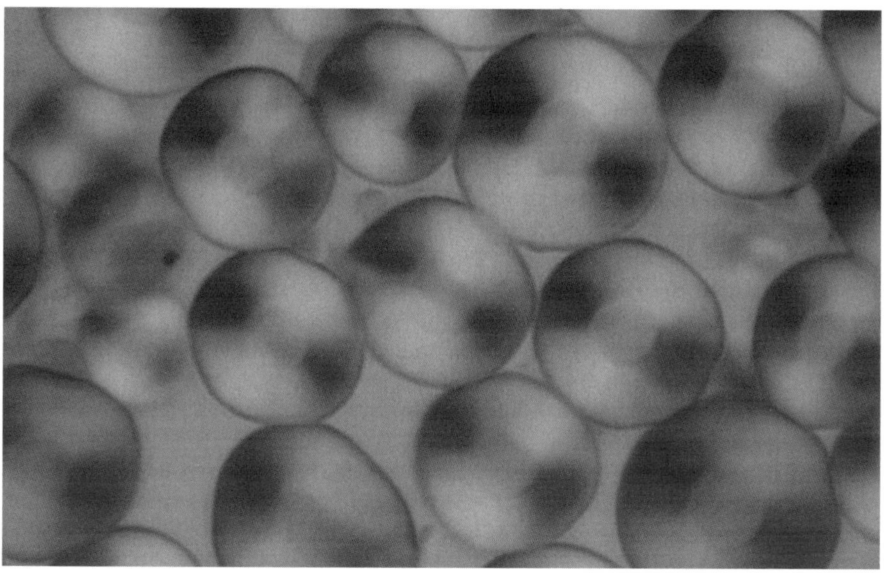

Fig. 5. *Emericella stellamaris* – Hülle cells.

are likewise adorned with beautiful natural patterns. The non-organic world is dominated by salt and gypsum crystals, while brine hides the diverse beauty of its living inhabitants.

Crystals are famous for their mathematical beauty of patterns, due to the submicroscopic symmetry of atomic arrangements. Pure polyhedron forms obtained *in vitro* can change due to nature's many irregularities. Salt, the essential requirement of life and one of the most abundant substances on Earth, always crystallizes in the shape of a simple cube. However, the size and final shape can vary. Francis Bacon said: "There is no excellent beauty that hath no some strangeness in the proportion," thus sometimes big, cube-shaped salt crystals can pile one on top of each other, resembling small Aztec pyramids (Picture). Ponds at lower salinity are characterized by massive amounts of precipitated gypsum crystals, shaped as tiny prismatic needles, resembling shattered glass. Beauty of a crystal is mathematical – it affects our sense of symmetry and complexity. Both are the essence of mathematics.

Brine also contains crystal-resembling microorganisms that bridge in shape the non-organic and organic world of the salterns. These salt-adapted Archaea (Fig. 6) are probably the only living beings, found nowhere else in nature, in which the cells are perfectly flat, square or rectangular, with straight angles and sharp corners (Walsby, 1980; Oren, 1999; Bolhuis et al., 2004; Burns et al., 2004).

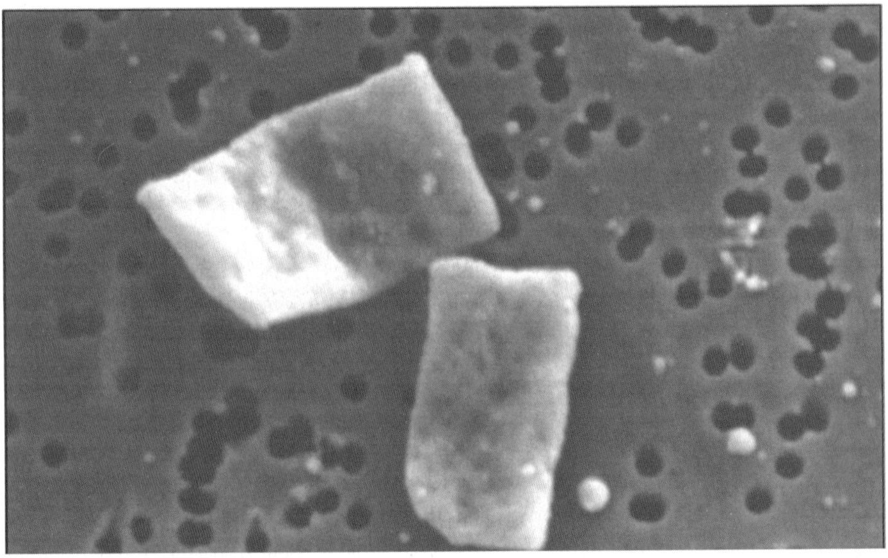

Fig. 6. Walsby's square haloarchaeon. (photo A: Oren)

Many other halophiles not only show highly unusual, but esthetically appealing shapes as well: from simple unicellular bacteria and multi-cellular threads of the filamentous cyanobacteria to the far more complex world of the fungi (Figs. 3, 5, 7, 8). As the squares, also these shapes could be described in terms of Euclidean geometry, extended to account for objects with a fractional dimension. Fractals possess structural self-similarity on multiple spatial scales, meaning that a piece of a fractal will often look like the whole (Barnsley, 1988). Although fractals are a mathematical idealized expression of beauty and symmetry, they come very close to capturing the richness and variety of forms found in nature. Some bodies have a more regular type of self similarity with little miniature parts as variations of the whole. In nature we observe and admire non-identical self similarity.

Small crystals of salt have the same random composition as piles of salt, big as mountains. Clouds, mountains, chains of fungal conidia on a conidiophore, coastlines, hypersaline ponds and lakes are statistically self similar, as well as many-fold repeated shapes of microorganisms, covering the water surface or microbial mats as biofilms. They give the impression of harmony, purely by the repetition of successfully reproduced forms.

Fig. 7. *Penicillium vulpinum* – conidiophores in bundles.

Nature does not display only fixed patterns, but also dynamic ones. When wind sweeps over the salterns, waves on brine and salt grains deposited elsewhere create the dynamic pattern of the moment. Besides physics also biological processes contribute their share. The form of the fungal colonies (Fig. 9) are the representation of the dynamic pattern of their life cycles. Chemicals that inter-react and diffuse through a tissue, can form spontaneous patterns, according to reactions and the speed with which they diffuse. The final picture is the combination of genetic instructions, physics and chemical dynamics.

We can extend this shape-directed, mathematical thinking even further, to include the evolutionary property of species adaptations. The neo-Darwinistic evolutionary distillation of ideas is often described by the combined action of variation, heredity and selection (Maynard Smith, 1975). This differs from strict Darwinism particularly in the crucial importance it gives to variation, which refers to how individuals in a population can differ from each other. Variations within a species, by definition, can be expressed only in terms of multiple individuals characterized by non-identical self similarity. Since evolution operates on no single individual but on the entire species, heredity can be seen as a form of temporal persistence of biological fractals. As two individuals are not the same, neither are two phylogenetic trees.

I should not continue any further since every scientist knows that good theories can be destroyed by facts. I will instead follow a simple scientific piece of advice, given by Albert Einstein: "Things should be as simple as possible, but not simpler." From now on pictures should speak for themselves.

6. Conclusions

After having decided to write this text on Beauty and Nature, I went through diverse evolutionary phases. As Ellen Sturgis Hooper said: "First I slept, and dreamed of that life was beauty, I woke, and found that life was duty." This duty made me think which type of beauty could be presented, which would not be already obviously noticed by everybody around. After much consideration my thoughts focused on the object of our scientific research, natural hypersaline environments and its inhabitants. It seemed a natural choice and obvious thing to do. After some thinking I felt, as George Bernard Shaw, when he wrote: "Beauty is all very well at first sight; but who ever looks at it when it has been in the house three days?" Is there really anything to show, is it not all too known, dull and obvious? It did not help to know, searching here and there for inspiration, what Stendhal thought: "One should only write when one has something important or profoundly beautiful to say, but then one must say it as simply as possible, as if one were trying one's best to prevent it being noticed." Not an easy task to describe a personal scientific fascination in simple and at the same time beautiful terms. Some encouragement in the end came from David Hume: "Beauty is no quality in things themselves. It exists merely in the mind which contemplates them." My mind contemplated my personal view on beauty. It is now up to you to contemplate whether this view can be shared.

Figures

Fig. 8. Most of the filamentous fungi found in the salterns reproduce asexually, by conidia or in the case of yeasts, by budding. Some fungi undergo sexual reproduction, forming ascopores or basidiospores. Some are very sturdy, while others compensate for the weaker nature of their spores by producing spores in extremely large numbers, ensuring survival in this way. Many spores become richly ornamented or have long appendages, helping them to float in the water or glide through the air. Most of them have beautiful shapes that speak for themselves.

-a- *Aspergillus versicolor*, conidia; -b- *Emericella stellamaris*, ascospores; -c- *Cladosporium* sp., conidia; -d- *Fusarium verticillioides*, conidia; -e- *Eurotium amstelodami*, ascospores; -f- *Phaeotheca triangularis*, conidia; -g- *Emericella appendiculata*, ascospores and Hülle cell; -h- *Epicoccum nigrum*, conidia; -i- *Hortaea werneckii*, yeast cells; -j- *Rhizopus oryzae*, sporangiospores.

Fig. 9. Colorful fungal colonies and dynamic patterns of life their cycles. Colonies are richly pigmented and display a variety of colors, from black to white, with shades of orange, red, green.
-a- Eurotium amstelodami; -b- *Epicoccum nigrum*; -c- *Mucor* sp.; -d- *Epicoccum nigrum*; -e- *Rhodotorula mucilaginosa*; -f- *Penicillium chrysogenum*; -g- *Epicoccum nigrum*; -h- *Eurotium sp.* contaminated with bacteria; -i- *Aureobasidium pullulans*; -j- *Emericella variecolor*; -k- *Aspergillus ochraceus*; -l- *Penicillium chrysogenum*; -m- *Aureobasidium pullulans*; -n- *Aspergillus flavus*; -o- *Aspergillus candidus*.

Fig. 10. Fractal beauty of fungal conidiophores.
-a, b- *Aspergillus candidus*; -c- *Emericella stellamaris*; -d- *Cladosporium sphaerospermum*; -e- *Emericella appendiculata*; -f- *Wallemia ichthyophaga*; -g- *Fusarium verticillioides*.

Fig. 11. Life is dynamic - adaptations and diversity of forms of fungal mycelium.
-a- *Trimmatostroma salinum*; -b- *Aspergillus candidus*; -c, d- *Phaeotheca triangularis*.

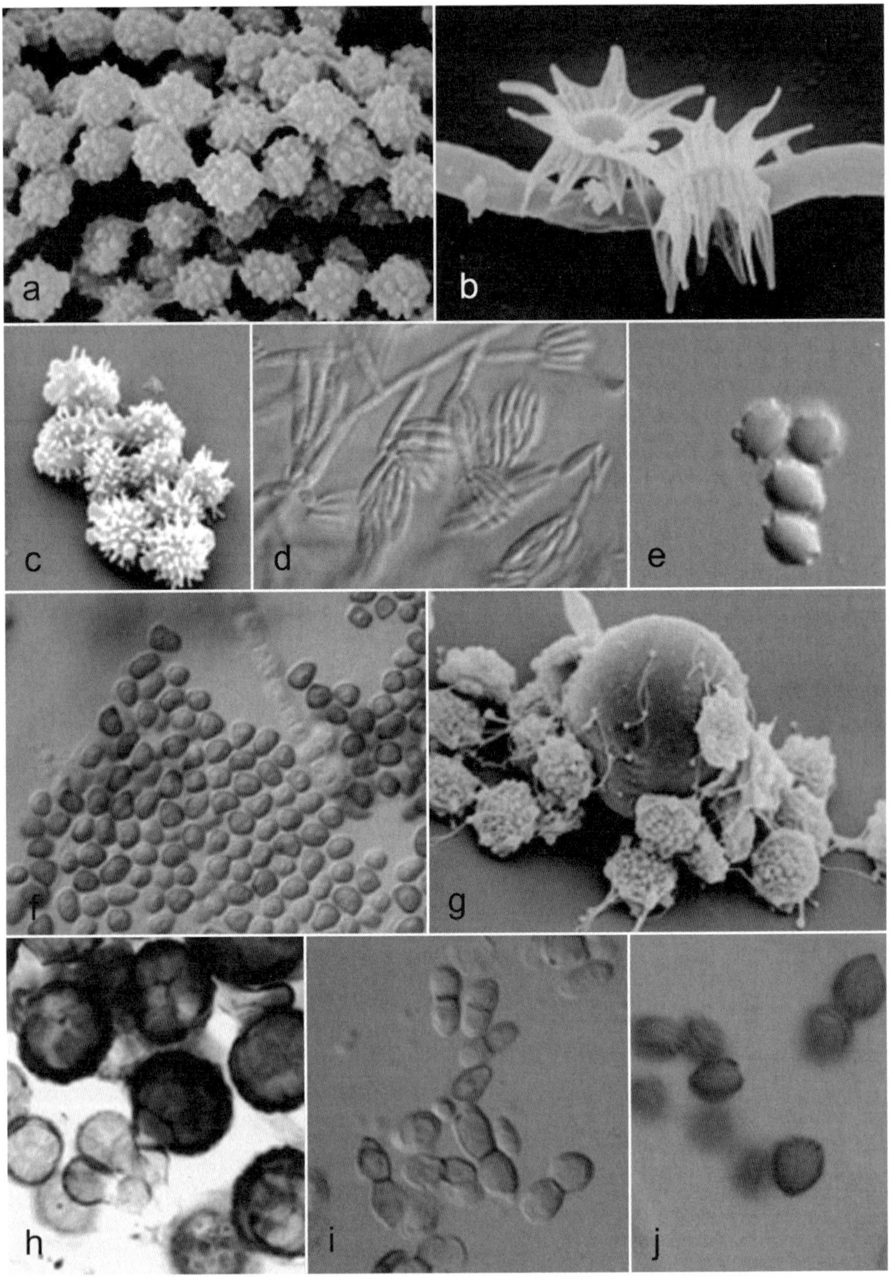

Fig. 8. Reproductive structures of fungi.

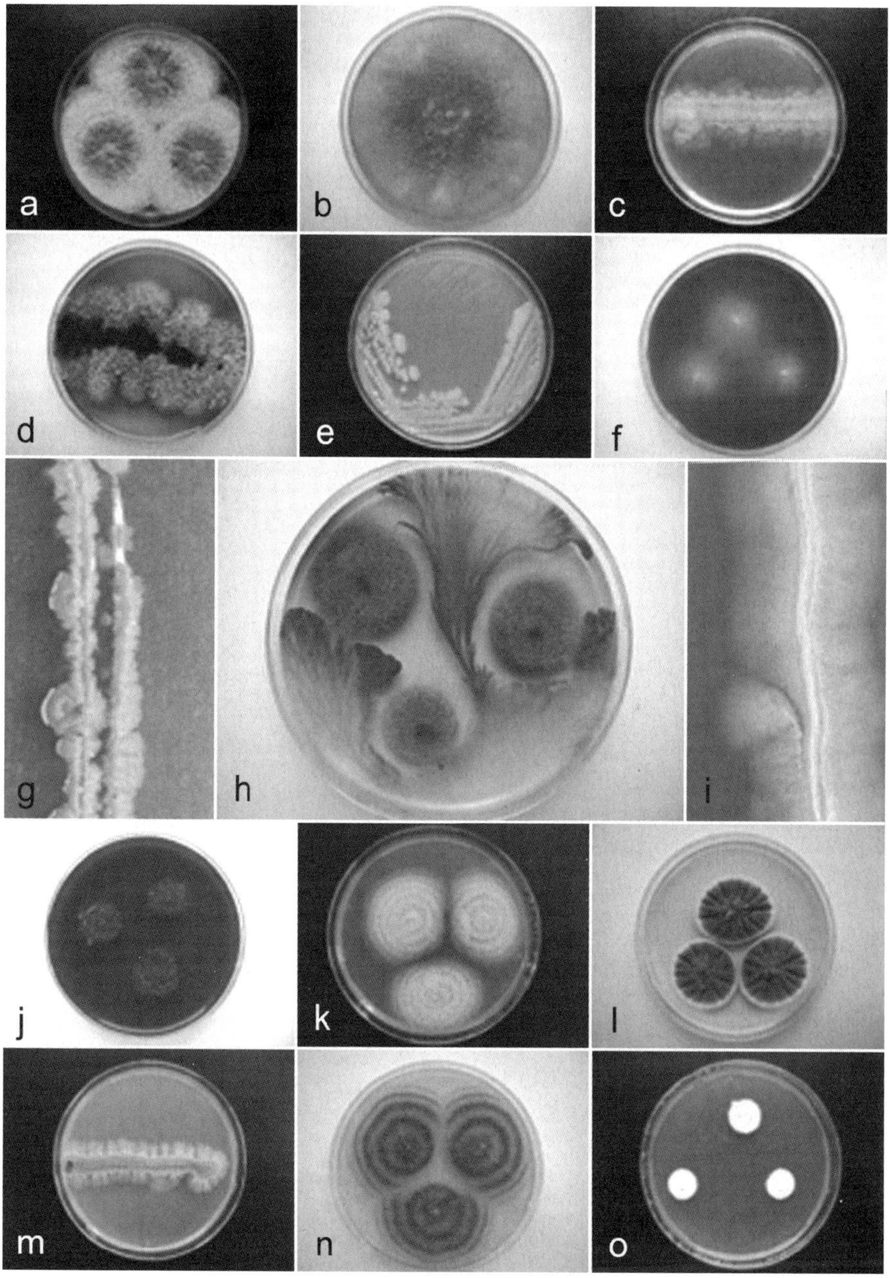

Fig. 9. Colorful fungal colonies.

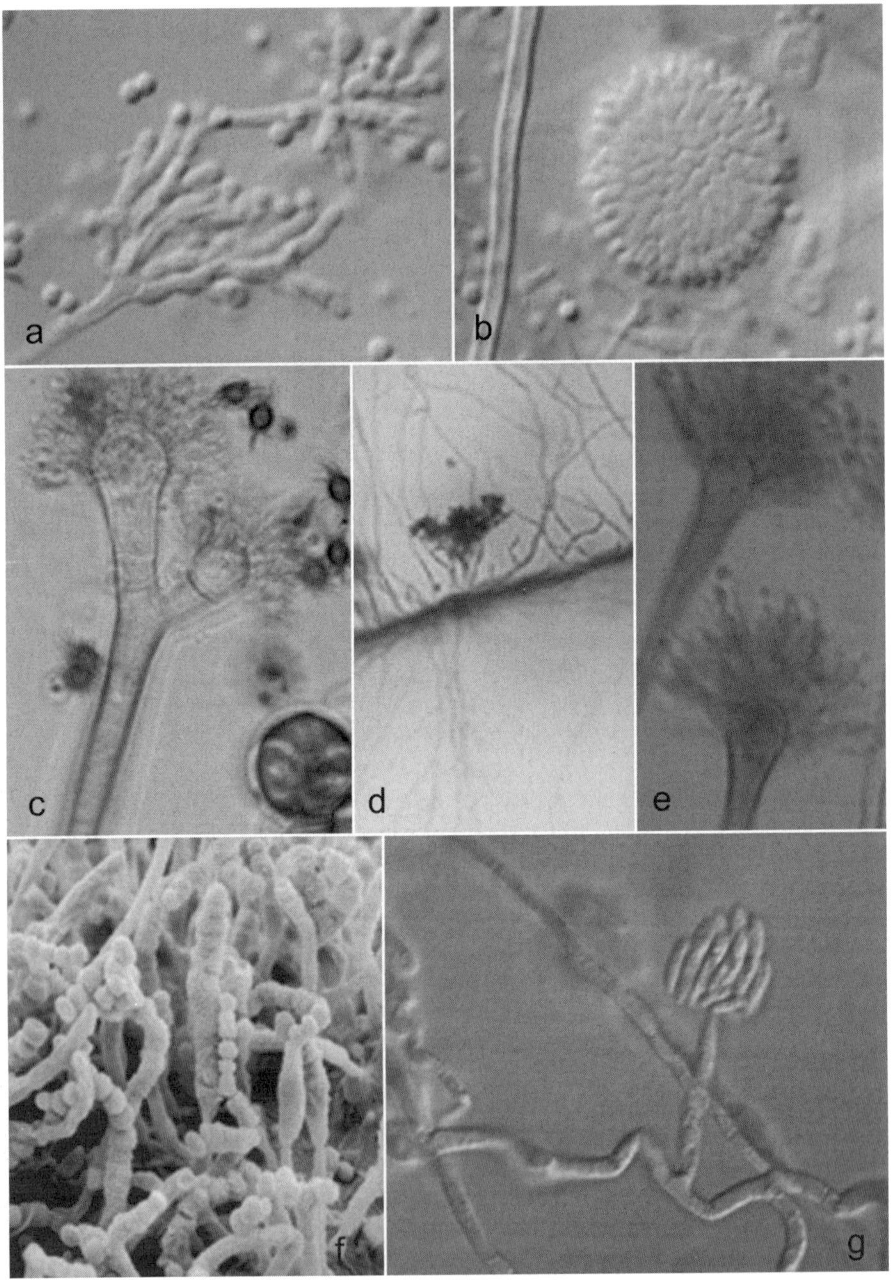

Fig. 10. Fungal conidiophores.

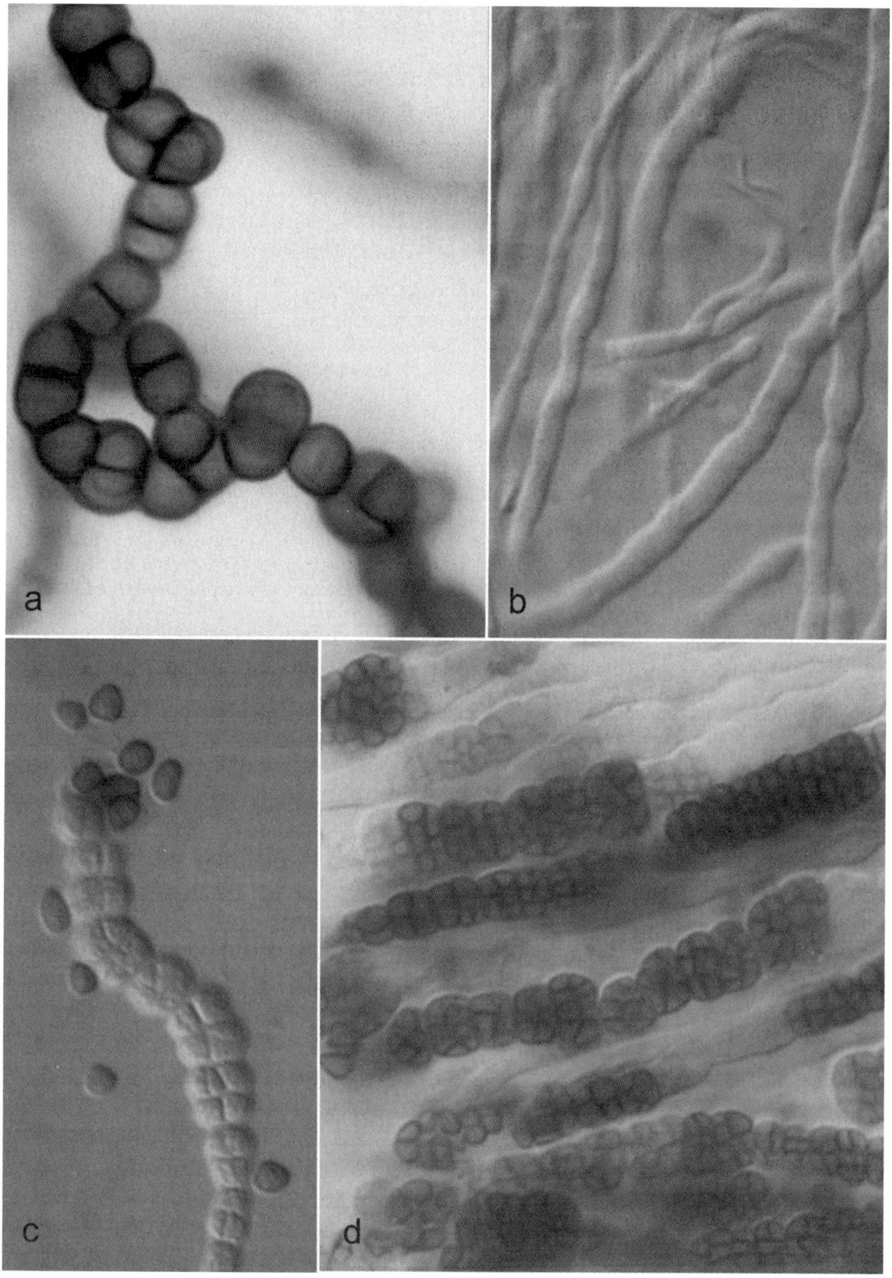

Fig. 11. Fungal mycelium.

7. Acknowledgements

We are grateful to prof. Aharon Oren for carefully reading the manuscript and to prof. Joseph Seckbach for his initial idea and friendly insistence.

8. References

Antón, J., Llobet-Brossa, E., Rodríguez-Valera, F. and Amann, R. (1999) Fluorescence *in situ* hybridization analysis of the prokaryotic community inhabiting crystallizer ponds. *Environ. Microbiol.* **1**, 517-523.

Barnsley, M. (1988) *Fractals everywhere*. Academic Press, New York.

Burns, D.G., Camakaris, H.M., Janssen, P.H. and Dyall-Smith, M.L. (2004) Cultivation of Walsby's square haloarchaeon. *FEMS Microbiol. Lett.* **238**, 469-473.

Bolhuis, H., te Poele, E.M. and Rodríguez-Valera, F. (2004) Isolation and cultivation of Walsby's square archaeon. *Environ. Microbiol.* **6**, 1287-1291.

Butinar, L., Sonjak, S., Zalar, P., Plemenitaš, A. and Gunde-Cimerman, N. (2005a) Melanized halophilic fungi are eukaryotic members of microbial communities in hypersaline waters of solar salterns. *Botanica Marina* **1**: 73-79.

Butinar, L., Spencer-Martins, I., Oren, A. and Gunde-Cimerman, N. (2005b) Yeast diversity in hypersaline habitats. *FEMS Microbiol. Lett.* **244**, 229-234.

Butinar, L., Zalar, P., Frisvad, J.C. and Gunde-Cimerman, N. (2005c) The genus *Eurotium* – members of indigenous fungal community in hypersaline waters of salterns. *FEMS Microbiol. Ecol.* **51**, 155-166.

Canfield, D.E., Sørensen, K.B., and Oren, A. (2004) Biogeochemistry of a gypsum-encrusted microbial ecosystem. *Geobiology* **2**, 133-150.

Caumette, P., Matheron, R., Raymond, N. and Relexans, J.-C. (1994) Microbial mats in the hypersaline ponds of Mediterranean salterns (Salins-de-Giraud, France). *FEMS Microbiol. Ecol.* **13**, 273-286.

Clavero, E., Grimalt, J.O., Merino, V. and Hernández-Mariné, M. (1994) Structure of the sediment at depositional saline environments, In: L.J. Stal and P. Caumette (eds.), *Microbial Mats. Structure, Development and Environmental Significance*. Springer-Verlag, Berlin, pp. 91-96.

Davis, J.S. (1974) Importance of microorganisms in solar salt production, In: A.H. Coogan (ed.), *Fourth Symposium on Salt.* Northern Ohio Geological Society, Cleveland, pp. 369-372.

Flake, G. W. (1998) The Computational Beauty of Nature. *Computer Explorations of Fractals,* Chaos,Complex Systems, and Adaptation, Bradford.

Gleick, J. (1987) *Chaos. Making a new science.* Viking, New York.

Gunde-Cimerman, N., Zalar, P., de Hoog, G.S. and Plemenitaš, A. (2000) Hypersaline waters in salterns – natural ecological niches for halophilic black yeasts. *FEMS Microbiol. Ecol.* **32**, 235-240.

Gunde-Cimerman, N., Sonjak, S., Zalar, P., Frisvad, J.C., Diderichsen, B. and Plemenitaš, A. (2003) Extremophilic fungi in Arctic ice: a relationship between adaptation to low temperature and water activity. *Physics and Chemistry of the Earth* **28**, 1273-1278.

Gunde-Cimerman, N., Frisvad, J.C., Zalar, P. and Plemenitaš, A. (2005a) Halotolerant and halophilic fungi, In: S.K. Deshmukh and M.K. Rai (eds.), *The Biodiversity of Fungi: Their Role in Human Life.* Oxford & IBH Publishing Cp. Pvt. Ltd., New Delhi, pp. 69-127.

Gunde-Cimerman, N., Oren, A. and Plemenitaš, A. (eds.) (2005b) *Adaptations to life at high salt concentrations in Archaea, Bacteria, and Eukarya.* Springer-Verlag, Berlin, in press. http://www. icbm. de/ pmbio/mikrobiologischer-garten.

Hall, N. (ed.). (1991) *Exploring chaos: A guide to the new science of disorder*. W.W. Norton & Co, New York.

Javor, B. (1989) *Hypersaline Environments. Microbiology and Biogeochemistry.* Springer Verlag, Berlin, pp. 163-175.

Kogej, T., Wheeler, M.H., Lanišnik Rižner, T. and Gunde-Cimerman, N. (2004a) Evidence for 1, 8-dihydroxynaphthalene melanin in three halophilic black yeasts grown under saline and non-saline conditions. *FEMS Microbiol. Lett.* **232**, 203-209.

Kohlmeyer, J. and Kohlmeyer, E. (1979) *Marine Mycology. The Higher Fungi.* Academic Press, New York.

Kirk, P.M., Cannon, P.F., David, J.C. and Stalpers, J.S. (2001). *Ainsworth & Bisby's Dictionary of the Fungi,* 9th. ed. CAB International, Oxon, U.K.

Kushner, D.J. (1978) Life in high salt and solute concentrations, In: D.J. Kushner (ed.), *Microbial Life in Extreme Environments*. Academic Press, New York, pp. 317-386.

Maynard Smith, J. (1975) *The theory of evolution* (third ed.). Penguin books, New York.

Oren, A. (1993) The Dead Sea – alive again. *Experientia* **49**, 518-522.

Oren, A. (1999) Microbiological studies in the Dead Sea: future challenges toward the understanding of life at the limit of salt concentrations. *Hydrobiologia* **205**, 1-9.

Oren, A. (1999) The enigma of square and triangular bacteria, In: J. Seckbach (ed.), *Enigmatic Microorganisms and Life in Extreme Environmental Habitats*. Kluwer Academic Publishers, Dordrecht, pp. 337-355.

Oren, A. (2000a) Salts and brines, In: B.A. Whitton and M. Potts (eds.), *Ecology of Cyanobacteria: Their Diversity in Time and Space*. Kluwer Academic Publishers, Dordrecht, pp. 281-306.

Oren, A. (2002) *Halophilic Microorganisms and their Environments*. Kluwer Academic Publishers, Dordrecht.

Poincaré, H. (1952). *Science and hypothesis*. New York, Dover.

Post, F.J., Borowitzka, L.J., Borowitzka, M.A., MacKay, B. and Moulton, T. (1983) The protozoa of a Western Australian hypersaline lagoon. *Hydrobiologia* **105**, 95-113.

Stewart, I. (2001) *What shape is a snowflake?* Ivy Press Ltd., Princeton.

Ventosa, A. (ed.) (2004) *Halophilic Microorganisms*. Springer-Verlag, Berlin.

Zalar, P., de Hoog, G.S and Gunde-Cimerman, N. (1999a) Ecology of halotolerant dothideaceous black yeasts. *Studies in Mycology* **43**, 38-48.

Zalar, P., de Hoog, G.S. and Gunde-Cimerman, N. (2001) Fungi from the salterns: halophilic versus eurytolerant species in the genus *Cladosporium*, In: A. Ventosa and D.R. Arahal (eds.), *Abstracts of the International Conference on Halophilic Microorganisms*, 23-27 September 2001, Sevilla, Spain, p. 17.

Zalar, P., de Hoog, G.S., Schroers, H.-J., Frank, J.M. and Gunde-Cimerman, N. (2005) Taxonomy and phylogeny of the osmophilic genus *Wallemia* (*Wallemiomycetes* and *Wallemiales*, cl. et ord. nov.). Antonie van Leeuwenhoek **87**, 311-328.

Walsby, A.E. (1980) A square bacterium. *Nature* **283**, 69-71.

Webster's encyclopedic unabridged dictionary of the English language (1996), *Gramercy books*, New York.

Biodata of **Gordon W. Beakes** author of ***"All Things Bright and Beautiful: The Hidden Cosmos of Microscopic Planktonic Algae"***

Dr Gordon Beakes is Reader in Developmental Mycology in the Division of Biology, School of Biology and Psychology, at the University of Newcastle upon Tyne, England. He obtained his Ph.D. from Imperial College, University of London, in 1974 and then spent time as a post doctoral demonstrator at both the Universities of Keele (1974-1976) and Newcastle (1976-81). After completing a one year Research Fellowship with Professor Salomon Bartnicki-Garcia, at the University of California, Riverside (1982-3) he returned to a lectureship in the then, Plant Biology Department at Newcastle, where he took over the post of the retiring phycologist. Dr Beakes research interests have revolved around the study of the structure and function of aquatic fungi (mainly oomycetes) and algae. He has championed rather esoteric, economically unimportant, but biologically fascinating organisms and has a long standing interest in searching for the evolutionary roots of the oomycete fungi. His proudest achievement is to have discovered several oomycetes with bald anterior flagella!

E-mail: **g.w.beakes@ncl.ac.uk**

J. Seckbach (ed.), Life as We Know It, 685-700.

ALL THINGS BRIGHT AND BEAUTIFUL:
The Hidden Cosmos of Microscopic Planktonic Algae

GORDON W. BEAKES
Division of Biology, Ridley Building, Newcastle University, Newcastle upon Tyne, NE1 7RU, United Kingdom

1. Introduction

"All things bright and beautiful, all creatures great and small" are the opening lines of a popular children's hymn, penned in the mid nineteenth century by Mrs Alexander. Little could the she have realised just what a remarkable hidden world of wonderful 'bright, beautiful and small' microscopic creatures were about to be shortly discovered. The world of microscopic algae is largely invisible to the human eye and only became apparent after the development of relatively affordable high quality light microscopes during the latter half of the nineteenth century. These extended the resolution of the human eye about a thousand fold, extending the size of objects we could see from around 0.2 mm to about 0.2 of a micrometer. The beauty and diversity of form amongst algal groups, such as the cyanobacteria (Figs. 1-8), dinoflagellates (Figures 9-14), diatoms (Figures 15-21, 35, 37) and desmids (Figures 28-32, 36), is awe inspiring and either provides evidence of a divine creator or illustrates the extra-ordinarily subtle powers of evolutionary selection pressures. Much of the beauty of microscopic algae comes from the aesthetic pleasure associated with the symmetry of form, as with the star-like colonies of the planktonic variant of the diatom *Tabellaria* (Figure 17) and the green algae *Pediastrum* (Figures 25, 26). Such symmetry can also be seen at the cellular level in the spoke-like bifurcating spines radiating from the diminutive central cell of *Bacteriastrum* (Figure 20) and from the perfect patterns of pores that decorate the valves of diatoms such as *Cyclotella* (Figure 21) and *Actinoptychus* (Figure 37) to the elaborately lobed plastids of *Zygnema* (Figure 23) and *Staurastrum* (Figure 31). By duplicating and rotating the elegantly curved cells of the desmid *Closterium*, I have replicated the aesthetic impact of this natural symmetry (Figure 36). One of the reasons why desmids such as *Micrasterias* (Figures 28, 29) and *Staurastrum* (Figure 30) are so visually captivating is that they predominantly occur as 'mirror-image' cells. Nature seems to be experimenting with symmetrical form in the same way that psychologists use mirrored ink-blot image to test their patient's perceptions. Replicating and mirroring the simple chains of *Anabaena* cells has added a mystical dimension to these images (Figures 33, 35).

Helically wound coils or spirals also have a universal visual appeal. The spiral elegance of a DNA molecule is duplicated at the macroscopic level by both the filaments of the cyanobacteria *Spirulina* (Figure 3) and *Anabaena* (Figure 33) and in the plastids of the common green pond alga *Spirogyra* (Figure 24). Conversely, complexity can also have an off-beat beauty of its own. The bizarre and often asymmetric forms of dinoflagellates such as *Ceratium* (Figures 9, 10), *Ornithocercus* (Figure 12) and *Dinophysis* (Figures 13, 14) are all strangely compelling, as are the complex stacked or twisted arrays of micro-algal filaments seen in colonies of *Aphanizomenon* and *Anabaena* (Figures 1, 2, 4, 5). The intricately sculpted punctuate shells of diatoms can also be used to generate images to great visual effect (Fig. 35). When contemplating the diversity and beauty of life, the images of these algae speak for themselves (Figures 1-37). However, it will be worthwhile to briefly reflect on some of the underlying science behind both how these images are produced and why these cells develop such fantastic forms.

2. Through the Glass Brightly – How Algal Cells are Examined

Once microscopes became widely available to biologists the systematic cataloguing of hitherto 'hidden' groups of organisms began in earnest and quickly led to the publication of many, frequently beautifully illustrated, accounts (diatoms - Van Heurck, 1885; desmids - Cooke, 1887). It was little wonder that the acid-cleaned silica (glass) shells of diatoms (Figure 15) soon became the specimens of choice for late nineteenth century microscopists to competitively demonstrate their prowess with these instruments. One of the earliest tricks developed to enhance what could be seen under the light microscope was dark field microscopy. Only light that interacts with the specimen is actually diffracted into the objective lenses. This results in specimens appearing as bright objects against a black background and transforms the dense, black-brown colonies of *Gloeotrichia* into something reminiscent of an exploding galaxy (Figure 7). In the twentieth century optical systems were developed to convert differences in thickness and refractive index of cytoplasmic compartments into optical contrast. Phase contrast systems (both conventional and differential interference contrast, DIC) not only enhanced the detail that could be seen but often greatly add to the aesthetic appeal of the micrographs. In the illustrations provided here, phase optics were used to examine samples of living algae, preserved algal specimens and cleaned diatom frustules (ordinary phase contrast - Figures 12, 14, 15, 17; DIC - Figures 5, 6, 8, 9, 13, 18, 22, 25, 28).

The widespread introduction of first transmission and then scanning electron microscopes in the nineteen sixties and seventies, extended the limits of what we could see (resolution) a further hundred fold and added a whole new dimension to our understanding of the complexity and micro-architecture of algae. In particular, they made the detailed study of the of the exquisite cell coats of dinoflagellates and diatoms much easier (Figures 10, 11, 15, 35, 37). However, such specimens have to be subjected to complex multi-stage preparation protocols (Beakes, 2003). These include chemical fixation and dehydration to enable them to withstand the vacuum conditions in the electron-microscope, followed by staining or coating in heavy metals to give them sufficient electron-contrast to deflect or reflect the electron beam.

As someone once, aptly, pointed out, electron microscopists deal only with autopsies!

One of the most exciting recent advances in microscopy has been the development of confocal laser scanning microscopes (Hepler and Gunning, 1998) which has enabled cellular detail in living cells to be resolved with often jaw-dropping clarity (Figures 2-4, 18b, 23, 24, 26, 27, 29-32).

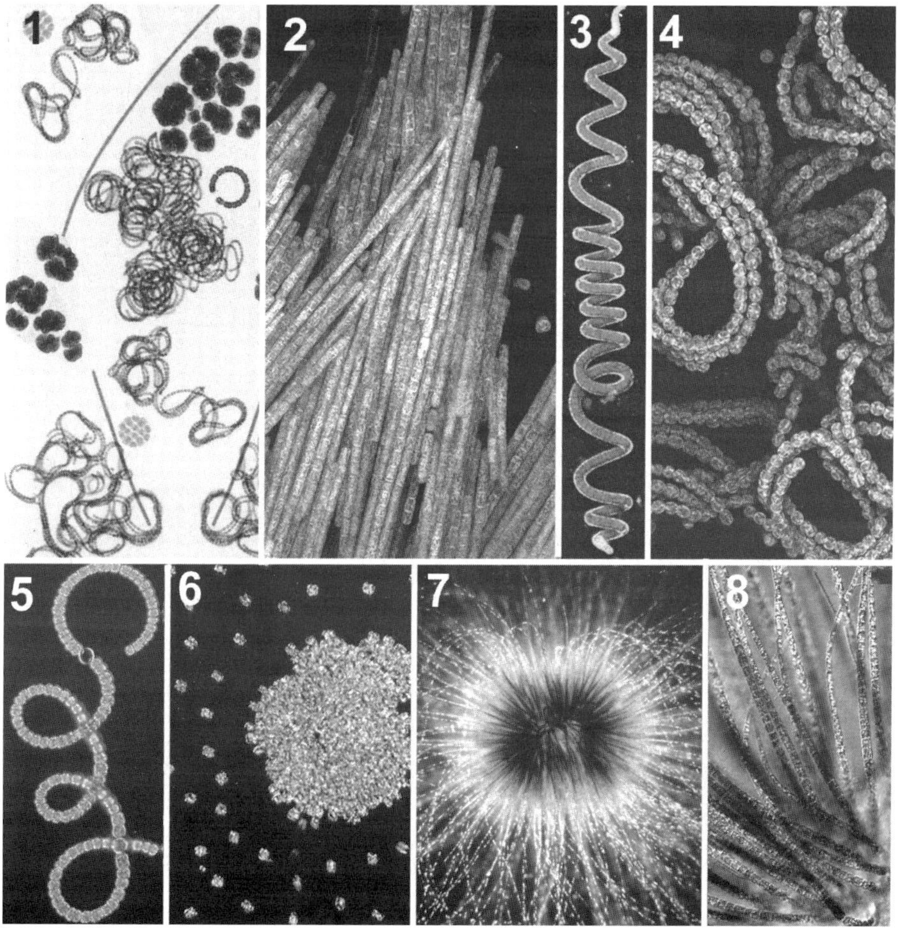

Figures 1-8. Cyanophyte Algae (Bacteria – Prokarya).

Figure 1. Micrograph of a bloom of phytoplankton in Eshwaite Water, Cumbria showing colonies of both cyanobacteria (*Anabaena, Aphanizomenon*) and green algae (*Eudorina, Botryococcus*).

Figure 2. Detail of a colony of the planktonic cyanobacterium, *Aphanizomenon*, from the above bloom showing dense clusters of architectural stick-like filaments, which make up the floating colonies of this, often toxin producing, species.

Figure 3. The cork tapered screw like filaments of the cyanobacterium *Spirulina* – looking like a stretched spring. This organism is harvested an eaten in some parts of the world.

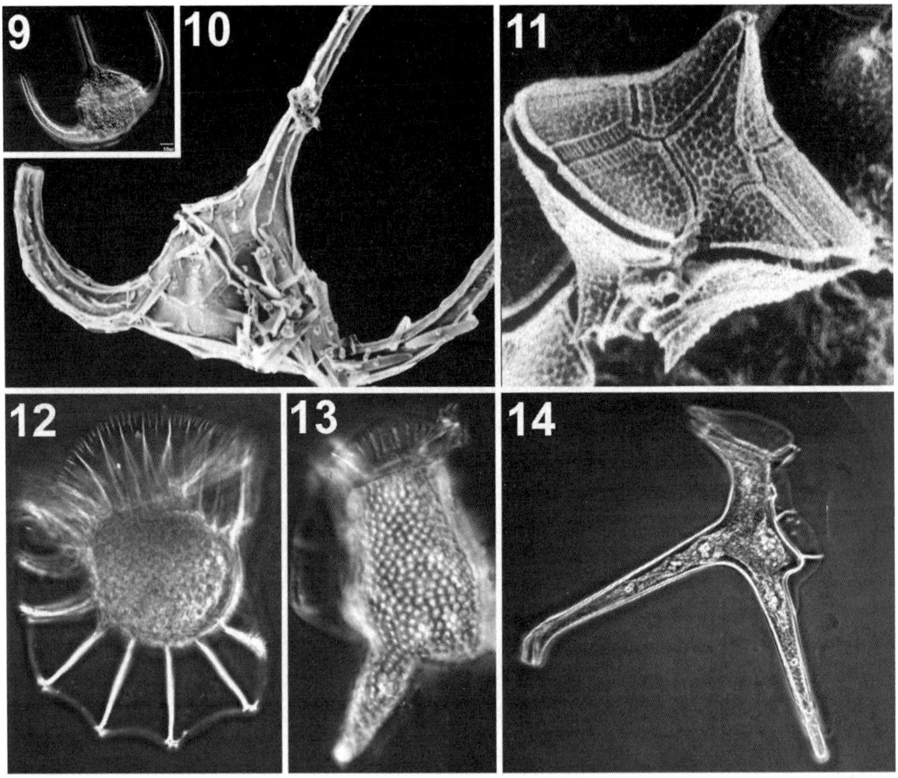

Figures 9-21. Chromalveolate Clade Algae: Dinoflagellates and Diatoms.

Figure 4. The great depth of focus achievable with the confocal microscope reveals the full beauty of the intricately coiled and knotted filaments of a colony of the cyanobacterium *Anabaena*.

Figure 5. The spiral form of *Anabaena* viewed under ordinary darkfield microscopy. The pressure of the cover slip has flattened it into an elegantly coiled ribbon. The slightly larger empty looking heterocyst cells which contain the biochemical machinery for nitrogen fixation.

Figure 6. The floating colonial cyanobacterium *Gomphosphaeria*. This species as a densely packed core of tiny ovoid cells, clusters of which disperse into the gelatinous matrix in which the cells are embedded.

Figures 7 & 8. The large colonial cyanobacterium *Gloeotrichia*, which at low power (7) looks like an exploding galaxy, but is made up of packed tapering filaments which terminate in translucent hair-like trichomes (8).

Figures 9-14 (above). A selection of marine armoured dinoflagellates from phytoplankton samples collected from British (10,11) and Indonesian (9,12-14) coastal waters.

Figures 9, 10. The distinctive triple-horned cells of the large genus *Ceratium*. In *C. symatrix* (9) the three horns are almost equal in length whereas *C. longipes* (10), has an elongated antapical horn.

Figure 11. SEM micrograph showing detail of the 'top-like' *Protoperidinium pentagonum*, showing the exquisitely sutured cellulosic plates which make up the cell coats of armoured dinoflagellates.

Figure 12. The amazingly elaborate cells of the alien-looking *Ornithocercus*. The large ribbed fin-like extensions to its body are characteristic of this tropical genus.

Figures 13, 14. The equally bizarre 'seahorse-like' genus, *Dinophysis* showing the variations in form. *D. caudata* (13) often occurs a twinned cells whereas another tropical species *D. miles* (14) has a long, almost trunk like, ventral extension and overall elongate form. Both these species occur in toxic blooms.

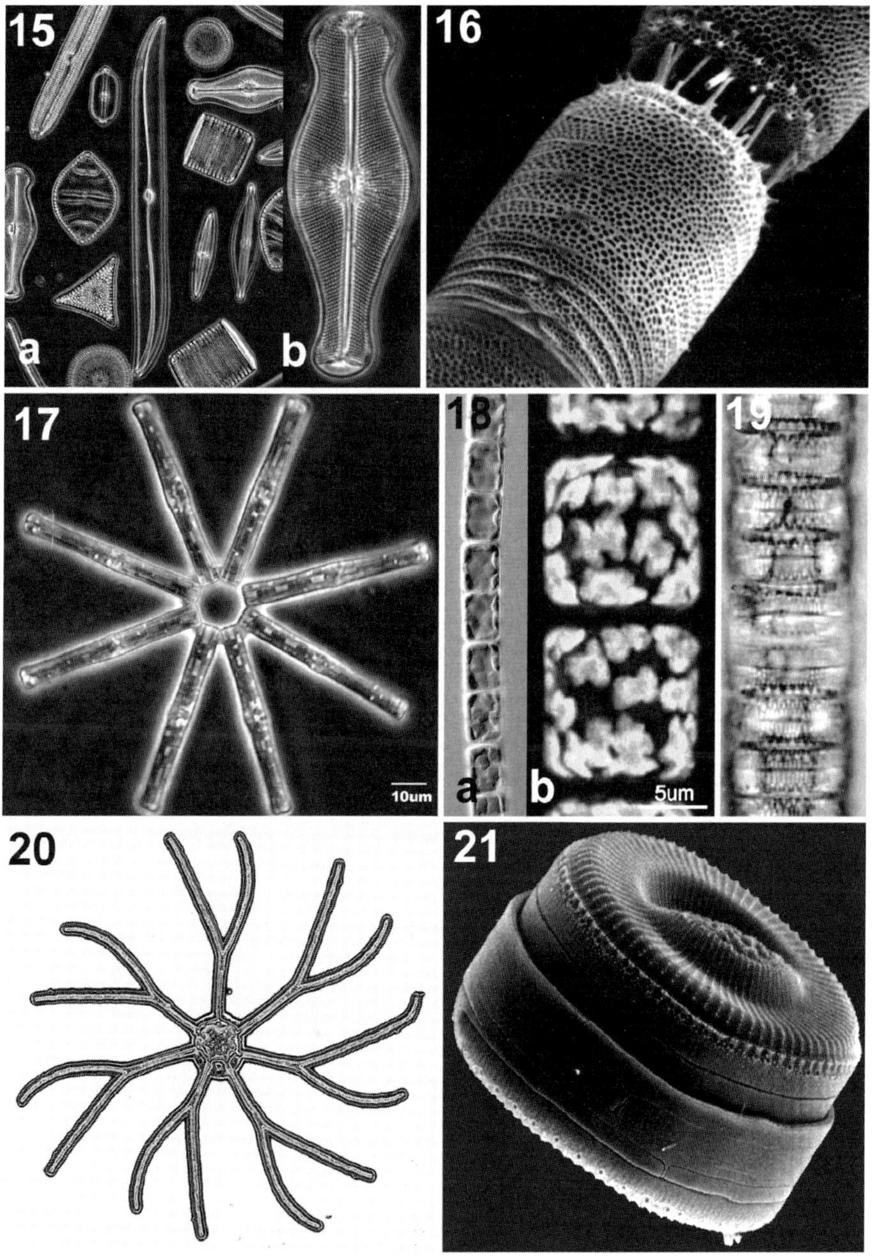

Figures 15-21 (opposite). A selection of diatoms from both marine (as above) and freshwater ecosystems.

Figure 15. A prepared slide of acid cleaned frustules planktonic diatoms (a) showing their exquisitely sculptured glass cases (frustules) the fine-detail of which is revealed using interference phase optics, as shown by the higher powered view of the jelly baby-like *Gomphonema* (b).

Figure 16. SEM of a tropical centric diatom species, whose filamentous cells are made up of concentric rings of filigree girdle lamellae (probably a *Detonula* sp.).

Figure 17. The pleasing radial symmetry of the colonial planktonic form of the freshwater diatom, *Tabellaria flocculosa*, more usually recognised by its distinctive zig-zag chains.

Figure 18. Phase (a) and confocal (b) images of the filaments of *Melosira*, showing peripheral plastids.
Figure 19. Surface view of another filamentous diatom, *Paralia*, showing spines which lock the cells together.
Figure 20. The extra-ordinary symmetry of the radiating appendages of the marine diatom *Bacteriastrum*.
Figure 21. The Petri-dish construction of another marine centric species, *Cyclotella*, is well shown here.

In such confocal micro-scopes specimens are illuminated with a narrow beam of laser light which is scanned across the specimen. The emitted fluorescence is then focussed through a pin-hole (confocal) aperture, which only permits light from the plane of focus to pass. This allows a series of thin optical sections to be acquired, which can then be assembled and manipulated by computer software. Whereas normal light microscope, the depth of field is often no more than a few microns or less, in focus reconstructions of 20-50 microns in depth can be achieved with the confocal microscope. This enables the three-dimensional beauty of colonies and cells to be fully appreciated (Figures 2, 4, 23, 23, 33). Although, confocal-microscopes have been most frequently used to examine fixed and stained specimens, they can equally well be used for examining living cells (Hepler and Gunning, 1998; Beakes and Cleary, 1998). Because chlorophyll is naturally autofluorescent, these algae make ideal subjects since their plastids can be directly observed in all their exquisite detail (Figures 2, 4, 18b, 23, 24, 26, 27, 29, 31). Natural chlorophyll autofluorescence is red but computer generated images can be colored to reflect the overall pigmentation seen in life (Figure 33).

Another dimension can be given to these cells by using vital stains to reveal specific cellular constituents. In a study of the oil producing green colonial alga, *Botryococcus*, the vital stains DIOC(6) and Rhodamine 123 were used, initially to reveal the endomembrane systems and mitochondria respectively (Beakes and Cleary, 1998). However, it turned out these stains were not particularly specific and that both also bound to the cell walls of some algae, particularly desmids and enables the full beauty of their cells to be fully resolved in stunning three dimensional glory (Figures 29, 30, 31). However, cell staining does with dyes such as Rhodamine 123 does vary between the main algal groups, as is shown in Figure 30, where the dye is clearly staining a cytoplasmic component in the colonial green algal *Kirchnieriella* but is mainly staining the walls of both the desmid *Staurastrum* (also Figure 32) and the filamentous green alga, *Chaetophora*.

3. Natural beauty – phylogeny and diversity

The entire diversity of terrestrial plant life from mosses to giant Sequoia trees is encompassed within a single side branch of one of the major three domains of life, the

Kingdom Plantae on the "Tree of Life" developed by comparative analysis of multiple genes (Van de Peer et al., 2000; Baldauf, 2003). In contrast, organisms defined as algae can be mapped onto no less than nine major side branches within three of the five, recently defined eukaryote "Kingdoms" – the Plantae (Chlorophytes, Glaucophytes and Rhodophytes), Chromalveolates (Dinoflagellates, Phaeophytes, Diatoms, Haptophytes and Cryptomonads) and Excavates (Euglenophytes). In addition to these, the so-called "blue green algae" (Fig. 1-8), have long been known to be oxyigenic photosynthetic prokaryotic cyanobacteria, closely related to the gram positive and purple bacterial groups. Many phylogenists believe that eukaryan organelles such as mitochondria and plastids have arisen from endosymbiosis although views as to the precise origins of these organelles are still subject to much debate. Based on the most recent molecular analyses, it is now thought that plastids within the Kingdom Plantae (i.e. green algae and red algae) evolved just once from the single endosymbiotic acquisition of photosynthetic ancestral cyanophytes (see review by Keeling et al., 2004). The chlorophyll a and c containing plastids found in members of the Chromalveolate clade, which encompasses both the dinoflagellates (Figures 9-14) and diatoms (Figures 15-22), are also thought to have arisen once, but from the acquisition of primitive photo-synthetic eukaryotes (Harper et al., 2005).

It is perhaps not surprising that this enormous evolutionary diversity is reflected in great structural diversity as well. This is well exemplified by huge variation shown in the structure and formation of algal cell coats (Figures 10, 11, 12, 13, 15, 16, 20, 21, 28, 32) and in their plastid morphology and biochemistry (Figures 18, 22-27, 29, 31). All land plants have cells delimited by an extra-cellular cell wall, consisting of an underlying cellulosic skeleton in which pectic substances and hemicelluloses act as amorphous fillers (Carpita and Gibeaut, 1993). In contrast the cell walls of many algae groups are considerably more varied. This is well exemplified by the complex coats of fused plates of the armoured dinoflagellates (Figures 9, 10, 11), the unique glass cases (frustules) of diatoms (Figures 16, 20, 21, 35, 37) and the elegantly lobed and ornamented walls of desmids (Figures 28, 29, 30, 32). In dinoflagellates the cellulosic wall plates are actually formed within vesicles which lie immediately beneath the plasma membrane (Bricheux et al., 1992; Hohfeld and Melkonian, 1992; Sekida et al., 2004) which are reminiscent of the alveoli of ciliates, to which the dinoflagellates are now known to be closely related (Schlegel, 2003; Harper et al., 2005). Dinoflagellate genera such as *Ornithocercos* (Figure 12) and *Dinophysis* (Figures 13, 14), have large fused punctuate plates, rather like the shells of almonds, but with elaborately veined extensions around their crown and dorsal regions, making them as strange as many alien creations dreamed up by the human imagination.

Most land plants have fairly uniform, rather small, ovoid-shaped chloroplasts. Some algal cells are also packed with small individual plastids, such as the filaments of the diatom *Melosira* (Figure 18a), although, they are irregular in shape (Figure 18b). Within the filamentous green algae (Fig. 22-24) the plastids take on a wide diversity of form, from simple spheres in *Zygogonium* (Figure 22), lobed stellate whorls in *Zygnema* (Figure 23) to elegant spirals of varying gyre as seen in *Spirogyra* (Figure 24a,b). In many colony forming (coenobial) green algae, such as *Pediastrum* (Figures 25, 26) and *Scenedesmus* (Figure 27), the chloroplasts fill the whole cell. It is also not surprising that the plastids in the elaborately lobed cells of desmids, the plastids take on a similarly complex morphology (Figures 29, 31). The ability to visualize in the confocal microscope autofluorescing plastids independently of the rest of the cell (Figures 18, 23, 24, 31) clearly reveals just how complex and elegant the plastid morphology actually is in species such as *Zygnema, Spirogyra* and *Staurastrum.*

4. Natural beauty – Form and Function

Nearly all the algal cells illustrated in this account are planktonic, which means they live suspended in the water column. In spite of their small size they contribute significantly to overall global carbon production as the main primary producers in freshwater and marine waters (Reynolds, 1984; Harris, 1986). Algae have adopted many strategies for maintaining themselves in suspension. Very small round or oval cells (below 5 micrometers in size) have low inherent sinking rates, but this does not exactly make them visually exciting, unless they aggregate together in colonies (Figure 6). Larger, heavier cells tend to form star-like colonies (Figures 17, 25, 30) or more

Figures 22-24. Variations in plastid organisation to be seen within the simple unbranched filaments of freshwater algae within the Zygnemataceae.

Figure 22. The filamentous algae *Zygogonium* showing pairs of globular plastids in each cell compartment.

Figure 23. Confocal reconstruction of a filament of *Zygnema*, which shows similar paired 'star shaped' plastids to above species, reveals an unexpectedly complex array of finely branched arms.

Figures 24. Spiral plastids in two (a, b) filaments of *Spirogyra*, showing variations both in the degree of plastid lobing and the tightness of the spiral coil.

Figures 25-27. Variations in organisation of planktonic coenobial colonies in Chlorophycean algae.

Figures 22-28. Chlorophyte algae: Filaments and colonies. (Plantae Clade).

Figures 28-32. Chlorophyte algae - Desmids (Plantae Clade):

Figure 25. Fenestrated plate-like colony of *Pediastrum duplex*, showing almost perfect symmetry.
Figure 26. *Pediastrum angulosum* forms plates of cells, arranged with jigsaw precision.
Figure 27. Confocal image of the rafts of plastids of *Scenedesmus* (probably *S. dimorpha*).
Figures 28-32. Planktonic freshwater desmids from Australia (28) and England (29-32). Desmids are immediately recognisable by the amazing symmetry of their 'twined' cells.
Figure 29. *Micrasterias hardyii.* This elegant Australian species, was collected from Lake Burley-Griffin opposite the Australian Parliament building!
Figure 30. Confocal microscopy reveals the full glory of cell symmetry in the desmid *Micrasterias rotata*. DIOC(6) staining (green channel) reveals wall and chlorophyll autofluorescence shows plastid.
Figure 30-32. Confocal images of the many appendaged desmid *Staurastrum arctiscon*, from a small upland tarn in the English Lake District.

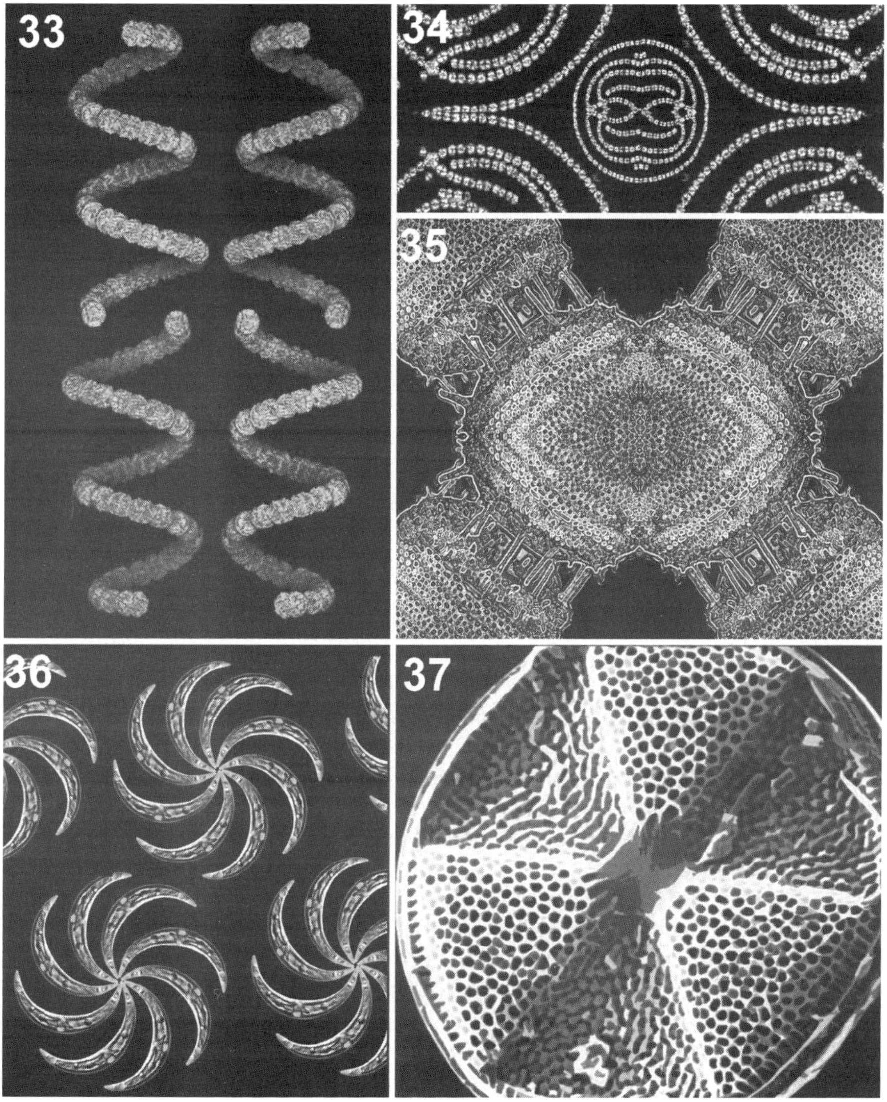

Figures 34-37. Digital art: Painting with algae!

Figure 30. A galaxy of freshwater algae – *S. arctiscon* in upper part of photograph, with the green colonial algae *Kirchneriella* (bottom left) and a branched filament of *Chaetophora* and some cells of *Ankistrodesmus* (bottom left).

Figure 31. *S. arctiscon* showing starfish like array of plastids when the cells are viewed end on.

Figure 32. The vital stain rhodamine 123, reveals details of the equisitely sculptured desmid wall.

Figure 33. *Space Spirals.* Based on confocal images of the spiral form of *Anabaena* (Fig. 4).

Figure 34. *Mystic Beads.* The more open filaments of another *Anabaena* species were the inspiration for this symmetrical composition.

Figure 35. *The X File* – composed from an image of a filigree shells of a marine diatom (*Detonula* sp?).
Figure 36. *War of the Whorls.* The elegantly curving sickle shaped desmid, *Closterium* was the starting point for this creation.
Figure 37. *Danger Within?* The shell of the diatom *Actinoptychus* is artistically rendered screen print style. This centric marine species is immediately recognised by its 'radiation warning' sculpturing.

elaborate shapes often with elongate appendages to increase turbulence and drag (Figures 9, 10, 12, 14, 20, 28, 30). Alternatively cells may link themselves together in filaments or chains (Figs 16, 18a, 19) or aggregate into flat, plate-like forms (Figs 21, 26, 29, 37) all of which have reduced sinking rates compared with spherical cells of similar volume (Reynolds, 1984; Harris, 1986; Hansen et al., 1996). In addition, these more elaborate forms may well show improved efficiency of nutrient uptake (Pahlow et al., 1997). In less dense tropical waters the larger and more spectacular the appendages tend to be (Figures 14, 20, 28). Such spines and appendages also have the additional advantage serve to deter predators (Harris, 1986). Cyanobacteria have acquired another strategy to keep them in suspension. They fill their cells with nitrogen-filled gas vesicles. This permits the large macroscopic rafts of filaments produced by genera such as *Anabaena* (Figures 1, 4, 5, 33), *Aphanizomenon* (Figure 2) and *Gloeotrichia* (Figures 7, 8) and the masses of spherical cells of *Gomphosphaeria* (Figure 6) to float near the water surface.

One downside of surrounding yourself with a protective silica case is that these still have to permit free gaseous exchange between the cytoplasm and external environment. For this reason, the glass shells of diatoms are delicately pierced with small holes (Figures 15, 16, 35, 37). This also improves the overall strength to weight ratio of these remarkably tough structures (Hamm et al., 2003). Exactly how diatoms have managed to create such exquisitely elaborate cell coats has long fascinated scientists and is still far from clearly understood (Sumper et al., 2005). It is known that the silica is deposited within an internal vesicle known as the silicalemma, and in simple centric species such as *Cyclotella* (Figure 21) the assembled valves are excreted as a single unit (Schmid and Schulz, 1979). However, valve formation in the centric species with elaborate spines, such as the *Bacteriastrum* shown in Figure 20, is by necessity more complex and involves a unique form of tip extension, in which spines seem to be teased out by actin-derived forces of rather than blown out by internal turgor pressure (Pickett-Heaps 1998). There is increasing evidence that species specific polyamines are able to directly influence the pattern of silica deposition but the precise underlying molecular control of this is still far from clearly understood (Kroger et al., 2000; Falciatore and Bowler, 2002; Sumper et al., 2005). Since complex silicon pattern generation is the basis of the vital semi-conductor industry, there has recently been much interest in unravelling how these simple organisms can achieve such wonders of nanofabrication (Tiffany, 2005).

The elegant symmetry of desmids (Figures 28, 29, 30) has also long fascinated cell biologists and the genus *Micrasterias*, in particular, has been taken up as a model system in which to study morphogenesis (Meindl, 1993; Holloway and Harrison 1999). There seems little doubt that in these algae also, actin plays a key role in the regulation of cell division and growth. The elaborate symmetrically lobed form of desmid cells (Figures 28-30), however seems pre-determined by changes in the cell plasma membrane which allows directed wall growth at specific points on the cell surface (Meindl, 1993; Holzinger and Meindl, 2001).

5. Concluding thoughts

For those wishing to explore further this extra-ordinary diversity many lavishly illustrated books of algae have been published of collections of both light (Hendey, 1964; Canter and Lund, 1995) and electron micrographs (Dodge, 1985; Pickett-Heaps, 1975; Round et al., 1990). However, it is also worth remembering that *en masse*, these organisms not only become apparent to the human eye but may even directly threaten our environment (Hallegraeff, 1993; Anderson, 1997). When the environmental conditions are favourable these cells multiply to such an extent, they form blooms which discolor water green or red. A sample taken from surface bluish-green scum from the edge of Esthwaite water is shown in Figure 1 is composed of a complex population of green and blue-green algae, many of which are potentially toxic. Indeed there are several references in the Bible (*Exodus 7, 17-25; Kings II 3 22)* to water either turning red or "to blood" that almost certainly allude to such blooms. Algal blooms, have achieved much notoriety of late, since many of the genera that are found in them, such as *Aphanizomenon* (Figure 2), *Anabaena* (Figures 4, 5, 33), *Peridinium* (Figure 11) and *Dinophysis* (Figure 13) can produce potent neurotoxins which accumulate in the human food chain (Anderson, 1997; Marasigan et al., 2001). So next time you fall into a slime-filled ditch or dive into the sea or lake, spare a thought for the fantastic hidden world of microscopic algae that is all around you.

6. References

Anderson, D.M. (1997). Turning back the harmful red tide. Nature **388**: 513-514.

Baldauf, S.L. (2003). The deep roots of eukaryotes. Science. **300**:1703-1706.

Beakes, G.W. (2003). Lower fungi: a review of microscopical techniques for the taxonomic and ecological the study zoosporic freshwater fungi. Fungal Diversity Research Series **10**: 51-79.

Beakes, G.W. and Cleary, A.L. (1998). Visualization of plastids and lipophilic components in living colonies of a wild strain of *Botryococcus* by laser scanning confocal microscopy. J. Appl. Phycol. **10**: 435-446.

Bricheux, G., Mahoney, D.G. and Gibbs, S.P. (1992). Development of the pellicle and thecal plates following ecdysis in the dinoflagellate *Glenodinium foliaceum*. Protoplasma **168**: 159-171.

Canter, H.M. and Lund, J. (1995). *Freshwater Algae. Their microscopic world explored.* Biopress, Bristol.

Carpita, N. and Gibeaut, D.M. (1993). Structural models of primary cell walls in flowering plants: consistency of molecular structure with the physical properties of walls during growth. The Plant Journal **3**: 1-30.

Cook, M.C. (1887) *British Desmids: a supplement to British Freshwater Algae*. London.

Dodge, J.D. (1985). *Atlas of Dinoflagellates.* Farrand Press, London.

Falciatore, A. and Bowler, C. (2002). Revealing the molecular secrets of marine diatoms. Ann. Rev. of Plant Biol. **53**: 109-130.

Hansen, J.L.S., Alldredge, A.L., Jackson, G.A., Passow, U., Dam, H.G., Drapeau, D.T., Waite, A. and Garcia, C.M. (1996). Sedimentation of phytoplankton during a diatom bloom: rates and mechanisms. J. Marine Res. **54**: 1123-1148.

Hallegraeff, G.M. (1993). A review of harmful algal blooms and their apparent global increase. Phycologia **32**: 79-99.

Harper, J.T., Waanders, E. and Keeling, P.J. (2005). On the monophyly of chromalveolates using six-protein phylogeny of eukaryotes. Int. J. of Systematic and Evol. Microbiol. **55**: 487-496.

Harris, G. (1986). *Phytoplanton Ecology: Structure, Function, and Fluctuation.* Chapman and Hall, New York.

Hamm, C.E., Merkel. R., Springer, O., Jurkojc P., Maier, C., Prechtel, K. and Smetacek, V. (2003). Architecture and material properties of diatom shells provide effective mechanical protection. Nature **421**: 841-843.

Hendey, N.I. (1964). *An Introductory Account of the Smaller Algae of British Coastal Waters. Part V Bacillariophyceae (Diatoms).* H.M. Stationery Office. Edinburgh.

Hepler, P.K. and Gunning, B.E.S. (1998). Confocal fluorescence microscopy of plant cells. Protoplasma **201**: 121-157.

Hohfeld, I. and Melkonian, M. (1992). Amphiesmal ultrastructure of dinoflagellates – a reevaluation of pellicle formation. Journal of Phycology **28**: 82-89.

Holloway, D.M. and Harrison, L.G. (1999). Algal morphogenesis: modelling interspecific variation in *Micrasterias* with reaction-diffusion patterned catalysis of cell surface growth Phil. Trans. Roy. Soc. Lond. Series **54**: 417-433.

Holzinger, A. and Lutz-Meindl U. (2001). Chondramides, novel cyclodepsipeptides from myxobacteria, influence cell development and induce actin filament polymerization in the green alga *Micrasterias* Cell Motil. and the Cytoskel. **48**: 87-95.

Keeling, P.J., Archibald, J.M., Fast, N.M. and Palmer, J.D. 2004. The evolution of modern eukaryotic phytoplankton. Science **306**: 2191.

Kroger, N. Deutzmann, R. Bergsdorf, C. Sumper, M. (2000). Species-specific polyamines from diatoms control silica morphology. Proc. Nat. Acad. Sci. **97**: 14133-14138.

Marasigan, A.N., Sato S., Fukuyo, Y. and Kodama, M. (2001). Accumulation of a high level of diarrhetic shellfish toxins in the gren mussel *Perna viridis* during a bloom of *Dinophysis caudata* and *Dinophysis miles* in Sapian Bay, Panay Island, Phillipines. Fisheries Science **67**: 994-996.

Meindl, U. (1993). *Micrasterias* cells as a model system for research on morphogenesis. Microbiol. Rev. **57**: 415-433.

Pickett-Heaps, J.D. (1975). *Green algae: Structure, reproduction and evolution in selected genera.* Sinauer Associates, Sunderland, Mass.

Pickett-Heaps, J.D. (1998). Cell division and morphogenesis of the centric diatom *Chaetoceros decipiens* (Bacillariophyceae) II. Electron microscopy and a new paradigm for tip growth. J. Phycol. **34**: 995-1004.

Pahlow, M., Riebesell, U. and Wolf-Gladrow, D.A. (1997). Impact of cell shape and chain formation on nutrient acquisition by marine diatoms Limnol. and Oceanography **42**: 1660-1672.

Reynolds, C.S. (1984). *The Ecology of Freshwater Phytoplankton* Cambride University Press, New York.

Round, F.E., Crawford, R.M. and Mann, D.G. (1990). *The Diatoms. Biology & Morphology of the Genera.* Cambridge University Press.

Schlegel, M. (2003). Phylogeny of eukaryotes recovered with molecular data: highlights and pitfalls. Eur. J. Protistol. **38**: 113-122.

Schmid, A.M.M. and Schulz, D. (1979). Wall Morphogenesis in Diatoms – deposition of silica by cytoplasmic vesicles. Protoplasma **100**: 267-288.

Sekida, S., Horiguchi, T. and Okuda, K. (2004). Development of thecal plates, pellicle in the dinoflagellate *Scrippsiella hexapraecingula* (Peridiniales, Dinophysceae) elucidated by changes in stainability of the associated membranes. Eur. J. Phycol. **39**: 105-114.

Sumper, M., Brunner, E. and Lehmann, G. (2005). Biomineralization in diatoms: characterization of novel polyamines associated with silica. FEBS Letters **579**: 376-3769.

Tiffany, M.A. (2005). Diatom auxospore scales and early stages in diatom frustule morphogenesis: Their potential use in nanotechnology. J. Nanosci. and Nanotech. **5**: 131-139.

Van de Peer, Y., Bauldauf, S.L., Doolittle, W. F. and Meyer A. (2000). An updated and comprehensive rRNA phylogeny of (crown) eukaryotes based on rate-calibrated evolutionary distances. J. Molec. Evol. **51**: 565-576.

Van Heurck, H. (1885). *Synopsis des Diatomees de Belgique.* Table Alphabetique (Index). J.F. Dieltjens, Anvers.

Biodata of **Lawrence Funderburke** author of ***"Life and Sport: Musings from a Pro Athlete"***

Lawrence Funderburke is currently working on an MBA through the University of Phoenix online program. He was born and reared in the Sullivant Gardens Housing Projects in Columbus, Ohio, in a single-parent home. Funderburke went to Indiana University and then transferred to Ohio State University, from which he graduated magna cum laude in 1994 with a degree in business finance. Currently he is finishing his Master degree. In 1992, Funderburke teamed with NBA star Jimmy Jackson to help Ohio State win the Big Ten Championship. He was drafted by the Sacramento Kings in 1994 and played three seasons overseas before returning to Sacramento, where he played from 1997 to 2004. He was nicknamed "Instant Offense," and a member of the famed "Bench Mob," Funderburke was a key reserve player for the Kings during his first five years in the NBA. In April of 2005 he signed with the Chicago Bulls.

Off the court, he was proud to work with inner city youth and honored to receive the first Hometown Hero of the Month Award in 2001. This award goes to NBA players for their dedication and commitment to the community. Funderburke founded the non-profit "Lawrence Funderburke Youth Organization" (LFYO) to help improve the social, moral, and academic standing of at-risk youth. Their motto has been, "Are you a positive one or a negative zero?" They aim to bridge the gap between corporate America and inner city youth (for more information on LFYO, go to www.lfyo.org).

Funderburke has many hobbies, including reading and writing, but his greatest joy is spending time with his family. He has recently (**2005**) authored the book, ***Hook Me Up, Playa,*** published by Wetherby Press. See: www.hookmeupplaya.com.

E-mail: **Lawrence@hookmeupplaya.com**

J. Seckbach (ed.), Life as We Know It, 701-706.

LIFE AND SPORTS:
Musings of a Pro Athlete

LAWRENCE FUNDERBURKE
12115 Magnolia Blvd. #113
North Hollywood, CA, 91607, USA

1. Introduction

Sports are a language and religion (shockingly to many people) that we can all relate to on some level or another. Our spirits rise or fall when our favorite sports teams triumph or go down in defeat. Avid fans paint their faces, wait in long lines for tickets to sporting events, and cheer fervently for their teams in an effort to influence the outcome of a game. Indeed, every fabric of their soul seems bent on being satisfied by what is euphemistically referred to as, "competitive entertainment."

Nevertheless we are tempted to ask "How did our culture arrive at this juncture, where sports have become the driving force behind many fans? Happiness or lack thereof?" Sports have satisfied the cravings of the human soul for fans from all walks of life. It does not matter if your are rich, poor, or middle class, it has the same influence. This otherworldly life that many people gravitate to for enjoyment allows them, however brief, to temporarily transport themselves inside the lives of pro sports as spectators.

Cheering hysterically for their favorite teams and players, whatever problems they had before the game begins have been all but forgotten, even the steep prices that they paid for their tickets to watch games in person. Some fans may even quip, "Who really cares about the drawbacks if our anticipated outcome materializes as expected, with a win?"

What insights can we leave behind for parents, high school, or college athletes who want to follow in this career path or fans who watch from a distance? Few outside the world of sports, and understandably so, comprehend the "game outside of the game" that pro athletes must regularly grapple with. "Who would have known?" Is the typical reaction when one peers behind the scenes into the lives of modern-day sports heroes?

2. Culture of Sports and Pro Athletes

Fans know (and rightly so) that they have a tremendous effect on the game's outcome. Whether positively influencing players to reach new heights in athletic competition or the referees officiating the game, many fans live and breathe sports. Their spiritual bliss is wrapped up in the euphoria that this form of entertainment provides. There is nothing wrong per se, and can certainly be encouraged, when fans on occasion use sports as an outlet from a hectic schedule and "microwave lifestyle." Spiritual bliss according to most pro athletes, however, is possible only in theory, especially once they leave their respective venues. Many pro athletes over the years have questioned the validity behind their title and its ability to provide happiness, peace of mind, and life-long financial security.

Do we place too much emphasis on sports as a society? Given the fact that roughly 20 million youth compete in sports annually, some pessimists might be inclined to emphatically argue, "Yes!" In terms of the residual benefits, including fame, fortune, and media attention, many people rarely take into consideration the common hindrances inherent to this field in spite of the manic obsession that many young American athletes or their parents have with professional sports.

Aspiring athletes, parents of athletically gifted children, and sports fans should be keenly aware of the less glamorous side of pro sports also, including but not limited to the following experiences of well-compensated athletes: get-rich-quick schemes; paranoia that wealth brings; loss of privacy; predators of thieves' interest; relatively short career span; vulnerability of life being scrutinized by the public; and the overwhelming expectations that close acquaintances place on athletes.

The past is a barometer of the future, if nothing else then for us not to repeat prior mistakes, whether ours or that of others. How often in America do the media report the failings of parents who push their children toward this supposedly lottery-winning lifestyle? Often. Since the late 80's and 90's this epidemic has gotten worse. Even famed television talk show hosts Oprah Winfrey and Dr. Phil have given lengthen dialogue to this troubling phenomenon. News broadcasts and other media outlets regularly report on this out-of-control problem. Who's to blame? Parents? Professional athletes? The media? Young people themselves? "Inner circle" influences? Your guess is as good as mine. But we hypothesize that the problem centers around two reasons: the desire of parents to live vicariously through their children's sports participation, and the monetary reward and fame that this life brings. Many parents, urban and city dwellers alike are pushing their children toward a career in pro sports because of the fame and fortune afforded to pro athletes. Instead of investing in their children's educational opportunities, many parents are defying conventional logic by investing in a possible pro career. Without question, a six- or seven-figure salary is a lucrative enough reason for a parent to steer a child in this direction. Some parents are taking second and third jobs or quitting their jobs entirely to care for their sport's prodigy. When they should be ensuring the child is also given a non-sports education. Certainly we don't discourage this as a career path since so many have graciously benefited from this profession, but the decision should fall squarely on the shoulders of the child, not his or her parents. Specialized coaches, mentors, and multiple sports camps are compelling parents to search out these and other beneficial avenues to boost their child's chances for sports success.

However, how many young athletes will reach the pinnacle of sports success, a career in pro sports? Very few. It has been said anecdotally that a young person stands a significantly greater chance of being struck by lightening then becoming a professional athlete. When it comes to money everyone should be careful how they spend it, especially professional athletes who believe they are protected from being ripped off. Reality shows they are wrong. Many pro athletes have lost considerable fortunes under the guidance of financial gurus and the "golden touch" mindset. Promises of doubling or tripling investment commitments, fortunes are lost just as quickly as they come. Lack of financial knowledge and giving carte blanche to unscrupulous financial advisors or business associates are to blame in most instances. Still other athletes are handicapped by their Vegas-style investment strategy (or frivolous lifestyle) while excluding due diligence from the equation.

Refusal to pay homage to due diligence and assess investment risks properly are additional hindrances that some pro athletes have succumbed to. Regardless of an athlete's or anyone for that matter talent level, he or she cannot always bend the odds in their favor, especially outside their "normal" sphere of expertise. The law of averages proves this to be true, although pro athletes have overcome incredibly high odds Wealth brings a great deal of paranoia, as many pro athletes can attest. Coupled with a loss of privacy, many pro athletes wonder why their spirits are vexed. The surreal dream of becoming a pro athlete that would one day come to fruition years later, at times can be an absolute nightmare. This paranoia increases with each subsequent financial and/or emotional betrayal at the hands of "insiders."

Thus, the athlete will oftentimes stereotype honest fans (and vice versa) when coming in contact with them. "So much for the fantastic dream," a pro athlete may lament while fans label the uncaring, spoiled, and overpaid stereotypical athlete, a jerk. Both groups are usually stereotyped by virtue of perception rather than reality. And on and on the vicious cycle goes.

Although no exact statistics exist on the number of pro athletes being on the receiving end of a crime, evidence supports the assertion that the yearly trend is tragically increasing. Personally, the author knows athletes who have been victims of crime, mainly robbery, and has heard numerous stories through the media or from casual acquaintances about other athletes. Whether at a nightclub, their old neighborhood, or just being at the wrong place at the incredibly wrong time, pro athletes attract criminals just as banks attract bank robbers. Why? Athletes have the material possessions and of course, money, and many criminals see them as a surprisingly soft, easy target. The majority of athletes will not risk their lives or livelihood over temporary possessions.

Another reason why many pro athletes become disenchanted lies in the average career span across four major sports that few players can manipulate in their favor. The National Basketball Association (NBA), the National Football League (NFL), the National Hockey League (NHL), and Major League Baseball (MLB). What may come as a surprise to many people is that the average career of a pro athlete in the "Big Four" is 3-5 years. That's it! An athlete may anticipate playing 10-15 years, but he could just as easily, which is likely to be the case, be out of work in as little as six months. When this happens to some athletes, they wrongly assume that their lives are over as well. Their once thriving marriage may begin to erode (usually ending in divorce), or in extreme cases, struggle with bouts of depression and suicide. No athletic career. No more attention. No more six or seven figure paychecks. No wife and no peace of mind. To some people, including media representatives, pro athletes are seen as "fair game." Every aspect of their lives is scrutinized and placed on public display. Being in the center of a controversial storm can hinder an athlete's image and endorsement possibilities, whether justified or not.

In the sports world, an athlete's private life is rarely, well, private. In essence, pro athletes belong to the world, or so the saying goes, and not to themselves. They must carefully find a way to sort of, take back their lives. But the public attention isn't necessarily a bad thing, especially when pro athletes are immediate role models to many disadvantaged youngsters. This privileged lifestyle affords us the opportunity to make a difference and touch countless lives, which we should graciously do. Because of sports, the author was able to created a youth foundation and donate $100,000 for a scholarship to students at Ohio State University.

In regard to close acquaintances, those in the inner circle of an athlete's entourage, many who fall in this category have dealt an incredible blow to the psyche and emotional stability of today's mega-paid athletes.

Open-ended monetary requests, business propositions, ticket demands, intervening in "crisis" situations, name dropping, etc. are common occurrences. The list is practically endless (although some requests are quite legitimate and noteworthy). Athletes are a peculiar breed. But what many of them want and yearn for are relationships where giving is reciprocal, not one-sided. The sting of betrayal from the hands of insiders can and, oftentimes does, leave a bitter taste in the mouth of a professional athlete. He realizes that he cannot regardless of his wealth, be all things to all people. A pro athlete's wealth is limited just as the next person is. Still, an athlete is left wondering, "What ever happened to that gigantic salary that was exhausted by others?"

3. Closing Comments

In closing, although the culture of sports and pro athletes are a way of life and religion to many people. For some of us at least, it cannot be the sum total of who we are as human beings. Blessed with physical and cognitive abilities, we have benefited greatly from a sport that has given us an opportunity to not only bless our families financially, but it has also allowed us to make a difference in the lives of others spiritually, emotionally, educationally, and of course, economically.

The Lawrence Funderbruke youth foundation works with inner city kids to improve their lives and teach them the value of education. From this perspective, the benefits far exceed the pulls and strains of being a professional athlete. It should not come as a surprise to pro athletes, that the awesome responsibility they have, has as many drawbacks as it has benefits. When they fail to grasp this, maybe the ancient and time-tested proverb that says, "To whom much is given much is required," will alert us to think beyond sports, whatever an athlete's interests may lie.

It is far too easy for any athlete to get swallowed up in the hype of pro sports while a member of "the show." Instantly granted celebrity status - but not necessarily as credible or reliable role models - can give us a bad case of the "entitlement bug," which has hindered many gifted athletes who have been mesmerized by the trappings of media attention, wealth, and troublesome conduct. Sports fans are drawn to the misconception that the life of grandeur this occupation allegedly exudes: peace, happiness, and satisfying materialism. In the past we have been bitten by the hypothetical and contagious, entitlement bug. And some of us are not too proud to admit this shortcoming. However, only by admitting one's mistakes can others be liberated in conjunction with my or our (and other pro athletes') mistakes.

More importantly, another world exists beyond sports where we would be wise enough to seek and obtain true peace and happiness - a loving relationship with our Creator. This spiritual true has no equal.

Biodata of **Moshe Sipper** author of ***"Artificial Life: From Life-As-We-Know-It to Life-As-It-Could-Be"***

Dr. Moshe Sipper received the B.A. degree from the Technion—Israel Institute of Technology, and the M.Sc. and Ph.D. (in 1995) degrees from Tel Aviv University, all in computer science. He is currently an Associate Professor in the Department of Computer Science, Ben-Gurion University, Israel. During the years 1995-2001 he was a Senior Researcher in the Swiss Federal Institute of Technology in Lausanne. Dr. Sipper's major research interests include evolutionary computation, bio-inspired computing, and artificial life; minor interests include cellular computing, cellular automata, and artificial self-replication, along with a smidgen of evolutionary robotics, artificial neural networks, and fuzzy logic. Dr. Sipper has published over 100 scientific papers, and is the author of two books: *Machine Nature: The Coming Age of Bio-Inspired Computing* and *Evolution of Parallel Cellular Machines: The Cellular Programming Approach.* He is an Associate Editor of the *IEEE Transactions on Evolutionary Computation* and an Editorial Board Member of *Genetic Programming and Evolvable Machines*. Dr. Sipper is the recipient of the 1999 EPFL Latsis Prize.

E-mail: **sipper@cs.bgu.ac.il**

J. Seckbach (ed.), Life as We Know It, 707-718.

ARTIFICIAL LIFE:
From Life-As-We-Know-It to Life-As-It-Could-Be

MOSHE SIPPER
Department of Computer Science,
Ben-Gurion University, POB 653,
Beer-Sheva 84105, Israel

1. Artificial Life

"He wanted to dream a man: he wanted to dream him with minute integrity and insert him into reality." This was the goal of the silent man who came from the South, in Jorge Luis Borges's short story *The Circular Ruins*. From Pygmalion, Frankenstein, and the Golem to Star Trek's Lieutenant Commander Data, the dream of administering the breath of life has fascinated mankind since antiquity (Sipper, 2002). This question moved from the realm of science fiction to that of science with the advent of the field known as *artificial life (ALife)*. The term was coined by Christopher G. Langton, organizer of the first artificial life conference, which took place in Los Alamos in 1987 (Langton, 1989).

"Artificial Life," wrote Langton (in the proceedings of the second conference), "is a field of study devoted to understanding life by attempting to abstract the fundamental dynamical principles underlying biological phenomena, and recreating these dynamics in other physical media—such as computers—making them accessible to new kinds of experimental manipulation and testing." (Langton et al., 1992) While biological research is essentially *analytic*, trying to break down complex phenomena into their basic components, artificial life is *synthetic*, attempting to construct phenomena from their elemental units, as such adding powerful new tools to the scientific toolkit. This is, however, only part of the field's mission. As put forward by Langton, "In addition to providing new ways to study the biological phenomena associated with life here on Earth, *life-as-we-know-it*, Artificial Life allows us to extend our studies to the larger domain of the 'bio-logic' of possible life, *life-as-it-could-be*, whatever it might be made of and wherever it might be found in the universe."

2. Key Challenges

One way to understand the workings and goals of a scientific endeavor is to consider the key challenges its practitioners are facing. Beadau et al. (2000) have recently published such a structured list of fourteen key open problems in artificial life. The challenges are classified under three broad categories: the transition to life, the evolutionary potential of life, and the relation between life and mind and culture.

A. How does life arise from the nonliving?

1. Generate a molecular proto-organism in vitro.
2. Achieve the transition to life in an artificial chemistry in silico.
3. Determine whether fundamentally novel living organizations can exist.
4. Simulate a unicellular organism over its entire lifecycle.
5. Explain how rules and symbols are generated from physical dynamics in living systems.

B. What are the potentials and limits of living systems?

6. Determine what is inevitable in the open-ended evolution of life.
7. Determine minimal conditions for evolutionary transitions from specific to generic response systems.
8. Create a formal framework for synthesizing dynamical hierarchies at all scales.
9. Determine the predictability of evolutionary consequences of manipulating organisms and ecosystems.
10. Develop a theory of information processing, information flow, and information generation for evolving systems.

C. How is life related to mind, machines, and culture?

11. Demonstrate the emergence of intelligence and mind in an artificial living system.
12. Evaluate the influence of machines on the next major evolutionary transition of life.
13. Provide a quantitative model of the interplay between cultural and biological evolution.
14. Establish ethical principles for artificial life.

As can be seen, the field of artificial life encompasses a wide range of problems, dealing with issues pertaining to the very basics of life, and all the way up to intelligence and societies, along the way passing through basic questions of organization, emergence, and structuring.

Since I cannot elaborate upon all of the above issues within the limited scope of this chapter, I shall try to focus on a small number of key issues, which underlie many of the above challenges. ALife pioneer Thomas Ray wrote, "I would consider a system to be living if it is self-replicating, and capable of open-ended evolution." (Ray, 1992) Indeed, self-replication and evolution are both crucial to ALife research.

3. Self Replication

The study of self-replicating structures in man-made systems began in the late 1940s, when John von Neumann—one of the twentieth century's most eminent mathematicians and physicists—posed the question of whether a machine can self-replicate (i.e., produce copies of itself). He wrote:

Living organisms are very complicated aggregations of elementary parts, and by any reasonable theory of probability or thermodynamics highly improbable. That they should occur in the world at all is a miracle of the first magnitude; the only thing which removes, or mitigates, this miracle is that they reproduce themselves. Therefore, if by any peculiar accident there should ever be one of them, from there on the rules of probability do not apply, and there will be many of them, at least if the milieu is reasonable. (von Neumann, 1966).

Von Neumann was not interested in building an actual machine, but rather in studying the theoretical feasibility of self-replication from a mathematical point of view. He succeeded in proving (mathematically) that machines can self-replicate, laying down along the way a number of fundamental principles involved in this process.

To conduct a formal mathematical investigation of the issue, von Neumann used a model conceived by his colleague, mathematician Stanislaw Ulam. The model, known as *cellular automata (CA)*, consists of a large grid of cells (similar to a checkerboard), each possessing a certain state at a given moment. The number of possible states per cell is finite and is usually small (in figures, each state is usually represented by a different color). All cells change state simultaneously such that the state of a cell at the next time step depends only on its state at the current time step and the states of its neighboring cells. The principle that guides state transformations is applied identically to all cells and is referred to as the *transition rule*. Note that this rule is entirely *local*, thus any global phenomena occurring throughout the grid are due to interactions of numerous cells.

A *machine* in the CA model is a collection of cells that can be regarded as operating in unison. Thus, one can observe simple "creatures" that are able to move within this austere universe, as demonstrated in Figure 1 for a well-known CA rule—the "game of life."

One of von Neumann's main conclusions was that the replicative process uses the assembly instructions in two distinct manners: as interpreted code (during actual assembly), and as uninterpreted data (copying of assembly instructions to offspring). During the following decade, when the basic genetic mechanisms began to unfold, it became clear that nature had "adopted" von Neumann's conclusions. The process, by which assembly instructions (DNA) are used to create a working machine (proteins), makes dual use of information: as interpreted code and as uninterpreted data, the former referred to in biology as *translation*, the latter as *transcription* (Figure 2).

A major problem with von Neumann's approach (as well as a number of his successors) is the complexity of the constructor, which requires hundreds of thousands of cells. In addition, each cell can be in one of 29 states rather than just two. In 1984 Christopher Langton observed that although the capacity for universal construction is a *sufficient* condition for self-replication, it is not a *necessary* one; furthermore, natural systems are not capable of universal construction. He devised a small CA "loop," capable solely of self-replicating, but which was much simpler than von Neumann's design, with but a few dozen cells (Langton, 1984; for a demonstration of its workings see http://necsi.org/postdocs/sayama/sdsr/java/).

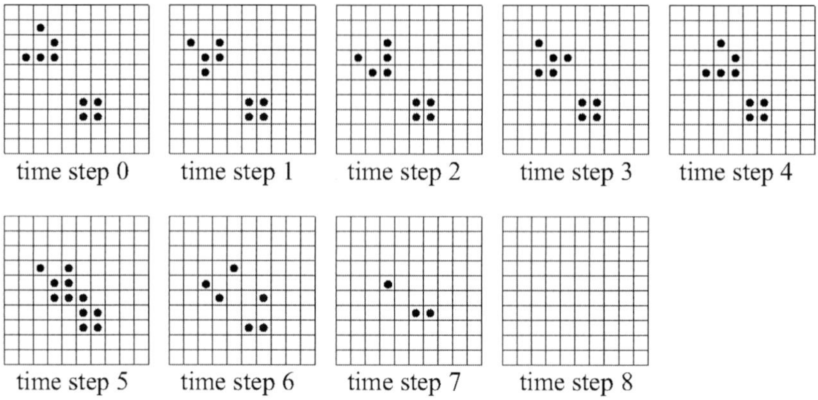

Figure 1. The "game of life" CA was defined by John H. Conway in 1968: "Life occurs on a virtual checkerboard. The squares are called cells. They are in one of two states: alive or dead. Each cell has eight possible neighbors, the cells which touch its sides or its corners. If a cell on the checkerboard is alive, it will survive in the next time step if there are either two or three neighbors also alive. It will die of overcrowding if there are more than three live neighbors, and it will die of exposure if there are fewer than two. If a cell on the checkerboard is dead, it will remain dead in the next time step unless exactly three of its eight neighbors are alive. In that case, the cell will be 'born' in the next time step" (Berlekamp et al., 1982). Shown above at time step 0 are two patterns, a stationary square, known as a "block," and a moving creature, known as a "glider." The latter displaces itself one square diagonally every four time steps. Upon meeting other creatures, such as the block, both are subject to this simple universe's basic rule, or "laws of physics," which cause their mutual annihilation in this case. Note that actions such as "movement" are purely in the eyes of the beholder—the most basic level consists of simple state transformations, with no movement at all. However, by considering higher-level views of the system, one can introduce new terms to describe the perceived phenomena. Such phenomena have often been termed "emergent," although this appellation is highly problematic (Ronald et al., 1999).

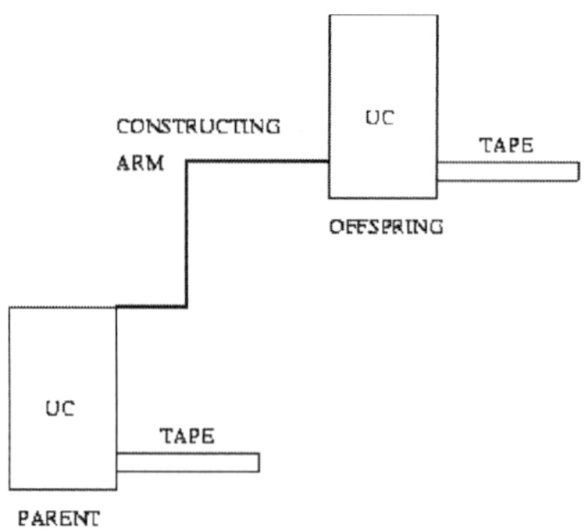

Figure 2. A schematic diagram of von Neumann's self-replicating CA. The machine is a universal constructor (UC) capable of constructing, through the use of a "constructing arm," any configuration whose description (genome) can be stored on its input tape. Thus, given its own description, the machine is capable of constructing a copy of itself, i.e., self-replicate. The parent UC first *interprets* the genetic information stored on the tape as building instructions to create an offspring UC, and then copies the genomic tape *uninterpreted* and attaches it to the offspring, so the latter can replicate in its turn.

The study of self-replication has been taking place now for more than half a century (Sipper, 1998; also http://www.cs.bgu.ac.il/~sipper/selfrep). This research might better our understanding of self-replication in nature, as well as find many technological applications. There is much talk today of nanotechnology, where self-replication is of vital import: we would like to be able to build one miniature machine, which would than sally forth and multiply.

As noted above self-replication is but one of two major characteristic of life, the second being evolution.

4. Evolutionary Algorithms

The only process currently known to have produced an ecosystem of living creatures, and in particular, of intelligent beings, is that of natural evolution. Darwin laid out the core of the currently accepted theory of evolution, its major elements being (Ray, 1994):

- Individuals vary in their viability in the environments that they occupy.
- This variation is heritable.
- Self-replicating individuals tend to produce more offspring than can survive on the limited resources available in the environment.
- In the ensuing struggle for survival, the individuals best adapted to the environment are the ones that will survive to reproduce.

The continual workings of this process over the millennia causes populations of organisms to change, generally becoming better adapted to their environment.

Having witnessed the study of self-replication in an artificial setting, we now do the same for evolution. The idea of applying the biological principle of natural evolution to artificial systems was introduced in the 1950s and the 1960s, when several researchers studied evolutionary systems with the idea that evolution could be used as an optimization tool for engineering problems. Central to all the different methodologies is the notion of solving problems by evolving an initially random population of candidate solutions, through the application of operators inspired by natural genetics and natural selection, such that in time "fitter" (i.e., better) solutions emerge (Sipper, 2002). Nowadays, these so-called evolutionary algorithms are ubiquitous, having been successfully applied to numerous problems from different domains, including optimization, automatic programming, machine learning, economics, operations research, immune systems, ecology, and population genetics. In particular, they are central to ALife research.

Based on (and inspired by) the workings of evolution by natural selection, the basic meta-algorithm is seductively (and, I might add, deceptively) simple, and can be expressed in a mere 8 lines of pseudocode:

1. produce an initial **population** of individuals, these latter being candidate solutions to the problem at hand
2. **evaluate** the fitness of each individual in accordance with the problem whose solution is sought

3. *while* termination condition not met *do*
4. **select** fitter individuals for reproduction
5. **recombine (crossover)** individuals
6. **mutate** individuals
7. **evaluate** fitness of modified individuals
8. *end while*

Over the past two decades evolutionary algorithms have proven their worth beyond a doubt, at times not only solving hard problems but indeed competing with their human designers (Koza et al., 2003).

A major reason for the success of this seemingly blind search is its not being blind at all (as in nature): though randomness and probability do play a part, selection is a function of fitness, and the recombination of good sub-pieces—or building-blocks—from two (or more) individuals enables the creation of offspring that outperform their parents (Figure 3).

ALife researchers often use some form of evolutionary algorithm to study the issues delineated in Section 2. For example, where the transition from the nonliving to the living is concerned, one can explore the evolution of self-replicating entities, as done, e.g., by Lohn and Reggia (1997). They used an evolutionary algorithm to automatically discover transition rules that govern emergent self-replicating processes in cellular automata. Given dynamically evolving automata, one of the most difficult tasks is that of identifying effective fitness functions for self-replicating structures. Simply ascribing a higher fitness to better replicators is of no use, since at the outset (generation 0) there are no replicators at all, i.e., all individual structures have a fitness of zero. And with no variability evolution grinds to a halt. Lohn and Reggia were able to ingeniously solve the fitness problem by assigning values to "partial" replicators, thereby demonstrating that self-replicating structures can be evolved.

5. Getting "Wetter": Lipidia

The presentation up until now may have seemed somewhat theoretical, and indeed much research in artificial life is, ipso facto, of an abstract nature. But in recent years ALife researchers have been attempting to get "closer" to biology, by building systems and models that attempt to answer biological questions through ALife methods. As an example I will briefly describe below the work my colleagues and I have undertaken in the area of Origin of Life: Lipidia.

Lipidia is a novel simulation system related to the "Lipid World" scenario for the origin of life. Lipidia allows conducting experiments with a population of assemblies containing lipid-like molecules on a two-dimensional grid (Naveh et al., 2004). The dynamics of the assemblies is modeled using the Graded Autocatalysis Replication Domain (GARD) model (Segre et al., 1998).

The "RNA World" is possibly today's most popular theory for the origins of life. Because RNA molecules can act as catalysts in addition to acting as templates, it is hypothesized they might have been able to do both: to store alphabet-based genetic information *and* to catalyze their own creation. Life, according to this theory, began when certain RNA molecules achieved the capability to replicate themselves. This scenario, despite its elegance, suffers from difficulties.

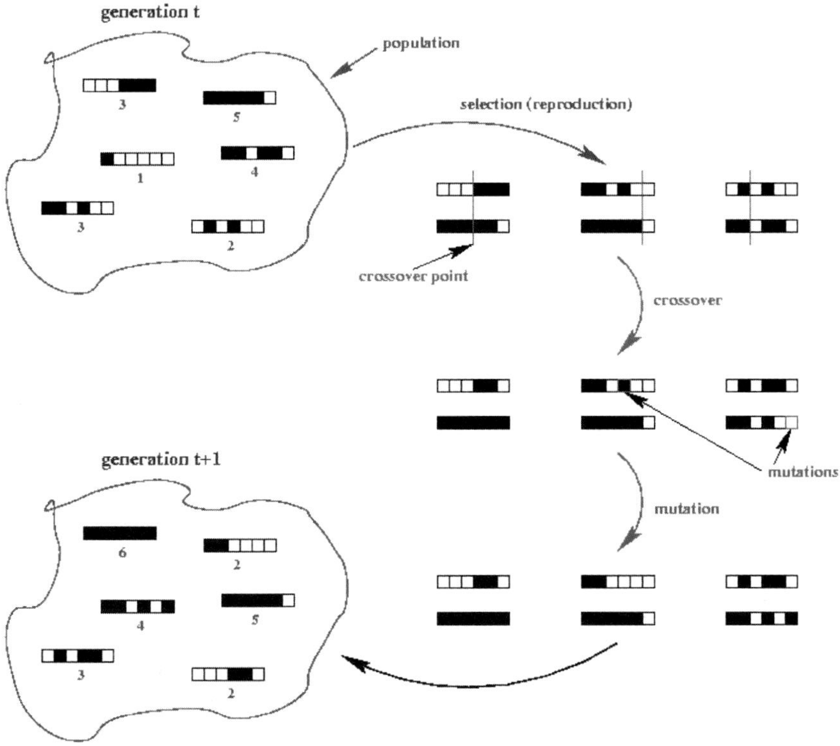

Figure 3. One *generation* (evaluation-selection-crossover-mutation cycle) of a simple evolutionary algorithm. Individuals are represented as bit strings (the so-called *genomes*). Fitness in this toy example is the number of 1 (black) bits. Selection is performed proportionately to fitness, so that high-fitness individuals are more likely to be selected. Recombination—or crossover—is performed on two individuals by selecting a crossover point at random and exchanging the chunks beyond this point. Mutation is performed by flipping a small number of bits (with low probability). As can be seen, this simple evolutionary scenario has produced a perfect solution by the next generation, in this case due to the crossover operation, which has glued a good sub-piece from one individual (3 1s on the right) with another good sub-piece from the second individual (3 1s on the left).

In an attempt to come up with a probable scenario, having observed that no known bio-molecule is capable of self replication in its naked form, it has been suggested that self replication might not have been achieved by a single molecule, but rather by a molecular ensemble (Kauffman, 1995). This work is based on "The Lipid World" scenario (Segre et al., 2001), which follows that line of thought. The scenario assumes that self-replication was initially achieved by non-covalent assemblies of lipid-like molecules that contained mutually catalytic sets. RNA according to this scenario, while possibly playing an important role, came later.

We used the GARD model to quantitatively model and simulate the developmental process of non-covalent assemblies of lipid-like molecules. Previous studies using the GARD model have mostly examined such assemblies in a one-at-a-time fashion. The behavior of assembly populations has been largely unexplored. In this work we expanded the model to a *population* of assemblies and obtained quantitative and qualitative results regarding its behavior. Also, previous studies assumed idealization of an infinite environment where the assembly's effect on the environment is negligible and "food" molecules are in infinite supply. In our work we introduced a finite environment to the model, which allows cross-interactions between assemblies via the environment. We also compared the effect of finite environments vs. infinite environments (Figure 4).

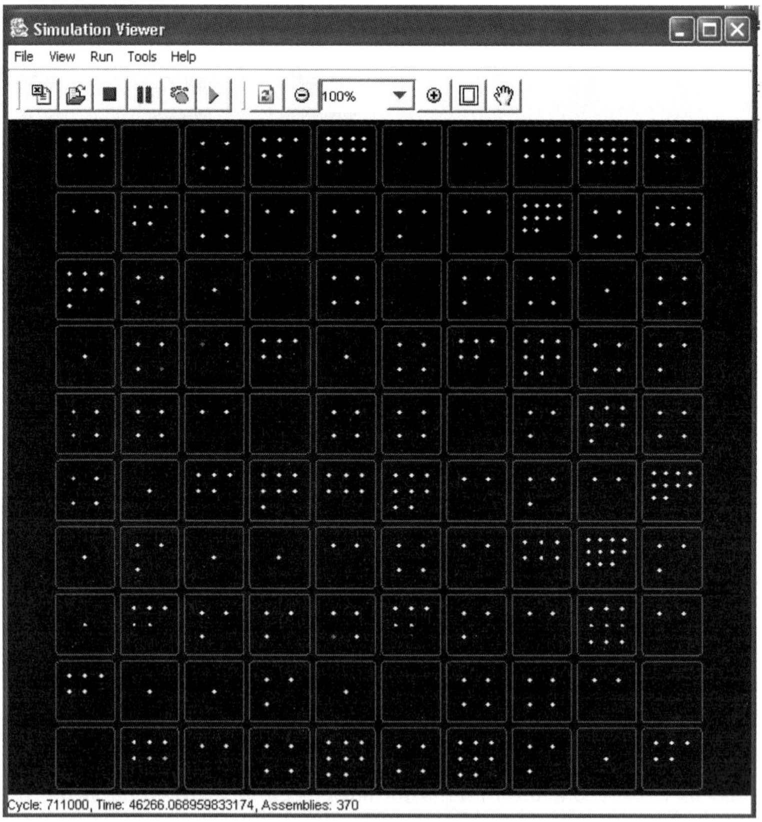

Figure 4. A Lipidia screen shot, typifying an A Life scenario. Lipidia is based on a two-dimensional interaction *grid*, as with cellular automata. Each square on the grid is called a *grid location*, or *location* for short. For each location there is a defined *environment* containing a variety of *molecules*. Each location may contain zero or more *assemblies* of molecules. The location's environment is common for all assemblies contained within it. Molecules from the environment may *join* an assembly, and molecules from the assemblies may *leave* their assembly back to the environment. "Matter" on the grid is therefore preserved—no matter is ever lost or created. The finite environment model can be replaced by an infinite environment model of fixed concentrations. For full details see (Naveh et al., 2004).

Our results showed that a finite environment produces more attractors (species), and faster, than an infinite environment. A finite environment allows more assemblies to occur in more attractors and in greater numbers (Naveh et al., 2004). Thus, diversity increases.

The results might be considered surprising. One might think that having an infinite supply of resources, in the form of "food" molecules, might help to "do more." According to our results, it only helps to "do more of the same." Diversity seems to spring when resources are limited. It is when resources for the "best solutions" run out that the race towards alternative solutions begins.

While the details are beyond the scope of this chapter, this work exemplifies an ALife research: the formation of digitally (but not carbonaceously) testable questions in a biological domain (in this case, Origin of Life), followed by the conception of a synthetic model, its implementation and the running of simulations, and the drawing of conclusions.

6. Strong Artificial Life?

The research described heretofore is often referred to as *weak* artificial life, namely, the application of synthetic—often digital—tools to the study of biological phenomena. But can we create bona fide life? This latter is the goal of *strong* artificial life, which for now remains mostly in the realm of theory and philosophy.

One way to create life, currently being explored by a number of groups, is to start at the very bottom—with bacteria. These have small genomes—on the order of a few hundred nucleotides—and we may be able to fabricate such small living organisms. A related research subjects simple bacteria to knock-out experiments, wherein genes are systematically taken out, in order to arrive at the very minimal organism which is still viable.

At the other end, we might employ *open-ended* evolution. As opposed to evolutionary algorithms described above, where a fitness function is imposed by the human designer, open-ended evolution—such as occurs in nature—has no such external imposition, but rather seeks out a trajectory on its own. Creating a set-up for such an open-ended process is a major challenge, probably still far in the future (a lengthy discussion of this challenge can be found in (Sipper, 2002)). Yet one cannot help but wonder, what if: What if we came up with the prodigious resources (computer power, "wet" power) necessary to mount an open-ended evolutionary process? What if we were able to *design* such an experiment—itself a fundamental problem? Would we then see—after a day, a month, or a decade—the emergence of living creatures? If so, would they really be artificial? After all, artificial life might in fact be an oxymoron—how can life be artificial? If something is truly alive—assuming we can somehow agree upon this fact, as no ironclad definition exists—then what is artificial about it?

Such speculative, futuristic science aside, artificial life today is providing a novel outlook on many scientific issues, and enabling the execution of experiments too costly or downright impossible in a "real" laboratory.

7. References

Bedau, M. A., McCaskill, M. A., Packard, N. H., Rasmussen, S., Adami, C., Green, D. G., Ikegami, T., Kaneko, K. and Ray, T. S. (2000) Open Problems in Artificial Life, Artificial Life, **6**: 363–376.

Berlekamp, E. R., Conway, J. H. and Guy, R. K. (1982) *Winning Ways for your Mathematical Plays*, Academic Press, New York.

Kauffman, S. A. (1995) *At Home in the Universe: The Search for the Laws of Self-Organization and Complexity*, Oxford University Press, UK.

Koza, J. R., Keane, M. A., Streeter, M. J., Mydlowec, W., Yu, J. and Lanza, G. (2003) *Genetic Programming IV: Routine Human-Competitive Machine Intelligence*, Kluwer Academic Publishers Boston. Langton, C. G. (1984) Self-reproduction in cellular automata, Physica D, **10**: 135–144.

Langton, C. G. (1989) *Artificial Life: Proceedings of an Interdisciplinary Workshop on the Synthesis and Simulation of Living Systems*, Addison-Wesley, Redwood City, CA.

Langton, C. G., Taylor C., Farmer J. D. and Rasmussen S. (1992) *Artificial Life II: Proceedings of the Workshop on Artificial Life,* Addison-Wesley, Redwood City, CA.

Lohn, J. D. and Reggia, J. A. (1997) Automatic discovery of self-replicating structures in cellular automata. IEEE Transactions on Evolutionary Computation, **1**: 165–178.

Naveh, B., Sipper, M., Lancet, D. and Shenhav, B. (2004) Lipidia: An Artificial Chemistry of Self-Replicating Assemblies of Lipid-like Molecules, In: J. Pollack, M. Bedau, P. Husbands, T. Ikegami, and R. A. Watson (eds.) *Artificial Life 9: Proceedings of the Ninth International Conference on Artificial Life*, pp. 466–471, The MIT Press, Cambridge, Massachusetts.

Ray, T. S. (1992) An approach to the synthesis of life, In: C. G. Langton, C. Taylor, J. D. Farmer, and S. Rasmussen (eds.) *Artificial Life II*, Addison-Wesley, Redwood City, CA, pp. 371-408.

Ray, T. S. (1994) Evolution, complexity, entropy and artificial reality, Physica D, **75**: 239–263.

Ronald, E. M. A., Sipper, M. and Capcarrere, M. S. (1999) Design, observation, surprise! A test of emergence, Artificial Life, **5**: 225–239.

Segre, D., Lancet, D., Kedem, O. and Pilpel, Y. (1998) Graded autocatalysis replication domain (GARD): Kinetic analysis of self-replication in mutually catalytic sets, Origins of Life and Evolution of the Biosphere, **28**: 501–14.

Segre, D., Ben-Eli, D., Deamer, D. W. and Lancet, D. (2001) The lipid world, Origins of Life and Evolution of the Biosphere, **31**: 119–145.

Sipper, M. (1998) Fifty years of research on self-replication: An overview, Artificial Life, **4**: 237–257.

Sipper, M. (2002) *Machine Nature: The Coming Age of Bio-Inspired Computing*, McGraw-Hill, New York.

von Neumann, J. (1966) *Theory of Self-Reproducing Automata*, University of Illinois Press, Illinois, Edited and completed by A. W. Burks.

Biodata of **Nakdimon Umiel** author of ***"Life-Cycles of Life"***

Dr. Umiel is currently (since 1977) senior research scientist in the Division of Ornamental Horticulture, Institute of Horticulture of the Agricultural Research Organization (The Volcani Center), Bet-Dagan, Israel. In addition he is teaching evolution in the Hebrew University Faculty of Agriculture in Rehovot, Israel. He Obtained his Ph.D. from the University of Wisconsin (Madison) in Plant-Breeding in 1970 and continued his studies and research as post-Doctoral (1970-1973) in the Zoology Dept. of the Univ. Wis. During 1972-1977 he held a faculty position at the Hebrew University Faculty of Agriculture, teaching genetics and carrying out genetic research in plants tissue cultures. Dr. Umiel's scientific interests are in the areas of: plant-breeding, genetics, and evolution, the domestication of plants from the wild, and also in the development, production, and marketing of horticultural products.

E-mail: vcumiel@int.gov.il

J. Seckbach (ed.), Life as We Know It, 719-733.

LIFE-CYCLES OF LIFE

NAKDIMON UMIEL
Division of Ornamental Horticulture, The Institute of Horticulture, Agricultural Research Organization, The Volcani Center, Bet-Dagan, Israel

1. Introduction

Little is known of how, when and where life on Earth had started or if it started on another planet and was transported to Earth by meteorites (Sleep et al., 1989; Nisbit and Sleep, 2001). Nevertheless, presently we know of only one type of life, the one that dominates our planet Earth. Knowing only one sample of life, we are unable to precisely define life (e.g., Nisbet and Sleep, 2001; Cleland and Chyba, 2002; Bains, 2004) but we can describe its main features as followed:

1. The basic feature of life is the existence of a "replicator" – a single or a collection of informational macromolecule(s) (e.g., DNA) that can store information and can be replicated to form new copies. The replication process has a very high fidelity but during the process, mistakes (mutations) do happen, resulting in changes of the stored information.
2. The replication process is carried by biochemical reactions governed by other macromolecules (e.g., RNA, proteins).
3. Life is built from basic units (e.g., cells or organisms), that contain the replicator(s) together with complex biochemical machinery that can carry the necessary biochemical reactions and processes which are essential and sustain the life units.
4. The basic units can grow and develop, adding mass, volume and complexity.
5. The basic units have the ability to reproduce, forming additional identical or somewhat modified (mutated) units.
6. Every basic unit (cell, organism, etc,) has a life cycle, which is terminated by the death of the unit.

An interesting and important question is how much "life" did exist in a given moment in a given place. The quantity of life itself can be measured in many ways. For example, in a given moment, it can be quantified by the total number of individuals of all species, or by the number of species. Alternatively, it can be quantified by totaling the fresh or dry weight of all living organisms. Another and preferable approach will be to base the quantification on the total quantity of the replicator (e.g., DNA) in all living organisms, in mass (in kg) or length (in light-years).

2. Life-cycles

A life-cycle is a measurement of quantity (e.g., weight, number, dollar value, etc.) versus time, as shown in the general description of life-cycle (Fig. 1). All life-cycles are started by birth (or for non-biological phenomena by introduction, initiation or appearance). The birth is followed by a growth period and then by maturation, then decline and finally by death (or termination, or disappearance). It is reasonable to say that the concept of life-cycle holds true for every thing in our Universe, from the life-cycle of the Universe itself which we only partially know (Adams and Laughlin, 1997), to the life-cycle of stars and galaxies (Gamov, 1945; Adams and Laughlin, 1997; Kauffmann and van den Bosch, 2002; Wakker and Richter, 2004), and to the life-cycle of each single insignificant cell or bacteria on Earth. We can also describe the life-cycles of non-biological (or chemical or physical) phenomena, such as the rise and fall of empires, cultures, economics processes, and many more.

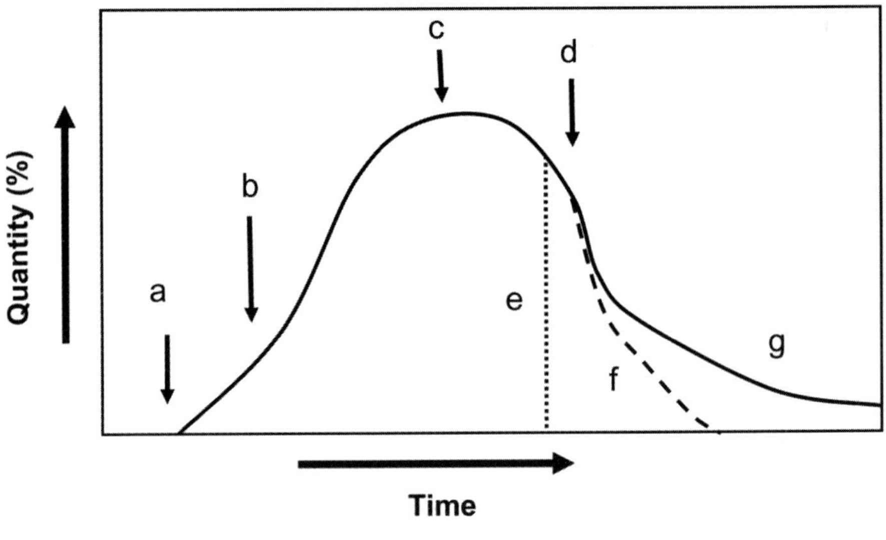

Figure 1. A general description of a life-cycle: a. birth, b. growth, c. maturation, d. decline, e. sudden death, f. slow death, g. small quantity (of units) surviving for a long time before final death.

In addition, we can describe the life-cycle of various species, including the succession of species in evolution (e.g., Fig. 2), when one form of life is replaced by another (or others) in a similar way that we can describe the life-cycles and succession of various products in the markets (Rymon et al., 1987).

If we agree that almost everything in the Universe has a life-cycle then we have to consider the possibility (or fact) that life itself has a life-cycle. Having a life-cycle means that at some point it had started and later on it will die and become extinct (Sleep, 2003). Jablonski (2004) have pointed out that "extinction is a fundamental part of nature". If so, it raises interesting questions on what are the main features (past, present and future) for this life-cycle of the only life we know, as well as possibly shedding some light on the life-cycles of other lives, which might exist elsewhere in the Universe (Bains, 2004), these that we presently do not know.

In order to graphically draw the life-cycle of life, we must have some kind of an estimate for its quantity during time. Presently lacking such data, we can alternatively use (in the present communication) for the quantity of life a rough estimate of percentage that life occupies, out of the total possible carrying capacity of each habitat (e.g., sea, land, space, other planets, etc.) that life can settle on Earth or elsewhere at a given time.

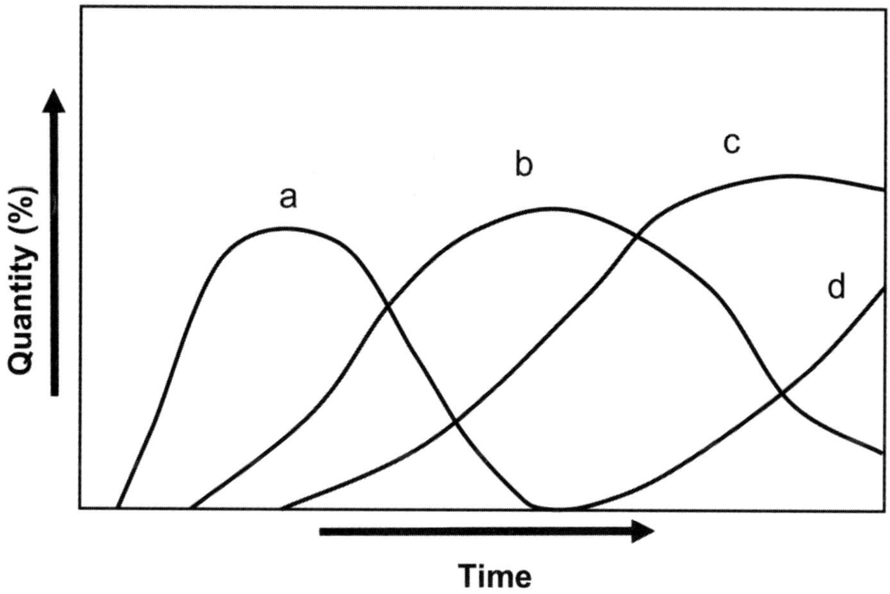

Figure 2. A series of life-cycles.

3. The life-cycle of replicators

Our solar system, the Sun with the planets, including Earth, were formed some 4.6 Bya (Gamov, 1945; Nisbet and Sleep, 2001). Life on Earth appeared about 3.8 Bya (Sleep et al., 1989; Nisbet and Sleep, 2001), perhaps out of the primordial soup in water (the sea), in which prebiotic chemical reactions took place some 4.0-4.2 Bya (Gilbert, 1986; Bartel and Unrau, 1999; Joyce, 2002; Dworkin et al., 2003; Orgel, 2004). It had been suggested and generally accepted (e.g., Dawkins, 1976; Gilbert, 1986; Joyce, 2002; Dworkin et al., 2003; Orgel, 2004; De Duve, 2005) that long before the appearance of cells life already existed in the form of free replicators which swam freely in the primordial medium. These replicators were naked macromolecules similar to RNA, which developed (by errors of replication) enzymatic capabilities, including the ability for self replication (Dawkins, 1976,

1982; Gilbert, 1986; Bartel and Unrau, 1999; Joyce, 2002; Dworkin et al., 2003). The ancient replicators mutated through mistakes in replication and had formed new versions, accumulating diversity in the population of molecules (Dawkins, 1976). At that stage, competition (e.g., for raw materials, and/or between predator and prey) between the various replicators started a Darwinian's chemical evolution for the survival of the fittest (Dawkins, 1976; Joyce, 2002; De Duve, 2005). Association of some replicators with other molecules (e.g., peptides, lipids, etc.) gave them relative advantages in their battles for survival later on (Dawkins, 1976). Slowly through many generations of replications and mutations useful for survival, a defense perimeter of a membrane was "invented" for their protection and the first primitive cell was born, some 3.6-3.8 Bya (see Dawkins, 1976; Bartel and Unrau, 1999; Orgel 2004; Joyce, 2002).

The DNA is now the replicator on which life (as we know it) is based in our DNA/protein-world. However, before the DNA-world there may have been other replicators which served as the base of the primordial life on Earth. Prior to the DNA-world was the RNA-world, some 3.8 Bya and before it probably was another (or other) replicator(s) in the pre-RNA-world some 4 Bya (Gilbert, 1986; Joyce et al., 1987; Joyce, 1989; 2002; Bartel and Unrau, 1999; Nelson et al., 2000; Dworkin et al., 2003; De Duve, 2005). Presently we do not understand and can only speculate about the processes of competition between the replicators and the ways that led to the replacement of one replicator by another, neither the exact time table of their domination (Joyce et al., 1987; Joyce, 1989; Poole et al., 1998; Dworkin et al., 2003). Nevertheless, we can suggest and graphically plot the possible estimated life-cycles of these replicators (Fig. 3).

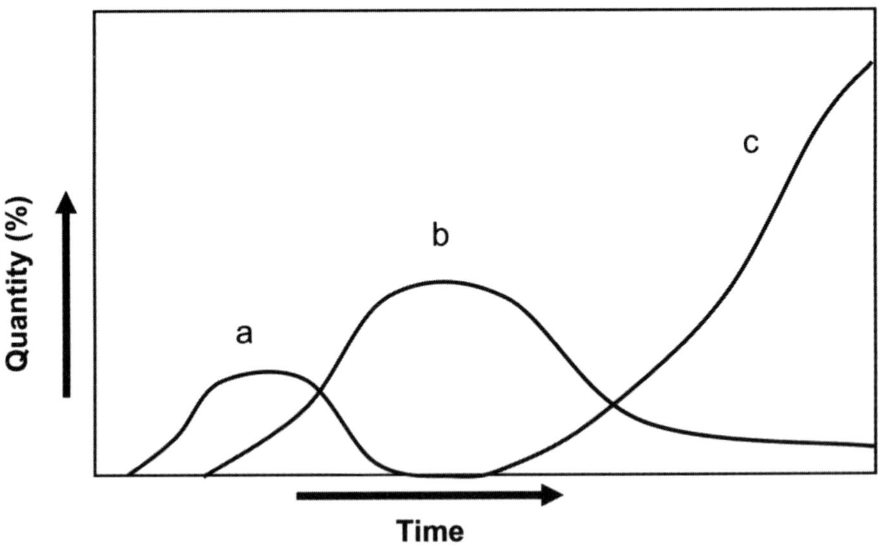

Figure 3. The possible succession of competing replicators and their life-cycles: a. pre-RNA replicator(s) in the pre-RNA-world, b. The RNA replicator in the RNA-world, c. The DNA replicator of the current DNA/protein-world.

Possible remnants of the RNA-world (such as ribozymes and the RNA-proteins machinery of proteins synthesis) were found as metabolic fossils in our DNA-world but nothing was found yet as remnants from the pre-RNA-world (Dworkin et al., 2003). The arguments for the existence of other replicators, which existed prior to the RNA-world, are based mainly on the long term instability of free RNA molecules and their components in the early Earth's environment (Joyce et al., 1987; Lazcano and Miller, 1996; Levy and Miller, 1998; Dworkin et al., 2003).

It was suggested (Maher and Stevenson, 1988; Sleep et al., 1989), that life had started on Earth more than once and it was terminated (again and again?) by impacts of giant asteroids followed each time by reformation. One can not exclude the possibility that during these prebiotic times various lives (representing different chemical elements and/or macromolecular replicators) were initiated in parallel, with only "ours" finally surviving.

4. The life-cycle – of life as we know it

The first living cells were most likely very simple prokaryotes like primitive bacteria and archaea and they had all the seas to inhabit (Bartel and Unrau, 1999). Shortly after the beginning of life, an explosive metabolic evolution took place (Lazcano and Miller, 1996). The primordial cells kept mutating and increasing their populations and the biodiversity through mutations, "inventing" new and useful (for survival) enzymes and chemical reactions, like the complex machineries of protein synthesis and photosynthesis. At that time, the primordial atmosphere of Earth did not have free oxygen but the vastly increasing populations of photosynthesizing organisms in the seas, produced vast quantities of oxygen that was released into the atmosphere (Knoll, 1992; Kasting, 1993; Lazcano and Miller, 1996). Some 2 Bya (Fig. 4c) there was considerable oxygen in the atmosphere (Knoll, 1992; Kasting, 1993; Bekker et al., 2004), but nevertheless, the air and land were still bare of life which then existed only in water.

Eukaryotic cells evolved again by mutations and selection, perhaps also by forming symbiosis with primitive bacteria and algae, resulting in cells that have mitochondria and chloroplasts (e.g., Nisbet and Sleep, 2001). By that time, some 2 Bya (Knoll, 1992), the habitable capacity of the seas was probably almost fully occupied by prokaryotic and protests, the unicellular eukaryotic organisms (Fig. 4c). Cooperation between cells in a compact colony, led some 1000-600 Mya (Knoll, 1992; Crawford, 2000), to the formation of multi-cellular organisms in which various cells differentiated into specific specializing organs within the individual organism. At that time, the seas began to flourish with many new and strange groups of organisms, many of which lost later on their Darwinian war for survival and were extinct, with only their remains left as fossils (e.g., Benton and Twitchett, 2003).

About 470 Mya, during the time that the seas were populated by vast biodiversity, some organisms (animals and plants) started to venture out of the sea to colonize the land (Shear, 1991; Kendrick and Crane, 1997). The move to life on land required adaptation to the then harsh environment (Shear, 1991; Kendrick and Crane, 1997). Mutations and natural selection led to changes, enabling the organisms to survive by breathing air and to reproduce on land without dependency on watery habitat. The biodiversity expanded on land, leading later on to the full occupation of the carrying capacity of each habitat (Fig. 4f, 5e).

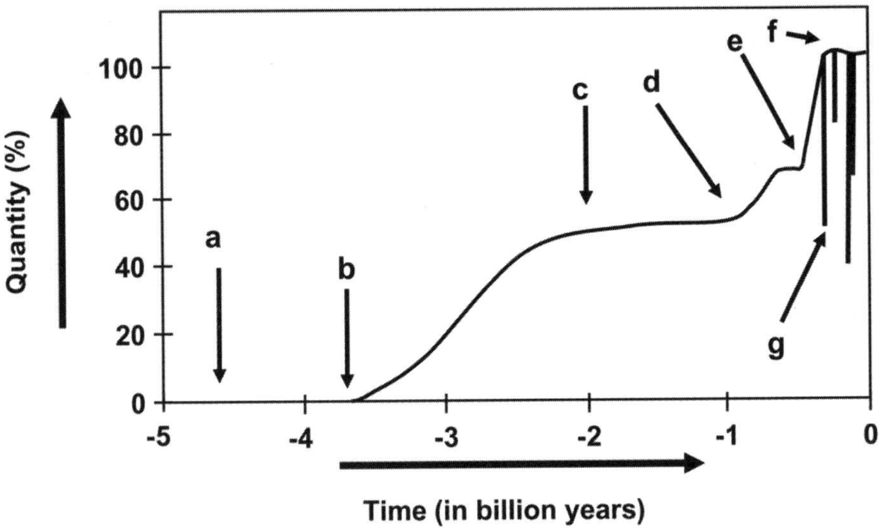

Figure 4. The life-cycle of life as we know it – A short history of "Earth" from the beginning to the present time: a. the formation of planet Earth, b. appearance of life (in the sea), c. considerable oxygen in the atmosphere, d. the appearance of multi-cellular organisms (in the sea), e. the beginning of life on land, f. the carrying capacity of the land is populated, g. mass extinctions. Zero value for time indicates the present. The quantity values are an estimate of accumulated percentages of the total carrying capacity for life on Earth.

Throughout the history of life on Earth, many catastrophes (e.g., climatic changes, vast volcanic eruptions, asteroids impacts, etc.) did happen, causing vast death and extinction for many individuals and species (Alvarez et al., 1980; Benton and Twitchett, 2003; Jablonski, 2004). Sometimes, more than 70% of the living species were totally killed, causing mass extinction (Benton and Twitchett, 2003). Each catastrophe decimated the populations and considerably decreased the total quantity of "life" (Fig. 4g, 5f-g). However, after each catastrophe the survivors flourished in the under-populated habitats, leading latter on to an increase in the quantity of life back to filling the carrying capacity (Fig. 5, between f and g). Mass extinctions have opened new opportunities for the surviving organisms and for newly formed (by mutations) biodiversity. The dinosaurs were the dominant group of creatures on land for more than 150 million years. Their extinction (Alvarez et al.,

1980; Sleep, 2003; Crawford, 2000), some 65 million years ago (Fig. 5g), has opened the way for the mammals to become the dominant creatures on Earth (Fig. 5h).

One might consider the fact that the only life we know was very lucky until now. Throughout the last 3.7 billion years, at any point it could have been exterminated by a giant catastrophe, but it did not, or did it (Maher and Stevenson, 1988; Sleep et al., 1989)? Collisions of Earth with an asteroid or a comet occurred many times in the past, causing mass extinctions (Alvarez et al., 1980; Maher and Stevenson, 1988; Sleep et al., 1989). But luckily, in the last 3.8 billion years the projectiles involved were not large enough to kill all life on Earth (Maher and Stevenson, 1988; Sleep et al., 1989). Some forms of life always survived and flourished after the catastrophe, until now.

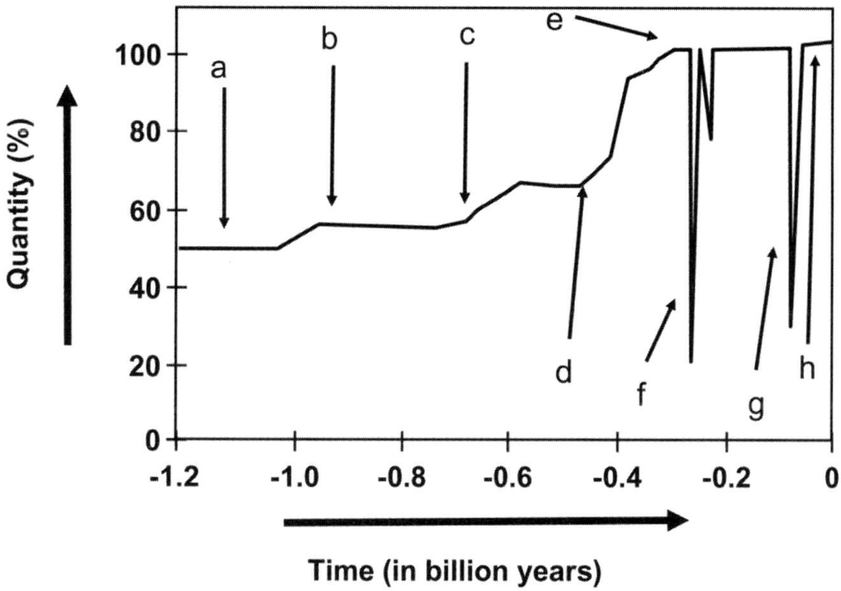

Figure 5. The life-cycle of life as we know it – From a world of prokaryotes to the present time (0 value of time): a. the only living organisms are unicellular prokaryotes, b. appearance and spread of unicellular eukaryotes, c. the appearance of multi-cellular eukaryotic organisms (in the sea), d. the beginning of life on land, e. the carrying capacity of the land is populated, f. mass extinctions, g. the mass extinction of the dinosaurs (65 million years ago), h. the age of mammals. Zero value for time indicates the present. The quantity values are an estimate of accumulated percentages of the total carrying capacity for life on Earth.

That brings us to the present time. We, *Homo sapiens*, the product of some 3.8 billion years of evolution, are mammals and the intelligent creature with technological culture, are only new-comers. The lineage of *Homo* was separated from the apes only a few million years ago (e.g., Stedman et al., 2004). We have discovered "evolution", electronics, flight, nuclear reactions and space flight, mainly in the last century.

Up to this point, this was a short description of life, the only one that we presently know. Based on this knowledge, it is possible to make some extrapolations about the future parts of the life-cycle of the life we know, as well as about other lives that existed, exist or will exist elsewhere in the Universe.

5. Alternative scenarios for the life cycles of life

Many scientists think that life on Earth is not the only one in the Universe (see Crawford, 2000). We can divide the various lives into two groups, depending on if they did (or will) develop intelligent creatures with technological culture.

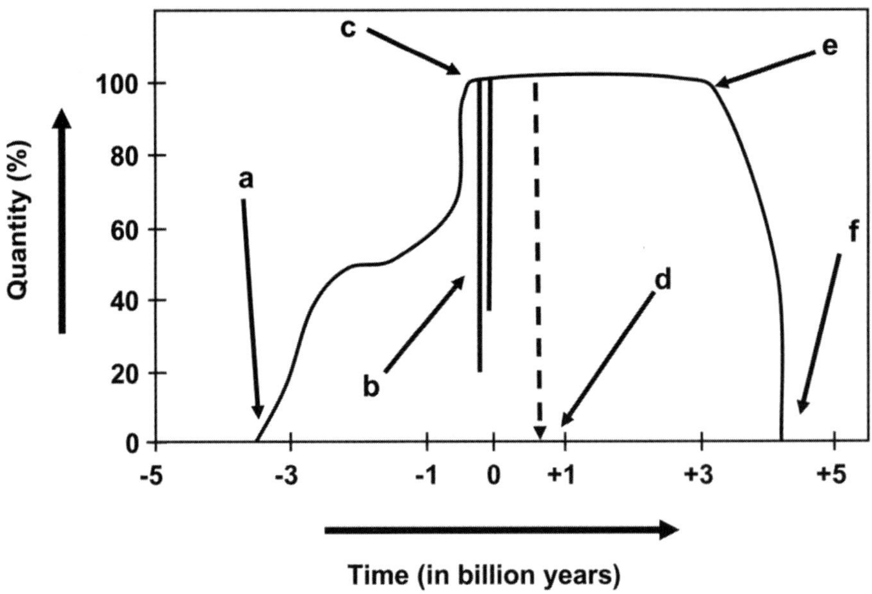

Figure 6. Future possible Scenarios for the life-cycle of life – without the appearance of intelligent creatures that have technological culture: a. appearance of life on the home planet, b. mass extinctions, c. the carrying capacity of the land is populated, d. total extinctions (by astronomical event), e. the local home star is turning into a red-giant, f. final and total extinction of life by overheating of the home sun or it becoming a red-giant. Zero value for time indicates the present. The quantity values are an estimate of accumulated percentages of the total carrying capacity for life on the home planet.

5.1. LIFE WITHOUT AN INTELLIGENT CREATURE THAT HAS TECHNOLOGICAL CULTURE

Let us consider the possibility of a planet that is totally covered by water and without any land. On such a planet (Fig. 6) life can evolve and even produce intelligent creatures, but it is unlikely that they will ever develop sophisticated technological culture, for which one needs fire, electricity, etc.

Even on a planet that has vast areas of land, life can thrive without ever developing intelligent creatures. Even if intelligent creatures were developed, they might never reach the stage of having sophisticated technological culture before their extinction. In such cases, the quantity of life will forever be restricted to the total carrying capacity of the home planet. On such a planet life will continue until life is exterminated by one of the following:

1. A major catastrophe, such as a collision with a giant astronomical projectile (Maher and Stevenson, 1988; Sleep et al., 1989), or a burst of a supernova in the close neighborhood of the home star (Fig. 6d). It is important to point out that such a catastrophe (Fig. 6d) can take place at any time during the life-cycle, earlier or later than indicated by Fig. 6d.
2. When the home star became old (Fig. 6e), it can overheat or turn into a red-giant (Gamov, 1945; Crawford, 2000; Brownlee, 2003; Lovelock, 2003), that will cook its planets and strip them of atmosphere and water leading to (Fig. 6f) the extermination of life (Gamov, 1945; Crawford, 2000; Brownlee, 2003; Sleep, 2003).

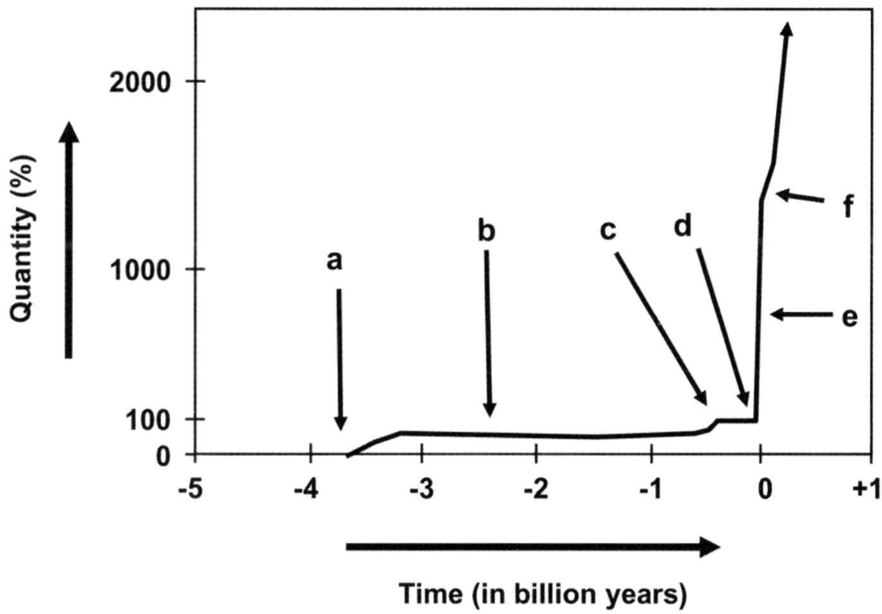

Figure 7. Future possible scenarios for the life-cycle of life – with the appearance of intelligent creature that developed technological culture: a. appearance of life on the home planet, b. the carrying capacity of the sea is populated, c. the carrying capacity of the land is populated, d. beginning of travel to space, e. settlement of space stations and planets within the boundaries of the home star system, f. beginning of settling space and planets of the neighboring star systems in the home galaxy. Zero value for time indicates the present. The quantity values are an estimate of accumulated percentages of the total carrying capacity for life on the home planet.

5.2. LIFE WITH AN INTELLIGENT CREATURE THAT HAS DEVELOPED TECHNOLOGICAL CULTURE

Unless an earlier total extinction by a major catastrophe (e.g., Fig. 6d), an intelligent creature, that has developed technological culture might sooner or later develop space flight (Fig. 7d), unless the use of technology by the intelligent creature will bring sooner a self-inflicted doom (Crawford, 2000; Brownlee, 2003).

Initiation of settlements in space within the boundaries of the home star system will enable life to increase its quantity far over the carrying capacity of the home planet. Even more, by forming settlements in space, life will be able to escape the possible total extinction caused by a giant asteroid or comet that might collide with the home planet (Maher and Stevenson, 1988; Sleep et al., 1989). Further development of space flight will enable life to leave the boundaries of the home star system to form colonies on planets of neighboring stars (Jones, 1976, 1981; Crawford, 2000). Such expansion will further and vastly increase the quantity of life (Fig. 7f) and save it from extinction by overheating associated with the stepwise turning of the home star into a red-giant (Gamov, 1945; Crawford, 2000). In such a case, life will probably have a very long life-cycle while it continues to spread within the habitable zone of the home galaxy and even beyond (Adams and Laughlin, 1997; Crawford, 2000; Gonzalez et al., 2001a).

Theoretically, any life in the Universe can continue for very long time, "in excess of 10^{50} – 10^{100} years, depending upon cosmological and biological issues" (Krauss and Starkman, 2000). One of the cosmological issues is the possibility that the Andromeda galaxy will encounter with our Milky-Way, in about 3 billion years, causing termination of all lives in both galaxies (Gonzalez et al., 2001b; Wakker and Richter, 2004). That, unless some lives managed previously to travel and settle in other galaxies in the Universe.

5.3. LIFE AS WE DO NOT KNOW IT

Other lives can exist in the Universe even in places that are prohibiting to the existence of the life we do know (Bains, 2004). Even more, some of such lives might be based on chemistry that is alien to us (Bains, 2004). Nevertheless, wherever any kind of life might exist it must have a life-cycle. At some point in time it was originated and at another point it will terminate. In between, it must have gone through various stages of its life-cycle. It is suggested here that the life-cycle of any life will basically follow along one of the general above mentioned alternative routes (Fig. 6, 7). The time scale and the physical chemical environment might be different, but the main important features will stay the same. One can prepare a check list of questions that will help to decide for a newly discovered life at what stage of its life-cycle it is. For example:

1. Are they still in the pre-cellular stage?
2. Did multi-cellular organisms developed there?
3. Did intelligent creatures evolved there?
4. Did they develop sophisticated technological culture?
5. Are they still bound to the home planet or to the space of the home star?

6. Have they managed to travel out of the home galaxy and to form settlements in other galaxies?

5.4. WHEN ONE REPLICATOR MEETS ANOTHER REPLICATOR

Any encounter between replicators will most likely result in various degrees of competition. The competition might be active or passive. Active competition will happen when one or both replicators are represented by intelligent creatures, expressing hostilities and/or ignorance to ecological issues, leading to the possible extinction of the native life (Crawford, 2000). An example of passive competition can be the case of one replicator causing an unwilling plague that will decimate or annihilate the populations of the other replicator. NASA and other such agencies are aware of such dangers and are taking various precautions to prevent contamination of space with our microorganisms as well as to prevent import of alien microbes by the returning astronauts and other space missions. The Fermi Paradox (Gonzales et al., 2001b) raise the question "if intelligent life is common in our galaxy and the Universe, where are they?" Astronomer Roger Angel (from the University of Arizona, Tucson) stated that "any civilization slightly more advanced than our own would know there was life on this planet" (see Pearson, 2002). One possible answer to the Fermi Paradox is that they know we are here, but they are aware of the dangers arising from direct contact and passive competition with other replicators, therefore they stay away.

6. Summary

Any kind of life, the one we know and others we do not know, by definition, each must have its own life-cycle. From the features of the life we know, we can fairly generalize on some important features of the life-cycle of other lives that might exist elsewhere in the Universe, those that we still do not know. The death of (any) 'life' can be caused by either one of:

1. By the over heating associated with the "death" of the home star.
2. Self inflicted death, by misuse of technology developed by the intelligent creature, mainly before forming settlements in the space of the home sun system.
3. A collision of the home planet with a giant asteroid or comet, before forming settlements in the space of the home sun system.
4. A supernova event in the neighborhood of the home star, before forming settlements in the space of other stars systems, located at a safe distant from the home star.
5. Encounter and competition between replicators, at any stage of their life-cycle, can end by the total extinction of one of them.
6. A major astronomical catastrophe affecting vast parts or the entire home galaxy (Gonzalez et al., 2001b; Wakker and Richter, 2004).
7. At the end of the Universe as we presently know it (Krauss and Starkman, 2000; Freese and Kinney, 2003).

Finally, it is worthwhile to cite Freese and Kinney (2003): "It is of course hubris to believe that humans can at this point foresee all the ideas that all future life forms will come up with to save themselves."

7. References

Adams, F. C. and Laughlin, G. (1997) A dying universe: the long-term fate and evolution of astrophysical objects. Rev. Modern Physics **69**: 337-372.

Alvarez, L. W., Alvarez, W., Asaro, F. and Michel, H. V. (1980) Extraterrestrial cause for the cretaceous-tertiary extinction. Science **208**: 1095-1108.

Bains, W. (2004) Many chemistries could be used to build living systems. Astrobiology **4**: 137-167.

Bartel, D. P. and Unrau, P. J. (1999) Constructing an RNA world. Trends in Genetics **15**: M9-M13.

Bekker, A., Holland, H. D., Wang, P. L., Rumble, III D., Stein, H. J., Hannah, J. L. and Coetzee, L. L. (2004) Dating the rise of atmospheric oxygen. Nature **427**: 117-120.

Benton, M. J. and Twitchett, R. J. (2003) How to kill (almost) all life: the end-Permian extinction event. Trends Ecol. Evol. **18**: 358-365.

Brownlee, D. (2003) A walk to the gallows. Nature **423**: 803-804.

Cleland, C. E. and Chyba, C. F. (2002) Defining 'life'. Origin Life Evol. Biosphere **32**: 387-393.

Crawford, I. (2000) Where are they? Sci. Am. **283** (1): 38-43.

Dawkins, R. (1976) *The Selfish Gene.* Oxford University Press, Oxford, UK.

Dawkins, R. (1982) *The Extended Phenotype: The Long Reach of the Gene.* Oxford University Press, Oxford.

De Duve, C. (2005) The onset of selection. Nature **433**: 581-582.

Dworkin, J. P., Lazcano, A. and Miller, S. L. (2003) The roads to and from the RNA world. J. Theor. Biol. **222**: 127-134.

Freese, K. and Kinney, W. H. (2003) The ultimate fate of life in an accelerating Universe. Phys. Lett. B558: 1-8.

Gamov, G. (1945) *The Birth and the Death of the Sun: Stellar Evolution and Subatomic Energy.* Penguin Books Inc., New-York.

Gilbert, W. (1986) Origin of life – The RNA world. Nature **319**: 618.

Gonzalez, G., Brownlee, D. and Ward, P. (2001a) The galactic habitable zone: chemical evolution. Icarus **152**: 185-200.

Gonzalez, G., Brownlee, D. and Ward, P. (2001b) Refuges for life in a hostile Universe. Sci. Am. **285**(4): 60-67.

Jablonski, D. (2004) Extinction: past and present. Nature **427**: 589.

Jones, E. M. (1976) Colonization of the Galaxy. Icarus **82**: 421-422.

Jones, E. M. (1981) Discrete calculations of interstellar migration and settlement. Icarus **46**: 328-336.

Joyce, G. F. (1989) RNA evolution and the origin of life. Nature **338**: 217-224.

Joyce, G. F. (2002) The antiquity of RNA-based evolution. Nature **418**: 214-221.

Joyce, G. F., Schwartz, A. W., Miller, S. L. and Orgel, L. E. (1987) The case for an ancestral genetic system involving simple analogues of the nucleotides. Proc. Natl. Acad. Sci. USA **84**: 4398-4402.

Kasting, J. F. (1993) Earth early atmosphere. Science **259**: 920-926.

Kauffmann, G. and van den Bosch, F. (2002) The life cycle of galaxies. Sci. Am. **286** (6): 45-58.

Kendrick, P. and Crane, P. R. (1997) The origin and early evolution of plants on land. Nature **389:** 33-39.

Knoll, A. H. (1992) The early evolution of eukaryotes: a geological perspective. Science **256**: 622-627.

Krauss, L. M. and Starkman, G. D. (2000) Life, the Universe, and nothing: life and death in an ever-expanding Universe. Astrophys. J. **531**: 22-30.

Lazcano, A. and Miller, S. L. (1996) The origin and early evolution of life: prebiotic chemistry, the pre-RNA world, and time. Cell **85**: 793-798.

Levy, M. and Miller, S. L. (1998) The stability of the RNA bases: implications for the origin of life. Proc. Natl. Acad. Sci. USA **95**: 7933-7938.

Lovelock, J. (2003) Gaia: the living Earth. Nature **426**: 769-770.

Maher, K. A. and Stevenson, D. J. (1988) Impact frustration of the origin of life. Nature **331**: 612-614.

Nelson, K. E., Levy, M. and Miller, S. L. (2000) Peptide nucleic acids rather than RNA may have been the first genetic molecule. Proc. Natl. Acad. Sci. USA **97**: 3868-3871.

Nisbet, E. G. and Sleep, N. H. (2001) The habitat and nature of early life. Nature **409**: 1083-1091.

Orgel, L. E. (2004) Prebiotic chemistry and the origin of the RNA world. Crit. Rev. Biochem. Mol. Biol. **39**: 99-123.

Pearson, H. (2002) They know we are here: Evolved space civilization will be eyeing Earth. *news@nature*, (18 Feb 2002) News. Nature Publishing Group.

Poole, A. M., Jeffares, D. C. and Penny, D. (1998) The path from the RNA world. J. Mol. Evol. **46**: 1-17.

Rymon, D., Umiel, N., Nakar, R., Spharim, I. and Fishelson, G. (1987) Marketing analysis of spray carnation: Product life cycle of the spray carnation (the product) and of single cultivar (the brand). Acta Hort. **216**: 165-175.

Shear, W. A. (1991) The early development of terrestrial ecosystems. Nature **351**: 283-289.

Sleep, N. H. (2003) Out, out, brief candle ? Nature **422**: 663-664.

Sleep, N. H., Zahnle, K. J., Kasting, J. F. and Morowitz, H. J. (1989) Annihilation of ecosystems by large asteroid impacts on the early Earth. Nature **342**: 139-142.

Stedman, H. H., Kozyak, B. W., Nelson, A., Thesier, D. M., Su, L. T., Low, D. W., Bridges, C. R., Shrager, J. B., Minugh-Purvis, N. and Michell, M. A. (2004) Myosin gene mutation correlates with anatomical changes in the human lineage. Nature **428**: 415-418.

Wakker, B. P. and Richter, P. (2004) Our growing, breathing Galaxy. Sci. Am. **290** (1): 38-47.

PART 4:
SUMMARY AND CONCLUSIONS

Seckbach, Raulin, Oren, Kolb, Chela-Flores

The Biodata of **Joseph Seckbach, Aharon Oren** and of **Julian Chela-Flores** are printed adjacent to their contributions in this volume.

Biodata of **François Raulin** co-author (with Seckbach et al.,) of the chapter "***What do We Call Life? A Brief Outlook of Life***"

Dr. **François Raulin** is Full Professor of Chemistry at University Paris 12, and head of PCOS (Space Organic Physical Chemistry) group of LISA, Laboratoire Interuniversitaire des Systèmes Atmosphériques. He is also Director of the CNRS federation of laboratories "Groupement de Recherche" in Exobiology. He got a diploma from the Ecole Supérieure de Physique et Chimie Industrielles de la Ville de Paris in 1969, and a Doctorat d'Etat (on the role of sulphur in prebiotic chemistry) from the Université Paris 6 in 1976. Dr. Raulin was research fellow at Carl Sagan's Laboratory, Cornell University (USA) in 1970-71, assistant Professor at University Paris Val de Marne until 1978 and CNRS/NSF post-doctoral fellow at Cyril Ponnamperuma's Laboratory of Chemical Evolution, University of Maryland (USA). Then he was Director of LISA for ten years until 2005. His scientific fields of Interest are related to planetology and exo/astrobiology: studies of organic chemistry in extraterrestrial environments (Titan, giant planets, comets and Mars) using complementary approaches: laboratory experiments, theoretical modeling; observational data and remote sensing and *in situ* space exploration.

Dr. F. Raulin is IDS (InterDisciplinary Scientist) of the Cassini-Huygens mission and Co-Investigator of the CIRS (Cassini), ACP and GC-MS (Huygens) experiments. He is Co.I of the COSAC and COSIMA experiments of the Rosetta European cometary mission. In addition he is Chair of COSPAR Commission F (Life Sciences), Vice Chair of COSPAR Planetary Protection Panel and 1rst Vice-President of ISSOL. Among his publication are 250 scientific papers (as an author and co-author), and 10 books related to the field of the origins of Life and Exobiology

E-mail: **raulin@lisa.univ-paris12.fr**

J. Seckbach (eds.), Life as We Know It, 737-743.

Biodata of **Vera Kolb** coauthor (with Seckbach et al.,) of the chapter ***"What do we call Life?"***

Dr. Vera Kolb is currently a Professor of Chemistry at the University of Wisconsin-Parkside, and interested in Astrobiology. She has received her PhD in organic chemistry from Southern Illinois University at Carbondale in 1976. Her training in exo-biology was received from the NASA NSCORT Center in San Diego, 1992-1994, where she worked with Leslie Orgel at the Salk Institute for Biological Sciences, and Stanley Miller, at the University of California San Diego. Dr. Kolb is the author of over 100 publications including patents and books. She is a recipient of awards for both research and teaching.
E-mail: **kolb@uwp.edu**

WHAT DO WE CALL LIFE?
A Brief Outlook on Life

JOSEPH SECKBACH[1], FRANCOIS RAULIN[2], AHARON OREN[3], VERA KOLB[4] AND JULIAN CHELA-FLORES[5]
[1]P.O.Box 1132, Efrat, 90435, Israel; [2]LISA, Universités Paris 12 and Paris 7, Créteil, France; [3]The Institute of Life Sciences, and The Moshe Shilo Minerva Center for Marine Biogeochemistry, The Hebrew University of Jerusalem, Jerusalem 91904, Israel; [4]Department of Chemistry, University of Wisconsin-Parkside, Kenosha, Wisconsin, 53141, USA, and [5]The Abdus Salam International Centre for Theoretical Physics, Strada Costiera 11; 34014 Trieste, Italy and Instituto de Estudios Avanzados, Caracas 1015A, Venezuela

1. Introduction

It is always frustrating to write about life itself, the answers are always unsatisfactory. Indeed to give a satisfactory definition for life is not easy. Many attempts to redefine life are bound to be unsuccessful. Round tables have recently been held during international conferences dedicated to this question, which ended with no clear conclusions or any consensus on the definition of life from the audience.

Many secrets of life have been unlocked in the past century thanks to the advancements of modern science. Examples include the discovery of the structure and function of DNA, mapping the genes of various species, and the discovery of organisms living under extreme harsh conditions (the "extremophiles").

In fact, whichever is the chosen definition strongly depends on the specific field of the scientist who is trying to give it. Thus, a biologist will speak about organisms' metabolism, reproduction, development and adaptation (in particular to the environmental constraints). In thermodynamics (see below), a living system is often considered as a system far from equilibrium, able to adapt and to reproduce by using a continuous energy flux. A philosopher would compare life to music, water or even fire, concepts which themselves are not easy to define. An exo/astrobiologist and a physical chemist would consider life as a phenomenon involving a high level of complexity, replication, and mutation, including optimization and adaptation. Thus we can conclude that life is the action, animation, organization, metabolism, information, perpetuation and multiplication of the organism.

We hope that this volume will eventually contribute to our attempts to provide an appropriate answer to the question "What is life?"

2. Origins, Evolution and What Kind of Life we are Searching?

The origin of life is still a mystery; however there are many hypotheses about the origin of life on Earth from abiotic matter. Pioneers like Oparin, Haldane, Miller, Orgel, Fox, Ponnamperuma and Oró are just several of the key scientists who advanced this field. All descriptions of life are still terra-centric. The need for a general, non-terra-centric definition of life arises as we are searching for life on Mars or in the giant planets and their moons. There is no *a priori* reason for extraterrestrial life to be substantially similar to ours. Moreover, life may not have fully emerged on Mars or Europa, but may be instead at the prebiotic stage, in the "chemical transition" zone. How we would recognize life that is different from ours is an open question. It maybe that extraterrestrial life is in the transition zone and has different chemical characteristics from those we imagine for the prebiotic stage on the primordial Earth. In this volume we concentrated mainly on terrestrial life as we know it and not on extraterrestrial life as we do not know it.

3. The Same Basic Principles Govern all Life on Earth

3.1. GENERAL FEATURES OF LIFE

The primary qualities of life as we know it, are: a) it is chemistry predominantly based on carbon atoms; b) its chemistry is based on macromolecules of polymeric nature, some of which can carry information by virtue of the monomer sequence of their particular three-dimensional features; c) the organization of its chemical system ensures preservation of its integrity and allows for its perpetuation; d) it is continuous, since life has not stopped existing since it emerged. The autonomous agents, whose life span is limited, carry out the continuity of life. It is essential to understand that, while the individual autonomous agents are considered alive, life as a phenomenon requires continuity that goes beyond an individual agent.

All organisms living on Earth, from the simplest prokaryote to the most complex eukaryotic representatives of the plant and the animal world, use the same basic biochemistry. All organisms possess DNA as genetic material, ribosomes (built of RNA and protein) to synthesize proteins, and protein- or RNA-based enzymes to catalyze chemical reactions. The lipid membranes of all organisms separate the intracellular space from the outer medium and (especially in eukaryotes) separate specialized compartments within the cell. Life on Earth, in all its diversity, thus functions according to very similar principles throughout. We can state that life is universal and ubiquitous on Earth.

3.2. MEMBRANES, THE CELLULAR GATES, ARE INDISPENSABLE FOR LIFE

Each cell is separated from its environment by a lipid membrane. Enzymatic activities within this membrane determine what goes in and out of the cell. Cells can thus regulate their intracellular environment to achieve optimal conditions for the functioning of their activities. Thus, intracellular pH and ionic concentrations can be kept constant also when the external environment changes. Some degree of homeostasis is a universal feature of life. Not for all environmental factors can homeostasis be achieved. Another universal function of membranes in all organisms (except for a few members of the domain *Bacteria* that obtain their energy only by fermentation) is the involvement of membranes in energy transduction.

3.3. LIFE MEANS ENERGY TRANSDUCTION

All cells known to us so far require energy for their existence and to function. In spite of the tremendous variety in life on Earth, the principles underlying energy conservation are markedly conserved, whatever the energy source and physiology of the organisms may be: phototrophic, heterotrophic, or chemoautotrophic. Cells have two forms of energy that can be used for energy-requiring processes: chemical energy – mainly in the form of high energy phosphoanhydride bonds in ATP, and electrochemical gradients across membranes – generally of protons, sometimes also of sodium ions. These two forms of energy are interconvertible through the activity of membrane-bound ATPase/ATP synthase.

3.4. LIFE IS SUBJECT TO THE LAWS OF CHEMICAL THERMODYNAMICS

The laws of chemical thermodynamics are valid in biology as well: Life cannot perform processes that are impossible chemically. No chemical process, including those performed by living organisms, enables a full conservation of energy. Energy losses as heat are inevitable. A constant input of energy is therefore essential to keep life on our planet going. Sunlight is the primary energy source that drives life on Earth. Geothermal energy, resulting in the formation of reduced compounds such as hydrogen, hydrogen sulfide or methane, can be locally important in specialized ecosystems such as the submarine hydrothermal vents. It is even possible that such environments were the first in which life originated, as the ability to use photons in the visible wavelength range developed relatively late in the course of evolution. However, nowadays all biological processes on Earth depend directly or indirectly on sunlight energy. This is even true for the dense hydrothermal vent communities that indeed rely on geothermally formed reduced inorganic energy sources. Some also need molecular oxygen as electron acceptor, and for this purpose they use oxygen that was ultimately derived from oxygenic photosynthesis on land or in the photic layers of the sea.

3.5. ALL CONCEIVABLE NICHES AVAILABLE ON EARTH ARE INHABITED

Life can be found almost anywhere on Earth, including environments hostile to most organisms. Extremophilic (micro)organisms can live up to the lowest and highest pH values, at temperatures below the freezing point of water and above its boiling point, at high pressures, at saturated salt concentrations, and at high levels of radiation. The prokaryotic world shows the greatest versatility when it comes to adaptation to environmental extremes. Especially the *Archaea* specialize in colonization of some of the harshest environments on Earth, not habitable by other forms of life.

4. Eukaryote Evolution Has Led to Increasing Intracellular Complexity

The evolution of the prokaryote world (*Bacteria* and *Archaea*) has never led to the formation of multicellular organisms with high levels of functional differentiation between cells. Only occasionally do we find cell differentiation within the prokaryotes, one rare example being the nitrogen-fixing heterocysts within filaments of cyanobacteria (domain *Bacteria*). The increasing complexity of the multicellular eukaryotic organisms has never resulted in development of metabolic possibilities not present in some prokaryotes. The opposite is true: there is not a single mode of assimilatory or dissimilatory metabolism that exists in the eukaryotes and that is not found in the prokaryote world. The metabolism of the eukaryotes is largely restricted to aerobic respiration and on oxygenic photosynthesis (performed by mitochondria and chloroplasts, which can be considered as prokaryotic “symbionts” of the eukaryotes) Most other options of energy generation known from the prokaryote world, such as anoxygenic photosynthesis, anaerobic respiration, most fermentations, and chemolithotrophy are not realized in the world of the eukaryotes.

5. Can We Anticipate Features of Living Organisms Elsewhere in Solar Systems?

When looking for forms of life on Mars or Europa we need to focus on the organic matter, to isolate and identify polymers, and to try to break their code of information transfer. This may be very difficult to do if we do not find live specimens from which we can learn the internal organization of their chemical components in the living system. In systems that are no longer alive (for instance, in the case of fossils) or which are in the prebiotic transition zone, we would only find the chemical pieces of the puzzle. The puzzle itself, which includes the organizational scheme for these pieces, would be very difficult to solve.

Space exploration has still failed to encounter living organisms that have evolved along a different evolutionary pathway from life on Earth. The discovery of an independent evolutionary line would provide us with significant deeper insights into the meaning of life. Can we anticipate some general features of living organisms that could possibly lie deep in the Martian surface, or in the subsurface oceans of some of the satellites of the giant planets?

If we assume that the science of biology is of universal validity, then its basic mechanisms such as natural selection and evolutionary convergence can orient us toward the general biological properties that we might expect to find in living organisms elsewhere. The assumption of the universality of biology has a significant implication, namely that we can expect natural selection to be working at the lowest stages of chemical evolution in terrestrial planets of the many extrasolar planetary systems that are now known to exist. Not only the basic biological processes may be going on now elsewhere in the cosmos, but we may also even enquire whether such processes may have been relevant before the birth of the Sun. By spectroscopic analysis of distant molecular clouds backlit by bright quasars, we may gather some information about the presence of biogenic elements heavier than hydrogen (the ‘metallic’ parameter) in the clouds themselves, and consequently in nascent solar systems that would originate from them after accretion. With this technique we may eventually learn whether the main mechanisms of biology could have been driving the evolutionary process well before the emergence of our own solar system, and possibly in other solar systems.

However, if biological processes are active now, or were active in the early stages of the evolution of the cosmos itself, the ingredients of a chemical evolution process in other solar systems should have been largely analogous to the same chemical elements that gave rise to life on Earth. This conjecture is shared implicitly by most researchers, since it is based on the remark that from interstellar spectroscopy, both in the infrared as well as in the microwave windows, the same chemical elements that are used by terrestrial life are indeed abundant in interstellar gas and dust.

Evolutionary convergence has been convincingly documented at various levels, namely at the molecular, cellular and multicellular levels. This aspect of Darwinian evolution argues in favor of different evolutionary pathways leading to largely similar biological properties. On the other hand, lineages are expected to be generally different in different solar systems. What is most relevant in our search for deeper insights into life as we know it is that experimental science is looking for reliable bioindicators that can be searched for during the exploration of our own solar system in the multiple missions that are now in progress or are currently in the planning stages.

We have emphasized convergence at the evolutionary level that forces us to consider general biological properties that are independent of specific lineages, such as anaerobic and aerobic metabolism, vision, the need to communicate, or the evolution of neurons, cerebral ganglions, brains and intelligent behavior. Amongst such general features of biology we may begin our search for the most reliable bioindicators of life and, consequently, we may eventually provide an appropriate universal answer to the question *"What is life?"*

6. Conclusions

We can not yet define life, but we are able to describe living organisms by their activities and features, and thus know in our own minds, and even perhaps to our own satisfaction what is life. Sometimes it is hard to delineate the border between living beings and non-living objects, such as viruses. If one of the definitions of life is the cellular feature of all organisms, then the viruses should be removed from this definition. We call an organism living if it has a cellular structure and fulfills all the functions of active metabolism (such as energy input, excretion output, respiration and production of energy, presence of the required intracellular ingredients – DNA and various proteins, and multiplication by cell division).

Life can be detected anywhere on Earth, from the "normal" to extremophilic organisms. Finding extraterrestrial life is one of our generation's greatest challenges. We hope to find "life as we know it." However it might be that there exists "life as we do not know it". In the Foreword of this volume it was pointed out that if extraterrestrial life is not based on carbon atoms, we should anticipate the publication of another volume, years in the future, entitled "life as others know it."

ORGANISM INDEX

Crenarchaeota, 93, 95, 138

SUBJECT INDEX

AUTHOR INDEX

Cellular Origin, Life in Extreme Habitats and Astrobiology

1. J. Seckbach (ed.): *Enigmatic Microorganisms and Life in Extreme Environments.* 1999 ISBN 0-7923-5492-3
2. J. Seckbach (ed.): *Journey to Diverse Microbial Worlds: Adaptation to Exotic Environments.* 2000 ISBN 0-7923-6020-6
3. J. Chela-Flores: *The New Science of Astrobiology: From Genesis of the Living Cell to Evolution of Intelligent Behavior in the Universe.* 2001 ISBN 0-7923-7125-9
4. J. Seckbach (ed.): *Symbiosis: Mechanisms and Model Systems.* 2001 ISBN 1-4020-0189-4
5. A. Oren: *Halophilic Micro organisms and their Environments.* 2002 ISBN 1-4020-0829-5
6. J. Seckbach (ed.): *Origins.* Genesis, Evolution and Diversity of Life. 2004 ISBN 1-4020-1813-4
7. J. Seckbach, J. Chela-Flores, T. Owen and F. Raulin (eds.): *Life in the Universe.* From the Miller Experiment to the Search for Life on Other Worlds. 2005 ISBN 1-4020-2371-5
8. J. Seckbach and E. Rubin (eds.): *The New Avenues in Bioinformatics.* 2004 ISBN 1-4020-2639-0
9. N. Gunde-Cimerman, A. Oren and A. Plemenitaš (eds.): *Adaptation to Life at High Salt Concentrations in Archaea, Bacteria, and Eukarya.* 2005 ISBN 1-4020-3632-9
10. J. Seckbach (ed.): *Life as We Know It.* 2006 ISBN 1-4020-4394-5